ENVIRONMENTAL MANAGEMENT IN PRACTICE: VOLUME 1

Methods of environmental management, and especially the 'tools' of environmental management, are increasingly being relied upon worldwide to deliver a degree of sustainability in all human activities. A thorough understanding of the nature, capabilities and limitations of these 'tools', as well as the conditions under which they can be best applied, is essential for students, researchers and practitioners within the field of environmental management.

Environmental Management in Practice presents three comprehensive volumes containing the most up-to-date research and practical applications in the field. Spanning the four main aspects of environmental management: instruments, compartments, sectors and ecosystems, this three-volume work contains over sixty contributions from leading specialists in each field and offers the first major source of contemporary international research and application within environmental management in practice.

Volume 1: Instruments for Environmental Management, focuses on the instruments and tools currently available to the environmental manager. A theoretical background to the instruments is given together with an overview of those instruments that are in common use today, with particular attention to the physical, economic, legislative and communication instruments.

Volume 2: Compartments, Stressors and Sectors, deals with the problems that occur in the three 'compartments' of the environment – namely, air, water and soil. The contributors also address the socio-economic sectors of industry, traffic, energy, agriculture and tourism.

Volume 3: Managing the Ecosystem, focuses on those ecosystems in which human intervention has been or continues to be predominant, specifically within cities and rural areas.

Packed with accessible and up-to-date information, these three volumes provide a comprehensive overview of environmental management for those studying, researching and practising in the field.

Bhaskar Nath is Director of the European Centre for Pollution Research, London. **Luc Hens** is Professor and Head of the Human Ecology Department at the Free University of Brussels. **Paul Compton** is an environmental and demographic consultant. **Dimitri Devuyst** is Co-ordinator of the programme of Environmental Impact Assessment in the Department of Human Ecology, Free University of Brussels

ENVIRONMENTAL MANAGEMENT IN PRACTICE: VOLUME I

Instruments for Environmental Management

EDITED BY

B. NATH, L. HENS,

P. COMPTON AND D. DEVUYST

London and New York

First published 1998
by Routledge
11 New Fetter Lane, London EC4P 4EE

Simultaneously published in the USA and Canada
by Routledge
29 West 35th Street, New York, NY 10001

Typeset in Galliard by The Florence Group, Stoodleigh, Devon
Printed and bound in Great Britain by T.J. International Ltd, Padstow, Cornwall

British Library Cataloguing in Publication Data
A catalogue record for this book is available from the British Library

Library of Congress Cataloguing in Publication Data
Environmental management in practice / edited by B. Nath ... [et al.].
p. cm.
Includes bibliographical references and index.
Contents: v. 1. Instruments for environmental management
– v. 2. Compartments, stressors and sectors – v. 3. Managing
the ecosystem.
1. Environmental management. I. Nath, Bhaskar.
GE300.E594 1998 98–11114
363.7'05–dc21 CIP

ISBN 0-415-14906-1

CONTENTS

FIGURES

TABLES

BOXES

CONTRIBUTORS

Henry van Asselt, LL.M., M.Sc., is an Assistant Professor of Technology at the Department of Technology and Innovation, Rotterdam School of Management, Erasmus University (The Netherlands) and before that was with the Dutch Ministry of Economic Affairs and a private consultancy. His main research interests are in the fields of environmental business management and innovation management.

Emmanuel K. Boon was awarded his Ph.D. in Economic Sciences by the Free University of Brussels (Belgium) in 1986. He is a Senior Lecturer in Management Studies at the University of Ghana, Legon-Accra, and a visiting Professor of Development Economics and Advanced Business Policy and Strategy at the Free University of Brussels.

David J. Briggs, Ph.D., is Research Director at Nene University College, Northampton, UK. He has compiled a number of state of environment reports at European, national, and local level, and is a regular consultant to the World Health Organisation, the European Commission and national agencies on both state of environment reporting and environmental indicators. His other research interests include environmental mapping, geographical information systems and environmental health. He has published extensively in all these areas.

Alfons Buekens, Ph.D., is a Chemical Engineer and professor at the Free University of Brussels. Over the years he has developed processes for the thermal treatment of waste and conducted numerous studies in the field of the environment and safety.

Bernard De Caevel is an engineer and a co-worker with Professor Buekens at the Free University of Brussels (Belgium). He is currently a private consultant on environmental engineering and counselling.

Liam Cashman studied Law at University College Cork in Ireland. Since 1988 he has worked for the European Commission as an Administrator in the legal unit of Directorate General XI – Environment, Nuclear Safety and Civil Protection.

Paul A. Compton, M.Sc., Ph.D., formerly Professor in the School of Geosciences at the Queen's University of Belfast, Northern Ireland, is an environmental and demographic consultant. His academic interests lie at the interface between the natural and social sciences.

Ana Cordova received her B.A. in biology from Harvard University and is currently pursuing an M.S./Ph.D. programme in natural resource policy at Cornell University (Ithaca, New York). She has been involved in studying agricultural practices and sustainability in rural Mexico and has assisted in drawing up a solid waste management plan for Chihuahua City, also in Mexico.

Aaron Cosbey holds a Masters degree in Development Economics from Dalhousie University (Canada) and is currently employed as an Environmental Economist and Program Manager at the International Institute for Sustainable Development, Winnipeg. He is a specialist on the relationship between international trade and sustainable development and has published widely on this topic.

Katrien Debeuckelaere studied law at the University of Ghent (Belgium) and received her Masters degree in International and Comparative Law from the Free University of Brussels. Mrs Debeuckelaere was an adviser to the Minister for Planning, Public Works and Transport in the Flemish Government and currently works at the Federal Planning Bureau, Task Force on Sustainable Development. She has also acted as consultant to the Commission of European Communities on environmental law and has worked with IJO in Rome on issues of International Environmental Law, in particular in connection with the Barcelona Convention and the preparation of the Draft Offshore Protocol.

Wouter T. de Groot, M.Sc., Ph.D., was trained as a civil engineer. On graduation he joined the Kenyan Ministry of Works before moving to the Centre for Environmental Science at Leiden University (The Netherlands) in 1977 where he is currently Professor. He has broad interdisciplinary research interests, embracing philosophy, biology and the social sciences. He is Deputy Director of the Centre where he is involved mainly in the Environment and Development Programme which focuses on desertification and deforestation in developing countries. He was awarded his Doctorate in Environmental Science in 1992 and was appointed Professor of Environmental Science at the University of Nijmegen in 1998.

Elizabeth R. DeSombre received her Ph.D. in Political Science from Harvard University and is now an Assistant Professor of Environmental Studies and Government at Colby College in Waterville, Maine. She specialises in international environmental law and policy and has undertaken work on ozone depletion, protection of endangered species, international fishery management and the relationship between free trade and environmental protection.

Dimitri Devuyst received his Ph.D. degree in Human Ecology from the Free University of Brussels (Belgium), as well as a Masters degree in Human Ecology and a Bachelors degree in Botany from the same university. His main interest is in the field of environmental impact assessment. Currently he is Postdoctoral Fellow at the Fund for Scientific Research – Flanders (Belgium) at the Department of Human Ecology, Free University of Brussels.

Walter Leal Filho, B.Sc., Ph.D., D.Sc., Eur.Biol., F.R.G.S., F.R.S.A., F.L.S. was trained as a biologist. Formerly a Senior Research Fellow at the University of Bradford and Director of the European Research and Training Centre on Environmental Education in England, he is currently a Professor at the University of Hamburg, Germany. He is the editor of *Environmental Management and Health* and a member of the editorial boards of *Environmental Awareness, Environmental Education and Information* and *Global Forum for Environmental Education.*

Edeltraud Günther studied management accounting and finance at the University of Augsburg (Germany) and also spent time at the Ecole de Traduction et Interprétation, Geneva (Switzerland). She is currently Professor at the Dresden University of Technology (Germany). Her academic interests are in the fields of environmental accounting and auditing. She also has direct practical experience of industrial applications through her work as Environmental Auditor for a German company.

Luc Hens graduated as a Biologist and later received his Ph.D. in Biology from the Free University of Brussels (Belgium) where he is at present Professor and Head of the Human Ecology Department. He also lectures at Antwerp University (Belgium), the Technical University of Sofia (Bulgaria) and Ankara University (Turkey). His specific area of research concerns the elucidation of interdisciplinary instruments for sustainable development. Professor Hens acts as an expert in environmental policy on several councils in Belgium. He is the European editor for the *International Journal on Environmental Pollution.*

Charles Vyvyan Howard, M.B., Ch.B., Ph.D., M.R.C.Path. is a qualified medical practitioner and Senior Lecturer in Foetal and Infant Toxico-Pathology in the Department of Pathology at the University of Liverpool. He is an expert in Quantitational Microscopy for 3-D measurement and has researched into the effect of dioxins and hormone disruption on the foetus. He is currently President of the Royal Microscopical Society and past-President of the International Society for Stereology. He was the General Editor of the *Journal of Microscopy* between 1985 and 1991.

Xuewen Huang is an Assistant Scientist in Agronomy at the Institute of Desert Research, Lanzhou, China with interests in sustainable agriculture, desertification, agroforestry, wind erosion and soil quality. Since 1995 he has been Visiting Researcher in the United States, first in the Agroecology Laboratory, College of Agriculture and Life Sciences, Cornell University with Professor David Pimentel and more recently at the USDA–ARS, Wind Erosion Research Unit, Manhattan, Kansas.

Paul Johnston obtained his Ph.D. at the University of London (UK) on 'Aspects of Selenium toxicity in *Daphnia Magna*'. In 1987 he established the research laboratories for Greenpeace International within the University of London, and has continued as principal scientist to the group since its move to the University of Exeter, Great Britain. Dr Johnston co-ordinates the research activities and represents the group at a national and international level.

William Luneburg received his J.D. from Harvard University and has been Professor of Law at the University of Pittsburgh (USA) since 1978. His first position was with the US Environmental Protection Agency, which he joined in 1971, and he has also practised as an attorney. He is an expert on environmental law and, in addition to his numerous reports and journal articles, has recently co-authored *The Clean Air Permitting Manual* which deals with the various permit and preconstruction review programmes for air quality under the US Clean Air Act.

Lars Fogh Mortensen is responsible for indicators, state of the environment reporting and OECD environmental performance reviews in the Danish Environmental Protection Agency, Ministry of Environment and Energy. From May 1994 to May 1996 he was the focal point on indicators of sustainable development in the Secretariat of the United Nations Commission on Sustainable Development (CSD). Mr Mortensen holds a M.A. in Economics from the University of Copenhagen.

Alain Neyt is an engineer specialising in safety techniques, risk analysis and disaster management. He received a Master's Degree in Human Ecology from the Free University of Brussels, specialising in environmental impact assessment.

Eddy Nierynck graduated as a molecular biologist from the Free University of Brussels. He joined the Human Ecology Department in 1990. As a scientific staff member of the Programme in Packaging Management, his research activities focused on life-cycle assessment. Currently, he is stationed in Hanoi as co-director of a European Union project, targeted at 'Capacity Building for Environmental Management in Vietnam', while remaining a staff member of the Human Ecology Department.

Tapio Pento holds a Professorship in the Department of Environmental Management, School of Business and Economics, University of Jyvaskya, Finland. He was awarded his doctorate by the University of Michigan and before his current appointment held positions in industry and at the University of Technology, Lappeenranta (Finland).

David Pimentel is Professor of Insect Ecology and Agricultural Sciences at Cornell University (Ithaca, New York) where he received his Ph.D. degree. His research interests cover a wide span, and include population ecology, genetics, economic and ecological aspects of pest control, energy use and conservation, genetic engineering, sustainable agriculture, soil and water conservation, natural resource management and environmental policy. Dr Pimentel has more than 470 scientific publications to his name, of which 19 are books. He has served on numerous national and governmental committees.

Marcia H. Pimentel is a Senior Lecturer in the Division of Nutritional Sciences, College of Human Ecology at Cornell University (Ithaca, New York). She received her university training in the College of Home Economics also at Cornell. She has published 30 scientific papers, including three books.

David Santillo is a marine and freshwater biologist who obtained his Ph.D. at the University of London, Great Britain. He is now a research scientist with the Greenpeace Research Laboratories at the University of Exeter, working on aspects of plankton ecology, industrial chemistry and environmental contamination. Dr Santillo has recently been involved in research projects in Ecuador, India and a number of Mediterranean countries.

Stefan Schaltegger, Ph.D., is an Assistant Professor at the Centre of Economics and Business Administration, University of Basel, Switzerland. His academic interests include corporate environmental management, environmental public policy, environmental and spatial economics. He is a member of the environmental accounting network of the US Environmental Protection Agency, the working group on Life-Cycle Impact Assessment for the European Society for Environmental Toxicology and was the head of the co-ordinated projects on eco-controlling, life-cycle assessment and environmental accounting of the Environmental Priority Programme of the Swiss National Science Foundation (1992–5).

Ruth Stringer is a senior scientist at the Greenpeace Research Laboratories, based at the University of Exeter, Great Britain. Along with Paul Johnston, she has been instrumental in establishing the laboratories under Greenpeace International, particularly the analytical capabilities of the group. Her expertise is in the environmental effects of chemicals, and in international policy.

Martin Tampier studied Environmental Engineering at the Technical University, Berlin (Germany). His contribution is based on his Masters thesis on the topic of *The UN and its Possible Future Role in Environmental Conflict Management* which he wrote for his Masters Degree in Human Ecology at the Free University of Brussels (Belgium).

Aviel Verbruggen received his training in engineering and economics at Louvain, Antwerp and Stanford universities. Currently, he holds a Professorship at the University of Antwerp (Belgium) and is President of STEM (Research Centre on Technology, Energy and the Environment). His main interest is in the field of energy economics and he has published extensively in such policy-related areas as the planning of power expansion, electricity pricing and market structures, energy conservation and environmental policy planning. He is the author of over two hundred reports, books and articles and is the project leader for the state-of-the-environment reports on Flanders.

Max Vojtísek, Ph.D., is a Senior Scientist at the National Institute of Public Health, Prague (Czech Republic), working in the field of experimental toxicology. By training he is a biologist with specific interests in animal physiology. He is a Visiting Professor at the Charles University of Prague where he lectures on Toxicology.

Hildegard De Weerdt holds a Doctoral degree in Human Ecology and a Bachelors degree in Botany, both from the Free University of Brussels (Belgium). Her doctoral research is in the field of 'the Application of Environmental Auditing in Environmental Protection'. She has a certificate of the Arthur D. Little Auditing Training Course and is the co-ordinator of the Environmental auditing module in the European Step Course. She is at present environmental consultant at Ecdos NV (Belgium).

Willy Weyns is a Professor and Environmental Consultant in the Department of Human Ecology at the Free University of Brussels. He holds a doctorate in Hydro-Geology from the University of Bordeaux and has interests in sustainable water management, environmental policy assessment, global change and cultural interactions.

James Winpenny obtained his M.A. in Economics from Cambridge University (UK) in 1962 and his M.Phil. from the University of East Anglia (UK) in 1970. With a career that has embraced academia, business and government service, at the time of writing he was Research Fellow at the Overseas Development Institute, London where he was engaged in research and related consultancy in the field of environmental economics. Mr Winpenny has an international reputation for his work on economic valuation, environmental policy and economic aspects of water management and has published extensively in these areas. He is a member of the WHO's Task Force on Tropical Diseases and the Environment.

Emiel Wubben, Ph.D., studied Economics and Philosophy at the Erasmus University and is currently Assistant Professor of Strategic Management and Business Environment within the Department of Strategic Management and Business Environment, Rotterdam School of Management, Erasmus University (The Netherlands). He was the winner of the Dorfman Dissertation Award, 1993. His main research specialisms are the management of uncertainty and environmental management.

PREFACE AND ACKNOWLEDGEMENTS

Environmental management draws its knowledge base from across the spectrum of disciplines – the natural, social and medical sciences, the humanities and engineering. It aims to maintain a harmonious relationship between the environment and human society, and in its approach to this adopts a holistic, interdisciplinary stance. Since value judgements are an integral part of environmental management, it is as much an art as a science in its methodology and application.

The growth of interest in environmental management is relatively recent. It reflects a widely held perception of accelerating environmental deterioration caused by the pressure of human activities, as evidenced by worsening problems of pollution and the destruction of natural landscapes and habitats. These concerns can be traced back to the 1960s, when the interconnectedness of nature was vividly demonstrated by the way in which seemingly benign activities such as the chemical control of pests could, by diffusing through the food chain, produce adverse environmental effects in regions ostensibly untouched by man's activities.

As our knowledge of the global environment has grown, other worrying effects have come to light. The emission of greenhouse gases is linked to global warming and climate change. Although we do not fully understand the probable effects of this, it may well result in greater temperate aridity and so jeopardise the world grain supply, with potentially disastrous consequences. Moreover, resultant changes in sea level could submerge major coastal sites of population.

There is also the well-established connection between CFC emissions, the depletion of upper atmosphere ozone, and increased ultra-violet radiation at the planet's surface. This has negative implications not only for human health but also for the well-being of other species. Likewise, the destruction of the tropical rain forest is seen as a grave threat to biodiversity and the world's gene pool. The fact that these hazards are the subject of internationally agreed measures of amelioration (albeit implemented with variable commitment by individual countries) testifies to the potential gravity of global warming, ozone depletion and loss of biodiversity.

These global issues also raise concerns at the level of ecosystems. The effects of modern agricultural practices on environmental quality are a case in point. Pesticide and fertiliser residues pollute the groundwater; animal and plant habitats are destroyed as hedgerows are removed and wetlands drained in the interest of intensive cultivation; soil structure is broken down, creating problems of soil erosion. Now, in addition, intensive rearing of plants and animals is even causing concern for the safety and wholesomeness of the food produced. Populations are no longer willing to accept assurances from experts that genetically engineered crops are safe, or that it is right to feed natural herbivores, such as cattle, protein supplements derived from the rendered remains of other animals.

Of course it is not only agriculture that is problematic. Urban living and its associated activities can be just as destructive of the environment; not least, the creation of built environments where residential, commercial and industrial areas and communications infrastructures either obliterate or radically change pre-existing landscapes and ecosystems. Moreover, urban systems depend upon the mobility of people and goods for their effective functioning, and so create the traffic problems associated with further detrimental effects on the environment. The problems caused by excessive use of energy and natural resources in production and consumption, and their implications for future generations, also have to be tackled. Measures to ensure effective waste disposal and the curbing of air and water pollution are vital for the maintenance of environmental quality.

When viewed over a longer time-scale, however, the environmental picture is somewhat less gloomy. For example popular coverage might lead one to suppose that human activity is the only cause of climatic change: the evidence does, after all, appear compelling, with the atmospheric content of the major greenhouse gas, carbon dioxide, having risen progressively since the start of industrialisation in the seventeenth century. But the record also shows that the world's climate has fluctuated markedly both in the recent geological and even historical past, and scientists still disagree on whether the rise now observed in global temperature should be attributed solely to greenhouse emissions. That this rise may be part of a natural progression readily absorbed by global systems cannot be ruled out at this stage.

It is also worth bearing in mind that ever since the domestication of plants and animals and the discovery of fire, human beings have moulded their natural surroundings to suit their purposes. During the medieval period, for instance, much of the forest that once covered the continent of Europe was destroyed (a process analogous with the present destruction of the tropical rain forest), without any apparent harmful consequences in the long term. Moreover, it was the agricultural and industrial revolutions that created those environmental conditions we consider 'natural' and with which we are comfortable. It is debatable, in other words, whether any truly natural landscapes and ecosystems remain – they have all to a greater or lesser extent come under human influence.

For most of human history our attitude to the environment has been purely exploitative: nature was there to be conquered, and the resource endowment to be used in the furtherance of human development. Little or no attention was paid to the possibility of detrimental environmental impacts – indeed, in most instances these were simply not appreciated because the complex relationships and linkages of environmental systems were not understood. It is only in this century that this attitude – of man as the conqueror of nature – has changed (indeed in Eastern Europe it persisted, with disastrous consequences, right up until the demise of communism). We now think more in terms of stewardship, whereby humans owe a duty of care to the environment, and in terms of sustainability. However it is still invariably the case that when choices have to be made, economic self-interest wins the day.

The broad scope of environmental management creates its own particular problems. The information on which it relies is scattered across disciplines isolated from one another by the traditional boundaries that demarcate major branches of academic endeavour. It follows from this that relevant advances in the natural sciences may not be appreciated by those working from a social science perspective, and so on. It is therefore a major objective of this book to bring together the expertise found within the diverse fields of environmental management, with the aim of providing an accessible overview of its content and methods. The treatment is biased towards environmental management as practised at the regional level – the so-called meso-scale. Global issues such as climate change and loss of biodiversity lie outside the scope of this book and so receive only incidental mention.

The idea for this book came initially from the involvement of the four editors in environmental training programmes in Eastern Europe, and a book was duly published by the Free University of Brussels Press in 1993. This publication is a revised, improved and extended edition of that earlier version and is presented in three volumes. The theoretical principles of environmental management are illustrated with the use of up-to-date examples and case studies, and self-assessment questions are included to aid students who may wish to use it as a textbook. It should also be of interest to policy-makers and researchers seeking information about the management of today's environmental problems.

Volume 1 considers the instruments for environmental management under four main headings – predictive and scientific instruments; economic instruments; legal instruments; and instruments for environmental communication and education. It covers not only relatively long-established instruments of management, such as environmental impact assessment and risk assessment, but also introduces more recently developed approaches such as material flow analysis and life-cycle assessment.

Volume 1 aims at up-to-date, comprehensive coverage and includes discussion of such important topics as the concept of sustainable development, environmental legislation in the European Union and the USA, and the management of environmental conflict.

Volume 2 is devoted to the environmental management of compartments, stressors and sectors. It covers the impact of population and the way environmental information is processed and interpreted through the filter of human culture. Soil, air and water are the environmental compartments referred to in this volume. They are subject to stress through overuse and pollution, and the manner in which these stressors should be managed constitutes a major strand of enquiry. The sectors referred to are agriculture, forestry, industry, transport and tourism, and how these should be managed in the interests of preserving environmental quality.

The theoretical and practical considerations involved in the appropriate management of major natural ecosystems types – wetlands, tropical forests, desert areas, the coastal margin, river and inland water environments – are discussed in Volume 3. These are, of course, 'natural' ecosystems only in a relative sense; the question of management arises precisely because of human impacts. Of equal importance are those environments created entirely by human activities, and in recognition of this Volume 3 also deals with rural and urban environments, as well as human ecosystems under threat and the management of archaeological sites.

The three volumes of this textbook contain over sixty chapters written by more than eighty authors. This was therefore a large project which would have been impossible without the support and active contributions of many colleagues whose names are not mentioned in the individual chapters. We would like to thank especially the secretarial staff of the Human Ecology Department, Free University of Brussels (VUB) for their excellent work; especially Mr Glenn Ronsse, who was responsible for the final formatting and overall secretarial management of the project. Sincere thanks are due to the peer reviewers of the chapters.

For Volume 1 these were: G.H. Addink, Universiteit Utrecht, Netherlands; M. Ahlberg, University of Joensuu at Savonlinna, Finland; G. Barbiroli, Università degli Studi di Bologna, Italy; H. Bergman, European Commission, Directorate General XI, Brussels, Belgium; M.H. Birley, Liverpool School of Tropical Medicine, Liverpool, United Kingdom; A.A. Brennan, University of Western Australia, Perth, Australia; M. Brusasco-McKenzie, European Commission, Directorate General XI, Brussels, Belgium; P. Calow, University of Sheffield, United Kingdom; R.A. Carpenter, Charlottesville, Va., USA; D. Ceuterick, VITO, Mol, Belgium; J.H.J. Copius Peereboom-Stegeman, University of Nijmegen, Netherlands; E. Corijn, Free University of Brussels, Belgium; M.A. Curran, Risk Reduction Engineering Laboratory, Cincinnati, O., USA; T.M. De Jong, Technische Universiteit Delft, Netherlands; N. De Sadeleer, Université Saint Louis, Brussels, Belgium; J. Delbeke, European Commission, Brussels, Belgium; F.J. Duijnhouwer, RMNO, Netherlands; H. Dumont, University of Ghent, Belgium; M.K. Garrett, Queen's University of Belfast, Northern Ireland, United Kingdom; P. Glasbergen, Universiteit Utrecht, Netherlands; L. Goeteyn, MINA-Raad, Brussels, Belgium; H. Greppin, Université de Genève, Switzerland; M. Hale, London Guildhall University, United Kingdom; P. Hardi, International Institute of Sustainable Development, Winnipeg, Manitoba, Canada; R. Harding, University of New South Wales, Kensington, Australia; S.A. Healy, Middlesex University, London, United Kingdom; R. Heremans, VN-Verbindingsbureau, Brussels, Belgium; R. Hillary, Imperial College, London, United Kingdom; D. Holemans, University of Ghent, Belgium; G. Huppes, Leiden University, Netherlands; A. Kiss, Centre National de la Recherche Scientifique, France; J. Kleinjans, Rijksuniversiteit Limburg, Maastricht, Netherlands; E. Kuerzinger, Deutsche Gesellschaft für Technische Zusammenarbeit, Bonn, Germany; B.L. Lamb, National Ecology Research Centre, Fort Collins, Colo., USA; W. Lang,

Embassador of Austria, Brussels, Belgium; N. Lee, Manchester EIA Centre, United Kingdom; R. Lesthaeghe and A. Libotton, Free University of Brussels, Belgium; M. Linddal, Royal Danish Ministry of Foreign Affairs, Copenhagen, Denmark; C.J. Lugtmeijer, Procter and Gamble, Brussels, Belgium; S. Lupkitaro, Kasetsart University, Bangkok, Thailand; J. Martinez Alier, Autonomous University of Bellaterra, Barcelona, Spain; R. Mitchell, Department of Political Science, Eugene, Oreg., USA; I. Moffatt, University of Stirling, Scotland, United Kingdom; T. O'Riordan, University of East Anglia, Norwich, Norfolk, United Kingdom; A.A. Olsthoorn, Vrije Universiteit Amsterdam, Netherlands; C. Perrings, University of York, Heslington, United Kingdom; J. Robinson, Institute for International Affairs, Directorate General XI, Brussels, Belgium; J. Schütz, Swiss Federal Institute of Technology, Zürich, Switzerland; L.P. Suetens, Catholic University of Leuven, Belgium; N. Sybesma-Knol, Free Univer-sity of Brussels, Belgium; J. Taylor, National Ecology Research Center, Fort Collins, Co., USA; E. Tellegen, Universiteit van Amsterdam, Netherlands; A. Tukker, TNO, Centre for Technology and Policy Studies, Apeldoorn, Netherlands; A. Vadineanu, University of Bucharest, Romania; N. Van Larebeke, VMM, St Genesius Rode, Belgium; P. van Meegeren, Universiteit Wageningen, Netherlands; P. Verbeeck, MINA-Raad, Brussels, Belgium; H. Verbruggen, Vrije Universiteit Amsterdam, Netherlands; F. Veroustraete, VITO, Mol, Belgium; H. Verschure, Catholic University of Leuven, Belgium; A. Vranckx, Free University of Brussels, Belgium; A. Wals, Agricultural University of Wageningen, Netherlands; D. Wilcock, University of Ulster, Coleraine, United Kingdom; G. Wouters, VITO, Mol, Belgium; C.W. Yang, Clarion University, Pennsylvania, USA.

We are deeply indebted to Mr K. Van Scharen, executive director of the VUB University Press, publishers of the first edition of this textbook in 1993, who was most supportive in the publication of this second edition. We thank Routledge for their guidance during the establishment of this project and in its final production.

The publishers have made every effort to contact copyright holders of material reprinted in *Environmental Management in Practice: Volume 1*. However, this has not been possible in every case and we would welcome correspondence from those individuals/companies we have been unable to trace.

LIST OF ABBREVIATIONS

AAC	Allowable Annual Cut
AADI	Adjusted Acceptable Daily Intake
ADB	Asian Development Bank
ADI	Acceptable Daily Intake
AE	Acidification Equivalents
AEDR	Alternative Environmental Dispute Resolution
ALARA	As Low As Reasonably Acceptable
ALCA	Average Life-Cycle Assessment
AMA	Accra Metropolitan Assembly (Ghana)
BACT	Best Available Control Technology
BATNEEC	Best Available Technologies not Entailing Excessive Cost
BOD	Biological Oxygen Demand
BP	British Petroleum
BS	British Standard
BSI	British Standards Institution
BUWAL	Bundesamt für Umwelt, Wald und Landschaft
BW	Body Weight
C & C	Command and Control
CAFE	Corporate Average Fuel Efficiency
CAG	Community Advisory Group
CBA	Cost–Benefit Analysis
CEAA	Canadian Environmental Assessment Agency
CEC	Commission of the European Communities
C.E.C.	Commission on Environmental Co-operation
CEE	Central and Eastern Europe
CEPS	Customs, Excise and Preventive Service (Ghana)
CEQ	Council on Environmental Quality
CERCLA	Comprehensive Environmental Response, Compensation, and Liability Act
CEU	Commission of the European Union
CFC	Chlorofluorocarbon
CFR	Code of Federal Regulations
CIA	Cumulative Impact Assessment
CITES	Convention on International Trade in Endangered Species (of Flora and Fauna)
CLEA	Centre Leo Apostel (Belgium)
CM	Contribution Margin
CMP	Co-operative Monitoring Programme
CNRS	Centre National de Recherches Scientifiques (France)
COD	Chemical Oxygen Demand
CoEC	Council of the European Commission
Com	Commission
COP	Conference of the Parties of the Convention
CSA	Canadian Standards Association
CSD	Commission for Sustainable Development (United Nations)
CSE	Centre for Science and Environment (New Delhi)
CVM	Contingent Valuation Method
DDT	Dichlor-Diphenyl-Trichlorethane
DEIA	Division of Environmental Information and Assessment
DG	Directorate General
DIP	De-Inked Pulp

DIT	Dual Inheritance Theory
DM	German Marks
DNA	Deoxyribonucleic Acid
DOA	Dioctyladipate
DOE	Department of the Environment
DPCSD	United Nations Department for Policy Co-ordination and Sustainable Development
DPSEA	Driving Forces–Pressures–State–Effects–Actions
DRIPS	Data-Rich-Information-Poor-Syndrome
DRS	Deposit Refund Scheme
DSB	Dispute Settlement Body
DSR	Driving Force–State–Response framework
EA	Environmental Assessment
EAF	Environmental Auditing Forum
EAR	Environmental Auditing Roundtable
EARA	Environmental Auditors Registration Association
EARP	Environmental Assessment Review Process
EC	European Communities
EC50	Median Effective Concentration
ECom	European Commission
ECE	Economic Commission for Europe
EcIA	Economic Impact Assessment
ECJ	European Court of Justice
ECLA	Economic Commission for Latin America
ECOSOC	Economic and Social Council
ECR	European Court Report
ECU	European Currency Unit
EDC	Ethylene dichloride
EE	Environmental Education
EEC	European Economic Community
EEZ	Exclusive Economic Zone
EI(s)	Economic Instrument(s)
EIA	Environmental Impact Assessment
EIS	Environmental Impact Statement
ELU	Environmental Load Units
EMAS	Eco Management and Audit Scheme
EMPEM	European Masters Programme on Environmental Management
EMS	Environmental Management Systems
ENRIN	Environment and Natural Resources Information Network
EP	European Parliament
EPA	Environmental Protection Agency
EPM	Eco-rational Path Method
EPS	Environmental Priority Strategies
ERA	Ecological Risk Assessment
ERC(s)	Emission Reduction Credit(s)
ETA	Event Tree Analysis
EU	European Union
EURATOM	European Community for Atomic Energy
EUROSTAT	Statistical Office of the European Commission
FAO	Food and Agriculture Organisation
FEARO	Federal Environmental Assessment Review Office
FISD	Framework for Indicators of Sustainable Development
FLPMA	Federal Land Policy and Management Act
FONSI	Finding Of No Significant Environmental Impact
FTA	Fault Tree Analysis
GAEC	Ghana Atomic Energy Commission
GATT	General Agreement on Tariffs and Trade
GDP	Gross Domestic Product
GEF	Global Environment Facility
GIS	Geographical Information System
GL	Guide Level
GNI	Gross National Income
GNP	Gross National Product
GRID	Global Resource Information Base
GSB	Ghana Standards Board
GSP	Green School Project
GV	Guideline Value
GWP	Global Warming Potential
Hazop	Hazard and Operability Studies
HDI	Human Development Index
HDPE	High Density Polyethylene
HMEMM	Human, Mechanical, Environmental, Material and Managerial
HUD	Housing and Urban Development
IARC	International Agency for Research on Cancer

ICC	International Chamber of Commerce
ICES	International Council for Exploration of the Seas
ICJ	International Court of Justice (The Hague, the Netherlands)
ICLEI	International Council for Local Environmental Initiatives
IEA	Institute for Environmental Auditing
IETO	Interagency Environmental Technologies Office
IGAE	Intergovernmental Agreement on the Environment (Australia)
IGBP	International Geosphere Biosphere Programme (ICSU)
IHE	Institute for Hygiene and Epidemiology (Brussels, Belgium)
IIA	Institute for Internal Auditing
IMF	International Monetary Fund
INGP	International Network of Green Planners
IPCC	Intergovernmental Panel on Climate Change
IS	Irish Standard
ISO	International Standards Organisation
IT	Information Technology
ITO	International Trade Organisation
IUCN	International Union for the Conservation of Nature
IWC	International Whaling Commission
JTP	Joined Time Projection
LAER	Lowest Achievable Emission Rate
LC50	Median Lethal Concentration
LCA	Life-Cycle Analysis
LCA-D	Life-Cycle Assessment with Delta Analysis
LCANET	European Network for Strategic Life-Cycle Assessment Research and Development
LCC	Life-Cycle Costing
LCI	Life-Cycle Inventory
LDCs	Less Developed Countries
LOAEL	Lowest (experimentally) Observed Adverse Effect Level
LOEC	Lowest Observed Effect Concentration
LOS	Law of the Sea Convention
LP	Linear Programming
LPG	Liquid Petroleum Gas
MAB	Man And the Biosphere (UNESCO)
M.A.C.	Marginal Abatement Cost
MAC	Maximum Admissible Concentration
MACT	Maximum Achievable Control Technology
MCA	Multi-Criteria Analysis
MDC	Marginal Damage Cost
MEA	Multilateral Environmental Agreement
MFM	Material Flow Model
MFN	Most-Favoured Nation
MGA	Mutual Gains Approach
MIUS	Material Intensity Per Unit of Service
MMPA	US Marine Mammal Protection Act
MP	Marketable Permits
MU	Monetary Unit
NAFTA	North American Free Trade Agreement
NEC	No Effect Concentration
NEMA	Network on Environmental Management and Auditing
NEPA	National Environmental Policy Act
NFMA	National Forest Management Act
NGO	Non-Governmental Organisation
NIMBY	Not-In-My-Back-Yard
NOAEL	No Observed Adverse Effect Level
NOEC	No Observed Effect Concentration
NOEL	No Observed Effect Level
NPV	Net Present Value
NSR	New Source Review
NZMOE	New Zealand Ministry of the Environment
ODP	Ozone Depletion Potential
ODS	Ozone-Depleting Substance
OECD	Organisation for Economic Co-operation and Development
OJ	Official Journal (of the European Communities)
OMB	Office of Management and Budget
OPEC	Organisation of Petroleum Exporting Countries

P & ID	Process & Instrumentation Diagram
PCA	Permanent Court of Arbitration
PCB	Polychlorobiphenyl
PCT	Polychlorotriphenyl
PET	Polyethylene Terephtalate
PiC	Problem-in-Context
PLA	Product-Line Analysis
PP	Precautionary Principle
PPA	Post-Project Analysis
PPMs	Process and Production Methods
PQLI	Physical Quality of Life Index
PSIR	Pressure–State–Impact–Response framework
PSR	Pressure–State–Response framework
PU	Physical Units
PVC	Polyvinylchloride
QMV	Qualified Majority Voting
QSAR	Quantitative Structure Activity Relationship
RA	Risk Assessment
R & D	Research and Development
RFPC	Reasonable Further Progress Curve
RIVM	National Institute of Public Health and the Environment (The Netherlands)
RMCL	Recommended Maximum Contaminent Levels
RONA	Return On Net Assets
SCOPE	Scientific Committee on Problems of the Environment
S.E.A.	Single European Act
SEA	Strategic Environmental Impact Assessment
SEEA	System of Integrated Environmental and Economic Accounting
SET	Synthetic Evolutionary Theory
SETAC	Society for Environmental Toxicology and Chemistry
SFA	Substance Flow Analysis
SGS	Société Générale de Surveillance (Ghana) Ltd.
SIA	Social Impact Assessment
SMR	Standardised Mortality Ratio
SNA	System of National Accounts
SoE	State of the Environment
SPA	Special protection area
SPOLD	Society for the Promotion of LCA Development
SPS	Sanitary and phytosanitary
SWOT	Strengths, Weaknesses, Opportunities and Threats
TA	Technology Assessment
T.B.T.	Technical Barriers to Trade
TBT	Tributyl Tin
TECM	Transnational Environmental Conflict Management
TEU	Treaty on European Union
TMC	Total Material Consumption
TOC	Total Organic Carbon
TRS	Total Reduced Sulphur
TSP	Total Suspended Particulates
TT	Technology Tribunal
TWA	Time Weighted Average
UF	Uncertainty Factor
UK	United Kingdom
UKEA	United Kingdom Environmental Agency
UN	United Nations
UNCED	United Nations Conference on Environment and Development (Rio de Janeiro, Brazil 1992)
UNCLOS	UN Convention on the Law of the Sea
UNCTAD	UN Conference on Trade and Development
UNDP	United Nations Development Programme
UNEP	United Nations Environment Programme
UNESCO	United Nations Educational, Scientific and Cultural Organisation
UNGA	UN General Assembly
UNICA	Association of Caribbean Universities
UNSD	United Nations Statistical Division
UNU	United Nations University
UPA	Units of Polluted Air
UPW	Units of Polluted Water
US(A)	United States (of America)
USEPA	United States Environmental Protection Agency

USGAO United States General Accounting Office
USSR Union of Soviet Socialist Republics
UVI University of the Virgin Islands
VCM Vinyl Chloride Monomer
VOC Volatile Organic Compound
WB World Bank
WCED World Commission on Environment and Development
WEED World Economy, Ecology and Development Association
WHO World Health Organisation
WICEM World Industry Conference on Environmental Management
WMO World Meteorological Organisation
WRI World Resources Institute (New York)
WTO World Trade Organisation
WWF World Wildlife Fund

LIST OF UNITS

Prefixes to the names of units

M	mega (10^6)
k	kilo (10^3)
d	deci (10^{-1})
c	centi (10^{-2})
m	milli (10^{-3})
μ	micro (10^{-6})
n	nano (10^{-9})
p	pico (10^{-12})
f	femto (10^{-15})

Units

Å	Ångstrom (0.1 nm)
atm	atmosphere
°C	degree Celsius or centigrade
cal	calorie
d	day
g	gram
h	hour
ha	hectare
J	joule
K	degree absolute (Kelvin)
l	litre
m	metre
M	molar (mol/litre)
min	minute
Pa	pascal (unit of pressure; 100kPa = 1 bar)
quads	measure of energy (1 quad = 10^{15} BTU or 337×10^{18} joules)
s	second
t	tons
V	volt
W	watt
yr	year

Other abbreviations

bn	billion (10^9)
kg_{bw}	kilogram body weight
ln	logarithm (natural, base e)
log	logarithm (common, base 10)
mn	million (10^6)
n or N	total number of individuals or variates
ppb	parts per billion
ppm	parts per million
s^2	sample variance

PART I

ANALYSING AND MANAGING ENVIRONMENTAL PROBLEMS

INTRODUCTION

Paul A. Compton

This book would be incomplete if it were to ignore the environmental impact of a growing world population and the manner in which environmental information is filtered through the medium of human culture. The middle of the seventeenth century marked the beginnings of the modern rise of population. Initially, growth was restricted to Europe and the overseas areas of European settlement, in particular North America, but has subsequently spread widely to affect the continents of Africa, Asia and Latin America this century. In the process, the population of the world has expanded from an estimated half a billion in 1650 to the current figure of 6 billion, equivalent to a twelvefold increase over a span of little more than thirteen generations.

The theory of demographic transition suggests that populations pass through three distinct stages of development. According to this model, in the premodern period numbers are essentially static with birth and death rates broadly equal; the second stage is associated with modernisation and is marked by a period of rapid population growth, while the third stage heralds renewed stability and low birth and death rates in the post-modern period. Viewing the world through the spectacles of this model helps explain the present variability in global population growth rates; the countries of the developed world (Europe, North America, Oceania, and the newly industrialised countries of Asia) have entered the post-modern phase and have low or even negative rates of population growth, whereas much of the developing world is still modernising and as a consequence rates of population growth, although decelerating in many countries, remain comparatively high.

However, the real significance of the demographic transition model is the implication that once modernisation is complete, the population of developing countries will also settle down as low rates of increase are re-established. In the process, the world population should also stabilise. The model is not sufficiently well defined to forecast when or at what level of population this might occur, but current trends suggest that stabilisation could well happen sometime during the second quarter of the next century when the world population is expected to be in the region of 10–12 billion.

By itself, a world population of this magnitude is bound greatly to increase the pressure on the environment; but it should also be borne in mind that population-induced pressure on the environment is as much a function of technology and living standards as of simple numbers. The affluence of the western world has been created by applying technology to the global resource base at the expense of fossil fuel exploitation and the creation of pollution. The per capita environmental impact generated by these countries is therefore many times greater than that generated by the population in the countries of the developing world. This poses a delicate problem for world environmental management because the people of the developing world also aspire to high living standards, and even though the environmental consequences could be grave, it is an aspiration that considerations of equity make impossible to deny. Hence, in future decades the environment will be in double jeopardy – jeopardy from a possible doubling of world population over the next fifty or so years and from the consequences of growing affluence in those countries where living standards are relatively low at the present time.

Pimentel and his co-authors paint an especially gloomy picture of the future in their chapter 'Impact of a Growing Population on Natural Resources'. They postulate that the world's environment is already overstretched to meet the demands of the present population, let alone those that will arise in future. Man already utilises about 50 per cent of all solar energy captured by photosynthesis, and all the fertile land suitable for arable farming is already under cultivation. On top of this, water resources have reached the limits of their utilisation in many regions while the proven reserves of fossil fuels have exploitation horizons that should be measured in decades and not centuries. In their view, there is little room for improving resource management because of the poor prospects of overcoming natural constraints with our present level of technical knowledge. The clear implication of their argument is that the planet only has a viable future, in the sense of providing an environment of acceptable quality together with adequate wealth and nourishment for all the people, if a drastic reduction in world population, say to between 1 and 2 billion, can be achieved. Although this conclusion may follow logically from their initial axioms, it is a much more demanding target than zero population growth and, given the human instinct to reproduce and survive, would not appear to be a prescription that could be put into practice, now or in the foreseeable future.

It is, of course, important that the world moves towards the creation of a sustainable environmental future, but the approach to this must be carefully balanced and the Pimentel standpoint may be overstated. History, indeed, is littered with the overblown predictions of the dire consequences of excessive population. As early as 1798, Malthus, a Church of England parson and early economist, postulated in his celebrated *Essay on Population* that, within a generation or so, population was bound to outstrip food supply if restraint was not exercised. By coincidence, he was writing at the exact time the population of Europe was entering its phase of explosive growth which, far from leading to crisis, was actually accompanied by an improvement in nutrition as grain flooded in from the New World, and rising living standards. Malthus failed to appreciate two things – the positive effects of the industrial and agricultural revolutions with the expansion of world trade, and the fact that population in itself can be a resource and not a liability.

Moving nearer to the present, the Report of the Club of Rome on the *Limits to Growth*, which was published in 1972, predicted the imminent exhaustion of a range of non-renewable resources and has been equally pessimistic as a forecast. The Report gained considerable initial credence because of the difficulties in the oil market at the time with shortage of supply leading to a quadrupling of the price. In some quarters this was interpreted as evidence that world oil supplies were on the point of exhaustion, but the gloom was misplaced. As is now well understood, the shortage was actually brought about by a political decision on the part of OPEC artificially to restrict output. The consequence of this was intensive international activity to discover and bring into production new sources of oil, including the North Sea. Energy conservation measures were also implemented and in a relatively short time the world moved from a situation of oil shortage to oil glut, with an inevitable collapse of the price.

The lesson from these examples is that the human ability to substitute raw materials and develop new technologies when the economic incentive is strong enough is immense and should not be understated. The danger with excessively gloomy scenarios is that, although they keep the conferences busy, the jets flying and the hotels full, they do so at the risk of being counter-productive. It is hard to reconcile the notion that world population is now nudging the environmental limits when global living standards continue to improve – including those in the two largest countries, China and India, with their 2.5 billion inhabitants. Moreover, the rising life expectancies revealed by United Nations statistics in the vast majority of countries are in contradiction to the notion that illness and disease is prevalent and widespread. It may be argued, of course, that these are only temporary improvements, that there is still much poverty and malnutrition in the world, and that in the longer term we shall have to face up to the reality of environmental limits and constraints. But pushing the 'green agenda' too hard clearly risks losing credibility; it is a message that may simply not be in sympathy with general perceptions.

Perceptions about the environment not only reflect objective scientific facts but also society's value and belief systems that come from our ideas about religion, morality, aesthetics etc.; it is these latter systems that constitute the cultural filter discussed by Weyns in his contribution. Because values and beliefs vary from society to society and also through time, perceptions about the environment similarly vary. Moreover, some would even argue that scientific findings themselves, although outwardly objective, are interpreted through the same cultural filter and are not value free. If we accept these arguments, it follows that effective management of the environment must be fully informed by these more intangible factors.

Examples of the importance of the *cultural filter* come readily to mind. In our type of society, for instance, the urban perception of the environment differs radically from that of country people. It is seen in the town dweller's desire to maintain a rural landscape that conforms with some traditional image of rural living. It is also exemplified in attitudes to wild life and animal welfare in general, with the anti-hunting and animal rights lobbies being essentially urban phenomena. In like manner, the work of anthropologists has demonstrated how much more in tune with their natural surroundings are primitive compared with modern societies. Perceptions, values and beliefs are, however, fluid and undergo change, and may therefore be moulded in a direction thought to be desirable through education. A case in point would be the way in which attitudes in western society have been transformed in recent decades, with the once dominant notion that the environment should be fully exploited as a natural resource giving way to the idea of stewardship that involves the nurturing and conservation of environmental quality. This does, of course, give rise to the crucial question of who should be the arbiter of desirable change; placing one's faith in the hands of governments and their expert advisers or in public opinion would appear to be equally hazardous as a prospect.

How might we integrate the scientific and society-related aspects of environmental management? Clearly, the scientific method which reduces a problem to its individual components cannot really help us as it is an interdisciplinary, holistic approach that is needed. Tackling this question through a systems methodology also has its disadvantages in the sense that boundaries may be drawn too narrowly. The strength of Wouter de Groot's 'Problem-in-context' framework lies in its ability to overcome both these drawbacks. As an approach, it encourages the investigator to extend the scope of an analysis until the context of a problem is fully defined. It is also holistic in explicitly unifying the *causes* of environmental problems, whose origins rest in society and therefore fall within the purview of the social sciences, and the *analysis* of environmental problems that lie with the physical and biological sciences. Values and perceptions are integrated into the methodology through the definition of a normative context which involves the investigation of whether a particular value perspective is justified.

1

IMPACT OF A GROWING POPULATION ON NATURAL RESOURCES

The challenge for environmental management*

David Pimentel, Xuewen Huang, Ana Cordova and Marcia H. Pimentel

SUMMARY

Rapid human population growth in the world today is a major challenge to environmental management. The current world population is nearly 6 billion – a quarter of a million people are being added daily – and populations in a few countries are doubling every seventeen years. As a result, pollutants are being produced practically everywhere in the world by human activities faster than they can be managed and/or degraded by natural biota and the environment.

In addition to the increasing pollution of the natural environment, land, water, energy, and biological resources are being rapidly degraded. Arable land, for example, is being eroded and abandoned at a rate of more than ten million hectares per year. To compensate for this an increasing amount of replacement land is being claimed from forests for agriculture, and it is the major cause of world deforestation. Some rivers, such as the Colorado River in the western USA, are being pumped dry before they empty into gulfs and oceans. Fossil energy resources, essential for food production, energy, services, and for the transportation of goods, are being depleted, and it is projected that most world oil reserves will be depleted in 30–50 years' time. Of equal concern is the loss of biodiversity, because natural biota are vital to a quality environment. About 150 species are being lost daily at a rate that is 1,000–10,000 times faster than the normal extinction rate.

However, despite these sobering statistics, humans have the intellect to reduce their numbers and to develop and implement effective environmental strategies so that people can live free from poverty and starvation in a sustainable and productive environment.

ACADEMIC OBJECTIVES

The objective of this chapter is to assess the impact of a rapidly growing world population on natural resources, and to determine what can be done to reduce pollution and to conserve land, water, energy, and biological resources. Because the per capita amounts of all essential resources are rapidly declining, there is an urgent need for adopting appropriate conservation measures and for developing sound environmental management strategies.

With regard to the above, various management practices are discussed for a target audience of students of environmental management. These include soil and water conservation, reduced pesticide and fertiliser use, reduced water and energy consumption for agriculture and other activities, and conservation of biodiversity.

The academic objective is twofold: first, to alert students to the unacceptable level of resource depletion and environmental degradation occurring everywhere in the world today in pursuit of a misplaced notion of the 'good life'; and secondly to inform them on up-to-date environmental management practices with which to address the problems and issues referred to above.

INTRODUCTION

A healthy and prudently managed environment is a major benefit to humans and other species. The natural environment is the sustainer of all life on earth. Indeed, humans and all living organisms rely for their very survival on the basic resources which earth's natural environment provides. Paramount among these resources are fertile land, fresh water, energy, and biodiversity. Already the human population is consuming more than 50 per cent of the solar energy captured by all the plant biomass on earth each year (Pimentel *et al.*, 1997a), and humans and other organisms obtain all their food and fibre from this photosynthetic activity.

Current world population is about 6 billion. It is projected to reach 10 billion by 2040, based on the present growth rate of 1.5 per cent per year (PRB, 1995). This large increase should signal a call for taking urgent and appropriate action on the use and management of resources because, clearly, erosion of farmland, overuse of both surface and groundwater, dwindling supplies of finite fossil fuels, and accelerating extinction of plant and animal species imperil the ability of earth's resources to meet the needs of such enormous numbers.

The world food situation is becoming critical. At present more than 2 billion humans are malnourished and live under unhealthy conditions (FAO, 1992a, 1992b; Neisheim, 1993; McMichael, 1993; Maberly, 1994; Bouis, 1995); the number of humans who are diseased is the largest ever; and about 40,000 children die each day from disease and malnutrition (Kutzner, 1991; Tribe, 1994).

Many of these obvious and growing problems emphasise the urgent need to reassess the status of environmental resources. Based on evidence, definitive plans must be developed to improve environmental management now and for the future. Of major importance in this is the control of human population growth in a way that establishes a balance between world population and the carrying capacity of earth's natural resources.

POPULATION AND CONSUMPTION OF RESOURCES

Human behaviour demonstrates a strong will to survive, to reproduce, and to achieve some level of prosperity and a reasonable quality of life. However, individuals as well as societies differ in their views of what they consider a satisfactory lifestyle. Contrasting some aspects of life in the United States, China, and the world reveals many disparities in lifestyles that are most often functions of the quantity of natural resources available per person. Furthermore, most of these basic resources (land, water, energy, and biota) are not unlimited in their supplies and many are finite. With a relentlessly growing human population, both prosperity and quality of life are expected to decline because limited resources must be divided among an increasingly greater number of people (UNFPA, 1991; RS and NAS, 1992).

A substantial population growth has occurred throughout the world in the last few decades (Figure 1.1). The population of the USA has doubled from 130 million to more than 260 million during the last 60 years (NGS, 1995); it is projected to double again to 520 million during the next 60 years, based on the current rate of growth of 1.1 per cent per year (USBC, 1994). In contrast, China currently has a population of 1.2 billion; but, despite the government policy of permitting only one child per couple, it too is growing at a rate of 1.1 per cent (PRB, 1995). China's desirable population is 650 million, which is roughly 50 per cent of her current population (Qu and Li, 1992). The number of people in India today is nearly 1 billion, living on about one-third the land area of the USA or China. However, India's current growth rate of 1.9 per cent per year means that her current population will double in the next 37 years (PRB, 1995). The combined populations of China and India today account for more than one-third of the total world population. Although these populations are projected to double

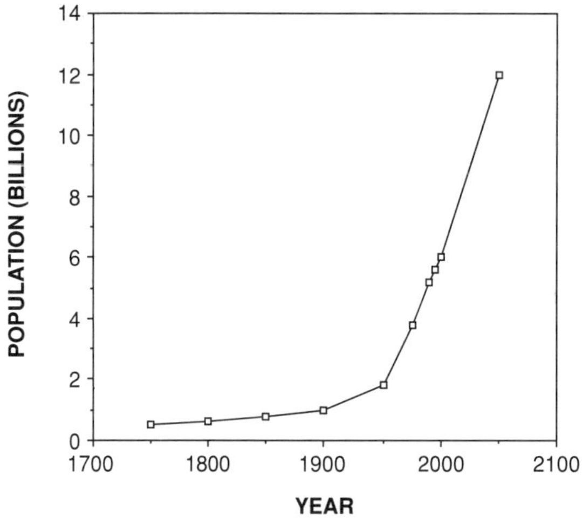

Figure 1.1 World population numbers for the past, present and projected future

Source: Pimentel and Pimentel, 1996

in the next 60 and 37 years respectively, it is doubtful that either population will actually double, because declining resources will put a brake on growth. Considering that most nations are subject to similar constraints of land, water, energy, and biological resources relative to population density, it is unlikely that world population will actually double in the next 50–100 years despite current projection (PRB, 1995).

In addition to constraints on resource availability, high consumption rates also put a limit on resources. For example, each American consumes about twenty-three times more goods and services than the average citizen of a developing country. In particular, an average American consumes about fifty-three times more goods and services than a Chinese citizen (PRB, 1995). Clearly therefore, based both on projections of future resource availability and population growth, it is unrealistic for developing countries to expect to achieve either the US standard of living or that of other highly developed countries. The excessive consumption levels of Americans have led to their dependence on imported natural resources, and this is reflected by the fact that the USA has the largest debt of any nation in the world (USBC, 1994). However, with declining resources throughout the world, the ability to import foreign resources will no doubt be curtailed in time.

Ever since the 1850s Americans have relied increasingly on diverse energy sources, rather than on human power, for their food and forest products. Typically, relatively cheap and abundant supplies of fossil fuel have been substituted for human and draft animal energy. Fossil-based fertilisers and pesticides, as well as fossil-run machinery, have enabled US farmers to minimise the need for human energy in farming activities. It is interesting to note, however, that at present Chinese farmers use more fertiliser and pesticides per hectare than American farmers (Wen and Pimentel, 1992). They also rely on about 1,200 hrs/ha/yr of human labour for grain production, compared with only 10 hrs/ha per year for the US farmer (Wen and Pimentel, 1984; Wen and Pimentel, 1992).

Industry, transportation, home heating, and food production account for most of the fossil energy consumed in the United States (DOE, 1991, 1995a). In China, on the other hand, most of the fossil energy is used by industry, although a substantial amount, approximately 25 per cent, is used for food production (Kinzelbach, 1983; Smil, 1984; Wen and Pimentel, 1992). Per capita consumption of fossil energy in the United States is 8,740 litres of oil equivalents per year, which is more than twelve times that of China (Table 1.1). Furthermore, per capita energy consumption in the USA is nearly seven times the world average – a clear demonstration of over-consumption.

So far the relative affluence enjoyed by most Americans has been possible because of abundant per capita supplies of fertile cropland, water, and fossil energy. But, if the population of the USA continues to grow, resource shortages similar to those now occurring in China and many other countries will become commonplace (Tables 1.1 and 1.2). In fact, there are already indications that the average standard of living in the United States began to decline during the last decade (Fuchs and Reklis, 1992), and that it will continue to decline if the population doubles during the next 60 years (Pimentel and Pimentel, 1996). Already measurable shortages of fertile land, water, and fossil energy are occurring in many regions of the world (Worldwatch Institute, 1992; WRI, 1994), and this should serve as a warning to all concerned.

Table 1.1 Resources used per capita per year in the USA, China and the world to supply basic needs

Resources	*USA*	*China*	*World*
Land			
Cropland (ha)	0.71[a]	0.08[c]	0.27[e]
Pasture (ha)	0.91[a]	0.33[c]	0.57[e]
Forest (ha)	1.00[a]	0.11[c]	0.75[e]
Total (ha)	2.62	0.52	1.59
Water (litres × 10^6)	5.1[b]	0.46[c]	0.64[c]
Fossil fuel			
Oil equivalents (litres)	8,740[b]	700[d]	1,500[f]
Forest products (kg)	1,091[b]	40[c]	70[g]

Notes: (a) USDA (1993), (b) USBC (1994), (c) PRC (1994), (d) Bennett (1995), (e) SSBPRC (1990), (f) Buringh (1989), (g) UNEP (1985)

Table 1.2 Foods and feed grains supplied per capita (kg) per year in the USA, China and the world

Food/feed	*USA*[1]	*China*[2]	*World*[2]
Food grain	100	228	171
Vegetables	105	86	69
Fruit	125	26	57
Meat and fish	137	40	45
Dairy products	247	7	70
Eggs	14	7	6
Fats and oils	28	5	11
Sugar and sweeteners	62	7	19
Total food	818	406	448
Feed grains	663	70	166
Grand total	1,481	476	614
Kcal/person/day	3,644	2,734	2,698

Notes: 1 USDA (1993)
2 Agrostat (1992)

Table 1.3 Land area (million ha) uses in major regions of the world

Region	*Total area*	*Cropland*	*Pasture*	*Forest*	*Other**
Africa	2,965	184	792	688	1,301
N. America	2,139	274	368	684	813
S. America	1,753	140	468	905	240
Asia	2,679	450	678	541	1,010
Europe	473	140	84	157	92
Total	10,009	1,188	2,390	2,975	3,456
	100%	12%	24%	30%	34%

Source: WRI, 1991
Note: * Land that is either too dry, too steep, or too cold to use for agriculture and forestry

STATUS OF WORLD ENVIRONMENTAL RESOURCES

The quantity and quality of arable land, water, energy, and biodiversity, balanced against the size of world human population, determine the current and future status of the very environmental resources which support and sustain human life.

Land resources

More than 99 per cent of human food comes from the terrestrial environment, and only 1 per cent from the oceans, lakes, and other aquatic ecosystems (Pimentel and Pimentel, 1996). The world's food and fibre crops are grown only on 12 per cent of the earth's total land area (Buringh, 1989). Another 24 per cent of the land is used as pasture to graze livestock for meat and dairy products, while forests cover an additional 31 per cent (Buringh, 1989). The percentage of forest and grassland, set aside as protected national parks to conserve biological diversity, amounts only to 3.2 per cent of the total terrestrial ecosystem (Reid and Miller, 1989). Most of the remaining land area (34 per cent) is unsuitable for crops, pasture or forests because it is too cold, dry, steep, stony, or wet; or because the soil is too infertile or shallow to support plant growth (Table1.3) (Buringh, 1989).

About 0.5 ha of cropland per capita is needed to provide a diverse and nutritious diet of plant and animal foods (Lal, 1989). At present there is slightly more than this amount in the USA. In China, on the other hand, there is only 0.08 ha of cropland per capita to feed the people, and this amount is fast declining because of both continuing population growth and rapid land degradation (Leach, 1995). The world average of cropland available per capita is only 0.27 ha, which is roughly one-half the amount needed (Table 1.1). This shortage of productive cropland is, in part, the cause of both food shortages and poverty in many parts of the world today (Leach, 1995).

Currently an average American consumes a total of 1,481 kg of agricultural produce annually. This compares with only 476 kg/capita/yr for an average Chinese and a world average of 614 kg/capita/yr (Table 1.2). Based on available data, each person in China eats essentially a vegetarian diet. By all measurements, the Chinese have reached the carrying capacity of their agricultural system. And their increasing use of fossil-based fertilisers and other inputs to compensate for severely eroded soils foreshadows problems in the future (Wen and Pimentel, 1992). They are also planning for huge imports of grains (Brown, 1995).

Escalating land degradation threatens most of the crop and pasture lands throughout the world (Lal and Pierce, 1991; Pimentel *et al.*, 1995). The major types of degradation include erosion by rainfall and wind, and salinisation and water-logging of irrigated soils (Kendall and Pimentel, 1994). More than 10 million hectares of the world's productive arable land are severely degraded and abandoned each year (Pimentel *et al.*, 1995). Moreover, each year an additional 5 million hectares of new land must be put into production in order to feed the nearly 100 million humans added annually to the world population. Most of the total of 15 million ha, needed for both replacement and expansion, is being claimed from the world's forests, and this urgent demand for agricultural land accounts for more than 60 per cent of the deforestation now taking place worldwide (Myers, 1990).

Soil erosion by wind and water is the single most serious cause of soil loss and degradation. Erosion rates are greater than ever before in history, averaging about 30 t/ha/yr worldwide (Pimentel and Hall, 1989; WRI, 1991; Pimentel *et al.*, 1995). Soil erosion on cropland ranges from about 13 t/ha/yr in the US to 40 t/ha/yr in China (USDA, 1994; Wen, 1993; McLaughlin, 1993). The rate of soil loss in Africa has increased by a factor of 20 in the last 30 years (Tolba, 1989). Wind erosion is so serious in China that Chinese soil can be detected in the Hawaiian atmosphere when planting starts in China (Parrington *et al.*, 1983). Similarly, soil eroded from Africa was detected in Florida and Brazil in 1992 (Simons, 1992). Worldwide soil erosion is about 30 t/ha/yr (Pimentel, 1993).

Current erosion rates throughout the world are of particular concern because of the slow rate of topsoil formation – it takes approximately 500 years for 2.5 cm (1 inch) of topsoil to form under agricultural conditions (OTA, 1982; Elwell, 1985; Troeh *et al.*, 1991; Pimentel *et al.*, 1995). Thus, topsoil is now being lost 13 to 40 times faster than it is being replaced.

Soil erosion has an adverse effect on crop productivity because it reduces water availability, water-holding capacity of the soil, and soil depth; it also depletes both nutrients and organic matter of the soil. And all these factors thwart plant growth. It is estimated that agricultural land degradation can by itself depress world food production by between 15 and 30 per cent during the next 25 years, unless new production strategies are implemented to reduce or arrest soil degradation by erosion (Buringh, 1989).

Unfortunately, the arable land currently in use for crop production already includes a considerable amount of marginal land which is highly susceptible to erosion. Also, since soil degradation depresses crop production, greater fossil energy inputs in the form of fertilisers, pesticides and irrigation are needed to offset the losses (OTA, 1982; Follett and Stewart, 1985; Pimentel, 1993; Pimentel *et al.*, 1995).

Water resources

The present and future availability of adequate supplies of fresh water is frequently taken for granted. Natural collectors of water, such as rivers and lakes, vary in their distribution throughout the world, and often they are shared within and between countries. Moreover, all surface water supplies, especially those in the arid regions, are subject to evaporation losses. In the case of reservoirs, for instance, the average annual evaporation loss is about 24 per cent (Meyers, 1962).

All vegetation needs and transpires massive amounts of water during the growing season. For example, a corn crop producing about 7,500 kg/ha of grain will take up and transpire about 5 million litres/ha of water during the growing season (Leyton, 1983). To supply this much water to the crop, not only must 8 million litres (800 mm) of rain fall per hectare, but a significant portion of it must fall during the growing season (Pimentel *et al.*, 1997b).

The greatest threat to the maintenance of fresh water supplies is the overdraft of surface and groundwater resources that are being used increasingly to supply the needs of the rapidly growing human population and of agriculture to provide its food. Agricultural production 'consumes' more fresh water than any other human activity (Falkenmark, 1989). About 87 per cent of the fresh water pumped worldwide is 'consumed' (non-recoverable) by agriculture (Postel, 1989), while in the United States this figure is about 85 per cent (NAS, 1989). An average person needs nearly three litres of water to drink per day, including the water in milk and other foods, in addition to a minimum of 90 litres per day for cooking, washing, and other domestic activities (Brewster, 1987). This compares with 400 litres per day for the domestic needs of an average American (USBC, 1994).

About eighty nations of the world are already experiencing significant water shortages (Gleick, 1993). For instance, more than 300 cities in China are short of water, and the problem is becoming increasingly more serious (WRI, 1994). Furthermore, in the arid regions where annual rainfall is low or very low, the future of agricultural production is grim, especially as their populations grow.

To meet the ever-increasing demand for water would require the use of all sources of water, both surface and underground, and this may entail important environmental consequences. For example, by the time the Colorado River reaches Mexico, it almost disappears before finally trickling into the Sea of Cortes. Excessive removal of its water, mainly by the states of California, Arizona, and Colorado, is the cause (Sheridan, 1983).

Surface water, in rivers and lakes, and groundwater constitute the freshwater resources of the world (Wolman, 1986; Falkenmark, 1989). Groundwater resources are renewed at various rates depending on hydro-geological conditions, but usually at the extremely slow rate of 0.1–0.3 per cent per year (UNEP, 1991; Covich, 1993). Because of their slow recharge rate, groundwater resources must be carefully managed to prevent overdraft. Yet there is no effective conservation of groundwater resources, and overdraft is now a serious problem worldwide. For example, in Tamil Nadu, India, groundwater levels declined by 25–30 m during the 1970s because of excessive, but necessary, pumping for irrigation (Postel, 1984; UNFPA, 1991). The groundwater table in Beijing, China, is being depressed at the rate of about 1 m per year; and by 4.4 m per year in Tianjin, also in China (Postel, 1984). In the United States aquifer overdraft averages 25 per cent higher than the recharge rate (USWRC, 1979). But in the vast Ogallala aquifer in the USA, annual overdraft is 130–160 per cent more than recharge (Beaumont, 1985). If this continues, this vital aquifer is expected to become non-productive in the next 40 years (Soule and Piper, 1992).

High rate of depletion of surface and groundwater resources is beginning to constrain the option of irrigating the arid regions. Furthermore, per capita irrigation area is also declining because of salinisation and water-logging, both deleterious effects of continual irrigation (Postel, 1989).

Another major threat to the maintenance of the integrity of fresh water resources is pollution caused by people and industry. Considerable water pollution is documented in the United States (USBC, 1994), but it is more serious in countries in which there are relevant regulations that are seldom enforced. In some Latin American countries, for example, untreated urban sewage is discharged into rivers and lakes (WRI, 1991), and this causes a serious health problem due to faecal-coliform bacteria counts that can be much higher than 1 million per ml (the acceptable level for US drinking water is zero). Carried by eroded soil, pesticides, fertilisers, and soil sediments also pollute water resources, while rivers and lakes are further polluted by untreated toxic chemicals discharged by industry (WRI, 1991). Pollution by sewage and disease organisms – as well as by the 100,000 different chemicals used worldwide – makes water unsuitable not only for drinking but also for application to crops (Nash, 1993).

Energy resources

Availability of some form of energy for performing essential human activities is vitally important to all societies, and it has been so throughout the ages. Historically, humans have relied on their own energy, animal power, wind power and water power – then turning gradually to wood, coal, gas, and oil for fuel and power. World consumption of energy today is about 381 quads per year from all energy sources (one quad is equal to 10^{15} BTU or 337×10^{18} joules). Increasing energy consumption is a direct reflection of the rapid growth in world population, and it is adding significantly to the degradation of the natural environment by human activities (Table 1.4).

Although about 50 per cent of all solar energy captured by photosynthesis is used by humans worldwide, this is proving to be inadequate for meeting the energy demands of food and forest production (Pimentel, 1989; Pimentel *et al.*, 1997a). To make up the shortfall, about 321 quads of fossil energy are being used worldwide each year (UNEP, 1985; IEA, 1991). Of this, 81 quads are consumed in the USA alone (DOE, 1995a, 1995b). American consumption of fossil fuel is about three times the 28 quads of solar energy harvested as crop and forest products each

Table 1.4 Fossil and solar energy use in the USA and the world (quads)

	USA	*World*
Total consumption	93.97[a]	381.13
Petroleum	33.71[a]	136.17[b]
Natural gas	20.81[a]	75.15[b]
Coal	19.43[a]	87.60[b]
Nuclear power	6.52[a]	22.10[b]
Biomass	6.80[a]	28.50[d]
Hydroelectric power	3.00[c]	23.81[b]
Geothermal and wind power	0.30[c]	0.80[b]
Biofuels (ethanol)	3.40[c]	7.00[e]

Notes: (a) DOE (1995a), (b) DOE (1995b), (c) DOE (1993), (d) Pimentel *et al.* (1994), (e) Pimentel and Pimentel (1996)

year. Thus, the US population consumes 40 per cent more fossil energy than all the solar energy captured by US vegetation (Pimentel and Pimentel, 1996).

Fossil fuels are used in agriculture and forestry to run machinery, pump water, and for the application of fertilisers and pesticides. Fossil energy is also used in a wide variety of other human activities including industrial production; fuel for cars, trains and trucks; highway construction; heating and cooling of buildings; and the packaging of goods, to mention but a few. In general, fossil energy is used for feeding an increasing number of humans and for improving their quality of life in many basic ways, protection against diseases for example. Typically, delivering clean water and treating sewage helps reduce a wide range of disease organisms transmitted by polluted water. It is interesting to note in this context that about 90 per cent of all human diseases are caused by water-borne organisms (Pimentel *et al.*, 1997b).

Developed countries consume about 80 per cent of the world's fossil energy annually, compared with only 20 per cent consumed by developing countries in which 75 per cent of the world's population lives (UNEP, 1985; DOE, 1991). The USA alone consumes about 25 per cent (81 quads) of the world's fossil energy output annually (Pimentel *et al.*, 1993). Fossil energy use in different US economic sectors has increased by a factor of between 20 and 1,000 in the last three to four decades, and this attests to the heavy reliance of Americans on this finite energy resource to support their affluent lifestyle (Pimentel and Hall, 1989; Pimentel and Pimentel, 1996).

Several developing countries with high population growth rates are now increasing their use of fossil fuels to augment their agricultural production of food and fibre. Typically in China, as mentioned earlier, there has been a hundredfold increase in the use of fossil energy in agriculture since 1955 (Wen and Pimentel, 1992), mainly for fertilisers, pesticides, and irrigation.

Overall projections of the availability of fossil energy resources are discouraging. A recent report, published by the US Department of Energy (DOE, 1991) and based on current oil-drilling data, shows that the estimated national oil reserve has plummeted. This means that, instead of the 35-year supply of US oil reserves projected about eight years ago, the current known and discoverable potential oil and natural gas reserves are limited to a 20-year supply based on present rates of pumping (DOE, 1990; Lawson, 1991; BP, 1994). Given that the USA is now importing more than half its oil, it is clear that a serious fossil fuel shortage already exists (Gibbons and Blair, 1991).

The world supply of oil is greater than that of the United States, and it is projected to last 30–50 years at current rates of pumping (Matare, 1989; BP, 1994). The world reserve of natural gas is adequate for the next 20–35 years, and coal for about a hundred years (Matare, 1989; BP, 1994; Bartlett and Ristinen, 1995). However, these estimates are based on current consumption rates and current population numbers. It is sobering to note that, if all people in the world enjoyed a standard of living similar to that of the USA, with a matching energy consumption rate, and if the world population continued to grow at a rate of 1.5 per cent, the world's fossil fuel reserves would last a mere 15 years!

At present, transportation represents 40 per cent, and electricity about 34 per cent, of total US energy consumption (nuclear energy contributes 18 per cent of the electricity need) (USBC, 1994). Electricity produced by nuclear technology has some advantages over electricity produced by burning fossil fuels because it requires less land than coal-fired plants, causes fewer human injuries and deaths, and it does not contribute to acid rain or global warming (Holdren, 1991; Meeks and Drummond, 1991). However, given the constraints to the further expansion

BOX 1.1 COMPARING THE USE OF SOLAR-ENERGY BIOMASS AS A WOOD-FUEL AND AS A CORN-ETHANOL FUEL IN THE USA

Biomass has been used as a woodfuel since humans first started to use fire for various purposes. It remains a dominant fuel, particularly in developing countries. Even in developed countries like the USA, biomass is still used by many people and is the dominant solar energy system in use.

The average annual harvest of fuel wood from a US forest could net about 3 t/ha per year, assuming a small amount of fertiliser (10 kg of nitrogen; 5 kg of phosphorus; 10 kg of potassium) is applied each year. The net yield of energy from a sustainable forest would be approximately 1.1 million kcal/ha (Pimentel *et al.*, 1994). Assuming that the wood were carefully harvested and no clear-cutting were practised, soil erosion would be minimal. The prime pollution problem would be associated with the actual burning of the wood. Up to a hundred different chemicals are produced in the wood smoke, several of which are carcinogenic (Pimentel *et al.*, 1994).

Producing ethanol using corn grain as the feedstock will produce 1 litre of ethanol per 2.7 kg of corn. After the energy inputs for growing the corn and inputs for fermentation and distillation are included, producing ethanol is a net loss. In fact, it takes 72 per cent more energy to produce a litre of ethanol than the energy that is contained in the ethanol itself (Pimentel, 1991). The only reason that ethanol is produced commercially in the US is that it is heavily subsidised.

Equally serious to these energetic and economic losses are the pollution problems contributed by the production and use of ethanol. Corn production results in severe soil erosion and also is the largest user of pesticides and fertilisers in the USA. Excess amounts of carbon dioxide are produced, thus contributing to the global warming problem. When burned as a fuel, it produces less carbon monoxide; however, the fuel adds large amounts of nitrous oxides and other air pollutants to the atmosphere. In addition, the use of ethanol results in aldehyde and alcohol air pollutants that are carcinogenic.

of nuclear fission technology, nuclear fusion technology would appear to hold greater promise for meeting the growing energy demand in the long term. But there are serious problems. First, the world reserve of uranium, needed for both, is limited and expected to last about a hundred years without nuclear fast-breeder reactors (fusion) (Hafele, 1991). Secondly, the risks of disposing radioactive wastes, and public hostility to their storage, may seriously impede the widespread use of both fission and fusion energy (Hafele, 1991). Finally, fusion technology would need many more years of research and development before it could be commissioned (Matare, 1989).

Both nuclear fission and fusion technologies produce enormous amounts of waste heat which causes serious environmental pollution (Bartlett, 1989, 1995). For example, it is estimated that if the number of nuclear power plants in the USA were increased from the 108 at present to 1,500, the temperature of US aquatic ecosystems would increase about 10°C (H. Kendall, Department of Physics, MIT, personal communication, 1992). Such heat pollution would seriously damage the environment and cause a major loss of biological diversity in both aquatic and terrestrial ecosystems. Also, it would alter existing climate patterns which could adversely affect agricultural and forestry production. Thus, it would not be possible to supply all US energy needs with fusion at the present rate of consumption, even if it were available now (Bartlett, 1995).

Biological resources

In addition to land, water resources, and crop and livestock species, humans depend on the existence and functioning of approximately 10 million other species present in agro-ecosystems and in nature (Pimentel *et al.*, 1992), and no technology can be substituted to perform the essential services they provide (e.g. the wild biota). In the United States alone there are approximately 500,000 species of plants, animals, and microbes which perform many essential activities, including the pollination of crop and wild plants, the recycling of manure and other organic wastes, the degradation of chemical pollutants, and the purification of water and soil (Pimentel *et al.*, 1992).

Clearly humans depend on these organisms, as they do on their livestock and crop species. For example, honey bees and wild bees play an essential role in pollinating about $40 billion worth of US crops annually, in addition to pollinating natural plant species (Pimentel *et al.*, 1997a). It is estimated that in a single bright sunny day in July in New York State, honey bees and wild bees pollinate 6 trillion blossoms (Pimentel, 1994). We have no technology to substitute for this natural service provided by wild biota.

In addition, a diverse group of species constitutes a vital reservoir of genetic material for the future development of agriculture and forestry, as well as for medicines to fight disease. Yet with each passing day an estimated 150 species become extinct because of increasing human numbers and human activities, including deforestation, soil and water pollution, pesticides, as well as urbanisation and industrialisation (Reid and Miller, 1989).

Resources and human diseases

Many of the diseases associated with the deteriorating quality of water, air, and soil rob humans of their health and eventually their livelihood. But there are profound differences between the causes of death in developed and developing regions of the world. Communicable, maternal, and/or prenatal diseases account for 40 per cent of the deaths in developing regions compared with only 5 per cent in the developed regions (WHO, 1994). While there is a host of reasons for this mortality and for disease and health problems in general, inadequate food and contaminated water and soil are considered to be the major contributors.

Disease and malnutrition are interrelated, and, as might be expected, parasitic infections and malnutrition co-exist where there is poverty and poor sanitation (Shetty and Shetty, 1993). Poverty and lack of sanitation can be as severe in some urban areas as in rural areas. Indeed, several studies point to different degrees of severity even in different parts of the same city (Atkinson, 1993). Urban environments, especially those without proper sanitation, are becoming a matter of growing concern in terms of their potential for the spread of disease as they become overcrowded (*Science*, 1995). The high density of people in urban environments provides no protection against pollution caused by the accumulation of city wastes in water, air, and soil, but creates favourable conditions for the rapid spread of infectious diseases which can easily reach epidemic proportions (WHO, 1992).

About 90 per cent of the diseases occurring in developing countries result from the lack of sanitary water (WHO, 1992). In the world today about 4 billion cases of disease are contracted from water, and each year approximately 49 million deaths are caused by diseases that are essentially water-borne (WHO, 1992). *Shistosomiasis*, which is common throughout the tropics, is a typical example of a parasitic disease associated with contaminated water, while hookworms, which infect nearly 1 billion people worldwide, are present in contaminated moist soils in the tropics. Malaria, spread by mosquitoes, infects 800 million worldwide (Walsch and Warren, 1979), while tuberculosis is estimated to infect about 1 billion people (Cohen, 1996). *Polyparasitism* is common in many parts of the world with poor sanitation, and it makes health problems even more difficult (Keusch and Migasena, 1982).

Intestinal parasites reduce the intake of nutrients in various ways: by causing nutrient loss through diarrhoea or dysentery, impairing nutrient absorption, lowering appetite for food, and through blood loss (Shetty and Shetty, 1993). Hookworms, for instance, can remove up to 30 cc of blood from a person in a day, leaving the person weak and susceptible to other diseases (Hotez and Pritchard, 1995). It is estimated that between 5 and 20 per cent of an individual's daily food intake is used for offsetting illnesses and for relieving stress of disease, thereby significantly lowering the nutritional status of the infected person (Pimentel *et al.*, 1997c).

At first glance human health would appear to be unrelated to resources, but it soon becomes apparent that it is affected not only by the quantities of resources (e.g. food and water) but also by the way in which they are handled and protected against contamination. As human populations grow, so do the risks to their health and productivity, especially where sanitation is inadequate.

CONSTRAINTS IN MANAGING RESOURCES

The primary resources – land, water, energy, and biological resources – function interdependently, but each can be manipulated to some extent to make up for a partial shortage or deficiency in one or more of the others. For example, desert land can be irrigated to bring it into agricultural production. However, this would be viable only if a sufficient quantity of groundwater or surface water is available, if sufficient fossil energy is available to pump and move the water, and if the soil is suitable for irrigation and sufficiently fertile to support crop growth.

The fertility of nutrient-poor soil can be improved by large inputs of fossil-based fertilisers. But this practice increases dependence on finite fossil fuel resources. A more basic problem, however, is that the natural reformation of topsoil to a depth of 15 mm, which is necessary for satisfactory crop production, takes about three thoudand years. And even with fertiliser use, soil erosion remains a critical problem for current agricultural production. On the other hand, crops can be grown hydroponically, but the cost of this in terms of energy, effort and inputs is approximately ten times that of conventional agriculture (Schwarz, 1995).

Although new technologies and environmental management practices are bringing improvements and efficiency in the use of resources, there are practical limits to the process. For example, fish production in oceans and other aquatic ecosystems has not increased with the availability of larger, faster ships or larger nets (Mitchell and Cleveland, 1993). In fact, fish production reached a peak in 1988, and since then it has been slowly declining because of poor management, over-fishing, and pollution. As a result, the cod fisheries on the east coast of Canada, for example, have been over-fished to the extent that there are now more than 80,000 fishermen on welfare and desperate for work (W. Rees, University of British Columbia, personal communication, 1996). Forest production has not increased either with the availability or use of large, efficient chainsaws. To the contrary, forests everywhere in the world are on the decline (WRI, 1994). Furthermore, no existing technology can increase the flow of the Colorado River, or enrich other surface and groundwater resources. The only hope is that, with better environmental management and effective conservation measures, it would be possible to make more efficient use of the remaining freshwater resources, aiming at some degree of sustainability. For example, drip irrigation in agriculture can reduce the use of water by nearly 50 per cent (Pimentel *et al.*, 1997b).

IMPROVING RESOURCE MANAGEMENT

Because the availability of all essential resources is fast diminishing, so are the options for their substitution. This emphasises the urgency for conservation and the development of alternative strategies. Certainly, soil erosion requires a major effort in order for agricultural production to be sustainable in the future. Numerous farm practices which conserve soil fertility are known, such as the use of cover crops, grass strips, ridge planting, and crop rotation (Pimentel *et al.*, 1995). But they are not used as much as in the previous decades. Neither is much attention given to the prevention of soil erosion which would conserve soil nutrients, increase water retention of soil, and improve overall soil quality.

In order to reduce dependence on commercial fertilisers, the under-utilisation and waste of livestock manure should be addressed (Pimentel, 1996; Pimentel *et al.*, 1995). Proper use of manure – which contains significant amounts of nitrogen, phosphorus, and potassium – will enrich soil nutrients and, at the same time, reduce demand on fossil fuels for agriculture.

Pests destroy growing crops and thereby reduce food and fibre supply. Yet, despite the annual use of 2.5 million tons of pesticides and other control chemicals worldwide, about 40 per cent of all potential crop production is lost to pests (insects, diseases, and weeds) (Pimentel, 1997). In the USA an average of about 0.5 of a ton of pesticide is applied per hectare, and yet pests destroy about 37 per cent of all potential crop production. Since 1945 the use of synthetic pesticides in the United States has grown by a factor of 33; yet crop losses to pests continue to increase (Pimentel *et al.*, 1991). For example, despite a thousandfold increase in the use of insecticides on corn, corn losses to insects have increased nearly fourfold (Pimentel *et al.*, 1991). Furthermore, residues of pesticides pollute the environment, destroy beneficial species, and create human health problems (Pimentel and Greiner, 1997).

Although large quantities of pesticides are being used, much is wasted because of inappropriate methods of application, or because they are applied when they are not really needed (Pimentel, 1997). Moreover, changes in agricultural technology, especially intensive agriculture, have led to increased pest populations. For some major crops like corn, the practice of crop rotation has been abandoned – in the USA nearly 50 per cent of the corn is grown continuously as a monoculture. These changes have caused an increase in the number of corn pests, necessitating the application of increased amounts of pesticides to protect the crop (Pimentel *et al.*, 1991). This is contrary to the principles of sustainable and environmentally sound agricultural practices, including a return to crop rotation, that would reduce soil erosion, conserve fertile land, reduce water demand for irrigation, reduce pesticide and fertiliser use, and save fossil fuel, soil, and water resources (Pimentel *et al.*, 1995).

There is no way of increasing the rainfall which replenishes surface and groundwater stores. However, more effective use of water is possible, for example by using drip irrigation instead of sprinkler irrigation. This would reduce agricultural water demand by 50 per cent (Pimentel *et al.*, 1997b).

Reduced water consumption by individuals, together with effective water conservation measures, is paramount for sustainability. If this is not done, crop production will decline, especially in the arid regions. Also, competition will intensify within regions and between countries as and when water becomes scarce.

The use of desalinated sea water is not a viable proposition for agriculture, because desalinisation is costly in terms of both energy and money. For example, it would cost $14,000 (excluding costs of transporting water from the sea to agricultural fields) to irrigate one hectare of corn with desalinated water, while all other inputs (e.g. fertilisers) would cost only $500 (Pimentel *et al.*, 1997b).

With the predicted decline of fossil fuels, it is clear that a transition will have to be made from fossil energy to alternative energy sources. High priority should be given to research aimed at converting solar energy into usable energy efficiently and economically, and to the development of new sources such as nuclear fusion. Although several promising solar technologies have already been investigated, including solar thermal receivers, photovoltaics, solar ponds, windpower, hydropower, and biomass (Pimentel *et al.*, 1994), most are of limited use at present.

If the available renewable energy technologies are exploited, an estimated 40 quads of energy could be produced in the USA using about 20 per cent of the available land area (Pimentel *et al.*, 1994). This development will affect agriculture and forestry to some extent, but not seriously, provided that US population is limited to about 200 million (Pimentel *et al.*, 1994). Clearly this is contingent upon the United States implementing a population policy which is commensurate with the availability of resources and the standard of living which it aspires to (Pimentel *et al.*, 1994).

As mentioned earlier, diverse species, large and small, perform vital functions in the environment, while plants remove carbon dioxide from the air and add oxygen to it. And there are no substitute technologies for most of these essential environmental and ecological functions. No one knows which or how many species can become extinct before the ecosystem is damaged irreversibly and ceases to function. Scientists hold the view that, in order to maintain sufficient natural biological diversity to ensure a quality environment, it is necessary to preserve about one-third of the terrestrial ecosystem as natural vegetation (Odum, 1971). Preservation of biomass at this level is considered to be essential for securing food and shelter for the present generation. It is also essential for protecting valuable species in order to ensure that adequate biodiversity is preserved for future generations (Heywood, 1995; Pimentel *et al.*, 1997a).

CONCLUSIONS

Improved environmental management, together with redistribution of food and wealth, would help relieve the problems of more than 2 billion malnourished people and another 1.5 billion currently living in poverty in the world today (Neisheim, 1993; Speth, 1996). The average income per family in the USA today is about $37,000 per annum, and, according to many, there is a good deal of over-consumption and waste of goods and resources in that country. Theoretically it would be possible to eliminate malnourishment with an equitable redistribution of food and wealth, whereby every family in the world

BOX 1.2 APPLICATION OF IPM (INTEGRATED PEST MANAGEMENT) PROCEDURES IN INDONESIA TO REDUCE PESTICIDE USE IN RICE, IMPROVE THE ENVIRONMENT, AND PROTECT PUBLIC HEALTH

In 1980, a new minister of agriculture who favoured the use of pesticides was appointed in Indonesia. The minister implemented a programme of subsidising the use of pesticides for farmers and generally encouraged farmers to use more pesticides in rice production. For the first year or two there were no major pest problems. However, starting about the third year pest insect populations began to increase, especially the brown planthopper pest of rice. By 1985 more than 80,000 ha of rice had to be abandoned because of severe outbreaks of the brown planthopper. Also yields of rice on other fields were significantly reduced. The cause of the problem was the destruction of the natural enemies of the planthopper by the widespread use of pesticides and the development of pesticide resistance in the planthopper itself.

The President of Indonesia sent his deputy to see Dr I.N. Oka and asked his advice on what should be done. Dr Oka recommended implementing an IPM programme that would rely primarily on natural enemies for insect pest control. He had already conducted research on how an IPM programme could be implemented in rice production. The President of Indonesia then appeared on television and banned 57 of the 64 pesticides that were currently being used on rice, as recommended by Dr Oka.

Within two years, Dr Oka's management policies had the Indonesian farmers using 65 per cent less pesticide while their rice yields increased 12 per cent. Subsidies for pesticides were reduced at the same time. Overall the farmers were making more money because they were using less pesticide while, at the same time, their yields were significantly higher. The environment improved, and there were fewer public health problems because of the new pesticide management programme.

Source: Dr I.N. Oka, Bogor Ford Research Institute, Bogor, Indonesia, personal communication, 1995

would receive an average annual income of about $6,700. For some families this would be an improvement, certainly a great improvement for countless others, but North American and European families will regard this as too little for maintaining the standard of living to which they have been accustomed or to which they aspire.

It is estimated that, with a world population of 1 to 2 billion, it would be possible for everyone to enjoy a relatively high standard of living, with access to ample and nutritious foods (Moffat, 1996). This level of global prosperity would be possible only if sound environmental management practices are adopted to ensure that water, energy, and biological resources are not degraded any further. However, even with sound environmental management practices and better food distribution, a greater number of people in the world than at present will be malnourished, hungry, and poor if the world population increases to 10 billion or thereabouts.

Unfortunately, the above is very much a 'utopian' scenario compared to the evolving situation in the real world. Fertile land, which is a primary resource, is not only deteriorating in quality through constant use and erosion, but available land is being subdivided into ever-smaller parcels among increasing numbers of people. A similar trend is also occurring with water supplies. As consumption increases, surface and groundwater sources are being used up faster than they can be replenished by rainfall. The problem is being exacerbated by increasing competition for dwindling supplies among individuals, agriculture, and industry, as well as between regions and countries.

At current and projected levels of consumption most oil, natural gas, and coal reserves will be used up during the next century because of population growth and rising consumer expectations. Since fossil energy resources are finite, new renewable energy sources are needed. Furthermore, plant and animal species are rapidly diminishing, both in numbers and diversity, as human activities expand throughout the earth's ecosystem. And, when lost, their benefits and services are no longer available to society or to the environment at large.

Based on their evaluation of available world natural resources, scientists of the Royal Society and the US National Academy of Sciences have issued a joint statement reinforcing their concern about the growing imbalance between world population and the resources needed to support it (RS and NAS, 1992). So far individuals and government leaders appear to be unaware, unwilling, or unable to deal with the growing imbalance between a rising population and available life-supporting resources. Indeed, the interdependencies among available life-supporting resources, individual standards of living, quality of the environment, environmental resource management, and population density are neither acknowledged nor understood. Sadly, despite their intelligence and inventiveness, humans have a disappointing record of effective environmental management to protect their essential resources and the environment from over-exploitation in the face of rapidly growing populations (Pimentel *et al.*, 1989).

Historically decisions have been made, and still are made, in response to isolated crises; and usually they are made when something approaching a catastrophe occurs. Such decisions have been *ad hoc*, designed to protect and/or promote a particular resource or aspect of human well-being instead of examining the problem in a holistic manner. Based on past experience, we are concerned that urgent decisions on the human carrying capacity of the world will not be made until the situation becomes intolerable, or worse still, irreversible (Pimentel *et al.*, 1994; Pimentel and Giampietro, 1994; Pimentel *et al.*, 1997a).

It should not be beyond human wit to develop a population control policy, within which basic rights of the individual are enshrined, together with sound policies on resource use and research for enhancing energy supplies in ways that do not damage the integrity of the environment. It is only in this way, in a true spirit of international co-operation and collaboration, that our fundamental obligations to future generations can be honoured. People would then be free from poverty and starvation and live in an environment that will sustain human life with dignity.

NOTE

* This chapter is based in part on the earlier article: Pimentel, D., Harman, R., Pacenza, M., Pecarsky, J. and Pimentel, M. (1994) 'Natural resources and an optimum human population', *Population and Environment* 15: 347–369.

REFERENCES

Agrostat (1992) *Agrostat Database*, Data Base, Rome: Food and Agriculture Organisation of the United Nations.

Atkinson, S.J. (1993) 'Urban health in the Third World: a guide to the literature', *Environment and Urbanization* 5: 146–152.

Bartlett, A.A. (1989) 'Fusion and the future', *Physics and Society* 18 (3): 11.

Bartlett, A.A. (1995) 'Fusion: an illusion or a practical source of energy?', *Clearing House Bulletin* 4 (1): 1–3, 7.

Bartlett, A.A. and Ristinen, R.A. (1995) 'Natural gas and transportation', *Physics and Society* 24 (4): 9–10.

Beaumont, P. (1985) 'Irrigated agriculture and groundwater mining on the high plains of Texas, USA', *Environ. Cons.* 12 (11 pp).

Bennett, G.M. (1995) *China Facts and Figures Annual Handbook*, Florida: Academic International Press.

Bouis, H.E. (1995) 'Breeding for nutrition', *Journal of the Federation of American Scientists* 48 (4): 1, 8–16.

BP (1994) *British Petroleum Statistical Review of World Energy*, London: British Petroleum Corporate Communications Services.

Brewster, J.A. (1987) *World Resources 1987*, a Report by the International Institute for Environment and Development and The World Resources Institute, New York: Basic Books, Inc.

Brown, L.R. (1995) *Who Will Feed China?*, New York: W.W. Norton.

Buringh, P. (1989) 'Availability of agricultural land for crop and livestock production', in D. Pimentel and C.W. Hall (eds) *Food and Natural Resources*, San Diego, Calif., USA: Academic Press, pp. 69–83.

Cohen, J.E. (1996) *How Many People Can the Earth Support?* New York: Rockefeller University.

Covich, A.P. (1993) 'Water and ecosystems', in P.H. Gleick (ed.) *Water in Crisis*, New York: Oxford University Press, pp. 40–55.

DOE (1990) *Annual Energy Outlook*, Washington, DC: US Department of Energy.

DOE (1991) *Annual Energy Outlook with Projections to 2010*, Washington, DC: US Department of Energy, Energy Information Administration.

DOE (1993) *Annual Energy Review*, Washington, DC: EIA, USDOE.

DOE (1995a) *Annual Energy Outlook with Projections to 2010*, Washington, DC: EIA, USDOE.

DOE (1995b) *International Energy Annual 1993*, Washington, DC: EIA, USDOE.

Elwell, H.A. (1985) 'An assessment of soil erosion in Zimbabwe', *Zimbabwe Sci. News* 19: 27–31.

Falkenmark, M. (1989) 'Water scarcity and food production', in D. Pimentel and C.W. Hall (eds) *Food and Natural Resources*, San Diego, Calif., USA: Academic Press, pp. 164–191.

FAO (1992a) *Food and Nutrition: Creating a Well-Fed World*, Rome: Food and Agriculture Organisation of the United Nations.

FAO (1992b) *Nutrition: The Global Challenge*, Proceedings of an International Conference on Nutrition Sponsored by FAO, Rome.

Follett, R.F. and Stewart, B.A. (1985) *Soil Erosion and Crop Productivity*, Madison, Wis., USA: American Society of Agronomy, Crop Science Society of America.

Fuchs, V.R. and Reklis, D.M. (1992) 'America's children: economic perspectives and policy options', *Science* 255: 41–46.

Gibbons, J.H. and Blair, P.D. (1991) 'US energy transition: on getting from here to there', *Am. Inst. of Physics*, July: 21–30.

Gleick, P.H. (1993) *Water in Crisis*, New York: Oxford University Press.

Hafele, W. (1991) 'Energy from nuclear power', *Scientific American*, September: 137–144.

Heywood, V.H. (1995) *Global Biodiversity Assessment*, United Nations Environment Programme, Cambridge: Cambridge University Press.

Holdren, J.P. (1991) 'Energy in transition', in J. Piel (ed.) *Energy for Planet Earth*, New York: W.H. Freeman Co, pp. 119–130.

Hotez, P.J. and Pritchard, D.T. (1995) 'Hookworm infection', *Scientific American* 272 (6): 68–75.

IEA (1991) *Energy Statistics of OECD Countries*, Paris: International Energy Agency.

Kendall, H.W. and Pimentel, D. (1994) 'Constraints on the expansion of the global food supply', *Ambio* 23: 198–205.

Keusch, G.T. and Migasena, P. (1982) 'Biological implications of polyparasitism', *Review of Infectious Diseases* 4: 880–882.

Kinzelbach, W.K. (1983) 'China: energy and environment', *Environmental Management* 7: 303–310.

Kutzner, P.L. (1991) *World Hunger. A Reference Handbook*, Santa Barbara, Calif., USA: ABC–Clio.

Lal, R. (1989) 'Land degradation and its impact on food and other resources', in D. Pimentel (ed.) *Food and Natural Resources*, San Diego, Calif., USA: Academic Press, pp. 85–140.

Lal, R. and Pierce, F.J. (1991) *Soil Management for Sustainability*, Ankeny, Ia., USA: Soil and Water Conservation Soc. in co-operation with World Assoc. of Soil and Water Conservation and Soil Sci. Soc. of Amer.

Lawson, R.L. (1991) 'The US should increase its use of coal', in C.P. Cozic and M. Polesetsky (eds) *Energy Alternatives*, San Diego, Calif., USA: Greenhaven Press, pp. 41–45.

Leach, G. (1995) *Global Land and Food in the 21st Century*, Stockholm, Sweden: International Institute for Environmental Technology and Management.

Leyton, L. (1983) 'Crop water use: principles and some considerations for agroforestry', in P.A. Huxley (ed.) *Plant Research and Agroforestry*, Nairobi, Kenya: International Council for Research in Agroforestry, pp. 379–400.

Maberly, G.F. (1994) 'Iodine deficiency disorders: contemporary scientific issues', *Journal of Nutrition* 124 (8 suppl.): 1473s–1478s.

McLauglin, L. (1993) 'Soil erosion and conservation in Northwestern China', in D. Pimentel (ed.) *World Soil Erosion and Conservation*, Cambridge, UK: Cambridge University Press, pp. 87–108.

McMichael, A.J. (1993) *Planetary Overload: Global Environmental Change and the Health of the Human Species*, Cambridge, UK: Cambridge University Press.

Matare, H.F. (1989) *Energy: Fact and Future*, Boca Raton, Fla., USA: CRC Press.

Meeks, F. and Drummond, J. (1991) 'Nuclear power can be environmentally safe', in C.P. Cozic and M. Polesetsky (eds) *Energy Alternatives*, San Diego, Calif., USA: Greenhaven Press, pp. 46–52.

Meyers, J.S. (1962) *Evaporation from 17 Western States*, Washington, DC: Geological Survey Professional Paper 272-D.

Mitchell, C. and Cleveland, C.J. (1993) 'Resource scarcity, energy use and environmental impact: a case study of the New Bedford, Massachusetts, USA fisheries', *Environmental Management* 17: 305–317.

Moffat, A.S. (1996) 'Ecologists look at the big picture', *Science* 273 (13): 1490.

Myers, N. (1990) 'Mass extinctions: What can the past tell us about the present and future?', *Global and Planetary Change* 2: 82.

NAS (1989) *Alternative Agriculture*, Washington, DC: National Academy of Sciences.

Nash, L. (1993) 'Water quality and health', in P. Gleick (ed.) *Water in Crisis: A Guide to the World's Fresh Water Resources*, Oxford: Oxford University Press.

Neisheim, M.C. (1993) 'Human nutrition needs and parasitic infections', in D.W.T. Crompton (ed.) *Parasitology: Human Nutrition and Parasitic Infection*, Cambridge, UK: Cambridge University Press, pp. s7–s18.

NGS (1995) *Water: A Story of Hope*, Washington, DC: National Geographic Society.

Odum, E.P. (1971) *Fundamentals of Ecology*, Philadelphia, USA: W.B. Saunders Co.

OTA (1982) *Impacts of Technology on US Cropland and Rangeland Productivity*, Washington, DC: Office of Technology, US Congress.

Parrington, J.R., Zoller, W.H. and Aras, N.K. (1983) 'Asian dust: seasonal transport to the Hawaiian Islands', *Science* 246: 195–197.

Pimentel, D. (1989) 'Ecological systems, natural resources, and food supplies', in D. Pimentel and C.W. Hall (eds) *Food and Natural Resources*, San Diego, Calif., USA: Academic Press, pp. 1–29.

Pimentel, D. (1991) 'Ethanol fuels: energy security, economics and the environment', *J. Agr. Environ. Ethics* 4: 1–13.

Pimentel, D. (ed.) (1993) *World Soil Erosion and Conservation*, Cambridge, UK: Cambridge University Press.

Pimentel, D. (1994) 'Insects and related arthropods', in *The Encylopedia of the Environment*, Boston, USA: Houghton Mifflin Co.

Pimentel, D. (1996)' Conservation of fertilizers and livestock manure: pollution prevention', manuscript to be published.

Pimentel, D. (ed.) (1997) *Techniques for Reducing Pesticide Use: Environmental and Economic Benefits*, Chichester, UK: John Wiley & Sons.

Pimentel, D. and Hall, C.W. (eds) (1989) *Food and Natural Resources*, San Diego, Calif., USA: Academic Press.

Pimentel, D. and Giampietro, M. (1994) 'Implications of the limited potential of technology to increase the carrying capacity of our planet', *Human Ecology Review* (Summer/Autumn) 1: 248–251.

Pimentel, D. and Pimentel, M. (1996) *Food, Energy and Society*, Niwot, Colo., USA: University Press of Colorado.

Pimentel, D. and Greiner, A. (1997) 'Environmental and socio-economic costs of pesticide use', in D. Pimentel (ed.) *Techniques for Reducing Pesticide Use: Environmental and Economic Benefits*, Chichester, UK: John Wiley & Sons, pp. 51–78.

Pimentel, D., Armstrong, L.E., Flass, C.A., Hopf, F.W., Landy, R.B. and Pimentel, M.H. (1989) 'Interdependence of food and natural resources', in D. Pimentel and C.W. Hall (eds) *Food and Natural Resources*, San Diego, Calif., USA: Academic Press, pp. 31–48.

Pimentel, D., McLaughlin, L., Zepp, A., Lakitan, B., Kraus, T., Kleinman, P., Vancini, F., Roach, W.J., Graap, E., Keeton, W.S. and Selig, G. (1991) 'Environmental and economic impacts of reducing US agricultural pesticide use', in D. Pimentel (ed.) *Handbook of Pest Management in Agriculture*, Boca Raton, Fla., USA: CRC Press, pp. 679–718.

Pimentel, D., Stachow, U., Takacs, D.A., Brubaker, H.W., Dumas, A.R., Meaney, J.J., O'Neil, J., Onsi, D.E. and Corzilius, D.B. (1992) 'Conserving biological diversity in agricultural/forestry systems', *BioScience* 42: 354–362.

Pimentel, D., McLaughlin, L., Zepp, A., Lakitan, B., Kraus, T., Kleinman, P., Vancini, F., Roach, W.J., Graap, E., Keeton, W.S. and Selig, G. (1993) 'Environmental and economic effects of reducing pesticide use in agriculture', *Agriculture, Ecosystems and Environment* 46 (1–4): 273–288.

Pimentel, D., Harman, R., Pacenza, M., Pecarsky, J. and Pimentel, M. (1994) 'Natural resources and an optimum human population', *Population and Environment* 15: 347–369.

Pimentel, D., Harvey, C., Resosudarmo, P., Sinclair, K., Kurz, D., McNair, M., Crist, S., Sphpritz, L., Fitton, L., Saffouri, R. and Blair, R. (1995) 'Environmental and economic costs of soil erosion and conservation benefits', *Science* 267: 1117–1123.

Pimentel, D., Wilson, C., McCullum, C., Huang, R., Dwen, P., Flack, J., Tran, Q., Saltman, T. and Cliff, B. (1997a) 'Environmental and economic benefits of biodiversity', *BioScience* 47 (11): 747–757.

Pimentel, D., Houser, J., Preiss, E., White, O., Fang, H., Mesnick, L., Barsky, T., Tariche, S., Schreck, J. and Alpert, S. (1997b) 'Water resources: agriculture, the environment, and society', *BioScience*, 47 (2): 97–106.

Pimentel, D., Huang, X., Cordova, A. and Pimentel, M. (1997c) 'Impact of population growth on food supplies and environment', *Population and Environment*, 19 (1): 9–14.

Postel, S. (1984) *Water: Rethinking Management in an Age of Scarcity*, Worldwatch paper no. 62, Washington, DC: Worldwatch Institute.

Postel, S. (1989) *Water for Agriculture: Facing the Limits*, Washington, DC: Worldwatch Institute.

PRB (1995) *World Population Data Sheet*, Washington, DC: Population Reference Bureau.

PRC (1994) *Year Book of 1994 of the People's Republic of China*, Beijing: PRC Yearbook, Ltd

Qu, G. and Li, J. (1992) *Population and Environment in China*, Beijing: China Environmental Science Press (Chinese).

Reid, W.V. and Miller, K.R. (1989) *Keeping Options Alive: The Scientific Basis for Conserving Biodiversity*, Washington, DC: World Resources Institute.

RS and NAS (1992) 'The Royal Society and the National Academy of Sciences on population growth and sustainability', *Population and Development Review* 18 (2): 375–378.

Schwarz, M. (1995) *Soilless Culture Managment*, Advanced Series in Agricultural Sciences 24, Berlin: Springer-Verlag.

Science (1995) 'Cities as disease vectors', *Science* 270: 1125.

Sheridan, D. (1983) 'The Colorado: an engineering wonder without enough water', *Smithsonian*, February, 45–54.

Shetty, P.S. and Shetty, N. (1993) 'Parasitic infection and chronic energy deficiency in adults', *Supplement to Parasitology* 107: S159–S167.

Simons, M. (1992) 'Winds toss Africa's soil, feeding lands far away', *New York Times*, October 29: A1, A16.

Smil, V. (1984) *The Bad Earth, Environmental Degradation in China*, Armonk, New York: M.E. Sharpe, Inc.

Soule, J.D. and Piper, D. (1992) *Farming in Nature's Image: An Ecological Approach to Agriculture*, Washington, DC: Island Press.

Speth, J.G. (1996) *The US and the UN: Back to the Future*, Charleston, S.C., USA: World Affairs Forum, 19 February.

SSBPRC (1990) *The Yearbook of Energy Statistics of China in 1989*, Beijing: State Statistical Bureau of PRC, Chinese Statistical Press.

Tolba, M.K. (1989) 'Our biological heritage under siege', *BioScience* 39: 725–728.

Tribe, D. (1994) *Feeding and Greening the World*, Oxford, UK: CAB International.
Troeh, F.R., Hobbs, J.A. and Donahue, R.L. (1991) *Soil and Water Conservation for Productivity and Environmental Protection*, Englewood Cliffs, N.J., USA: Prentice-Hall.
UNEP (1985) *Energy Supply Demand in Rural Areas in Developing Countries*, Report of the Executive Director, Nairobi, Kenya: United Nations Environment Programme.
UNEP (1991) *Freshwater Pollution*, Global Environment Monitoring System, Nairobi, Kenya: United Nations Environment Programme.
UNFPA (1991) *Population and the Environment: The Challenges Ahead*, New York: United Nations Population Fund.
USBC (1994) *Statistical Abstract of the United States 1991*, Washington, DC: US Bureau of the Census, US Government Printing Office.
USDA (1993) *Agricultural Statistics*, Washington, DC: US Department of Agriculture, Government Printing Office.
USDA (1994) *Summary Report 1992 National Resources Inventory*, Washington, DC: Soil Conservation Service.
USWRC (1979) *The Nation's Water Resources 1975–2000. Second National Water Assessment*, Washington, DC: US Water Resources Council.
Walsch, J.A. and Warren, K.S. (1979) 'Selective primary health care: an interim strategy for disease control in developing countries', *New England Journal of Medicine* 301: 967–974.
Wen, D. (1993) 'Soil erosion and conservation in China', in D. Pimentel (ed.) *World Soil Erosion and Conservation*, Cambridge: Cambridge University Press, pp. 63–86.
Wen, D. and Pimentel, D. (1984) 'Energy inputs in agricultural systems of China', *Agriculture, Ecosystems and Environment* 11: 29–35.
Wen, D. and Pimentel, D. (1992) 'Ecological resource management to achieve a productive, sustainable agricultural system in northeast China', *Agriculture, Ecosystems and Environment* 41: 215–230.
WHO (1992) *Our Planet, our Health: Report of the WHO Commission on Health and Environment*, Geneva: World Health Organisation.
WHO (1994) *Global Comparative Assessments in the Health Sector: Disease Burden, Expenditures, and Intervention Packages*, Geneva: World Health Organisation.
Wolman, M.G. (1986) 'Consensus, conclusions, and major issues in water resources', in J. Cairns and R. Patrick (eds) *Managing Water Resources*, New York: Praeger, pp. 117–126.
Worldwatch Institute (1992) *State of the World*, Washington, DC: Worldwatch Institute.
WRI (1991) *World Resources 1990–91*, New York: Oxford University Press.
WRI (1994) *World Resources 1994–95*, Washington, DC: World Resources Institute.

SUGGESTED READING

Giampietro, M. and Pimentel, D. (1993) 'The tightening conflict: population, energy use, and the ecology of agriculture', *The NPG Forum*, October, Teaneck, N.J.: Negative Population Growth Inc.
Grant, L. (ed.) (1992) *Elephants in the Volkswagen*, New York: W.H. Freeman and Co.
Kendall, H.W. and Pimentel, D. (1994) 'Constraints on the expansion of the global food supply', *AMBIO* 23 (3): 198–205.
Pimentel, D. and Pimentel, M. (1996) *Food, Energy, and Society*, Niwot, Colo., USA: University Press of Colorado.

SELF-ASSESSMENT QUESTIONS

1. What is the current world population and how many people are added daily?
2. How much food per year is required to feed a person?
3. To produce food, how much land, water, and energy is required?
4. How rapidly is arable land being lost and what management strategies are available to conserve land?
5. How much water is required to produce a hectare of a food crop? Can this amount be reduced?
6. How serious is water as a source of human disease, and what can be done to reduce this problem?
7. How soon will fossil energy resources like oil be depleted?
8. Will solar or renewable energy be able to replace fossil energy consumption completely?
9. Solar energy production requires land. How will this requirement be met without interfering with food and forest production?
10. Approximately how many species are there, and how rapidly are they being lost?
11. What role does biodiversity play in maintaining a quality environment?
12. Human diseases appear to be associated with the diminishing quality of water, air, and soil resources. How do diseases affect health and the livelihood of humans?
13. What are some management strategies that can be utilised to reduce environmental pollution and resource degradation?
14. If humans do not control their rapid growth in numbers, what role will nature and the environment play in limiting human numbers?

2
PROBLEM-IN-CONTEXT
A framework for the analysis, explanation and solution of environmental problems

Wouter T. de Groot

SUMMARY

All environmental problems share a common conceptual structure that is composed of two parallel causal chains: (1) a chain of *facts* of human activities, their effects on the environment and their further impacts on health, nature, etc., and (2) a chain of *values*, i.e. assertions about the desired state of health, nature etc., the desired state of the environment and the desired human activities. Problem analysis consists of research into this structure. Environmental problems are set in a causal context, i.e. the chains of *causes* of the problems. Problem explanation is research into these causal chains, which branch 'outward' from the environmental problem. Explanations are (1) normative, connected to the values that co-constitute the environmental problem, (2) physical-scientific, connected to the carrying capacity of the environment, and (3) social-scientific, explaining the human activities from which the problem originates. Social-scientific explanation starts with the actors carrying out the problematic activities, their options and motivations, and proceeds to further actors and factors influencing these options and motivations. By these means, one identifies the target groups and the social options for solving the environmental problem. These, jointly with the technical options identified in the problem analysis, can go into the process of designing and evaluating environmental policies and projects. Problem-in-Context is a methodological framework which brings all this together in an interdisciplinary whole.

ACADEMIC OBJECTIVES

The aim of this chapter is to establish familiarity with the common conceptual structure of all environmental problems in their social, physical and policy contexts, and the approach to solving them. On completion of this chapter the student should be able:

- to see, roughly but systematically, the interconnection of core concepts such as carrying capacity, environmental values, norms derivation, secondary actors, etc;
- to use this structure for the analysis, explanation and solution of environmental problems by integrating knowledge from the physical, social and normative sciences.

INTRODUCTION: FRAMEWORKS AND CAUSAL CHAINS

'Environmental Impact Assessment' (EIA) is a legal procedure to evaluate proposed activities. At the same time, EIA is a *methodological framework*; that is, a structure of concepts that prescribes what questions to ask, what data to gather, what models to apply, and so on. Another example of such a framework is 'Life-Cycle Assessment' (LCA), which is a methodology for studying the impacts 'embodied' in industrial products. Being methodological in nature, these frameworks do not assert how the world *is*, but how to *discover* it. This chapter will also be devoted to a framework of this kind. It differs from EIA, LCA and all the others, however, in that it is fully comprehensive. It is applicable to all kinds of environmental problems, and spans all disciplines that relate to them: the physical sciences, the social sciences and the normative sciences (e.g. ethics). It is an instrument for research that runs the whole cycle of the analysis of an environmental problem, of its causes in society and of its possible solutions (the keys for which often lie in society, too, rather than in the environment itself). Relatively great emphasis will be placed on the social-scientific explanation of the causes of environmental problems, because this area is usually unfamiliar to environmental scientists, weakly treated by social scientists and neglected in textbooks on environmental management. The framework is called 'Problem-in-Context' (PiC). This name itself discloses some of its structure: the environmental problem stands in the middle and holds everything together; the 'context' is where the problem comes from, i.e. its roots in human values and social institutions. More information on all aspects of the framework is to be found in de Groot (1992).

One issue is basic before we can discuss the practicalities. This issue may be called 'systems versus causal chains'. Interdisciplinary research is often associated with following a systems approach, meaning that we have to conceptualise the physical and social world as being composed of systems. However, while it obviously cannot be denied that many systems do exist (ecosystems, farming systems, the 'world system', etc.), plenty of relevant research may not be necessary, or may even be downright inadequate, if the causal relationships are conceptualised only in terms of systems. What is a system? A system is a whole of interconnected elements with a well defined boundary and with system-level characteristics of its own. One example is the shallow lake summarised in Figure 2.1. The shallow lake is obviously a system, with boundaries and many interconnected elements (in water, sediments, biota), and system characteristics such as the seasonal dynamics of the energy flow. But let us take another example: let us say a country is obliged by the IMF to undergo a Structural Adjustment Programme. As a result, the agricultural agencies have to change their policies towards farmers. The farmers are also affected by the Programme directly. Because of these influences, the farmers may decide to start growing more cotton. This may result in soil degradation and in less space being available for cattle and subsistence crops. This in turn changes gender relations and undermines the resource base for future generations. All this is easy to follow, but is it a *system*? It is not. It is a causal story, a causal chain. It may be perfectly sensible for researchers to study this chain without necessarily studying the world system, the national agricultural system, the 'value system' and so on (Vayda, 1996). Generally speaking, systems often do not exist to any relevant degree (or they may be inaccessible to study in a given research project). Causal chains always exist, however, and the Problem-in-Context framework is therefore constructed from causal chains. If we encounter a system 'along the research road', we take it up as a part of an ongoing causal chain. Figure 2.1 is an example. Going from the bottom upwards on the left side, we see that a human activity such as boating causes phosphates to enter the lake. The lake itself is a system, visible in the figure as the interconnected circle of broad-lined relationships. From the lake onwards, however, the causal chain continues to further effects ('impacts') on humans, which in turn may have other effects, and so on. We can study these without introducing system assumptions.

The broader lines in the figure indicate the interrelationships in the lake ecosystem. Starting with the human activity of boating and going straight upwards, this activity causes emission of phosphate to the lake. This causes rising phosphate levels ('immissions'), which in turn cause higher levels of algae in the water. Because of this, the transparency becomes lower, and the lake loses its attractiveness for swimming. This affects the economic welfare of recreationists and owners of amenities, which in turn gives rise to further effects. Many other causal lines are linked to this one. For instance, boating is only an arbitrary starting point; it has underlying causes in society (economy, culture, etc.). Also, the lake ecosystem creates complex feedbacks within the lake. Because of the decreased plant cover and prey visibility, for instance, the pike stock declines, which in turn gives rise to higher whitefish levels, lower levels of water fleas, therefore less feeding on algae and hence even higher algal levels. These phenomena are also connected to economic welfare (e.g. through the fishing attractiveness of the lake). Moreover, the algae can be a threat to human health, the cultural value of the lake is affected, and so on. Overall, we see the lake ecosystem as a part of a longer, ongoing causal chain. Note that the facts and the values sides of the figure are each other's exact mirror images, but with a reversed causal direction.

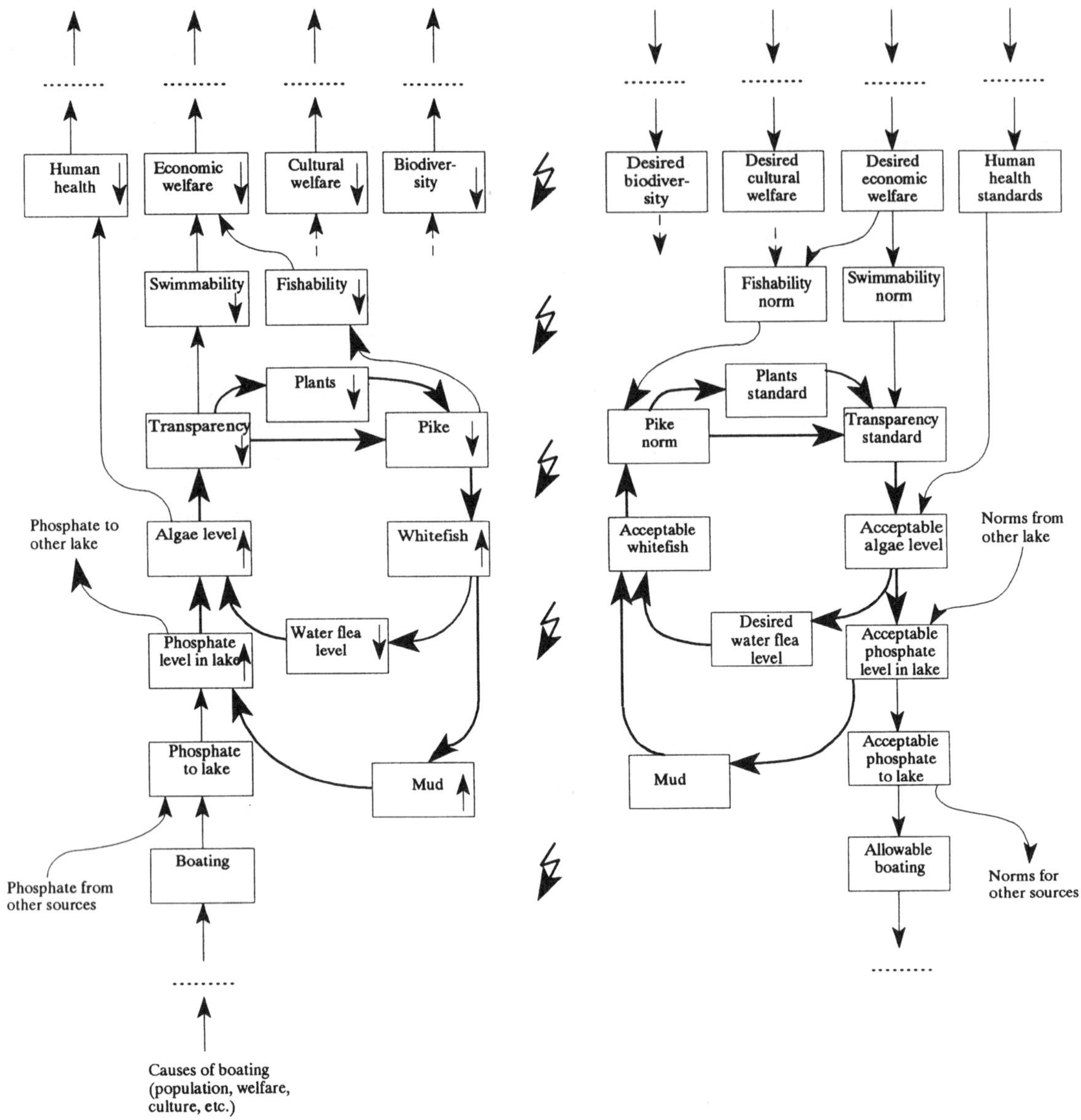

Figure 2.1 Effects chains and norms chains of a shallow lake eutrophication problem

Systems have boundaries, but causal chains do not. They are like causal stories that never end. In Figure 2.1, this is visualised by the dots at the bottom and at the top. Downwards, the human activities have causes, which in turn have other causes, and so on. Upwards, the effects have other effects, which in turn have other effects, and so on. Figure 2.1 represents only a *cut-off* of the endless chain. Hopefully, this cut-off is well chosen, because (for example) it is the core of an environmental problem (see next section). The 'causal chain' concept also marks another difference between the 'PiC' and 'systems' approaches. In systems research, the elements of the system are usually pre-defined, and the research then serves only to quantify the elements and relations. In many parts of PiC, however, the elements become identified during the research process. This is the principle of 'progressive contextualisation', to be discussed later.

ANALYSIS OF ENVIRONMENTAL PROBLEMS

Generally stated, 'analysis' means investigating how something is built up of various elements. Thus, the analysis of an environmental problem entails investigating the nature of the components. The analysis of any environmental problem will be greatly facilitated if the researcher knows beforehand how environmental problems are built up in general, conceptual terms. Getting to know this general structure of (all) environmental problems is the first objective of this section. After that, the *process* of analysis will be briefly discussed.

The structure of problem analysis

Is soil erosion an environmental problem? Looking at a country such as Nepal, the answer would seem to be affirmative. In the coastal dunes of Holland, however, biologists are happily welcoming the wind that blows sand from specially arranged erosion sites, where rare plants now have a new opportunity to grow. Erosion *per se* is merely a *fact*. It becomes problematic (or welcome) only in confrontation with *values*. For 'values' in this sentence we may also read 'norms', 'objectives', 'normative statements', 'standards', 'preferences' or any other term that stands for assertions of how the world *should be* (or should become, or should have been). 'Facts', then, are their empirical counterparts – that is, assertions of how the world *is* (or will be, or has been). In drawings of the Problem-in-Context framework, such as Figure 2.1, facts are on the left and values on the right. Note that in Figure 2.1, every fact has its perfect mirror on the values side. There is a factual (e.g. predicted) level of algae, mirrored on the values side by the normative 'acceptable' or 'desired' level of algae. There is a factual and a desired level of 'swimmability' of the lake, and so on for every variable. Generally speaking, we say that facts and values are expressed *in the same terms*; e.g., facts and norms in mg of algae per litre of water, or factual and desired incomes in dollars per year, and so on. Hence, PiC does not say that facts and values follow from each other (the 'naturalistic fallacy'). Neither does PiC assume the opposite extreme ('Cartesian') position, saying that facts and values are totally incommensurable, the former being 'objective' and the latter 'subjective'. Facts and values are both interpretations of reality, 'subjectivity objectified' as de Groot (1992) puts it, or 'social constructs', as many sociologists say. The simple essence of these difficult words is that in PiC facts and values can be *compared* so as to assess an environmental problem. In Figures 2.1 and 2.2, these comparisons are depicted as the flash-like 'tension arrows' that denote the facts/values discrepancies. Thus, the core of PiC looks like a zipper: chains of facts and values, joined by the 'point-by-point comparisons of the real and the ideal worlds' (von Wright, 1986).

PiC is drawn such that the direction of cause-and-effect relations in the effects chain runs upwards. This is merely conventional; the whole PiC picture may be turned upside down without changing its meaning. It is important to note, however, that the causal direction on the values side is *reversed*; the chains of norms run from the 'higher' values downwards. This is not conventional, but is truly logical. In Figure 2.1, for instance, we say that if we want a shallow lake with a good 'swimming value', then the transparency of the water should be more than 1 metre, so that the level of algae should be less than so-and-so many mg/l, and so on downwards, until we can specify a norm in terms of human activities such as the acceptable number of boats on the lake. In general terms, this may be called a process of *norms derivation*. It is currently an important type of research in environmental science, because it makes the crucial interconnection between effect-oriented policies ('We want our lakes to be swimmable') and source-oriented policies ('The maximum number of boats, therefore, is so-and-so many').

Figure 2.2 is the overall picture of the Problem-in-Context framework. It is easy to see that its centre – that is, the environmental problem – is only a generalisation of Figure 2.1. Although complicated problems such as climate change have quite complex effects and norms chains, they have the same basic structure. 'Problem analysis', in general terms, thus consists of

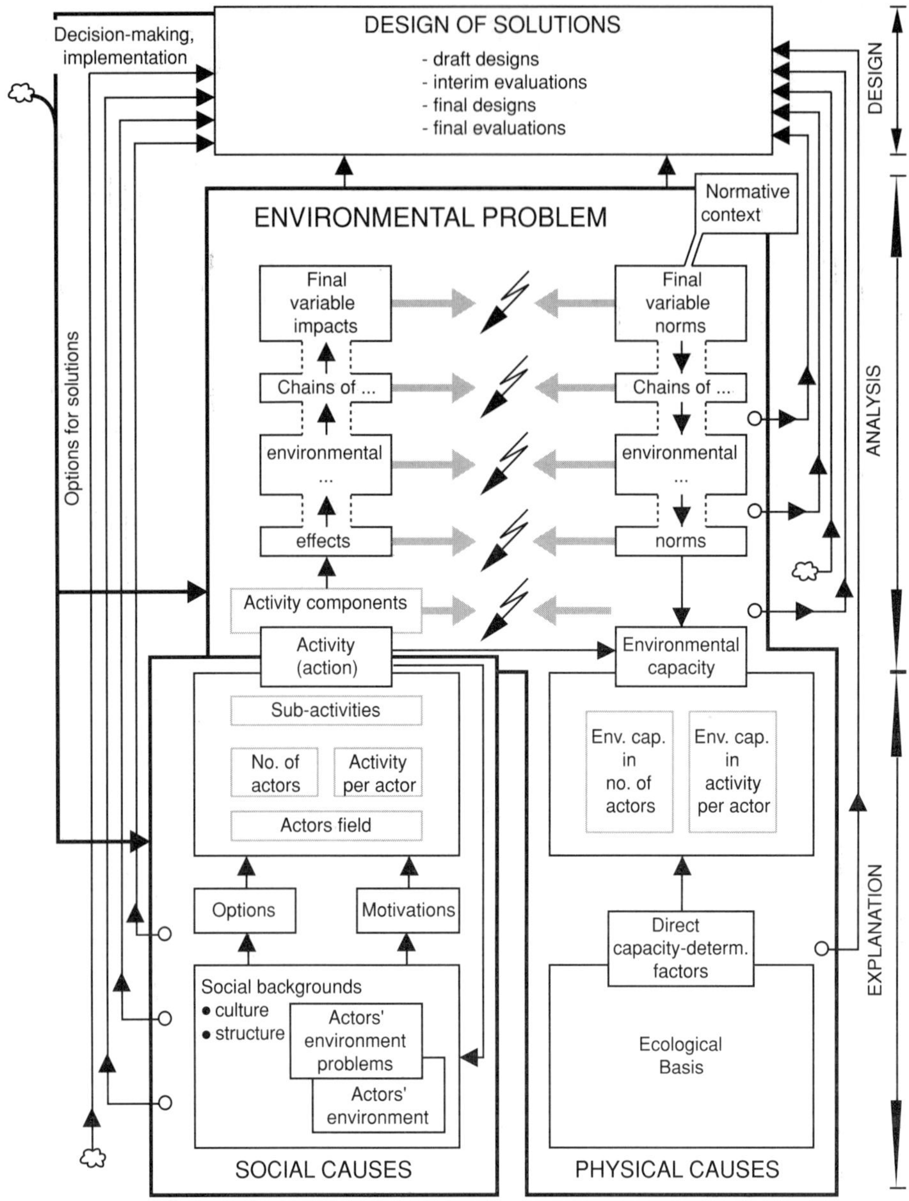

Figure 2.2 The summary picture of the Problem-in-Context framework: the environmental problem at the core, the three causal contexts and the feedback of the normative observer, through the design of solutions

performing research to specify the whole 'problem block' in Figure 2.2 to the degree of detail and certainty necessary for one's specific case. A number of core concepts of problem analysis will now be discussed one by one.

Final variables

At the top of the central 'problem block' of Figure 2.2 stand the facts and values in terms of 'final variables'. What are these? How can variables be called 'final' if causal chains in fact never end? The answer is

BOX 2.1 'STANDARD' FINAL VARIABLES FOR ENVIRONMENTAL SCIENCE

'Standard' final variables for environmental science are facts and norms in terms of:

- *Human health and security;*
- *Material welfare* (goods and services, efficiency and equity);
- *Non-material welfare* (cultural heritage, expressive landscape, nature close to home, etc.);
- *Intrinsic values of nature* (diversity and naturalness of, and within, populations and ecosystems);
- *Harmony in the relationships between humans and nature.*

All of these apply in the short term as well as for future generations (*sustainability*).

not that we now suddenly assume that causal chains do end; just that final variables are a privileged place for a cut-off of causal chains. If we say, for instance, that some polluting activity may result in 100 deaths per year, we do not usually have to follow that chain any further. Nobody will ask us: 'So what? Please indicate the further effects of these 100 deaths.' Therefore, the formal definition of 'final variables' is: *the variables that do not usually require further normative justification.* Obviously, from this example, human health is such a variable. Where should we look to arrive at a full list? One natural place to look is in the policy documents that lay down the objectives of environmental policies, e.g. documents of the UN or national governments. We then find a list such as that given in Box 2.1. The list has a logical structure. A first major distinction is between (1) three types of human values, (2) the worth of nature as an 'intrinsic' value in itself, and (3) the values in the relationships between humans and nature. All these, then, are relevant not only in the short term but also in the long term, thus adding the value of sustainability, i.e. the care for future generations. A further discussion is to be found in de Groot (1992, Chapter 4) and, in one formulation or another, in other chapters of this volume.

Whose values?

The set of final variables used to analyse an environmental problem may be called the 'value perspective' of the analysis. Value perspectives have a special relevance because they determine much of what an environmental problem is constructed to be. This immediately poses the question: why should we use the list of Box 2.1? Why not take a few specific values we personally like best? Or why not adopt the value perspective of, say, poor female farmers? The first, methodological answer is that for Problem-in-Context itself it makes no difference what we choose. The framework always works, and we can make a perfect analysis from *any* perspective. The second, substantive, answer is that for the content and ethics of our work it does of course make a lot of difference which values are taken to generate the problem analysis. From this angle, we can say that Box 2.1 represents a 'standard' set for integrated studies designed to lie close to normal policy work. In other circumstances, it may make much sense to narrow down the value set and make a more one-sided, say a 'Save the Whale!' analysis; this is sometimes called 'advocacy planning'. A second important notion here is the distinction between insiders' and outsiders' problem analyses, related to conflict resolution and participation (See Chapter 3). An insiders' analysis is an analysis adopting the values of the people 'inside' the problem, hence usually the local community. Thus, Box 2.1 is typically the standard outsiders' list. Participatory approaches, often applied in developing countries, usually try to find some middle ground between the insiders' and outsiders' perspectives. In conflict resolution, it may sometimes be advisable to make two entirely separate problem analyses, one generated by the insiders' value perspective and one generated by that of the outsiders, and then negotiate over the differences. Finally, the term 'local knowledge' is often used to denote analyses in which the knowledge of the community is applied instead of official science. Note however that for true participation, adopting community *values* is more important than applying community knowledge.

Activities and environmental capacity

In Figure 2.1, the human activity that causes the eutrophication problem is 'boating'. This activity, of course, has further impacts down the causal chain. Conventionally, however, we say that the activity is the first element of the environmental problem. Thus, the problem is that people have too many boats, causing too much phosphate in the water, and so on. Everything causally below this, then, is not the problem itself but the problem's causes ('social backgrounds', 'economic driving forces', etc.). Roughly speaking, the problem analysis is physical-science work, while going into the problem's causes is social-science work. Arriving at a well-structured identification of the problematic activities connected to an environmental problem is therefore always very important, because through the pivotal position of the activities element, the physical and social disciplines can be properly interconnected. In Figure 2.2, we see the activities at their borderline position between the problem analysis and the social-scientific problem explanation. We also see that the normative counterpart of the human activities has a special name: 'environmental capacity'. The concept goes under many other names, too, such as 'carrying capacity'. The concept always boils down to the same, however – namely, the *desired content and level of human activities*, derived from the desired levels of final variables such as human health, protection of nature and sustainability. Depending on the environmental problem under study, the environmental capacity takes on a new appearance each time. For the problem in Figure 2.1, it is the acceptable number of boats on the lake. In other cases, it is the maximum allowable emission (emission standard) of a toxic substance, or the optimum sustainable farming system, or the desired level of technologies that prevent carbon dioxide emissions on a national scale, and so on. In all cases, these are the end-points of norms derivation research. Although we may use the same physical-science models to move down from the final variables along the norms chain, each new vision in terms of final variables (e.g. not caring about the basic needs of the poor, or wanting very stringent health protection) results in new levels of environmental capacities.

Activity components and sub-activities

'Boating' may be such a simple activity that it does not need to be analysed further before the effects can be assessed. Other activities, however, may be so complex that they need to be subdivided into various components first, as shown in Figure 2.2. A drinking water supply system may be such an activity, for instance, which has to be specified as a complex of different activity components at different locations. Note that we do this with a view to the subsequent impacts analysis. If we look in the other direction, i.e. that of the *causes* of the activity, there may be no need for subdivision, e.g. if there is only one actor, such as the drinking water company, that decides on the activity as a whole. This may again be different for the 'boating' activity. Although the phosphate emissions of all types of boats may be equal, the causes may be quite different, e.g. if one type is used for commercial fishing and the other type for recreation. Then we may separate the activity into 'sub-activities' as they are called in Figure 2.2, i.e. 'boating = boating for fishing + boating for recreation', connected to two different actor categories. The same is done in the example of Figure 2.3, where the actors field is identified by going from the primary activities 'outwards'; hence the sequence of primary, secondary, etc. actors. The actors are connected not directly ('social network') but through their options and motivations.

Levels of problem formulation, and functions of the environment

In Figure 2.1, we may say that the environmental problem is 'too many boats on the lake'. We then formulate the problem as a fact/value discrepancy in terms of human activities; 'too many' is the difference between the actual and desired levels of boating (carrying capacity). The problem can also be formulated in terms of environmental variables, e.g. 'too high levels of algae' or 'undesired fish stock composition'. And, finally, the problem can be formulated on the level of the final variables, e.g. 'unacceptable economic and biodiversity loss'. Confusions often arise when it is not noted that this is the *same* problem all the time, only formulated on different levels along the facts/values zipper.

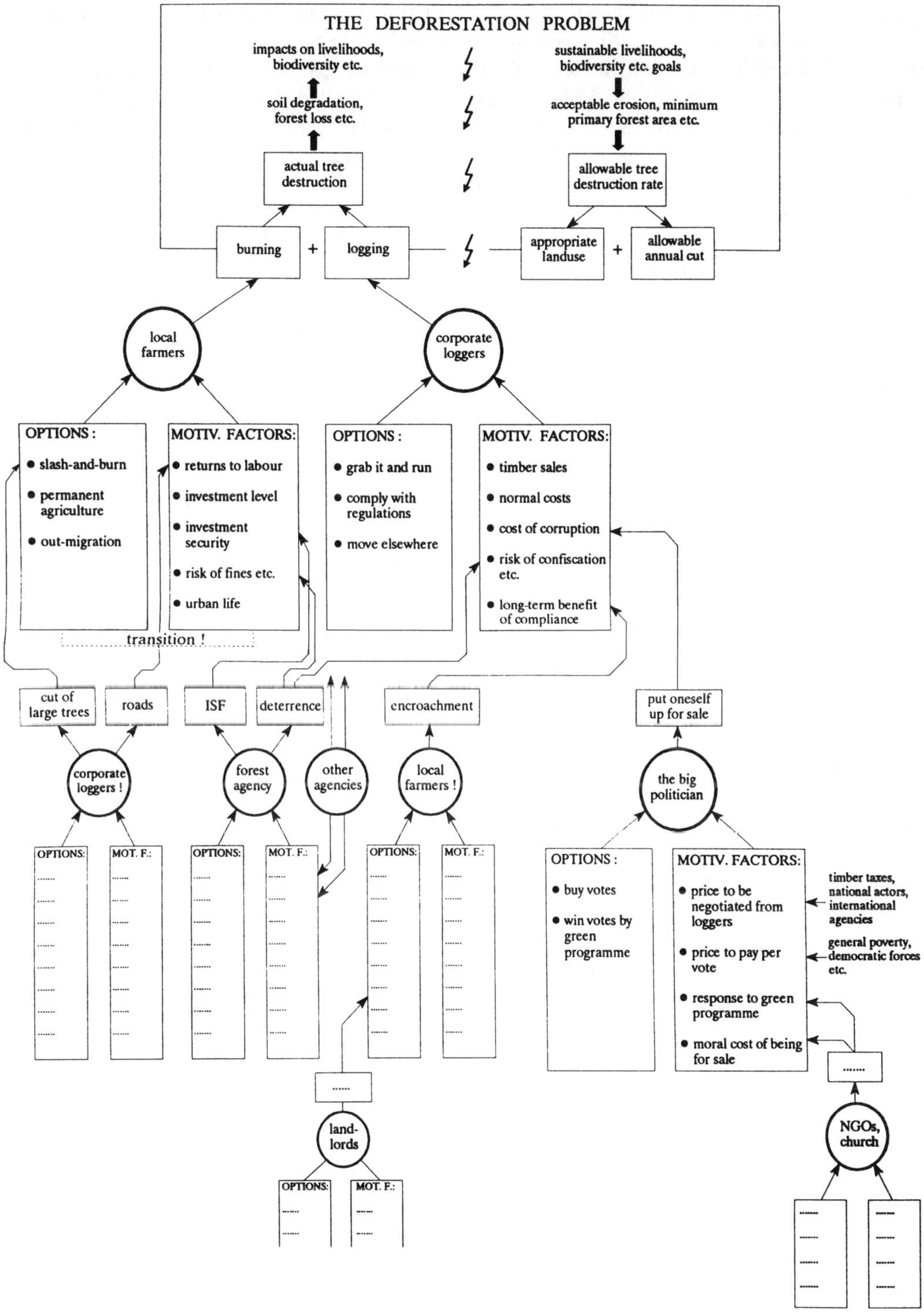

Figure 2.3 A simplified PiC of tropical deforestation, with the emphasis on the actors field

One level of formulation is not shown in Figure 2.2; it is often called that of the *functions of the environment*. This level lies between the level of the environmental variables and that of the final variables; the functions of the environment can be defined as 'all the ways in which the environment is linked to the welfare of people and nature'. Typically, functions of the environment are presented as systematic checklists, comprising items such as various 'carrying functions', various 'signification functions', etc., which can be of great help in organising problem analyses, especially in complicated cases such as the environmental problems analysis of a whole region. Lack of space prohibits a listing here, but de Groot (1992: 229) gives a good example. Economists often use the functions concept for quantifying the value of the environment to society.

People in three positions

Figure 2.2 shows that environmental problems run 'from people to people'. There are 'causing people' (e.g. in Figure 2.1, the people who go boating), and there are 'impacted people'; these are hidden in the final variables but are quite concrete nevertheless. In Figure 2.1, they may be the ones who can no longer go swimming, or whose fishing pleasure is lost. Then there are the people whose value perspective (final variable norms) sets the definition and analysis of the problem. In our case, this might be the water management agency. The general PiC term for this position is the 'normative observer'; this abstract term has been chosen because PiC can be applied irrespective of what agency, group or person 'sets the values'. Thus, PiC distinguishes between (1) actors, (2) victims and (3) normative observers, but it is important to note that these are *abstract positions* which assume a concrete aspect only in the analysis. If, for instance, I spray pesticides in my garden that may impact my own health, and if I start wondering whether this might be a problematic thing to do, then in such an analysis, I am actor, potential victim and normative observer at the same time! Or, to take another example, I may adopt the normative perspective of the water management agency in the case of the shallow lake problem of Figure 2.1; then, the agency is the normative observer, who may identify recreationists as victims as well as actors, and possibly also the agency itself as another actor contributing to the problem it is analysing. All this can be handled in a problem analysis without confusion if we keep a clear view on the three positions, the PiC way.

'Branching out' and 'weaving in'

In Figure 2.1 we see causal arrows on the left and right that branch out from the picture to other causal places, or weave into the picture from outside. For instance, water from our lake also flows out of it, so that the phosphate may also have impacts on a lake downstream, outside our problem picture. The first thing to note here is a technical rule: *what branches out on the facts side weaves in on the norms side*, and vice versa. For instance, the impact branching out to the downstream lake on the facts side generates a norm for our lake that weaves into our problem analysis on the values side. In fact, the downstream lake may be more vulnerable than the one in our analysis, so that the norm that will be set for the boats on our lake is in fact generated by the norm weaving in from the downstream lake! After all, if two norms 'compete', the lower standard has to be chosen (Ortolano, 1984). Conversely, it may be noted that on the facts side the level of algae may be determined not only by the boating, but also by sewer effluents entering our lake. In such a case of weaving in on the facts side, the norms branch out on the values side; a single standard with respect to the algae generates a norm for the boats as well as a norm for the sewer effluents. A strict norm for the boats gives more room for the effluents, and vice versa. Setting an appropriate balance between these two is a political matter that becomes part of the norms derivation process. On a global scale, we encounter this type of discussion in the well-known 'global equity' debates between the North and the South on who should bear the strictest norm of carbon dioxide emission abatement.

The process of problem analysis

A process of problem analysis may start out from any element in the 'problem block'. The starting point may be a proposed activity, for instance, such as the construction of a new road. In other cases, algal blooms or other environmental phenomena may trigger the analysis. Alternatively, a government agency may wish to know what the consequences would be of adopting a new normative perspective in its field; the analysis then starts out from the 'final values' corner of the problem block. Whatever the point of departure, however, the major rule to keep in mind is always to first try to get a grip on the whole of the problem before jumping into all kinds of more monodisciplinary details. In an interdisciplinary team, the project leader is typically the one who should keep that grip on the whole at all times, in order to steer the various monodisciplinary contributions effectively. The second thing to keep in mind is that the same Problem-in-Context structure works on all levels of sophistication. In a situation of conflict mediation, for instance, it may already be very useful to apply the framework (without using the abstract terms, of course) to clarify the conceptual confusions that the various parties may be locked into: what is normative and what is empirical, what are the problem boundaries, what is the influence of normative perspectives on environmental capacity norms, and so on. Examples of the other extreme are the ongoing research and debates on global climate change. Taken as a whole, this process is a huge problem analysis that may last far into the next century, with very large and competing models, constant political inputs, shifting paradigms, new empirical data and surprises that have to be accommodated into the analysis along the way, and so on. The most usual level of sophistication of problem-analytical research is somewhere between the two extremes. The grip-on-the-whole rule implies that problem analyses are often cyclical. They start out on an exploratory level, then seek to gain more budget and more sophistication on the basis of the initial results, and so on. The PiC structure can guide the work on all these levels of sophistication.

Types of partial problem analysis

Often, a partial problem analysis in set in motion after sufficient grip on the whole has been gained. These types of research focus on a specific area of the 'problem block'. A grip on the whole always remains necessary in order to guarantee the relevance of the partial work. Norms derivation research, for instance, focuses on the values side of the problem, but this makes sense only if we also keep an eye on the facts side. In Figure 2.1, for instance, what would be the use of deriving norms for boating if the lake were too shallow for boating anyway, or had already been declared a nature reserve, or if the people were too poor to buy boats? Some types of partial problem analysis have been worked out such as to become frameworks in their own right, usually much more detailed and strict than is the all-embracing PiC. Characteristically, too, these frameworks are becoming institutionalised as standard models for applied research. Some have already been mentioned in the introduction. In Environmental Impact Assessment (EIA, Chapter 10), the point of departure is some proposed activity, and the prescriptions then specify which causal routes to take 'upward' and what are the norms and criteria on the values side with which to compare the impacts in the final assessment. In Life-Cycle Assessment (LCA, Chapter 11), the 'activity' is an industrial product, which is then analysed in great detail, specifying all its intermediary stages 'from the cradle to the grave' (these are 'activity components' in PiC terms). Emissions of substances are then connected to all these stages, added up and assessed against various norms, in order to gain insight into the total impact of the product and its alternatives. Finally, the Land Evaluation procedure of the FAO (1989) is an elaborate model for norms derivation connected to land use, starting out from final variable values such as economic optimality and sustainability, and working its way downwards to arrive at a map of desired land use types in a region ('environmental capacities' in PiC terms).

EXPLANATIONS: ENVIRONMENTAL PROBLEMS IN TRIPLE CONTEXT

Problem analysis, roughly, is physical and normative science work. Translated into policy recommendations, problem analysis logically leads to prescriptions in physical terms: what purification plant to build, which forest to protect, how much fertiliser to apply, and so on. In some cases, especially if there is a strong government and much money around, this may be sufficient for a solution of the problem. Many of these solutions have been exhausted by now, however. Increasingly, environmental policies will have to be based on more than problem analysis alone. With this in mind, the present and the subsequent sections will focus on the methodology for explaining environmental problems, that is, identifying their *causes*, as a further basis for designing solutions. The causes of a problem do not lie in the problem itself but outside, in the problem's context. Environmental problems have three such contexts, i.e. the types of causes of the problem. These can be discovered by enumerating the three essential preconditions for the existence of environmental problems. There would be no environmental problems (1) if the normative observer had no values (no norms, no policy targets), (2) if the capacity of the environment were infinite, and (3) if people did not carry out problematic activities. The explanation of an environmental problem therefore implies posing questions in three contexts:

1 *the normative context*: on what grounds can the normative observer maintain that his value perspective is justified?;
2 *the physical context*: why does the environment react the way it does? Why can it not produce or process more?;
3 *the social context*: why do people undertake the problematic activities?

These three contexts can be seen in Figure 2.2, each connected to one corner of the 'environmental problem' block. The normative context is connected to the final variables (values side), the physical context to the environmental capacity, and the social context (social causes) to the activities. The basic way to do explanatory research is to take one of these corner elements as the point of departure, and then proceed step by step, going further and further into the context. Vayda (1983) terms this approach 'progressive contextualisation'. In this section we shall touch briefly on the three routes of explanation. In the following section we look at the social context in greater depth, because this context is usually the most relevant for solving the problem.

(1) Normative contextualisation is not empirical research, but a type of (applied) ethics research. The question is not what other people happen to value, but what the normative observer *himself should* value. Examples of this type of question include:

- If a factory has to lay off labourers because of an environmental regulation, this is often regarded as a negative impact on economic welfare. But why should this position be justified? Economic policies have objectives in terms of *net* effects on the *regional* employment, not gross effects on specific factories.
- We usually take the value of humans (health, welfare, etc.) and the intrinsic value of nature as final variables. But is there not also a value or disvalue in the way that humans and nature *relate to* each other, e.g. the intensity and harmony of these relationships? If we say 'yes' to this question, it obviously has many practical consequences for the analysis of nature conservation problems and buffer zones, the design of good farming practices, and so on. In nature protection, for instance, it leads us to focus not only on endangered species, but also on the places where children in nature play. We may also go a step further in the normative context and ask: what could be the ground for this idea of valuing people–nature relationships? Are there other ethics where 'relationship values' can be found? This leads us into ongoing discussions between traditional ethics (often called the 'ethics of rights and obligations') and new ways of ethical thinking, such as the 'ethics of care' (Cheney, 1987). Besides emphasising the values in relationships, the latter ethics makes a plea for doing away with abstract criteria, abstract cost-benefit analysis and so on, and instead letting problems and people tell their own story, in their own voice. And this, turning back to the practicalities of applied environmental science work, leads to entirely new, 'post-modern' ways of analysing problems and relating to people in the design of solutions.

In applied research one often comes up against questions of this kind, but they generally remain unanswered; as defined, the final variables are those variables that *usually* require no further argumentation. Nevertheless, the examples show that normative-contextualising questions can be very practical for the content and legitimacy of environmental policies. Environmental philosophy is not for philosophers only!

(2) Physical-scientific contextualisation poses the question 'why?' with respect to all the dose-effect relationships used in the problem analysis; theoretically, it therefore lies 'below' the whole 'problem block' in Figure 2.2. Because this is too difficult to draw, the physical context is shown as a 'leg' in the figure, hooked on to environmental capacity. Immediately below the environmental capacity (and, consequently, below all physical-science models that have been used in the derivation of the environmental capacity) we find the 'direct capacity-determining factors'. The nutrients in a soil are an example, determining, for instance, the minimum rotation time of a shifting cultivation system. Another example is the minimum foraging area of a predator, determining, for instance, the minimum size of a nature conservation area. Beneath these factors lies the 'ecological basis', denoting the fundamental patterns and processes in the environment, connected to basic research in geography, oceanography, biology, etc.

(3) Social-scientific contextualisation is research into the social causes of an environmental problem. The structure of the relevant part of Figure 2.2 is as follows. First, there are the *actors*. Actors are defined as the social entities (individual people and organisations) with a significant decision-making capacity regarding the activity concerned. Actors make choices on the basis of the available alternatives for action (their *options*) and the pros and cons of these options, here called the *motivations*. Options and motivations are usually co-determined by other actors. A government agency, for example, that offers a subsidy on fertiliser co-determines the motivations of many farmers. A large landowner who buys up ever more land restricts the space, and hence the options available, for small farmers. The agency and the landowners thus co-generate the environmental problem. This is the meaning of the term '*actors field*' in the figure. The actors field is a form of social *structure*, which, together with the *culture* in which the actors are embedded, forms the social basis of the options and motivations. Research that starts out from a problematic activity (action), then identifies the primary actors, and works its way from there, progressively contextualising towards other actors and towards structure and culture, is called Action-in-Context research in de Groot (1992), and forms part of the Problem-in-Context framework as a whole. By taking the action-in-context approach, we connect the decisions of farmers, consumers, firms, agencies and other actors to the underlying power relationships, culture and markets in society (Blaikie, 1985).

As has been stated above, the activities stand at the pivotal junction where physical-scientific problem analysis connects to social-scientific problem explanation. Crossing that interface, the researchers must make a switch in perspective. In order to understand why actors act the way they do, one must proceed from *their* knowledge, values and ideas, not primarily from those of oneself or those embodied in the natural sciences. In other words, we must at this point drop the value perspective of the normative observer and put ourselves in the place of the actor whose action we seek to explain. This switch in perspective is why an *'environment of the actors'* is included in Figure 2.2; it is defined as the environment that co-determines the actors' decisions. We find an example of this in the inspiring story of Mishra and Sarin (1987). These Indian researchers did a problem analysis that had started out from the silting up of a reservoir close to a city. During this analysis, they travelled upstream along the river (and upstream along the causal effect chain), in search of the areas from where the silt was eroding into the river feeding into the reservoir. Thus, they identified upstream villagers as the actors causing the erosion, by keeping many more goats on steep slopes than the environmental capacity allowed for. The environment that had played a role in the researchers' problem analysis thus consisted of the reservoir, the river and the slopes, seen from the normative and physical-science point of view of the researchers, who here acted as normative observers. Discussing the causes of over-grazing with the villagers, the same slopes (but now viewed from the point of view of the villagers) proved to play a role in the explanatory accounts of the villagers too. Moreover, the villagers explained the presence of the goats by referring to other places where they might graze, and to the decline of the yields of arable fields around the village, and so on. Thus, the environment of the problem analysis turned out to be very different from this 'environment of the actors', which became part of the problem explanation and, with that, the core of its solution. Emphasising this, Figure 2.2 states that actors have their own environmental problems that co-explain the problem as defined by the normative observer. Actors who carry out problematic activities not only *are* problems, but also *have* problems! These problems may be analysed by doing a mini PiC as part of the big PiC. (In mathematical terms, PiC is a recursive framework.)

The final element in the 'social causes' part of Figure 2.2 is the separation of the activity into 'number of actors' and 'activity per actor'. This is a multiplication which is trivial in itself, but which may be helpful in the explanation. For instance, with regard to the rain forest, we may say: 'hectares of forest burned each year = (number of farmers) × (hectares burned each year per farmer)'. The first factor then relates to demography and migration, and the second factor to the farmers' land-use decisions. These may be explained separately, running along different action-in-context routes (and later giving rise to different policy options). The same goes for the carrying capacities side: we may say that given the farming system, the acceptable number of farmers is so-and-so many (this is often called 'volume policy' in general terms), or, given the number of farmers, that the acceptable farming system is such and such. The number of people that can live off the land sustainably, given a certain farming system and satisfying a certain income standard in a given context of markets, is often called the 'human carrying capacity' of a region (e.g. Daly, 1991). All this, in turn, is part of the well-known 'I = PAT paradigm' (e.g. Hardin, 1991). This formula says: environmental Impact = Population × Affluence × Technology.

SOCIAL-SCIENTIFIC EXPLANATION OF ENVIRONMENTAL PROBLEMS

The objective of this section is to deepen and to illustrate the social contextualisation following the action-in-context principles introduced in the previous section. To start with, two methodological principles should be noted:

1. *Actor models.* If we know the options an actor can choose between, and if we know what the actor sees as the pros and cons of these options, do we then know what the actor will in fact choose? We often do to a sufficient degree of precision by following the basic rule of putting oneself in the place of the actor: what would *you* do if you were these farmers, this firm, these boat owners, this politician? It may sometimes be necessary to apply a more explicit and theory-connected actor model. One way to do this is to use a model of 'moral domains' (e.g. de Groot, 1992: section 5.7), which assumes that actors have various more or less consistent 'calculi' of decision-making, such as the 'homo economicus', 'homo communalis' and 'homo honoris' modes of reasoning. Another way is to use the narrow but powerful actor model of microeconomics, which assumes that people are 'homo economicus' all the way (e.g. Ellis, 1993); 'rational choice theory' is another term for this approach.
2. *Guidance on the route of progressive contextualisation.* As we have said, the explanation of environmental problems is an open search along causal chains that run far and deep without end. Which one to take, and when to stop? If we identify a factor, should we go on and search for the actors behind this factor? How deep into history and culture should we go? Should we divide an actors category such as 'farmers' into rich and poor farmers, or men and women? Besides using one's own common sense and following a cyclical approach, the most general advice is to keep tabs on the problem relevance of these choices, i.e. whether they make a difference to the final variables. In very applied research projects that are tied to a specific agency, this advice may be narrowed down to 'solution relevance', i.e. a focus on the actors and factors that are within the reach of control of this agency. For more free types of research, Vayda (1983) stresses the 'surprise principle', i.e. an open eye for, and a subsequent focus on, the unexpected things one may discover. For this type of research, it is noteworthy, too, that causal insights are enhanced by contrast in the data, e.g. by also including actors who do unexpected things, even though they may be few in number.

As an example for this section, we take a deforestation problem in a hypothetical tropical country. Figure 2.3 shows a large part of the PiC, with the problem itself very simplified because we here focus on the social explanation. We see that the activity has been divided into two sub-activities, connected to two categories of primary actors: migrant farmers and logging companies. They do the forest burning (to create fields) and the forest cutting, respectively. These we call 'primary' activities and 'primary' actors, because we encounter them first on the road of progressive contextualisation from the environmental problem downwards (hence not because they might be more important or more responsible than the 'secondary' and 'tertiary' actors behind them). From the primary actors onwards, an action-in-context analysis can be made in two steps:

- first the identification of the *actors field*, seeking horizontally, as it were, the secondary and tertiary actors behind the primary ones
- and then a *deeper analysis*, vertically as it were, for relevant actors in the actors field.

The actors field

The actors field is defined as the structure of how actors are connected to other actors in the causation of a problem. The notion of 'actors connected to other actors' is usually translated directly by social scientists into 'social network', which is the structure of how actors are connected to other actors through communication. If we start out with a migrant farmer in Figure 2.3, for instance, his social network will consist of his neighbours, his friends and so on, i.e. other migrant farmers. But do these determine why the migrant farmers burn the forest once every few years to create new fields, or might do something else? They hardly do; all migrant farmers are in the same boat, and in a *causal* analysis we would prefer to know who pushes and pulls this boat! In other words, we want to establish the lines of social influence (the 'power lines') that run from the whole category of migrant farmers to other actors, irrespective of whether the farmers know these actors or not. Finding these is basically very simple. Actors make certain decisions because they have certain options that have certain pros and cons in terms of certain motivations. Therefore, the actors behind this are the ones *who have an influence on either these options or these motivations.* This principle generates the whole actors field.

Figure 2.3 is simplified in many respects. One is that the actors have only 'homo economicus' motivations, for instance. But it is detailed enough to explain the essentials of the deforestation actors field. The first thing to note is that the picture is built up as a repetition of a single triangular image: action/actor/options/motivations. Starting out with the migrant farmers, we see that these have three basic options: (1) the slash-and-burn agriculture they are practising at present, (2) investment in more sustainable agriculture and (3) to migrate. The motivations concern short-term benefits, long-term benefits, investments and risks. Influencing these options and motivations, we find the logging corporations in a role of secondary actors. Because they cut down large trees, they open up the forest for the farmers' slash-and-burn option, and because they make logging roads, they connect the farmers to markets, thus creating their cash benefits. Alongside the farmers we discover the loggers as primary actors. Their options are (1) 'grab-it-and-run' (that is, cut as much as possible in a short time), (2) 'comply with regulations' (that is, sustainable logging by not cutting more than the 'Allowable Annual Cut' (AAC)), and (3) give up and re-invest elsewhere. Remarkably, we find the migrant farmers as secondary actors here; they encroach in the loggers' concession areas and, in doing so, they take away the loggers' long-term benefit of compliance. If the loggers were to apply a sustainable rotation and come back to the cutting site thirty years later, they would not find any trees left to cut! Behind this we may identify government actors and politicians who deny secure land tenure to both farmers and loggers; these tertiary actors certainly have their own reasons for doing so, but they are not included in Figure 2.3.

Every log cut by the logging company above the AAC norm stated in its permit is an illegal log. In some countries this would rule out the 'grab-it-and-run' option, but in many others 'illegal' in fact only means that you have to pay up to get the log to the harbour, e.g. pay the forestry officials and the police.

This is the 'cost of corruption' mentioned in the figure. Corruption is a kind of market, in which officials offer for private sale the powers vested in their public office. In Figure 2.3, the loggers are the buyers and the supply side of the market is summarised in one secondary actor, the 'big politician', who is the political boss without whom nothing moves in the region. To try to explain the big politician's action we have to put ourselves in his place. What are his options and motivations? Figure 2.3 gives these. The basic mechanism is that in order to remain in power in a 'patronage' context, you need a lot of money because, one way or the other, you have to buy votes, either in blocks or by direct trading; a usual price in poor regions of the Third World is around $5 per vote. And often the forest and the loggers are the only sources around from which this money can be extracted. Connected to this phenomenon are many tertiary actors, e.g. democratic movements that teach the people what in fact they are selling. Many more tertiary and quaternary actors could be drawn in Figure 2.3, e.g. national governments and the World Bank, but the actors field principle has by now been sufficiently illustrated. The final thing to note is that *all* the actors now identified are potential target groups for policy-making. Quite often, it will turn out that working with the secondary actors may be more effective than (only) doing local projects with farmers!

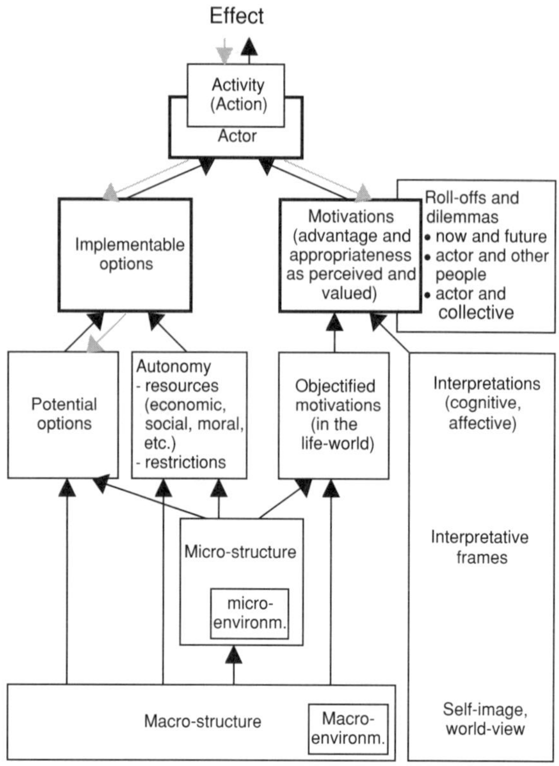

Figure 2.4 The schema of the single-actor deeper analysis

Note The solid arrows are the causal direction, the other arrows indicate the research sequence of progressive contextualisation

The single-actor deeper analysis

By analysing the actors field, we have identified *who* may become involved in the solution of the problem. In order to understand better what to *do* with these target groups it is often worth looking at the backgrounds of their decisions in more detail. We do such an analysis separately for each of the most relevant actors in the actors field. Figure 2.4 has been constructed for this 'single-actor deeper analysis'. We simply go through this figure element by element, going from the top downwards, which is the normal research route in a progressive contextualisation. At the top, in the thick-line boxes, we see the familiar triangle of action, actor, options, motivations. With respect to the options, it is now specified that we focus first on the options implementable for the actor in question. With respect to the motivations, it is now stated explicitly that we are interested in the 'advantage and appropriateness' of the options 'as perceived and valued' by the actor. The term 'advantage' relates to the 'homo economicus' mode of reasoning, the term 'appropriateness' to the others. Visible, too, in the figure is that a brief list has been added to the motivations element. This refers to a number of social mechanisms which are often identified as causes of environmental problems. A well-known one here is the 'tragedy of the commons', whereby the benefits from a certain activity accrue to the actor alone, while the costs are spread over a larger collective. Often in such cases the activity is not stopped even if it means over-exploitation of the environment, e.g. overgrazing or overfishing. It is important to note that the term is in fact a misnomer;

the tragedy pertains largely to open-access resources rather than to real commons, which have a governing body.

One step further down in the figure we find the element of 'potential options', denoting all the things the actor might conceivably do (including the implementable options). The next element, 'autonomy', is defined as the degree to which the actor can actually implement the options of the potential options list. Autonomy means capital in the widest sense, i.e. economic, social and cultural (Bourdieu, 1986), but (as Figure 2.4 states) diminished by governmental or cultural restrictions (prohibitions, permit conditions, taboos, 'things a real leader just doesn't do', etc.). In the deforestation example of Figure 2.3, the logging corporation has enough capital to counteract the AAC restriction on its autonomy and implement the 'grab-it-and-run' option. 'Empowerment' is the general term used for the policy instrument of raising the autonomy of actors in the financial, social or cultural field, so that more potential options become implementable. With regard to the logging corporation of our example, it is easy to see that the reverse is advisable. This does not mean, however, that empowerment is a policy option that should be confined to the poor; what counts for its application is the balance of motivations for or against the environment.

The next elements focus on the motivational side. First, 'objectified' motivations are defined as the motivations which we may express in simple quantitative terms, e.g. hours of work or hours of waiting, dollars of income, calories of food, risk percentage of a certain event and so on. Identifying these is a tool with which to distinguish between the structural and cultural aspects of the actor's motivation. As drawn in the figure, objectified motivations are often connected to structural elements of society, such as markets, transport systems, regional rates of unemployment and so on. The next element, 'interpretations', is then defined as the difference between the objectified motivations and the ones above these in Figure 2.4, i.e. those that really count for the actor in question. An example is that half an hour of passive waiting for a bus may have a very different value from losing half an hour in a traffic jam. Another example are the many books written by anthropologists about the difference between food as calories and 'proper' and 'tasty' food in various cultures. The Dutch eat half-rotten raw fish, but of course only if it is herring; a half-rotten raw mackerel would be downright disgusting, although it is the same in objectified terms. The word 'interpretation' implies that we assume the actor has richer and more active mental capacities than suggested by the frequently used term 'perception' or the even more passive term 'information filter' (Gold, 1980). In Figure 2.4, the element of 'interpretations' extends downwards to indicate that the interpretations arise from 'deeper' and more encompassing 'frames' of interpretation. The three modes of moral reasoning are an example of what is meant here. Next in the figure, the interpretative frames are seen as embedded in a subsequent 'deeper' level, which are the self-images of what it is to be a good farmer, a real manager, a member of the community, a person in nature.

In the element of objectified motivations, the words 'in the life-world' have been added. The actor's life-world is defined as everything of immediate significance to the actor. Its elements are parts of the micro- and macro-structure of community and society, found at the bottom of Figure 2.4. The distinction between 'micro' and 'macro' is not a distinction of geographical scale but of social scale. A micro-structure is any social system where the actor has a co-determining role, he is missed when he is not there, be it in the classroom, the family or the Internet discussion group. In a macro-structure, however, the actor does not make a difference. Consequently, the oil market is macro for most actors but micro for OPEC member states. The distinction between micro- and macro-structures is important for many social phenomena. If you make a difference, you act differently. The distinction thus also has a bearing on the instruments for environmental policies: macro-causes require macro-instruments (e.g. a change in the tax system), or the rehabilitation of a micro-structure (a commons) that may manage the environment.

DESIGN AND EVALUATION

The focus of this section is the top block in Figure 2.2, called 'design of solutions'. As a basic sequence, design follows problem analysis and problem explanation;

it is the research activity that aims to turn all the knowledge acquired in the analysis and explanation into practical action. The result of a design process can be a national policy plan for forest protection or acid rain abatement, a programme for local environmental projects, a plan for a green tax system, a regional soil restoration strategy or any such thing. The large arrow originating from the design block in Figure 2.2 depicts the implementation of the design in the real world, initiated by the normative observer. We see that the arrow is directed towards the physical elements that have been researched in the problem analysis, and towards the social elements that have been researched in the problem explanation. This is to say that designs often contain a mixture of physical and social elements. As said before, the social policy elements, reaching deeper into the roots of a problem, are often the more effective ones in the longer term. Also visible in Figure 2.2 is that the design process is fed from various sources. First, there are many facts from the problem analysis and explanation that are re-used in the design. Then there are the final variable values, usually the same ones that have also driven the problem analysis; they are often summarised in the 'terms of reference' of the design assignment. Most importantly, there are the many 'thin' arrows visible in Figure 2.2 that originate from all over the problem analysis and explanation. These are the 'options for solutions', or, put more formally, the *potential plan elements*. Put together, these potential plan elements form a long and loose list of everything that might help to alleviate the problem. We often encounter such long loose lists, under the heading of 'policy recommendations', on the final pages of research reports. These recommendations are potential building blocks for a solution, but they are not the solution itself. For a solution to arise, the options have to be selected and combined in a design process, in order to form the higher system level of the plan as a whole.

A logical first step in a design process is to take full stock of all the potential plan elements. How to arrive at this long loose list? The basic rule is that every causal linkage in the problem or its context might be intervened in so as to alleviate the problem. In other words, as Figure 2.2 shows by the thin lines rising from all over the picture: *every causal element or linkage of the problem analysis and explanation generates one or more options* for a solution. In Figure 2.1, for instance, going from the bottom up, options are: (1) reduce the number of boats, (2) accept the number of boats, but see to it that each boat results in less phosphate emission, (3) accept the number of boats and the phosphate emission per boat, but try to let that result in lower phosphate levels, e.g. by adding iron to the water so that the phosphate flocculates and sinks, (4) accept the phosphate level but try to have that result in less algae, e.g. by . . . , and so on over the whole picture. Such an exercise will be the starting point of a design that may include, for instance, catching almost all whitefish in order to break the self-reinforcing feedback loop in the system, combined with the regulation of boats, planting of trees along the banks to reduce the wind that whirls up phosphate-rich and transparency-reducing sediments, and so on. Several general insights can be distilled from this example. One is that the closer to the final variables the options are generated, the more they take on the character of symptoms abatement. A second insight is the general rule that, in principle, options for solutions should never be rejected because they look somewhat strange *as such*; after all, the plan will later be evaluated with reference to the criteria of cost, biodiversity, etc. of the plan *as a whole*, not of the separate elements. The tree planting element, for instance, might work well in combination with others, and be of negligible cost if it could be integrated into a landscaping programme.

The generation of social policy options follows the same principle as for physical ones, but taking the problem explanation as the generating field. As stated, the actors field generates the potential target groups. To find out what to *do* with these actors, the generating principle says that we should seek to influence their options and/or motivations; for instance, with regard to the big politician in Figure 2.3, trying to raise the average price per vote. A more detailed source of social policy options is the 'deeper analysis' schema of Figure 2.4. For each actor, we can go through this schema on the level below the level of options/motivations, and look systematically at what might be changed in these factors of 'potential options', 'autonomy', and so on. Doing so, in general terms, we arrive at a well-known type of list usually called the list of 'policy

BOX 2.2 TYPES OF OPTIONS FOR SOLUTIONS ('POLICY INSTRUMENTS') FOLLOWING THE PROBLEM-IN-CONTEXT LOGIC

1 *Physical policy options* (identified in the problem analysis): insulation, clean-up, treatment, recycling, conservation, physical zoning, etc.
2 *Policy options focusing on actors' options*
- increasing the number of potential options (technological research, environmentally sound agricultural systems, public education);
- increasing the autonomy of actors (their economic, social and other resources);
- decreasing the autonomy of actors ('direct regulation'): prohibitions, permit conditions, regulatory zoning, etc.

3 *Policy options focusing on actors' objectified motivations*
- with main emphasis on macrosphere: financial and legal internalisation
  - levies, subsidies, deposit money, etc. (incentives),
  - legal liability and security;
- with main emphasis on microsphere: social internalisation
  - turn macro (free access) into micro (commons),
  - intensify the visible rightness (honour and care aspects) of right action (product information, publication of pledges, public honour for caretakers); participatory policy design; covenants.

4 *Policy options focusing on actors' culture*
(interpretation schemata, world-views): make visible the consequences of environmental problems; stimulate the theory and practice of environmental ethics; tell the story, show the beauty and encourage the experience of what is valuable, e.g. nature close to home.

instruments', with items such as 'direct regulation', 'economic instruments', and so on. Box 2.2 gives this list on the basis of Figure 2.4, as a general support of design studies. Because of the systematic character of Figure 2.4, the list of Box 2.2 is the only fully comprehensive one in the literature.

After the inventory of the potential plan elements, the next step in a design process is the selective combination of these options into a number of possible plans. It is usually advisable first to make several of these, each designed to be optimal for a different set of 'weights' on the various final variables, e.g. one draft design that is more 'nature-friendly', one more 'people-oriented', and one more 'cost-conscious'. These draft designs can then be subjected to an informal 'pre-evaluation'. The general definition of evaluation is to take a systematic look at the degree to which various plans satisfy the various final values (cf. Box 2.1). The pre-evaluations in a design process do not yet serve to select one best plan, but to check if the plan designed to be most nature-friendly really is so, or might be improved by including elements from other plans to make it less costly, and so on, improving each draft design one by one. These improved designs may then be re-evaluated, and so on. Typically, therefore, design is a cyclical process; the sequence written in the design block of Figure 2.2 is only a simplified example. All the way, the design process has to be steered by the abstract image of the final variables (that is, the *aims*) and not be focused on the concrete means. De Groot (1992) gives a number of design methods. Ranging from simple common sense to the rather exotic or heavily computerised, they all are meant to facilitate and systematise the creative jump from the separate potential plan elements to the system level of the plan as a whole.

Methodologically, policy design is not essentially different from the design of, say, a house. The architect, too, has final variable values (e.g. the house's beauty, efficiency and cost) that are used both to steer the design process and to evaluate its result; the architect also starts out with a long loose list of potential plan elements (e.g. all possible attractive doors, effective types of roofing, efficient room patterns, etc.). And s/he, too, will select and combine these in the design of the higher system level of the plan (the designed house) as a whole. But the architect is trained for this, and we are not; design requires a type of creative rationality that is actively 'out-trained' in the education of most scientists. But, as stated by Alexander (1982) and Tjallingii (1996), you may analyse, explain and evaluate as much as you like – the quality of what will finally be implemented in the real world will never be greater than the quality of what was designed!

Plan evaluation plays an informal role in various stages of a design process, but is most visible at the end when the plan is handed over to the decision-makers. Environmental Impact Assessment is a formalised plan evaluation process. In PiC terms, plan evaluation is a repeated problem analysis. What you do to evaluate a plan is to 'feed' the elements of the plan into the PiC structure, and then assess the impacts, going upwards until you meet, on the value side of Figure 2.2, the appropriate standards to compare the impacts with. Especially in the more official plan evaluations, we need a methodology to aggregate the multitude of impact-versus-standard comparisons into a smaller number of final variables. The two methods most widely used for this purpose are multi-criteria analysis (MCA) and cost–benefit analysis (CBA); literature under these two key terms is easy to find. In MCA, effects are grouped according to a set of final variables (often approximating those of Box 2.1), called 'criteria' in this type of analysis. Thus, one arrives at a table with, on one axis, the different plan alternatives, and on the other the indications of how these plan alternatives score in terms of criteria designed to express the values of health, economy, nature and so on. The analysis then consists of manipulations of these indications, e.g. by assigning different weights to the different criteria, or by pairwise comparison of the plan variants. In CBA, as many impacts as possible are expressed in monetary terms, e.g. by calculating the construction costs, future benefits of production, loss of recreation possibilities, and so on, so that money becomes the single, overall final variable. Then, the costs and benefits are 'discounted': that is to say that future costs and benefits count less than those in the present. Usually, a discount rate of 5 or 10 per cent per year is chosen, so that one dollar next year counts for only 95 cents today (note that benefits and costs accruing to future generations now count for almost nothing). Then, the 'net present value' (NPV) is calculated by adding up all discounted costs and benefits. The plan with the highest NPV is chosen.

MCA and CBA are beset by serious problems. Both, for instance, have an objective appearance but are very susceptible to the conscious or unconscious biases of the researchers, who have to make many decisions on how to aggregate the data and how to present them. The traditional rarity scale of plant species, for instance, puts so much emphasis on the rarest of plants that plans containing isolated nature reserves automatically come out better than plans containing more integration of people and nature. Another example concerns the 'willingness-to-pay' methodology often used in CBA (de Groot, 1992, section 4.10). A third problem with MCA and CBA has already been touched on (see p. 32): both are *abstract* methods that stand in the way of really listening to the voices emanating from the concrete problematic situation. However, decision-makers are often more interested in that concrete account of the situation. Finally, the most fundamental criticism of MCA and CBA is that they both are *efficiency* instruments, bypassing the morality of *equity*. Both methods trade everything off against everything else, trying to establish a total, overall 'best way'. Many authors (e.g. Page, 1991 and Tjallingii, 1996) therefore make a plea for separating equity decisions from efficiency (cost–benefit) decisions, in a 'two-tier' evaluation system. Hence we first concentrate on the equity issue, that is, the *protection of the weak* (the basic needs of the poor, the protection of nature), and sustainability (the protection of future generations); only then do we assess the overall efficiency by applying MCA or CBA. In the 'Safe Minimum Standards' approach, for instance, the three equity conditions are incorporated into every plan. Thus, no unsustainable plans leave the hands of the designers, to be fed into the cost–benefit evaluations of the economists. The other way to arrive at a two-tier evaluation is to first assess the 'equity acceptability' of a plan in a public or legal discussion (Sagoff, 1988). The great advantage of two-tiered evaluation is that the cost–benefit analysis itself is freed of the duty to incorporate equity arguments. For instance, if a project is designed so that it is in itself sustainable and poses no sustainability risks elsewhere, evaluating this project in a CBA with *any* discount rate can no longer be a source of unsustainability.

The end result of a design process is a plan or a set of alternative plans, usually with some kind of evaluation attached, to be fed into political or social decision-making. Making a set of alternative plans gives more formal freedom of choice to the decision-makers. In practice, however, the outcome of their choice can be almost fully manipulated by the designers and evaluators. The 'democratic quality' of a design is therefore largely determined by the degree to which people (problem victims, problem actors, the public, the politicians, etc.) have been involved in the design and evaluation process itself. 'Participation' is the general term for this involvement. Participation can vary in several dimensions, e.g. the width of the circle of participants and the intensity of their inputs (de Groot 1992, ch. 6). Sometimes, these inputs are confined to a few specific moments in the process, e.g. discussing the terms of reference and discussing the draft designs. In other cases, for instance in participatory projects in developing countries, people are included over the full duration of the design process. Roughly speaking, participation makes design studies technically more difficult (for instance, because we have to also put our participants in the relatively loose and abstract 'design mode' of thinking), but also more rewarding. This is not only because it satisfies the aim of democratic quality, but also because participation commits participants to the plan. It becomes their 'own', and that may be crucial for the plan's implementability. Thus, participation is an aim but at the same time a means (a 'policy instrument', cf. Box 2.2).

CONCLUSIONS

This whole chapter has focused on a straightforward understanding of PiC as an integrative instrument for applied 'whole-problem' research. In this final section we will look at PiC in a less straightforward and less applied way, so that the reader will be able to use the framework with a maximum of flexibility in the maximum of situations.

A more sociological look: PiC as explained here is made for the environmental scientist, i.e. somebody in a position of doing the problem analysis, explanation and design. It is also possible to look at this matter more from the outside – more sociologically, studying how others do problem analysis, etc. PiC is helpful here too, but some of its elements shift. In particular, the normative contextualisation shifts away from ethical study to sociological study of how various interests mobilise to establish their normative perspective as the dominant one.

What is in the clouds?: PiC as explained here is a purely 'problem-driven' approach. If there is no problem, you cannot apply PiC. Although environmental problems will not be in short supply in the years to come, it is a definite limitation that PiC is conceptually blind to what may be called environmental *opportunities*, i.e. potential improvements that are not directly linked to a problem. A framework designed to handle both problems and opportunities would become much too abstract for a book like this, however. In Figure 2.2, this limitation of PiC is acknowledged by the two lowest clouds drawn on the left and right. They symbolise the non-PiC world of opportunities, out of which options for solutions may arise as if out of 'nothing'. In this way, the picture reminds us to stay alert for this world of opportunities, e.g. when consulting experts, farmers, firm managers and other sources of inspiration. A third cloud is visible at the top of Figure 2.2: this one symbolises the hidden world of what researchers often call the 'irrationality' of outsiders, which comes to the fore especially at the stage of plan implementation. Sometimes, an analysis is applauded, an explanation is endorsed and the ensuing plan is even adopted, but nothing further happens! The powerful often have other agendas, or the local people may have politely 'adopted' our plan purely because they did not want to disappoint us after all our work. Participation, of course, is the key to minimising such surprises, but they can never be fully mastered.

Whole-PiC cyclicality: It has been said in the previous sections that analysis, explanation and design are often cyclical types of research. In practice, it is usually advisable to go one step further, and include the whole basic sequence of analysis, explanation and design in a cyclical approach. We then first follow the sequence on an exploratory level, then do it again with more sophistication and more focus on the areas that appeared to be the most relevant, and so on. Two interesting extremes here are 'rapid prototyping' and 'learning-by-doing'. In 'rapid prototyping', we start an exploratory design phase almost directly at the beginning of a research project. We accept that this design will most probably end up in the waste basket later, but we use it to overcome the 'design fear' often present in scientific teams, and to prepare for a well-focused problem analysis and explanation. The approach of 'learning-by-doing' means also including the implementation phase in the research cycle. A 'pre-implementation' is then started very early in the process; analysis, explanation and re-design continue during the pre-implementation, incorporating the experiences gathered during that work, for instance, in a flexible programme of soil and water conservation (Zanen and de Groot, 1991). This is of course the perfect way to address the 'cloud of surprises' mentioned above, in a permanent 'dialogue of action' between researchers, local people and nature. 'Learning-by-doing' is feasible only if the learning can be organised without risks or high cost, however. Finally, one thing which must *not* be done in terms of PiC cyclicality is to skip the phase of social-scientific problem explanation. Although a physical/technological science approach is still prevalent in most textbooks on environmental management, it tends to lead to shallow policies that do not address the causes of problems.

Other PiC applications: PiC is not only applicable to applied research on whole problems. It can also be used for 'strategic' research, for instance, that is also problem-oriented but with a longer-term objective. In strategic research, more attention than has been possible here has to be paid to the many linkages between PiC and scientific theories, from economics, geography, political science, biology and so on. PiC can also be used to provide context to monodisciplinary endeavours. Soil research, political studies, etc. usually focus on a small area of the PiC image; the framework can serve to set that research in a whole-problem overview that clarifies its linkages to other disciplines and to society.

Everybody with his own PiC: Like any other framework, Problem-in-Context is not something to be applied rigidly as if it were a mathematical method with only one good way of doing it. In Slootweg *et al.* (1996) we find an example of a PiC adapted to become a kind of big-step tool for natural resource assessment. The steps, roughly, are (1) resource inventory, (2) functions and problems of these resources, (3) primary actors connected to these functions and problems and (4) secondary actors connected to the primary ones. A second example is Kessler *et al.* (1997), in which PiC is simplified in structure but more elaborate in process terms, with the outputs of one research step explicated as the inputs of the next. I do not propose, however, that every student should now rush out and invent his or her very own framework. The better route seems to be that of the two examples, i.e. first to master PiC as well as one's own field of environmental expertise, and only then to make one's own well-grounded adaptation.

REFERENCES

Alexander, E.R. (1982) 'Design in the decision-making process', *Policy Sciences* 14: 279–292.

Blaikie, P. (1985) *The Political Economy of Soil Erosion in Developing Countries*, New York: Longman.

Bourdieu, P. (1986) 'The forms of capital', in J.G. Richardson (ed.) *Handbook of Theory and Research for the Sociology of Education*, New York: Greenwood Press.

Cheney, J. (1987) 'Ecofeminism and deep ecology', *Environmental Ethics* 9 (2): 115–145.

Daly, H.E. (1991) 'Elements of environmental macroeconomics', in R. Costanza (ed.) *Ecological Economics*, New York: Columbia University Press, pp. 32–57.

De Groot, W.T. (1992) *Environmental Science Theory*, Amsterdam: Elsevier Science Publishers.

Ellis, F. (1993) *Peasant Economics* (revised edition), Cambridge: Cambridge University Press.

FAO (1989) *Guidelines for Land Use Planning*, Rome: Food and Agriculture Organisation.

Gold, J.R. (1980) *An Introduction to Behavioural Geography*, Oxford: Oxford University Press.

Hardin, G. (1991) 'Paramount positions in ecological economics', in R.Costanza (ed.) *Ecological Economics*, New York: Columbia University Press, pp. 47–57.

Kessler, J.J., Toornstra, F.H., Van Wetten, J.C.J. and Heringa, A.C. (1997) *Strategic Environmental Assessment (SEA); A framework for planning and integration of environmental care in SNV policies and interventions*, The Hague: Organisation of Netherlands' Volunteers.

Mishra, P.R. and Sarin, M. (1987) 'Social security through social fencing', *Papers Congress 'Only One Earth'*, London: Internat. Instit. of Envir. and Devel. (summarised in Conroy, C. and Litvinoff, M. (eds) (1988) *The Greening of Aid*, London: Earthscan).

Ortolano, L. (1984) *Environmental Planning and Decision Making*, New York: John Wiley.

Page, T. (1991) 'Sustainability and the problem of valuation', in R. Costanza (ed.) *Ecological Economics*, New York: Columbia University Press, pp. 58–74.

Sagoff, M. (1988) 'Some problems with environmental economics', *Environmental Ethics* 10 (1): 55–74.

Slootweg, R., Maresch, S. and De Vries, D. (1996) *Terms of Reference voor het opstellen van geintegreerde profielen van departementen in Bolivia*, Haarlem: Ecotec Resources.

Tjallingii, S.P. (1996) *Ecological Conditions; Strategies and structures in environmental planning*, Wageningen: DLO Institute for Forestry and Nature Research (IBN Scientific Contributions 2).

Vayda, A.P. (1983) 'Progressive contextualisation: methods for research in human ecology', *Human Ecology* 11 (3): 265–281.

Vayda, A.P. (1996) *Methods and Explanation in the Study of Human Actions and their Environmental Effects*, CIFOR/WWF Special Publication, Jakarta: Centre for International Forestry Research.

Von Wright, G.H. (1986) 'Is and ought', in M.C. Doeser and J.N. Kraay (eds), *Facts and Values*, Dordrecht: Nijhoff, pp. 31–48.

Zanen, S.M. and De Groot, W.T. (1991) 'Enhancing participation of local people; some basic principles and an example from Burkina Faso', *Landscape and Urban Planning* 20, 1, 151–158.

SUGGESTED READING

The emphasis here is on practical but non-physical science knowledge, methods and theory for analysing and solving environmental problems, because these are often under-represented in textbooks on environmental management. Of the books mentioned in the references list, then, the ones of general importance are:

Ellis (1993). Although focusing only on farming households, this is a good book for gaining an economic understanding of other actors, too, because it does not narrow down human decision-making to profit-seeking alone.

Blaikie (1985). Following a relaxed kind of action-in-context approach, this gives good insights into the linkages between the decisions of actors and the surrounding society.

De Groot (1992). The general reference of this chapter, but also provides material on interdisciplinarity, problem-oriented modelling and environmental philosophy.

Costanza (1991). An inspiring compilation resulting from the need to build up an environmentally conscious discipline of economics.

Tjallingii (1996). Although the applications focus on the relatively narrow field of suburban planning, this book gives many noteworthy general design principles.

Among books not mentioned in the list of references, the following are suggested reading:

Bernard, H.R. (1994) *Research Methods in Anthropology*, London: Sage – a practical and balanced overview of methods for basically all types of social science work.

Elster, J. (1989) *Nuts and Bolts for the Social Sciences*, Cambridge: Cambridge University Press – a sharp, wise and practical introduction to the social sciences, also for social scientists.

Rolston, H. (1986) *Philosophy Gone Wild*, Buffalo: Prometheus Books – still my favourite among the many good books on environmental philosophy.

Engel, J.R. (ed.) (1990) *Ethics for Environment and Development*, London: Belhaven Press – remarkable for its breadth, also containing contributions from Hindu dharma and Islam.

Ness, G.D., Drake, W.D. and Brechin, S.R. (eds) (1993) *Population–Environment Dynamics*, Ann Arbor: University of Michigan Press – theory and case studies on an issue of rapidly growing importance, linked also to the upcoming debate on 'environment and security'.

SELF-ASSESSMENT QUESTIONS

1 Let us assume you have a vegetable garden. Like everybody else in your neighbourhood, you have used pesticides to control weeds and insects. But recently you have read an article about the possible consequences. Hence, there may be an environmental problem in your garden. Analyse this problem, and its triple context, with the PiC framework.

2 Suppose you are an environmental scientist employed by the Public Works department. Your agency plans the construction of a new highway to develop a certain area. By law it is obliged to make an Environmental Impact Assessment. The EIA will be written by a firm of consultant engineers, but it is your job to write a memorandum of guidance to that firm, so that the experts will focus on the relevant issues. How would you go about doing this, with the help of the PiC framework?

3 Severe water pollution is observed in the lakes of a certain area. One of its main causes are the nearby wood pulp factories. Waste water is discharged into the lakes causing algal blooms and so on. Design a strategy for the solution of this environmental problem, surveying all the types of options identified in the Box 2.2 (p. 39).

3
THE CULTURAL FILTER

Willy Weyns

SUMMARY

In recent years there has been a growing public awareness of global environmental challenges throughout the world. Unprecedented in human history, this awareness is fostering a new sense of urgency to protect and preserve the natural environment. In response, both the emerging academic discipline of human ecology and the practice of environmental management are now striving to reintegrate humans, with all their complexities and ambivalencies, into nature.

In this chapter the first objective is to overview recent scientific developments that are relevant for humankind in its endeavour to conceive global environmental strategies in various disciplines such as:

- physics (thermodynamics, quantum-physics, general relativity and chaos theory);
- mathematics (Gödel theorems), (para-consistent) logic, epistemology;
- anthropology, cultural history and comparative philosophy;
- biological sciences (ecology, bio-history, etc.).

In a very general way, the common denominator of many of the findings in those disciplines now poses a substantial challenge to the old assumptions and provides the impetus for developing an alternative new paradigm of what may be called 'reflexive equilibrium'. This new world-view, which is tending to replace the old model, highlights the multiple interactions and dialectical relationships between:

'Hard' material culture	↔	'Soft' ideas, world-view, cultural filter
Bio-evolution	↔	Cultural evolution
Determinism	↔	Freewill philosophies, choices open to humankind
Facts	↔	Values
Scientific expertise	↔	Public decision-making

A number of methodological needs have so far been identified for acknowledging and taking full advantage of these new scientific findings, notably the need for interdisciplinarity, the need for interculturality, and the need to develop arguments in a public debate, both about facts and values. All these are considered in this chapter which ends with an appeal for thinking the unthinkable: is there anything left for humans to manage in the emerging new paradigm of 'humans-in-nature'?

ACADEMIC OBJECTIVES

1 To advance convincing arguments for, and hints on, implementing interdisciplinary programmes for environmental management.
2 To bridge the gap between natural sciences and human sciences by advocating the theory and practice of public debate on facts, values and norms, and by defining accurately the role of scientific expertise in public decision-making on issues of environmental management and environmental policy.
3 To view local environmental problems in the global context, and to make the concepts of sustainable development (WCED, 1987) and Agenda-21 (UNCED, 1992) more operational than at present.
4 To link the present and future generations in a new and environmentally ethical way with respect to the numerous global environmental challenges, and to deal with the 'unity-in-diversity' paradox in a positive way.

5 To take advantage of cultural diversity in developing robust environmental strategies, involving peoples from all around the world in spite of their current socio-economic status.
6 To construct realistic environmental management strategies simultaneously with bridging the gaps in human development.
7 To examine the importance of the cultural filter in developing multicultural consensus for environmental management strategies at the global level.

ENVIRONMENTAL MANAGEMENT, CULTURE AND SUSTAINABILITY

Environmental management can be defined as 'the art that seeks to balance human demands upon the Earth's natural resource base, with the natural environment's ability to meet those demands on a sustainable basis' (Colby, 1990). One can draw upon this definition to show that the practice of environmental management is necessarily a part of the overall human–nature relationship in any culture at any moment of its history. This is certainly the case with the very broad anthropological definition of culture, which includes everything that is transmitted from one generation to the next through teaching and imitation (Boyd and Richerson, 1985; Boyden, 1990; Pinxten, 1994). In this way anthropologists include in the cultural realm much more than the so-called 'high Culture', with a capital 'C', which is restricted mainly to the fine arts, religion, philosophy and science. Their definition includes beliefs and values, institutions, artefacts (material culture) and practices, as well as other factors which affect human behaviour (Figure 3.1).

The wide perspective of anthropologists' definition of culture, out of more than one hundred and sixty definitions surveyed by Kroeber and Kluckhohn (1952), is the most adequate for environmental management for a number of reasons:

First, it overcomes the conventional dualism upheld by both neo-liberals and traditional Marxists. Neo-liberals believe that only economic growth can provide the means to address both environmental and social problems (the so-called 'trickle-down' theory); while orthodox Marxists separate reality in terms of a 'hard' economic foundation and a 'soft' realm of (high) culture – the latter placed as a 'superstructure' on top of the first. In contrast to both these ideologies, the operational definition of culture we will use encompasses all the layers and segments of society, stressing the multiple interactions which

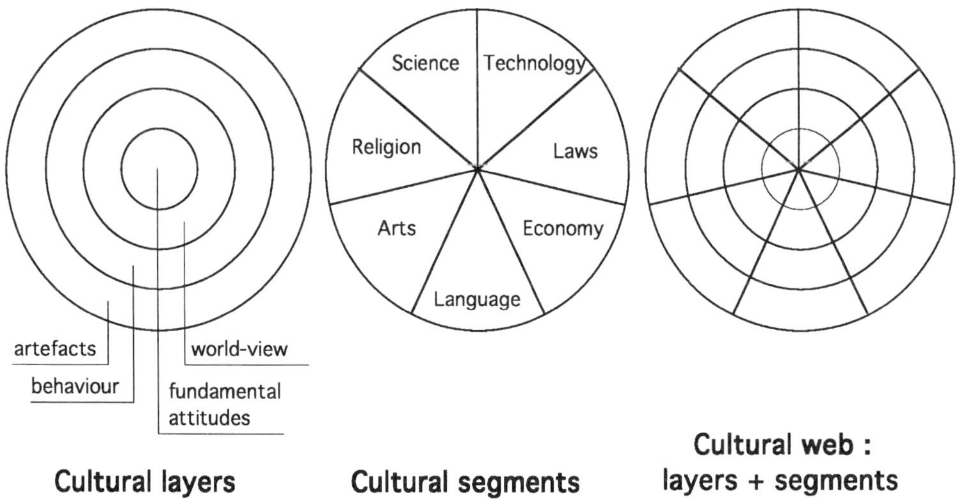

Figure 3.1 An anthropological definition of culture
Source: Weyns, 1995a

occur between humans and the ways in which they impact upon and transform nature, and vice versa.

Second, the social context, which can be seen as being intermediate between the above-mentioned symbolic and material poles, is also intertwined with them both, because the social position of any person inevitably constrains to a certain extent the way he or she views the human–nature relationship (see Table 3.1). The following are germane in this context:

- Entrepreneurs generally perceive nature as cornucopian (i.e. blessed with plenty), benign and highly resilient, capable of making good any damage that humans could inflict upon it. They claim an unlimited freedom of action.
- Hierarchists – the socially well integrated – perceive nature as being tolerant. In general they respect thresholds of irreversibility and favour imposing limits on environmental risks that can or should be taken.
- Egalitarians see nature as ephemeral and extremely fragile. They prescribe rigorous accountability as a responsibility for preventing further damage to the environment.
- Fatalists – the silent majority – view nature as rather capricious and uncontrollable. Management strategies do not really make any difference according to them, because people have no real influence in society.

Social change, resulting from important development projects, can also have important environmental impacts. Aware of this, the World Bank has included societal analysis as a component of its Environmental Assessment Guidelines (Box 3.1).

In spite of the multiple interactions which occur between the different realms, the adjustment of one with respect to the others is not always swift or simultaneous. Significant or even substantial time-lags often occur. Today an institutional framework, based on the dated concepts of modernism and European Enlightenment, still prevails in many parts of the world, in spite of the overwhelming challenges of a new paradigm originating from twentieth-century science, post-modernism and radical ecology. Here 'paradigm' means a world-view, or a model, around which perception of reality is organised. This socially expanded concept of a paradigm is broadly derived from the original Kuhnian definition which restricts scientific enterprise, meaning 'a criterion for choosing problems that can be assumed to have solutions. Other problems are rejected as metaphysical' (Kuhn, 1970).

Third, culture, thus conceived, includes 'hard' material artefacts such as buildings and road infrastructure, in addition to mobility and industrial activities. Although the anthropogenic transformation of the earth's surface is inevitable, its magnitude has been changing dramatically over historical times. Hence the recent emergence of the new disciplines of bio-history and environmental history which focus on the history of environmental load exerted by humankind (Arrow *et al.*, 1995; Hughes, 1995). According to Boyden (1990), in approximately the twelve millennia separating the Neolithic revolution from the present high-energy phase, human population has increased by a factor of 1,000 and its energy use by a factor of 15,000! Furthermore, the environmental space, defined as annual material throughput of the actual techno-metabolism expressed as per capita global average, can be quantified as shown in Table 3.2.

The last figure in Table 3.2, which refers to the most important greenhouse gas responsible for global

Table 3.1 Four social positions related to four concepts of nature and environmental management

Social position	*Views nature as*	*Concept of environmental management*
Individualist	Nature benign (cornucopian)	Unlimited liberty
Hierarchist	Nature tolerant	Respect thresholds!
Egalitarian	Nature ephemeral	Accurate accountability
Fatalist	Nature capricious	Management is illusory

Source: After Thompson *et al.*, 1990

BOX 3.1 SOCIAL AND CULTURAL ISSUES RELATED TO ENVIRONMENTAL REVIEW

- *Indigenous peoples* are dependent on ancestral land subject to customary rights to use natural and environmental resources. They tend to be divided in gender, age, occupational and ethnic minority groups, that are quite often in fierce competition for their use. Concretely the incidence of contrasting the socio-economic position of women and men can play a major role in the use of water and other environmental resources.
- *Cultural property* includes archaeological remains and structures of high cultural, religious, and historical value as well as natural sites of esteemed aesthetic value that deserve conservation measures.
- *Involuntary resettlement* is quite a common drama of watershed inundations as a consequence of many dam construction programmes. In case this cannot be avoided, an active participation of the people involved proves to offer the highest chance for avoidance of social and environmental destruction.
- *New land settlement* has to take both the physical characteristics of the new agricultural land and the social aspect of land allocation into account, in order to be socially and environmentally sound.
- *Induced development* such as increased population size may enhance environmental destruction through the increased competition for fuelwood, water, building materials and waste disposal that it entails. The project should provide adequate compensation to prevent those nuisances.

Source: World Bank, 1991

Table 3.2 The global technometabolism

	T / capita / year
Iron	0.5
Gravel + stones	7.3
CO_2–exhaust	20.0

Source: Boyden, 1990

climate change, has increased by a factor of one hundred. Comparing it to gas production by biological respiration alone, Boyden estimates the actual load of humans on earth as 5 per cent of the ecological load of the entire biosphere. If we carry on according to the 'business-as-usual' scenario, this percentage is estimated to increase by a factor of about ten and reach 50 per cent by the year 2100, although such predictions are subject to a plethora of assumptions and uncertainties. At the end of this chapter we will be somewhat better equipped to proceed a little further in this speculative direction.

For the time being we will return to the definition of environmental management, given its clear linkage with the concept of sustainability. In so far as it relates to environmental management, we will abandon the original consensus definition of sustainable development (WCED 1987) in favour of one that is more operational. Such a definition is given by Daly (1985), who is an environmental economist, according to whom 'sustainable development is a process in which qualitative development is maintained and prolonged, while growth in the biophysical scale of the economy becomes increasingly constrained by the capacity of the socio-economic system to perform two essential functions in the long run: to regenerate raw material inputs and to absorb waste outputs of the human economy'. In short, this is what environmental management is all about.

THE CULTURAL FILTER

Within the all-encompassing framework of any culture, we have so far stressed the multiple interactions between socio-economic and physical realities *vis-à-vis* the prevailing ideas. These ideas tend to be structured, coherent in part and contradictory on the whole, into a 'cosmology' or a 'world-view' (Worldviews Group, 1994). Together with explicit assumptions, such a social construct of reality implicitly integrates unconscious mental habits. And the construct and its attributes, both quintessentially ever-changing and provisional, function as a model or a map. It is thus a prerequisite for any rational action, as it is for environmental management. And it is precisely the mechanism we call the 'cultural filter', on the basis of which humans decide how to act according to their perception of the environment and reality, and which shapes their belief in nature (including themselves) and in the 'world-out-there' (Apostel, 1994;

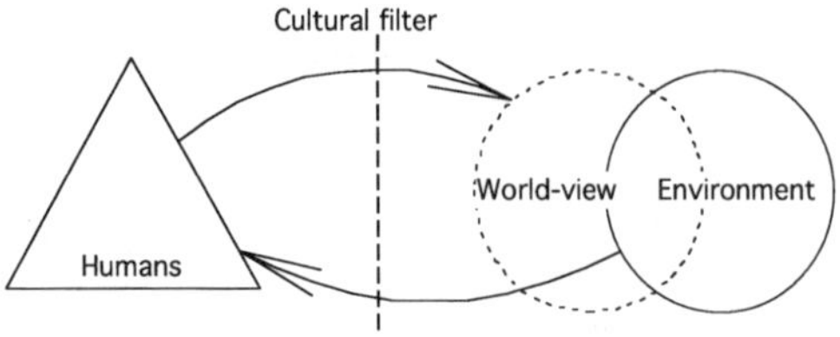

Figure 3.2 The world-view and the cultural filter

Naess and Rothenberg, 1989; Pepper, 1984). Thus, because people wear differently coloured sun-glasses so to speak, an average Chinese views the environment rather differently than a North American, a European or an African (Figure 3.2). And through education and imitation, their world-view is heavily indebted to the specific cultural context in which they are embedded.

At this point it is necessary to mention the perennial philosophical dispute between the advocates of determinism and the free-will philosophers. Determinists tend to stress the inescapable laws of nature, such as laws of economics, influence of the genes and the like. Politically this often paves the way to an attitude which is overtly conservative or even fatalistic. On the other hand, defenders of free-will philosophy, like existentialists and phenomenologists, proclaim that humans are free to choose and act either against a supposed alien nature, or in favour of it. The 'deep ecologists', who respect nature for its intrinsic values, belong to the latter category.

Since environmental management is concerned with voluntary and desired effects of human intervention, it has, at least implicitly, greater affinity with the free-will philosophies. But the application of science and its inherent causality, in an attempt to predict the impact of its action, inevitably imports a good deal of natural science's determinism at the same time.

To conclude, Apostel (1994), Georgescu-Roegen (1971), Ost (1995), Pepper (1984) and others have advanced convincing arguments in favour of what is called 'dialectical thinking' in the sense of a continuing search for a 'reflexive equilibrium' between extreme determinism and free will (freedom of choice).

FACTS AND VALUES

Can science be objective?

Empirical science relies on the collection of facts through observation and experimentation. Facts are descriptions of events or states of affairs. Thus facts have a dual nature: an *objective* one, in that they come from the world, and, at the same time, a *subjective* one too because they depend on human observation and description. The question then is, is this a sufficient guarantee for objectivity? Often the more information a fact yields, the less certain it is. Experiments can only be controlled by a very limited number of peers. Moreover, the art of experimentation is *implicitly* value-laden; and *values* are qualities or attributes of a person, a group, an object or a process, and ideally they transcend individual or group interests or preferences. Experimentation is by definition a manipulation of nature (Bacon). The *de facto* decontextualisation carried out in experiments, as well as the extreme simplification of reality in constructing models, is in contradiction with the complex web of inter-relationships of any subject and its environment. But this nexus is precisely the subject-matter of environmental sciences. Furthermore, the very act of making a choice, which for example favours the ideals of objectivity and academic autonomy, is in itself totally subjective, and it can hardly be justified on rational grounds.

Such considerations led Laudan (1984), an epistemologist, to conclude that proper scientific methodology is both value-laden and norm-laden, including methodologies that are commonplace in environmental management such as those of environmental impact assessment (EIA) and cost–benefit analysis (CBA) (Schrader-Frechette 1985). In the case of EIA, for example, the initial 'scoping' or listing of a (necessarily) limited set of alternatives is not, and cannot be, value-free. Secondly, the idea of a common unit to make trade-offs between incommensurable qualities can hardly be justified, because commonly used monetary units are subject to market imperfections. Moreover, neither the EIA nor any other efficiency-oriented, utilitarian tool has the ability to promote distributive equity. All such arguments clearly show that neither EIA nor CBA is in any way objective or value-free. But their bias to a specific world-view is rarely admitted by their proponents (Weyns 1995b).

Notwithstanding the above, it should be made clear that facts, by their very nature, are distinct from values, although this distinction is seldom respected or even acknowledged. And this is precisely where lies the root of what is called the 'naturalistic fallacy'.

The naturalistic fallacy

The Roman Catholic Church tried to prevent Galileo Galilei from stating scientific facts; Nazism abused Darwinism in the form of social Darwinism; and state socialism has dictated the laws of biology (Lysenko) – to name but a few of the best-known cases of 'values directing facts'. An emancipatory reaction against those manifest historical aberrations has been both necessary and vigorous. 'Scientific facts should be endorsable by scientists of multifarious belief-systems', is a famous dictum ascribed to Max Weber. And empiricist David Hume formulated his weak form of non-naturalism in the following terms: '"ought" can in no way be logically derived from "is"'.

But the pendulum does not stop halfway, so to speak. A stronger variant of non-naturalism was defended by the logical positivists in this century. This prompted Moore (1929) to coin the phrase 'naturalistic fallacy' which defines the distinction between 'natural' facts and 'non-natural' values. For logical positivists of the 'Vienna Circle', authentic scientific research *is* simply value-free. And even today this is the dominant belief held by most scientists and laymen alike, for fear of repeating the aberrations that occurred when the pendulum was on the other side. But, in these times of complex global change, is it still defensible, adequate or desirable to maintain such simplistic and absolute distinctions?

Science in the global age

As a result of the ideology of rigorous fact–value separation, which is the trademark of conventional neo-positivistic science, ethical questions and matters of human well-being that are most important in human ecology and environmental management tend to remain labelled as 'subjective', 'irrational', or even 'meaningless'. But, nearly a century after the Vienna Circle, both problems and science have changed a great deal. Human ecology has reintroduced *Homo sapiens* into nature. Recent findings in epistemology, in logic and in mathematics all support the proposition of a gradual transition from facts to values, from natural to human sciences, instead of an unbridgeable gap separating them (Batens, 1978). Clearly, more research on values is absolutely necessary to build bridges by working prudently on the real problems of a new (weak) naturalism that seems to be inescapable in the global age. More specifically, progress has to be made on the problem of universality of values and on how to arrive at ethical conclusions.

We have learned from modern astrophysics to perceive the cosmos in a permanent state of expansion since the 'big bang'. This cosmology is consistent with a world-view of successive levels or layers of increasing complexity, each evolving from those before it – from quarks, elementary particles, atoms and molecules, cells, multi-cellular organisms (e.g. humans) and populations (e.g. human societies including their cultural aspects), to ecosystems and finally the biosphere as a whole. Higher levels evolve from the lower levels, and, therefore, knowledge of the qualities of the lower levels helps us to understand those of the higher levels. Also, something novel is added as the level increases. There is a preponderance of general theories (e.g. in physics) at the lower levels where facts are of paramount importance, whereas in higher levels we encounter contingency, singularity and complexity which necessitate description, values and evaluation.

The proponents of what is called 'post-normal science' have opened a promising avenue with regard to the above. By integrating humans, human populations, ecosystems and the biosphere as a whole, the disciplines of environmental management and human ecology have also integrated the higher levels of complexity and uncertainty into the process. Thus, the more global the environmental problem in question is, the greater the integration, complexity and uncertainty. Traditional scientific methodologies are no longer appropriate for addressing problems of such complexity. In a situation 'where facts are uncertain, values in dispute, stakes high and decisions urgent', Funtowicz and Rawetz (1995) propose a 'second-order science' or 'post-normal science' (Figure 3.3).

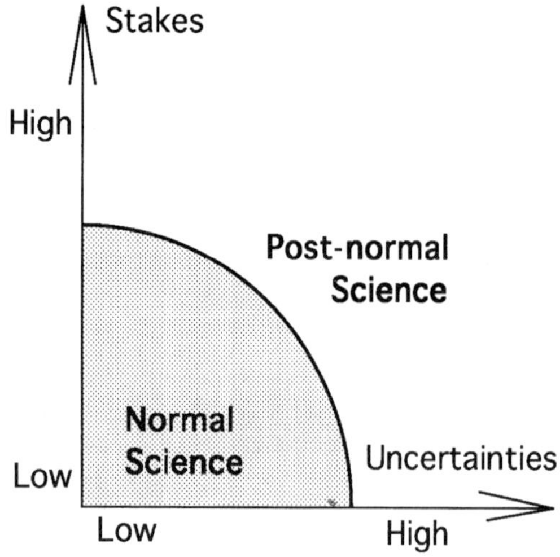

Figure 3.3 Uncertainties and stakes at hand

Source: After Funtowicz and Rawetz, 1995 in Westra, L. and Lemons, J. (1995) *Perspectives on Ecological Integrity*, Dordrecht: Kluwer. Reproduced with kind permission from Kluwer Academic Publishers.

These concepts also validate and require the systematic involvement of educated laymen in promoting greater democracy for making hard policies based on soft scientific inputs.

The delicate relationship between scientific expertise and public policy decisions will be explored further in the next section.

INTERDISCIPLINARITY

The environment being the interface between nature and humans, it is clear that contributions from both natural and human sciences are needed in environmental management for addressing or managing environmental problems. To this end multidisciplinarity, which is the mere juxtaposition of research findings from a number of monodisciplinary investigations, can be considered as the first step. But proper interdisciplinarity is a more ambitious objective. It aims at a real integration of individual disciplines into an overall dynamic process. Everyone embarking on an interdisciplinary project has to know the other participants personally. And, in a truly interactive process, a common framework (a common language, mathematical models, translation mechanisms) needs to be constructed progressively. At the other extreme lies transdisciplinarity as the total merging of the constituent disciplines into a single, all-encompassing one. In what follows we will focus on interdisciplinarity – its motivations and advantages, handicaps, historical antecedents and some examples, and its typical internal dynamics.

For interdisciplinarity

Apostel and Vanlandschoot (1988) from the Brussels-based interdisciplinary CLEA/Worldviews Group enumerate the following important motivations for interdisciplinarity:

1 Reality itself is not subdivided into disciplines, but science is.
2 Almost everyone – laymen and scientists alike – have a psychological need to integrate their experience and perception of nature and culture into a 'world-view' or a 'total view'.
3 Environmental issues are extremely complex and constitutively transboundary in nature.
4 There is a new and evolving relationship between science and society – an emerging trend towards applied, action-oriented science which is in synergy with society at large.

In comparison with their monodisciplinary counterparts, interdisciplinary task forces are far better equipped to face the combined challenges of (3) and (4) above under mounting urgency for taking strategic action.

Against interdisciplinarity

The complexity of reality itself has caused the historical proliferation of monodisciplines, and the history of science bears testimony to the rapidity with which specialist monodisciplines have multiplied. But a growing debit side of this hitherto success story is now emerging, revealing the deficiencies inherent in the extreme fragmentation of both institutionalised research and education. Typically, almost all scientists are trained in a strictly (mono)disciplinary tradition. And, because of this exclusively disciplinary context, interdisciplinary programmes are almost always very difficult to fund, and can often be developed only at the expense of ingenious administrative complexity.

Frequently these impediments deny an understanding of how the environment works in its true modality as an 'organic whole'.

Antecedents and current international programmes

Some disciplines such as anthropology, geography, civil engineering and ecology have been from the outset more biased towards interdisciplinarity.

The subject-matter of geography, for example, comprises both the natural and the social environment, thus making geographers the natural bridge-builders between the natural and social sciences (Jollivet, 1992).

Typically, the systems-analysis methodology, developed by civil engineers for problem-solving, has proved to be a very useful tool for transdisciplinary study and research. Probably the best-known transdisciplinary programme, for which credit is due to engineers, is the Apollo Programme of the NASA which successfully landed humans on the moon and brought them back to earth.

The interdisciplinary process

Given below is a typical chronology of the successive phases encountered by an interdisciplinary project team (Apostel and Vanlandschoot 1988):

(a) Inertia, self-confirmation and self-repetition by the monodisciplinary participants. Such power and prestige 'games' seem to be inescapable.
(b) Team members with backgrounds in different disciplines naturally tend to adopt fundamentally different approaches, languages and methodologies to address the same problem. Often they are not qualified for what is needed in an interdisciplinary endeavour; or if they are, their relevant qualification often proves to be 'superficial'. Some members tend to leave the team at this stage.
(c) Creative search for a common ground which is often to be found in common historical roots. In practice this can turn out to be an unproductive search for what is essentially an escape route.
(d) Definition of terms and concepts. Although necessary, this phase is not risk-free.
(e) Search for common fields of application suitable for evolving new plans.
(f) *Disillusion*: the new interdisciplinary instruments do not seem to add value compared to the familiar monodisciplinary approach, and disillusion sets in as a result.
(g) *Total loss*: the participants drop-out in large numbers. Hopefully some *deus ex machina* (e.g. the obligation to make a progress report) would save the project at this stage.
(h) *The grand transformation*: both the remaining participants and the drop outs have been thoroughly transformed by the process. The remaining team members are motivated to invest more time and energy in continuing what now appears to them to be a promising experiment.
(i) Translation of the fundamentals of one's own discipline into one's own words. The rationale has to be made comprehensible to experts in other disciplines. Only highly qualified scholars prove to have sufficient self-confidence to surmount the boundaries of their own discipline.
(j) *The real start*: from now on participants start to be really productive and begin to function in a truly interdisciplinary way.

If anything, the above shows that interdisciplinarity is a very time-consuming process, and no one would take the trouble to engage in it unless the rewards were commensurately high. Thus continuity of funding and other support systems is an indispensable condition of success. Also, positive external stimuli can often help to revitalise the process at its critical junctures.

Human ecology and environmental management

The historical evolution of concepts in scientific ecology, as described by Di Castri (1981), is shown in Table 3.3.

Unlike environmental sciences, the new academic discipline of human ecology clearly belongs to (e) in Table 3.3 which is the latest stage (Hens and Devuyst, 1994). Although human ecology has its historical roots in urban planning in the USA in the 1930s and the ideas of Burgess, human ecologists have gradually enlarged its scope of application. As we have already

Table 3.3 Historical stages in scientific ecology

Type	*Focus*	*Scale*
(a) Auto-ecology	One single species	Anywhere
(b) Syn-ecology	Different species	Anywhere
(c) Ecosystems	All species except humans	One single location
(d) The biosphere	All species except humans	The whole earth
(e) Humans in the biosphere	All species including humans	The whole earth

Source: Di Castri, 1981

mentioned, in human ecology 'the environment' is defined in a holistic way as the overlapping of humans in their socio-economic context, with nature as its subject-matter. Thus it includes, among many others, complex value-laden human problems of poverty, hunger, unemployment and health. Being action-oriented, human ecology offers an interdisciplinary approach to environmental management and policy-making in accordance with the imperatives of sustainable development (WCED 1987) and the spirit of Agenda-21 (UNCED 1992). At the same time, human ecology is in a privileged position to promote the implementation of advanced public decision-making procedures on important environmental issues.

PUBLIC DECISION-MAKING

In almost all democracies there is a trend towards increasing public involvement in environmental policy and management issues. As amply argued in Chapter 2 of this volume, entitled 'Problem-in-Context' (de Groot), environmental problems are becoming ever more difficult to address with traditional technocratic and purely efficiency-oriented methods such as cost–benefit analysis (CBA), environmental impact assessment (EIA) and life-cycle analysis (LCA). What is at stake here is an effective and productive articulation of technical expertise versus public values.

The most promising proposals advanced to date in this regard are 'discursive decision-making tools' in developments such as 'citizen's juries' (Jacobs, 1995), 'commissions of inquiry' (Funtowicz and Rawetz, 1995) and 'technology tribunals'. The last has been proposed by Schrader-Frechette (1985) as an amendment to the Science tribunal experiment proposed by the US Presidential Task Force (1976) (Box 3.2).

Tuler and Webler (1995) define discursive decision-making tools as 'structured discourse settings for the exchange of both *knowledge* (facts and uncertainties) and often conflicting *values* (world-views and life-styles)'. We have already pointed out that the ability of the public to argue and debate on issues of norms and values can be acquired through education (Naess, 1966; Naess and Rothenberg, 1989).

A high standard of public debate on important environmental issues demands a shift from the narrow aspirations of individuals, or of vested interest groups, to considerations of the greater common good of society as a whole (Sagoff, 1989). The hope that an

BOX 3.2 TECHNOLOGY TRIBUNAL

Technology tribunal is a procedure of adverse public proceedings, inspired by court trials. A jury of 'intelligent' citizens – not scientists – is invited to decide on collective environmental stakes. The jury relies on testimonies of independent scientists, competent in the relevant aspects of the problem.

Arguments in defence of the TT:

- *Balance and objectivity:* hearing all the sides of the controversy is the only possible remedy to experts' discord (uncertainties).
- *Democracy:* weakening the threat of technocratic decisions above the heads of concerned citizens.
- *Education:* educative effect upon intelligent laymen throughout the procedure.
- *Political realism:* implementation of any policy cannot be successful until the public has fully endorsed it.
- *Avoidance of the naturalistic fallacy:* no ethical conclusions (how to manage, what policy to implement) can be derived from natural sciences exclusively.

Source: Schrader-Frechette, 1985

open dialogue would promote mutual respect and understanding, and even a wide consensus on a growing number of environmental issues, is also based on the same premise.

INTERCULTURALITY

Unity in diversity

In the last decades an increasing number of environmental problems have grown in scope from local to regional, then to international, and are finally reaching the global or even the planetary scale. This unprecedented historic fact obviously makes the implementation of environmental strategies contingent upon the success of international agreements. At the same time, the anthropological definition of culture underlines the multiple interactions which link environmental management practices to their specific cultural context.

To take but one example from Chinese culture, Wang and Ye (1991) observe that, 'with 7,000 years of history, Chinese agriculture has been nourishing nearly one-fourth of the world population with only seven percent of the arable land of the world, while maintaining sustainable production and soil fertility. It is considered that the holistic ecological worldview of the Chinese has been playing an important role in these achievements.'

In spite of powerful unifying trends at work in financial, economic, environmental (e.g. global change) and other sectors at the close of the millennium, prevailing cultural contexts of countries and regions are still characterised by contrasts and diversity. Comparative cultural and philosophical studies seek to find key variables with which to compare contrasting cultures (Callicott and Ames, 1989; Engel and Engel, 1990; Libbrecht, 1995; Nakamura, 1986; Pinxten and Balangadhara, 1993). To give but a flavour of the findings of those studies, we give below a tentative list of the key variables for some important religious and geographic regions (Weyns, 1995a):

- Taoism: harmony with nature (e.g. deep ecology).
- India, Africa and Latin America: intertwining of social and environmental concerns.
- Japan and the Buddhist traditions: characterised by a more ambivalent attitude to nature in which aesthetic fascination seems to be incapable of preventing savage ecological damage and destruction.
- Jewish, Christian and Islamic traditions: ranging from domination (management?) over nature to its human trusteeship.

In Box 3.3 we summarise the methodology developed by Hofstede (1991) for comparing management skills in relation to the cultural nexus, by taking the example of the subsidiaries of one multinational corporation in more than fifty countries. In this cultural differences in attitudes to management have been studied with regard to the five key parameters described below:

- *Small versus large power gap:* the degree of acceptance of unequal power distribution. A small power gap means prevalence of values of equality, and easy delegation of power and harmony, overdependence (on a superior), hierarchy, status differences and centralisation.
- *Collectivism versus individualism:* preference for a tight social framework and unlimited loyalty of the members of an in-group; and interdependence above a social setting where individuals are expected to look after themselves.
- *Femininity versus masculinity:* femininity stands for caring social roles, whereas masculinity stands for more assertive and performance-oriented roles.
- *The long-term perspective:* perseverance, pragmatism (many truths) and acceptance of change versus the short term, stability, and acknowledgement of one absolute truth. In Asia the first perspective is known as 'Confucian dynamism'.

The main conclusions to emerge from this and similar surveys (e.g. the Chinese Value Survey) are these:

(a) Cultures are well contrasted.
(b) Cultural contrasts are reflected in different management skills.
(c) The linkage between cultural settings and management skills is not simple or causal (i.e. culture is not sufficient ground). Culture functions more as a trigger or a constraint (a necessary condition) in its influence on management practices. Nevertheless, it is in no way the case that one

BOX 3.3 FIVE DIMENSIONS FOR STUDYING THE INFLUENCE OF CULTURAL DIFFERENCES ON MANAGEMENT AROUND THE WORLD

0 (Small)	(Scale 0–100)	100 (Large)
	POWER DISTANCE	
Relative importance of:		*Relative importance of:*
Equality		High dependence
Delegation		Hierarchy
Equality		Status
Harmony		Centralisation
(e.g.: Scandinavia, Israel, South Africa, Switzerland)		(e.g.: Arab world, Malaysia Singapore, Guatemala, Philippines)
COLLECTIVISTIC	↔	*INDIVIDUALISTIC*
We-consciousness		I-consciousness
Social system		Self-actualisation
In-group		Universalism
Organisation		
(e.g.: Latin America, Africa)		(e.g.: Australia, Canada, Netherlands)
FEMININE	↔	*MASCULINE*
Quality of life		Performance, wealth
Small and slow		Big and fast
People and environment		Growth
Interdependence		Independence
(e.g.: Scandinavia, Africa, Thailand, Costa Rica)		(e.g.: Japan, Austria, Switzerland)
	UNCERTAINTY AVOIDANCE	
Competition		Consensus
Tolerance		Security
Willingness to take risks		Legal system
Citizen competence		Expertise
(e.g.: Scandinavia, Hong Kong)		(e.g. Portugal, Greece, Uruguay, Philippines, El Salvador)
	LONG-TERM PERSPECTIVE (cfr. Confucianism/Asia)	
Tradition		Perseverance, initiative
Personal stability		Status
		Working hard, investment
(e.g.: Pakistan, Africa, Philippines)		(e.g.: Hong Kong, Japan, Taiwan)

Source: After Hofstede, 1991

cultural background is definitely more successful in management than others.

The acculturation process

So far attention has been given to the synchronic contrasts between different cultures. But there is another more dynamic way in which to consider the same, in terms of one culture in a process of acculturation with another cultural unit (Figure 3.4).

In fact, communication with other cultural systems often acts as a catalyst for evolution through acculturation. The potential for acculturation depends on the degree of permeability or openness to information from outside of the cultural province under consideration. Both awareness of commonality and of differences may increase during the acculturation process, and enhanced awareness of cultural differences can ease the acceptance of partial compromises and foster tolerance of the remaining differences.

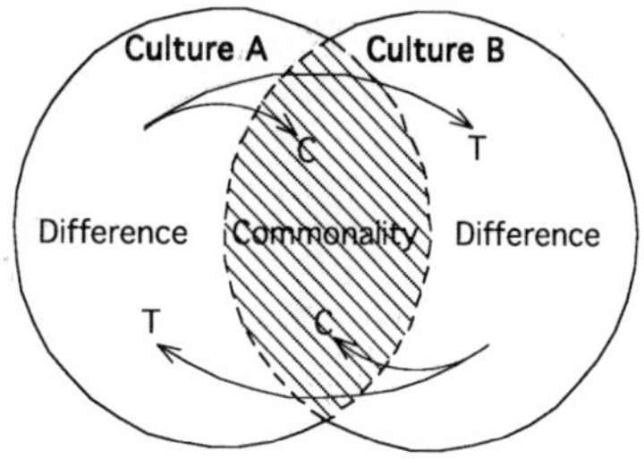

Figure 3.4 The acculturation process

In what follows we will discuss two specific strategies for reaching agreement on global environmental issues which, in spite of or because of their co-existence within the same intercultural context, require intercultural negotiations governed by the principles of overlapping consensus.

Intercultural negotiations

Cultural anthropologists, such as Pinxten and others, see the evolution towards a limited number of important cultural provinces as more likely than the advent of a unique and homogeneous world culture in the near future. Intercultural negotiations hold the promise of a great future (Fischer and Ury, 1981; Pinxten, 1994) for developing a common strategy for survival and for addressing global environmental challenges and other world problems. But this diplomatic art has to be developed and mastered. Global environmental negotiations are bound to be infinitely more complex than the very complex disarmament talks during the cold war era, for example, because in global environmental matters everything is intertwined with almost everything else.

A crucial prerequisite for success seems to be that all parties should be thoroughly aware of, and well informed about, each other's world-views, beliefs and values, before negotiations can begin. No doubt a wealth of useful hints can readily be found in Gandhi's non-violent communication (Gandhi, 1948; Naess and Rothenberg, 1989). Nevertheless, an adequate theory of intercultural negotiations is yet to be developed. To that end the rudimentary principle of overlapping consensus might be a useful starting point.

Overlapping consensus

The principle of overlapping consensus, advanced by Naess and Rothenberg (1989) in the deep ecology debate, appears to hold much promise for intercultural negotiations. Naess distinguishes the four levels of understanding, shown in Figure 3.5, that are central to it.

Naess's 'apron' diagram (Figure 3.5), showing the four levels of communication and understanding, offers a promising model for articulating pluralism at the deeper level (Level 1), while shared principles lie at Level 2 which is more superficial. Cosmologies, world-views and cultural-religious backgrounds belong to Level 1. They tend to be all-encompassing, coherent or contradictory in part, and probably mutually incommensurable. An overlapping consensus on a carefully negotiated number of priority actions belongs to Level 2. Levels 3 and 4 are the concrete implementation levels of environmental management. Although they are broadly derived from Levels 1 and 2, new divergences are common at these levels.

Naess acknowledges the difficulty of communicating ultimate views contained in religious, philosophical and scientific traditions, and he accepts it as a precondition for preserving a highly valued cultural diversity (Level 1). But this does not inhibit his picture from being an extremely dynamic one (i.e. progressive in time). He is constantly exhorting everyone to construct, as coherently as possible, a total view (Level 1) and to improve the linkages

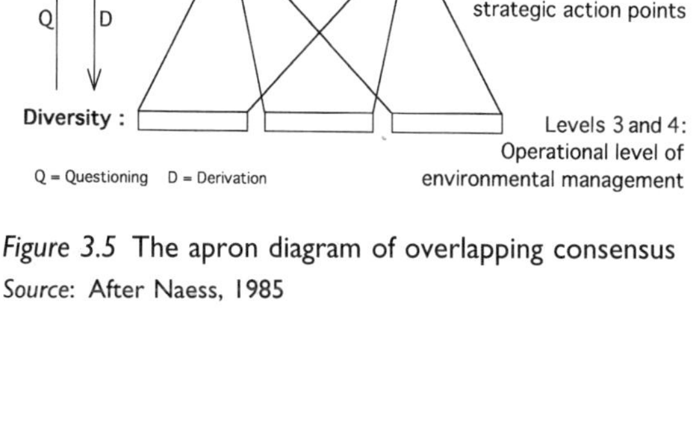

Figure 3.5 The apron diagram of overlapping consensus
Source: After Naess, 1985

between all four levels through deep questioning (bottom–up) and logical derivations (top–down).

The following example typically illustrates the influence of culture on world-view. During the preparation for the UNCED Conference in Rio de Janeiro (1992), the Centre for Science and Environment (CSE) at New Delhi severely criticised the world-view and value-system implicit in the strategic proposals, prepared by the World Resource Institute (WRI) based in New York, for combating the anthropogenic greenhouse effect. Agarwal and Narain (1991) from the CSE elaborated alternative proposals based on a different set of assumptions. The contrasting picture is shown in Box 3.4.

It seems highly probable that cultural diversity should be cherished, just like biodiversity, because it has the potential to enhance the overall chances of survival of the human species as a whole, and to improve the quality of life. Unquestionably the nature–culture relationship lies at the heart of cultural

BOX 3.4 TWO CULTURALLY CONTRASTING ALTERNATIVE PROPOSALS FOR COMBATING THE ANTHROPOGENIC GREENHOUSE EFFECT (AFTER A GLOBAL MAXIMUM HAS BEEN AGREED UPON) AT UNCED, 1992

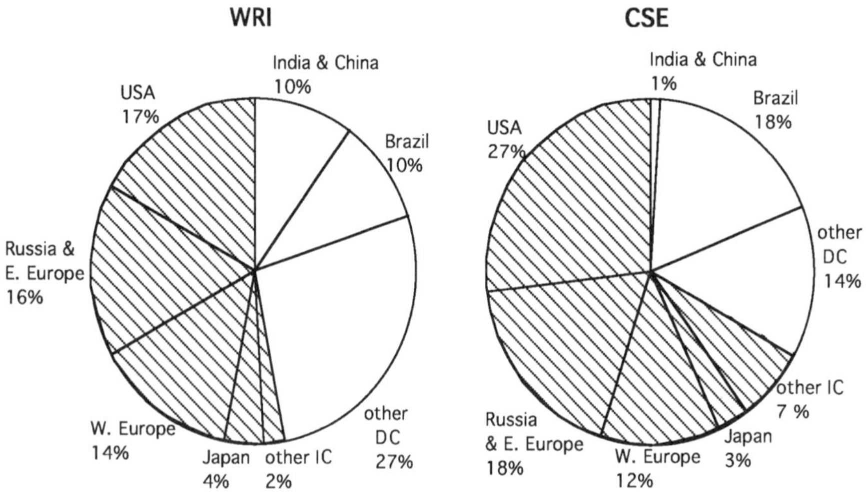

Contrasting world-views produce contrasting environmental data

World Resource Institute (New York)	*Centre for Science and Environment (New Delhi)*
Basic idea:	
The greenhouse effect is a global problem, hence all nations should contribute to its solution. forests	All inhabitants have an equal per capita share of global commons, inclusively absorbing sinks such as oceans and all over the world.
Strategic headlines:	
Reduce national emissions of greenhouse gases by an equal percentage of present emissions.	Only emissions in excess of a per capita share of absorbing capacity of natural sinks should be annihilated.
Practical consequences:	
Net emitter: Japan Net absorbers: USA, Brazil, China, India Nearly zero emitters: none	Net emitters: USA, Brazil Net absorber: Japan Nearly zero emitters: China, India, Indonesia
Critical remarks:	
Greatest polluters are favoured.	Most populous nations and lowest polluters are favoured.

specificity. In the next section we will examine in greater detail the tension between the biological and cultural aspects.

BIOLOGICAL AND CULTURAL EVOLUTION

An extension of earlier considerations of the processes of dynamic acculturation to a longer time perspective raises more philosophical questions about the future evolution of humankind and about life on earth (Sattler, 1986). And here we have to wrestle with the duality of the position of the human species: humans have been, and still are, a part of nature, but, at the same time, they have also emerged in part from the natural sphere through culture. With reference to the determinism versus free-will debate (p. 48), our concern here is with this ultimate question: 'what is there for humankind to manage in the long-term evolutionary process of life on earth'?

The time-arrow

As a background to the concluding section, we elaborate below the omnipresent 'time-arrow' which pervades all three spheres – physical, biological and cultural:

(a) *The entropy law:* contemporary astrophysics describes the cosmos as one that is constantly changing, expanding and unfolding following the big bang. The irreversibility of the time-arrow is thus introduced, and it is echoed in thermodynamics by the second law of Clausius which is an expression of the fatally increasing entropy or disorder (in closed systems). Recently this has inspired further and new considerations of evolutionary processes at higher levels. For example, the systems ecologist Margalef (1986) has linked the rise in entropy, in and around living organisms, to gains in information, while the forerunner of environmental economics, Georgescu-Roegen (1971), has described the human economic system as a substantial accelerator of entropy generation.

(b) *The Synthetic Evolutionary Theory:* long before the advent of human ecology, Charles Darwin began the process of rooting humans firmly back into nature, thus making them fully-fledged participants in the process of biological evolution. Remarkably, in spite of the fierce opposition of creationists and ideological deviations in the form of social Darwinism, Darwin's scientific theory has remained intact and lies at the heart of the Synthetic Evolutionary Theory (SET) developed about fifty years ago (Mayr, 1982). Even today scientists from various disciplines are further exploring and building on Darwin's theory of evolution. Wilson (1975) for example, who is a bio-sociologist, has made the first serious but highly controversial attempt to study the biological framework of cultural development. And, drawing on the Game Theory, Axelrod (1984) has discovered the evolution of co-operation even in very simple biological systems.

(c) *Dual Inheritance Theory:* with culture humans have added a supplementary and highly geared layer into the global evolutionary process, shown in Table 3.4. Boyd and Richerson (1985) have adapted Darwin's theory of evolution to cultural organisms (humans). Their Dual Inheritance Theory (DIT) rests on the premise that culture can also be most profitably viewed as a system of inheritance. According to them, parallels between genetic and cultural evolution are sufficiently profound to make a common conceptual framework adequate, in spite of important differences (Box 3.5).

Ultimately, the human sciences are bound to be unified with the physical and biological sciences anyway. Indeed, the emerging disciplines of biohistory (Boyden 1987) and environmental history (Hughes 1995) are manifestations of the same trend linking biological and cultural evolution.

Table 3.4 Biological and cultural layers

Layer	*Scope*
Individual	One individual
Cultural	One cultural province
Biological	The human species

BOX 3.5 DIFFERENCES AND PARALLELS BETWEEN CULTURAL EVOLUTION AND BIO-EVOLUTION

Differences

- Mating system: cultural offspring different from biological offspring.
- Culture transmitted horizontally within generations: 'epidemiological' instead of genetic transmission.
- Culture acquired throughout life cycle.
- Time-scale much shorter for cultural than for genetic evolution.

Parallels

- Random cultural variation is a cultural analogue of genetic mutation.
- Cultural traits also can get entirely lost.
- Majority opinions can be compared to biased genetic transmission.
- Natural selection of cultures can operate on human survival and reproduction (both genetic and cultural).

Source: Boyd and Richerson, 1985

Uncertainties

No doubt planet earth will survive for a long time into the future, with or without humans. The main parameters humans can hope to manage, or at least influence, are the time period and the nature of transition towards the next state of dynamic equilibrium. Humans will either manage to negotiate that transition earlier and quickly to reach a richer state (defined below), or the changes will occur anyway with humans as mere spectators – probably the latter in possibly an abrupt way, as determined by resource depletion and environmental degradation. The two scenarios, described below, are germane in this context:

1 What is called the 'successful management scenario' builds on the emerging concepts of 'ecosystem integrity' (Goodland, 1995), 'ecosystem health' and 'ecosystem resilience'. These concepts aim at predicting the magnitude of disturbances that can be absorbed by the (global) ecosystem before its capacity for sustaining human life is diminished dramatically and irreversibly. Important indicators of ecosystem resilience are:
   - biodiversity;
   - heterogeneity of ecological functions;
   - biological productivity;
   - options for present and future generations.

2 The 'abrupt degradation scenario', on the other hand, tentatively explores the process of ecosystem degradation leading to ecosystem breakdown (Simmons, 1989). Recent empirical evidence has confirmed the following key indicators (Kareiva, 1996) for this:
   - drastic lowering of primary productivity;
   - very low quality of energy in biomass;
   - very low ecosystem complexity.

The non-linearity of complex dissipative systems, which occurs when they are far from equilibrium, makes the potential fluctuations unpredictable (Prigogine and Stengers, 1984), as illustrated in Figure 3.6.

The only rational way to face those huge uncertainties about the very future of humans on earth

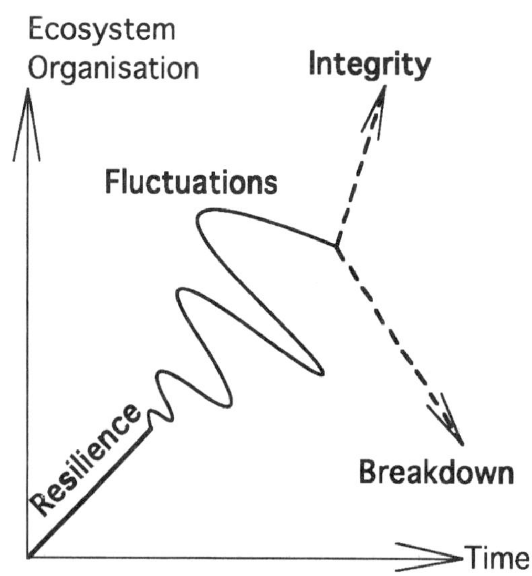

Figure 3.6 Ecosystem unpredictability

would appear to be by adhering to what is called the Precautionary Principle (PP) (Dovers and Handmer, 1995). This principle is defined in the 1992 Australian Intergovernmental Agreement on the Environment (IGAE) as follows: 'Where there are threats of serious or irreversible environmental damage, lack of full scientific certainty should not be used as a reason for postponing measures to prevent environmental degradation.' In our judgement these measures should include qualitative human development, adequate environmental management, and less or no action at all elsewhere.

CONCLUSIONS

In this chapter we have documented the view that scientific findings of the twentieth century and the global environmental challenges have reintegrated humans into nature. Consequently, the subject-matter of human ecology and the core of environmental management are bound to be of the highest complexity. How could it be otherwise, as they are integrating both the 'soft' sphere of world-views, human attitudes and so on, and the 'hard' economic and technological spheres, human history and natural evolution, (human) values and scientific facts.

Hence human ecology is a privileged locus for interdisciplinarity, a cross-over between the humanities, the social and the natural sciences. Procedures for public decision-making bringing into practice an adequate articulation between scientific objectivity and enlightened citizenship are emerging here and there. They will be further developed by human ecology and lead to refined environmental management practices.

As an integral part of all human activities, environmental management, the art of interaction between humans and nature, cannot but be culturally determined. Asians have different management practices than Europeans, North Americans or Africans. In the age of global environmental challenges this state of affairs demands intercultural negotiations, with the aim of reaching step by step an ever-changing overlapping consensus about global strategies of the type of Agenda 21 (but better).

In case the growing awareness of this cultural diversity would foster mutual respect and instil a sustained intercultural negotiation process, it could be that the threats of anthropogenic climate change, soil loss, the so-called ozone hole and other global challenges will be viewed retrospectively as the greatest opportunities humankind has ever faced to develop novel ways of environmental management. At the brink of the third millennium, cultural diversity and biodiversity alike could be the best guarantees for humans to survive and develop a better life on earth in the age of 'constructive post-modernism'.

REFERENCES

Agarwal, A. and Narain, S. (1991) 'Global warming in an unequal world', New Delhi: Centre for Science and Environment (CSE).

Apostel, L. (1994) *Wereldbeelden, ontologie en ethiek*, Antwerpen: Humanistisch Verbond.

Apostel, L. and Vanlandschoot, J. (1988) 'Interdisciplinariteit, wereldbeeldenconstructie en diepe verspreiding als tegenzetten in een kultuurcrisis', Brussels: Centrum Leo Apostel (CLEA)/VUB.

Arrow, K., Bolin, B., Costanza, R., Dasgupta, P., Folke, C., Holling, C., Janssen, B., Levin, S., Mäler, K., Persings, C. and Pimentel, D. (1995) 'Economic growth, carrying capacity and the environment', *Science* 286: 520–521.

Axelrod, R. (1984) *The Evolution of Cooperation*, New York: Basic Books.

Batens, D. (1978) 'Rationality and ethical rationality', *Philosophica* 22: 23–45.

Boyd, R. and Richerson, P. (1985) *Culture and the Evolutionary Process*, Chicago: University of Chicago Press.

Boyden, S. (1987) *Western Civilization in Biological Perspective*, Oxford: Clarendon Press.

Boyden, S. (1990) 'Human ecology and societal change', *J. Hum. Ecol.* I (1): 7–19.

Callicott, J. and Ames, R. (eds) (1989) *Nature in Asian Traditions of Thought*, New York: State University.

Colby, M. (1990) *Environmental Management in Development*, World Bank Discussion Paper 80, Washington, DC: World Bank.

Daly, H. (1985) 'Steady-state versus growth economics', Paper presented to the Conference on Population and Environment, Stanford, Calif.: Hoover Institute.

de Groot, W. and Stevers, R. (1993) 'Problem in context', in B. Nath, L. Hens, P. Compton and D. Devuyst (eds) *Environmental Management Volume I*, Brussels: VUB Press.

Di Castri, F. (1981) 'L'Écologie: naissance d'une science de l'homme et de la nature', *Le Courier de l'UNESCO* 34, Paris: UNESCO.

Dovers, S. and Handmer, J. (1995) 'Ignorance, the precautionary principle, and sustainability', *Ambio* 24 (2): 92–97.

Engel, R. and Engel, G. (eds) (1990) *Environment and development*, Tucson, Ariz.: University of Arizona Press.
Fischer, J. and Ury (1981) *Getting Yes*, Harmondsworth: Penguin.
Funtowicz, S. and Rawetz, J. (1995) 'Science for the post normal age', in L. Westra and J. Lemons (eds) *Perspectives on Ecological Integrity*, Dordrecht, The Netherlands: Kluwer Academic Publishers.
Gandhi, M.K. (1948) *Nonviolence in Peace and War*, Ahmedabad, India: Navajiwan Press.
Georgescu-Roegen, N. (1971) *The Entropy Law and Economic Process*, Cambridge, Mass.: Harvard University Press.
Goodland, R. (1995) 'Universal environmental sustainability and the principle of integrity', in L. Westra and J. Lemons (eds) *Perspectives on Ecological Integrity*, Dordrecht, The Netherlands: Kluwer Academic Publishers.
Hens, L. and Devuyst, D. (1994) 'Human ecology: the European perspective', *J. Hum. Ecol. Rev.* 1: 326–332.
Hofstede, G. (1991) *Cultures and Organisations: Software of the Mind*, London: McGraw-Hill.
Hughes, J. (1995) 'Ecology and development as narrative themes of world history', *Env. Hist. Rev.* 19 (1): 1–16.
Jacobs, M. (1995) 'Environmental valuation, deliberative democracy and public decision-making institutions', in J. Foster (ed.) *Valuing Nature*, London: Routledge.
Jollivet, M. (ed.) (1992) *Sciences de la nature, sciences de la société*, Paris: CNRS.
Kareiva, P. (1996) 'Diversity and sustainability on the prairie', *Nature* 379: 673–674.
Kroeber, A. and Kluckhohn, C. (1952) *Culture: A Critical Review of Concepts and Definitions*, New York: Random House.
Kuhn, T. (1970) *The Structure of Scientific Revolutions* (2nd edition), Chicago: University of Chicago Press.
Laudan, L. (1984) *Science and Values*, Berkeley, Calif.: University of California Press.
Libbrecht, U. (1995) *Inleiding comparatieve filosofie*, Assen, The Netherlands: Van Gorcum.
Margalef, R. (1986) *Perspectives in Ecological Theory*, Chicago: University of Chicago Press.
Mayr, R. (1982) *The Growth of Biological Thought*, Cambridge, Mass.: Harvard University Press.
Moore, G. (1929) *Principia Ethica*, Cambridge, UK: Cambridge University Press.
Naess, A. (1966) *Communication and Argument*, Oslo, Norway: Universitetsforlaget.
Naess, A. (1985) 'Ecosophy T', Appendix A in B. Devall and G. Sessions (eds) *Deep Ecology*, Salt Lake City, Ut.: Peregrine Books.
Naess, A. and Rothenberg, D. (1989) *Ecology, Community and Lifestyle*, Cambridge, UK: Cambridge University Press.
Nakamura, H. (1986) *A Comparative History of Ideas*, London: Routledge.
Ost, F. (1995) *La Nature hors la loi*, Paris: La Découverte.
Pepper, D. (1984) *The Roots of Modern Environmentalism*, London: Croom Helm.
Pinxten, R. (1994) *Culturen sterven langzaam*, Antwerpen: Hadewijch.
Pinxten, R. and Balangadhara, S. (1993) 'Comparative anthropology and rhetorics in culture', in Berrill (ed.) *Argumentation and Written Language*, New York: Springer.
Prigogine, I. and Stengers, I. (1984) *Order Out of Chaos*, London: Bantam Books.
Sagoff, M. (1989) *The Economy of the Earth*, Cambridge, UK: Cambridge University Press.
Sattler, R. (1986) *Biophilosophy*, Berlin: Springer.
Schrader-Frechette, K. (1985) *Science, Policy, Ethics and Economic Methodology*, Dordrecht, The Netherlands: Reidel.
Simmons, I. (1989) *Changing the Face of the Earth*, Oxford, UK: Basil Blackwell.
Thompson, M., Ellis, R. and Wildavsky, A. (1990) *Cultural Theory*, Boulder, Colo.: Westview Press.
Tuler, S. and Webler, T. (1995) 'Process evaluation for discursive decision-making in environmental and risk policy', *J. Hum. Ecol. Rev.* 2: 62–71.
UNCED (1992) *Agenda 21*, New York: United Nations.
US Presidential Task Force, Advisory Group on Anticipated Advances in Science and Technology (Chair: Kratowitz, A.) (1976) 'Science court proposal', *Science* 193 (4254): 653–656.
Wang, R. and Ye, Q. (1991) 'Human ecology in China', in S. Suzuki, R. Borden and L. Hens (eds) *Human Ecology – Coming of Age: An international overview*, Brussels: VUB Press.
WCED (1987) *Our Common Future*, Oxford, UK: Oxford University Press.
Weyns, W. (1995a) *Het gras onder onze voeten. Over interculturaliteit in tijden van global change*, Brussels: VUB Press.
Weyns, W. (1995b) 'Scientific objectivity and worldviews in environmental impact assessments', in *Proceedings Int. Sympos. E.I.A. for water management* (Bruges, 5–17 May), Brussels: AMINAL/GTE/VMM.
Wilson, E. (1975) *Sociobiology. The New Synthesis*, Cambridge, Mass.: Harvard University Press.
World Bank (1991) *Environmental Assessment Sourcebook* (Vol.1), World Bank Technical Paper 139, Washington, DC: World Bank.
Worldviews Group (1994) *World-views. From Fragmentation to Integration*, Brussels: VUB Press.

SUGGESTED READING

Agarwal, A. and Narain, S. (1991) *Global Warming in an Unequal World*, New Delhi: Centre for Science and Environment (CSE). [Culturally contrasting views about global environmental strategies.]

Boyden, S. (1987) *Western Civilization in Biological Perspective*, Oxford, UK: Clarendon Press. [Human history and economics in ecological context.]

Choucri, N. (ed.) (1993) *Global Accord*, Cambridge, Mass.: MIT. [International dimension of global environmental challenges.]

Colby, M. (1990) *Environmental management in development*, World Bank Discussion Paper 80, Washington, DC: World Bank. [Typology of environmental paradigms.]

Funtowicz, S. and Rawetz, J. (1991) 'A new scientific methodology for global environmental issues', in R. Costanza (ed.) *Ecological Economics*, New York: Cambridge University Press. [Second order or post-normal science for global environmental challenges.]

Jacobs, M. (1995) 'Environmental valuation, deliberative democracy and public decision-making institutions', in J. Foster (ed.) *Valuing Nature*, London: Routledge. [Public decision-making.]

Naess, A. and Rothenberg, D. (1989) *Ecology, Community and Lifestyle*, Cambridge, UK: Cambridge University Press. [Argumentation about values, logico-normative systems, objective axiology and Gestalts, overlapping consensus.]

Pepper, D. (1984) *The Roots of Modern Environmentalism*, London: Routledge. [Determinism–free will controversy, scientific and societal backgrounds for environmental management, dialectical thinking.]

Schrader-Frechette, K. (1985) *Science, Policy, Ethics and Economic Methodology*, Dordrecht, The Netherlands: Reidel. [The technology tribunal for public decision-making.]

Westra, L. *et al.* (eds) (1995) *Perspectives on Ecological Integrity*, Dordrecht, The Netherlands: Kluwer Academic Publishers.

Zimmerman, M. (1994) *Contesting Earth's Future*, Berkeley, Calif.: University of California Press. [New conceptual challenges to the mainstream world-view, dialectical thinking.]

SELF-ASSESSMENT QUESTIONS

1 What are the concrete implications for environmental management with an all-inclusive anthropological definition of culture?
2 Explain what you understand by 'the historic root of the naturalistic fallacy'.
3 What are the arguments against the factual and desirable objectivity of science at the threshold of the global age?
4 Explain how the transition from facts to values and norms can be made in a gradual way.
5 Describe the strengths and weaknesses of the interdisciplinary process in addressing environmental issues and problems.
6 Describe the modalities for promoting public involvement in making policies and decisions in environmental management.
7 Explain how cultural diversity translates into differences in management skills.
8 What is the purpose of intercultural negotiations and overlapping consensus in global environmental management?

PART II

INSTRUMENTS FOR ENVIRONMENTAL MANAGEMENT

INTRODUCTION

Luc Hens

SUMMARY

This section unravels the structure of this volume. It starts from a broad definition of instruments for environmental management, which entails both effect-directed and source-directed approaches. This volume deals with source-directed instruments, which are subdivided into legal, economic and communicative approaches. The legal section is the most established. It entails the genesis and analysis of the environmental legal systems in the US and the EU, but also describes in different chapters the main instruments. Both the directive (e.g., standards, permits, etc.) and the preventive (e.g., environmental impact assessment, life-cycle analysis) groups of instruments are included. They are complemented by sections on economic and communicative instruments. The aim of this volume is not only to review and systematise the main instruments for environmental management, but also to describe them in the context of their workability, realisability and efficiency which qualify their use.

INTRODUCTION

There are two main reasons for the editorial decision to have a separate volume on instruments for environmental management in this series of textbooks. The first has to do with the key role of these instruments, which is often overlooked. Many environmental plans worldwide provide clear analysis and description of environmental problems, and often put forward ambitious targets, but fail to clarify adequately the pathways and instruments they will use to meet those targets. This lack of attention to instruments often accounts for the limited implementation of the plans. The second reason is that the development, application and evaluation of environmental management instruments is a rapidly growing field, both scientifically and from the policy point of view. Over the past thirty years the number of instruments for environmental management has grown fast and their specificity has increased substantially. In this context, this volume can be seen as a systematic overview of a *capita selecta* of the available information.

What are instruments for environmental management? There is no unequivocal definition available in the literature. For the purpose of these volumes, we will understand by 'instruments for environmental management' all things which are used or can be used by or on behalf of a particular target group to reach or to promote one or more targets in its environmental policy (based upon Bressers, 1994). According to this definition, different target groups use instruments for environmental management: not only the authorities, but also business and industry, unions, environmental NGOs, etc. use instruments to reach their environmental targets. Moreover, these instruments are characterised by a continuously increasing degree of specificity.

According to the definition given above, a whole series of elements qualify as instruments for environmental management: plans and planning procedures, organisations, resources such as money or manpower, etc. In this book, however, instruments for environmental management are more specific.

EFFECT-DIRECTED AND SOURCE-DIRECTED INSTRUMENTS

Instruments for environmental management can be classified according to different characteristics. A first subdivision can be made according to the effect-directed versus source-directed character. Instruments aiming to mitigate or solve *effects* are, e.g., water treatment plants, filters to purify air, soil decontamination technology, catalytic converters or noise insulation walls. If correctly applied, these instruments change the quality of the environment without moving consumers and producers towards changing their behaviour. When it comes to *source-directed* instruments, behavioural changes are central. Examples of these are education, information, (voluntary or imposed) agreements, subsidies, taxes. This group of instruments is discussed in this volume. Instruments, strategies and procedures used to manage the quality of the different environmental compartments are found in Volume 2, and those aiming to influence the quality of ecosystems are collected in Volume 3.

Source-directed instruments may be subdivided and ordered in different ways. The one used in this volume is partially based on Van der Doelen (1993). He describes three main groups: legal, economic and communicative; however, there are no clear borders between them. Not only standards and permits, but also subsidies and taxes have a legal basis. In countries where green product labels exist as an instrument for environmental communication with consumers, they are often regulated by law. Moreover, both standards and taxes can be set up in such a way that they have an important communicative component.

LEGAL INSTRUMENTS

There are different reasons why authorities choose to use *legal instruments* in environmental management. It is an approach compatible with the health and safety regulations which predate more modern environmental legislation. From this point of view, establishing and using laws is almost the natural management reflex of policy-makers. Moreover, in many OECD countries, environmental policy was established during a period in which the steering capacity of the authorities was held in high esteem. The limited results in the field of environmental legislation, characterised by increasing complexity, the results of evaluations and the general discussion about a withdrawal of authority, have led to more attention being paid to the limitations of legal instruments.

This volume deals with the genesis of and the philosophy behind two models of environmental legislation: that of the United States and that of the European Union. This analysis shows that the great majority of environmental legislation aims to control the effects of environmental problems by addressing environmental sectors, or tries to conserve nature and its values. Most legal regulations were established when problems came to the public's attention on a so-called *ad hoc* basis. In many countries, this resulted in a system of rules which were complex, non-integrated and difficult to maintain, and usually not the best way to manage environmental quality efficiently. Those systems of environmental legislation which are much more recent are more integrated.

This volume also pays special attention to a selection of instruments which are pillars of environmental management, as much in an *ad hoc* legal system as in a more integrated one. Much has been achieved in environmental management by planning and its prerequisite: 'State-of-the-Environment' studies. Of equal importance is the use of standards, still the main instrument for protecting human health against environmental pollution. Risk analysis and hazard assessment aim to develop protection measures in a broader context. More recent is the use of instruments such as environmental care systems which include environmental auditing. Most of these instruments aim to manage an existing situation, preferably without any fundamental changes to the situation, and as such they have a *retroactive* character. There are only a few instruments in environmental management which are essentially *proactive*: the most important of these are environmental impact assessment (EIA) and life-cycle analysis (LCA). EIA is intended to support decisions which prevent or minimise environmental damage resulting from projects, and this will be soon be extended in the form of strategic EIA to programmes, plans and policies. Basically, LCA does the same for products.

All the instruments discussed above share the characteristic that authorities have an important influence on their functioning. Environmental care systems (of which environmental auditing is part) and environmental accounting systems are instruments for improving environmental housekeeping in organisations, whether industrial or in the non-profit sector.

For each of these interdisciplinary instruments in environmental management, this volume contains an individual chapter in which both the fundamentals and the prospects for application are discussed.

The picture which emerges from this section is illustrative of some trends and changing views on environmental management during the last few decades. It all started by using plans, standards, permits, obligations and prohibitions. This quite *directive approach* can be successfully applied in controlling environmental excesses, such as very high SO_2 emission values or heavy metal pollution. On the other hand, it is also characterised by focusing on the problems at the so-called 'end of the pipe', it is relatively flexible, and it entails some major difficulties in control and maintenance. These are the main reasons why, even in societies which use these instruments in an effective way, environmental degradation continues. Therefore while these directive instruments are necessary, they are insufficient alone and need to be complemented.

A first addition is provided by the *preventive instruments* environmental impact assessment and life-cycle analysis. Evaluation studies (see, e.g., Devuyst, 1994; Lee and Colley, 1992; Lee and Wood, 1993) have shown however that the use of these instruments is not easy and could be substantially improved in most countries where they have been used, that they do not cover the whole of the environmental problem and that, on the whole, prevention instruments have only limited capability to lessen the problems they address.

ECONOMIC INSTRUMENTS

A second addition is a *market-oriented approach*. The philosophy underlying the use of economic instruments in environmental management is basically twofold. First there is the *financial revenue* aspect: in their search for money to finance mainly the effect-directed instruments, authorities decided to tax different aspects of waste, air and water pollution. Most of these taxes not only have financing capacity, but in a more or less pronounced way are also intended to change the attitude of the taxpayer to creating pollution. To this *regulatory* character may be added the element that environmental quality (next to safety, user-friendliness, etc.) might become a *competitive advantage* (Agenda 21). Essentially, underlying this market-oriented approach is the use of money and/or the creation of an artificial market (of taxes, subsidies, permits, etc.). The basic hypothesis is that companies, farmers and consumers will find the way which provides them with maximum utility. The authorities have to adapt conditions, using economic instruments in such a way that the actors will find the most utility by behaving in the way that the authorities want them to.

The section in this volume on economic instruments enters into three aspects of the use of these elements of environmental management. The first chapter in the section explains the basic rules of economic instruments as applied to environmental problems. The second chapter illustrates these concepts and shows how they can be applied in practice. The third chapter is concerned mainly with the limitations of the system, and points to the difficulties an open free-market system faces in attempting to integrate the environment and social fairness into a worldwide trade system. It shows that (the lack of) autoregulation, and a too far-reaching belief in an open, deregulated, free market system are the main limiters of the application of economic instruments.

INSTRUMENTS FOR COMMUNICATION

A third section concerns the use of *communication*. It is of the utmost importance for all environmental management target groups to have access to accurate and up-to-date information, and to be aware of existing problems and their context. This awareness should lead to involvement and active participation in environmental management by all actors, each at his or her most appropriate level. Therefore, education and training are of central importance.

Communication can be seen either as (a set of) instruments or as a condition to make legal, economic or physical instruments for environmental management to function properly. Instruments themselves are 'mechanical' in nature, because their use depends on the willingness of the public, government, responsible agencies and organisations to use them. Even the very best instruments are of no consequence if there is no willingness or only half-hearted willingness to use them. The most efficient way to increase willingness among the public at large, and thus the willingness of the government, is by raising the level of public knowledge, awareness and commitment to the environment.

Environmental discussions are still characterised by the idea of conflict. Two chapters discuss how to deal with environmental conflicts, both at the local and at the international level.

Among the different groups of instruments discussed in this volume, communication is the most recent. The chapters here essentially show the need for communication between the different target groups, and the methods which are used in this context.

CONCLUSIONS

The different models for environmental management – legal, economic and communicative – show three distinct approaches. All of them have a clear-cut capacity to improve environmental quality but also have obvious limitations. In the everyday practice of environmental management they are handled together, as a mix of instruments. In using this mix (and to some extent the individual instruments also) three processes are fundamental:

1 During the phase of policy formulation, the primary questions concern *feasibility* issues. Is the implementation using the available and indicated instruments workable, given the target groups of the environmental policy?
2 During the phase of policy implementation, the questions to be asked will concern elements of *achievability*. Can the application of a particular tax be achieved or will it be hampered by administrative or technical difficulties?
3 During the phase of policy execution, the way the target groups handle or respond to the instrument is of key importance. In this phase the *effectiveness* of the instrument is important.

Until now only limited scientific information has been available relevant to questions of which instrument, or combination of instruments, is the most appropriate for a particular situation, i.e., will give the best results when applied to a particular issue in environmental policy. But whatever policy is set out by any of the target groups involved, feasibility, achievability and efficiency are the best parameters for evaluating the use of various instruments. It is the explicit aim of this volume to enable students to weigh up in practice the use of different instruments or their combination for environmental management according to these criteria. Moreover, the lack of scientific information on integrated systems reflects at once the strength and the limitations of this volume: it provides an up-to-date analysis and systematic review of instruments for environmental management, but can offer only preliminary guidance for their use in combination.

REFERENCES

Bressers, J.M.A. (1994) 'Beleidsinstrumenten in het beleidsproces', in P. Glasbergen (ed.) *Milieubeleid* (4th edition), 's-Gravenhage, The Netherlands: VUGA Uitgeverij, BV.

Devuyst, D. (1994) 'Instruments for the evaluation of environmental impact assessment', Ph.D. thesis, Human Ecology Department, Vrije Universiteit Brussel, Brussels, Belgium.

Johnson, S.P. (1993) *The Earth Summit: The United Nations Conference on Environment and Development*, London, UK: Graham & Trotman.

Lee, N. and Colley, R. (1992) *Reviewing the Quality of Environmental Statements*, Occasional Paper 24 (2nd edition), EIA Centre, Department of Planning and Landscape, University of Manchester, Manchester, UK.

Lee, N. and Wood, C. (1993) 'Five year review of the implementation of the EIA directive', in *EIA Centre Leaflet Series 14*, EIA Centre, University of Manchester, Manchester, UK.

Van der Doelen, F.C.J. (1993) 'De gereedschapskist van de overheid: een inventarisatie', in *Beleidsinstrumenten bestuurskundig beschouwd*, Maastricht, The Netherlands: Assen.

SUGGESTED READING

Ryding, S.-O. (1994) *Environmental Management Handbook*, Boca Raton, Fla.: Lewis Publishers.

PART II.1

PREDICTIVE AND SCIENTIFIC INSTRUMENTS

4

ENVIRONMENTAL PLANNING

Emmanuel K. Boon

SUMMARY

Improving the quality of life of citizens is the main policy objective of socio-economic development in all countries of the world. Hitherto most countries have attempted to achieve this by formulating and implementing development plans without any serious effort to integrate environmental considerations into the process. However, in order to ensure that human activities interfere only to an acceptable degree with the functioning of life support systems or natural ecological processes, environmental planning is now becoming an increasingly important imperative in the rapidly growing liberalised world economy.

This chapter contains a review of the literature on this emerging topic with a view to exploring what it actually means, together with an overview of the scope and application of environmental management. Fundamental problems concerned with environmental health and environmental planning are first identified. The complex concept of the environment is then briefly explained, followed by a review of the trends in development planning. The focus then shifts to environmental planning, with a discussion of strategic environmental planning and management.

ACADEMIC OBJECTIVES

On completion of this chapter, the reader should be able to:

- understand the concept of the environment;
- appreciate how the rapid evolution of environmental problems, together with greater public awareness and concern, brings pressure to bear on governments to develop more credible environmental planning and management systems;
- acquire an overview of both development planning and environmental planning;
- establish a link between development, environmental planning and environmental management;
- understand the need for the various actors and the public to participate in both environmental planning and management of environmental health, and gain an insight into their respective roles in the participatory process.

INTRODUCTION

Although the extractors of various natural resources hold different views on the justification of their activities, continuing depletion of the earth's resources at an increasing rate remains a fundamental and increasingly serious problem. The various actors who exploit and use these resources often look myopically at the environmental system and see only those parts of it that are within their own sphere of activity and/or interest. Indeed, it is because of this myopic viewpoint that various human interventions in the environment often tend to work against each other. Clearly, given the fact that environmental issues are complex, plagued with uncertainty and highly political in nature (Bardwell, 1991), more attention than hitherto is needed in planning and implementing human activities in order to ensure a move towards sustainable development.

Introduction of environmental management systems at the national level in many countries, and the requirement that such systems should go beyond merely restoring environmental damage to actually reducing or eliminating pollution, clearly shows the importance of environmental planning in forestalling further damage to the natural environment. Not surprisingly, economic policies that are not integrated with environmental planning, or do not address environmental protection issues adequately, tend to exacerbate environmental pollution. The extensive damage that has been caused in many countries through the exploitation of resources such as forest products, minerals, agricultural land and water bodies is symptomatic of the lack of proper environmental planning. Furthermore, planning for socio-economic development and health of the environment are inextricably linked. For example, if the landscape loses its productivity, standards of living are bound to decline; and, when it loses its aesthetic appeal, property prices may fall. Clearly therefore, environmental planning needs to be made and implemented in an integrated manner at the level of the local community, as well as at regional, national and international levels. Rosenbaum (1973) has identified a number of obstacles which hinder effective environmental planning:

> in the past, national governments have been potent environmental degraders because their planners usually treated the environmental consequences of policy as a 'secondary effect' – unnoticed, unplanned, and seldom emphasised. . . . Whether through indifference, ignorance, or political expediency, public officials, by disregarding ecological values, established a highly stable formal and informal policy process that circumvented the environmental problems they were generating.
>
> (Rosenbaum, 1973: 253)

Another increasingly serious problem facing the world today arises from the fact that environmental pollution does not respect national boundaries. Increasingly each nation is both creator and victim of pollution created by others and, consequently, it is becoming impossible for nations to protect their own environment without regional or international collaboration. A key political problem for most governments, especially those of the developing world, is concerned with the heavy 'sunk costs' which governmental agencies incur by investing in environmental management. Rosenbaum (1973) provides a convincing example of this:

> the White House and Congress view any proposal for environmental planning as a major political issue – a multitude of private interests involved in such planning must be weighed, their stake in present and future policy carefully assessed. The electoral costs of supporting or failing to support comprehensive plans must be pondered.
>
> (Rosenbaum, 1973: 255–6)

Furthermore, many of the developing countries are seriously disadvantaged by their limited planning and administrative structures. Often their ability to plan effectively is also thwarted by poverty (sometimes extreme poverty), low level of public awareness, little or no public participation, and civil unrest (OECD, 1995a). The absence of effective legislative and administrative procedures for incorporating ecological data into the decision-making process constitutes another hindrance to effective environmental planning and management. According to Stohr and Taylor (1981), due in the main to these reasons, many of the developing countries have so far failed in their planning effort to achieve their stated policy objectives of intended development.

The approach to developmental planning in most of these countries has been 'top-down'; this led to excessive centralisation of the planning machinery, concentration of development efforts on selected sectors of the economy without any appropriate linkages among the sectors or with the rest of the economy, and concentration of socio-economic activity in a few of the urban centres. Slocombe (1993) adds that while the negative environmental and social impacts of mega-projects, suburbs, and urban growth have long been known, only recently have environmental problems been framed in planning terms. Devuyst (1993: 151) rightly emphasises that 'environmental concerns in developing countries should be marked by a determination to achieve sustainable development in an environmentally sound manner'.

Political and organisational obstacles constrain the ability of governments in developing countries to formulate more credible environmental management systems (Carley and Christie, 1992). Also, the prevailing acute shortage of indigenous management skills at nearly all levels constitutes a major, if not the major, constraint to the achievement of socio-economic development. The inability of current managerial and planning systems to cope either with the dynamics of rapid change or with the interaction between numerous organisations, entrepreneurs and the environment, is referred to as 'limits to governance'. These interactions among the various actors produce enormous turbulence in the environment which further erodes the ability of governments and organisations to achieve sustainable development.

Two other critical problems, which hinder effective sustainable environmental planning in both developed and developing countries, are fragmentation of policy and institutions, and tensions which can occur between central, regional and local governments. Governments often pursue contradictory policies. But both politicians and the public appear to be impotent and lack the necessary will or motivation to force those in power to undertake the difficult task of mediating between conflicting social, economic and environmental goals for protecting the environment.

A further difficulty in environmental planning is caused by international rivalry. It creates tensions in international relations and makes collaboration between and among nations difficult to achieve. A typical example of this is the wide and growing socio-economic disparity between the rich and poor nations which is often a critical issue in all discussions on environmental planning and management at the international level. In fact, the developing countries have identified a number of problem areas, listed below, which they claim impede international co-operation in environmental planning and management:

- the growing fear of under-development in the South;
- the fear that current environmental concerns in the developed countries will affect the developing countries adversely in trade, aid and technology transfer;
- the fear of 'neo-protectionism', meaning that the affluent nations may discriminate in international trade against the products of developing countries because they do not implement acceptable environmental protection measures;
- fear that developed countries may divert foreign aid funds to their own ecological projects;
- fear that the donor countries may attach the implementation of acceptable environmental protection measures in developing countries as a necessary condition for receiving aid;
- fear of undesirable technologies being pushed to the developing countries.

Similar problems of both developmental planning and environmental planning also occur at the organisational level, as they do at national and international levels. Daneke (1982) illustrates the situation thus:

> The incentives for corporate planners to interact with social and environmental policy analysts are certainly present in terms of reducing investment uncertainties, but an atmosphere of mutual co-operation to achieve viable trade-offs between economic and environmental concerns may be difficult to achieve.
>
> (Daneke, 1982: 127)

Clearly, therefore, ensuring a good environment, and a move towards sustainable development, requires a complete understanding of the total environmental system as well as of the political, organisational and managerial functions in modern society. In other words, learning to understand the interactions between the environment and society is an important prerequisite of effective environmental planning and management.

UNDERSTANDING THE ENVIRONMENT

It would be instructive at this stage to focus on the complex concept of the environment itself. The environment comprises many subsystems, actors and processes. The main environmental subsystems are physical, biotic, ecological, social, political, economic, and technological. These subsystems include all subsurface elements, all man-made and natural surface elements, as well as those of the atmosphere. The local community, public agencies, professional associations, property developers, property owners and the general public are the key actors in the environmental decision-making process.

Often environmental decision-making is significantly influenced by both behavioural and cultural dimensions, and to this extent such decisions can be nation-specific or region-specific. There are, on the other hand, increasingly serious global environmental problems that can be addressed only through international co-operation and collaboration. Stratospheric ozone depletion, the greenhouse effect, and diminishing biological diversity are today the main global environmental problems (Field, 1994). A number of problems concerning human health and agricultural crop damage have been traced to elevated levels of surface ozone. The greenhouse effect, which raises the earth's surface temperature, is said to be responsible for changing the world's climate, and climate change in turn is causing a whole range of adverse impacts. The earth's rich biodiversity is being destroyed at an alarming rate, and this is putting at risk the proper functioning of global ecosystems that is so vital to both a healthy environment and human well-being. Biodiversity gives the system the means with which to adapt to change, while biological uniformity, resulting from lack of (or reduced) biodiversity, makes the system inflexible and weakens its ability to respond to change.

Given the complexity of the environment, the complex and hitherto unknown interactions that occur among its numerous subsystems, and the increasing environmental problems being created by human activities, it is clear that a holistic approach must be taken to study and understand it. To that end, knowledge of the environmental subsystems and their respective elements should be structured in a way that facilitates easy communication and understanding by all the actors and stakeholders involved. This means that formats, which permit intelligent debate and decision-making and minimise confusion and conflict between and among the actors, must be adopted. The approach to be taken to understand the environment as a system, together with the interactions among its subsystems, thus needs to be based on an analysis of the cause and effect relationships between the various constituent parts as well as the participants. In addition, a methodology is needed which permits simplification and eliminates duplication or overlap within the existing system. It should be recognised, however, that the final outcome of the environmental decision-making process depends on the values which the community attaches to the environment. It is to be noted in this context that each community is unique, with its unique set of values which have evolved from its unique perception of and attitude to the environment. The uniqueness of a given community should therefore be the principal determinant of how its environment should be preserved or modified.

TRENDS IN PLANNING

Definition and importance of planning

The elusive nature of planning is acknowledged in the literature (Smith, 1993; Wildavsky, 1973). It is argued that planning can be all things to all people – the ability to control the future by current acts, the application of causal knowledge, a form of political power, and an act of faith. Benveniste (1989: 45) sees planning as 'the elaboration of a set of related programmes designed to achieve certain goals . . . [and the] . . . planning process is the set of interventions and other actions undertaken during the elaboration of a plan'. Planning may be said to be a rational methodology for approaching the future. It involves the definition of the problem to be addressed, establishment of planning objectives, identification of the assumptions on which planning is to be based, search for and evaluation of alternative courses of action, and the selection of a particular action to be followed (Compton, 1993).

What is probably more important is the fact that development, or human well-being, is a primary concern of planning. Development itself is increasingly defined in social and environmental terms. It is seen as enlarging the range of people's choices, including access to income and employment opportunities, education, health and a clean and safe physical environment. Beer (1990) adds that the need for food, clean water, clean air and shelter is fundamental to human survival. Carley and Christie (1992) also recognise the prevalence and importance of developmental planning:

> Most . . . countries have . . . economic plans, forest plans, plans for tourism, industry, services, ranching etc. – not to mention plans for rural agricultural development. All these plans and development efforts impact on one another; and can conflict with one another . . . in an attempt to harmonise economic development with sectoral objectives, a land use planning overview is essential . . . a planning framework within which public bodies, the private sector and individuals can all operate.
>
> (Carley and Christie, 1992: 159)

Baldwin (1985: 4) prefers to consider planning within a wider context of resource management as 'the initiation, transformation, distribution and disposal of resources in a manner capable of sustaining human activities, with a minimum disruption of physical, ecological, and social processes'. For Slocombe

(1993), on the other hand, planning is the collection and analysis of information to serve the public interest by performing a wide range of human economic and other developmental activities.

Another significant trend in planning is the need to maintain cultural diversity, while at the same time aiming to provide people with an environment that supports their way of life. People from different cultural backgrounds have different ideas about what constitutes a satisfactory environment. Planners should therefore have a proper understanding of how people react to their environment, otherwise unnecessary social costs may be incurred through the unintentional creation of environments that alienate people.

Approaches to planning

There are many approaches to planning, including substantive planning and procedural planning. Substantive planning is concerned with the subject matter of planning, or the object of planning. Procedural planning, also called the classical model of the planning process (Faludi, 1973; Cayer and Weschler, 1988), is a means for promoting economic growth through the application of rational procedures of thought and action. Faludi argues that the planning process should be separated from the substantive content of the plan itself. Procedural planning consists of the following inter-related steps (Cayer and Weschler, 1988):

- identification of needs;
- specification of goals and objectives;
- development of alternative means to attain each goal;
- estimation of the costs of each alternative;
- selection of the most promising alternative(s).

Darke (1983) advocates the integration of substantive planning with procedural planning. He argues that procedural planning theory is only a partial approach to the explanation of planning, because it is inherently confined to a conservative political philosophy that views social action from a strictly functional perspective. Moreover, the role of the state as an agent of change is not taken into account in procedural planning, neither is an analysis based on class, gender or political ideology. Procedural planning theories are also said to have limited methodological bases and are constrained by their adherence to the notion of rationality. These weaknesses of the approach prevent it from addressing adequately the larger social milieu of a planning endeavour. It is argued, therefore, that the substance of planning should not be separated from the planning process.

Hudson (1979: 45) has proposed yet another approach to planning, based on five schools of planning thought. For him planning is 'foresight in formulating and implementing programmes and policies'. He identifies the five schools of planning thought listed below:

- synoptic planning;
- incremental model of planning;
- trans-active approach to planning;
- advocacy planning;
- radical planning.

Steiner and Miner (1977) have identified four different approaches to formal planning: the top-down approach, the bottom-up approach, a mixture of the top-down and bottom-up approaches, and the team approach. At the micro level planning generally entails the setting of the enterprise objectives and establishing the policies, procedures, and programmes for their realisation. It involves determining what needs to be done, by whom, and by when, in order to discharge one's assigned responsibilities. In other words, the exercise is to establish a predetermined course of action within an agreed policy framework. Customarily defined as strategic, tactical, or operational, it can be at the individual level, at the level of a working group, or at the functional level of an organisation. At the 'individual level' planning is required to deliver assigned tasks; at the 'working group' or the 'functional' level it must include agreement on purpose, assignment and acceptance of individual responsibilities, co-ordination of activities, increased commitment to group goals, and lateral communication.

Mintzberg (1994) gives an excellent review of the various definitions of planning and their interpretation. For some it is thinking about the future, or simply taking the future into account. To others planning is about controlling the future – not just thinking about it but acting on it or enacting it. More comprehensively, planning is decision-making – the conscious determination of courses of action designed to accomplish purposes – a formalised procedure for delivering an articulated result in the form of an integrated system of decisions.

There are many reasons why organisations need to plan. One is to ensure the co-ordination of their activities, taking the future into account by preparing for the inevitable, pre-empting the undesirable, and controlling the controllable. Organisations must plan to be 'rational'. The planning process also determines the position of the planner and indicates how, and to what extent, the general public should be involved (Slocombe, 1993).

CONCEPT AND APPROACHES

The concept of environmental planning

It is unfortunate that while developmental planning is largely the work of economists and mainstream urban and regional planners, environmental planning remains the exclusive and separate realm of environmentalists, ecologists, and resource managers of various kinds. Beer (1990) sees the role of the planner as thinking more widely about how environments can be maintained or created to ensure human well-being within the socio-economic constraints and conditions of the community in question and its perception of the environment. In other words, planners do need to translate the concerns of the community for their immediate habitat into strategies and policies in order to guide the planning process. Growing concern for the quality of the environment, and for the preservation and conservation of nature, is thus a fundamental objective of environmental planning. It underlines the need for integrating both environmental and health aspects into the decision-making process of national planning. Rosenbaum (1973) argues the central role of environmental planning thus:

> No goal is more central to the environmental movement or more politically contentious than comprehensive environmental planning . . . Planning is becoming the central axis about which evolve the multitude of individual conflicts over particular environmental difficulties.
> (Rosenbaum, 1973: 179)

Faludi (1987: 252) sees environmental planning as 'the sum total of all public environmental measures which an environmental authority can take'. According to him an environmental plan – which is a document (or documents) used in guiding operational decision-making on public environmental measures – therefore stands for the outcome of deliberations in environmental planning. Westman (1978) views environmental planning in terms of measuring the value of the earth's resources and predicting changes caused by disturbances. It is concerned with the analysis of social, economic and environmental relationships resulting from human activities, and with the design of programmes and action plans for improving both social well-being and environmental health. He stresses the importance of environmental planning in these words:

> It seems clear that if we are to preserve for future generations some semblance of the biological order of the world of the past and hope to improve on the deteriorating standards of urban public health, environmental science and technology must quickly come to play a dominant role in designing our social and industrial structure for tomorrow . . . The prediction of impacts may form part of the tasks in environmental planning and design, resource management, and applied ecology.
> (Westman, 1978: v)

The objectives of environmental planning thus include measures to be taken under a variety of circumstances, some of which cannot be known in advance. The environmental planning process also forces public officials to aggressively assert governmental power in vast areas of the economy that have been left largely to private interests. Rosenbaum (1973) maintains that national governments must establish priorities for environmental protection, must calculate and create the 'trade-offs' to be made between environmental protection and other national goals, and must plan resource use and protection over several generations. In addition, the planning process should give equal emphasis to both national socio-economic development and environmental quality.

It is very important for national governments to be more sensitive to the environmental impacts of their own policies, and more willing to sacrifice other objectives to environmental protection when good ecology so dictates. Nations will bear not only severe environmental costs for delaying such planning, but will also incur high political and economic costs. Delaying of environmental planning may lead to severe environmental costs; it may also lead to near-crisis situations whose management would entail high or very high social costs. The design and implementation of appropriate and effective environmental plans depend on the ability to prepare accurate Environmental Impact Assessments (EIAs), which are discussed elsewhere in this book (see Chapter 10).

Approaches to environmental planning

The approach to environmental planning should be broad, based on both monetary and non-monetary considerations. Good interdisciplinary skills are needed in the integrated planning approach. From the very outset, the environmental planner needs to understand the distinction between environmental 'assessment' and 'evaluation'. The purpose of Environmental Impact Assessment is twofold: first, it attempts to identify what functions of the affected resources would be modified by each alternative proposal; second, it attempts to predict reliably the direction and magnitude of any modification likely to result from each alternative decision or action as compared to when that action is not taken. Evaluation, on the other hand, is the subsequent examination of both the patterns of modifications caused by the action(s) taken and the consequences thereof.

Faludi (1987) suggests that planning should depend on the context in which it takes place (the 'planning environment'), while Ryding (1994) proposes two different but closely related environmental planning and decision-making approaches:

1 The institutional approach, used in the social planning process (including agricultural, forestry, municipal and transportation sectors), in which remedial measures are often in the form of legislation, regulations, taxes and charges.
2 The commercial approach, used for planning in the business sector (including energy production, and industrial, trade and retail sectors), in which remedial measures can be in the form of non-regulatory initiatives for improving existing treatment processes, developing new environmentally friendly processes and products, reducing waste through waste minimisation programmes, and/or for improving the logistics of transportation, storage and sale of products.

A generalised planning and decision-making model for all aspects of environmental protection, developed by Ryding (1994), is shown in Figure 4.1. Three pertinent features of environmental planning can be deduced from this figure. The first is the multidisciplinary nature of planning which, for example, involves economic, social, ecological, land-use and technical considerations. The second and third features refer to public involvement in the planning process, and to the systematic approach of the process starting from the setting of objectives through to the implementation of remedial programmes, assessment of results and a permanent feedback mechanism.

Input to the general planning and decision-making model derives from the separate priorities set in the institutional and commercial approaches, as well as from the outcome of analyses of potential consequences based on user and cost/benefit elements. The

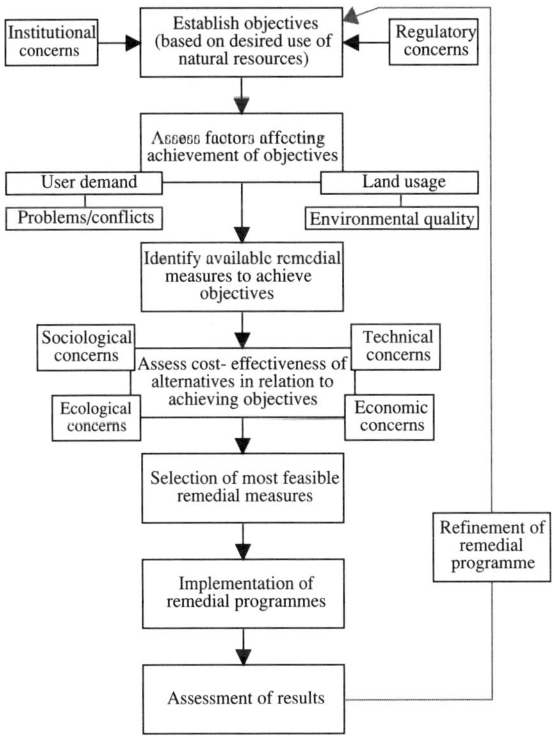

Figure 4.1 General sequence of planning and decision-making aspects in implementing remedial programmes
Source: Ryding, 1994: 681

following must be considered during the planning process:

- environmental effect-oriented issues should focus on the minimisation of ecological threats *vis-à-vis* the input of the mixture of pollutants released;
- technologically oriented issues should focus on access of all countries to best available technologies;
- economically oriented issues should focus on potential financial consequences before decisions are made; also, both financial aid programmes and modalities for transferring necessary technology should be decided upon;
- user-oriented issues should focus on information campaigns to alert all those likely to be affected by the decision.

The planning and decision-making approach may be based on the following:

- initial evaluation of the environmental areas under threat;
- a priority list of actions;
- a plan of action;
- a procedure for remedial actions;
- regular auditing.

In order to ensure that all the various issues are properly addressed and that nothing important is forgotten or omitted, each step in developing a common strategy needs to be comprehensively assessed and checked with rules of thumb. The acronym ACTION, explained below, provides a suitable set of rules of thumb for this:

A Anticipate the participation/involvement of all the stakeholders and the general public in the environmental planning and decision-making process as well as in the different environmental, technical and economic consequences.
C Conclude the most plausible economic and environmental development.
T Tabulate the possible advantages and disadvantages of all potential decisions.
I Initiate a thorough evaluation of the advantages and disadvantages of alternative decisions.
O Omit all less useful alternatives and highlight the preferred ones.
N Negotiate with all relevant parties for the acceptance of the preferred action.

In general, the following characterise an environmental plan:

- all stakeholders and the general public should be involved in setting and discussing environmental priorities;
- the plan should be relevant and capable of meeting the stated environmental objectives;
- the plan should be operational and cost-effective at both regional and global scales (this implies the allocation of environmental investments beyond national and regional boundaries);
- the plan should be realistic and feasible.

STRATEGIC ENVIRONMENTAL PLANNING AND MANAGEMENT

The application of the concept of strategic planning and management to environmental planning is discussed in this section, together with how the long-term success of such plans can be ensured. In this context the relevant elements of strategic planning and management include the following: the setting of clear environmental objectives and the definition of appropriate policies and programmes for their achievement; effective environmental education and training; organisational and institutional development; participation; co-operation and co-ordination; harmonisation of policies; and, finally, evaluation, monitoring and control.

Objectives, planning and management

The setting of clear environmental objectives is important in the planning process, because the objectives help determine how specific actions and policies should be developed for their achievement. Environmental management is about the achievement of particular environmental goals and objectives which may be broadly defined in terms of the protection and/or improvement of environmental quality (Compton, 1993). Environmental objectives may be set as precise quantitative targets, or expressed as broad aspirations. As previously pointed out (see pp. 74–76), planning is central to any attempt at managing the environment effectively, and it involves the identification of objectives and the selection of the means with which to achieve them. Faludi (1987) and Wyatt (1989) suggest that planning practice must be capable of adapting to both the changing political climate and the rising level of uncertainty.

Strategic planning is an effective tool which organisations and nation-states can use to estimate the degree of risk of their policies and actions, thereby helping to facilitate the realisation of their environmental goals and objectives. But the top managers of an organisation must have a sound conceptual understanding of long-term planning in order for them to apply this tool effectively. Daneke (1982: 123) sees strategic planning as 'a process for distilling social goals and objectives, and design of alternative scenarios or "road maps" that become the basis for meaningful collective choice'. According to Steiner and Miner (1977) the essential features of strategic planning are that:

- it deals with the future impact of current decisions and actions;
- it examines the cause-and-effect relationships of an actual or intended decision or action over time;
- it examines the alternative courses of action that could be taken in the future, and, when choices are made, how they should form the basis of current decisions;
- it considers future opportunities and threats together with other relevant data, thereby creating a management regime to improve current decisions for exploiting future opportunities and for avoiding future threats;
- it is a process which begins with the setting of objectives and the definition of strategies and policies to achieve those objectives; detailed plans should be developed to make sure that appropriate strategies are implemented to achieve the stated objectives;
- it is a continuing process of deciding in advance what is to be done, when it is to be done, how, and who is going to do it;
- it is a philosophy, an attitude and a way of life. It is an understanding that planning requires dedicated action based on the contemplation of the future; a determination to plan constantly and systematically as an integral part of management;
- it may be defined as a structure of plans – a structure which integrates strategic plans with short-term operational plans.

In the absence of comprehensive strategic environmental planning and management, most countries, especially the developing countries, will face mounting social costs in the long term. Studies made by the UNDP (1995) on the experience of a number of developing countries in formulating and implementing their national conservation strategies, national environmental action plans and national Agenda-21 reveal the following weaknesses:

- the strategy was not a priority, either at the government level or at the local level;
- the process was externally driven, without domestic resources and without prioritisation;
- emphasis was on the product (i.e. strategy or action plan as a document) rather than on the process;
- there was lack of stakeholder involvement, and no ownership or implementation;
- each report or strategy was treated in isolation; there was no coherence, no information gathering and no historical antecedent which could facilitate subsequent work;
- locally available management skills were insufficient, sometimes grossly so.

An effective strategic environmental management system can serve as a basis for overcoming the above deficiencies. In this context Ansoff (1977: 30) defines management as 'the creative and error-correcting activity that gives the firm its purpose and cohesion, and assures a satisfactory return on investment'. Strategy, on the other hand, refers to a well co-ordinated set of action programmes aimed at securing a long-term sustainable advantage. Strategic management therefore refers to systematic identification of, and response to, opportunities and threats, both of which can and do originate from a firm's domestic and/or international market. As a continuous iterative process founded on the reality of good management practice, strategic management thus focuses on the present strategy of the firm, supported by the resources and strengths generated in the past. It also focuses on the design and implementation of a strategy that will carry it successfully into the future (Stahl and Grigsby, 1992). The analysis of strengths, weaknesses, opportunities and threats, referred to as SWOT analysis, usually covers both the internal and external environments of the firm.

Wheelen and Hunger (1995) have proposed a strategic management model, shown in Figure 4.2, which separates the various activities that must be linked together in making strategic decisions for an organisation. As can be seen from this figure, strategic management involves the formulation, implementation, evaluation and control of strategic plans of the organisation in question. Strategic management is also concerned with how an organisation goes about translating its policies, missions, objectives and goals into results. This, however, can only be done if the organisation carries out a proper assessment of its internal and external environments.

Environmental education and training

Probably the most demanding task in environmental management is that of educating and training people in skills for predicting the likely outcome of human activities on society and the environment itself. For this the concept of strategic environmental management, discussed earlier, must be integrated into environmental education (EE) programmes. Indeed, enhanced public awareness of environmental issues and problems is a critical and over-riding concern in successful environmental planning and management (Buckley, 1991):

> In the longer term this may be done through the normal school education system. Unless public environmental awareness is improved in the short-term, however, long-term improvements will be too late. More effective means to communicate the importance of environmental concerns to those in business, government and the general electorate are urgently needed.
>
> (Buckley, 1991: 29)

Clearly, therefore, any effort to develop an innovative and productive environmental planning system is unlikely to succeed unless public officials, charged with a nation's environmental management, cultivate an ecological conscience themselves (Rosenbaum, 1973). As discussed elsewhere in this book, environmental

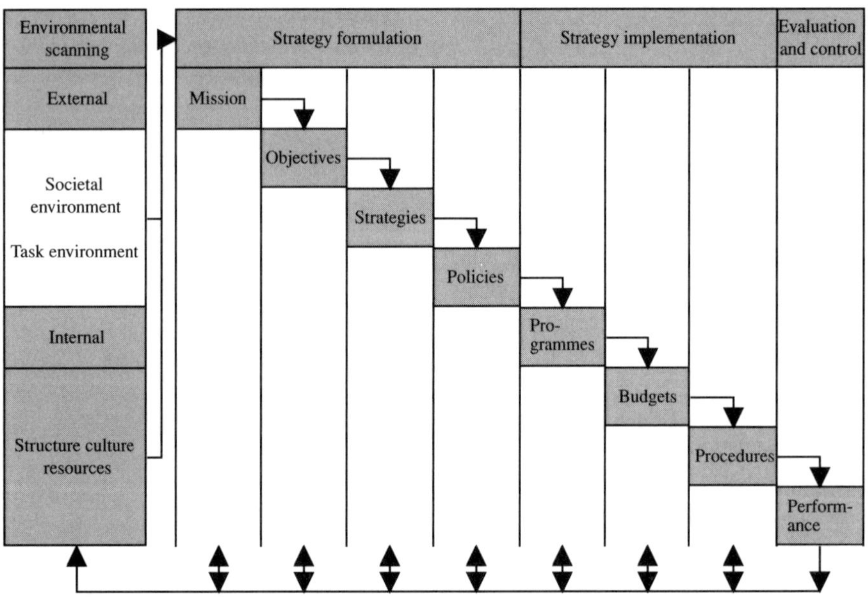

Figure 4.2 The Strategic Management Model
Source: Wheelen and Hunger, 1995: 114

education and training are most appropriate for equipping such officials and other environmental actors with essential skills in environmental planning and management. In particular, environmental education would help environmentalists in developing countries to acquire essential skills for understanding and using basic analytical techniques and tools such as environmental impact assessment (EIA), social impact assessment (SIA) and economic impact assessment (EcIA) (Clark, 1989; Jain and Hutchings, 1978; Welford and Gouldson, 1993).

Apart from helping to redress the lack of skills referred to above, environmental education and training will also equip environmental scientists, especially in the developing countries, with expertise in the appropriate and 'integrating' styles of environmental management, as well as in institutional and organisational development through action research. However, the approach must be multidimensional, as shown in Figure 4.3. Single issues must be analysed in their wider environmental context. And this would require the acquisition of wider and deeper knowledge of the environment which should be documented and disseminated in a simple but effective manner to all the social sectors concerned. Dissemination of information would facilitate both the setting of environmental priorities and international co-operation in environmental planning and management.

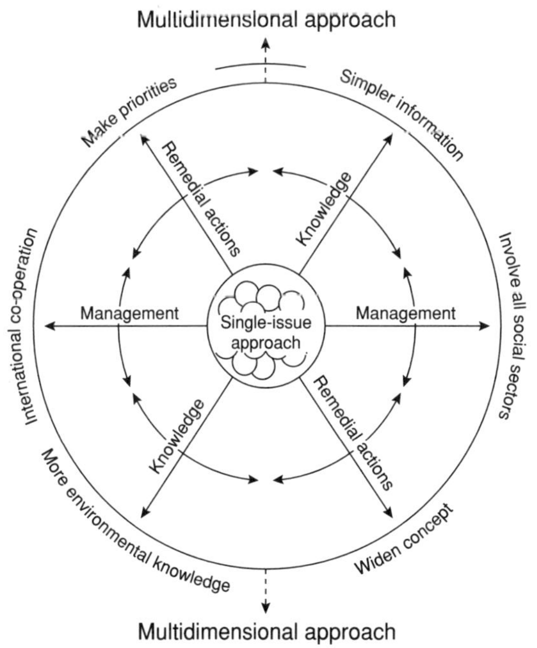

Figure 4.3 Multidimensional approach to planning
Source: Ryding, 1994

Institutional and organisational development

Institutional and organisational development is also critical for effective environmental planning. Institutional development addresses legal and organisational constraints and limitations in human resources, while legal institutions underlie the fabric of development and environmental control (Carley and Christie, 1992). For example, property systems and land-use planning controls are very important factors in any rural or agricultural development or conservation initiative, yet very often programmes are initiated without an analysis of these fundamental systems. Legal systems include the necessary laws, regulations and environmental standards, and the government framework to implement them. The legal system also embodies formal arrangements for centralisation or decentralisation of planning and implementation, as well as the institutional arrangements of government.

In some countries environmental plans are made by a number of agencies. Such 'multi-planning' agencies are clusters of related planning organisations which permit a wider input of specialisation as well as co-ordination of tasks (Faludi, 1987). The aim of interorganisational development should be to forge new, formal and informal linkages among agencies of the government, and beyond government to business and the voluntary and community sectors.

There is now an urgent need in most of the developing countries for institutional capacity building for sustainable development. For this, it is suggested that a critical element in strengthening institutional and legislative structures is the selection of environmental policy instruments (Buckley, 1991). There are three main types of instruments for putting environmental policies into effect – namely, technological, regulatory and economic. Training of professionals and strengthening of institutional structures are the two critical areas in which aid donors can assist the developing countries (Buckley, 1991). The OECD (1995a), for example, regards co-ordination and harmonisation of donor approaches to the formulation and implementation of sustainable development activities as key factors for improving the effectiveness, efficiency and quality of development co-operation. Measures proposed for improving co-ordination include setting realistic goals, better information networks, the recognition that 'people count', training of relevant donor and recipient personnel, and the assessment of co-ordinated activities as part of professional evaluation processes.

A key challenge in harmonising environmental plans and planning requirements is to find ways of mutually acknowledging the substance of different plans and strategies, or even common points or areas between different plans. It is also necessary to develop a better understanding of how sectoral strategies relate to the overall planning framework. The 'translation' of plans to suit different national requirements on the basis of experience is an important measure for promoting harmonisation. For example, donors, whether bilateral or multilateral, need to be careful so that they do not accentuate the existing problems of the recipient countries.

Interorganisational development and networks help to facilitate both exchange of information and resources needed for environmental problem-solving. Acting as non-hierarchical social systems, networks permit co-operation and collaboration in implementing sustainable environmental projects. For example, research by Lee (1994) on community participation in the planning process shows that, with necessary support, citizens at neighbourhood and community levels can manage lanes, waterways, and waste collection and disposal systems, and that they can also develop and enforce access rules and reciprocal obligations among the members of the community.

According to Carley and Christie (1992) there are four main types of networks serving different functions. They include policy networks that are based on the major functional interests of the government. Such networks are characterised by stable relationships, continuity of restricted membership, shared responsibilities for service delivery, and insulation from other networks. Issue networks are considerably less integrated by comparison. They have a large number of participants but a limited degree of interdependence. Some are based on the need to share technical information, others on shared concern over issues such as conservation of the rain forest. Professional networks cut across policy and issue networks, while producer networks are concerned with economic functions and the relationship between the private and the public sector.

Investment in individual and team skills is one of the keys to organisational innovation for improving the management process, not only in terms of both personnel and professional development but also for upgrading all other supporting organisational, institutional and legal structures. Incentives are an important means of generating commitment. At higher levels both networking and consensus-building must be made so stimulating and essential to good decision-making that senior officials would not like to be left out of the process.

Increasing participation

For Stohr and Taylor (1981) the most appropriate approach to environmental planning is planning 'from below'; that is, planning for the people and with the people. As Carley and Christie (1992) point out, environmental issues must be fully integrated into socio-economic planning and involve public participation at all levels – individual, community and global. Faludi (1987) also refers to planning as teamwork, in which discussions should not be limited to the chief planners and politicians but should involve all sections of the community. He sees participation as a means of furthering criticism, thereby expanding the range of options considered. Ryding (1994) also suggests that, at the individual level, people should assist in environmental protection work through participation in environmental organisations. At the community level, environmental considerations should be integrated into all sectors of society, while the responsibility of safeguarding the environment at the global level should be shared by all countries, both rich and poor. Planning without participation is ineffective, undemocratic, paternalistic, and doomed to failure.

Co-operation, co-ordination and harmonisation of policies and programmes

Chechile and Carlisle (1991) cite a number of new environmental challenges which compel the various stakeholders in the environment to change their attitudes to environmental issues. These challenges include the high degree of uncertainty attaching to the long term, complex inter-relationships of numerous variables and time constraints, shrinking distances, widening scope, and cultural diversity. The complexity of these challenges implies that environmental policies, plans and programmes of individual countries need to be effectively co-ordinated and harmonised.

The OECD (1995b) also recognises that environmental planning needs to be effectively co-ordinated at the international level, because planning requirements are proliferating and becoming increasingly more complex. Also, given that environmental plans are made at the local, regional, national and international levels, and that the planning process often involves several agencies, it is clear that co-operation between and among the different agencies is paramount. To this end Morell (1995) suggests that the planning process must have two characteristic features: first, a mechanism for wide dissemination of decisions and information; second, established structures enabling cross-functional activity. Effective modalities of communication are critical to both and should be developed to ensure wide access to relevant information. For example, videoconferencing and teleconferencing facilities can help to reduce the cost of interaction and, at the same time, increase the frequency of communication among the experts. Also, networks should be developed to facilitate both understanding and consensus-building among people with diverse expertise, outlook and cultural background.

Both governments and environmental groups often play an important role in ensuring that environmental policies, programmes and actions are co-ordinated and harmonised. Box 4.1 gives a typical example of co-operation in environmental policy and planning by the International Network of Green Planners (INGP). A successful implementation of environmental plans requires co-ordination at all levels. In the case of developing countries, a high level of co-operation between aid donors and the recipient countries is needed to make sure that all these actors understand each other and that they do not work against each other (OECD, 1995c). National governments must be at the centre of co-ordination at the following levels:

- between sectoral ministries and their related organisations;
- between local, regional and national authorities;
- between various external actors such as UN agencies, bilateral donors, and international organisations.

Some current attempts to improve co-ordination include the following:

- better information networks;
- sensitivity to national characteristics (culture, tradition, etc.);
- multilayer co-ordination; institution-building and tools;
- co-ordination at the sectoral level as a practical starting point;
- providing incentives for co-ordination at the individual career level;
- training experts in multisectoral planning and implementation;
- indicators of sustainability; agreed indicators to measure changes in both natural and social environments to determine whether sustainability of activities is being achieved.

Evaluation, monitoring and control

Evaluation, monitoring and control are important elements in developing and implementing environmental plans. They enable planners to keep track of planned actions and to monitor and compare the actual performance of plans with their anticipated performance. This process permits corrective measures to be taken as necessary. Taking the case of the USA as an example, McAllister (1986: 4) has shown how the evaluation and monitoring of environmental actions can be inadequate; he argues that 'a disappointingly small proportion of the decisions by local governments is made on the basis of systematic evaluations of alternatives. Evaluations are often *ad hoc*, quick, and impressionistic.' It is important for monitoring and evaluation to pervade the whole planning process. In simple terms, the process should involve the following inter-related steps:

BOX 4.1 AN EXAMPLE OF CO-OPERATION IN ENVIRONMENTAL PLANNING: THE INTERNATIONAL NETWORK OF GREEN PLANNERS

The International Network of Green Planners (INGP) was founded in December 1992, following the call at UNCED for states to 'co-operate in a spirit of global partnership to conserve, protect and restore the health and integrity of the Earth's ecosystem'. The INGP was set up by a number of environmental policy experts drawn from the UNDP, UNEP, OECD, the Federation of Malaysia, and the Environment Ministries of Canada and the Netherlands (INGP, 1994).

The INGP was established in response to the emerging 'new profession' of environmental practitioners; that is, people actively involved in developing and implementing environmental policy with the long-term objective of achieving sustainable development. Many nations now have some kind of environmental policy/resource management objectives – legislation to establish a process for developing long-term environmental objectives, or a more comprehensive development plan which includes a strong environmental component. These strategies vary widely in their scope, approach and objectives, and this variety is matched by the wide variety of practitioners.

The INGP is an informal network; its aim is to provide a forum for environmental policy-makers from all over the world to exchange ideas, share information and experience, and benefit from the pooled expertise of their fellow members. It develops concepts and provides practical support to members responsible for developing and implementing long-term environmental policy. The network employs a multidisciplinary approach and extends both its membership and topics to include economics and the social sciences. The INGP advocates greater integration of policy-making in these areas, together with co-ordination of objectives, for the achievement of sustainable development. INGP's key concerns refer to the following:

- goals and processes are important elements in environmental planning;
- donor organisations need to take greater account of national circumstances and priorities;
- participation is important, but approaches must depend on national circumstances. It is important to work through key actors and other groups in government and society;
- successful environmental policy must take account of those responsible for implementation; priority-setting is essential, and so is greater transparency and clarity in decision-making on priorities;
- monitoring and review of environmental plans: monitoring of capacity, procedures and check points should be incorporated from the beginning of the policy process;
- environmental plans and monitoring systems should be designed, as far as possible, so that they can be easily compared with each other;
- state of the environment (SoE) reports are valuable;
- political leadership is necessary for promoting both development and implementation of environmental and sustainable development plans.

The INGP believes that many of the problems encountered by green planners are universal, national differences notwithstanding. It recognises the importance of the environmental planning process, and these aspects in particular: planning and implementation go hand in hand; plans and procedures should be kept simple; the environmental strategy process should be country-driven; environmental policy-makers must not become too specialised; it may be useful to listen to 'subversive ideas'; and development should be both open and flexible.

- identification of the problem to be addressed;
- design of alternative solutions to the problem;
- evaluation of the alternatives;
- decision on actions to be taken through the appropriate political process, and the implementation of those actions;
- monitoring of the results.

Figure 4.4 shows how a proposed activity can be evaluated in two phases (McAllister, 1986). Phase one comprises the analysis which defines and estimates the various impacts of the proposed action. This phase facilitates a good understanding of the consequences of the action, but at the same time it poses a dilemma with regard to achieving coherence of the many diverse parts of an environmental action programme. The second phase is basically a synthesis which attempts to solve the dilemma by integrating the impacts into an overall view so that a judgement can be made as to whether the action should or should not be undertaken.

The evaluation process should include the following: performance monitoring to control the use of

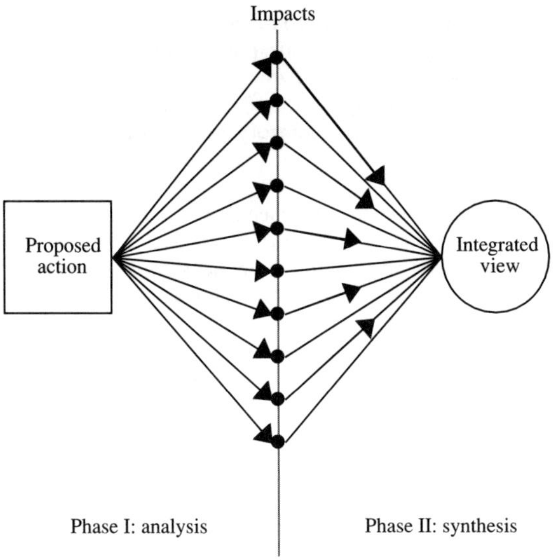

Figure 4.4 The phases of evaluation
Source: McAllister, 1986: 7

inputs and the production of outputs; process monitoring to evaluate the efficiency and effectiveness with which the plan is implemented; impact evaluation to estimate the quantitative effects of the project on the socio-economic conditions of the target population; and cost–benefit analysis to compare alternative projects in terms of their costs and benefits (Bamberger and Hewitt, 1986). The responsibility for monitoring and evaluating environmental plans lies with the environment ministry (or its equivalent) and related agencies. Plans may be evaluated in-house, or through the services of consultants for an independent perspective. In order for the evaluation to be operationally useful, senior management must participate actively in the planning and review of the evaluation programme. Taking the investigation of the Parliamentary Committee, Government of Ghana, into an illegal importation of toxic plastic waste materials into that country as an example, Box 4.2 highlights the importance of evaluating, monitoring and controlling environmental actions; it also provides an insight into some of the fundamental problems environmental authorities face in the process.

BOX 4.2 EVALUATION, MONITORING AND CONTROL: INVESTIGATION INTO THE DISPOSAL OF PLASTIC WASTE MATERIALS IN GHANA BY METALLOPLASTIC (GHANA) LIMITED

Metalloplastic (Ghana) Limited is a registered company in Ghana which manufactures plastic goods from imported raw materials. The company was alleged to have illegally imported into Ghana a quantity of cytotoxic materials from an American company, Petroco Americas Uniondale of New York, for disposal in Ghana. On receipt of a quotation from the supplier, Metalloplastic (Ghana) Ltd established irrevocable letters of credit at the Bank of Ghana through the National Investment Bank to pay for the 'raw materials' to be imported. Société Générale de Surveillance (Ghana) Ltd (SGS) confirmed its inspection of the cargo prior to shipment from Houston, USA, on 9 March 1992. The consignment arrived at the Port of Tema, Ghana, on 15 April 1992, and was cleared under a 'certificate of urgency'.

The arrival of the waste material caused a stir in the country and had to be referred to the Parliamentary Committee on Environment, Science and Technology for investigation. The Committee's terms of reference were to determine the following: the circumstances under which Metalloplastic (Ghana) Ltd imported plastic waste material into the country and cleared it from the Tema Harbour; whether any post-shipment tests were made on the toxicity or otherwise of the material; who approved the disposal of the waste locally and under what circumstances; whether there was any evidence of toxicity prior to its disposal; whether a second batch of plastic waste material imported by Metalloplastic (Ghana) Ltd was re-exported to its supplier and, if so, under what circumstances; why was the first batch of the plastic waste material, imported by the company, not re-exported to the supplier; why was the Ghana Standards Board not involved in the laboratory examination of the plastic waste material; and any other issues incidental to the foregoing.

The following gave evidence to the Committee: 43 witnesses from Metalloplastic (Ghana) Ltd, the Ghana Police Service, the Environmental Protection Agency (EPA) of Ghana, the Accra Metropolitan Assembly (AMA), the Ghana Atomic Energy Commission (GAEC), the Ghana Standards Board (GSB), Société Générale de Surveillance (Ghana) Ltd (SGS), the Customs, Excise and Preventive Service (CEPS) of Ghana; and the Ministry of Foreign Affairs of

→

Ghana. In addition to the above evidence, the Committee visited the Accra Metropolitan Waste Management Department's landfill site to see where the material had been dumped by Metalloplastic (Ghana) Ltd.

Inspection of the cargo, which was stored in the company's warehouse, revealed that only 1,063 of the bags (11 per cent), out of a total of 9,334, contained the imported material. This implied that 89 per cent of the cargo was not meant for the manufacture of plastic products. Although the results of laboratory tests did not establish the presence of radioactivity, the material was found to contain heavy metals like lead and copper. Part of the cargo was also found to be cytotoxic.

In spite of these revelations, and instructions from the Ministry of Environment, Science and Technology that the material should not be disposed of in Ghana, the EPA issued Metalloplastic (Ghana) Ltd a licence to dump the waste in AMA's landfill site. Surprisingly, however, the company could not provide the Committee with any record to prove that the waste was actually disposed of at that site. Since it had no expertise for determining the toxicity or otherwise of the material, the National Toxic Waste Task Force recommended the re-exportation of the waste to the supplier. But Metalloplastic went ahead regardless and dumped the waste at AMA's landfill in Accra.

A second consignment of plastic waste material arrived at the Port of Tema on 10 May 1992, but it was re-exported to the supplier on 4 June 1992. A relevant question to ask here is, why was the first consignment not re-exported to the supplier even though 89 per cent of it did not comply with the specifications of the import order? It was established, furthermore, that Ghana did not have any need for such waste, neither did it have any facility for treating it. Moreover, no prior consent was given by the Government of Ghana for accepting such wastes, as required under the Basel Convention.

A reason given for dumping the waste material in Ghana, contrary to instructions, was that attempts to return it to the US exporter through the American Embassy in Ghana and Greenpeace International had not been successful. Also, while investigations were going on in Ghana into the circumstances leading to the illegal importation of the waste, Petroco Americas Uniondale could no longer be contacted in the USA.

This case shows, clearly and typically, that the evaluation, monitoring and control of environmental actions in Ghana and many of the developing countries is far from efficient. Co-ordination of the roles of the various organisations in the country, responsible for protecting the environment, was rather weak and poor. Indeed, the fact that the National Investment Bank and the Bank of Ghana agreed to issue irrevocable letters of credit to Metalloplastic (Ghana) Ltd for the import is ample evidence of the deficiency in evaluating, monitoring and controlling environmental actions in Ghana.

The lack of effective communication between and among the various organisations involved in the episode has been a major contributory factor. For example, the Ghana Standards Board was not informed about the arrival of the material in the country, neither was it asked to take samples for testing. The directive issued by the EPA, authorising Metalloplastic (Ghana) Ltd to go ahead with the disposal of the waste in spite of instructions from the Ministry of Environment, Science and Technology not to do so, is symptomatic of the total breakdown of the environmental decision-making process; it also illustrates vividly the lack of effective co-operation, collaboration and co-ordination among the environmental agencies in the country.

Furthermore, the capacity of these agencies and their staff to ensure effective environmental planning is questionable. For example, the National Toxic Waste Task Force did not have the expertise to interpret SGS laboratory test results sent from Belgium. Also, different laboratory test results were obtained, each capable of different interpretations, variously by the SGS, the Nogouchi Memorial Institute for Medical Research at the University of Ghana, and the Ghana Standards Board. This points to a lack of uniformity in international environmental standards, as well as to lack of capacity in Ghana for evaluating, monitoring and controlling environmental actions. Finally, the reluctance of both the American Embassy in Accra and Greenpeace International to help re-export the first consignment of the waste to the US supplier highlights the problems of international co-operation and co-ordination in matters of environmental protection.

Source: Report of the Select Committee on Environment, Science and Technology, 1994.

CONCLUSIONS

The crucial need for integrating the environment into developmental planning in a holistic manner has been highlighted in this chapter. Such an approach, it is argued, would help arrest the rapid decline of the earth's natural resources and restore the health of the environment in the long term. In order to do this effectively, a deep understanding of the environment as a system is needed, together with access to essential information, basic methodologies, and appropriate instruments of environmental planning and management.

In addition, the application of strategic planning and management to environmental planning and management is proposed as an appropriate conceptual framework for enhancing environmental management practices. The conceptual framework includes environmental education and training; greater participation of the key actors and the general public in planning and management processes; facilitating effective co-operation and co-ordination between and among the countries of the world; and harmonising policies and plans at all levels.

The planning process needs to guarantee the right of all to participate. Environmental information should be reduced to a form in which it is easy to understand, and it should be easily accessible. Future environmental plans should be made more relevant and cost-effective on a national, regional and global scale. Thus, effective environmental planning, especially to deal with transboundary problems, would require investment in the environment beyond national and regional boundaries.

In order to ensure this, those responsible for planning should have access to effective environmental education and training in order for them to acquire the necessary scientific and multidisciplinary skills. Environmental planners must be skilled in integrating new environmental fields into the planning process. They should also develop their capacity for integrating the different contributions to the process in terms of general points of departure, time spans, ways of formulating hypotheses, and uncertainties which inevitably attach to environmental problems. It is equally important for them to be able to present the conclusions of their plans in a simple and succinct way for the attention and scrutiny of both the public and decision-makers. Strategic environmental planning is a very useful analytical tool in this regard. Finally, environmental plans would become effective instruments for improving the environment only if they are subjected to continuing and systematic evaluation, monitoring and control. This process should include performance monitoring, process monitoring, impact evaluation, and cost–benefit analysis.

REFERENCES

Ansoff, H.I. (1977) 'The changing shape of strategic problems', in D.E. Schendel and C.H. Hofer (eds) *Strategic Management: A View of Business Policy and Planning*, Boston, Mass.: Little Brown.

Baldwin, J.H. (1985) *Environmental Planning and Management*, Boulder, Colo.: Westview.

Bamberger, M. and Hewitt, E. (1986) *Monitoring and Evaluating Urban Development Programmes: A Handbook for Managers and Researchers*, Washington, DC: The World Bank.

Bardwell, L.V. (1991) 'Problem-framing: a perspective on environmental problem-solving', *Environmental Management* 15 (5): 603.

Beer, A.R. (1990) *Environmental Planning for Site Development*, London: Chapman and Hall.

Benveniste, G. (1989) *Mastering the Politics of Planning*, San Francisco: Jossey-Bass.

Buckley, R. (1991) *Perspectives in Environmental Management*, Berlin: Springer-Verlag.

Carley, M. and Christie, I. (1992) *Managing Sustainable Development*, London: Earthscan Publications Ltd.

Cayer, N.J. and Weschler, L.F. (1988) *Public Administration: Social Change and Adaptive Management*, New York: St Martin's Press.

Chechile, R.A. and Carlisle, S. (1991) *Environmental Decision Making: A Multidisciplinary Perspective*, New York: Van Nostrand Reinhold.

Clark, B.D. (1989) 'Environmental assessment and environmental management', in *Proceedings of 10th International Seminar on Environmental Impact Assessment and Management*, Aberdeen: University of Aberdeen, 9–22 July 1989.

Compton, P. (1993) 'Introduction', in B. Nath, L. Hens and D. Devuyst (eds) *Environmental Management, Vol. 1: The Compartmental Approach*, Brussels, Belgium: VUB Press.

Daneke, G.A. (1982) *Energy, Economics, and the Environment: Towards a Comprehensive Perspective*, Lexington: D.C. Heath and Company.

Darke, R. (1983) 'Procedural planning theory', in I. Masser (ed.) *Evaluating Urban Planning Efforts*, Aldershot, Hampshire: Gower.

Devuyst, D. (1993) 'Environmental impact assessment', in B. Nath, L. Hens and D. Devuyst (eds) *Environmental Management, Vol. III: The Compartmental Approach*, Brussels, Belgium: VUB Press.

Faludi, A. (1973) *Planning Theory*, Oxford: Pergamon Press.

Faludi, A. (1987) *A Decision-centred View of Environmental Planning*, Oxford: Pergamon Press.

Field, B.C. (1994) *Environmental Economics: An Introduction*, New York: McGraw-Hill.

Hudson, B.M. (1979) 'Comparison of current planning theories: counterparts and contradictions', *Journal of the American Planning Association* 45: 387–398.

International Network of Green Planners (INGP) (1994) *Report*, First Meeting of the Network, held in Maastricht, The Netherlands, 30 March–1 April 1994.

Jain, R.K. and Hutchings, B.L. (1978) *Environmental Impact Analysis: Emerging Issues in Planning*, Urbana, Ill.: University of Illinois Press.

Lee, Yok Shui F. (1994) 'Community-based urban environmental management: local NGOs as catalysts', *Regional Development Dialogue* 15 (2), Autumn.

McAllister, D.M. (1986) *Evaluation in Environmental Planning: Assessing Environmental, Social, Economic, and Political Trade-offs*, Cambridge, Mass.: MIT Press.

Mintzberg, H. (1994) *The Rise and Fall of Strategic Planning*, Hemel Hempstead: Prentice-Hall International Ltd.

Morell, J. (1995) 'Integrating technological change into planning: the case for an interdisciplinary perspective in socio-economic planning sciences', *The International Journal of Public Sector Decision-Making*, 29 (3): 224–225.

OECD (1995a) Room Document No. 7, Joint Workshop on National Planning for Sustainable Development: Donor Co-ordination and Harmonisation, OECD Development Assistance Committee, Working Party on Development Assistance and Environment, Paris, 26 October 1995.

OECD (1995b) Room Document No. 1, Special Session on National Plans for Sustainable Development, OECD Development Assistance Committee, Working Party on Development Assistance and Environment, 12th Session, 27–28 April 1995.

OECD (1995c) Room Document No. 3, National Sustainable Development Plans and Strategies, Information Note by UNDP, Paris.

Report of the Select Committee on Environment, Science and Technology, on the Investigation into the Importation and Disposal of Plastic Waste Materials by Metalloplastic (Ghana) Limited, Parliament House, Accra, November 1994.

Rosenbaum, W.A. (1973) *The Politics of Environmental Concern*, New York: Praeger Publishers.

Ryding, S.O. (1994) *Environmental Management Handbook*, The Netherlands: IOS Press.

Slocombe, D.S. (1993) 'Environmental planning, ecosystem science, and ecosystem approaches for integrating environment and development', *Environmental Management* 17, 3: 289.

Smith, L.G. (1993) *Impact Assessment and Sustainable Resource Management*, Harlow, Essex: Longman Group UK Ltd.

Stahl, M. and Grigsby, W. (1992) *Strategic Management for Decision Making*, Boston, Mass.: PWS-Kent Publishing Company.

Steiner, G.A. and Miner, J.B. (1977) *Management Policy and Strategy*, New York: Macmillan Publishing Co.

Stohr, W.B. and Taylor, D.R.F. (1981) *Development from Above or Below? The Dialectics of Regional Planning in Developing Countries*, Chichester: John Wiley & Sons Ltd.

UNDP (1995) National Sustainable Development Plans and Strategies, Information Note by UNDP, OECD/DAC Working Party on Development Assistance and Environment, 27–28 April 1995.

Welford, R. and Gouldson, A. (1993) *Environmental Management and Business Strategy*, London: Pitman Publishing.

Westman, W. (1978) *Ecology Impact Assessment, and Environmental Planning*, Chichester: John Wiley & Sons.

Wheelen, T. and Hunger, J. (1995) *Strategic Management and Business Policy*, New York: Addison-Wesley.

Wildavsky, A. (1973) 'If planning is everything, maybe it's nothing', *Policy Sciences* 4: 127–153.

Wyatt, R. (1989) *Intelligent Planning*, London: Unwin Hyman.

SUGGESTED READING

Baldwin, J.H. (1985) *Environmental Planning and Management*, Boulder, Colo.: Westview.

Faludi, A. (1987) *A Decision-centred View of Environmental Planning*, Oxford: Pergamon Press.

McAllister, D.M. (1986) *Evaluation in Environmental Planning: Assessing Environmental, Social, Economic, and Political Trade-offs*, Cambridge, Mass.: MIT Press.

Mintzberg (1994) *The Rise and Fall of Strategic Planning*, Hemel Hempstead: Prentice-Hall International Ltd.

Ryding, S.O. (1994) *Environmental Management Handbook*, The Netherlands: IOS Press.

SELF-ASSESSMENT QUESTIONS

1 The term 'environment' means?
   (a) The natural resources of the planet.
   (b) Air.
   (c) All man-made and natural surface and atmospheric elements.

2 Which of these is not a core concern of environmental planning?

(a) The need for including environmental health aspects in national planning.
(b) Public environmental measures which an authority takes.
(c) Promoting rapid economic growth.

3 Indicate which of the following is not an approach to environmental planning.
(a) Substantive planning.
(b) Sideways planning.
(c) Process planning.

4 The elements of the five schools of planning include which of the following?
(a) Radical planning.
(b) Gender planning.
(c) Synoptic planning.

5 According to the UNDP which of the following are weaknesses in the formulation and implementation of national conservation strategy, national environmental action plans and national Agenda-21?
(a) The strategy is not a priority locally or for the government.
(b) The emphasis is on the process and on the product.
(c) Each report is treated in isolation.

6 Which of the following factors tend to hinder effective environmental planning?
(a) Governments regard environmental consequences of national policies as a 'secondary effect'.
(b) Lack of comprehensive planning.
(c) Refusal by nature to satisfy human needs.

7 Effective international co-operation in environmental planning and management is influenced by:
(a) The growing fear of under-development in the South.
(b) God's anger with human society.
(c) Lack of effective communication.

5

STATE OF ENVIRONMENT REPORTING

David J. Briggs

SUMMARY

State of environment (SoE) reporting has developed rapidly in recent years in response to improvements in information technology and increased demands for information by both public and politicians. The adoption of Agenda-21 at the Earth Summit in Rio de Janeiro in 1992 has given added impetus to this process. As a result, almost all countries in Europe now publish information on the state of environment, and a number of international reports are also produced.

The character of SoE reports is nevertheless varied, reflecting the constraints and influences of institutional factors (e.g. the type of the institution producing the report), the audience at which it is aimed, and the availability and quality of data.

SoE reports are the products of long and complex data streams, involving the collection of primary data (e.g. by survey or monitoring techniques) and many, often repeated stages of aggregation, processing, selection and presentation. Errors, biases and inconsistencies, which may develop at all points in this process, tend to accumulate down the data stream. Particular problems tend to occur because of one or more of the following: differences in the definitions and classifications used to describe environmental phenomena; limitations of the sample design used during monitoring and measurement; errors and inconsistencies in the methods of measurement used; and the effects of aggregating and selecting data for presentation. As a result, much of the information presented in SoE reports must be interpreted and used with extreme care if environmental problems and policy needs are not to be misunderstood. In order to help users in this respect, environmental information is often presented in the form of 'indicators'. These are designed to summarise the information in a simple and readily applicable form. Designing effective indicators is nevertheless difficult, especially if they are to take account of the conflicting interests and concerns of different sections of society, as recommended by Agenda-21.

ACADEMIC OBJECTIVES

The aims of this chapter are to introduce readers to the role, history and processes of state of environment reporting in Europe.

The chapter should provide a basis for readers to understand the conceptual framework within which state of environment reporting occurs, the methods used for data compilation and presentation, and the various problems and pitfalls involved. On completion of this chapter the reader should thus be able to use and interpret state of the environment reports critically, and be aware of their functions and limitations.

THE NEED FOR INFORMATION

The rapid increase in the demand for, and use of, information has been one of the defining features of the last 20–30 years, and one which perhaps sets the present apart from the past. We live in an information age. We depend on information to guide our decisions and actions, to help monitor and evaluate the effects of the decisions and actions we take, to exchange and share experience, and as a means of facilitating participation in the affairs of society.

Information on the environment is no exception in this respect. Like other forms of information, environmental information has become increasingly important in recent years. Many factors have contributed to this trend. One is the increasing awareness of the role which such information can play, especially in a world which is itself complex and in which many different stakeholders and actors are involved. We need environmental information, for example:

- to highlight environmental problems and issues of concern;
- to identify 'hot-spots' and areas in need of special attention;
- to give an early warning of future environmental problems and trends;
- to help set sustainability targets and goals;
- to evaluate and compare policy options likely to affect the environment;
- to contribute to environmental policies and plans;
- to monitor policy performance and effects;
- to monitor compliance with policy requirements;
- to meet statutory requirements for environmental reporting;
- to help incorporate sustainability principles into other policies and areas of activity;
- to raise awareness about the environment;
- to inform the public and other actors;
- to provide a baseline for debate;
- to promote and justify policy initiatives.

Another factor has been the growing availability of information on the environment. In part this derives simply from the accumulation of data over the years. It is also due to the improvements which have been made in environmental survey and monitoring, especially through the use of remote-sensing techniques. Equally, however, it is a product of the major advances which have been made in computer technology – advances which allow us to develop more complex environmental models, to map the environment more effectively (using GIS for example), and to handle larger datasets. Together, these developments have meant that environmental information has become more deeply embedded in our society – from decisions by individual consumers about what to buy through to major questions of international and global policy.

THE DEVELOPMENT OF STATE OF ENVIRONMENT REPORTING IN EUROPE

The development of state of environment reporting is but one component of this growing dependence on, and use of, environmental information. In so far as any evolutionary process can be traced back to a single cause, its origin may be sought in the adoption of the National Environmental Policy Act (NEPA) in the USA in 1969. Prior to this, many countries were already collecting information on the environment routinely, and several – including the USA and Canada – were developing environmental information systems to support policy. In establishing the Council on Environmental Quality (CEQ), however, Article 201 of NEPA defined for the first time the principle of publishing regular state of the environment reports. Starting in July 1970, it required the CEQ to produce an annual report which would:

- describe the state of the natural, managed and artificial environment, including in particular the air, water and land (especially forests, arid zones, wetlands, high plains and urban, suburban and rural areas);
- show the current and foreseeable trends in environmental quality and the state and use of resources, and indicate their potential social and economic implications;
- assess the balance between available natural resources and the economic and social demands placed on them;
- describe the policies, programmes and actions being taken at different levels, with special reference to their effects on the environment, conservation, development and use of natural resources; and
- outline the actions needed to remedy the weaknesses in current policies.

In practice these ambitious aims could not be achieved in the time available, neither has any of the subsequent reports managed to address all these issues. The CEQ report nevertheless acted to focus the attention of the various environmental organisations in the western world on the concept of public SoE reporting as a means of public communication, and on the potential role of such reports in policy formulation and monitoring.

In Europe further impetus to the concept of SoE reporting was given by two developments: the Stockholm Conference in 1972, and the adoption by the European Community of a formal environmental policy in the following year. Each stressed the need for concerted action to tackle the growing environmental problems at an international level. Each also emphasised the need for preventive action which, as far as possible, avoided environmental damage rather than simply attempting to control or ameliorate damage after it had occurred. Within this context each recognised the need for information in order to direct and underpin policy action as well as to inform the public.

More recently, another crucial episode in the development of SoE reporting has been the United Nations Conference on Environment and Development (the so-called 'Earth Summit') in Rio de Janeiro in 1992. One of the main outcomes of this conference was the adoption of Agenda-21 which sets out a series of principles and actions for achieving sustainable development (Hens, 1996). Providing relevant information in a readily accessible form is seen as one of the keys to this objective. Chapter 40 of Agenda-21, for example, states that 'Countries and international organisations should review and strengthen information systems and services in sectors related to sustainable development at the local, national and international level. Special emphasis should be placed on the transformation of existing information into forms more useful for decision-making and on targeting information at different groups.' To support this, Chapter 8 calls for 'Monitoring and evaluating the development process systematically, conducting regular reviews of the state of human resources development, economic and social trends, the state of the environment and natural resources.'

At the same time, Agenda-21 introduces a range of other principles which impinge on SoE reporting. One is the concept of sustainable development itself. By recognising the way in which society is embedded within the environment, and the strong social and economic dimensions of environmental management, this extends the scope of SoE reporting from the environment *per se* to wider cultural, social and economic issues. It also emphasises the large number of different actors and stakeholders concerned – not just politicians, scientists and environmental managers but also local and regional planners, business and industry, education, voluntary groups and the public. All these groups need to participate in the processes of both decision-making and environmental action. SoE reports thus need to address them all and to take account of their various interests and concerns.

Stimulated by these developments, SoE reporting has spread widely throughout Europe in recent years. SoE reporting, however, may take many different forms and generate different responses depending, among other things, on the motives involved, institutional structures and functions, national cultures and availability of data. One response has been to expand the environmental component of existing statistical compendia. A second, often as an offshoot of these, has been to compile new compendia dedicated wholly to the environment. A third has been to publish full state of the environment reports in the mould of the CEQ report. A fourth has been to develop even broader-based reports on sustainable development, which consider not only the environment but also its associated social and economic implications. As part of these trends, many different groups of actors have become involved in SoE reporting, including national agencies such as ministries of the environment, intergovernmental and non-governmental organisations (e.g. the Commission of the European Union, United Nations, WHO, World-Wide Fund for Nature), regional and local authorities, and voluntary organisations.

Because of this divergence of mechanisms, it is not easy to chart precisely the development of state of environment reporting in Europe over the last two decades. In many cases SoE reports have tended to evolve gradually from the mutation of existing organisms rather than arising in a newly created form. The question of what exactly constitutes state of environment reporting (as distinct, perhaps, from a formal state of environment report) is one which could be debated at length. Several reports have also appeared in an experimental form prior to formal publication, while others have been produced informally by private organisations. An attempt is made in Table 5.1 to summarise some of the main national reports on the state of the environment (in their various guises) which have been published in Europe. As this table shows, one of the first European countries to establish a regular reporting mechanism was

Table 5.1 Selected national reports on the state of the environment

Country	*Publisher*	*Title*	*Comment*
Belgium	IHE	*L'Environnement en Belgique*	Irregular
Denmark	Danmarks Statistik	*Tal om natur og miljú*	First published 1990
Finland	Central Statistical Office for Finland	*Environmental statistics*	Published 1973, 1977, 1981, 1987
France	Ministère de l'Environnement	*L'État de l'environnement*	Annual since 1978
Germany	Umweltbundesamt	*Daten zur Umwelt*	Biannual since 1984
Greece	National Statistical Service for Greece	*Environmental statistics*	Published 1987, 1988
Hungary	Hungarian Academy of Sciences	*State and protection of the Hungarian Environment*	First published 1990
Ireland	An Foras Forbartha	*The state of the environment*	Published 1986
	An Foras Forbartha	*Irish environmental statistics*	Published 1986
Italy	ISTAT	*Statistische ambientali*	Published 1984, 1991
	ISTAT	*Annuario statistico Italiano*	Contains statistics on the environment
	Ministry of Environment	*Report on the state of the environment*	Published 1989
Netherlands	RIVM	*Concern for tomorrow*	Biannual since 1989
	CBS	*Environmental statistics of the Netherlands*	Biannual since 1973
Norway	CBS	*Environmental statistics*	Published 1976, 1978, 1983, 1988
Poland	State Inspectorate for Environmental Protection	*State of the environment in Poland*	First report published 1993
Portugal	Ministerio do Planeamento e de Administraçao do Territórlo	*Relatório do estado do ambiente e ordenamento de ambiento*	Published 1987, 1989
Spain	Dirección General de Medio Ambientale	*Medio ambientale en España*	Annual since 1985
Sweden	Sveriges Officiella	*Naturmiljön in Siffor*	Published 1984, 1987, 1990
UK	DoE	*Digest of environmental protection and water statistics*	Annual since 1978
	DoE	*The UK environment*	First published 1992
	DoE	*Indicators of sustainable development for the United Kingdom*	First published 1996
USSR	USSR State Committee for the Protection of Nature	*Report on the state of the environment in the USSR*	First published 1989

Germany, which published its original environmental compendium (*Daten zur Umwelt*) in 1972. Finland and the Netherlands each produced their environmental compendia in the following year, and Norway followed suit in 1976. The UK published the first *Digest of Environmental Protection and Water Statistics* in 1978, while France produced the first edition of the formal SoE report, *L'État de l'environnement*, in the same year. Today, as the example of the UK shows, a wide variety of national reports on the state of the environment is being published.

By the late 1980s, almost all European countries were reporting on the state of the environment in some form, although the character of the reports, their frequency of publication, and the audience at which they were aimed varied. By then several international reports had also begun to appear. OECD began to publish a series of state of environment reports in 1979 with a periodicity of 5–6 years, and in 1984 it initiated the *Environmental Data Compendium* which is published biannually. UNEP, UN–ECE, the Worldwatch Institute, and the World Resources Institute have also produced reports which include coverage of the European area (Table 5.2). In the European Union the picture is somewhat more confused, however. The Commission of the European Union (CEU) produced its first report in 1977, followed by the publication of three further editions in 1979, 1987 and 1992 – the last only in draft form. Eurostat, the Statistical Office of the European Union, has also published a number of compendia of environmental statistics. In addition, and in response to an agreement by the European environment ministries at the Dobris Castle, a pan-European report was completed in 1995 (the so-called *Dobris Report*). To what extent these various reports will be continued on anything like a regular basis is not clear. It is likely that the European Environment Agency, established in Copenhagen in 1994, will take over the responsibility for future SoE reporting. Meanwhile, the World Health Organisation (WHO) has also compiled a pan-European report, *Concern for Europe Tomorrow*, which sets environmental issues alongside information on the state of human health.

Table 5.2 International reports on the state of the environment

Publisher	*Title*	*First published*	*Periodicity (years)*
CEC–DGXI	*The state of the environment*	1977	Irregular
CEC–Eurostat	*Environmental statistics in the European Community*	1989	Irregular
CEU	*Europe's environment*	1995	?
DocTer	*European environmental yearbook*	1984	2
Gale Research	*European environmental statistics yearbook*	1993	1
OECD	*The state of the environment*	1979	6
	OECD environmental data compendium	1984	2
	Environmental indicators	1991	2
UN–ECE	*Environmental statistics in Europe and North America: an experimental compendium*	1987	Irregular
UNEP	*State of the environment report*	1974	1
	Environmental data report	1987	2
Worldwatch Institute	*The state of the world*	1984	1
WRI	*World resources*	1986	1

At the same time, SoE reporting has begun to descend to the regional level. In Germany, for example, the first report for Baden-Württemberg was published in 1985, followed by that for Nordrhein-Westfalen in 1986. In France, the Délégation Régionale à l'architecture et à l'environnement and the Conseil Régionale have produced a series of *Tableaux de bord de l'environnement* for most regions and departments of France. In Belgium, the Vlaamse Milieu Maatschappij has been producing a report for the Flanders region every two years since 1984. In the UK, the Scottish Office produces regular statistical compendia, while the first state of environment report for Scotland was published by the Scottish Wildlife and Countryside Link in 1992 (Dargie and Briggs, 1992). The Kirklees Metropolitan Council was the first local authority to publish an SoE report in 1989, but since then many district and county councils have also produced reports. In addition, SoE reporting is emerging as a crucial component of environmental auditing – introduced in the UK by the government White Paper (*Our Common Heritage*) and now being extensively adopted at the regional level.

THE CHARACTER OF SoE REPORTS

The form and format of SoE reports

Traditionally SoE reports have been produced as textual documents. In this form they have tended to vary substantially, although three broad types of report can be identified:

1 Statistical compendia, comprising tables of environmental statistics with supporting notes (e.g. the *OECD Environmental Data Compendium* and Eurostat's *Environmental Statistics*).
2 Environmental digests, comprising sets of statistics on the environment in tabular, graphic or map form with accompanying explanatory commentaries

(e.g. Umweltbundesamt's *Daten zur Umwelt* and the UK Department of the Environment's *Digest of Environmental and Water Protection Statistics*).
3 Interpretative reports, discussing and analysing specific problems and themes, supplemented by statistical appendices (e.g. the OECD *State of Environment* reports and RIVM's *Concern for Tomorrow*).

With recent developments in Information Technology (IT), however, the reliance on textual documents is rapidly diminishing and a wide range of alternative media and formats is being used. Some reports, for example, are now supplemented by diskettes or CD-ROMs containing the statistical and graphical materials relating to the textual document. Some are available wholly in computer-readable form – and increasingly in a form which allows interactive access to the underlying data. Summaries or selections of the reports are also being made available via the Internet. An example of a computer-driven SoE approach is described in Box 5.1.

BOX 5.1 BUILDING CAPACITIES IN ENVIRONMENTAL INFORMATION AND REPORTING (BY OTTO SIMONETT AND NICKOLAI DENISOV, UNEP/GRID, ARENDAL, NORWAY)

One of the primary objectives of the United Nations Environment Programme (UNEP) is to catalyse co-operative action between nations to address environment and development issues of international significance. Part of that process is the undertaking of authoritative and coherent global assessments of the scope and magnitude of environmental issues. It is clear, however, that the authority of such assessments depends much on the ability to integrate national and sub-national data, rather than on assumptions about the national and regional significance of findings based on global data.

Chapter 40 of Agenda-21, which is on information for decision-making, outlines how the capacity for managing information could be improved to provide better access to environmental information at all levels. To meet these needs, UNEP's Division of Environment Information and Assessment (DEIA) is providing preparatory assistance to developing countries, and to countries with economies in transition, with the aim of strengthening their environmental data and information management capacities through its Environment and Natural Resources Information Network (ENRIN) programme.

National and sub-regional partner institutions and agencies are being enabled through ENRIN to collate, store, manage and disseminate environmental data and information, and to use modern information technologies to assess environment and development issues for decision-making, policy-setting and planning. Although the main objective is to improve the availability of environmental data and information for international action, this process also benefits national decision-makers and the general public, thus contributing to the implementation of Agenda-21 at the national level.

Since 1994 UNEP has been implementing ENRIN in the transition countries of Central and Eastern Europe (CEE). By 1997 the programme will have involved 18 out of the 27 countries in the region. ENRIN's approach involves the compilation of assessments of national and regional environmental information needs and capacities, establishment of co-operative agreements with governments and international organisations, development of strategies and proposals to strengthen national and regional information networks, co-ordination of implementation projects, provision of direct assistance and training in information management technologies, and wide dissemination of information products.

ENRIN operates on the basis of a network comprising international bodies, national environmental authorities, and collaborating institutions such as sectoral agencies, academic and research organisations, and public groups. In each participating country the process consists of UNEP-funded assessment (6 months), feasibility study (1 year), and externally funded implementation (3–5 years). After that a national network becomes operational, self-sustaining, and fully capable of producing internationally harmonised information products.

Since ENRIN aims at reaching the widest possible audience of the general public and decision-makers in a cost-efficient way, particular attention is given to the production of easily understandable SoE reports on the Internet. The first electronic indicator-based Norwegian SoE report, prepared by UNEP/GRID-Arendal in 1992 and regularly updated since then, is now used as a model by the partner countries. The Georgian national SoE report, officially launched in 1996, was the first. Six other CEE countries have completed their prototype reports on the Internet in accordance with UNEP's methodology.

These changes in format are not cosmetic. The way in which SoE reports are produced fundamentally affects their accessibility and use. In this respect many traditional SoE reports have been far from successful. Analysis of the distribution and sales of reports at all levels, from the regional to international, shows that they have often been used by a very limited range of people; typically the main users have been staff within the organisation which produced them, and staff in closely related agencies and universities. The many other actors and stakeholders who might be regarded as key audiences for SoE reports – such as industry, commerce and the public – have only been rare users. Clearly, the SoE reports must reach a much wider audience if they are to fulfil their role of providing a pathway for the dissemination of environmental information and a means of encouraging participation in environmental decision-making. Whether the use of new technology will help to achieve this remains to be seen.

The structure of SoE reports

In whatever form it is produced, an SoE report is an attempt to condense a set of extremely complex and multidimensional systems (the 'environment') into a much simpler and more linear entity (the 'report'). The task is far from easy, and it is becoming increasingly more difficult with the expanding scope of SoE reports and as wider issues of sustainable development come into consideration. Much thought is thus being given to ways of organising SoE reports in a manner which reflects both the underlying structure of the environment and the inter-relationships within it, on one hand, and the needs of users, on the other.

One of the earliest structures to be proposed for this purpose was the so-called P–S–R (Pressure–State–Response) model. This was initially suggested by the United States Environmental Protection Agency (USEPA) and subsequently adopted by the OECD for its reports. Since then it has been widely copied in many other SoE reports, including the third report of the European Union and the Dobris Report.

A central tenet of the P–S–R model, illustrated in Figure 5.1 and as used in the 1991 OECD report, is that

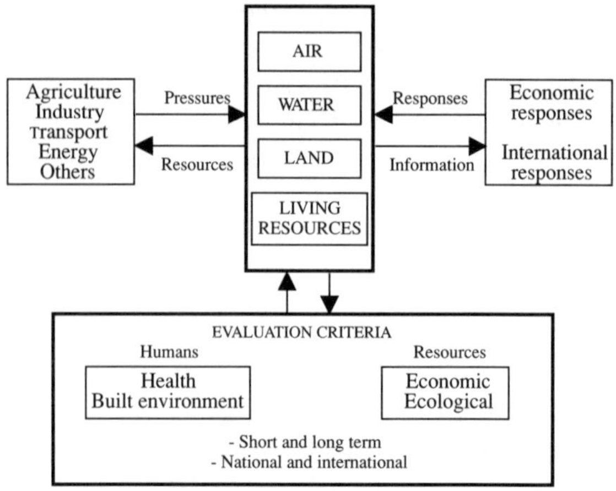

Figure 5.1 Conceptual framework for the OECD state of environment reports
Source: From OECD, 1991

> human activities impose pressures on the environment (e.g. pollution, waste, environmental restructuring) and depend on it for natural resources inputs (e.g. water, marine, forest, land resources). The state of the environment also depends on its management through responses from economic and environmental agents to environmental changes. Environmental changes can be evaluated with two types of criteria relating to man and natural resources (from an economic and ecological point of view) . . .
>
> (OECD, 1991: 14)

The P–S–R framework certainly has much to commend it in that it helps to emphasise the role of human actions both as sources of pressure on, and means of protection of, the environment. Nevertheless, it is also very limited in a number of respects, especially in giving a relatively simple, linear view of these relationships. In recent years, therefore, a number of alternative models of the environment have been developed which are now beginning to find their way into SoE reporting. Possibly one of the most useful of these is the so-called 'Deepsea' (DPSEA) model.

The origins of this model are somewhat obscure in the sense that it appears to have been developed independently (or semi-independently) by a number of different parties, albeit in slightly different forms. One form of the framework, illustrated in Figure 5.2, comprises the following elements:

- *Driving forces* (D) represent the underlying or exogenous factors which lead to pressures on the environment, such as by economic development, population change, global warming.
- *Pressures* (P) comprise the actions, infrastructure and immediate products (e.g. wastes, emissions) associated with human use of the environment.
- *State* (S) refers to the measurable condition of the environment *per se*; for example the extent and quality of its various resources and media.
- *Effects* (E) represent the wider consequences of changes in the state of the environment; for example on human health, social conditions and the economy.
- *Actions* (A) relate to the measures taken to protect and manage the environment. As Figure 5.2 shows, these may be directed at each of the other four links in the DPSEA chain.

To date the DPSEA model has not been formally used in SoE reporting, though it has been suggested as a basis both for integrated environmental assessment and for developing indicators on environmental health and the countryside. As a framework for SoE reporting, however, it certainly has a number of advantages. One is its recognition of the wider, and often less immediate, factors which drive changes in the environment (the aptly named 'driving forces'). Another is its acknowledgement that changes in the state of the environment do not represent the end of our concern. Often of far greater importance – and ultimately our real motive for action – are the subsequent social and economic impacts, since these affect the quality of human life. Indeed, these wider effects are especially significant in the context of sustainable development. Moreover, the model recognises the very different types of action which can be taken to protect and manage the environment, and the ways in which these feed back at different points within the overall system. This helps to emphasise the fact that actions aimed at subsequent links in the DPSEA chain are essentially mitigative, while those aimed at earlier links can be more preventive.

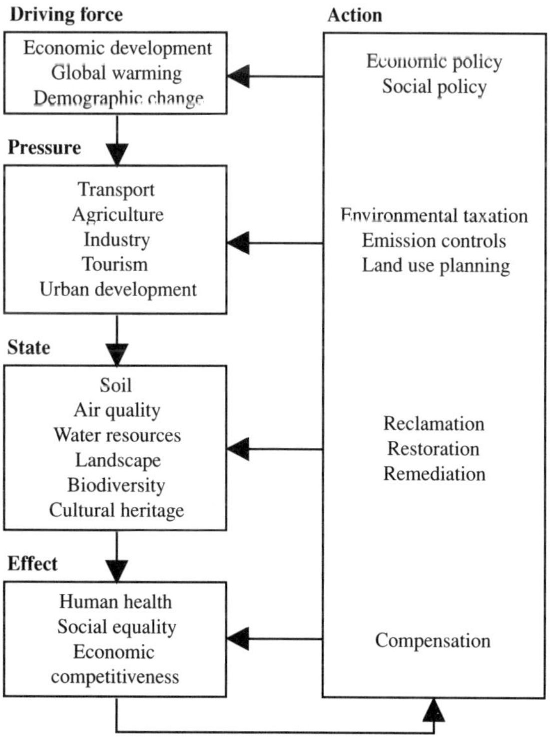

Figure 5.2 The DPSEA framework

Scope and content of SoE reports

The choice of what to include in a state of environment report is influenced by a number of factors, not least who is producing it, the area for which it is being produced, the intended audience, and the particular environmental issues that are of concern. In addition, the availability of data is often a major constraint. As a result of these influences and through their evolution over time, SoE reports vary widely in both scope and content. Indeed, the evolution of SoE reports for the European region illustrates the point. The first two reports, published in 1977 and 1979, were relatively poor in information, concentrating as they did on policy actions rather than on describing in any quantitative manner either the prevailing conditions or trends in the state of the environment. In part this reflected the lack of environmental data available at a European level at the time, or at least the lack of institutional mechanisms for data collection. But it was a consequence also of the relatively recent emergence of the European Community's environmental policy, so that attention was still focused on reviewing and promoting the actions and successes achieved. In contrast the 1986 report, published in 1987, was far more quantitative in its approach. It attempted to draw together data describing the state of the environment at national and regional levels as a means of defining spatial patterns and temporal trends, assessing the pressures acting on the environment, and highlighting gaps in knowledge. This change in approach derived from a number of factors, including:

- the increased availability of environmental information in the EU member states (e.g. through the publication of national SoE reports);
- the increased emphasis being placed on data gathering by the CEC, both in environmental Directives (e.g. on surface freshwater quality, bathing water quality, air quality) and through the CORINE Programme (CEC, 1985);
- the specific background and philosophy of the consultants commissioned to undertake the work.

It is to be noted that these reports cover a gradually increasing geographic area, and thus a greater range of environmental problems and issues, as a result of the expansion of the European Union.

The Dobris Report has taken the process much further forward. It not only covers a much wider area – the whole of Europe excluding Russia – but also explicitly considers issues of sustainability in relation to Europe's environment (Stanners and Bordeau, 1995).

This trend towards greater quantification and increasing scope of SoE reports is widespread. To a large extent it is a natural process of evolution in that, in order to show progress, each report needs to be seen to be bigger and better than the one before it. As a result there is a tendency for many SoE reports to become 'State of Everything' reports rather than State of Environment Reports! Whether or not this process is improving their effectiveness is, however, a moot point. It is a trend which makes them more encyclopaedic in nature, with the undoubted benefit of bringing together more information in a consistent form, thereby emphasising the interconnectedness of environmental and other issues. It also improves their role as works of reference. At the same time, however, it may make them less accessible to many users, especially policy-makers who typically do not have the time or levels of specific expertise to read such lengthy and detailed documents, and the public for whom a large and complex report may appear too costly and intimidating. Furthermore, as SoE reports become bigger and more encompassing, they become correspondingly more difficult and costly to maintain.

INDICATORS

The problem of DRIPS

As we have pointed out, the development of SoE reporting in recent years owes much to the increased availability of environmental information. Strangely as it may seem, however, this has in many ways made the process of SoE reporting harder rather than easier, because, typically, it means that data for inclusion in the report must be selected, often from a large pool of detailed statistics – a process which is itself subject to bias and likely to influence the message which the report conveys. It also means that commonly used environmental statistics need to be reduced to simpler and more easily interpretable form in order to ensure that they are user-friendly. In many ways, therefore, SoE reports suffer from what some observers have termed DRIPS: Data-Rich-Information-Poor-Syndrome.

Environmental indicators

Indicators are one solution to this problem. This is because they are seen as a way of adding value to data by converting them to a more usable and meaningful form. As such, indicators offer a means of:

- *simplification*, because they provide a way of expressing knowledge in a simpler and more easily interpretable form;
- *data reduction*, because they provide a way of condensing large quantities of data into smaller and more manageable amounts;
- *targeting*, because they help to make the information more directly relevant to the needs of the user;
- *surrogation*, because they offer a means of describing or measuring conditions or processes which are otherwise not directly amenable to measurement;

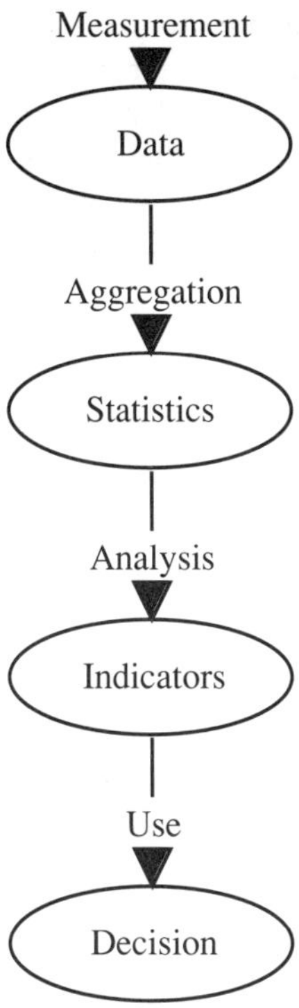

Figure 5.3 The place of indicators in the information chain

- *communication*, because they offer a common language for communicating information to others in an understandable form.

Indicators represent an important link and point of convergence within the information chain: measurements produce data; data are compiled to produce statistics; statistics are translated into indicators; and indicators are used in decision-making (Figure 5.3).

The concept or use of indicators is not new. Historically they have been used in disciplines such as economics (e.g. GNP, GDP, unemployment rate), resource management (e.g. indicators of land suitability), and ecology (e.g. the use of 'indicator species'). In recent years, however, there has been a marked growth in interest in the use of indicators in many other fields. The use of social indicators (e.g. of deprivation, poverty) is now widely accepted, while performance indicators are increasingly being used to monitor the activities of industry and public services.

In recent years environmental indicators have also become well established to the extent that they are now accepted as important tools in SoE reporting.

An environmental indicator can be defined as:

> a measure which describes the state of, or trends in, the environment and the factors affecting it, in terms which are relevant to and readily usable by the decision-maker.

Criteria for environmental indicators

An environmental indicator is intended to provide information on the environment concisely, simply, unambiguously, effectively and clearly. Achieving this is, however, far from easy because good environmental indicators need to satisfy a number of challenging and often contradictory criteria, as the following list shows:

1 Environmental indicators must be scientifically valid; that is
  - they must be based on a valid relationship between the characteristics which are measured and the conditions which they are intended to indicate;
  - they must be consistent and comparable across the area and over the time for which they are used;
  - they must be unbiased and representative of conditions within the area concerned;
  - they must be robust and unaffected by minor changes in the methodology or data used to construct them.

2 Environmental indicators must be fitted to the purpose for which they are intended; that is
  - directly relevant to the specified objectives;
  - sensitive to changes in the conditions concerned;
  - timely and up to date;
  - available at an appropriate spatial scale or resolution.

3 Environmental indicators must be practicable; that is
  - they must be based on data which are available at an acceptable cost–benefit ratio;
  - they must be compilable using readily available methods and technology.

A crucial requirement is that indicators must be scientifically valid and reliable. In this context, and as indeed their name implies, indicators must *indicate* something which is not easily and clearly expressed otherwise. This means that a known and consistent relationship must exist between the indicator and the conditions it is intended to portray. However, in practice such relationships are often extremely complex and may vary from one area, or one time period, to another. As a result many indicators are difficult to interpret reliably. A typical example is the use of levels of air pollution as an indicator of health risk. In reality the links between air pollution and health are so subtle, so varied and so dependent upon other factors (such as lifestyle, socio-economic status, quality of the health system) that even quite major variations in air pollution levels cannot be relied upon to indicate differences in health impacts.

The validity of environmental indicators also depends on the details of how they are designed and compiled. Indeed, indicators may tell very different stories, for example, depending on:

- the level of aggregation used – for example, whether they are assessed at the level of individual sites, regions, or at the national level;
- the period to which the indicator relates – for example, indicators of pollution levels may vary depending on whether they relate to mean annual levels, mean summertime levels, or to peak daily levels;
- the denominator used to express the indicator – for example, indicators of waste generation give a different picture depending on whether they are expressed as total waste produced, or waste produced per unit of land area or per head of population;
- the threshold or target level to which they are related – for example, indicators of ozone pollution may vary substantially depending on whether they are related to EU levels for health impact or vegetation damage.

A second important criterion is that indicators must be fitted to the purpose for which they are intended. This means that they must be relevant to the needs of the user, and sensitive to the characteristics of the particular prevailing conditions. An implication of this is that most indicators are use-specific, meaning that different indicators need to be developed for different purposes. Thus an indicator developed in one area, or for a given purpose, cannot easily be transferred to another. Also, indicators of interest to one set of users may not be relevant to others. Sadly, this means that attempts to develop core indicator sets, which can serve the needs of a wide range of users, are likely to meet with only limited success.

The third crucial criterion is that it should be practicable to compile and use the indicators. Cost is a vital consideration in this respect, for environmental surveys and monitoring work are intrinsically expensive and can seldom be justified simply for the purpose of SoE reporting. In most cases, therefore, indicators need to be based on data which already exist and which can be compiled by existing methods and institutions. But inevitably this often limits the range of indicators which can be used in SoE reports.

THE QUALITY OF ENVIRONMENTAL INFORMATION

As we have pointed out in the previous section, issues of data availability and quality are of fundamental concern in SoE reporting. In fact, availability of data predetermines the range of indicators which can be included in SoE reports, while the quality of data determines the validity of these indicators and their ability to detect real variation or changes in the environment.

The information used in compiling SoE reports can be seen as the end product of a lengthy and complex data stream (Figure 5.4). Errors, inconsistencies and biases may arise at any point in the stream, and they may tend to accumulate downstream. Thus the information contained in SoE reports may suffer from a wide variety of quality problems, including:

- inconsistencies in the definition of the measured variables; in the way in which municipal wastes are defined for example;
- inconsistencies in the actual variable (the 'determinant') which is measured; for example, whether nitrogen concentrations in stream waters are measured as total oxidisable nitrogen or as nitrates;
- inadequacies or inconsistencies in sampling frequency; for example, stream water quality is

measured no more than once per month at most of the monitoring stations;

- inadequacies and differences in sample layout and design; for example, air quality monitoring stations tend to be biased towards specific types of environment;
- errors and differences in measurement and analytical technique; in the methods used for water quality analysis for example;
- differences and inadequacies in modelling; as in the models used for assessing atmospheric emissions;
- inconsistencies and uncertainties in the classifications used; in classifying protected habitats and species for example;
- bias and differences in the selection of sites or data values for reporting; as in the selection of air quality or water quality sites;
- errors and uncertainties in data aggregation, averaging and processing; ambiguities in the questionnaires used for collating information at the international scale for example.

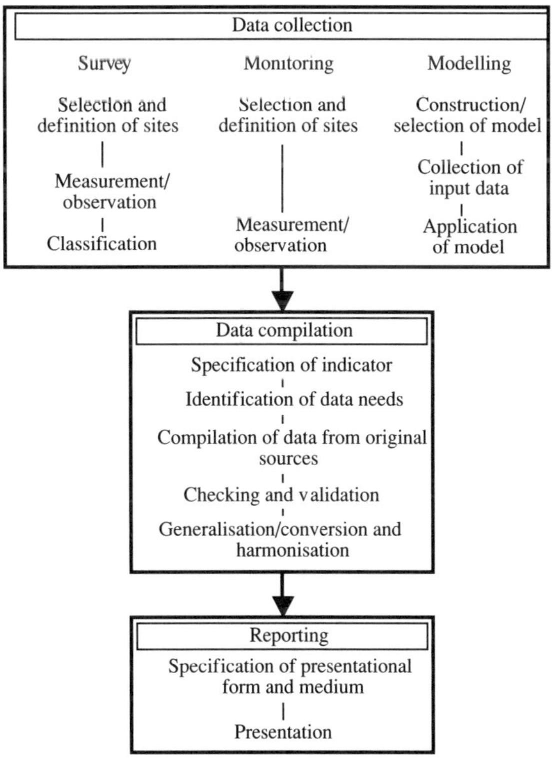

Figure 5.4 The SoE data stream

In view of the problems listed above, it has been suggested that most statistics probably have a minimum margin of error of at least 25 per cent, and that in many cases (as in statistics on wastes for example) it is likely to exceed 100 per cent (Briggs, 1995). In the following sections we illustrate some of these problems in relation to environmental statistics on atmospheric emissions, air quality, wastes, water quality, and nature and wildlife.

Atmospheric emissions

Statistics on atmospheric emissions are generally derived from emission models. In the European Union many different models are used at the national level, although the CORINAIR system, devised as part of the CORINE Programme in association with OECD, may eventually become the standard. These models typically take the following general form:

$$Ek = \sum (f_{ij} \cdot e_{ijk})$$

in which f_{ij} is the fuel consumption factor; that is, the amount of fuel of type i used in combustion process j, and

e_{ijk} is the emission function; that is, the unit rate of emission of substance k from fuel type i in combustion process j.

The quality of the results is probably constrained mainly by differences and errors in the emission function. Indeed, these emission functions have had to be revised on many occasions as new information became available on unit rates of emission (from manufacturers' information and experimental studies among others). In the UK, for example, the emission function for NO_2 was adjusted every year between 1984 and 1991 such that estimates of NO_2 emissions were revised upwards by more than 42 per cent. Over the same period, estimates of CO emissions in the UK were reviewed downwards by more than 43 per cent, while estimates of dust emission in France were increased by almost 75 per cent.

Data on fuel consumption are also difficult to obtain. Thus it is often necessary to estimate them from secondary sources such as production or employment data. Significant errors may arise in such cases, however. In addition, differences often occur in the classification of the sectors of activity to which the emission estimates relate, so that comparisons between different countries, or between different published sources, must be made with care. Because of these various errors and uncertainties, CBS (1987) suggests margins of error for emissions from stationary sources in the Netherlands of 10 per cent for sulphur dioxide, 25 per cent for nitrogen oxides and suspended particulates, and up to 100 per cent for CO and hydrocarbons.

Air quality

Statistics on air quality are derived primarily from permanent monitoring sites. Most countries run several different monitoring networks for different substances and purposes: for example, to monitor compliance with EU Directives or national legislation; as part of the EMEP programme on long range transboundary pollution; or to provide an early warning system for possible health effects. The major substances covered by these networks (and thus included in SoE reports) are particulates in various forms, sulphur dioxide, nitrogen dioxide, hydrocarbons, carbon monoxide, ozone and lead.

However, various sources of error and inconsistency occur in the data obtained from these networks. Differences occur, for example, in the methods of measurement used, and in the detection levels and accuracy of these methods. For example, the strong acidity method, traditionally used to assess SO_2 in the UK, Ireland, France and Luxembourg, is known to under-predict high concentrations and over-predict low concentrations. Inter-calibration studies have similarly shown significant differences in laboratory performance. In a study covering nine laboratories in seven EC countries, Apling *et al.* (n.d.) found measurement errors for NO_2 of up to 25 per cent. In addition, major variations may occur in the definition of the measured variables. For example, the size range of particulates measured in national networks varies from $<7\ \mu m$ in Belgium to $>20\ \mu m$ in Dublin and $<60\ \mu m$ in some towns in Italy.

The most serious problems with reported statistics on air quality, however, relate to the geographic distribution of monitoring sites and the way in which they are selected for reporting. Almost all networks are designed to monitor air pollution within specific types of environment and localities (high pollution urban sites close to emission sources for example); they do not, therefore, give an indication of more general patterns or levels (at the city, regional or country scale, for example). Statistics derived simply by averaging data from these sites – whether at the national or city level – inevitably tend to give a false impression of ambient conditions, and in most cases they are likely to produce significant overestimates. Studies in the USA, for example, have indicated that simple arithmetic averages of air quality data from monitoring sites overestimate actual city averages by up to 12 per cent compared to area-weighted techniques (Muschett, 1981). Analysing data from Antwerp and Brussels, Briggs (1995) found overestimates of 20–40 per cent for particulates, but underestimates of about 20 per cent for SO_2. Perhaps more significantly, neither method gave results similar to those reported by OECD (1991) for these two cities. For Brussels, for example, OECD data were six to seven times higher than those derived from the nationally published data; more than twice as high for SO_2. The causes of these discrepancies are not known.

Wastes

Statistics on wastes are notoriously unreliable. Since direct measurements of waste generation, disposal and treatment are rarely made, reported statistics are generally derived from a wide range of indirect estimation methods. Considerable difficulties occur even when measurements are made, because wastes vary substantially in their moisture content, degree of dilution and composition, etc. Thus direct measurements of either wet weight or volume reveal little about the actual quantities of wastes involved, especially in the case of hazardous wastes in which the hazardous component may be a small proportion of the total waste volume. Classifications of waste materials are also rarely consistent because of the following reasons, among others:

- municipal wastes may comprise household wastes alone, or may include a range of commercial, public and light industrial wastes;

- industrial wastes may include some commercial, construction and demolition wastes, as well as sewage sludge, waste-water sludge and fly-ash;
- agricultural wastes may include or exclude straw, forest products and livestock wastes.

As a result of these discrepancies, statistics on wastes reported in different SoE reports vary substantially. Briggs (1995) compared reported estimates for both municipal and industrial wastes from four main sources – the CEU, Eurostat, OECD and UN–ECE. Major inconsistencies were apparent. In the case of industrial wastes, for example, estimates for Luxembourg ranged from 135,000 tons to 1,961,000 tons; those for Portugal from 662,000 tons to more than 11 million tons; those for Spain (which were in any case remarkably low) from 5.1 million tons to 12 million tons; and those for France from 56 million tons to 206 million tons. The uncertainty attached to such waste statistics is clear.

Surface freshwater quality

Like statistics on air quality, those on water quality typically relate to permanent monitoring sites that are operated as part of national networks. Results are normally reported as annual average (and sometimes extreme percentile) concentrations of selected pollutants or water quality indicators such as dissolved oxygen, biochemical oxygen demand, nitrate and phosphate. The number of sites, sample design (both in terms of site location and frequency of sampling), the range of pollutants measured and methods of measurement used all vary from one country to another, and in some cases even from one region to another. Several sources of error and bias also occur. As a result, the statistics on water quality included in SoE reports are difficult to interpret or compare in any meaningful way.

One of the main problems relates to sampling frequency. Most networks still rely on manual sampling methods, and, because such methods are relatively costly, they are seldom conducted more than once or twice a month. Moreover, for various technical and administrative reasons sampling may not be done regularly, and the number of samples taken may vary from station to station or from year to year. Consequently, the number of available observations, from which to derive annual averages, is both variable and low – generally not more than 12 observations/year and often considerably fewer. Such sporadic measurements are thus likely to miss major, short-term pollution events, such as the Sandoz incident on the Rhine or the Camelford incident in the UK. Clearly, therefore, annual averages derived from such data give no more than broad order-of-magnitudes at best, and are generally likely to underestimate actual pollution levels.

The selection of sites for reporting, and the way in which the data are reported, may also be a source of considerable bias and misunderstanding. Most international agencies request data measured at monitoring sites located at the mouths of major rivers or at their exit frontiers. Whilst these sites could clearly provide useful information on the transfer of pollutants across national borders, or into the sea, and thus could give a broad measure of the quantities of pollutants added to the environment by each country, unfortunately the data are normally reported only as concentrations without accounting for stream discharge. Such data thus do not indicate the total pollution load and tend only to provide a weak and partial picture of water quality status.

Wildlife and nature

Two main sets of information tend to be provided on wildlife and nature in most SoE reports: those relating to wildlife protection, such as the number and total area of protected sites; and those relating to biodiversity, for example the total number of known species, species status, trade in threatened species, and extent of selected habitats. Both tend to suffer from serious problems, however, due largely to problems of definition and classification. Data on species may also be subject to large margins of error because of limited knowledge about species numbers and population status.

The problem of classifying protected areas is well known. It derives from the circumstance that protection may take many different forms and that the nomenclature used for classifying protected sites is far from consistent. For example, national parks, as defined at the national level, may range from essentially wilderness areas in which almost all human activity is strictly controlled or even banned, to landscapes within which many forms of farming, recreation, mineral extraction and manufacture are permitted or even promoted. In order to provide a basis for comparison, IUCN (1984) have developed a common classification which recognises eight major types of protected area. This classification is used for many international SoE reports. Even so, considerable differences in interpretation may arise, and significant discrepancies may occur between statistics published in different SoE reports. The IUCN classification also deliberately excludes many less strictly protected categories of designation; thus it tends to ignore the vast bulk of protected sites, many of which provide an important source of contact with nature for local people. In the Netherlands, for example, OECD (1995) report only 68 of about three thousand sites which are recognised at the national level as under some form of protection; in Belgium it reports only two out of more than three hundred sites, and in Luxembourg none out of about one hundred and forty sites.

Similar problems also occur in the case of wildlife species. Again, in order to provide some consistent basis for comparison, IUCN have devised a classification of species status according to their degree of threat. Eight categories of threat are recognised: extinct, endangered, vulnerable, rare, indeterminate, out of danger, insufficiently known, and not threatened. This classification is widely used to report on species status at both national and international levels, often by reference to national 'Red data books'. However, different SoE reports have tended to combine different IUCN classes in order to provide a single estimate of the number of species under threat or at risk. Reported statistics thus appear to be inconsistent. Moreover, data on wildlife populations are typically derived from the work of amateur or professional groups who undertake field surveys. Countries with strong amateur interests in natural history, such as the UK and France, tend to have reasonably good data on wildlife species as a result, at least for mammals, birds, reptiles, amphibians and plants. Data are more scarce and less reliable in other countries, however, including many Mediterranean countries where there are fewer volunteer recorders. Consequently the status of many species, especially fish, invertebrates and lower plants is still inadequately known. Therefore, data on these, as given in both national and international SoE reports, must be treated with the utmost care.

CONCLUSIONS

State of environment reporting is a young science (or art?) which is still evolving. The agencies concerned are still seeking the most appropriate form for SoE reports and the most effective means of compiling them. To date a number of competing or complementary models of state of environment reporting have emerged. But which of these will eventually become paramount or the standard remains to be seen. Clearly, however, further consideration will need to be given to the possibility of defining agreed sets of environmental indicators, as well as to the potential for using new technologies for SoE reporting.

Whichever approach is taken, it is clear that availability and quality of essential information are the main limiting factors for many state of environment reports. The problem is that available data tend to be scattered across different institutions; they also vary in both quality and character and are often incomplete. Data availability thus tends to restrict both the content of many SoE reports and the depth at which specific problems can be considered and addressed. There is a great need, therefore, to improve data availability, quality and consistency, and clearly this calls for action to be taken at almost all points in the data stream. This need is increasingly being highlighted by SoE reports.

Yet the role of state of environment reporting is not merely to provide information. Its ultimate aim is to improve environmental management and protection by helping to inform and direct policy development. The extent to which this aim is being achieved at present is far from clear. Also, the relationship between SoE reports and policy practice remains vague and enigmatic: are SoE reports guiding policy, or are they merely serving to reflect policy concerns? Little is known about how SoE reports feed back into policy mechanisms. Also there is little evidence to suggest any detectable policy effect.

The European Environment Agency has a major role to play in providing a link between SoE reporting and policy action in the European Union. Other international agencies, such as UN–ECE and OECD, are also important in this respect. If state of environment reporting is to become an effective part of the policy-making process, however, its role within the decision system must be much more explicitly addressed.

Nevertheless SoE reporting is no longer the exclusive domain of national and international agencies. As argued in the Fifth Environmental Action Plan, and as the principles of Agenda-21 have forcefully shown, a wide range of stakeholders and actors need to be involved in environmental protection and management. Whereas SoE reports have traditionally been produced in a 'top-down' fashion, there is now a need to develop a more democratic process – a 'bottom-up' approach – in which the wider public can be involved. This calls for far-reaching changes to be made to the mechanisms of state of environment reporting, and requires the agencies concerned to act in a more responsive and participatory manner. It also means that all those with interests in the environment should be actively consulted to identify the issues which they think the SoE reports should address, to define the indicators which should be used to describe them, and to help collect the necessary data. This, in turn, requires the use of appropriate consultative mechanisms – for example formal and informal meetings (such as focus groups, environmental forums), questionnaire surveys, environmental 'watches' (in which members of the public are encouraged to report environmental problems), and different types of awareness-raising and training events using leaflets, newsletters, information packs, etc. This shift, which has already been adopted in a number of local SoE reports, often motivated by local versions of Agenda-21, is indeed an important trend which deserves to be encouraged and promoted. However, it represents a change which, once initiated, is likely to be self-sustaining. This is because the very act of participation breeds and nourishes a wider understanding of the environment; it also instils both a recognition of the role we each have for environmental protection and a commitment to be involved.

REFERENCES

Apling, A.J., Peperstraete, J.J. and Rudeolf, W. (n.d.) *CEC Harmonisation of Methods for Measurement of NO_x: Intercalibration of Measuring Equipment*, CEC Report EUR 7865, Brussels.

Briggs, D.J. (1995) 'Environmental statistics for environmental policy: genealogy and data quality', *Journal of Environmental Management* 44: 39–54.

CBS (1987) *Luchtverontreininging emissies door verbranding van fossiele brandstoffen in vuurhaarden 1980–1985*, 's-Gravenhage, The Netherlands: CBS.

CEC (1985) 'Council Decision of 27 June on the adoption of the Commission work programme concerning an experimental project for gathering, co-ordinating and ensuring the consistency of information on the state of the environment and natural resources in the Community (85/338/EEC)', *Official Journal of the European Communities* L176: 14–17.

Dargie, T.C.D. and Briggs, D.J. (1992) *State of the Scottish Environment*, Perth: Scottish Wildlife and Countryside Link.

Hens, L. (1996) 'The Rio conference and thereafter', in B. Nath, L. Hens and D. Devuyst (eds) *Sustainable Development*, Brussels, Belgium: VUB Press.

IUCN (1984) 'Categories and criteria for protected areas', in J.A. McNeely and K.R. Miller (eds) *National Parks, Conservation and Developments: The Role of Protected Areas in Sustaining Society*, Washington, DC: Smithsonian Press, pp. 47–53.

Muschett, F.D. (1981) 'Spatial distributions of urban atmospheric particulate concentrations', *Annals of the Association of American Geographers* 71: 552–565.

OECD (1991) *Environmental Indicators: A Preliminary Set*, Paris, France: OECD.

OECD (1995) *The State of the Environment*, Paris, France: OECD.

Stanners, D. and Bordeau, P. (eds) (1995) *Europe's Environment: The Dobris Assessment*, Copenhagen, Denmark: European Environment Agency.

SUGGESTED READING

The best way of understanding state of environment reporting is to read different examples of published reports. New reports appear annually. The Dobris Report (see Stanners and Bordeau, 1995) is perhaps currently the definitive report for Europe. In addition to this and the national reports relating to your own country, however, the following international reports are likely to be of particular value (though look out for updates and new editions).

OECD (1991) *Environmental Indicators: A Preliminary Set*, Paris, France: OECD.

OECD (1995) *The State of the Environment*, Paris, France: OECD.

UNEP (1993) *Environmental Data Report*, 1993–4, Oxford: Basil Blackwell.

World Resources Institute (1994) *World Resources. A Guide to the Global Environment*, New York: Oxford University Press.

SELF-ASSESSMENT QUESTIONS

The whole thrust of this chapter has been to emphasise that state of environment reports are not merely technical entities which give a factual description of reality. Instead, they must be viewed both within the cultural and institutional context in which they are produced, and within the framework of data availability and quality. The understanding which you should have obtained from this chapter cannot therefore be tested by standard self-assessment questions which imply answers that are 'right' or 'wrong'. Instead, you should have developed a wider appreciation of the complexity and uncertainty involved in SoE reporting. The following questions are thus more open-ended than those you will have met elsewhere in this book. Their aim is to make you assemble and expand your ideas by thinking about what you have read.

1 The following table shows data on concentrations of nitrogen dioxide in three cities as reported by national SoE reports and in the OECD Environmental data compendium, 1991. Suggest reasons for some of the differences which exist, both between cities and between reports. All data refer to the same year.

Mean nitrogen dioxide concentration ($\mu g/m^3$)

City	*National report*	*OECD Compendium*
Copenhagen	74.0	64.0
London	62.8	32.0
Milan	86.4	75.3

2 State of environment reports are the product of a long and complex data stream. The nature of this data stream varies considerably, however, according to the particular phenomenon under consideration. Construct diagrams outlining the data streams necessary for reporting information on the following phenomena at the national level:
(a) stream water quality;
(b) number of mammal species 'at risk';
(c) loss of natural habitats;
(d) rates of soil erosion.
In each case, identify the main sources of error which are likely to occur.

3 State of environment reports vary in their content depending upon the agency involved in compiling the report and the audience at which it is aimed. Prepare a list of the main topics which you think should be included in a state of environment report compiled by your local community as a basis for promoting environmental awareness.

4 Good environmental indicators are difficult to compile. They need to be carefully designed to match the issue of concern and the specific conditions. Develop indicators for the following issues:
(a) pressures from agriculture on groundwater in Europe;
(b) pressures on the environment from domestic wastes in Europe;
(c) biodiversity in Europe;
(d) air quality in a city of your choice;
(e) health risks from road transport in your country;
(f) actions being taken to reduce reliance on fossil energy sources in your country.
Note: In devising indicators, you will need to consider the following questions:
- What specific feature(s) should you measure to provide an indicator for this topic?
- At what geographic scale should the indicator be applied (i.e. what are the smallest geographic units for calculating the indicator)?
- How should the indicator be scaled (e.g. should it be calculated as a simple total, on a per capita basis, or per unit area)?

- What averaging time should be used (i.e. should air quality be measured as an annual average, seasonal average or daily average)?
- Are data likely to be available for the indicator as you have designed it?

5 It is sometimes suggested that, in order to summarise the state of environment simply and effectively, we need to develop a composite index of the state of the environment (rather like the retail prices index) which can be easily compared over time.

(a) List a maximum of 12 key environmental indicators which you think should be included in such an index. (*Hint*: commonly quoted examples are 'average atmospheric SO_2 concentration' and 'average bathing water quality'.)

(b) List the main problems you envisage in developing an index of this kind.

6

THE ESTABLISHMENT OF HEALTH AND ENVIRONMENTAL STANDARDS

Luc Hens and Max Vojtísek

SUMMARY

New industrial processes and intensive agricultural practices have resulted in the introduction of numerous chemical substances into the environment. Increased awareness of the possible adverse effects of this 'chemification' has led to scientific, social and political pressures to control the effects of these substances. A common environmental management response to this pressure has been the introduction of standards to protect human health and environmental quality.

The methods of establishing standards are described in this chapter. After separating the scientific and political components of standards, the conceptual and laboratory elements used in developing guidelines for carcinogens and non-carcinogens have been described. Underlying theory is illustrated by the example of establishing standards for pollutants in drinking water. Boxed texts provide additional examples on air and food quality guidelines and standards.

Given that, apart from the protection of human health, standards can also be used to improve environmental quality, this chapter overviews a wide array of current environmental quality standards, ranging from behavioural and product standards on emission, zoning and immission values to environmental quality and exposure levels. The underlying principles of setting environmental standards are explained, together with their links with the establishment of human health guidelines, and their advantages and limitations.

ACADEMIC OBJECTIVES

In this chapter several aspects of human health and the setting of norms for environmental quality are explained. After working through the text the reader should know the following:

1 Definition of standards, guidelines and alarm levels.
2 The strategy and conceptual background to the elements necessary for establishing a health standard.
3 The problem in context approach of ADI, AADI, NOAEL, UF, RMCL, and guideline values.
4 The difference between carcinogenic and non-genotoxic substances.
5 The advantages and limitations of the use of health standards.
6 Types of environmental standards and areas of their application.
7 Background to the setting of environmental quality standards – their application, advantages and limitation.

GUIDELINES, STANDARDS, LIMIT VALUES AND ALARM VALUES

Standards are one of the oldest instruments of environmental management. Their purpose is to protect both the environment and human health by defining maximum permissible pollution levels. Techniques and methods of defining maximum permissible concentrations were originally developed by occupational hygienists in order to protect those in the working environment. Similar methods were subsequently applied to identify maximum permissible concentrations in air, surface water, soil, drinking water and food.

The basis on which a standard is set is a 'guideline', and the 'guideline value' is the estimated maximum permissible concentration that would still guarantee a reasonable health or environmental quality. This estimation is based on a critical review of all available and relevant toxicological information, including data on the fate and behaviour of substances in the human body and in the environment. In addition to toxicological data, guidelines also incorporate a 'safety factor' which reflects scientific uncertainty attaching to the establishment of a guideline when scientific knowledge on the subject is incomplete. Typically, the safety factor accounts for risks inherent in the extrapolation of results obtained for certain individuals to the whole population, or when results of laboratory tests on animals are extrapolated to humans. But safety factors can differ substantially from pollutant to pollutant, and, consequently, it is not possible to quantify the immediate health risk of a pollutant when its guideline value is exceeded.

In general, guidelines are established by international panels of experts acting for and on behalf of relevant international organisations. The World Health Organisation (WHO) has established guidelines for air pollutants (WHO, 1987) and for drinking water (WHO, 1984). In the United States the Environmental Protection Agency (USEPA) is a leading authority on establishing guidelines (Stara and Erdreich, 1984).

The 'standard' is a value 'fixed by authorities'. In general, it is enshrined in a legal regulation which determines either the maximum permissible concentration of a substance in food, water, air or soil, or the maximum permissible stress resulting from exposure to an environmental factor such as noise or radiation. Although a standard is established on the basis of a 'guideline', it also takes into account technical, economic, social or political considerations as deemed necessary.

An example of a technically influenced standard is the European standard for pesticides in drinking water. The basic idea behind this standard is that people have the right to drink pesticide-free water. This zero-level exposure standard coincides with the smallest quantity of pesticide which could be measured in the 1970s when this particular European regulation was formulated.

The European Community standard on the concentration of nitrates (NO_3) in drinking water is a typical example of a standard influenced by economic and political considerations. Although the 'natural' concentration of nitrate in water is less than 1 NO_3 mg/l, it is substantially increased by pollution caused by both manure and inorganic fertilisers used in agriculture. The WHO has established a guideline of 45 mg NO_3 per litre (NO_3 mg/l). But, despite substantial evidence suggesting an even lower concentration for safety (Speyers *et al.*, 1987), European regulation, based on the EC Directive of 17 July 1982, specifies a higher value of 50 NO_3 mg/l. The main justification for this higher guideline is economic, because nitrate pollution in several regions of the European Community is dramatically high, and much of the drinking water in those regions has NO_3 concentrations of 50 NO_3 mg/l and higher. But reduction of nitrate concentration from this level is not an economically feasible proposition, and hence the EC guideline of 50 NO_3 mg/l. On the other hand, the Community recommends bringing the level down to 25 NO_3 mg/l. The 50 NO_3 mg/l value is therefore recommended as the 'Maximum Admissible Concentration' (MAC) and the 25 NO_3 mg/l value as a 'Guide Level' (GL) limit. Thus the guide limit is the 'ideal' to be reached in order to avoid damage to health or the environment.

Another example of socio-political influence on standards is given in Box 6.1. Here a comparison is made between the values of a selection of standards before and after the political changes which took place in the Czech Republic following the 'Velvet Revolution' in 1989.

For a number of pollutants, however, the use of guidelines and standards aimed at avoiding negative impacts on human health and the environment has become unrealistic. For example, tropospheric ozone concentrations in many countries often exceed the WHO guideline of 200 micrograms of ozone per cubic metre of air ($\mu g/m^3$) for 1 hour, or 65 $\mu g/m^3$ for 24 hours. There is no safety factor built into these guidelines because they are based on data obtained from human populations. As a result, actual ozone concentrations during summer months regularly impact on human health. It is for this reason that standard regulators advocate the establishment of what are called 'alarm levels'. For 'Los Angeles type' smogs (i.e. those with high concentrations of ozone) the recommended alarm level values should be based on:

BOX 6.1 ENVIRONMENTAL STANDARDS IN FORMER CZECHOSLOVAKIA AND THE CZECH REPUBLIC

Neither the establishment of standards nor environmental awareness can be isolated from the social context. An interesting example of this is provided by the effect of the transition of Central and Eastern European countries towards a free market economy. The situation in the Czech Republic, before and after the Velvet Revolution of autumn 1989, is discussed here as a typical case study. Critical elements in the discussion to adapt environmental standards are the following:

The social aspects
In spite of the fact that information on the catastrophic state of the environment was freely accessible after the changes of 1989, transformation of the economy focusing on employment rates and rising income became top priority in the Czech Republic.

The 'Rainbow' programme (Moldan *et al.*, 1991)
The former anthropocentric approach to the environment is criticised in this document, intended as an environmental recovery programme for the Czech Republic. It recommended the following:

1 Laws to protect the environment, e.g. regulations to gradually reduce and prevent pollution.
2 A state administration system for the effective enforcement of laws and other measures.

Reactions and the situation today
1 Development of strategies to deal with the limitations of the pre-1989 anthropocentric approach and the problems which it created. In air quality, for example, the former dispersion principle for the atmosphere (the 'high chimney' policy) was modified with new limits for immission, emission and deposition.
2 Feed-back is provided by the newly formed Czech Inspectorate for the Environment and by other monitoring institutions.

Harmonisation
Another reason for changing the environmental standards is because of the Czech Republic's membership of the OECD which initiated a process of harmonisation with European standards.

The above explain the difference between standards before and after 1989, as shown in the following table. This table shows the standards for selected air pollutants before and after 1989 in Czechoslovakia and the Czech and Slovak Republics. The main difference lies in the number of pollutants for which standards have been established. For pollutants such as CO and NO_2, the standard has become less stringent.

Comparison of standards for selected pollutants before and after the Velvet Revolution in former Czechoslovakia

Pollutant Name	*Symbol*	*Allowable concentration before 1989*[1] C_d [μg.m^{-3}]	C_{max} [μg.m^{-3}]	*Allowable concentration*[2] C_d [μg.m^{-3}]	C_{max} [μg.m^{-3}]
Arsenic[3]	As	3	–		
Lead[4]	Pb	0.7	–		
Sulphur dioxide	SO_2	150	500	150	500
Carbon monoxide	CO	1,000	6,000	5,000	10,000
Nitrogen dioxide	NO_2	100	100	100	200
Airborne particles[3]		150	500		
Airborne particles with lead[5]	Pb			0.5	
Airborne particles with cadmium[6]	Cd			0.01	

[1] Guidance notes of Ministry of Health of CSR, Vol. 51/1981.
[2] Immission limits for pollutants from Supplement 1: Collection of Laws of Czech and Slovak Federative Republic, 1991, Part 84.
[3] Inorganic substances except AsH_3.

→

[4] Except tetraethyl lead.
[5] Airborne particles without biologically active components.
[6] Median concentration observed on the site during one year.
C_d = Median 24 hrs concentration observed on the site.
C_{max} = Median concentration observed on the site during 30 minutes.

In 1996, the philosophy underlying the setting up of standards in the Czech Republic put greater emphasis on emission than on immission values, because the former can be monitored and controlled more effectively. New limits are currently being considered jointly by the National Institute of Public Health, the Ministry of Health, and the Ministry of Environment.

Source: Collection of Laws of the Czech and Slovak Federative Republic, 1991; Ministry of Health of the CSR, 1981, 1986

1 Specification of what effects on human health and the environment the authorities definitely wish to avoid, and
2 specification of pollutant concentrations causing these effects.

These alarm values will generally be higher than guideline or standard values. The establishment of alarm levels takes into account an evaluation of the seriousness of damage to health or environment. It is clear that such a choice is ethically loaded and, therefore, much more difficult to make objectively than a 'no effect' level. Moreover, at present there is some experience of establishing guidelines and standards, but almost none for defining alarm levels. Figure 6.1 shows the different control level values and their key characteristics.

HEALTH STANDARDS

Strategy for health standards establishment

The purpose of health standards is to control human exposure to environmental substances or agents in such a way that no, or only negligible, risk occurs. As a rule, exposure to new chemicals, introduced either as food additives or as medication, is only permitted if it can be shown that their use is effective and that they have no adverse effects on human health. In the case of some drugs the 'no adverse effect' concept is replaced by that of 'relative benefit' which the use of those drugs brings. In all instances, and especially in relation to environmental pollutants, particular attention is paid to lifetime exposure and to groups at special risk such as children, elderly persons and individuals with specific sensitivity.

The toxicology of chemical substances is commonly divided into two broad classes: (1) acute or chronic toxicity, and (2) carcinogenicity. However, teratogenic and mutagenic risks should also be considered. This duality arises mainly from the fact that there are dose thresholds for chronic and acute toxicity effects, but not for the majority of carcinogens. When there is a dose threshold, the ideal goal is to set a guideline whereby a total daily dose during lifetime exposure to the substance remains, with all practical certainty, below the level at which any injury would occur to any individual of the population. A realistic assessment of human health hazards, however, calls for a distinction between absolute safety and acceptable risk. Very often relevant data for determining safety levels are scarce and the quantitative relationships uncertain. Scientific judgement and consensus then play an important role in establishing acceptable levels of population exposure.

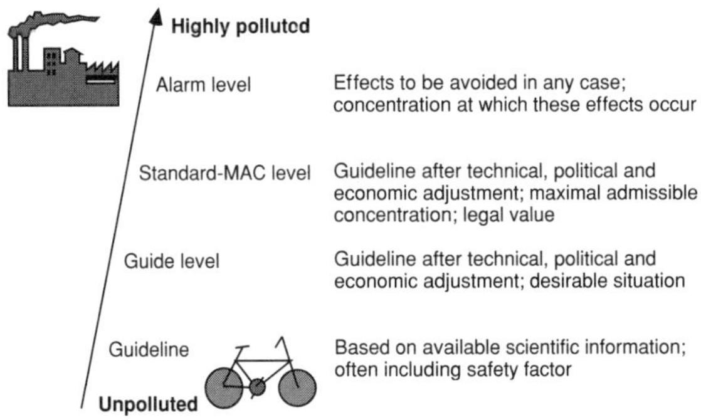

Figure 6.1 Control level values in ascending order; the right side of the figure shows the key characteristics of the corresponding value

As a rule, there are no thresholds for carcinogens or mutagens. It follows therefore that there may be finite risks from any non-zero dose exposure level. For such substances guidelines are established in terms of a lifetime exposure level resulting in one cancer per million, or per one hundred thousand, of exposed persons. Since only less than 5 per cent of the carcinogens do not interact directly with the genetic material, it is necessary in establishing guidelines to have a clear insight into how they act within the human body.

Criteria used in establishing guideline values

Criteria common to carcinogens and non-carcinogens

The first aspect of this is concerned with the evaluation of sources, levels and routes of exposure. For sources of pollution and exposure levels, special attention is paid to highly polluted areas. For example, air pollution requires not only an evaluation of the rural areas, but also of urban and indoor atmospheric concentrations. As for routes of exposure, an evaluation of uptake by inhalation, ingestion from water and food, and dermal contact has to be made. For many chemicals, however, data on total human exposure are incomplete at present.

The second aspect is concerned with the evaluation of the kinetics and metabolism of the chemical in question. Here the main aims are to identify the active metabolites, determine the distribution of the chemical throughout the body, and to calculate tissue doses to the target organs.

Criteria for non-carcinogenic substances

The 'acceptable daily intake' (ADI) procedure is widely used for setting standard concentration levels when appropriate data are available in the epidemiological and toxicological literature. The ADI is the daily dose considered to be without risk to humans for lifetime intake.

The ADI is usually derived from a detailed toxicological analysis. The 'no observed adverse effect level' (NOAEL) is determined for the most sensitive adverse effect observed in tests, usually carried out on animals. Occasionally NOAEL is also determined from epidemiological or even experimental data obtained from tests on humans, and a 'safety' or 'uncertainty factor' is applied to the NOAEL so derived to determine a level which is believed to be safe for the general human population.

The ADI is computed by multiplying the experimental NOAEL (in milligrams per kilogram per day) by the weight of a typical adult (70 kg), or a child (10 kg), and then dividing it by a safety factor. Thus, for an adult,

$$\text{ADI (mg/person/day)} = \text{NOAEL (mg/kg/day)} \times 70 \text{ (kg/person)/safety factor}$$

Table 6.1 lists some of the safety factors which can be used in the calculation of the ADI. The data show that the selection and application of these uncertainty factors is important in the calculation of guidelines, because the limit set can vary by one or two orders of magnitude when these factors are applied. In most cases an overall safety factor of 100 (inter-species variation × intra-species variation = 10 × 10 = 100) is used, even though it can sometimes be 1,000 or more.

Figure 6.2 shows the process by which the ADI for a particular substance is calculated. The solid line to A is the dose-response curve, as determined in the laboratory by a multiple-dosing experiment performed under conditions prescribed by 'good laboratory practices'. Point A represents the highest 'not observed' level in mg per kg per day for the most sensitive adverse end point, determined in general from animal studies, while A″ represents the lowest (experimentally) observed level for the same effect (LOAEL). Points B, D and E are the presumed thresholds of adverse effect in the human population, provided that the extrapolated dose-response curves AB, AD or AE are correct. Point C would then represent the ADI

Table 6.1 Derivation of safety (uncertainty) factors if laboratory experiments are indicative of this attitude

Source of uncertainty	*Factor*
Inter-species variation	10
Intra-species variation	10
Nature and severity of effect	1–10
Adequacy of study	1–10
Interactive effects	1–5

concentration, determined by applying the selected safety factor to the dose at point A. The real dose-response curve in the low concentration area of the figure is unknown; therefore, any of the extrapolated curves, AB, AD or AE, could be correct. The guideline setter would hope that AB is the true curve, because the 'no response' dose would then be greater than the calculated ADI value at C. In this case the safety (or uncertainty) factor will be chosen appropriately. If, however, AD or AE were the true dose-response curves, the calculated ADI would be too large and the safety factor too small. Consequently, some persons in the population might suffer adverse effects. AE represents a non-threshold dose-response. The size of the gap between C and B is also interesting because, if it is excessively large, over-regulation can result in excessive control entailing expenditure without any benefit.

The ADI, even when it is rigorously determined, does not guarantee absolute safety because the assumption that the same threshold applies to all individuals in a large population is simplistic and patently not true. Consequently, in spite of the use of safety factors, there are likely to be individuals in a given population with genetic predisposition or susceptibility to specific environmental health impacts (stressors). It is possible, therefore, that certain individuals

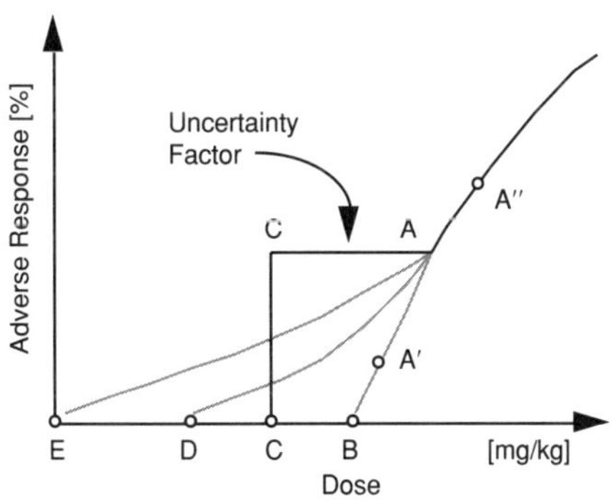

A: NOAEL (experimentally derived)
A′: NOAEL (theoretically possible)
A″: LOAEL (lowest observed adverse effect level)
B: Presumed threshold for any effect (not adverse)
C: ADI or RMCL 'no effect level'
D: Another possible presumed threshold for any effect (not adverse)
E: Non-threshold end point of toxicity

Figure 6.2 General illustration of the process for calculating the ADI of a substance

in a given population would be at high or very high risk while others would be at low or very low risk. It is to be noted that the concept of the ADI is difficult to apply to heavy metals and lipophilic substances because they tend to bio-accumulate, and it is difficult to mimic lifetime human exposure with bio-accumulation in laboratory experiments.

The distinction between adverse and non-adverse effects also creates considerable difficulties. Both the USEPA and the WHO define an adverse effect as 'any effect resulting in functional impairment and/or pathological lesions that may affect the performance of the whole organism or which contributes to a reduced ability to respond to an additional challenge'. But even with this definition a significant degree of subjectivity and uncertainty remains.

Substances with malodorous properties at concentrations far below those at which toxic effects occur present a particular problem. Odour annoyance does not constitute an adverse health effect, but it clearly affects the quality of life. Determination of odour threshold levels has its own method which is distinct from the toxicological and epidemiological approaches to determining ADI. As a rule, odour threshold levels are therefore given separately from the NOAEL-based values.

Criteria for carcinogenic substances

The guiding principle is that exposure to genotoxic substances should be prevented. Genotoxins or mutagens are substances which in themselves, or after metabolic conversions, have the ability to change genetic material. Sometimes exposure to genotoxic substances cannot be avoided. But some of the mutagenic substances occur naturally in the environment, while others were introduced into it before there was evidence of their mutagenic properties. Still others are allowed because of fears that the use of their substitutes might be more dangerous than the limited presence of mutagens in the environment. In such cases an indicative risk analysis may be performed to assess the adverse health effects which exposure to mutagens could cause.

The majority of genotoxic substances are carcinogenic. This means that long-term exposure of laboratory animals to these substances causes malignant tumours. A small minority of carcinogens, such as asbestos, are non-genotoxic.

The carcinogenicity of a substance is determined by expert groups. The guidance of two agencies are accepted internationally as bench-marks – the International Agency for Research on Cancer (IARC) and the United States Environmental Protection Agency (USEPA). Both these agencies have established guidelines for weighing epidemiological and animal toxicological data, on the basis of which the strength of evidence of carcinogenicity is assessed (Table 6.2).

The main difference between the two approaches is that while the IARC does not distinguish between those chemicals for which there is inadequate evidence of carcinogenicity and those with no evidence at all, the USEPA does. The list, given in Table 6.3, of carcinogenic substances identified so far is limited.

Table 6.2 IARC and USEPA classification of chemicals according to their carcinogenic potential

1 IARC classification	
Group 1:	the agent is carcinogenic to humans (sufficient epidemiological evidence)
Group 2A:	the agent is probably carcinogenic to humans (sufficient evidence from animal studies plus limited evidence from epidemiological studies)
Group 2B:	the agent is probably carcinogenic to humans (sufficient evidence in animals plus inadequate evidence from epidemiological studies)
Group 3:	the agent is not classifiable as to its carcinogenicity to humans
2 USEPA classification	
Group A:	human carcinogen: sufficient evidence in epidemiological studies to support causal association between exposure and cancer
Group B:	probable human carcinogen: limited evidence in epidemiological studies (group B1) and/or sufficient evidence from animal studies (group B2)
Group C:	possible human carcinogen: limited or equivocal evidence from animal studies and inadequate or no data in humans
Group D:	not classified: inadequate or no human and animal evidence for carcinogenicity
Group E:	no evidence of carcinogenicity for humans: no evidence of carcinogenicity in at least two adequate animal tests in different species or inadequate epidemiological and animal studies

Source: After IARC, 1982 and USEPA, 1986

If toxicological and/or epidemiological evidence leads to the classification of a substance as a known or probable human carcinogen (group A or B), mathematical models are employed to calculate and predict the risk of exposure to it. The data used in these estimates usually come from lifetime exposure studies in animals. Two general approaches have been proposed for assessing risk: a strictly linearised estimate used by the USEPA (Anderson, 1985), and the linearised multistage model. Non-linear relations, resulting in two-hit, logit, probit (see Glossary) and multistage models, have been proposed by numerous authors. Since the mechanism of how cancer is caused is not completely understood, there is no evidence that one model predicts the risk more accurately than another. The USEPA therefore uses one of the more conservative models for carcinogenic risk assessment. That model fits linear dose-response curves for low doses and is consistent with a no-threshold model for carcinogenesis. This means that, according to this model, exposure to even a very small amount of the substance can increase the risk of cancer by a finite amount. Consequently, no safe level can be specified for carcinogenic substances, and there is always a risk from exposure to them. In its guidelines the WHO (1987) estimates carcinogenic potency in terms of 'incremental unit risk'. For an air pollutant, for example, the incremental unit risk is defined as 'the additional lifetime cancer risk occurring in a hypothetical population in which all individuals are exposed continuously from birth throughout their lifetimes to a concentration of 1 $\mu g/m^3$ of the agent in the air they breathe'. In practice this means that risk from exposure to a carcinogen, benzene for example, would be determined as follows: at an air concentration of 1 μg benzene per m^3, the estimated lifetime risk for leukemia is 4×10^{-6}.

The determination of exposure level to an 'unavoidable' genotoxic substance requires the assessment of incremental risk, defined as the exposure level resulting in a risk of one cancer in *x* persons during their lifetime exposure. The value of *x* can be 10^4 or 10^5, as is generally the case in Central European countries such as Czechoslovakia, or 10^6 as in the EU countries generally.

Table 6.3 Human carcinogens so far identified

Industrial process causally associated with human cancer exposure	*Chemicals and groups of chemicals causally associated with human cancer for which exposure has been mostly occupational*
Aluminium production	4-aminobiphenyl
Auramine (manufacture of)	Arsenic and arsenic compounds
Boot and shoe manufacture and repair	Asbestos
Coal gasification	Benzene
Coke production	Benzidine
Furniture and cabinet making	Bis (chloromethyl) ether and chloromethyl methyl ether (technical grade)
Underground haematite mining, with exposure to radon	Chromium compounds, hexavalent
Iron and steel founding	Coal-tars
Isopropyl alcohol manufacture, strong acid process	Coal-tar pitches
Magenta (manufacture of)	Mineral oils, untreated and mildly treated
Painters (occupational exposure as)	Mustard gas (sulphur mustard)
Rubber industry	2-naphtylamine
	Nickel and nickel compounds
	Shale oils
	Soots
	Talc containing asbestiform fibres
	Vinyl chloride

Drugs causally associated with human cancer exposure	*Environmental agents and cultural risk factors causally associated with human cancer*
Analgesic mixtures containing phenacetin	Aflatoxins
Azathioptine	Alcoholic beverages
N N Bis(2-chloroethyl)-2-naphtylamine (Chlornaphazine)	Betel-quid with tobacco
1,4-Butanediol dimethanesulfonate (Myleran)	2,3,7,8-TCDD (dioxin)
Chlorambucil	Erionite
1-(2-Chloroethyl) 3-(4-methylcycloexyl)-nitrosourea Methyl-ccnu	Radon and its decay products
Cyclophosphamide	Tobacco products, smokeless (chewing tobacco, oral snuff)
Cyclosporine	Tobacco smoke
Diethylstilbestrol	Hepatitis B virus
Oestrogen replacement therapy	Human T-Cell leukaemia virus
Oestrogen, nonsteroidal	Ionising radiation
Oestrogens, steroidal	Ultraviolet radiation
Melphalan	
8-Methoxypsoralen (Methoxsalen) plus UV radiation	
MOOP and other combined chemotherapy including alkylating agents	
Oral contraceptives, combined	
Oral contraceptives, sequential	
Thiotepa	
Treosulfan	

Source: After Tomatis *et al.*, 1989

It must be noted that this approach, based on the linearised multistage model, is subject to different assumptions. In practice these assumptions appear to be valid or acceptable in a wide variety of situations. There are nevertheless known situations in which they are not valid, as may be in the case with some non-mutagenic carcinogens. These substances cause malignant tumours in animals, but the cellular and biochemical mechanisms underlying cancer induction and progression might differ substantially from the genetically mediated mechanism which is present in genotoxic carcinogens. A possible example illustrating this point is 2, 3, 7, 8-tetrachlorodibenzo-dioxin, which is the most toxic member of the dioxin family. There is evidence (Roberts 1991) to suggest that a molecule of this substance can only interact with the DNA after it has been linked to molecules from the nuclear membrane. On the other hand, these molecules can 'absorb' a certain amount of dioxin before the DNA-damaging substances are released. On the basis of this evidence the USEPA has decided to review its risk calculations based on the linear multistage model and to refine it according to the new insight into how the mechanism works. Whether this revision will result in higher or lower risk figures for dioxin is not known at present.

Other elements, which may influence risk and which are not considered in this model, are specific toxico-kinetic properties such as the relatively higher excretion rate of a carcinogen at higher doses of exposure, or a relatively lower production rate of carcinogenic metabolites at lower doses.

Health guidelines in practice: establishment of standards for drinking water quality
(Premazzi *et al.*, 1989)

Non-carcinogens

Recommended maximum contaminant levels (RMCLs used by USEPA and equivalent to EC guidelines) are calculated as follows:

- identify the highest no-observed adverse effect level (NOAEL) based on an assessment of human or (usually) animal data;
- determine the ADI. The NOAEL is divided by an appropriate safety (uncertainty) factor (UF) to accommodate extrapolation data (usually $10 < UF < 1{,}000$);
- adjust the ADI to mg/litre (AADI) by factoring in the weight of the consumer (70 kg for an adult) and the amount of water consumption per day (2 litres for an adult):

$$\text{AADI} = \frac{\text{NOAEL} \times \text{BW}}{\text{UF} \times \text{litre/d}}$$

where AADI = adjusted acceptable daily intake,
NOAEL = no-observed-adverse-effect-level,
BW = body weight (70 kg for an adult; 10 kg for a child),
UF = uncertainty factor,
litre/d = assumed daily water consumption of protected individual (2 litres for an adult; 1 litre for a child);

- calculate the RMCL by multiplying the AADI by the estimated contribution of drinking water to the total daily intake (20 per cent for organics; 10 per cent for inorganics). Thus
  $\text{RMCL} = \text{AADI} \times 0.2 \text{ (or } 0.1) \times 1{,}000$ (μg/litre)
  where 1,000 is the conversion factor from mg/l to μg/l.

The WHO (1984) approach is slightly different in that the guideline value (GV) is calculated as

$$\text{GV} = \text{AADI} \times \text{A} \times 1{,}000 \ (\mu\text{g/litre})$$

in which A = relative source contribution of drinking water: 1 per cent for compounds which 'bioaccumulate', 10 per cent for compounds which do not 'bioaccumulate'.

As an example, Table 6.4 illustrates the different guidelines established by the WHO for drinking water quality. Not only are the different guidelines for the main contaminants indicated, but the physico-chemical characteristics are also shown. It should be noted that the WHO also provides guidelines for many other compounds, mostly organic, in drinking water that are not listed in Table 6.4.

Carcinogens

The principle of 'zero-level' exposure is adopted by most international agencies for carcinogens. But the argument against it is that the zero-level cannot often be achieved, neither can it be measured. The only scientifically acceptable alternative is therefore not to

Table 6.4 Guidelines for drinking water quality issued by WHO (1984)

Parameter	*Guideline*	*Nature of guideline*	*Remarks*
Hg	0.001	HR	a
Pb	0.05	HR	a
Se	0.01	HR	a
As	0.05	HR	a
Cr (VI)	0.05	HR	a
CN	0.1	HR	a
Cd	0.005	HR	a
Ba	–	NAR	–
F	1.5	HR	a*
NO_3	45	HR	a
PAH	0.00001	HR	a–2
Total solids	1000	AQ	a
Colour	15	AQ	a–3
Turbidity	5;<1	AQ	a–4
Taste	not offensive for most consumers	AQ	–
Odour	–	–	–
Fe	0.3	AQ	a
Mn	0.1	AQ	a
Cu	1	AQ	a
Zn	5	AQ	a
Ca	–	NC	–
Mg	–	NC	–
SO_4	400	AQ	a
Cl	250	AQ	a
pH	6.5–8.5	AQ	a
Ph.Sub.	0.001–0.04	HR–AQ	b–5
EEC	–	NC	–
ABS	0.2	AQ	a
Mineral oils	–	NC	–
Total hardness	500	AQ	1

Categorisation of the parameters
HR Health-related substances/chemicals: organic–inorganic constituents/contaminants that can cause severe effects after different types of exposure (acute, prolonged, chronic).
NAR No action required.
AQ Substances which affect aesthetic and organoleptic quality of the water.
NC Not considered.

Type of standard
a Guideline value.
b Limits considering both organoleptic and health consequences.

Remarks
* Depending on climatic conditions.
1 Such as $CaCO_3$ mg/litre.
2 Only BαP (benza(α)pyrene).
3 True colour units.
4 Nephelometric turbidity units; preferably < 1 for disinfection efficiency.
5 Only mono-dichlorinated phenols.

specify a safety level, but to quantify the risk of exposure to carcinogens. To this end the WHO employs the concept of the 'unit lifetime risk' (UR), defined as the risk associated with a lifetime exposure to 1 $\mu g/m^3$ of carcinogen. It is calculated as:

$$UR = \frac{Po(R - 1)}{X}$$

in which Po = background lifetime risk, which is taken from age/cause-specific death, or incidence rates, found in national statistical tables of a matched control population;

R = relative risk, defined as the ratio between the observed (O) and the expected (E) number of cancer cases in the exposed population. Relative risk is sometimes expressed as the standardised mortality ratio, SMR, which is equal to (O/E) × 100 times lifetime average exposure (standardised lifetime exposure for the study population based on a lifetime of continuous exposure).

X = lifetime average exposure (standardised lifetime exposure for the study population on a lifetime continuous exposure basis). In the case of occupational studies, X represents a conversion from the occupational 8-hour, 240-day exposure over a specific number of working years and can be calculated as X = 8 hour TWA × 8/24 × 240/365 × (average exposure duration (in years))/(life expectancy (70 years)), where TWA is the time-weighted average ($\mu g/m^3$).

Where scientific information only allows for the quantification of risk, official regulations (standards) which take both costs and feasibility into account, generally fall within the lifetime risk range of 10^{-4} to 10^{-6}.

Another practical problem of dealing with carcinogens, which can be more difficult than problems of acute toxicity, is concerned with access to information on the state-of-the-art on epidemiological and/or laboratory evidence for a particular substance or a given exposure episode. The only help in this respect comes from the use of databases, some of which are accessible on-line (Gold *et al.*, 1995).

ENVIRONMENTAL STANDARDS

Standard classification (Ragas *et al.*, 1994)

Although the protection of human health against environmental pollution has been the primary aim of and motivation for environmental standards, the concept can be extended to the preservation of environmental quality, too.

From this point of view there are as many (types of) environmental standards as there are aspects to the environment that are taken as targets of environmental policy. One way of systematising environmental standards is by considering what is called the 'environmental chain effect' (Udo de Haes and Van der Voet, 1987), which is essentially a variant of the 'pressure–state–response' model (OECD, 1993).

The 'environmental chain effect', illustrated in Figure 6.3, starts with perceived human needs. Societal activities in every society are aimed at meeting those needs. But these activities, and the consequences that flow from them, create what is called an 'environmental load'. An excessive environmental load produces negative effects for humans, animals, plants, ecosystems and materials. It is only when these effects are recognised by society that one begins to deal with environmental problems.

The traffic problem, especially in cities, provides a typical illustration of the environmental chain effect. In this case the chain starts with the need for mobility. The motor car meets this need in part, but its negative effect is the release of polluting gases and particulates into the environment. They are harmful to the environment as well as to human health, especially in cities where people are exposed to high

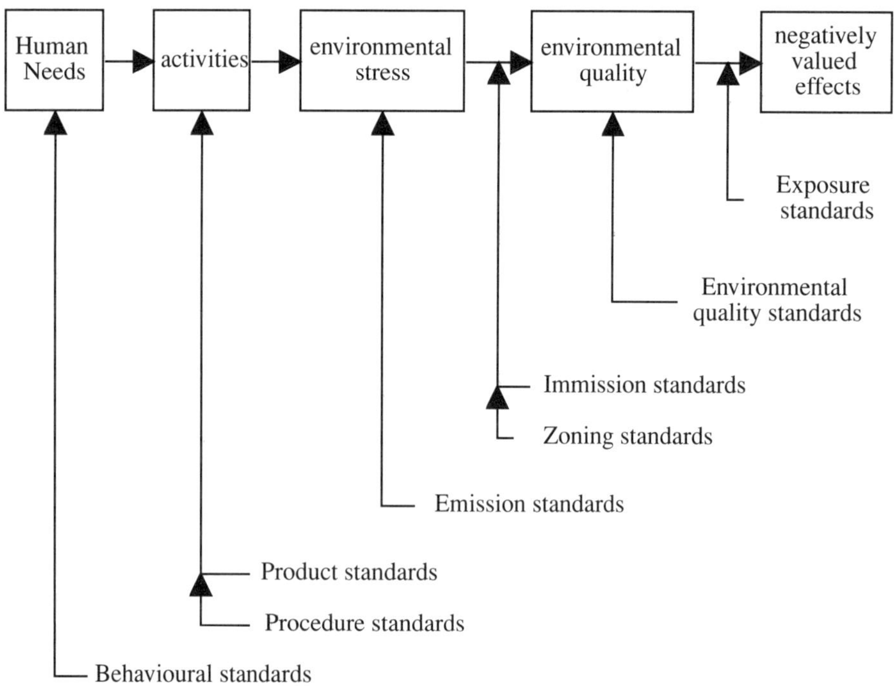

Figure 6.3 Standards in relation to the environmental chain effect
Source: After Udo de Haes and Van der Voet, 1987

concentrations of polluting substances in the air they breathe. But clearly, the problem will be seen as an environmental problem to be addressed only if the health of a sufficient number of persons is affected by the polluting traffic.

Standards can be established for each element of the chain, as well as for the junctions between elements of the chain shown in Figure 6.3. The main standards are these:

1 *Behavioural standards*: These standards indicate how people should respect the environment and behave in order to protect it. As a rule, they are not enshrined in legal regulation, but are part of education and information aimed at enhancing public awareness of the environment. The advice to turn off the motor car engine while waiting in front of a crowded elevated bridge can be interpreted as a typical behavioural standard.
2 *Product standards*: These standards concern the requirements for formulating a particular product. They give guidance either on the composition of the product, or on emissions generated by its use. The lead or benzene content of car fuel is a typical example of a product standard.
3 *Process standards*: They stipulate the maximum permissible emission concentrations for a particular installation, or specify the conditions to be taken into account in designing, building or exploiting a facility. Prescribing catalytic converters for cars and specifying their performance is a typical example of a process standard.
4 *Emission standards*: These standards are to determine which emission concentrations should not be exceeded, either in absolute terms or in terms of a particular frequency and/or during a specified time period. Emission standards are often regulated by permits. The maximum concentration of CO which can be emitted by a car is a typical example of an emission standard.
5 *Zoning standards*: These standards relate environmental quality to the distance of a source of emission. Their purpose is to specify the safe distance between a pollution source and the receptor. The minimum prescribed distance to be maintained between a highway and a residential area is a typical example of a zoning standard.
6 *Immission standards*: These standards specify, sometimes in terms of probability, which concentrations or exposure levels should not be exceeded in a particular area. A typical example is the amount of lead deposited on soil per square metre per day.
7 *Environmental quality standards*: These standards specify the amount, concentration or intensity of a harmful chemical, or a physical or biological agent, which should not be exceeded in an environmental compartment. A typical example is the stipulation that 1 kg of soil should not contain more than 1 mg of cadmium.
8 *Exposure standards*: These standards specify the maximum amount of chemicals, or physical or biological agents, to which ecosystems, people or materials can be exposed. All the health standards, discussed earlier in this chapter (pp. 111–118), are based on exposure guidelines.

Standards can be systematised according to different classification systems as follows:

- The environmental compartment they address, e.g. water-, air-, and soil-related standards.
- The effect they address, e.g. health, environment and materials.
- Nature of the pollutant, e.g. chemical substance, radiation, noise, bacteria, viruses, odour and genetically modified organisms.

Moreover, almost all of these main groups can be subdivided. For example, water standards can be subdivided into those for surface water, groundwater, coastal water, sea water, etc.

Establishing environmental standards

In contrast to health standards, there is less consensus on how to establish standards aiming to protect the environment and cultural heritage. There are basically two different policy approaches for establishing environmental standards. The first is substance-related which, to a large extent, can be compared with the approach we have already discussed for establishing health-related guidelines. Analogous to the standard experiments toxicologists use to determine NOAE-levels, eco-toxicologists have developed experiments to determine no (adverse) effects in a number of laboratory organisms. It is thus possible to establish guidelines for a number of environmentally important substances on the basis of their effects on important indicator organisms. However, the limitations of this classical substance-related approach, often called the 'technological approach', are much more important than those concerned with the setting of health guidelines because effects on *Daphnia* (or the 20 other regularly used laboratory organisms) provide only partial information on impacts on the ecosystem. In practice the method has only been used for a limited number of substances and requires priority rules. Also, the substances studied are usually known to have already been introduced into the environment and to cause damage – a non-preventive approach. Moreover, analogous to the health approach, the environment is not stressed by only one substance at a time. Actually, the real stress results from a cocktail of substances. But the resistance of an ecosystem is limited, and, once it is exceeded, the negative effects become evident in a pronounced way. Another limitation is the finding that environmental quality should not be defined only in terms of concentrations of substances. An ecologically healthy river system is not characterised only by the absence of heavy metals or excessive nutrients or hydrocarbons. Its biological quality is also determined by the biological diversity of the species present in the water and in the river bed.

A more holistic approach is thus needed to deal with the limitations of the substance-related method. An integrated risk assessment of the complex stresses affecting the global environment constitutes such a holistic approach, shown schematically in Figure 6.4. This figure shows how the definition of risk results from a process, taking into account the relevant laboratory and field data. It will also be seen that once a risk is defined, it is possible to develop and adopt policies for appropriate environmental management aimed at minimising or eliminating that risk. Monitoring enables verification of whether the management aims are being realised.

This holistic method, however, needs fundamental scientific data to support it and to make its practical application possible. As the conservation of important ecological functions is one of the aims of this approach, there is some criticism that the dynamic of the method is too explicitly directed towards a past utopian situation and that it does not take into account the socio-economic realities of life.

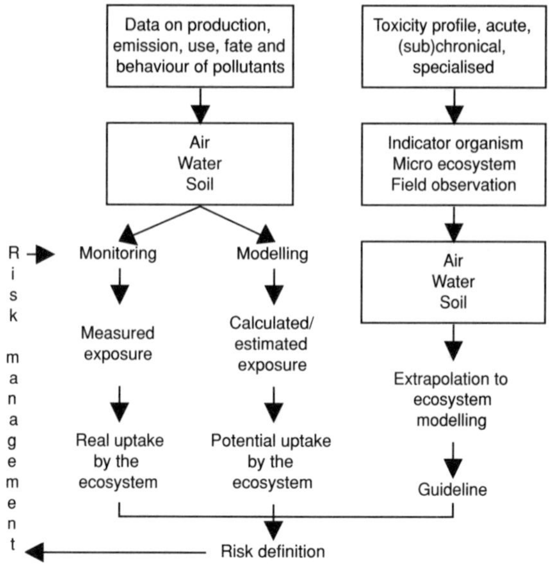

Figure 6.4 Risk assessment for environmental objectives
Source: After Kroes, 1989

Further strategies would probably be needed to combine the substance-related approach with the holistic approach. The main aim of this 'marriage' is to replace the restorative attitude, which is characteristic of the substance-related approach, with an effective and efficient approach to prevent further degradation of the environment.

GUIDELINES FOR SUSTAINABLE DEVELOPMENT?

The use of health and environmental standards has so far proved to be a reasonable policy instrument, especially for controlling the excesses of environmental pollution. For example, the high SO_2 immissions in Europe during the 1960s are no more. Although the actual values in the major cities of the Northern hemisphere are still above the WHO guidelines, they are certainly not as high as those of the notorious London smog of 1952. Thus, standards in general appear to be a useful instrument with which to curb the excesses of pollution, but in practice they seem to be less efficient when it comes to preventing chronic negative impacts on health or the environment.

One might wonder if the useful character of standards, which curbs excesses of pollution, might also be applied to achieve a minimum level of sustainable development. This would entail the specification of critical levels for socio-economic activities which create pollution. The strategy chosen in this context is however less directive, less prescriptive and softer compared to the use of standards. It is explained in the other chapters in this volume that, at least for the time being, indicators of sustainability are developed with the aim of quantifying sustainable development. One of the major objectives of the indicators is to clarify trends in development, but they do not help in defining absolute or minimum levels of poverty, illiteracy, or unsustainably high energy consumption or resource utilisation. Neither is there any indication in current literature of how this could be done (Goeteyn, 1996).

CONCLUSIONS

Health standards for individual substances are still one of the most widely used instruments for framing legislation on environmental hygiene. The development of such standards is basically a two step process in which the first step, in which the guidelines are defined, takes into account both scientific information and expert judgement. Feasibility, costs and policy are considered in the second step.

An analysis of the methodology shows that, without a detailed record of the data used for establishing standards, it is impossible to evaluate a standard on its health implications alone. Moreover, guidelines are, as a rule but not always, established in such a way that acute danger to health is not caused when they are slightly exceeded. It is thus important to note that at present the environmental manager faces a situation in which exposure often exceeds the guideline or standard. Therefore, before assessing a health risk, it is prudent for the professional to look into the nature of the most sensitive adverse end points (and other known or presumed effects), and to determine if a safety factor has been used, and, if so, the type of uncertainty it covers.

The use of standards has distinct limitations too. Many real-life exposures, often to cocktails of substances, have dynamic and chronic patterns that do not have counterparts in laboratory systems and are not covered by standards for individual substances. Alternative approaches need to be developed to deal with these aspects (Stara, 1984; Mumtaz *et al.*, 1993).

Other limitations are the lack of transparency in socio-economic discussions, on the basis of which standards are established, the long list of substances for which standards are still to be established, and the lack of sufficient information on the effects of those substances. Standards tend to provide safety, but at the same time they conceal the actual degree of uncertainty. Also, because they are enshrined in legal regulations, it is sometimes difficult and time-consuming to update them. Furthermore, to date standards have been established for a limited number of products and substances. Environmental episodes are, on the other hand, often characterised by exposure to mixtures of substances, many of which are not currently covered by standards. The use of standards thus leads to a fragmentary and non-systematic policy.

There is, moreover, a growing conviction that standards for environmental hygiene should not only protect human health, but they should also guarantee the quality of ecosystems, including those to which humans are integral. Methods of establishing environmental standards to meet these aims are now being developed (Lindhurst *et al.*, 1995). The basic data needed for this holistic approach are derived from eco-toxicological laboratory experiments, ecosystem models, field observations and environmental risk assessment (see Chapters 8 and 9). Theoretically, an intelligent combination of these data can lead to environmental norm-setting, and the realisation of their objectives can be controlled by monitoring systems. At present, though, the main problem with this approach is the substantial lack of essential basic data. Clearly, important and comprehensive scientific and intellectual inputs are needed to solve this problem so that future generations can inherit an environment of acceptable quality.

ACKNOWLEDGEMENTS

The authors wish to express their thanks to the following colleagues: V. Bízek, P. Jílek, B. Moldan and O. Velek.

REFERENCES

Anderson, E.L. (1985) 'Quantitative approaches in use in the United States to assess cancer risk', in V.B. Vouk *et al.* (eds) *Methods for Estimating Risk of Chemical Injury in Human and Non-human Biota and Ecosystems*, New York: John Wiley & Sons.

Collection of Laws of the Czech and Slovak Federative Republic, Supplement 1, Part 84, Prague, 1991.

Goeteyn, L. (1996) 'Measuring sustainable development at the national and international level', in B. Nath, L. Hens and D. Devuyst (eds) *Sustainable Development*, Brussels, Belgium: VUB Press.

Gold, L.S., Manley, N.B., Slone, T.H., Garfinkel, G.B., Ames, B.N., Rohrbach, L., Stern, B.R. and Show, R. (1995) 'Carcinogenic potency database', *Env. Health Perspectives* 103 (Supplement 8): 3–122.

IARC (International Agency for Research on Cancer) (1982) *IARC Monographs on the Evaluation of the Carcinogenic risk of Chemicals to Humans*, Suppl. 4, Lyon, France: IARC.

Kroes, R. (1989) 'Normstelling voor chemische verbindingen', in A.R.J. Stumpel and R. van de Doel (eds) *Medische Milieukunde*, Utrecht/Antwerpen: Bohn, Schultema and Holkema.

Lindhurst, R.A., Bourdeau, P. and Tardiff, R.G. (1995) *Methods to Assess the Effects of Chemicals on Ecosystems*, Chichester, UK: Wiley & Sons.

Ministry of Health of the CSR (1981) 'Guidance Notes of Ministry of Health of CSR', Vol. 51/1981, Directive No. 58 on basic hygiene demands, maximal allowable concentrations of the most serious air environment pollutants and on the evaluation of the degree of contamination, Prague.

Ministry of Health of the CSR (1986) 'Guidance Notes of Ministry of Health of CSR', Vol. 61/1986, Directive No. 69, changing and and fulfilling Directive of Ministry of Health of the CSR – chief hygienist No. 50/1978 Coll. Guidance notes on contaminants in food, respecting Directive No. 63/1984 Coll. Guidance notes, Prague.

Moldan, B. *et al.* (eds) (1991) *Rainbow Programme*, Ministry of Environment of the Czech Republic, Prague: Academia Press.

Mumtaz, M.M., Sipes, I.G., Clewell, H.J. and Yang, S.H. (1993) 'Risk assessment of chemical mixtures: biologic and toxicologic issues', *Fundamental and Applied Toxicology* 21: 258–269.

OECD (1993) *Core Set of Indicators for Environmental Performance Reviews*, Paris, France: OECD.

Premazzi, G., Chiaudani, G. and Ziglio, G. (1989) *Scientific Assessment of EC Standards for Drinking Water Quality*, Luxembourg: Office for Official Publications of the European Communities.

Ragas, A.M.J., Leuven, R.S.E.W. and Schoof, D.J.W. (eds) (1994) *Milieukwaliteit en normstelling*, The Netherlands: Boom, Meppel.

Roberts, L. (1991) 'More pieces in the dioxin puzzle', *Science* 377: 347.

Speyers, G.J.A., van Went, G.F., van Apeldoorn, M.E., Montizaan, G.K., Jamus, J.A., Canton, J.H., van Gestel, C.A.M., van der Heyden, C.A., Heyna-Merkus, E., Knaap, A.G.A.C., Luttik, R. and de Zwart, D. (1987) *Integrated Criteria Document Nitrate*, Appendix to Report 758473007, Bilthoven, The Netherlands: Natl. Inst. Public Health Env. Prot.

Stara, J.F. (ed.) (1984) *Selected Approaches to Risk Assessment for Multiple Chemical Exposures*, Cincinnati, O. 45268: USEPA – 600/9–84–014a.

Stara, J.F. and Erdreich, L.S. (1984) *Approaches to Risk Assessment for Multiple Chemical Exposures*, Report No. 600/9–84–008, Cincinnati, O.: USEPA.

Tomatis, L., Amo, A., Wilbourn, J. and Shuker, L. (1989) 'Human carcinogens so far identified', *Jpn. J. Cancer Res.* 80: 795–807.

Udo de Haes, A. and Van der Voet, E. (1987) 'Richtlijnen voor ammoniakemissie van intensieve veehouderijbedrijven. Een voorbeeld van afgeleide ecologische normstelling. Landschap', themanummer *Ecologische Normstelling*.

USEPA (1986) 'The Guidelines for carcinogenic risk assessment', *Federal Register* 51: 33992–34003.

WHO (World Health Organisation) (1984) *Guidelines for Drinking Water Quality*, Geneva.

WHO (1987) *Air Quality Guidelines for Europe*, Copenhagen: Regional Office for Europe.

SUGGESTED READING

Calabrese, E.J. and Kenyan, E.M. (1991) *Air Toxics and Risk Assessment*, Chelsea, Mich.: Lewis Publishers Inc.

Gorids, L. (ed.) (1988) *Epidemiology and Health Risk Assessment*, Oxford, UK: Oxford University Press.

SELF-ASSESSMENT QUESTIONS

1 A 'guideline' differs from a 'standard' because (indicate one or more that apply)
  (a) a guideline is based on both scientific evidence and political decisions, while a standard is set only on the basis of a political decision;
  (b) a guideline includes a safety factor to account for scientific uncertainty, but a standard does not;
  (c) a guideline results from scientific evaluation, while costs and feasibility aspects are taken into account in formulating a standard.

2 An ADI value (indicate one or more that apply)
  (a) is usually derived from a detailed analysis of the toxicology of a chemical substance;
  (b) indicates the value at which the most sensitive adverse effect, caused by a chemical, occurs;
  (c) is computed by taking into account the body weight of the subject and a safety (uncertainty) factor.

3 A genotoxic substance is characterised (indicate one or more that apply)
  (a) as a rule, by an exponential dose–effect relationship;
  (b) by its capacity to interact with the genetic material;
  (c) by a threshold dose which allows the determination of a no-effect level.

4. Suppose that the lifetime mortality rate for *myelogenous leukemia* caused by exposure to benzene is 0.3 per cent (that is, three out of 1,000 persons die from this cancer of the white blood cell). Also suppose that four cases of *myelogenous leukemia* are found in a survey of workers exposed to benzene, while only 0.9 cases were expected. Assuming that the calculated average exposure was 2.66 mg/m^3, indicate which of the following is the incremental unit lifetime risk for this group (in your calculation disregard the fact that the number of cases observed in such a small group may vary widely):
  (a) 3.88×10^{-6}
  (b) 3.88×10^{-3}
  (c) 1.16×10^{-6}

5 A holistic risk assessment for setting environmental guidelines (indicate one or more that apply)
  (a) is uniquely based on laboratory experiments and results of ecological modelling;
  (b) shows that the resistance of an ecosystem is limited, and that the negative effects become evident once this resistance is exceeded;
  (c) is in urgent need of fundamental scientific data.

6 Environmental quality standards (indicate one or more that apply)
  (a) are less frequently used in environmental hygiene laws than health standards are;
  (b) provide better protection against exposure to a combination of chemicals than health standards;
  (c) are unsuitable for combining with health standards, when environmental protection is the objective.

7
MEASURING SUSTAINABLE DEVELOPMENT

Lars Fogh Mortensen

SUMMARY

Since the report of the World Commission on Environment and Development (WCED, 1987) and the United Nations Conference on Environment and Development (UNCED) in Rio de Janeiro in 1992, sustainable development has been widely recognised as a common goal. Ability to measure progress towards sustainable development is a critical factor in achieving that goal. The need to measure sustainable development had led to a large number of initiatives aiming at providing information.

This chapter begins with a brief introduction to the concept of sustainable development. The need for information on sustainable development for decision-making at different levels – local, regional, national and global – as well as in different phases of decision-making, is then described. This is followed by an explanation of the 'information pyramid' comprising primary data, analysed data, indicators and indices.

In addition, criteria for choosing measures of sustainable development are discussed together with an assessment of the state-of-the-art in their development. Examples of initiatives in this regard at local, regional, national and global levels are given, highlighting recent important initiatives of the United Nations Commission on Sustainable Development (CSD) and the World Bank. Frameworks for measuring sustainable development, based on static and dynamic approaches, are introduced and analysed. Some problems in the development of measures of sustainability, such as availability and access to data, development of indices and the need for integration, are also identified and analysed.

Finally, an overall assessment is made of the major steps to be taken for the further development of measures of sustainability.

ACADEMIC OBJECTIVES

The aim of this chapter is to introduce the concept of sustainable development, and to discuss and analyse how to measure it. On completion of the chapter the reader should have sufficient information on, and insight into, the following aspects:

- what sustainable development is;
- why information on sustainable development is useful;
- spatial scales at which this information is useful;
- which types of information are useful;
- the relevant criteria for choosing measures of sustainability;
- the state-of-the-art in the development of measures of sustainability;
- the major initiatives at the international level;
- the advantages and disadvantages of using different frameworks;
- the important gaps in developing measures for sustainable development;
- which next steps are of particular importance.

INFORMATION ON SUSTAINABLE DEVELOPMENT FOR DECISION-MAKING

The concept of sustainable development

The United Nations Conference on Environment and Development in Rio de Janeiro in June 1992 (UNCED) was a global political effort to address the issue of sustainable development. Agenda 21, which is the report of the UNCED (United Nations, 1993a), is a globally agreed political document which gives recommendations for the integrated treatment of the environment and development aiming at sustainable development. However, Agenda 21 does not define the concept of sustainable development, the underlying assumption being that a successful implementation of Agenda 21 would in time lead to significant progress towards it.

Numerous attempts have been made to define sustainable development. The definition which has achieved widest global recognition is that of the Bruntland report which defines it as 'development that meets the needs of the present generation without compromising the ability of future generations to meet their own needs' (WCED, 1987: 43). However, there are many other definitions of sustainable development deriving from different perceptions and disciplinary backgrounds. For example, it has been defined and interpreted from economic, ecological, ethical, cultural, social and system-oriented viewpoints. But an element which is common to all these definitions is the focus on intra-generational and inter-generational equity as a goal. Intra-generational equity is about ensuring a certain level of equity in the distribution of income and goods within one generation, while inter-generational equity focuses on equity between present and future generations.

Our purpose here is not to add one more definition of sustainable development, but rather to discuss and analyse how progress towards it could be measured. We will therefore refrain from judging the usefulness or relevance of the various definitions of sustainable development. Instead, we will assume that, in order to achieve sustainable development, it is important to focus on intra-generational and inter-generational equity as well as on the integration of its various aspects such as the environmental, the economic, the social (including human and cultural) and the institutional. Due in the main to the proliferating and often competing definitions, and the complexity of the concept itself, it is difficult to arrive at an operational definition of sustainable development. In this regard the development of a methodology to measure progress towards sustainable development may help by focusing attention on its aspects that are of particular importance.

Information on sustainable development at local, regional, national and global levels

The recognition of sustainable development as a shared goal since UNCED has led to a need for information on it. Following the agreement on Agenda 21 (United Nations, 1993a), there has been an increasing demand for information on sustainable development for use at all levels of decision-making. In fact, Chapter 40 of Agenda 21, entitled 'Information for Decision-making', specifically calls for such information and emphasises that 'indicators of sustainable development need to be developed to provide a solid basis for decision-making at all levels' (United Nations, 1993a: 473).

Information on sustainable development is important for making decisions that are based not only on economic conditions and trends but also on the environmental, social and institutional aspects of sustainability. Such information is needed to guide the process of development in a sustainable manner, for example by assessing the magnitude of problems that may arise, defining possible counter-measures, and by monitoring the implementation and effectiveness of decisions at all levels.

Furthermore, information on sustainable development is needed at all levels – local, regional, national and global. Information at global and regional levels is needed when decisions affecting the present and future of the planet and its population are to be made. For example, information on the ozone layer affects decisions to be made for reducing emissions of ozone-depleting substances. At the national level information on sustainable development is needed to guide both national decision-making and policy-making, focusing on issues that are of particular importance to the country in question. In addition, such information can provide mechanisms to guide policy implementation, as well as to indicate whether trends are in line with national priorities, goals and targets.

Furthermore, information at the national level can be used by international organisations to analyse national situations and trends and to compare the performance of different countries. At the local level information on sustainable development can assist local governments, organisations and the public in moving towards a development pattern that is locally sustainable, and this can contribute to the achievement of sustainable development at national, regional and global levels. In addition, at the local level companies can use the information in their in-house decision-making so that their activities make a useful contribution to the achievement of sustainable development.

As will be seen in a later section (pp. 126–128), different types of information on sustainable development may be needed for different levels (e.g. local, national, etc.).

The decision-making phases

Information on sustainable development is useful in all phases of the decision-making process. In general, decision-making includes problem identification and raising of awareness, policy formulation, policy implementation, and policy evaluation (Table 7.1) (UNEP/DPCSD, 1995). These processes take place at all levels – local, regional, national and global. It is pointed out that sustainable development is not the primary goal of all decision-makers. Nevertheless, in order for society to make progress towards it at all levels, information on its various aspects should be used by all decision-makers as an important input to the decision-making process.

Table 7.1 The decision-making phases

Phase	*Use of information on sustainable development*
1 Problem identification and awareness raising	Problem identification and awareness raising
2 Policy formulation	Identification of policies and strategies
3 Policy implementation	Identification of goals, targets and objectives
4 Policy evaluation	Assessment of whether goals, targets and objectives have been met

Source: UNEP/DPCSD, 1995

Problem identification and awareness raising is the first phase of the decision-making process. Problems are identified in this phase, and the attention of decision-makers is drawn to them. Information on sustainable development can contribute to awareness raising, and to the identification of problem areas within the integrated issues of environment and development, including environmental, economic, social and institutional aspects.

Policies and strategies to address problems are identified in the policy formulation phase, which is the second phase. The policies and strategies can be projects, legislation, economic instruments, plans, management systems, etc. In this phase information on sustainable development can contribute to the identification of policies and strategies with which to address the problem or problems in question.

The policies and strategies, identified in the policy formulation phase, are implemented in the implementation phase – the third phase – in which information on sustainable development plays a useful role in setting goals, targets and objectives.

In the policy evaluation phase – the fourth phase – the effectiveness of policies and strategies implemented are evaluated. Information on sustainable development can provide a very useful tool for assessing whether, and the extent to which, the goals, targets and objectives of the policies and strategies have been or are being met.

The four phases described above should be seen as a conceptual illustration of the process. While they are used by most countries and regions, some may choose to use other phases of decision-making that are more or less specific. The phases may also be seen in terms of a cycle, because, once a set of policies has been implemented and evaluated, the process starts once again as new problems are identified and new policies are formulated (or old policies updated or upgraded), implemented and evaluated to address them.

The 'information pyramid'

Information on sustainable development can be provided with different levels of detail depending on the spatial level (local, regional, national or global), phase of decision-making, and the particular needs of decision-makers and other users. Different kinds of information which may be useful in measuring sustainable development are described in this section, together with an assessment of their usefulness in different decision-making phases and at different spatial levels. Figure 7.1 shows the schematic of the different kinds of information in the form of what is called the 'information pyramid'.

Primary data, sometimes referred to as raw data or statistics, give very detailed information on aspects of concern in a given situation. Detailed data and statistics on the use of pesticides in agriculture, by type of pesticide and by location, are a typical example of primary data on sustainable development. Such information is often compiled by national statistical services and used by decision-makers and others concerned with the specific issue to which the information refers. Often primary data is used in the problem identification and awareness raising phase, and also in the implementation phase. But such data can be relevant in other phases of decision-making too, especially when very detailed information is needed. Primary data is used at all spatial levels, but they can be particularly useful at local and national levels when

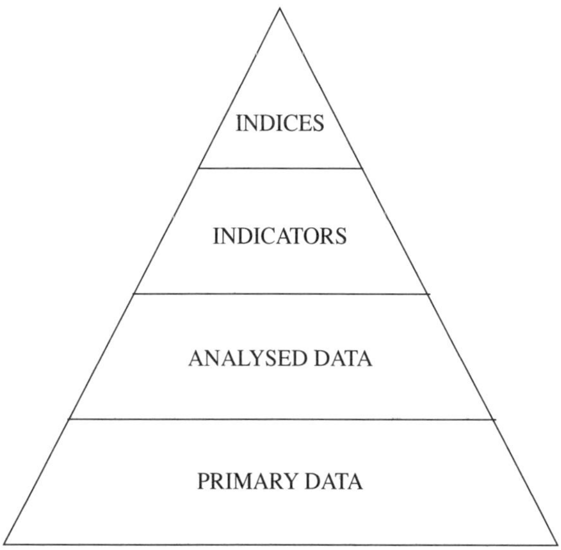

Figure 7.1 The information pyramid
Source: World Resources Institute, 1995

specific information is needed to identify problems and implement policies. However, since this kind of information is often very highly detailed, a need may arise for analysing or aggregating the information to give a better overview of complex problems in the context of sustainable development.

Analysed data is information processed from primary data, as explained above. And the purpose of analysed data is to produce information that is more easily accessible to decision-makers and others who need it. As with primary data, analysed data can be useful in all phases of decision-making when rather detailed information is needed, especially at local and national levels. Since the level of detail is also high in analysed data, a need may arise for further condensing the information, as in the case of indicators for example.

Indicators generally contain information based on both primary and analysed data. Often primary data is both analysed and aggregated in order to provide decision-makers with condensed information contained in indicators which is more useful to them. The term 'indicator' comes from the Latin verb *indicare*, which means 'to disclose or point out, to announce or make publicly known, or to estimate or put a price on' (World Resources Institute, 1995: 1). Indicators are used to quantify information, to make its significance apparent, and to simplify information on complex phenomena such as sustainable development.

There is a variety of indicators, of which the 'descriptive indicators' are the most common. In general, they summarise sets of individual measurements of different issues and communicate the most relevant information to potential users, primarily the decision-makers (UNEP/DPCSD, 1995). These indicators are based on technical and/or scientific values, or statistics, and they are in general simple to use and easy to understand. The 'descriptive indicators' are widely used in the problem identification and awareness raising phase of decision-making, and less so in the other phases. What are called 'performance indicators' incorporate a descriptive indicator and a reference value or a policy target. Thus they provide decision-makers with information on performance undertaken to meet national or international goals, targets and objectives (UNEP/DPCSD, 1995). 'Performance indicators' are used at all spatial levels and, in particular, in the policy evaluation phase of the decision-making process. There are also other kinds of indicators to which we will not allude here. Basically they differ from one another in terms of level of aggregation, level of detail of information, and whether or not they relate to any goals, targets or objectives. The choice of an indicator to be used in a given situation depends on the needs of decision-makers, as well as on the spatial level and phase of the decision-making process. An 'index', which is a specific type of indicator, presents information in a highly aggregated form.

An 'index' is a single aggregated indicator obtained by aggregating and weighting a number of indicators. The Gross Domestic Product (GDP), used by decision-makers all over the world, is a typical example of an index. It provides information on the total production of a country in a single index. Another well known example is the Human Development Index (HDI) which is an index of human development based on three equally weighted elements of human development – namely, longevity (life expectancy at birth), knowledge (adult literacy and mean years of schooling), and income (purchasing power) (UNDP, 1995).

Indices are especially useful to high-level decision-makers (e.g. ministers and heads of agencies) who, depending on the issue in question, may ask for between one and four indices that are easy to understand and on the basis of which decisions can be made. But often the indices they need are difficult to develop and use, because a major problem with indices is that they are derived from complex processes of aggregation and weighting that can be highly subjective, such as expert judgement, the Delphi technique, multicriteria analysis, public opinion polls, and modelling experiments (RIVM, 1994). In order to develop an index, the different indicators on which it is based need to be weighted according to their relative importance. In the case of the GDP, for example, the weighting to be given is in terms of the monetary value of the goods produced. However, when considering environmental, social and institutional aspects, valuation becomes problematic since such goods cannot presently be given a monetary value. Natural resources, on the other hand, can be valued by estimating their potential value to the world population, or to the population of a region or a country as the case may be. Other experts argue that an economic value should not be estimated in this way. Instead, they maintain, the different elements should be weighted according to available scientific information. For example, the impact of different greenhouse gases could be weighted according to their individual impacts. However, we will not pursue this topic any further here because it is somewhat tangential to our discussion. Interested readers are referred to the literature cited under 'Suggested Reading' (p. 141–142).

Another major shortcoming of indices arises from the loss of analytical power which inevitably occurs when less detailed information is used, as in the case of indices. This can often put a significant question mark against whether, or the extent to which, an index takes into account the reality of the situation in question or measures what it is actually supposed to measure.

The kinds of information described above are used for measuring sustainable development. The particular kind to be used depends on the needs of users, phase of decision-making and the spatial level. Analysed data or descriptive indicators are frequently used for measuring specific aspects of sustainable development, air pollution for example. They give information in sufficient detail on the issue in question, and, at the same time, the information is analysed, thereby making it easier for decision-makers to understand and use it. Descriptive indicators can be useful in measuring overall progress towards sustainable development, at the national level for example. But often there is a need for a greater aggregation of indicators or indices to give decision-makers a better overview of the issue under consideration.

Henceforth the term 'measures' will mean the different kinds of information mentioned above, including primary data, analysed data, indicators and indices.

CRITERIA FOR CHOOSING MEASURES

When attempting to measure sustainable development, or progress made towards it, it is important to focus on aspects which effectively capture the essence of the issue. The multitude of aspects contained in this very broad concept makes it necessary to address only those aspects that are of particular importance. Furthermore, the measurement of sustainable development is obviously contingent upon the availability of necessary data.

Various criteria have been developed by organisations, governments and others in an attempt to measure sustainable development. In recent years the following criteria have been used by various international organisations and governments (United Nations, 1996b; OECD, 1994):

- relevant: the phenomenon measured should have direct relevance to sustainable development;
- understandable: the measures should be simple, clear and unambiguous;
- conceptual: the measures should be conceptually well-founded;
- limited: the measures should be limited in number, remaining open-ended and adaptable to future developments;
- data available: necessary data should be readily available or available at a reasonable cost–benefit ratio, adequately documented, of known quality, and updated at regular intervals.

The criteria to be adopted in a given situation depend on the specific purpose of the measurements to be made within the context of that situation. Thus some measures may be used at the national level, others at the local level, and yet others in different phases of decision-making. However, the criteria listed above give good and comprehensive guidance for developing measures of sustainable development.

However, in order to make the above criteria easier to understand and apply, we will rephrase them slightly and add a few more as follows: first it is important to know for which phase or phases of decision-making the measures are to be developed, and for which spatial level, because, as we have pointed out earlier, this would determine the kind of information needed. Second, it is important to consider a minimum number of measures in order to be able to capture those elements that are most relevant to sustainable development. Third, the underlying methodology of the measure (e.g. how to calculate it) and the data needed to produce it should be readily available. Fourth, it is important that the measures are easily understood by the users, and that conclusions can be drawn from the information revealed by the measures. Finally, there must be an underlying framework from within which measures are developed; this will help decision-makers to identify problems and to seek ways in which to address them for improving the situation under consideration.

STATE-OF-THE-ART IN THE DEVELOPMENT OF MEASURES

Although significant strides have been made in developing sustainability measures since UNCED in 1992, measures have been developed for particular aspects of sustainable development rather than for the relationship between the various aspects. An assessment is made in this section of the progress achieved to date in the development of measures for the economic, social, institutional and environmental aspects of sustainable development. This is followed by an assessment of measures developed specifically for measuring sustainable development.

Economic measures

Economic measures are routinely used by decision-makers and others everywhere in the world. The best-known example of such a measure (index) is the Gross Domestic Product (GDP) which, together with its variants such as the Gross National Product (GNP) and the Gross National Income (GNI), is widely used as the primary macroeconomic measure of a country's economic performance. National strategies and goals focus on achieving a continuing high annual growth in per capita GDP, and the overall economic performance of the country is measured using GDP as a yardstick. Other economic measures, such as rate of inflation, money supply, etc., are also routinely used to check the economic health of a country everywhere in the world.

Economic measures, such as those mentioned above, are derived from the System of National Accounts (SNA) which gives a qualitative interpretation of the economic process at the national level, based on the monetary value of economic flows (United Nations, 1993b). For example, the GDP measures the total production of goods and services in a country in monetary terms. This qualitative interpretation represents a step from the traditional statistical description to a purely economic measurement based on monetary values.

It is often argued that economic measures, such as the GDP (which only measures the economic wealth of a society and fails to measure its environmental, social and institutional wealth), give an incomplete picture of welfare and development. The international debate has thrown up many proposals and ideas for including the elements not measured by the GDP, for example, into the SNA. However, the main problem which makes it very difficult to include them is their quantification (Van Dieren, 1995): whereas all transactions in the SNA are measured by market prices and/or real flows of money, it is very difficult to put monetary values on environmental, social, and institutional wealth or endowment.

The United Nations has proposed a System of Integrated Environmental and Economic Accounting (SEEA) as a satellite system of the SNA (United Nations, 1993c). The SEEA tries to address the omission of environmental aspects in the SNA by costing fixed and natural capital consumption and by introducing broader concepts of capital and capital accumulation (Bartelmus, 1995). In the SEEA environmental assets and their changes, notably from depletion and degradation, are added to economic assets. This system is still very controversial in many countries, but progress is being made at the international level by organisations such as the United Nations Statistics Division (UNSD), the United Nations Environment Programme (UNEP), and the World Bank. At the national level some countries, especially those in the industrialised world, are beginning to use methods such as the SEEA in their national accounting.

The Genuine Progress Indicator (GPI), which subtracts from the GDP the costs of pollution clean-up and adds the value of unpaid household work (Cobb and Halstead, 1994), represents another attempt at improving the GDP. Furthermore, the Wuppertal Institute in Germany has been attempting to develop an indicator of the Total Material Consumption (TMC) of a region which measures all primary material input to the economy and its output to the environment (Bringezu, 1993).

In addition to their traditional usage, economic measures are now beginning to make a major input to the measurement of sustainable development.

Social measures

The development of social measures began in the 1960s and gained momentum in the early 1970s. These measures were developed to provide alternative measures of development and welfare to the widely used economic measures. Some social measures, such as measures of population, health and education, are relatively well developed and used all over the world. However, adequate measures for other social aspects of sustainable development are still to be developed and used on a wide scale. They include poverty (income does not cover the issue adequately), civic life, human well-being, and cultural values. It is pointed out that some of the measures, referred to as social measures, are sometimes known as 'human' or 'cultural' measures.

Social measures, developed and used in the past, include the Physical Quality of Life Index (PQLI) developed in the United States, as well as other measures used by the World Bank and published annually in its World Development Report (World Bank, 1995). The best-known social measure is probably the Human Development Index (described on p. 128) which has been published annually since 1990 in the Human Development Report (UNDP, 1995). Due to its reliance on data which is often readily available, the HDI is frequently used for measuring social progress in developing countries and in the countries in transition (i.e. former socialist countries of Eastern and Central Europe). Other well-known social measures include those used by national governments such as employment indicators, indicators of population growth, health indicators, and educational indicators (e.g. school enrolment ratios).

Social measures, such as those mentioned above, are increasingly being used as important elements in measures to assess sustainable development.

Institutional measures

By contrast, hardly any progress worthy of note has been made by the international community towards the development of institutional measures (Gouzee, 1996). This is due in the main to the difficulties in accessing data, statistics or measures that say anything about changes or progress made in this field.

The institutional aspects of sustainable development can be monitored, but they are very difficult to measure. A typical example of an institutional measure is the mandated Environmental Impact Assessment (EIA), which is an indicator developed by UNEP (United Nations, 1996a). It is relatively easy to monitor whether EIA is being mandated, but very hard to develop a measure of its success. In practice the measurement thus becomes a so-called yes/no indicator. It is probably better than having no measure at all, but certainly not satisfactory in the long term. Many of the institutional measures only exist as yes/no indicators, if they exist at all. Clearly, therefore, there is an urgent need to develop measures to monitor the institutional aspects of sustainable development.

An attempt is now being made to further the development of institutional measures of sustainable development. In November 1996, the government of Belgium, in co operation with the United Nations Department for Policy Co-ordination and Sustainable Development (DPCSD) and others, organised an international workshop with the aim of making progress in this field.

Environmental measures

Significant progress has been made in the 1990s towards developing environmental measures, especially in the industrialised countries. Growing concern for our common environment has led to increased data collection and development of measures at both national and international levels. Environmental aspects of sustainable development include the following among others: the atmosphere, oceans, fresh water, forests, agriculture, land resources, biodiversity, chemicals and natural resources. All these are key environmental issues with implications for the ability of human societies to move towards a more sustainable development pattern. Examples of environmental measures are numerous; some of those used at the national level are concerned with pesticides and fertilisers, protected forests, land-use change, emissions of greenhouse gases, waste generation, and protected species (United Nations, 1996a).

Efforts of organisations such as the UNEP, the Organisation for Economic Co-operation and Development (OECD) and the European Environment Agency have to a large extent furthered the development of environmental measures. Perhaps the most important contribution to such developments is OECD's Pressure–State–Response (PSR) framework, described below (see pp. 134–137) which continues to inspire many.

The Dutch attempt to develop indicators, based on environmental themes and target groups, is another important contribution to the development of environmental measures (Adriaanse, 1993). Also important is the initiative by Adriaanse, Hammond and others (World Resources Institute, 1995), who use an ecosystem approach for developing environmental indicators by distinguishing between source indicators, zinc or pollution indicators, life support indicators, and human impact indicators. The long-term purpose of the initiative is to develop four key aggregate indicators, one for each of the issues mentioned above. Another initiative is that of Schmidt-Bleek who developed the resource use index, called the Material Intensity Per Unit Service (MIPS) index, which measures the total material consumption from cradle to grave per unit of service or function (Schmidt-Bleek, 1993).

The Ecological Footprint, which is an index measuring the environmental impact of a person, a city, an industry or a country in terms of the amount of land the entity needs for its sustainable maintenance (Wackernagel and Rees, 1996), is one among several other important recent initiatives. The Ecological Footprint shows how much more (or less) land such an entity would need for its sustainable maintenance than it actually occupies.

The environmental aspects of sustainable development are often said to suffer from lack of information. But the industrialised world has made considerable progress in this direction during the last decade in most of the environmental issues. As a result, environmental measures are increasingly being used in OECD countries such as Canada, Denmark, the Netherlands, Norway, Sweden and the United States of America. But the fact that such measures are being developed and used almost exclusively in the developed countries, and seldom in developing countries or in the transition countries mainly because of scarcity of national data, is a problem in the international context with serious implications. Unfortunately, the economic problems of many of these countries are making it impossible for them to develop or use such measures in a significant way in the near future, although the situation would appear to be more promising in a few of the growing economies of South-East Asia.

Measures of sustainable development

The term 'measures of sustainable development', which we will use throughout, is used to mean those measures in which economic, social, institutional and environmental aspects are combined in an attempt to measure progress towards sustainable development. Measures of sustainable development are not directly aimed at assessing whether or not progress towards sustainable development is being made, or has been made. Rather, it is an attempt to measure progress towards sustainable development by measuring progress made in a number of selected individual fields. Measures of progress (or lack of it) made in the individual fields are then combined to assess the overall progress made towards sustainable development.

Since the UNCED in Rio, there have been various efforts at local, regional, national and global levels to develop measures of sustainable development. Following UNCED recommendations, a large number of international governmental and non-governmental organisations, national governments and scientific institutions have been co-operating to further the development of these measures.

One of the most important international initiatives in this regard is the work programme on indicators of sustainable development of the United Nations Commission on Sustainable Development (CSD), which is being implemented in co-operation with national governments and a large number of international organisations. If implemented successfully, this programme may lead to a wide use of sustainability measures as a basis for decision-making at national, regional and global levels. This initiative is described in the first case study (Box 7.1). Another important initiative, described in the second case study (Box 7.2), is that of the World Bank which has developed measures for various aspects of sustainable development, as well as an index of wealth (World Bank, 1996).

Another initiative, called the Barometer of Sustainability, attempts to develop an index of sustainable development (Moldan and Billharz, 1997: 133–7). It consists of two components – namely, ecosystem well-being and human well-being. Parallel progress in both is necessary for achieving sustainable development. In this the values of ecosystem well-being and human well-being at the national or local level are plotted on a two-dimensional system of co-ordinates to produce a rating on the Barometer of Sustainability.

In different regions of the world, Latin America, the Caribbean, and Asia and the Pacific for example, efforts are being made to develop measures of sustainable development for use at the regional level. For the Latin American region, including the Caribbean, a report on indicators of the environment and sustainable development was published in 1995 (Winograd, 1995). Also, the Approximated Sustainability Index, developed at the University of Costa Rica, represents a Latin American attempt to develop a highly aggregated sustainable development index based on the four attributes of productivity, equity, resilience and stability (Baldares *et al.*, 1993). Organisations active in Asia and the Pacific region, such as the Asian Development Bank, have made attempts to develop sustainable development indicators for use in that region.

At the national level various governments in many parts of the world have begun the process of developing their own measures of sustainable development. Among the countries active in this field are Bhutan, Brazil, Canada, China, Costa Rica, the Czech Republic, Germany, Japan, South Africa, the United Kingdom, and the United States of America (United Nations, 1996c). For example, a report on sustainable development indicators for the UK was published in March 1996 and presented to the Commission on Sustainable Development (CSD) in April 1996 (Department of the Environment, United Kingdom, 1996).

At the local level initiatives such as that of Sustainable Seattle (in the city of Seattle in the USA),

BOX 7.1 THE WORK PROGRAMME ON INDICATORS OF SUSTAINABILITY (UNITED NATIONS COMMISSION ON SUSTAINABLE DEVELOPMENT)

The United Nations Commission on Sustainable Development (CSD) was established in 1993 to monitor the follow-up to UNCED and the implementation of Agenda 21. At its third session, in April 1995, the CSD approved a work programme on indicators of sustainable development (United Nations 1996b). The primary purpose of this programme is to make indicators of sustainable development accessible to decision-makers at the national level. A secondary purpose is to promote the use of these indicators in national reporting to the CSD and other inter-governmental bodies, and thereby provide these bodies with a tool for assessing progress made towards sustainable development.

The work programme includes elements such as the development of methodology sheets for 130 indicators of sustainable development; testing of the indicators in a number of countries; development of aggregated indicators; identification of linkages between the indicators; and the further development of indicators for aspects for which suitable indicators do not yet exist (United Nations 1996b). The work programme is now being implemented by approximately twenty-five international organisations in co-operation with national governments. Implementation is being co-ordinated by the Division for Sustainable Development in the United Nations Department of Economic and Social Affairs (formerly Department for Policy Co-ordination and Sustainable Development). Implementation is well on its way and 130 methodology sheets, describing policy relevance, underlying methodology and the availability of data for each indicator, have been developed by lead organisations and made available to national governments (United Nations 1996a). The indicators are presented in what is called the Driving Force–State–Response framework described below (see pp. 134–137). These indicators attempt to cover social, economic, environmental and institutional aspects of sustainable development and all chapters of Agenda-21. A number of countries, including developing countries and countries with economies in transition, have volunteered to test the indicators at the national level, and this process started in late 1996. Although the indicators included in this work programme are intended for use by all interested countries, they may be particularly useful to developing countries, and countries with economies in transition, which may have limited experience in this field.

The following are typical examples of indicators included in the CSD set (United Nations 1996a):

- Chapter 5 of Agenda-21: Demographic Dynamics and Sustainability: population growth rate; net migration rate; total fertility rate; and population density.
- Chapter 9 of Agenda-21: Protection of the Atmosphere: emissions of greenhouse gases; emissions of sulphur oxides; emissions of nitrogen oxides; consumption of ozone-depleting substances; and ambient concentration of air pollutants in urban areas.

The CSD work programme on indicators may be seen as an attempt at operationalising sustainable development by defining and testing a large number of indicators and the links between them. It is important to note that none of these indicators measures sustainable development as such. Rather, the 130 indicators and the links between them give an indication of whether progress is being made towards it. The critical value of each indicator depends on the overall targets and goals set by the country in which the indicators are used.

in which the number of salmon in local rivers is taken as one of the indicators of local sustainable development, show how much local effect such initiatives can have on both awareness raising and measurement of sustainable development.

In recent years there has been an enormous increase in the number of initiatives for developing measures of sustainable development at all levels. It is obviously not possible to refer to them all here, and so we have outlined above only those that are considered to be most important. It is clear, however, that with increasing regional and international co-operation greater progress is expected to be made in the near future.

FRAMEWORKS

The focus of this section is on the discussion and analysis of the various types of frameworks employed for developing and using measures of sustainable development. The purpose of frameworks is to place the measures of sustainable development in an analytical context, with the aim of being able to develop, present and use them in a way that combines or

BOX 7.2 WORLD BANK REPORT ON MONITORING ENVIRONMENTAL PROGRESS

The World Bank report, *Monitoring Environmental Progress: Expanding the Measure of Wealth* (World Bank 1996, and Dixon and Hamilton 1996), addresses the measurement of sustainable development. The chapter on measuring the wealth of nations is particularly important in this regard. The World Bank assumes that the wealth of each nation (or region) consists of the produced assets (conventionally measured in terms of the GNP), natural capital and human resources. Another important assumption, made and used in calculations, is a social discount rate of 4 per cent.

The value of the produced assets (including machinery, equipment, buildings, construction, etc.) in a country is calculated using a perpetual inventory model, major inputs to which are investment data and an assumed life table of assets. The calculation of the value of produced assets is well defined and generally accepted.

The value of natural capital is assumed to consist of agricultural crop land; pasture lands; forests; protected areas; non-timber forest benefits; metals and minerals; and oil, coal and gas. The natural capital is valued in the following way, using world market prices: the value of agricultural crop land is based on average production multiplied by the world market price for rice, wheat and maize. In addition, a crop-specific adjustment factor of 30–50 per cent is applied to the value of each hectare. The expected future value is discounted at a rate of 4 per cent. Pasture lands are valued in a similar manner whereby the national output of meat, wool and milk is valued. The value of forest land is based on the annual production of round-wood, valued at timber rents (i.e. price minus production costs). An opportunity cost approach has been used to calculate the value of protected areas. Values of pasture lands have been used as a proxy for a minimal value of protected areas. The reasons are that agriculture, forestry and pasture are the most likely alternative uses, and that the protected areas must have at least the value of the pasture land; otherwise they would not be protected. For non-timber forest benefits, one-tenth of the forest area in each country is multiplied by a per hectare estimate of non-timber forest benefits. The value of non-renewable resources, metals and minerals is estimated as a function of production and reserves, average extraction rate, and the economic rent accruing from their extraction. Annual production of oil, coal and gas is valued using estimates of resource rents over time.

The value of human resources is measured residually, combining many elements including the impact of investments in education and health, as well as social capital. The wealth value of returns on labour and capital is measured as the discounted present value of the following: non-agricultural GNP, plus agricultural wages, minus rents on minerals and fossil fuels, and minus depreciation of produced assets. The present value is taken as the mean productive years of the population minus the mean age of the population. Subtraction of produced assets from this present value gives the value of human resources at current exchange rates. It is then revalued using the purchasing power parity rate to obtain the final value of human resources.

Based on the methods described briefly above, the World Bank has produced a ranking of countries by their per capita wealth in 1994. In this the OECD countries, and the oil exporting countries of the Middle East, are ranked considerably higher than the developing countries, the top three countries being the United States of America, Switzerland and Canada. The large gap between the rankings of developed and developing countries is due mainly to human resources and produced assets. The value of natural resources does not vary a great deal between countries. However, it should be obvious to the reader from what has been said above that the method used by the World Bank for calculating the wealth of nations is based on a large number of assumptions that are simplistic and controversial. Even so, this work is undoubtedly a significant contribution to the development of measures of sustainable development.

integrates the various elements of sustainable development.

Static approaches

Frameworks based on the static approach are most commonly used in developing measures of sustainability. In this context the work of the OECD on environmental indicators has heavily influenced subsequent work on measures of sustainability. It recommends a core list of about seventy-five environmental indicators to be used by the OECD countries in their environmental performance reviews (OECD, 1994). This core list is presented within a framework, called the 'Pressure–State–Response' (PSR) framework, based on the concept of causality. The conceptual foundation of the PSR framework is that human activities exert pressure on the environment which changes both the quality and quantity of the environment and its natural resources (state). Society responds to these changes through environmental, economic and sectoral policies (response). This framework can be seen as a static framework because it focuses on pressure, state and response at a given point in time.

In the PSR framework the indicators are structured vertically by environmental issue, and horizontally by pressure, state and response, as shown in Table 7.2. For each cell in this table, the OECD has developed between one and eight indicators (a few examples are given in the table). As the table shows, OECD distinguishes between three types of issues. The first eight issues are the 'sink-oriented' issues concerned with environmental quality. The next four issues are 'source-oriented', focusing on the quantity of natural resources. Issues which cannot be put into either of these two categories, such as those concerned mainly with social and economic aspects, are placed in a general category.

The CSD work programme on indicators of sustainable development of the United Nations (described in Box 7.1) can be seen as an initiative which has been heavily influenced by the OECD's work in this area. Indeed, the CSD framework (United Nations, 1996a), shown in Table 7.3, can be considered a further development of OECD's PSR framework. The concept of what is called a 'Driving Force–State–Response' (DSR) is used in the CSD framework, and it differs from the OECD's framework in a number of aspects (United Nations, 1996b): first, there is no implied causality among the indicators between cells, either horizontally or vertically. This means that the indicators shown in the rows or columns are not necessarily directly linked through a causal relationship. Second, the term 'pressure' has been replaced by 'driving force' in an attempt to accommodate more accurately the social, economic and institutional indicators. In the CSD framework driving force indicators indicate human activities, processes and patterns which impact on sustainable

Table 7.2 The OECD framework for environmental indicators (including examples of indicators)

Issues	*Pressure (indicators of environmental pressure)*	*State (indicators of environmental conditions)*	*Response (indicators of societal response)*
Climate change	• Index of greenhouse gas emissions • CO_2 emissions	• Atmospheric concentrations of greenhouse gases • Global mean temperature	• Energy efficiency • Energy intensity • Economic and fiscal instruments
Ozone layer depletion			
Eutrophication			
Acidification			
Toxic contamination			
Urban environmental quality			
Biodiversity/landscape	• Habitat alteration and land conversion from natural state	• Threatened or extinct species as a share of total species known	• Protected areas as % of national territory and by type of ecosystem
Waste			
Water resources	• Intensity of use of water resources	• Frequency, duration and extent of water shortages	• Water prices and user charges for sewage treatment
Forest resources			
Fish resources	• Fish catches	• Size of spawning stocks	• Regulation of stocks (quotas)
Soil degradation			
General indicators	• Population growth and density • Growth of GDP		• Environmental expenditures

Source: OECD, 1994

Table 7.3 The CSD framework of indicators of sustainable development (including examples of indicators)

Category	*Chapters of Agenda 21*	*Driving force indicators*	*State indicators*	*Response indicators*
Social	Chapters 3, 5, 36, 6, 7	• Unemployment rate • Population growth rate • Adult literacy rate	• Measures of poverty • Population density	• GDP spent on education • Contraceptive prevalence
Economic	Chapters 2, 4, 33, 34	• GDP per capita • Annual energy consumption • Net resources transfer/GDP	• Debt/GNP • Lifetime of proven energy reserves • Intensity of material use	• Environmental protection expenditures as % of GDP
Environmental				
Water	Chapters 18, 17			
Land	Chapters 10, 12, 13, 14			
Other natural resources	Chapters 11, 15, 16			
Atmosphere	Chapter 9			
Waste	Chapters 21, 19, 20, 22	• Generation of industrial and municipal solid waste • Generation of hazardous waste	• Chemically induced acute poisonings • Area of land contaminated by hazardous waste	• Waste recycling and re-use • Municipal waste disposal • Expenditure on hazardous waste treatment
Institutional	Chapters 35, 37, 8, 38, 39, 40, 23–32			

Source: United Nations, 1996a

development; the state indicators indicate the state of sustainable development, while the response indicators indicate policy options and other responses to changes in the status of sustainable development. The CSD framework is also a static framework, like the OECD framework, because it looks at the situation at a given point in time. It should be noted, however, that the OECD and CSD frameworks differ in scope, because, while the OECD framework focuses almost strictly on environmental issues, the CSD framework tries to cover all environmental, economic, social and institutional aspects of sustainable development.

There are other examples of static frameworks apart from the PSR and the DSR. One such is the 'Framework for Indicators of Sustainable Development' (FISD) developed by the United Nations Statistical Division (UNSD). It is based on these four categories (Bartelmus, 1995): (A) socio-economic activities and events; (B) impacts and effects; (C) responses to impacts; and (D) inventories, stocks, and background conditions. Vertically it divides into Agenda 21 issues in much the same way as the DSR framework, but differs from it mainly because the state indicators are divided into categories (B) and (D) as defined above. Categories (A) and (C) in the FISD are similar to those in the DSR framework. The 'Pressure–State–Impact–Response' (PSIR) framework, developed and used by UNEP (UNEP/RIVM, 1995), is very similar to the FISD framework.

The static frameworks can be seen as ordering mechanisms which structure and organise measures of the complex phenomenon of sustainable development in an analytical format. An advantage of using static frameworks, such as those described above, is that they facilitate both the development of measures of sustainable development and their use in the decision-making process.

Static frameworks have limitations, however. First, the fact that they are static implies that their output (analysis, observations, conclusions, etc.) is limited to sustainable development in the past and the present, and that they are not capable of making projections about the future. And second, they are not very good at analysing linkages between the different aspects that are so crucial to sustainable development. In spite of these limitations static frameworks are being widely used today, partly because they are simple and easy to use and understand, and partly because there is a considerable lack of knowledge about more advanced frameworks.

Dynamic approaches

More advanced frameworks are increasingly being employed for the development and application of measures of sustainable development. Their common characteristic is that they are all based on integrated dynamic models of sustainable development. A major problem in measuring sustainable development is that there is a variety of linkages between measures at different levels, as well as between levels, sectors and themes. It may therefore be an advantage to apply a systemic approach to analyse these linkages. What are called integrated dynamic models provide the means for systematically structuring the interlinkages between and among measures.

The development of dynamic models has been in response to the need to acknowledge the dynamic nature of sustainable development. Based on certain assumptions, these models try to reflect the various components of sustainable development and their interlinkages. They are capable of making scenario analyses (i.e. descriptions of alternative futures) on the basis of an assumed logical framework of how the real world works (UNEP/RIVM, 1995). Scenarios are images of the future created from mental maps or perspectives of the past and the present. The role of such scenarios is to provide decision-makers with projections of possible future effects of policy interventions.

The main advantages of linking the measures of sustainable development within an integrated dynamic framework are the following: it shows how the various measures are inter-linked within each issue (cause–effect) and between different issues; it provides insight into the dynamic behaviour of measures; it enables long-term projections of sustainable development; and it identifies critical system variables and provides guidance on how the measures should be selected and aggregated. However, integrated dynamic frameworks have disadvantages and limitations too: because they are more advanced than the static frameworks they are of necessity based on a greater number of assumptions on how the different elements of sustainable development have been inter-related in the past, are inter-related in the present, and will be inter-related in the future. Due to the enormous complexity of the real world which these frameworks try to model, it is very difficult to make assumptions that would be strictly valid or truly representative of the real world. Clearly therefore, the projections which these models make are only as good as the assumptions on which they are based; far too often the underlying assumptions are simplistic and do not conform to the workings of the real world. Due caution should therefore be exercised when using such a framework as a decision-making tool. However, with growing knowledge and experience the assumptions would hopefully converge more and more to the modalities of the real world; the projections made by the dynamic models would then be correspondingly more reliable.

Most of the more advanced integrated dynamic models of sustainable development currently available are based on, or variants of, the static frameworks. Often they combine models of different environmental, economic, social and institutional aspects of sustainable development and try to establish the linkages and interactions among the various aspects of these models.

The TARGETS 1.0 Model, developed jointly by the Dutch National Institute of Public Health and Environment (RIVM) and UNEP (UNEP/RIVM, 1995), is an example of an integrated dynamic model. It consists of sub-models that are representations of the sub-systems of the real world. The sub-models are, for example, a stratospheric ozone model, a population/health model and a carbon-cycle model. With the PSIR framework as its basic starting point, the model distinguishes between the human system and the environmental system and focuses on the linkages between the two. It combines the models of pressure, state, impact and response for both human and environmental systems as shown in the two-dimensional integration matrix of Figure 7.2. The rows in this matrix integrate the interlinkages among various pressures, states and impacts, while the columns integrate the cause–effect chain of an issue or a theme. Since it is only meant as an illustrative example, we will not dwell on it any further.

Other integrated models of sustainable development are being developed. The most important recent initiative is by DPCSD, UNEP and the United Nations University (UNU). They have established the Global Modelling Forum on Sustainable Development whose aim is to enhance the compatibility of model projections used throughout the United Nations system. This forum also supports periodic reports on long-term social, economic and environmental trends by producing trend analyses. One of the intended activities of the forum is to link short-term macroeconomic models (with a time horizon of less than ten years) with long-term integrated assessment models (with a time horizon of more than one hundred years). The Threshold 21 model, developed

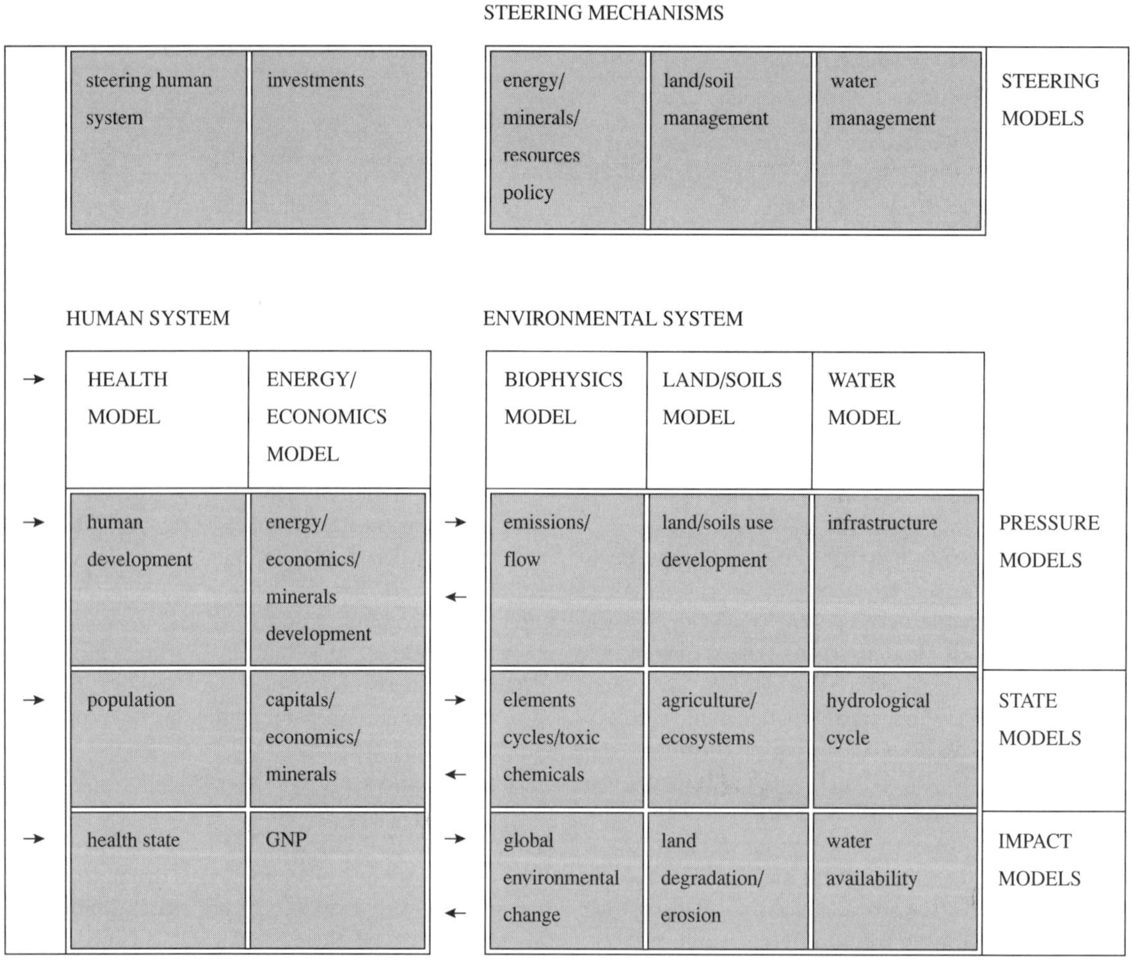

Figure 7.2 The TARGETS 1.0 Model: a modular perspective

Source: Rotmans, J. and de Vries, H.J.M. (1997) *Perspectives on Global Change: the Targets Approach*, Cambridge University Press

by the Millennium Institute, is yet another example of an integrated model. This national development simulation model assesses the long-term impacts of alternative tax, subsidy and other relative price factors on national development indicators. Social, economic, environmental and security implications can also be analysed by the model, in which 15 different sectors are included.

Existing models are numerous, especially for analysing the economic and environmental aspects of sustainable development. The challenge for the international community is to further integrate them into frameworks of sustainable development, and to use measures for making projections and analysing scenarios for the measurement of sustainable development.

SOME PROBLEMS IN DEVELOPING MEASURES OF SUSTAINABILITY

In this section we will identify some of the problems in developing measures of sustainable development, but we will not repeat those to which we have already referred, such as that concerned with the development of institutional measures (pp. 130–131).

Availability of data

Perhaps the most important problem in the development of measures of sustainability is the scarcity or unavailability of essential data in many countries, and often the poor quality of what is available. In many countries, especially developing countries and the countries in transition, availability and quality of essential data is so poor that reliable measures of sustainable development are difficult to develop.

Lack of data, or the poor quality of available data, is a particular problem for the institutional and environmental aspects of sustainable development. In many countries these issues have only recently become a priority, and so effective methods and dedicated institutions for collecting and processing such data are not yet in place. Moreover, financial problems impose significant and even severe constraints on the ability of many countries of the world to undertake such activities.

Training and capacity-building programmes are urgently needed at all levels for improving both availability and quality of data, and some initiatives are beginning to emerge to this end; but, given the global decline in development aid, future prospects do not appear to be too promising.

Indices

Development of indices of sustainable development can be controversial because, in order to develop a given index, agreement is first needed on elements that are judged by all parties to be sufficiently important to be included in that index. Weightings to be given to the different elements of an index, and especially diverse expert opinion on such matters, can be controversial, as we have pointed out earlier (see pp. 126–128). Even so, a number of regional and global indices of sustainable development are now being developed. For example, the Statistical Office of the European Commission (EUROSTAT) is working on the development of Pressure Indices for Europe (Jesinghaus, 1995), while the project on indicators of the Scientific Committee on Problems of the Environment (SCOPE) have made attempts to aggregate some of the indicators of the CSD into four specific indices (World Resources Institute, 1995). The work of the World Bank, described in Box 7.2, is another example of an international initiative to develop an index for measuring sustainable development.

Integration

An important problem in developing measures of sustainability is the integration of the different aspects and measures of sustainable development. In this regard the international community has come a long way in its attempt to identify the elements that are important for measuring sustainable development, and for selecting the most appropriate measures for the purpose. Current knowledge of how the different aspects of sustainable development should be integrated is, however, poor. This deficiency is expected to be addressed, at least in part, by the effort now being made to develop reliable dynamic models to which we have referred earlier.

Another problem is that present knowledge of the interlinkages among the various aspects of sustainable development, such as environmental, economic and social aspects, is quite inadequate. A great deal of work remains to be done in this area.

THE NEXT STEPS

The problems, to which we have referred in the previous section, in fact identify most of the next steps to be taken for the further development of measures of sustainable development. The major next steps to be taken are:

1 Effective and appropriate mechanisms should be put in place for the collection of reliable data necessary for developing measures of sustainable development. The need for this is greatest in developing countries and in the countries in transition.
2 Single compound indicators should be developed for use by high level decision-makers, based on or aggregated from a range of indicators, aspects or elements.
3 Greater effort should be made to understand the interlinkages which inevitably occur among the various aspects of sustainable development. Progress towards true sustainability cannot be measured without a proper understanding and consideration of these interlinkages and interactions.
4 Further work is needed on the development of measures for the various environmental, economic, social and institutional aspects of sustainable development – especially the last, on which there is very little available data at present.
5 Greater and more effective international co-operation is needed at all levels (e.g. political, scientific, etc.) for the development, testing and implementation of national, regional and global measures of sustainable development.
6 Emphasis should be given to training and capacity-building in the development and proper use of indices and indicators of sustainable development. They are no more than tools and, as such, they will not give a true indication of progress towards sustainability unless they are developed and used properly.

CONCLUSIONS

Information on sustainable development is useful to decision-makers at all levels – local, national, regional and global; it is also useful in all phases of decision-making for measuring, promoting and achieving sustainable development. Different kinds of information, such as primary data, analysed data, indicators and indices are useful to decision-makers and others for measuring progress towards sustainable development.

Important criteria for choosing measures of sustainable development include the following: the measure should be relevant; it should be conceptually well founded and easy to understand; only a limited number of measures should be considered; essential and reliable data should be available for developing the measures; and the measures should be based on a robust analytical framework. Economic and social measures of sustainable development are widely used all over the world, while environmental measures are used mainly in the industrialised countries. However, useful institutional measures are still to be developed. Measures of sustainable development have been developed in the last few years by national governments and international organisations, notably the United Nations and the World Bank.

Static frameworks have been widely used for developing measures of sustainability. On the other hand the integrated dynamic models, which are considered to be more realistic for the purpose, are still to be developed for wider application. Major problems in the further development of measures of sustainable development are concerned with the development of institutional indicators, availability of reliable data, development of single compound indices, and the integration of the various aspects of sustainable development in a meaningful way. Ways of overcoming these problems have been elaborated in this chapter.

REFERENCES

Adriaanse, A. (1993) *Environmental Policy Performance Indicators*, The Hague: Ministry of Housing, Physical Planning and Environment of the Netherlands.

Baldares, M., Gutierrez, E., Alvarado, A. and Brenes, L.G. (1993) *System Development of Sustainability Indicators for the Agriculture and Natural Resources Sectors of Latin*

America and the Caribbean Countries, San Jose: University of Costa Rica.

Bartelmus, P. (1995) 'Indicators of sustainable growth and development – linkage, integration and policy use', Background paper for Scientific Workshop on Indicators of Sustainable Development, Wuppertal, 15–17 November 1995.

Bringezu, S. (1993) *Towards Increasing Resource Productivity: How to Measure the Total Material Consumption of Regional or National Economies*, Wuppertal: The Wuppertal Institute.

Cobb, C. and Halstead, T. (1994) *The Genuine Progress Indicator: A Measure of the Socio-Economic Well-Being and Sustainability of the Nation 1950–1992*, San Francisco, California: Redefining Progress.

Department of the Environment, United Kingdom (1996) *Indicators of Sustainable Development for the United Kingdom*, London: HMSO.

Dixon, J. and Hamilton, K. (1996) 'Expanding the measure of wealth', *Finance and Development*, December, 15–18.

Gonzee, N. (ed.) (1996) *Indicators of Sustainable Development for Decision-making*, Report of Workshop of Ghent, Belgium, 9–11 January 1995, Brussels: Federal Planning Office of Belgium.

Jesinghaus, J. (1995) *The Pressure Indices Project: Theory and Structure*, Brussels: Commission of the European Union.

Moldan, B. and Billharz, S. (eds) (1997). *Sustainability Indicators: Report of the Project on Indicators of Sustainable Development*, Chichester: SCOPE/Wiley.

OECD (Organisation for Economic Co-operation and Development) (1994) *Environmental Indicators*, Paris: OECD.

Prescott-Allen, R. (1993) *The Barometer of Sustainability: A Method of Assessing Progress Toward Sustainable Societies*, Washington, DC: PADATA and IUCN (World Conservation Union).

RIVM (1994) *Global Change and Sustainable Development: A Perspective for the Next Decade*, RIVM report no. 461502004, Bilthoven: Dutch National Institute of Public Health and Environmental Protection.

Schmidt-Bleek, F. (1993) 'A universal ecological measure', *Fresenius Environmental Bulletin* 2(8) 529–534.

UNDP (1995) *Human Development Report 1995*, New York: Oxford University Press for the United Nations Development Programme.

UNEP/DPCSD (1995) *The Role of Indicators in Decision-Making*, Discussion paper for the Workshop on Indicators of Sustainable Development for Decision-Making in Ghent, Belgium, January 1995, jointly prepared by the United Nations Environment Programme and the United Nations Department for Policy Co-ordination and Sustainable Development, Brussels: Federal Planning Office of Belgium.

UNEP/RIVM (1995) *Scanning the Global Environment: A Framework and Methodology for Integrated Environmental Reporting and Assessment*, UNEP/EATR.95-01; RIVM 402001002, Nairobi: Environment Assessment Sub-Programme, UNEP.

United Nations (1993a) *Agenda 21 – Report of the United Nations Conference on Environment and Development*, Rio de Janeiro, 3–14 June 1992, Volume 1: *Resolutions Adopted by the Conference*, A/CONF.151/Rev.I (Vol. 1), Sales No. E.93.I.8, New York: United Nations.

United Nations (1993b) *System of National Accounts*, Series No. 2, Rev. 4, Sales No. E94.XVII.4, New York: United Nations.

United Nations (1993c) *Integrated Environmental and Economic Accounting*, Series F, No. 61, Sales No. E.93.XVII.12, New York: United Nations.

United Nations (1996a) *Indicators of Sustainable Development: Framework and Methodologies*, New York: United Nations.

United Nations (1996b) *Work Programme on Indicators of Sustainable Development of the Commission on Sustainable Development*, prepared by the Division for Sustainable Development in the Department for Policy Coordination and Sustainable Development, New York: United Nations.

United Nations (1996c) *Report of the Secretary General on Chapter 40 of Agenda 21, Information for Decision Making, to the Fourth session of the Commission on Sustainable Development*, New York: United Nations.

Van Dieren, W. (1995) *Taking Nature in Account – A Report to the Club of Rome*, New York: Springer-Verlag, Inc.

Wackernagel, M. and Rees, W. (1996) *Our Ecological Footprint – Reducing Human Impact on Earth*, Gabriola Island: New Society Publishers.

WCED (1987) *Our Common Future*, Oxford: World Commission on Environment and Development/Oxford University Press.

Winograd, M. (1995) *Environmental Indicators for Latin America and the Caribbean – Toward Land-use Sustainability*, Washington, DC: World Resources Institute.

World Bank (1995) *World Development Report 1995*, Washington, DC: Oxford University Press.

World Bank (1996) *Monitoring Environmental Progress: Expanding the Measure of Wealth*, Washington, DC: The World Bank.

World Resources Institute (1995) *Environmental Indicators – A Systematic Approach to Measuring and Reporting on Environmental Policy Performance in the Context of Sustainable Development*, Washington, DC: World Resources Institute.

SUGGESTED READING

Anderson, V. (1991) *Alternative Economic Indicators*, London and New York: Routledge.

Bartelmus, P. (1994) *Environment, Growth and Development: The Concept and Strategies of Sustainability*, New York: Routledge.

Billharz, S. and Moldan, B. (eds) (1995) *Report of Scientific Workshop on Indicators of Sustainable Development, Wuppertal, Germany, November 15–17, 1995*, Prague: Scientific Committee on Problems of the Environment.

Dixon, J. and Hamilton, K. (1996) 'Expanding the measure of wealth', *Finance and Development*, December, 15–18.

Goodland, R., Daly, H., El Serafy, S. and von Droste, B. (1991) *Environmentally Sustainable Economic Development: Building on Brundtland*, Paris: UNESCO.

Gouzee, N. (ed.) (1996) *Indicators of Sustainable Development for Decision-making*, Report of the Workshop of Ghent, Belgium, 9–11 January 1995, Brussels: Federal Planning Office of Belgium.

Hodge, T., Holtz, S., Smith, C. and Baxter, H. (eds) (1995) *Pathways to Sustainability:Assessing our Progress*, Noranda.

Serageldin, I. and Steer, A. (eds) (1994) *Making Development Sustainable – From Concepts to Action*, Environmentally Sustainable Development Occasional Paper Series No. 2, Washington, DC: The World Bank.

Trzyna, T. (ed.) (1995) *A Sustainable World: Defining and Measuring Sustainable Development*, Washington, DC: IUCN.

UNEP/RIVM (1994) *An Overview of Environmental Indicators: State of the Art and Perspectives*, UNEP/EATR.94–01; RIVM/402001001, Nairobi: Environmental Assessment Sub-Programme, UNEP.

UNEP/RIVM (1995) *Scanning the Global Environment: A Framework and Methodology for Integrated Environmental Reporting and Assessment*, UNEP/EATR.95–01; RIVM 402001002, Nairobi: Environmental Assessment Sub-Programme, UNEP.

Van Dieren, W. (1995) *Taking Nature into Account – A Report of the Club of Rome*, New York: Springer-Verlag, Inc.

WCED (1987) *Our Common Future*, Oxford: World Commission on Environment and Development/Oxford University Press.

World Bank (1996) *Monitoring Environmental Progress: Expanding the Measure of Wealth*, Washington, DC: The World Bank.

World Resources Institute (1995) *Environmental Indicators – A Systematic Approach to Measuring and Reporting on Environmental Policy Performance in the Context of Sustainable Development*, Washington, DC: World Resources Institute.

SELF-ASSESSMENT QUESTIONS

1 Information on sustainable development is needed to:
   (a) provide the World Bank with a tool to be used for evaluating countries with regard to Structural Adjustment Programmes;
   (b) provide information for decision-makers;
   (c) provide a legal instrument to enforce environmental regulation.

2 Indicators of sustainable development contain:
   (a) more detailed information than primary data;
   (b) more detailed information than analysed data;
   (c) more detailed information than indices.

3 Which of the following criteria is not particularly relevant for developing indicators of sustainable development:
   (a) information should be contained in a database;
   (b) information should be simple and unambiguous;
   (c) information should be relevant for decision-makers.

4 Discussion on the measures of sustainable development:
   (a) started in the 1960s and is now at a stage where further progress is very difficult to make;
   (b) has not really started yet;
   (c) gained momentum in the early 1990s and considerable progress is now being made at many levels.

5 What is common between the PSR, the DSR and the PSIR?:
   (a) they are all static frameworks for measures of sustainable development;
   (b) they are all based on the concept of causality;
   (c) they have all been developed in an attempt to model sustainable development.

6 Which of the following is considered an advantage of integrated dynamic models for sustainable development?:
   (a) easy for decision-makers to use;
   (b) easy for decision-makers to understand;
   (c) enables long-term projections;
   (d) enables static descriptions of the state-of-the-art.

7 The work programme on indicators of the Commission on Sustainable Development attempts to cover the following aspects:
   (a) mainly environmental aspects of sustainable development;
   (b) environmental, social and economic aspects of sustainable development;
   (c) economic, social, environmental and institutional aspects of sustainable development.
8 Which of the following is not considered a problem in developing measures of sustainable development:
   (a) integration of the different aspects of sustainable development;
   (b) availability of economic indicators;
   (c) development of indices.
9 Which of the following countries has recently developed sustainability measures?:
   (a) Norway;
   (b) Denmark;
   (c) United Kingdom.
10 Which of the following is considered to be particularly important for the further development of measures of sustainability?:
   (a) development of new social and economic measures;
   (b) training and capacity-building programmes;
   (c) development of additional environmental measures.

8
RISK ANALYSIS

Alfons Buekens, Bernard De Caevel and Alain Neyt

SUMMARY

Attention is focused in this chapter on risk, industrial risk in particular, and the consequences that flow from it to affect both people and the environment. The nature of risk is explored after a general introduction, followed by a discussion of how risk is perceived subjectively by both people and societies with respect to their norms, values and attitudes. Formal definitions of risk and 'acceptable risk' are then given, together with methods of ecological risk assessment and the uncertainties inherent to them.

The role and impact of the 'human factor' in the management of both risk and its consequences are then discussed, pointing out that risk management works most effectively and efficiently only when it is integrated into the general corporate management of organisations. This is followed by a discussion of both qualitative and quantitative risk analysis, along with the limitations of the latter and how an 'acceptable level of risk' is perceived. Finally, the roles of governments, the public, industry and scientific research in risk analysis and management are elaborated with the future perspective in mind.

It is argued that further progress will require careful study of the consequences of decisions on the value of unknown parameters and the narrowing of confidence intervals in risk analysis. And that, in spite of its limitations, risk analysis can greatly contribute to the identification and evaluation of various hazards, to the prevention of accidents, and to the attenuation of damage, whether caused by a sudden or permanent exposure.

ACADEMIC OBJECTIVES

After completing the chapter the reader should be familiar with, and have a working knowledge of, the following:

- what constitutes a risk, how it occurs, and its consequences;
- what are the current methods of analysing and managing risk, how they are applied, their limitations, and ways in which they can be improved for greater reliability;
- human error, its causes, and how it can upset best laid risk management plans supported by state-of-the-art technology;
- contributions that governments, the public and industry can make to improve the industrial landscape with regard to risk and the consequences that flow from it, and the role of scientific research in the process.

THE CONCEPT OF RISK

Introduction

Industrial activity is a welcome source of employment, wealth and useful products, but at times it is also a source of accidents and is dangerous to human life, industrial plants and entire ecosystems. Accidents vary both in frequency and the extent of damage they cause. Although technically it is possible to reduce both these factors markedly, a completely accident-free society remains a utopian ideal none the less.

There are two distinct approaches to an ideal, accident-free society: a *technical* approach and a *social–educational* approach. Technical conditions influence both the probability of occurrence of an accident and the amount of damage it can cause. But ultimately, the provision of technology and how it is used – whether in a positive way by preventing accidents and limiting damage, or in a negative way by creating dangerous situations – is determined by the so-called 'social factors' – namely, individuals, industries and governments. The issue of safety may thus be seen as a problem of *distributing social resources.* A good example of a technical choice, involving large potential damage but with a low probability of its occurrence, is nuclear power generation (McCormick, 1981).

The willingness to reduce a particular type of danger in one sector by making sacrifices in other sectors, depends on social attitudes, technical conditions and events. Contrary to popular belief, risk should not be reduced to the lowest possible level, but to a level 'As Low As Reasonably Acceptable' (ALARA) in order to spend social resources in ways that are most beneficial to society. Furthermore, since social parameters also influence the *perception* of a given danger, this subjective assessment is pertinent to a society's relative willingness to reduce danger. Numerous strict codes in the field of safety, fire protection and the environment have been passed in the wake of disasters that have shaped, and continue to shape, public opinion.

Both accidents and acute toxicity may cause violent death. But other effects of industrial emissions or products on humans and the environment are more insidious because of their gradual or imperceptible release, chronic exposure and bio-accumulation, as well as long-term effects such as carcinogenicity, mutagenicity, reproductive toxicity and teratogenicity, many of which are still scarcely documented either for humans or for a myriad of other species. Some hints are given in this chapter for the further study of these and other even more challenging aspects of *ecological risk assessment and management.*

The nature of risk

The focus of this section is on industrial hazards and risks, but the methodology elaborated herein can easily be extended to other types of hazard too.

What is called *voluntary or speculative risk* occurs when an action taken may lead to advantages as well as disadvantages; buying shares on the stock market is a typical example of this. Non-speculative risk, on the other hand, is related to a danger. It may lead to an undesirable occurrence (an *accident*) with a certain chance (*probability*) of causing negative consequences (*damage*) to the parties involved, as well as to parties not involved, in the hazardous activity. Risks may be encountered in various social spheres such as industry, traffic, business, household and recreation, some of which are listed in Table 8.1.

Material damage, such as loss of equipment, materials or production is fairly easy to estimate in monetary terms, unlike *personal damage* such as death, injury, temporary or permanent incapacity and mental

Table 8.1 Some technical, social, marketing, juridical and financial hazards

Technical hazards	fire explosion loss of containment → release of flammable, toxic or polluting chemicals air pollutants water pollutants soil contaminants radiation chronic exposure to toxic substances unsafe appliances
Natural hazards	lightning hurricane flooding volcanic eruption earthquake
Human hazards	operating error sabotage vandalism
Social hazards	social consequences of an accident unsatisfactory working conditions poor work relationship poor job motivation stress from job insecurity
Marketing hazards	negative publicity negative consumer attitudes negative effects on the workforce
Juridical hazards	neglect of mandatory procedures damage to third parties
Financial hazards	contractual shortcomings bankruptcy fiscal error fraud theft

strain. For example, what is the monetary value of the life of a girl of 6, a man of 65, or a rich or a poor victim who may be an entrepreneur, an employee, a neighbour, or a chance passer-by?

Estimation of the *probability* of occurrence of a certain event is an important part of risk assessment in industry, as it is in book-making and insurance policies. In fact, risk assessment underpins the entire insurance industry and, consequently, insurance companies have accumulated a wealth of actuarial data on many types of activities and accidents. Typically, Figure 8.1 shows the probability of the natural death of a person as a function of his or her age. Clearly, for children, the probability is lowest in the age ranges of 5–9 and 10–14 years, after which it increases continuously with age. In the same vein Figure 8.2 shows that the probability of accidental death from a traffic accident is highest for young adults (15–19 years) as well as for the elderly. An objective estimate of the probability of a Belgian citizen dying in or after a traffic accident was 198 per million per year in 1990, compared with 225 per million per year in 1981. However, these risks should be compared with those of natural but improbable events, such as being struck by lightning (about 1 in a million killed per year in Belgium) or killed as a result of floods, earthquakes, etc.

Perception of risk (the risk–benefit relation)

The capacity to live with risks is an intrinsic and basic human characteristic. Almost everyone fully accepts the danger inherent in driving a car. The degree of difficulty with which we accept, say, accidental death, is influenced by a host of subjective factors that are very difficult to quantify. A child, struggling for life after having fallen into a well, is likely to raise greater public concern than 200 miners asphyxiated in a mine shaft, even though a much larger number of children may rely on the mining victims for their subsistence. A soldier killed in action, a fatally injured racing driver, a young mother dying from cancer, and a criminal cornered and shot by the police, are examples of four deaths that will be perceived differently by people everywhere.

Most people seem to ignore the multitude of dangers surrounding them, or assign varying degrees of importance to them, and concentrate only on a selected number of hazards which they tend to magnify or exaggerate. Interesting questions in this

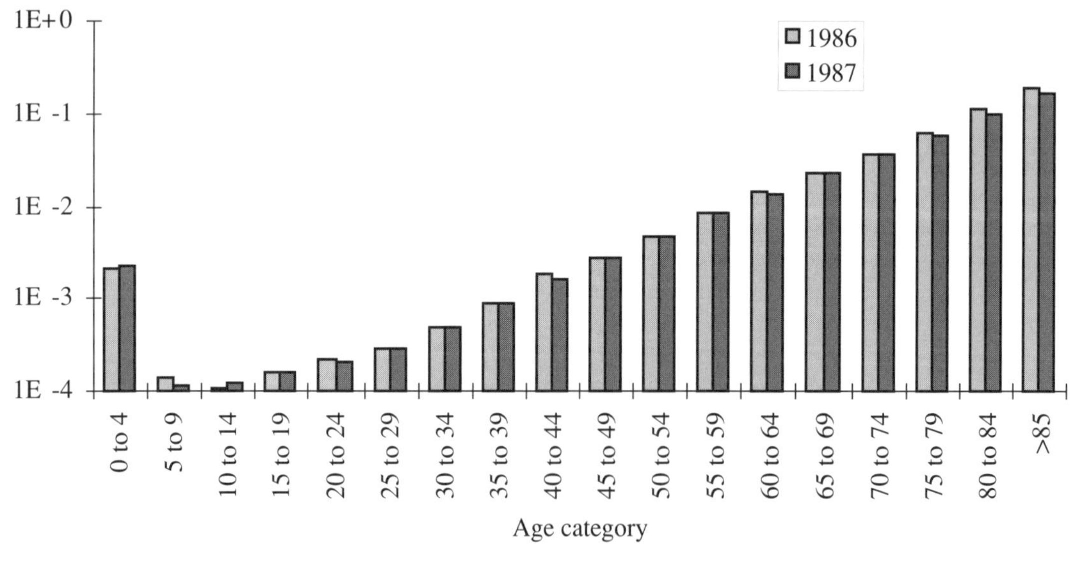

Figure 8.1 The probability of natural death of a person as a function of age in Belgium (logarithmic scale)

Source Cl. Vancayseele, Ieper, Belgium, personal communication

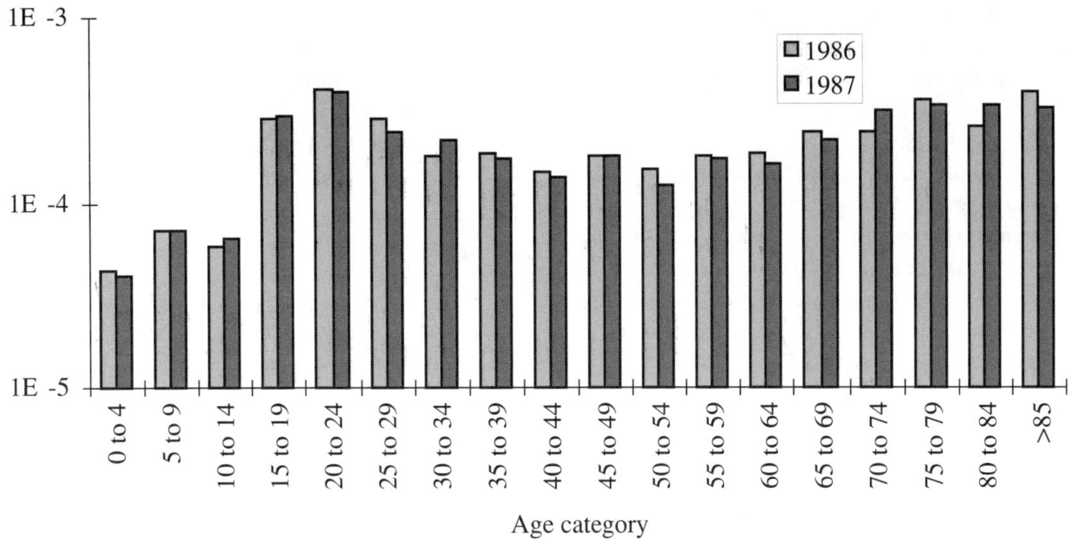

Figure 8.2 The chance of accidental death from a traffic accident in Belgium (logarithmic scale)
Source Cl. Vancayseele, Ieper, Belgium, personal communication

regard are: what are the psychological mechanisms that make objectively tolerable risks seem unbearable? Why are some low-risk situations considered to be unacceptable whereas other, more hazardous (according to objective standards) situations are accepted as normal or inevitable facts of life (e.g. smoking, drinking alcoholic beverages, giving birth to children, racing, mountain climbing)? Some of the factors which influence the perception of risk are:

- Is the risk taken voluntarily?
- Were advantages gained by taking the risk?
- Is the risk avoidable?
- What are the alternatives?
- Were people aware of the risk?
- What is the likelihood of an accident?
- What are the consequences of the hazard?
- Could the hazard and its consequences be perceived by the exposed group?
- Can the consequences be alleviated in the long run or are they irreversible?

Thus these rules-of-thumb would seem to apply:

(a) the higher the perception of risk, the more likely society would be to take measures to reduce its frequency of occurrence and/or the gravity of its consequences;
(b) the less society is capable (technically and/or psychologically) of taking action, the more inclined it would be to legitimise hitherto unbearable situations by reducing or even masking the perception of risk;
(c) the further away an accident occurs, the lower the perception of its magnitude;
(d) the psychological impact of an accident killing a thousand people is greater than that of daily traffic accidents causing one or more deaths;
(e) the more an individual or a society benefits from an action, the higher the level of risk it would accept for that action. It is a general principle that dangerous work should be rewarded appropriately, but a bonus payment is not really proportionate to the risk taken.

Sometimes people prefer not to be informed about a given risk, trusting that all those responsible will do their best to keep it within reasonable limits. Obviously, people accept greater risks by participating in a potentially hazardous activity or situation because they derive some satisfaction by doing so:

(a) in some cases people take risks just for fun, for the exhilaration, excitement and satisfaction they derive from the potentially hazardous activity (e.g. driving a car at 200 km/h, practising martial sports, etc.);

(b) in other cases attitudes are either positively or negatively correlated with the perception of risk, for example:
- *negative correlation*: a brand of cigarettes in a black packet, with a skull and cross-bones vividly printed on it, may sell better than other brands despite the morbid nature of the advertisement it carries;
- *positive correlation*: a driver may enjoy speeding until the moment the car goes into a skid, or until he or she is pulled over by the police.

According to Figure 8.3, people will not participate in a hazardous activity when the perceived risk lies above a certain threshold. To illustrate this point, consider rock climbing for example. People who do not enjoy rock climbing (low benefit) will either not participate in it because they find it too hazardous (group 1), or they will take part without great enthusiasm, only because they are not aware of the hazards (group 3). People in group 2 will persist in rock climbing, even when they are fully aware of the dangers, simply because they enjoy it (high benefit).

On the other hand, even people who accept rational decision-making may experience irrational fear if they have no feeling for a very low probability or relative values. For instance, while the difference between one chance out of two and one chance out of ten can be understood easily, differences between low probabilities, such as one in a million or one in a billion, are not meaningful to most people.

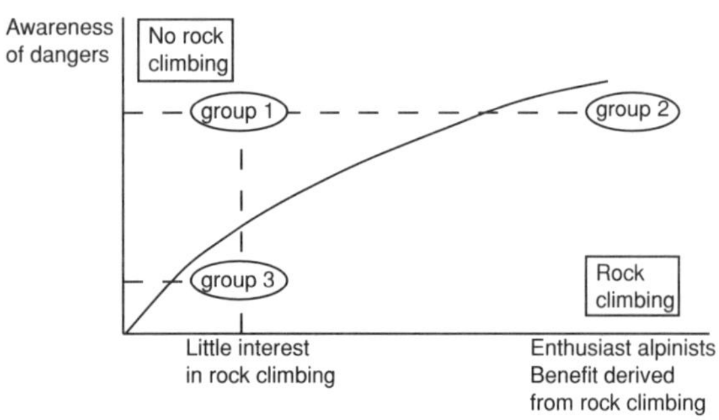

Figure 8.3 Rock climbing: danger/benefit evaluation

Moreover, social acceptance of a hazard is all but static. The quality of the environment is a good example of this, in which the acceptance of risk has declined dramatically.

Definition of risk

Any risk (R) can be expressed as a product of the probability (P) of the occurrence of some undesirable or even disastrous event, and the amount of damage (D) which that event could cause. Thus, for a given accident x, the risk is given by:

$$R(x) = P(x) \cdot D(x).$$

For a group of *n* discrete accidents, the risk is given by:

$$\text{Risk} = \sum P(x_i)\ D(x_i) \qquad i = 1,2.\ldots.n$$

in which x_I = i-th. accident of a given type
$P(x_i)$ = the probability that accident x_i will occur, and
$D(x_i)$ = damage caused by accident x_i.

In some cases accident x may be regarded as a continuous function, with both probability of its occurrence and its effect varying with x. In such a case the risk function is expressed in the integral form as:

$$\text{Risk} = \int_0^\infty P(x)D(x)\ dx$$

The formula $R = P \cdot D$, which accords with common sense, is derived from the concept of mathematical expectation as used in the calculus of probabilities; it is merely an expression of common sense rather than a tool for fundamental investigation. In daily life a situation is perceived as being hazardous only if both P and D are present and recognisable; if one or both is negligibly small, or unheard of, most people will not be inclined to perceive any risk at all.

For example, if a student prepares well for an examination and if the examination is not important, then no significant risk would attach to failing that examination. The risk would be little or insignificant if the student is well prepared and if the examination is important, or if the student does not prepare well and the examination is not important. But the risk will be high if the examination is important and the student does not prepare well for it.

Clearly, the *probability* of everything going well would depend on how well the student prepares. In this case the importance of the examination determines the *damage* that would be caused by failing it.

The perception and interpretation of P and D may differ from person to person, or vary with cultural and circumstantial factors. P is a matter of probability, even though the probability has to be seen as an outcome of the systemic variables. D is a function of more or less interconnected circumstances and consequences that are largely unpredictable and vary both in nature and context, and can therefore be calculated only on the basis of more or less arbitrary (but preferably standardised) assumptions. Also, there is no functional relationship such as:

$$\text{Damage} = f\,(\text{initial event})$$

Thus the expression $R = P \cdot D$ is only a rough approximation in which P and D are influenced by each other. In the case of a high probability (P) (e.g. a car accident), measures will be taken to reduce the damage (D). Conversely, when damage (D) is high and perceived as such, all conceivable measures will be taken to reduce the probability (P) (e.g. an accident in a nuclear plant). There are many well-known examples of such feed-back mechanisms, primarily at the local level (e.g. safety procedures in industry, fire prevention, etc.). At the macro-level, however, the hyperbolic relationship shown in Figure 8.4 may be distorted by unquantified factors, complexity of the subject-matter, the political structure, imperfection of decision-making processes, etc. Typically, the hyperbolic function of Figure 8.4 depicts the human desire to minimise the occurrence of high risks, while accepting low risks when their avoidance is excessively expensive. Since different risks call for different types of control and investment, the curve in Figure 8.4 will never be unique, but only one member within a disparate family of curves.

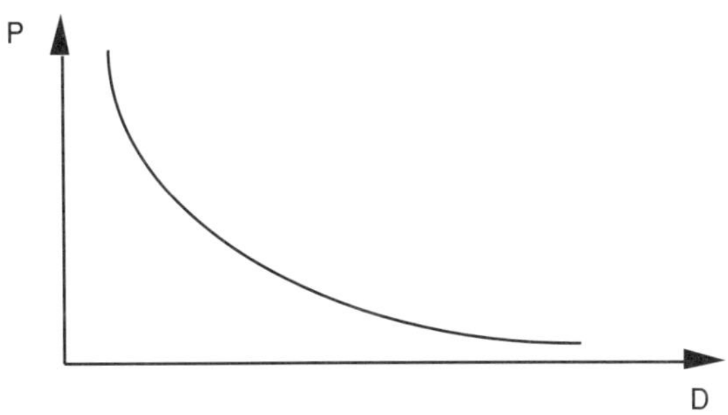

Figure 8.4 Relationship between P and D for a given system: maximum acceptable probability (P) of occurrence of an accident which causes a certain damage (D)

Acceptable risk

The Dutch government has developed legislation in which the concepts of *risk* and *acceptable risk* are clearly defined. The death of a person, or of a group of people simultaneously, is an event of primary importance in these definitions of *risk*, *damage*, and *chance*. Other damage, such as material or psychological damage, is not taken into consideration at present in the risk criteria.

Some key definitions in this context are:

(a) *individual risk* is defined as the probability that a person will be subjected to a certain negative effect (death, injury, burns, etc.) resulting from exposure to an agent (explosion, fire, toxic gas cloud, etc.). Individual risk is calculated for a person subjected to it in a defined area. In practice this refers to a person who is exposed to hazardous effects or activities 24 hours each day;

(b) *group risk* is defined as the annual probability that an entire group of a given size will fall victim to an accident. Group risk is used for evaluating the probability that people outside the site will be killed as a result of an accident occurring at the site. The calculation of group risk is based on the probability of survival of a single individual and the number of persons present at the time within a given risk area. The result is expressed in terms of the number of deaths per accident, and not in terms of the probability of individual survival or death;

(c) *maximum acceptable risk* is defined as the level of risk above which the activity responsible for the risk must either be forbidden or restricted;

(d) *negligible risk level* is usually set at 1 per cent of the maximum acceptable risk. The difference between the maximum acceptable level and the negligible level must be sufficiently large so that the consequence of multiple exposures to a negligible risk, as well as the uncertainties in its evaluation, does not have a significant effect on the comparison.

For serious accidents, the following values for the above limits have been set by the Dutch government:

(a) For *individual risk* the limit is for a single person to die as a result of exposure to an agent:

> The maximum acceptable risk corresponds to a probability of 10^{-6} per year.
> Negligible risk corresponds to a probability of 10^{-8} per year.

(b) There are also maximum levels for *cumulative risk*, defined as a risk which is linked with several different activities. Cumulative individual risk, which is the sum of risks attaching to all activities resulting in a fatality, leads to the following risk levels:

> Maximum acceptable risk is 10^{-5} per year.
> Negligible risk level is 10^{-7} per year.

(c) *Group risk* is considered to be less acceptable. If the number of casualties (D) is multiplied by factor n, then the probability of occurrence (P) must be divided by n^2, and consequently:

The maximum acceptable risk is:	10^{-5} per year for $n = 10$
	10^{-7} per year for $n \geq 100$
The negligible risk level is:	10^{-7} per year for $n = 10$
	10^{-9} per year for $n \geq 100$

All these limits for individuals, groups and cumulative risks must be adhered to. Industrial safety can then be ascertained by using suitable quantitative risk models. If the estimated risk is found to be above the limit, the activity should be forbidden.

Diffuse risk: Ecological Risk Assessment

In Dutch legislation risk is defined on the basis of probability, or the number of casualties likely to be caused by a disaster. Some industrial risks don't involve accidents, but result from continuous exposure either to polluted air, water or soil, or to toxic, carcinogenic, mutagenic, teratogenic agents (whether anthropogenic or naturally occurring) in food, drinking water or air. Inventories of emission sources (flue gas, particulates, genetically modified organisms, radioactive emissions, leachate, noise, etc.) are being prepared, and immission values (ambient concentration levels) are being monitored routinely. Under adverse circumstances, such as atmospheric inversion or accidental contamination, specific measures can be taken to control the level of exposure of the population by legal, administrative and technical means that are comparable to those used in the case of a disaster or a catastrophe.

Risk assessment is also making substantial progress in these cases, although a number of specific problems remain, notably the following:

- it is much easier to establish the cause of violent death than to assess the risk of developing malignant tumours after a dormant period which can often last 20 years or more;
- the effect of exposure on humans varies considerably, as do their individual lifestyles;
- synergistic and antagonistic effects are still very poorly documented.

The consequences of, or damage from, an emission can be assessed by means of epidemiological methods, or in the context of an environmental impact assessment.

Ecological Risk Assessment (ERA) is currently evolving as a discipline in its own right, graduating from the assessment of adverse impacts of new chemicals or old contaminants on single species, to one of forecasting and evaluating the impact of multiple chemicals, and physical and biological stressors, on an entire aquatic or terrestrial ecosystem which can be small-scale, an entire watershed, or even an entire geographical region. The phases of the ERA are given in Box 8.1.

Ecological risk assessment follows a classical methodology to identify pathways and mechanisms

BOX 8.1 ECOLOGICAL RISK ASSESSMENT (ERA): A TOOL FOR DECISION-MAKING

Ecological Risk Assessment (ERA) is useful because it provides a framework giving information needed to make management decisions. ERAs help to identify existing problems, anticipate the risks of planned actions, establish regulatory programmes and research priorities, and provide a scientific basis for regulatory actions. Most importantly, ERAs provide the basis for prioritising management and regulatory efforts, because objective comparisons of a variety of situations can be made. Also, the use of risk as a common denominator allows the decision-maker to select the most effective management actions. Following the implementation of management actions, ERAs can also help to measure the effectiveness of those actions.

Before any data are collected, or any research undertaken, the researcher must consider how the ERA will be carried out and what specific questions it must answer. The process of conducting an ERA requires a careful identification of the assessment objectives (environmental values to be protected), deciding on the appropriate scale and level of biological organisation, assembling multidisciplinary data collection and assessment teams, rigorously interpreting results using both quantitative and qualitative methods, and communicating the results in a manner that facilitates risk management. The components of a framework for ERA are illustrated in the figure below. This risk assessment framework is intended to facilitate the integration of human health and ecological risks of toxic chemicals and other environmental stresses. As far as the USA is concerned, it was developed by combining the essential features of the National Academy of Sciences/National Research Council framework for health risk assessment, as well as those of the US Environmental Protection Agency framework for ecological risk assessment. This framework comprises five phases: (1) hazard identification; (2) exposure assessment; (3) exposure–response assessment; (4) risk characterisation; and (5) risk management, as described below.

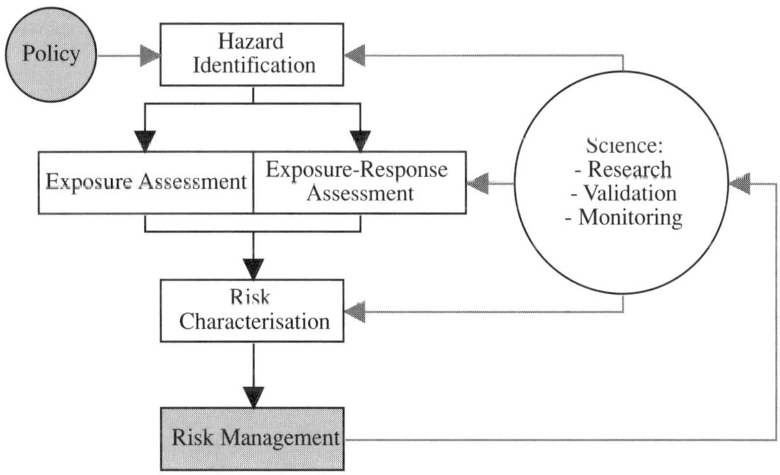

Risk assessment framework which integrates human risk and the ecological risk of chemicals

1 *Hazard identification*

This phase determines whether a particular danger exists, if the effects associated with the hazard are sufficiently significant to warrant further study or immediate management action, and the kinds of data required to determine the level of risk. This systematic planning phase establishes the goals and focus of the assessment, and identifies the major factors to be considered. A major factor to be considered in this phase is the selection of ecologically based end-points relevant to the ultimate decisions to be made. Information used for hazard identification includes short-term or screening toxicity tests, and reviews of existing information which characterise the potentially affected ecosystems and the contaminants in question.

2 *Exposure assessment*

This phase is concerned with the determination of exposure to the hazardous agent in question. The process includes the measurement or prediction of movement, fate, and partitioning of chemicals in the environment. This

→

phase is typically accomplished through the chemical analysis of site media or ecological receptors and/or mathematical modelling.

3 *Exposure–response assessment*
The process in this phase is concerned with the determination of the relation between the magnitude of exposure and the probability of occurrence of the expected effects. Information useful in this phase includes toxicity data (chronic toxicity, mode of action, sensitivity of particular species), mesocosm or field test data, field surveys comparing exposed and unexposed sites, and population or ecosystem modelling.

4 *Risk characterisation*
This phase involves describing the nature and magnitude of risks, including the inherent uncertainties, expressed in terms understandable to decision-makers and the public. Information from the previous phases is integrated in this phase and communicated to the decision-makers in a manner relevant to the decisions being made. Because the purpose of risk assessment is to support decision-making, communication with decision-makers is a critical step in an environmental risk assessment.

5 *Risk management*
This process is the one in which decisions are made about whether an assessed risk needs to be managed, and the means for accomplishing it, for the protection of public health and environmental resources. Managing risks involves making decisions based on the information collected in the previous phases of the risk assessment, along with a consideration of the social and cultural values, economic realities, and political factors. Monitoring and research are used to provide the data necessary to conduct an ERA. Important aspects to consider when planning monitoring for a risk assessment include the assumptions involved in determining which environmental values are to be protected by the risk management decisions; which levels of biological organisation must be studied to obtain the data necessary to answer the pertinent questions; and the temporal and spatial scales relevant to the study.

Source: Linthurst *et al.*, 1995

of exposure to agents with a detrimental impact, such as mortality in 10 per cent of the exposed population, or a reduction in reproductive success, etc.

Efforts are now being made to develop tools, procedures and norms required to establish ERA as a formal discipline. But, since the ERA is a new and complex discipline, and because it relies on professional experience and judgement, it is generating considerable controversy at present (Renner, 1996; Suter, 1993; EPA, 1992, 1995).

Uncertainties

The study of uncertainty normally involves the analysis of complex systems consisting of numerous parts, events and decisions. The evaluation of the probability of a given chain of events, or of their combinations, leading to an undesirable event is based on Boolean algebra.

Rational management of safety can only be pursued when there is objective information as well as the willingness to deal with it. The term *objective information* implies that people responsible for providing the information have no personal interest in any decision that may be based on it. But, since there is no guarantee that the objective information provided will also be precise and accurate, any reasoning or calculation derived from it will not necessarily lead to optimal or well-founded decisions. Uncertainties are indeed numerous: data on the probability that an entire system, or a certain part of it, will malfunction are very imprecise; damage prediction models are still under development; the level of *acceptable risk* is not yet generally agreed upon and varies according to circumstances and public perception; and the *weight* to be assigned to a given safety measure in a global strategy has to be resolved in the context of a complex resource allocation problem.

THE HUMAN FACTOR

Humans as an important factor of risk

According to Heinrich (1931), 88 per cent of all accidents can be attributed to unsafe human behaviour, and the remaining 12 per cent to unsafe conditions

– 10 per cent to purely technical defects and only 2 per cent to the disregard of design conditions. A modern variant of this concept is reflected in ongoing publicity campaigns aimed at persuading workers to use their personal protection equipment.

However, Heinrich's theory has been challenged on the ground that most accidents in the workplace result from a combination of several forms of negligence and error, and that the characteristics of the situation also play an important part in promoting accidents.

In the 1970s *total safety* was believed to be a realistic goal, which could be pursued with technical progress on the one hand and systems analysis on the other. The underlying thesis was that human behaviour is predictable and can be controlled, more or less like any other part of the system. Accordingly, emphasis was placed on automatic system control equipped with foolproof measurement, and on control and alarm systems. These automated production systems were expected to make production processes almost totally reliable and independent of human judgement. So the need for human intervention, it was thought, would become the exception rather than the rule.

In the meantime, the limitations of this theory have been revealed. For example, it is thought that some recent major aeroplane accidents, especially during take-off and landing, can be attributed to the excessive automation of measurement and control systems. The crash of an A320 Airbus at Mont-Saint-Odile on 20 January 1992, in which 87 people were killed, is cited as a typical example of this. Pilots no longer *feel* the plane because they have nothing to do under normal flying conditions. Thus they lack familiarity with the way the plane would react to their commands, lose their touch, and become less attentive and motivated as a result.

It can be argued, therefore, that treating human beings as predictable and controllable pieces of mechanical 'equipment' is a restricting and peculiar concept which may even cause accidents. Furthermore, human behaviour under stress or abnormal conditions is still fairly unpredictable, and this makes the implementation of safety procedures in hazardous plants a very complex exercise. Given that for every case there is probably an optimal degree of automation, a careful consideration of the 'human factor' is thus paramount in the technical management of complex and potentially dangerous plants. In such plants, such as nuclear power plants that are highly automated, every attempt is therefore made to minimise the probability of malfunction due to the human factor by:

1 offering regular training to staff on simulated reactor accidents;
2 keeping their knowledge at a high level with regular courses and training;
3 stimulating their attention with frequent reminders of the disastrous consequences of an accident.

Today similar procedures are also in place in large chemical plants in which safety management policies are developed and monitored by top management. Development of personnel, who are highly aware of safety issues and well trained to deal with process upsets as well as normal operation, is a key objective of safety management. The current trend in legislation is to ensure that this responsibility rests with industry.

Human reliability and human error

A quantitative approach to human reliability has been derived from empirical data relating to the chances of errors people make in performing simple operations. More complex tasks are then composed from these simple operations, and the probability of making errors in them are calculated. Swain and Guttman (1980) have compiled tables giving the probability data on such failures. But the validity of such data is questionable, especially when they are subsequently introduced into complex systems, or extrapolated to conditions of emergency, distraction, or apathy. The data in Table 8.2 are interesting to note in this context.

Table 8.2 Probability of occurrence of some common human failures

	Probability per command
Operator does not observe a signal	10^{-4}
Operator does not notice an alarm	3×10^{-4}
Operator fails to take action	3×10^{-4}

The most frequent human errors are caused by (Roos, 1985):

1 Disregarding instructions because:
   - the instructions are wrong or unclear (35 per cent);
   - the operators do not know or understand the instructions sufficiently well (30 per cent);
   - nonchalance, most often due to fatigue, stress or lack of time (33 per cent).
2 Reduced attention, particularly in jobs involving vigilance, when:
   - the hours of work are too long;
   - incidents which occur infrequently have to be detected;
   - signals are difficult to differentiate from background noise, or when a specific noise must be identified.
3 Incorrect diagnosis:
   - the less frequently a phenomenon occurs, the less experienced the operator is at diagnosing it. This is, of course, especially true when very serious incidents occur. Operators may not be trained for a proper reaction in the case of a catastrophe.

Some of the factors that are reputed to hamper a task, or which act as stress-inducers, are given in Table 8.3 (Van de Kerckhove, 1990).

Table 8.3 Psychological and physiological stress-inducing factors

Psychological stressors
Sudden events
Over-supply of information
Threats
Events occurring in a rapid sequence
Distraction
Monotonous and mind-numbing work
Physiological stressors
Lack of sleep, fatigue
Pain
Thirst
Work stress
Surroundings
Climate

Humans and their environment

Human behaviour and human reliability cannot be isolated from the complex environment in which humans may be, and often are, both actors and victims at the same time.

Sources of accident can be classified into five groups of factors: human, mechanical, environmental, material and managerial (HMEMM). Table 8.4 shows the possible causes of an accident which can occur from the use of a ladder, corresponding to the HMEMM factors.

In general, reference to *human reliability* or *human factors* implies that the components of a more complex human–machine–environment system must still be considered independently, and that insufficient attention has so far been given to the relevant processes and their interactions. Clearly, an efficient and safe human–machine–environment system must be organised according not only to the logic and rationality of techniques but also to the existing measures and the physical and intellectual possibilities of humans. Nowadays managers and system planners no longer try to convert humans into pieces of technical equipment, but, on the contrary, make efforts to humanise technical systems. Indeed, people are neither robots nor biological machines, and each person has unique and specific limits and capacities. This implies that humans:

- have limitations and, to a certain extent, will always be prone to failure;
- are creative and able to discover problems and solve them;
- are autonomous, with their own individual needs, objectives, hopes and desires which they pursue at all times, consciously or subconsciously.

Table 8.4 Analysis of an accident occurring with a ladder

Source	*Shortcoming*
Human	Did not set up the ladder properly
Mechanical	The ladder had no anti-slip devices
Environmental	The floor was slippery
Material	Polish was on the floor
Managerial	The employee was pushed to work too fast

In order to involve people in complex systems, it is clear therefore that all these peculiarities of human nature must be taken into consideration. Furthermore, it is essential to acknowledge that human behaviour will never be predictable or computable, because it varies with personality, training, motivation, and distraction by external factors (radio, visitors, personal problems, etc.). Of course this does not mean that people cannot be relied upon, but merely that factors such as work stress, training, motivation and personal problems should be acknowledged and given due consideration by management.

Trends in human reliability assessment

Much progress has been made in equipping technical facilities with internal safety systems, and technical errors are increasingly less likely to occur as a result. Consequently, human error is becoming increasingly more important in the residual risk. It is thus something of a paradox that people look for technical solutions to problems that are more likely to have been caused by human error. In part, at least, this may be because a clear distinction between human, technical and managerial mistakes is difficult to make.

Three major trends can be recognised in the assessment of human reliability:

1 *Technical evolution.* This has profoundly changed the duties and behaviour of plant operators. Routine activities have largely been taken over by computer-controlled technical systems. As a result, operator activity has shifted towards the supervision of systems, which increasingly requires an in-depth knowledge of system structures, and their possible reactions and interactions. Also, the more complex the job-specification of the operator, the more difficult it becomes to analyse, structure and model his or her behaviour.
2 *Psychological analysis.* The behaviourist theories were abandoned in the 1960s in favour of cognitive psychology, with the recognition that human behaviour is more than just a stimulus–response relationship. Emphasis is now on a well thought out and meaningful relationship between people and the environment. People act autonomously and plan actively for achieving their personal goals, and they do not just react passively to external stimuli. Employees can no longer be regarded as an extension or part of the equipment.
3 *Evolution in management and theory of organisation.* Earlier theories held a rather mechanistic view of what constitutes a successful organisation. The prevailing scientific view of management was one in which the workers acted as 'cogs' within a more complex machine (organisation). Requirements, aims and expectations of the workforce were largely ignored. In contrast, current thinking is that successful management should be based on the concord of the organisation's objectives and workers' aspirations. This calls for a permanent dialogue between the workforce and the management for maintaining harmony, and for promoting attitudes conducive to the achievement and maintenance of the highest possible level of safety in the workplace.

The 'integrated safety' approach

The accident with the ladder of Table 8.4 occurred as a result of at least five quasi-independent factors – namely, the worker, ladder, floor, polish and workload. But possibly the true initiating event of the accident should be sought further upstream. To identify the culprit, further questions should be asked, such as:

- Why was an unsafe ladder available for use?
- Was it inspected and approved?
- Did the supervisor allow the use of that ladder?
- Was the worker well trained, experienced and familiar with all the safety codes?
- Was the worker motivated to apply all the safety rules?

It follows from the above that safety should be integrated into all other activities within industry, and that it should not be perceived as a physical and financial burden or an optional extra but as a long-term strategy towards saving materials, equipment and possibly even human lives.

The concept of integrated safety should be periodically reviewed by management, and responsibility for safety and the environment must rest with the top level of management. A permanent dialogue

Table 8.5 The 10-Point Code of Professional Practice on Risk Issues

1 Professional responsibility	Exercise reasonable professional skill and care
2 Law	Know about and comply with the law
3 Conduct	Act in accordance with the codes of conduct
4 Approach	Take a systematic approach to risk issues
5 Judgement	Use professional judgement and experience
6 Communication	Communicate within your organisation
7 Management	Contribute effectively to corporate risk management
8 Evaluation	Assess the risk implications of alternatives
9 Professional development	Keep up-to-date by seeking education and training
10 Public awareness	Encourage public understanding of risk issues

Source: Reproduced by courtesy of the Engineering Council

should be established to promote a positive attitude to safety for maintaining good working relations (see the 10-Point Code of Professional Practice on Risk Issues, given in Table 8.5).

METHODS OF RISK ANALYSIS

Survey

Risk assessment is still a young and rapidly evolving science. The first quantitative study on reliability and risk was published by the Aeronautical Research Council of London, which in 1939 and 1942 issued reports recommending measures to lower the frequency of air crashes to below a threshold value of 11×10^{-5}/h, taking into account all conceivable causes of accidents (Olivier and Van Overmeire, 1990). The probability of a crash caused by the structural failure of a plane had to remain below 10^{-7}/h, which is less than 0.1 per cent of all other causes combined. In 1960 the first *fault tree analysis* (FTA) was developed by a telephone company at the request of the US Air Force. In the FTA an accident is broken down into its basic causes which are then depicted by means of diagrams for identifying the logical interrelationships between them. From the diagram it is possible to generate a list of the various combinations of equipment and human failures that are sufficient to cause the accident. A major strength of the method lies in the systematic and logical development of the many possible equipment failures and human errors that might result in an accident (AIChE/CCPS, 1985).

The longest experience with reliability and risk has been acquired in the field of labour protection. Views on the role of humans in their work environment have evolved, and continue to do so, and so have the techniques of risk analysis. The latter have also been extended to other domains such as quality assurance, efficiency improvement and environmental protection.

The three major tasks of risk assessment are (also see Table 8.6):

1 *identification* of hazards, with the aim of developing and completing the concept of safety (AIChE/CCPS, 1985);
2 evaluation of hazards according to the respective *probability* of their occurrence with a view to optimising safety systems, with particular reference to availability and reliability (AIChE/CCPS, 1989);
3 evaluating the *consequences* of hazards, with the aim of minimising hazard potential and adopting optimum preventive, remedial and protection measures.

Qualitative analysis

Principles

Qualitative methods have now been developed to discover and identify hazards, and earlier methods have been adapted to make them more powerful. But none of these offers a guarantee that it will unearth all possible errors or malfunctions under all conceivable circumstances. It is therefore very important to develop a methodology which permits the location and identification of all causes that may lead to damage on the one hand and an estimation of the extent of the damage on the other. The methodology must be applicable in the different stages of the definition and development of both technical (preliminary and final design, construction and

Table 8.6 Schematic classification of risk-assessment methods

Application	*Use*	*Basic principles*	*Methods*
1 Identification of hazards	Completing safety concepts	Systematic and organised mental search using questionnaires, keywords, etc.	Checklists Error analysis Operating error analysis
2 Hazard control by reducing the probability of their occurrence	Design optimisation to enhance reliability and availability	Representation of cause–effect relationships of interactions between system components, including chance estimation	System analysis Fault tree analysis Event tree analysis
3 Controlling hazards by limiting the consequences of accidents	Design optimisation to reduce the potential consequences of undesirable events	Mathematical modelling of physical phenomena, e.g. dispersion of flammable or toxic clouds	Effect and damage models

Source: Pitblado and Turney, 1996

operation) and non-technical systems. A procedure for the application of risk management in this context is given in Figure 8.5.

It is helpful to identify the possible *initiating event* which triggers an accident, the *intermediate events* in the accident event sequence, and the consequences of each hazardous accident. Some of the procedures used for hazard identification are also useful for this purpose. Other procedures, such as fault tree analysis (FTA) and the event tree analysis (ETA), can be used to identify the intermediate events. These methods are discussed in detail in the specialised literature (see 'Suggested Reading', p. 167).

The principal methods are based on:

1 The *process/system checklist* which specifies those aspects of a plant which require attention for safe design. They address known hazards derived from industrial codes, regulations, past accidents and judgement, but they are not effective for novel technologies or new hazards. Checklists are also useful during:
2 Safety review.
3 The relative ranking of hazardous plants can be based on the 'Dow and Mond Hazard Indices' (see 'Suggested Reading', p.167).
4 Preliminary hazard analysis which considers the adverse consequences of unexpected events revealed during a creative, brainstorming examination of a process or procedure in which questions such as the following may be raised:

- what if the raw material is of the wrong concentration?
- what if the cooling ceases?

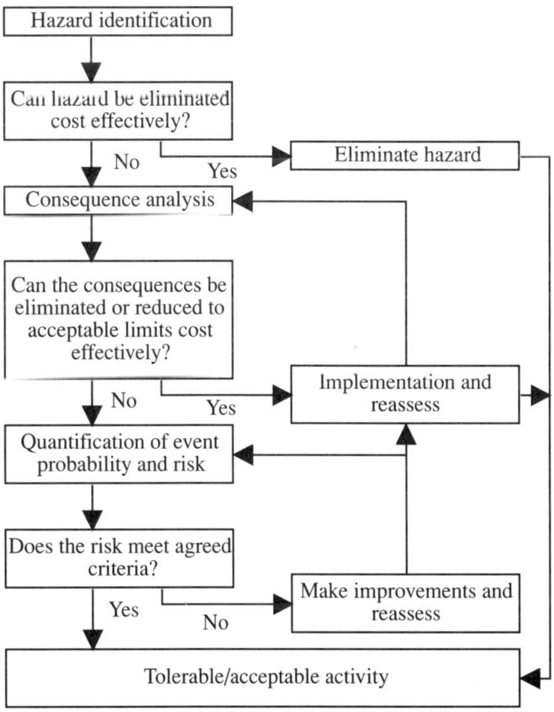

Figure 8.5 Procedure for the application of risk assessment
Source: Pitblado and Turney, 1996

Table 8.7 A list of guide words

Guide words	*Meaning*	*Comments*
NO or NOT	The complete negation of intentions	No part of the intentions is achieved, but nothing else happens
MORE	Quantitative increase	These refer to quantities and properties such as flow rates and temperature, as well as to activities such as 'HEAT' and 'REACT'
LESS	Quantitative decrease	
AS WELL AS	A qualitative increase	All the design and operating intentions are achieved, together with some additional activity
PART OF	A qualitative decrease	Only some of the intentions are achieved, and some are not
REVERSE	The logical opposite of the intention	This applies mostly to activities such as reverse flow or chemical reaction. It can also be applied to a substance such as 'POISON' instead of 'ANTIDOTE'
OTHER THAN	Complete substitution	No part of the original intention is achieved. Something quite different happens

Source: Pitblado and Turney, 1996

- what if the vessel agitation stops?

5 The *what if* analysis.
6 The hazard and operability study (Hazop), in which a set of 'guide words' such as *no, more, less* or *reverse* are methodically used to generate possibly unforeseen deviations from the designer's intention. It is the most widely used technique applied to batch or continuous and new or existing processes, plants, procedures and projects. A list of 'guide words' is given in Table 8.7, together with their meaning.

A useful method of identifying hazards is based on the systematic questioning of internal procedures and organisation, including the views of, and approach to, the 'inner circle' of management with respect to:

- the behaviour of employees and operators;
- the causes of the accident;
- the managerial implications of the accident, and
- the role and limitations of the technical plant in question.

The relevant findings can then be placed in the well-known human–mechanical–environmental–material–managerial (HMEMM) context.

It is also very important that the results of each analysis are correctly interpreted. For this it is necessary first to be aware of, and duly consider, the limitations of each of the methods. An advantage, which is common to almost all the methods, is that they initiate a process of reflection on hazards, their nature, origins and potential effects. During this process the analyst becomes aware of all kinds of potential hazards, including at least a fraction of those that would normally be discovered only after an accident.

Application of qualitative analysis

During the design, construction and operational stages of a plant, different priorities should be given to the management of activities in these consecutive stages. The same also holds for safety, as well as for environmental assessments that are often related to it.

Clearly, the earlier hazards can be detected, identified and evaluated, the easier it would be to adapt a project to necessary safety requirements. In practice, however, much of the more detailed information needed for this becomes gradually available only as the project takes shape. For example, the chances of a valve, a flange, or a gasket malfunctioning can be evaluated only after a decision has been made to use them in the first place and their technical specifications have been stipulated. Table 8.8 lists the factors relating to risk and their investigation during the successive stages of developing a chemical plant.

In a given project it is not always clear as to which

Table 8.8 Qualitative analysis of successive stages in plant development

R & D work (research and development)
Safety in the storage and use of hazardous, toxic, flammable and irritating raw materials, intermediates, products, catalysts and utilities (risks caused by material properties).
Study of reaction kinetics and equilibria with special attention to dangerous side-reactions, such as the formation of unstable compounds.
Experimental or computational assessment of potential run-away conditions (risks caused by reactions – reaction matrix – operating conditions).

Process
Establishing a list of processes, plants and machinery required in the project (risks connected with the layout chosen and with process implementation).
Definition of the required level of technical safety management – analysis of potential consequences (release of gases, vapours or boiling liquids, dispersion of gas clouds, explosion, fire, toxicity, estimation of *worst case* consequences).

P & ID (process and instrumentation diagram)
Measurement and monitoring of temperature, pressure, flow, level, etc. in all equipment (risks associated with excessive temperature, pressure, high or low flows or levels, etc.).
Technical and material specification of reactors, vessels, columns (risks associated with design decisions).

Construction
Construction and assembly strictly according to plans (no deletion of safety or instrumental features due to budget restrictions).
Site organisation and inspection, collective and individual safety precautions, etc.

Start-up/shut-down
Risks associated with non steady-state conditions.

Operation
Organisation of training, inspection, preventive maintenance, calibration, quality control and emission monitoring.

Plant de-bottlenecking and modification
Ensure that all plans are regularly up-dated.
Review all procedures.

Source: Pitblado and Turney, 1996

of the aforementioned hazard identification methods can or should be used. The best approach is therefore to subdivide the project, and then to choose and apply the method that is most suitable to the nature and status of each sub-project with regard to available information. Also, the results of previous stages of risk analysis can be used as input for subsequent and probably more detailed risk analyses, possibly involving different methods.

As the hazard evaluation process proceeds, it may become necessary to set up a specific safety system, or a particular modification to the process, in order to achieve the desired level of risk. But termination of *hazard evaluation* at this point would be premature and can even be dangerous. At the very least the hazard identification process should be completed, since elimination of the risk of the first identified hazard will not necessarily reduce the risk of the remaining hazards which may still exist. Also, a 'worst case scenario' estimate of the consequences of accidents involving the identified hazards should be prepared, both prior to and after the changes have been made, in order to reduce their risks. Finally, since a change, which reduces the risk of one particular process hazard, may actually increase the risk of another hazard or introduce a new danger, interactions of this type must be considered before the hazard evaluation is terminated.

Risk analysis is virtually incapable of dealing with wilful or criminal action such as sabotage, disregard of compulsory procedures, or suicidal interventions. The only way to deal with these delicate problems is through an analysis of what is called 'worst case hypotheses'. It should be mentioned in this context that a sizeable fraction of all fire accidents in industry are caused by malicious acts.

Quantitative analysis

Principles

A quantitative analysis of risk must answer two specific questions:

1 What is the probability (P) of a particular accident occurring?
2 What is the damage (D) which the accident would cause?

Possible sources of knowledge for responding to these questions are personal experience, consulting someone with relevant experience, similarity of the hazard in question to one that happened in the past, codes and practice, and databanks.

Probability data should first be derived or calculated on the basis of real accidents, which have occurred in the same type of industry worldwide, and on information on near-misses or failures in the plant in question. Accident data are available in public databases such as MHIDAS (AEA, UK), FACTS (TNO, the Netherlands), Loss Prevention Bulletin Index (IChemE, UK), etc. (also see AMINAL (1994), Belgium, CEC Major Accident Reporting Systems). What are called 'effect and damage models' can be used to calculate the damage. These are mathematical models for evaluating physical phenomena.

When no historical data are available on the process in question, the following methods can be used to evaluate the probability of an accident.

1 *Fault Tree Analysis* (FTA) is a graphical representation of a variety of combinations of failures which produce an undesirable *top event* (an explosion in a reactor for example). FTA is a process of deductive reasoning which attempts to identify the root causes of top events, as well as alternative outcomes to help identify further top events.
2 *Event Tree Analysis* (ETA) is an inductive process of evaluating the potential for an accident as a result of equipment failure or a process upset, called the *initiating event*. Starting from the initiating event, the possible sequences leading to a potential accident are analysed. A typical example is given in Figure 8.6.
3 *Cause–Effect Analysis* is a method which combines the ability of fault trees to show how the various factors may act together to cause a hazardous event indicating the possible consequences. The symbolism used is similar to that of fault trees, and sequences and time delays can be indicated in the ETA part of the analysis. The method is suitable for quantitative study, but it is used less frequently than single fault or event trees (Pitblado and Turney, 1996).
4 *Human Error Analysis*: Human error is often an important factor in the logic chain leading to an undesirable event. Identification of the role of human error provides insight into the process of failure. But data on the probability of human error quoted in the literature, though based on a number of studies, are still arbitrary and uncertain (Pitblado and Turney, 1996).

The problem of the human element is primarily one of identification. But, once identified, the chance of the error occurring is reduced. Moreover, some types of human error are implicitly catered for in the reliability of components by making them undergo a series of human interventions including design, installation, inspection, calibration and maintenance.

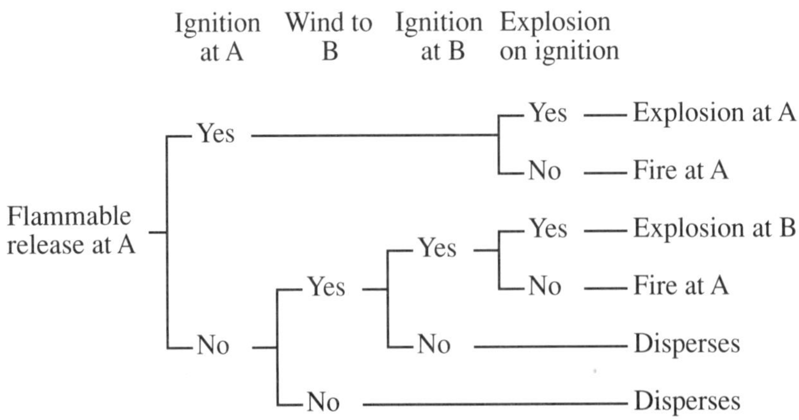

Figure 8.6 Example of an event tree
Source: Pitblado and Turney, 1996

Other types of human error, for example operational error such as failing to carry out a certain task or performing the wrong task, are considered in accident data (AIChE/CCPS, 1989; Lees, 1980; SRD Association, 1995; Kirwan and Ainsworth, 1992).

A general description of these methods will be found in the literature (AIChE/CCPS, 1989; Pitblado and Turney, 1996).

Damage evaluation models

There are many possible kinds of accidents, and a multitude of consequences which flow from them, shown schematically in Figure 8.7. Of these we will only consider three main types here:

1 emission of toxic compounds leading to acute poisoning;

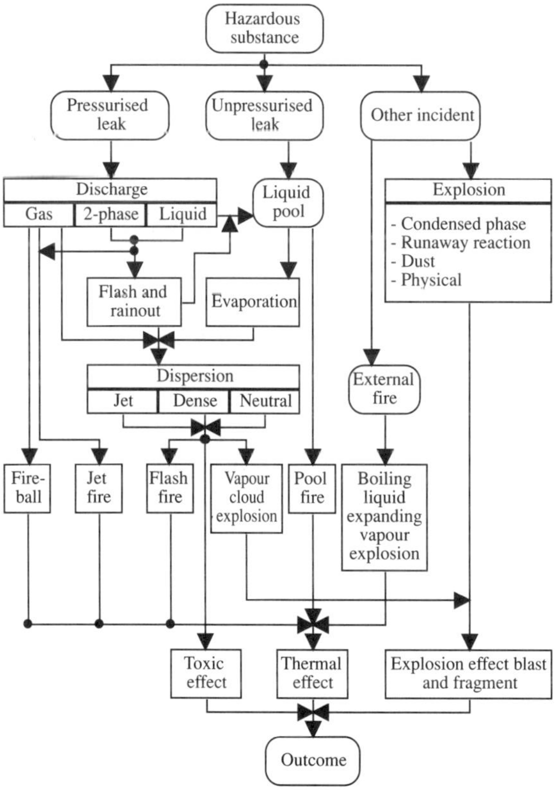

Figure 8.7 Logic diagram for consequence analysis
Source: Pitblado and Turney, 1996

2 explosion, including flying fragments and effects of pressure blast on personnel and buildings;
3 fire, and exposure to heat radiation.

The generation or release of toxic chemicals can be lethal if inhaled, ingested or absorbed through the skin. The case of a toxic gas cloud, moving with the wind and being dispersed by it, is well documented in the literature. In this case the chances of a victim's survival can be assessed with regard to the toxicity of the compound, the degree and duration of exposure, and the rhythm and volume of inhalation.

This methodology was directly applied in the Bhopal disaster in India. It is thought that there would have been fewer casualties if the exposed victims had fled upwind, even through the toxic cloud, rather than downwind which lethally prolonged the exposure of their lungs to the toxic gas.

Explosion, which is the second type of accident, includes events such as detonation and deflagration in which death can be caused by the impact of a mechanical blast, or by flying debris. Empirical data show which kinds of impact (over-pressure, mass, velocity) a building or a victim is likely to survive, and when the eardrum will be damaged.

Explosions and fires are amenable to the following reasoning:

1 Every human being present within a fireball, or in a strong fire, will perish.
2 Outside the fire the victim will be exposed to heat radiation. His or her chances of survival will be determined by the prevailing heat flux, the surface area of the body irradiated, age of the victim, and, obviously, whether his or her clothes catch fire.

Attempts have been made to model these phenomena and to quantify their statistical effects on humans as a function of the various operating parameters such as distance from the place of the accident, wind direction and velocity, presence of physical obstacles, position of the body, duration of exposure, shielding, etc.

Sometimes several of these phenomena may act together. Typically, toxic compounds may be released by an explosion, or generated by the combustion of other materials, or a fire may result from an explosion and vice versa. This illustrates the difficulty of developing a model which is both computable and realistic. The so-called TNO 'Yellow Book' (TNO, 1988) lists the best available models to describe such situations, pertinent data for which include the following:

- physical and chemical properties of the materials in the process;
- toxicological and biological properties (biodegradability, bio-accumulation, persistence, etc.);
- reaction parameters (enthalpy, onset of exothermal reactions, etc.);
- thermal stability, including the properties of decomposition products;
- reactivity of process chemicals with materials of construction;
- flammability and explosion limits for fuel/oxidant/inert gas mixtures, including dust.

Structured procedures have been developed to ensure that data on the above, and other information necessary for an effective study, are collected and reviewed from the earliest stage of a project.

Limitations of quantitative analysis

The objective of predictive hazard evaluation is to identify and evaluate the hazards and the unintended events which, if associated with a hazard, could cause an accident. Predictive hazard evaluation procedures have been developed for the analysis of certain processes, systems and operations that are so far removed from real situations that adhering to established codes of good practice may not be adequate. Nevertheless, they can be used for evaluating an accident of which there is little or no experience, and the probability of whose occurrence is very low, but the consequences can be very serious if it occurred. Predictive procedures can provide semi-quantitative or quantitative estimates of the probability of occurrence of an accident and of the consequences flowing from it. Often these procedures can also point out ways in which to reduce the risk, if it is considered necessary so to do.

There are still several sources of imprecision in risk analysis, however, arising notably from available or estimated probability data, physical representation of the disaster in question, damage evaluation models used, and the level of acceptable risk.

Probability

Probability data on accidents can differ from actual values by at least two orders of magnitude. This very great uncertainty arises from several causes:

- technology evolves very quickly, and it takes time before reliable failure data for new equipment become available;
- processes and control systems are continuously being improved, and this lowers their chances of malfunctioning. However, new conditions may increase their chances of failure;
- probabilities are derived from a very limited number of accidents, and this reduces their statistical significance;
- the behaviour or reliability of human beings is very difficult to predict, and it is strongly correlated with many other factors such as training, culture, etc.

Damage

Theories aspiring to model accidents, and the damage they cause, do not describe precisely the reality of a given situation because of the following reasons:

- models have not been extensively tested due to the lack of a sufficient number of well-documented accidents, or because of scarcity of data on them;
- only a few models have so far been developed, so that it is not possible to compare the results predicted by alternative models;
- models are based on different hypotheses and parameters which limit both their fields of application and the precision of their predictions;
- models do not take into consideration the possible interactions among several phenomena occurring simultaneously. They are based on idealised geometric and physical conditions, while in reality local factors may just as well attenuate or exacerbate the negative effects of an accident.

What constitutes an acceptable level of risk?

Natural risks, such as those of earthquakes, volcanic eruptions, hurricanes, floods, avalanches, meteorites, and epidemics are largely beyond human control. The level of such risks, which are implicitly accepted by the people at risk, may therefore be taken as the 'datum' with respect to which the level of an *acceptable risk* may be defined.

Growing focus of the media on catastrophes and accidents makes people aware, possibly too aware, of risks which they perceive as being threatening to their lives. Television instantly transforms people into spectators of every disaster anywhere in the world almost as it happens. This creates the impression that one day perhaps catastrophes may also befall us, our family or our business. The combination of a growing urban concentration of population and relentless sophistication of technology creates an ideal breeding ground for potential catastrophes, as well as for the fear surrounding them.

As a matter of fact, while almost everyone agrees on the need for essential industrial plants, few are willing to have them close to their homes. This is the so-called NIMBY (Not In My Back Yard) syndrome. The upshot of this is that, when licensing an undesirable but essential plant, government officials must make unpopular decisions which may result in officials or politicians postponing decisions until someone else comes into power. This is known as the *Not In My Term* mentality.

Even a simple and transparent formula, such as $R = P \cdot D$ which defines risk, does not make much sense to decision-makers or the population at large who consider it too abstract or too formal. In general, people are only willing to recognise hazards on the basis of perception, not analysis. In this sense the equation $R = P \cdot D$ may be construed as a relativistic concept which enables comparisons between different social or societal situations. Indeed, it obliges us to acknowledge notions of totally different levels of risk within a given society, and, as such, it has some heuristic value.

RISK MANAGEMENT

The task of governments

Risk evaluation and management should be conducted independently, but, ultimately, both should be the responsibility of governments. The search for, and identification of, hazards plays an essential role in developing preventive policies which governments are responsible for implementing. Clearly, a greater number of hazards will be discovered as search strategies become more sophisticated, and, provided that these strategies do not entail undesirable side-effects, this would go a long way to reducing people's feelings of insecurity. It is important, therefore, for risk analyses also to point out less controversial alternatives for reducing risk. It is in this spirit that governments must try to identify and locate hazards, and thus minimise people's feelings of insecurity, especially of those living close to large industrial plants.

However, two major problems remain to be addressed in risk analysis. The first refers to *decision-making*. How should decision-makers account for risks and include their *possible costs* in the cost–benefit analyses? How can *standards* be developed for routine application to enable the classification, assessment and comparison of hazardous situations as such? In short, the question is how to make analyses more transparent and transform them into instruments of a more responsible policy. The study of such problems pertains to the definition of *acceptable risk*.

A second problem, which is germane to the first and yet very different in nature and scope, is concerned with *risk acceptance*. Decision-makers, as well as the public at large, are often unwilling to accept that a hazardous situation has been sufficiently analysed and controlled, even when robust objective data are made available to support the proposition. The question is, why are some low-risk activities or situations still considered to be unacceptable, while others that are much more risky by objective standards are accepted?

Many of the mental mechanisms involved in the second problem are well known in cognitive psychology, some of which are concerned with the study of personality and sociological irregularities. Thus, the study of risk acceptance is more a matter of reshuffling and applying existing insights, rather than developing new theories or methods as would be the case in the first problem.

It is also the responsibility of governments to protect their citizens against catastrophic hazards which may threaten their very existence. Work is therefore needed on the prevention of disasters (*disaster prevention*) as well as on their curative aspects (*disaster fighting*). Taken together, these two measures constitute what is called *disaster management*.

Disaster management should be introduced at each of these three consecutive levels:

1 the *industrial plant level*: adoption or adaptation of internal working procedures and technical measures (this process is typically a matter of a few weeks);
2 the *sector level*: adaptation of codes of *good practice* (a process which may take a couple of years);
3 the *societal level*: adaptation of legislation which pursues determined safety objectives with a more general purpose (a process which can last five to ten years).

The accident at the Icmesa plant at Seveso, Italy, in 1976 serves as a good example of reaction at the societal level. This accident occurred when a part of the contents of a synthesis reactor was vented through a safety valve, thus creating a toxic cloud over the surrounding area. Six years later, in June 1982, the European Communities issued a Directive (82/501/EEC) obliging the EEC member states to legislate for reaching certain safety objectives in large industrial plants (Box 8.2).

Role of the public

Risk management must take into account social (risk acceptance) and political factors as well as economic ones. Regulations and measures must be clear and easy to understand in order to satisfy people engaged in hazardous activities, as well as those who live in exposed surroundings.

Governments, on their part, must ensure that industry can still be located in specific regions selected through a process of zoning and urban planning, because, clearly, adoption of unrealistic norms for safeguarding the environment can threaten the very existence of industry, and the resulting loss of jobs can lead to social problems, instability and even strife that can actually be more dangerous than an accident. It is obviously essential that risks are reduced to below an acceptable minimum, but a requirement for zero risk is as unreasonable as it is technically and economically impossible.

These problems become especially poignant in densely populated areas where hardly any hazardous activity is possible without causing potential danger to human life and the environment. A traffic accident, in which a truck load of a toxic chemical is released in a densely populated area, is a typical example of this.

It is also the responsibility of governments to ensure that risks are evaluated objectively, and to stipulate a limit above which a given risk is unacceptable (maximum allowable level) and below which it is negligible (acceptable risk). Also, people in hazardous surroundings should be allowed to choose different resource allocations in order to achieve and maintain a certain level of risk. But, in return, this should accrue some benefits to their community. For example, imagine that a certain company has to spend one million dollars each year for ensuring extra safety for its neighbours by lowering the level of risk from 5×10^{-7} to 2×10^{-7}. The neighbours would probably prefer the company to save 0.5 million dollars and donate the other 0.5 million dollars to a local charity or activity, such as a school or a festival, instead of lowering the risk by the proposed amount, because, in their view, this would improve their social resources. But, when there is a lack of communication, they would probably demand that the risk be lowered to below the ALARA level. However, when a suitable compensation is offered, they may accept a more reasonable level of risk.

The role of industry

In the context of safety, the role of industry is primarily to prevent accidents through the application of sound, good practices and standards for the design, construction, operation and maintenance of plants. This is fundamental and the best approach to minimising risks. Accordingly, an important safety management objective is to offer regular training to personnel in industry in safety awareness and in the application of safe work practices to the operation and maintenance of plants.

In addition, safety systems should be used to prevent the propagation of an accident. But this strategy will be successful only if such systems are well designed and function properly in the first place, and this brings us back to the probability factor. Passive mitigation systems, which do not affect the

BOX 8.2 THE SEVESO ACCIDENT

The Seveso accident was the result of an ordinary and perfectly controllable chemical reaction at a temperature of 140–180°C. But it had potential for two other consecutive exothermal reactions. After a certain higher temperature was reached, an uncontrollable level of heat was generated, leading to the following:

Primary reaction: The formation of 2,4,5-trichlorophenol sodium salt.

Exothermic secondary reaction: The formation of adducts of trichlorophenol with ethylene glycol.

Second exothermic secondary reaction: The combination of two molecules of 2,4,5-trichlorophenolate, forming the highly toxic substance called 2,3,7,8-tetrachlorodibenzo-p-dioxin through the elimination of sodium chloride occurring at temperatures above 230°C.

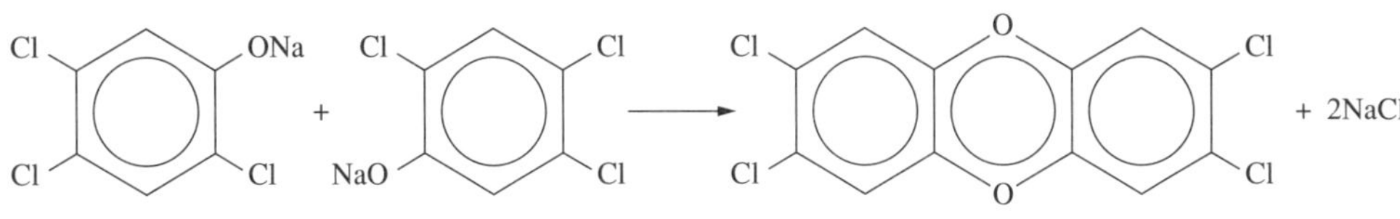

'TCDD'

2,4,5-trichlorophenol is produced by the partial alkaline hydrolysis of 1,2,4,5-tetrachlorobenzene. Temperatures of 140 to 180°C are necessary for this reaction.

TCDD, whose chemical structure is shown above, is known from runaway or decomposition reactions as a contaminant of herbicides. It is also a trace contaminant formed in numerous thermal and metallurgical processes. A very toxic and dermatitic compound, it is extremely resistant to normal chemical decontamination procedures. In the first edition of the *Handbook of Reactive Chemical Hazards*, which appeared before the Seveso incident, Bretherick had already demonstrated the hazards connected with the above reaction; seven serious accidents had occurred before the Seveso incident.

The reaction enthalpy of the hydrolysis of 1,2,4,5-tetrachlorobenzene is −108 kJ/mol (essentially equal to double the heat of neutralisation). Two processes have mainly been used for the hydrolysis reaction. In the earlier process methanolic alkali was used under autogenous pressure to effect the hydrolysis. Four incidents have so far occurred with this type of reaction. The second process, in which ethylene glycol is used as solvent, operates at atmospheric pressure. At a reaction temperature of 140–180°C, several hours are needed for the hydrolysis. A solvent (dichlorobenzene or xylene) is usually added for the subsequent solvent distillation. Three other incidents had occurred with this type of hydrolysis before the Seveso incident.

In Seveso, a town in northern Italy, glycol was used as a solvent in hydrolysis. At the beginning of the operation, 3,235 kg of ethylene glycol, 603 kg of xylene, 2,000 kg of 1,2,4,5-tetrachlorobenzene, and 1,100 kg of sodium hydroxide were charged into the 10 m^3 reactor. The incident occurred on Saturday, 10 July 1976. The reactor was charged during the afternoon of 9 July. In the early morning the hydrolysis was completed. The water formed during the reaction, together with the xylene and about 500 kg of glycol, had been removed by distillation. Heating by steam was discontinued, and vacuum was terminated by connecting the reactor to the atmosphere. Stirring was maintained for 15 minutes, after which all operations were stopped and the workers left the plant. At that time the temperature of the reaction mixture was 158°C. The wet part of the reactor wall was at the same temperature as the liquid, while the temperature of the upper dry part of the reactor was, at least partially, about 300°C. Steam at a pressure of 12 bar was used for heating, but it arrived superheated to about 300°C at the upper part of the coil welded to the outside of the reactor. The bursting disc, rated at 4 bar, ruptured 7.5 hours later at midday on 10 July when the reactor was unattended. The entire reaction mixture was dispersed into a large area of the factory's environment. Although the reactor had not been damaged, it was one of the most expensive incidents in the chemical industry, because the cost of cleaning-up a large polluted area was very high.

The incident was followed by many experimental and theoretical investigations carried out by scientists from the company which owned the plant, by other Italian experts, and by a scientist recruited by the company. A comprehensive list of publications on the incident will be found in the paper by Cardillo *et al.*, cited by Grewer (1994).

An important lesson of the Seveso incident is that a venting system without a collection device is not sufficient if toxic material is discharged. Kletz said about this incident, 'If a "catchpot" had been installed at Seveso, the name would still be unknown'.

Source: Grewer, 1994

probability of the event sequence occurring but alleviate the consequences, are often a cost-effective way of reducing risk.

We have already pointed out that reducing either the probability or the consequences of a potential accident will also reduce the risk of its occurrence. Reducing the inventory of a hazardous material, which could be involved in an accident, is thus a suitable, inexpensive and practical way of reducing the consequences, and thereby the risk.

Furthermore, often some of the organisational aspects of the system itself may be implicated in risk when a search is made to identify the basic causes of accidents. In this sense an accident may be seen as a symptom of the dysfunctioning of the organisational system itself.

Risk management can only be successful when it is fully integrated into the overall management of the organisation. This is what is referred to as *integrated safety*. Management of safety should not be limited to technical prevention alone, but must be part of the general trend towards efficiency, effectiveness and quality improvement of both managerial and industrial processes. People pay attention to safety not only for preventing accidents at work, but also as part of a more positive way of thinking, and indeed as a way of promoting the health and well-being of workers and of the organisation as a whole. In the wider context, risk analysis can thus improve the global economic situation of industry, and should *not* be seen as waste of money that has to be minimised.

The role of scientific research

As industry continues to develop at an accelerated pace, it is becoming increasingly more difficult to rely on past experience. As a result, new and hitherto unknown risks appear periodically, such as stratospheric ozone depletion, global warming, etc. This lack of experience obliges society to develop new and effective tools for predicting and analysing risks.

Scientific research can contribute significantly or even substantially to the identification and evaluation of risk. The study of the consequences and importance of damage is also the subject of scientific research. Even industrial disasters serve at least one useful purpose in the sense that their meticulous *post factum* analysis helps to identify the initiating events, as well as the subsequent chain of events, from which lessons can be learned. Other scientific issues relating to risk analysis are:

- development of emission models for evaluating impacts on humans and assessing environmental damage;
- research on how materials behave in fire, explosion, etc.;
- modelling of accidents, such as outflow of gas as a turbulent jet, heat radiation, explosion of a gas cloud, boiling of a liquid with expanding vapour, etc.;
- how humans and the environment respond to exposure to some (toxic) agents, heat radiation, shock waves, etc.

REFERENCES

AIChE/CCPS (1985) *Guidelines for Hazard Evaluation Procedures*, New York: Center for Chemical Process Safety, American Institute of Chemical Engineers.

AIChE/CCPS (1989) *Guidelines for Chemical Process Quantitative Risk Analysis*, New York: Center for Chemical Process Safety, American Institute of Chemical Engineers.

AMINAL (1994) *Handboek kanscijfers*, Brussels.

Engineering Council (1993) *Engineers and Risk Issues – A Code of Professional Practice and Guidelines on Risk Issues*, London.

EPA (1992) *Framework for Ecological Risk Assessment*, Risk Assessment Forum, US Environmental Protection Agency, Washington, DC, EPA/630/R-92/001.

EPA (1995) 'Draft Proposed Guidelines for Ecological Risk Assessment', US Environmental Protection Agency, Washington, DC, EPA/630/R-95/002.

Grewer, T. (1994) *Thermal Hazards of Chemical Reactions*, Industrial Safety Series, Vol. 4, Amsterdam: Elsevier, Chapter 7, pp. 299–342.

Kirwan, B. and Ainsworth, L.K. (1992) *A Guide to Task Analysis*, London: Taylor & Francis.

Lees, F.P. (1980) *Loss Prevention in the Process Industries*, London: Butterworths.

Linthurst, A., Bourdeau, P. and Tardiff, R.G. (eds) (1995) *Methods to Assess the Effects of Chemicals on Ecosystems*, New York: John Wiley & Sons Ltd.

McCormick, N.J. (1981) *Reliability and Risk Analysis: Methods and Nuclear Power Applications*, San Francisco: Academic Press.

Olivier, P. and Van Overmeire, M. (1990) *Betrouwbaarheids- en risicoanalyse van industriële systemen*, Brussels: Vrije Universiteit Brussel.

Pitblado, R. and Turney, R. (eds) (1996) *Risk Assessment in the Process Industries* (second edition), Rugby, UK: Institution of Chemical Engineers.

Renner, R. (1996) 'Ecological Risk Assessment Struggles to Define Itself: Practitioners are debating how to turn this young field into a rigorous discipline', *Environmental Science & Technology* 30 (4): 172A–174A.

Roos, A.J. (1985) *De kans op menselijk falen: Methoden voor het bepalen en verwerken van kansen*, Arbeidsinspectie, Voorburg, 7.3–7.21.

SRD Association (1995) *Human Reliability Assessor's Guide, Report SRDA-R11*, Harwell, Oxon: AEA Technology.

Suter, G.W. II (1993) *Ecological Risk Assessment*, Chelsea, Mich.: Lewis Publishers.

Swain, A. and Guttman, H. (1980) *Handbook of Reliability Analysis with Emphasis on Nuclear Power Plant Application*, NUREG/CR-1278.

TNO (1988) (The Yellow Books) *Methods for the Calculation of the Physical Effects of the Escape of Dangerous Materials: Liquids and Gases*, 2 volumes, PO Box 342, 7300 AH Apeldoorn, The Netherlands.

Van de Kerckhove, J. (1990) *Is het menselijk gedrag berekenbaar? Enkele kanttekeningen i.v.m. Human Reliability*, Leuven, Belgium: Katholieke Universiteit Leuven, Permanente Vorming.

SUGGESTED READING

The Dow Chemical Company (1994) *Dow's Fire and Explosion Index Hazard Classification Guide and Chemical Exposure Guide*, AICHE Technical Manual, 2 vols, LC 80–29237 (AIChE).

Grewer T. (1994) *Thermal Hazards of Chemical Reactions*, Industrial Safety Series, Vol. 4, Amsterdam: Elsevier.

ICI (1985) *The Mond Index* (second edition), London: Imperial Chemical House.

Pitblado, R. and Turney, R. (eds) (1996) *Risk Assessment in the Process Industries* (second edition), Rugby, UK: Institution of Chemical Engineers.

van Leeuwen, C.J. and Hermens, J.L.M. (1995) *Risk Assessment of Chemicals – An Introduction*, Dordrecht, The Netherlands: Kluwer Academic Publishers.

SELF-ASSESSMENT QUESTIONS

These questions are open and require some thought. The answers serve only to illustrate the line of thought and the main points which should be covered.

1 Consider a car accident.
   (a) Which of the following factors would influence the *probability* and *severity* of the accident? Consider for instance:
   - the driver (age, sex, religion, drinking habits, etc.);
   - the car (make, model, year, horsepower, condition of the tyres, etc.);
   - the weather (rain, fog, sleet, snow);
   - others.

   (b) If you work for an insurance company, which kinds of data would you request on cars and car drivers and which data would you collect on traffic accidents?
   (c) Under what circumstances should a particular driver or a particular car not be allowed on the roads? Can you develop rational criteria for making this decision?

2 Are you afraid of engaging in certain activities or behaviours? List the undesirable consequences they may have and try to assess the probability of their occurrence. Use the factors listed on pp. 146–148 and bear in mind that the perception of risk, as experienced by various social groups, may differ from yours (e.g. groups of a certain age, sex, economic or social status, people with certain hobbies, etc.).

3 Do you think that the probability of a single death, or the total number of casualties in an accident, are sufficient criteria in risk assessment? If not, propose criteria in which other forms of damage are taken into account. How would you assess the relative values of material damage and human life?

4 How would the accident with the ladder (Table 8.4) be treated in your organisation? (Try to list the elements needed for short- and long-term safety and their respective economic benefits):
   - without further interest;
   - in an administrative manner;
   - leading to disciplinary measures;
   - analysis by an *ad hoc* working group.

   What are the pros and cons of each approach?

5 What are the actual limits in quantitative risk assessment? What is your personal view of the usefulness of collecting more or better data?

6 Which of the following aspects of risk analysis would you personally prefer to deal with? (Explain the reasons for your preference and the strategy you would follow in pursuing your task):
- administrative tasks for the authorities;
- measures to be taken by industry;
- disseminating information to the public.

9

HAZARD, EXPOSURE AND ECOLOGICAL RISK ASSESSMENT

Paul Johnston, Ruth Stringer, David Santillo and Charles Vyvyan Howard

SUMMARY

The regulation of chemicals entering aquatic environments is based largely on the level of hazard they pose, as assessed using an iterative framework. Hazard assessment is carried out using toxicity tests to gauge the magnitude of any likely problem together with a determination of the key physico-chemical parameters. This information can be used in two ways. First, it may be used to set an environmentally 'safe' level. Second, it may be used in conjunction with an exposure assessment to prepare a probabilistic risk assessment. In theory this process leads to a probability value indicating the likelihood of unacceptable impacts in the receiving environment.

Although ecological risk assessments are being used more widely, there is a large number of assumptions inherent in the process along with a larger number of uncertainties. A critical overview of the processes involved is given in this chapter, together with an examination of the state-of-the-art in hazard assessment of chemicals, exposure assessment of target systems, and their joint relationship to ecological risk assessment. Application to aquatic systems is emphasised, since the methods are particularly well developed in this discipline

ACADEMIC OBJECTIVES

On completing this chapter the reader should gain:

- a clear overview of the elements involved in the hazard assessment of individual chemicals, the procedures used for applying the derived results to the wider environment, and the assumptions involved therein;
- an understanding of how hazard and exposure assessment data are applied in the risk assessment process, and the inherent uncertainties at each stage;
- knowledge of the difficulties in applying predictive models to study the behaviour and impact of complex mixtures of chemicals released into the environment, even when supporting toxicity data are available.

HAZARD, EXPOSURE AND ECOLOGICAL RISK ASSESSMENT

For the purposes of this discussion, hazard assessment and ecological risk assessment may be regarded as two branches of environmental assessment. Other branches include resource management and environmental impact assessment. Taken together, these constitute perhaps the most diverse assessment dealing with all the environmental aspects of a project at the planning stage. Although these broad fields are interrelated, there is a considerable diversity in the way that environmental problems are defined and analysed under each discipline. Hazard assessment may be considered broadly as a methodology for analysing and investigating the effects of chemicals on the natural environment (Suter, 1993). It has primarily been developed as a discipline by aquatic toxicologists, and it is in this field of environmental protection that the techniques are most highly developed. The expected concentration of a given chemical in the environment is compared with the estimated toxic threshold using a tiered toxicity testing system. This comparison is then used to determine whether a given release or discharge is safe, hazardous or whether the implications are unknown or uncertain. Hazard assessment does not by itself provide information which can be used to predict the magnitude or nature of likely impacts, nor for making comparisons between the relative impacts of different chemicals and discharges in the wider environment.

Having evolved from toxicological studies focusing on the impacts of individual chemicals on single species of test organisms, ecotoxicology has now become an established discipline in its own right concerned with the wider consideration of the effects of chemicals on whole ecosystems. As a developing science it constitutes the foundation of the evolving field of ecological risk assessment. The latter has been described by Suter (1993) as the 'application of ecotoxicology to public policy'. In practical terms, risk assessment is a means by which a derived probability can (in theory) be assigned to the possible environmental impacts of a chemical released or discharged into the environment. Risk assessment embodies an array of methods based on the actuarial techniques which have been applied for many years in the insurance industry. More specifically, therefore, ecological risk assessment is intended to provide an objective and probabilistic assessment of the likelihood of undesirable changes taking place in an ecosystem exposed to chemical inputs. As pointed out by Power and McCarty (1996), whereas some ecological risk assessments do provide true probabilistic estimates of risk, there is no presumption that they must do so. Indeed, frequently a very high degree of uncertainty remains in risk assessments used for legislative purposes. None the less, probabilistic risk assessment is seen as a potentially powerful tool for gaining deeper insight into stressor-response distributions and, consequently, their use is likely to increase markedly in the future.

Input data needed for ecological risk assessment procedures are derived from two defined areas of analysis: characterisation of ecological effects and characterisation of exposure. Information from these two areas is combined to characterise the ecological risk, whereupon a risk management strategy is devised. Figure 9.1, adapted from USEPA (1989), shows the basic elements of a risk assessment framework designed for humans exposed to chemicals through the consumption of contaminated aquatic resources. Although the emphasis here is clearly anthropocentric, this diagrammatic representation incorporates all the major elements of an ecological risk assessment framework which has been used in a number of recent studies (see: Kendall *et al.*, 1996; Solomon *et al.*, 1996).

In the estimation of both hazard and exposure, and hence in any subsequent ecological risk assessment, there are several sources of uncertainty and imprecision that are seldom identified explicitly in any given scheme. In addition, these procedures incorporate a number of assumptions which may not be explicitly defined, despite the fact that they are of considerable importance in the application of the derived results to environmental protection. These important areas are discussed in greater depth in the broad procedural descriptions given below.

HAZARD ASSESSMENT PROCEDURES

The elements of information required for the hazard assessment of a chemical are shown in Figure 9.2 (adapted from Cairns, 1989). In practice these elements are derived from iterative laboratory testing procedures to determine the toxicity of the chemical (toxicity testing), and from laboratory and other investigations of the properties of the chemical relating to its environmental fate and occurrence. These may include, among others, the speciation of the chemical, its resistance to degradation, and a measure of its potential to accumulate in biological systems.

Information thus obtained is pooled, and the hazard associated with the chemical assessed. The final stage is the formulation of criteria specifying concentration limits which, if met, will protect the aquatic environment. Thus the purpose of hazard assessment is to predict (and ultimately to provide the means to regulate) possible damage to natural systems and to the organisms which inhabit them. Hazard assessment is generally conducted as a tiered routine, or as a series of progressive stages, and may typically comprise the following protocol:

TIER 1: Range-finding toxicity tests whereby the effective concentration range of the toxicant is established by determining 100 per cent and 0 per cent response concentrations.

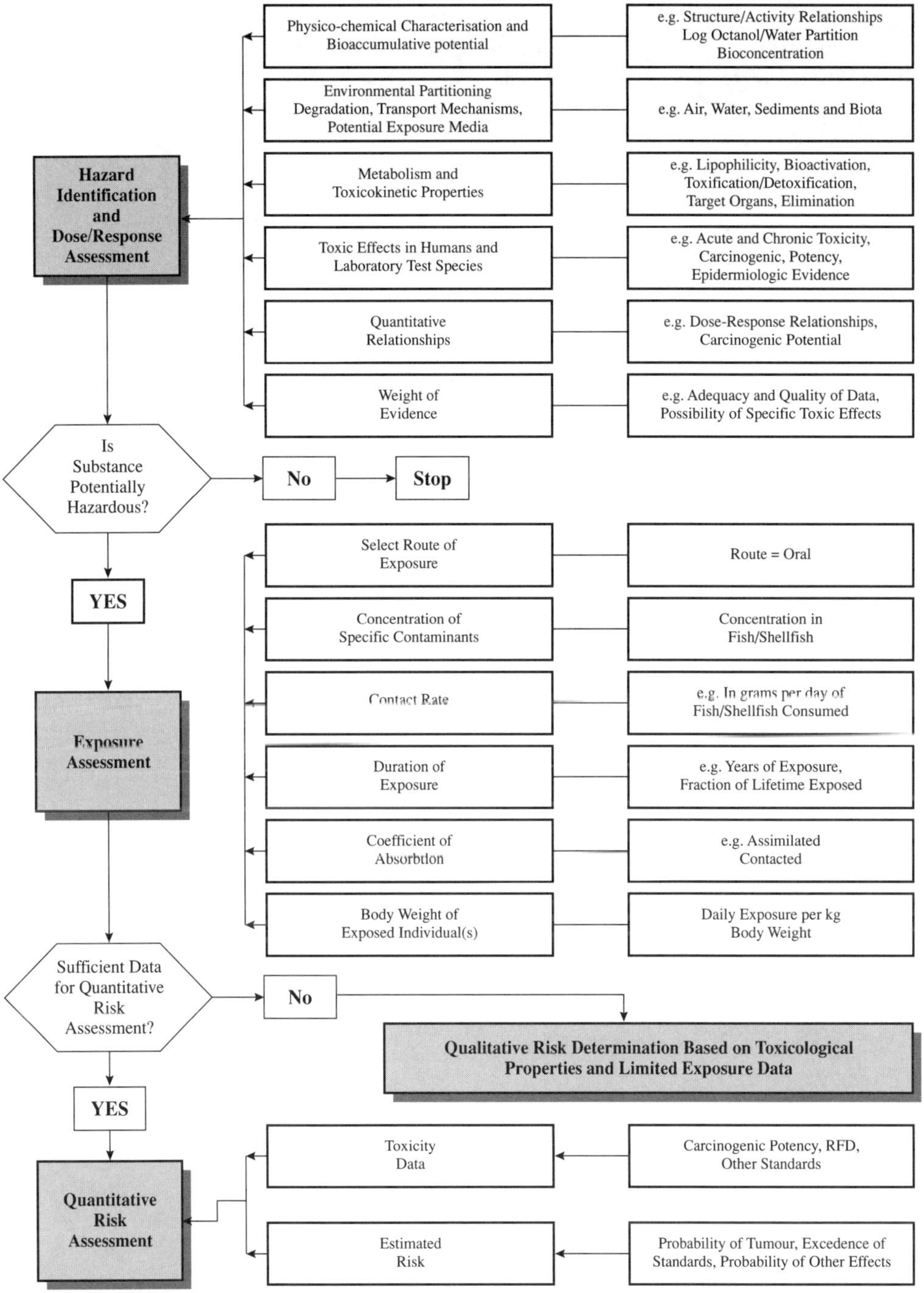

Figure 9.1 Flow diagram of a risk-assessment framework designed to address the consumption of contaminated aquatic resources by human populations. Despite this being a relatively simple exposure scenario, the system requires substantial input of information

Source: Adapted from USEPA, 1989

TIER 2: Acute, sub-acute and chronic toxicity tests to determine the LC50 and/or EC50 and other intermediate responses. This tier gives greater precision over results of the range-finding tests in Tier 1.

TIER 3: Validation of the results in a natural or surrogate system (e.g. field manipulations, mesocosm experiments).

TIER 4: Emplacement of a biological monitoring system to establish that predetermined conditions are being met.

In general, as pointed out by Cairns (1989), Tiers 1 and 2 are commonly applied, whereas the use of validation tests and biological monitoring schemes, Tiers 3 and 4, is relatively uncommon.

Toxicity testing

The first stage in a hazard assessment is to design and execute test procedures to determine the toxicity of a chemical to organisms representative of the ecosystem likely to be impacted. Aquatic toxicity test systems need to be highly standardised. In an ideal experimental set-up, the only variables will be the concentration of the toxic chemical present and the time of exposure. Parameters such as oxygen concentration or pH can change due to the metabolic activities of the test organism. Also, the concentration of a toxicant under test may be difficult to predict if its solubility is poor. Concentration may vary considerably if the chemical is volatile, or subject to adsorption by the walls of the test vessel. The results of a test can be influenced considerably by the design of experiments, while the selection of a particular test protocol would depend on the information required from the test, the test species and the test chemical. Standardisation, considered to be very important, has been widely reviewed and practical guidance documents for the conduct of tests have been issued by various bodies (ASTM, 1991; OECD, 1995). Working definitions of terms, used in toxicity testing and in the subsequent application of results, are given in the Glossary (pp. 476–488) (adapted from Sprague, 1990; Suter, 1993; UKEA, 1996).

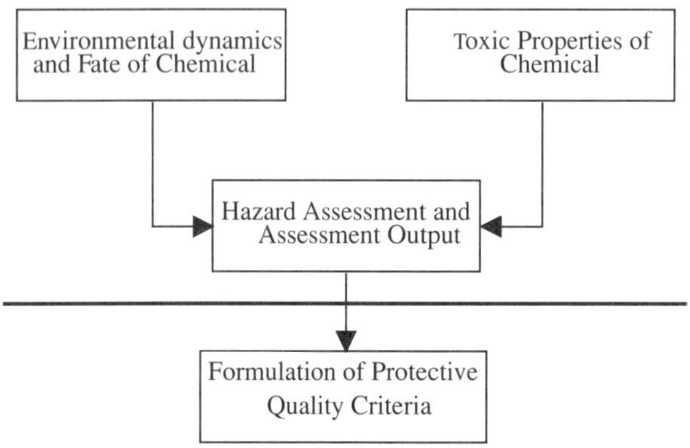

Figure 9.2 Components of a hazard assessment protocol showing the input information required to perform the task. Protective criteria are generally formulated through the use of 'application factors' based on toxicity test output data and knowledge of the properties of the chemicals concerned. Expert judgement is widely used as a surrogate for empirical data where this is required

Not all individuals within a population respond identically to a given concentration of a toxicant. The tolerance of individuals within a single species to a toxic chemical can vary within very wide limits. The distribution of tolerance within the population may be markedly asymmetrical when the individuals are extremely tolerant to high concentrations, or when a small number of individuals are in poor physiological condition. Tolerance cannot be measured directly, and so toxicity tests are designed to measure the proportion of subjects, in random samples of the population, responding to different concentrations of the test chemical. Data (i.e. the distribution of the response) can generally be normalised by using the logarithm of concentration, and this simplifies their statistical analysis.

The cumulative frequency of the response is recorded where mortality at several points is measured in tests over a given period of time. When plotted as a log of concentration against response, the resulting curve is found to be *sigmoidal*. It is customary to transform the response data into units derived from probability, called 'probits'. Such an analysis is known as a 'probit analysis' (Finney, 1962). A detailed description of the derivation and treatment of toxicity test data in this way is given by Sprague (1969, 1990), and the application of such techniques by Johnston (1987). The LC50 or the EC50 then represents the median lethal concentration, i.e. the concentration producing a response in 50 per cent of the population. The slope of the line reflects the variability of response.

Often the consequence of the distribution of tolerance values is not accorded the appreciation it deserves. For example, not all individuals in a given population will be equally affected by a toxic chemical present in the environment at a concentration that is sufficiently high to elicit a detectable response. This raises an interesting question about the low numbers of the most susceptible members of the population affected, as well as ethical issues when the population is human. This is of particular interest in the case of chemicals, described in greater detail later (pp. 183–184), which disrupt endocrine systems.

The output data from toxicity tests are generally in terms of an LC50 or EC50 value to which statistical confidence limits can be attached. Furthermore, the 'lowest observed effect concentration' (LOEC) and 'no observed effect concentration' (NOEC) can be determined from tests in which the tested effect is sub-lethal. Both these parameters are used in regulation, although recently the use of the NOEC for such purposes has been the subject of considerable criticism on the ground that it lacks precision and reproducibility; there is also some doubt about its statistical validity (Chapman *et al.*, 1996). The NOEC, increasingly considered to be a poor estimator of the true 'no effect concentration' (NEC), is also prone to extreme variability between tests. It is for these reasons that there is now an apparent preference for the use of the EC50. The outcome of this ongoing debate has considerable implications for regulatory toxicology, since the use of the NOEC is mandated by several national legislative bodies.

In order to arrive at a notional biologically 'safe' concentration of a chemical, an application factor, which is a multiplying factor ranging from 0.1 to 0.0001, has traditionally been applied to the LC50 or EC50 value. As pointed out by Cairns (1989), the values of the application factors have generally been selected on a 'trial and error' basis incorporating some empirical evidence, although they have been entirely intuitive in some cases. The application factor focuses on species protection rather than on community and ecosystem protection in the sense that toxicity tests are generally carried out on single species. Such factors have underpinned the formulation of water quality criteria used in the regulatory system of both the UK and the USA (Train, 1979).

Formalised protocols for toxicity testing are increasingly accommodating different ecosystem components through a battery of tests involving different organisms. True multi-species toxicity tests in a simulated natural environment (e.g. a mesocosm experiment) are as complex to design as they are to interpret (Smith, 1995). They may, for example, reach different equilibria even before the commencement of tests. But, on the other hand, they can provide more information than single-species toxicity tests. Such tests generate an effective concentration value, based on some predetermined effect on the species composition of the mesocosm or on the energy flowing through it. Alternatively, statistical methods may be deployed to detect differences between the exposed and control systems. But such tests can be expensive to carry out, and they are therefore unlikely to be used routinely (Cairns, 1989).

Selection of test species

In addition to the standardisation of test conditions, the organisms under test should also be standardised. The purpose of this is to ensure that test data on each species, or chemical combination, can be reasonably reproduced between laboratories and over time. The derived LC50 and EC50 information would then assume the status of basic information which can be used as part of the information base for deriving quality criteria. Such information can also be used for ranking toxic chemicals in prioritised groups, based on the degree of hazard attaching to their use and transportation as well as to their use in the development and marketing of new chemicals (Mount, 1995).

In view of the above, it has been common practice to devise a battery of tests using a number of different species representing different levels or components of the ecosystem. A battery of tests, proposed by the UKEA (1996) for the toxicity testing of a whole range of complex effluents listed in Table 9.1, is a recent example of this. This battery of tests is designed to generate data on acute lethal and acute sub-lethal modes of action. Any subsequent

Table 9.1 Battery of toxicity tests using test species at different trophic levels (primary producer, primary consumer or herbivore, secondary consumer or carnivore) within an ecosystem adapted from UKEA (1996)

Type of test	*Environmental application*	
	Freshwater	*Marine*
Rapid	Chemiluminescence (enzyme)	Chemiluminescence (enzyme)
Rapid	Bioluminescence (Bacterium)	Bioluminescence (Bacterium)
Trophic Level 1 (primary producer)	*Selenastrum* (algal)	*Skeletonema* (algal)
Trophic Level 2 (primary consumer)	*Daphnia magna* (crustacean)	Oyster embryo larvae
		Acartia tonsa (crustacean)
Trophic Level 3 (secondary consumer)	Trout (fish)	Turbot/plaice (fish)

Note: All tests at trophic levels 1–3 are designed to derive information about acute lethal and acute sub-lethal effects

processing of data generated by this scheme would be based on the most sensitive response identified. The scheme thus illustrates what Cairns (1989) describes as two pivotal assumptions in single-species toxicity test regimes. First, the selection of organisms for toxicity testing based on the assumption that, if the most sensitive species present in a natural system are protected, then all other species will be protected too. The second assumption is that, as a result of such protection, deleterious effects on the ecosystem will simply not be possible.

Under ideal circumstances, organisms to be used in laboratory tests need to fulfil the following criteria:

- they should be available throughout the year;
- they should be easy to collect and culture in large numbers;
- they should tolerate laboratory handling;
- they should be available as juveniles or larvae (generally the most sensitive life stages);
- they should be amenable to culture through several generations in order to enable chronic testing;
- they should give reproducible test results;
- they should be present in, or related to organisms present in, the ecosystem of interest;
- their ecology and distribution should be well understood;
- they should be ecologically important.

In practice, this means that relatively few organisms would qualify as routine and standard test species. The overriding limitation is the amenability of the species – from standard breeding, or culture stock, to laboratory culture under standard conditions.

The scheme of Table 9.1 highlights another aspect of ecotoxicology which has so far received relatively little attention; it is that of organism interaction at different trophic levels. Taub (1989) describes it as the 'third leg' of ecotoxicology – a discipline which had until then largely rested on the two 'legs' of chemical fate and single-species toxicity tests. To date it still remains an apt description of the state of development. Clearly, further studies are needed on species interactions, contaminant metabolism and the transmission of chemicals between compartments of the trophic web for a deeper and wider understanding of ecotoxicology.

Environmental dynamics and the fate of chemicals

The behaviour and fate of a given toxic chemical comprises the second element of information required as input into a hazard assessment. In practice, much of the existing information has been gathered in order to prioritise chemicals for regulation and control. The need for, and indeed the inevitability of, such prioritisation arises from the fact that most regulatory instruments are permissive in nature because they define acceptable levels of discharge. But there is a vast multitude of other chemicals that are also being discharged into the environment. It has been estimated that currently some 63,000 chemicals are in common use worldwide, of which about 3,000 account for 90 per cent of the total production; and that between 200 and 1,000 new synthetic chemicals enter the market each year (Shane, 1994). Other figures suggest that 50,000 substances are in common use in the EU alone, of which 4,500 could merit priority regulation on the basis of their chemical and biological properties (Edwards, 1992).

Three properties are generally selected for the evaluation of chemicals: toxicity, persistence in the

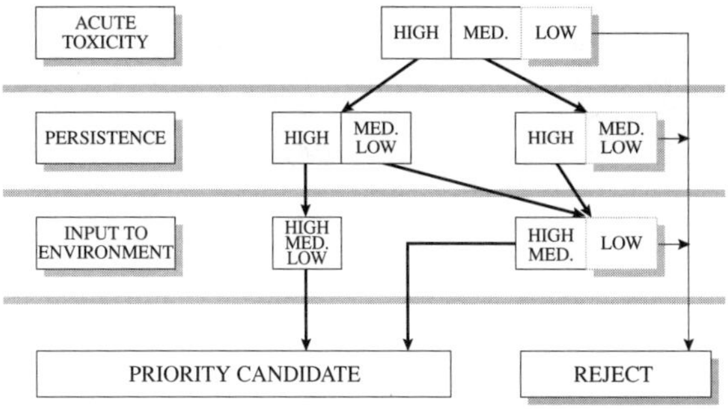

Figure 9.3 Decision tree for the evaluation of priority chemicals for regulation on the basis of the short-term scenario detailed in Agg and Zabel (1990). The threshold values for the above parameters, and those used in the long-term and bio-accumulation scenarios, are given in Table 9.2

environment, and tendency to bio-accumulate. These three properties are also perceived as being of greatest use in defining a potential hazard, not only to aquatic ecosystems but also to terrestrial systems. Accordingly, a number of schemes have been devised on the basis of these properties. A typical example is the UK 'Red List' in which the production volume of the chemical concerned is also considered. One of the basic schematics for evaluation is given in Figure 9.3, adapted from Agg and Zabel (1990). The numerical values to be used for each property of the chemical in this ranking scheme are given in Table 9.2. The evaluation procedure is based on three scenarios. Scenario 1, which is the short-term scenario, is based on acute toxic effects. Persistence and chronic (long-term) toxic effects are considered in Scenario 2; while Scenario 3, which is called the 'food chain scenario', is concerned with the potential for bio-accumulation and effects on higher organisms. The threshold values used in each decision tree are, by admission, arbitrary. In the case of chronic toxicity, an application factor of 0.01 is assumed in the absence of empirical data in order to derive a value from acute lethal toxicity data. In the case of bio-concentration, provision is made for the use of the log octanol/water partition coefficient as a surrogate value, because it is considered to be a good indicator of whether the chemical will associate with lipid tissues. The Red List framework is based on a very limited sub-set of the universe of data, shown typically in Figure 9.1, which could be regarded as being important under ideal circumstances. In particular, the characterisation of environmental partitioning, degradation and transport mechanisms is a complex task. Contaminants are subject to partitioning between sediments, water, atmosphere and biological materials, thus creating a complex pattern of interfacial movements. Establishing quantitative values, even for single contaminants, can therefore be a highly costly and time-consuming procedure. Some insight into these problems can be gained from the work of Swackhamer and Eisenreich (1991) in which details are given of the behaviour of various organic chemicals in the North American Great Lakes. None the less, such data are a critical component of the exposure assessment which is required for a full risk assessment.

The assessment of chemical hazards is an ongoing process, and at present adequate assessments are available only for a relatively small number of the vast multitude of chemicals in use today. Within legislative forums, such as the European Union, non-assessed chemicals present enormous problems for regulation. These chemicals, which represent the vast majority of existing substances and preparations, may escape all forms of discharge control simply because there are no data on their toxicity or fate, or because such data are inadequate if available, or indeed because there are no data on their existence or usage. There are clearly enormous risks inherent in delaying regulation until such time as adequate data become available.

In an attempt to address this issue, a group of experts from the Danish Board of Technology have recently submitted a proposal to the EU outlining a scheme for the temporary classification of non-assessed chemicals on a group basis, such that:

> any chemical which is without data or inadequately documented within a group, block or cluster, shall be classified and labelled for hazards similar to those of the most hazardous chemical represented in the group.
>
> (Bro-Rasmussen *et al.*, 1996)

The intent of this proposal is clearly to provide some degree of interim environmental protection until such time as adequate data become available for a more complete hazard assessment. Should such a proposal be adopted, and given the large proportion

Table 9.2 Threshold parameter values for use in the decision tree to select UK Red List chemicals

Property or input of chemical	*Assessment parameter and units*	*Application to scenario(s)*			*Numerical thresholds used in ranking*		
		1	*2*	*3*	*High*	*Medium*	*Low*
Persistence	Half-life in water (days)	x	x	x	≥ 100	> 10 < 100	≤ 10
Acute aquatic toxicity	96-hour LC50 (mg/l)	x			≤ 1	< 100 > 1	≥ 100
Chronic aquatic toxicity	EC50 or 0.01 × 96h LC50		x		≤ 0.01	< 1 > 0.01	≥ 1
Toxicity to higher organisms	Oral LC50 in the rat (mg/kg body weight)			x	≤ 50	< 500 > 50	≥ 500
Bioaccumulative potential	Bioconcentration factor **			x	≥ 1,000	> 100 < 1,000	≤ 100
Production volume (UK or European)	Tonnes per annum	x	x	x	≥ 10,000	> 1,000 < 10,000	≤ 1,000
Source identification	Point, diffuse or mixed	x	x	x			
Solubility in water	mg/l	x	x	x	≥ 1,000	> 1 < 1,000	≤ 1
Volatility	Pascals (Pa)	x	x	x	≥ 0.133	> 0.000133 < 0.133	≤ 0.000133

** The given values are essentially arbitrary. Note the use of the application factor/safety factor in the evaluation of chronic toxicity, and the use of the log/octanol water partition coefficient as a surrogate for the bio-concentration factor

of chemicals that are at present not assessed, it is likely that the majority would be classified on a group basis for some considerable time to come.

Output data from hazard assessment

The output data from a hazard assessment consist of the toxicity values determined from tests, together with statistical confidence intervals where a sufficiently robust protocol has been devised. In addition, the assessment generates (or should generate) information on the physico-chemical properties of a given substance and on its likely inputs to the environment. These, in turn, form the basis for a number of subsequent actions to be taken, including the prioritisation of regulatory action and the setting of safe concentrations for the environment. Although the intuitive basis for assessing hazards in this way appears to be robust, using formalised frameworks such as that adopted in drawing up the UK Red List, in practice there are a number of limitations which attach varying degrees of uncertainty to the derived data.

HAZARD ASSESSMENT, UNCERTAINTY AND LIMITATIONS

Hazard assessment procedures, based on the tiered framework, are historically the single most important basis for understanding the effects of chemicals in the wider environment. Much of the procedural basis for this has been developed in the field of aquatic toxicology. While similar iterative processes can be identified in, for example, workplace hazard assessment for chemicals, the science has not yet achieved anything like the degree of sophistication of the frameworks applied to aquatic systems. This makes the study of the protocols applied to aquatic systems useful in identifying both the shortcomings of the methodology and the factors which give rise to uncertainty. Generally, the degree of uncertainty attaching to chemical effects will tend to increase in non-aquatic systems, including those in which humans may act as target organisms through the consumption of aquatic resources.

Figure 9.4, based on NZMoE (1996), illustrates the relevance of the specific effects of a toxic chemical to a given level of biological organisation. A

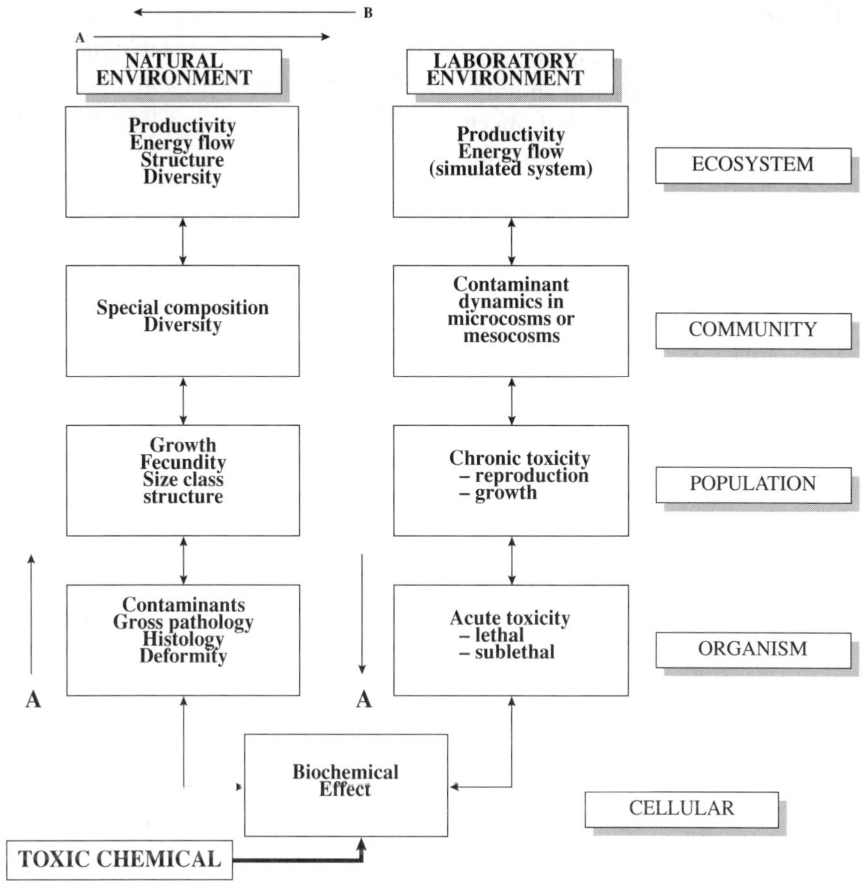

Figure 9.4 Representation of relevance relating to the toxicological understanding of levels of biological organisation. Relevance to natural systems increases in the direction of arrows labelled 'A'; levels of understanding increase in the direction of arrows labelled 'B'. Endpoints of interest at each level of biological organisation are shown

Source: NZMoE, 1996

comparison is also given in this figure of the overall level of understanding of toxicology at different levels of organisation (e.g. cellular, population, ecosystem). From an environmental protection viewpoint, however, the understanding of long-term effects at the whole ecosystem level is of paramount importance, but, sadly, there is at present a great lack of understanding at this level. Broadly, our understanding of laboratory systems, which by themselves are less relevant to environmental protection, is greater than that of natural systems. Similarly, although our understanding of the lower levels of biological organisation is greater than that of the higher levels, its relevance to environmental management and protection is low. In short, levels of our understanding with regard to both field and laboratory studies are greatest for individual species over relatively short time-scales, and this provokes the question as to the degree to which an understanding of individual species is, or could be, relevant to whole ecosystems. This paradox – that the organisational levels of greatest relevance are also the least understood – is discussed in more detail by Suter (1994) in the context of 'endpoints' applied to hazard and ecological risk assessment.

Assessment endpoints

The search for suitable endpoints in toxicity test procedures, to be applied in hazard assessment and subject to iterative verification, has provoked much discussion in the literature (Cairns, 1995). The discussion has extended to ecological risk assessment since the two forms of assessment are interdependent. Suter (1994) has drawn attention to the substantial difference between test endpoints, derived from toxicity tests (such as LC50 and EC50 values), and assessment endpoints which nearly always refer to effects on higher levels of ecosystem organisation. Much larger spatial and time frames are applied in assessment endpoints compared to even the most sophisticated simulated systems used in tests. Also, interactions are much more subtle and complex in real ecosystems than in simulated systems. In general, however, there are constraints which limit the predictive power of test endpoints. These constraints, or the way in which they operate, may differ markedly between simulated and real-world systems. Obviously, it would be very convenient if test and assessment endpoints coincided, but this would be highly unusual. Endpoints themselves fall into four broad categories as shown in Figure 9.4.

What are called sub-organismal endpoints involve biochemical, physiological or histological parameters, referred to as bio-markers. Although important, they are by themselves not usually interesting to environmental managers and, moreover, there are few predictive models capable of using such data as input. Organismal assessment endpoints, which measure death, life-span, reproductive vigour or behavioural response, are most often generated from standard toxicity test data (LC50 or EC50), but they cannot predict ecosystem effects. Population responses are favoured by environmental managers, while toxicity tests are rarely designed to generate data at the population level. Also, such responses are generally difficult to relate to specific chemicals.

Even so, population level responses are frequently used to assess response at the next level of organisation, i.e. response of whole ecosystems. Assessment at this level is fraught with difficulties, however, because both assessment and test endpoints are difficult to define. Therefore, a general strategy is to select ecosystem endpoints on the basis of defined management goals. Inevitably, the use of ecosystem endpoints in the assessment of ecosystems has been simplistic as experience shows, mainly because there is no realistic way of assigning causality by using what can be described as ecological epidemiology. With each layer of ecosystem organisation, there is thus a progressive move away from endpoints which could be useful in the assignment of causality.

Moreover, many ecosystem endpoints may take a long time before they can be expressed overtly and detected. Furthermore, changes in low-abundance species may be difficult to detect with standard statistical methods, even though they may well be the most sensitive to a particular environmental stressor. It is interesting to note that these problems are not unique to aquatic toxicology. They are equally apparent in human epidemiology (Taubes, 1995) too, where the issue of protecting the most sensitive individual, rather than the median, becomes essential. The difficulties of establishing causality in human populations imply a high degree of uncertainty. In fact, there are relatively few well-defined endpoints – and therefore a high level of dependence on carcinogenic effects as an index (Crouch, 1996) – outside the universe of other potential health problems that can be attributed to environmental or occupational exposure to toxic chemicals.

Any toxicity test is limited in its output to the endpoints essayed. Rather than demonstrating whether a chemical is or is not toxic, a test serves to answer specific questions about biochemical or physiological responses to a particular chemical under the established conditions and duration of the test. In this way a positive response may be much more informative than a negative one in the context of setting priorities for regulation and control.

Appraisal of single-species toxicity testing

Single-species toxicity testing has been the subject of numerous reviews and critiques, especially with regard to its relevance to wider ecosystem protection. The underlying uncertainty is that the test species selected may not adequately predict effects in the wider system. Cairns and Pratt (1989) point out that extrapolation of test results can be acceptable when the degree of environmental realism of the system, as simulated in the laboratory, is high. But such extrapolation becomes indefensible on scientific grounds when the natural environment is poorly simulated in laboratory systems. According to Cairns (1989) the most important deficiencies of the single-species toxicity test, where the test species is regarded as being the most sensitive, are:

- the most sensitive species may not be culturable under laboratory conditions and, therefore, of little use in toxicity testing;
- the relative sensitivity of a species to different toxicants is likely to vary with exposure to different toxicants;
- the response of a species to a toxicant does not reflect the effects of that toxicant on holistic ecosystem functions.

Some further general observations (NZMoE, 1996) may be added to the above list relating to the utility of the test species as predictors of wider effects:

- it may be difficult to detect non-lethal endpoints which are none the less significant at an ecosystem level;
- rare or endangered species, which may be highly vulnerable, are not considered in toxicity testing;
- it may not be possible to test all the life stages of an organism in the laboratory;
- the majority of toxicity data are acute data rather than chronic data;
- laboratory tests tend to operate over short timescales and are of little relevance to long-term prediction;
- the highly standardised test conditions do not reflect real environmental conditions.

On the whole, these observations indicate that there is a considerable (but unquantifiable) degree of uncertainty in the use of toxicity test data as predictive tools for wider environmental effects.

Appraisal of chemical dynamics and fate data

As described above, hazard assessment is generally based on tests of single chemicals on single species. In view of this and given the large number of chemicals currently being manufactured and used, it is clear that such assessments are fundamentally inadequate because they are also based on a highly simplified and stylised representation of what actually occurs in the natural environment. Although chemicals enter the environment as mixtures, relatively few of the toxicity tests have attempted to address the problem of chemical mixtures. However, recently there has been a general move towards conducting toxicity tests on whole effluents for regulatory purposes, involving the testing of mixtures (Wharfe and Tinsley, 1995; Bergman *et al.*, 1986). While these initiatives purport to address the issue of complex chemical mixtures, the test protocols are essentially based on the single-species format with little power to elucidate interactions between the components of such mixtures. Chemicals may interact toxicologically in a number of ways. Such interactions are shown conceptually in Figure 9.5 which illustrates the potential interactions between two toxic chemicals mixed in various proportions.

Analysis of complex effluents using gas chromatography, coupled with mass spectrometry, shows that they may contain a large number of chemicals (Johnston *et al.*, 1996). But only a relatively few of the various chemicals isolated can be readily identified. This has two important implications. First, if it is not possible to identify a chemical, then it would not be possible to carry out a hazard assessment on it. Second, the potential interactions are far too numerous to be determined using currently available methods. This is a little-appreciated limitation of the hazard assessment process which, as pointed out by Johnston *et al.* (1996), impinges directly on risk assessment procedures. Even when the chemicals can be identified, there may be little information available on their toxicological properties. A typical example of this is given in a recent study on the complex chemical mixture present in the River Tees in the UK; in this case no toxicological information could be found for 75 per cent of the chemicals isolated. In order to attempt a hazard assessment, it was therefore necessary to infer the toxicological properties from the structure of the chemical, using quantitative structure activity relationships (QSARs) (Matthiessen *et al.*, 1993). This method is gaining in popularity because of the large number of chemicals present in natural

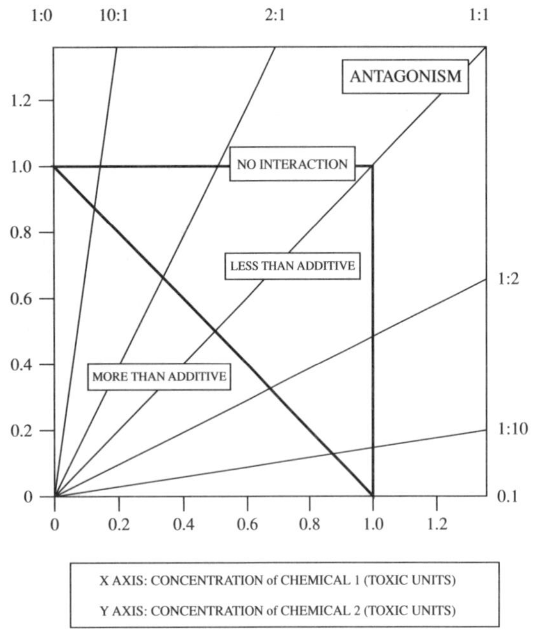

Figure 9.5 Isobologram showing potential interactions and lines of equal effect for a test species exposed to two toxic chemicals (LC50 or EC50 used as test endpoint). The radial lines denote fixed concentration ratios of the two chemicals. Other lines indicate the set of concentrations of both chemicals at which the response occurs. There is no standard biometric method for combinations that are more than additive, or which are antagonistic in their mode of action

Source: Adapted from Suter, 1993

environments and because current methodologies vary so widely (Bol *et al.*, 1993). The use of such methods represents a distinct move away from the traditional use of empirical data in that particular component of hazard assessment in which chemical properties are needed. This feature is mirrored in the use of surrogate values for the determination of bioconcentration factors, and also in the use of application and safety factors in the results of toxicity tests.

An overview of hazard assessment

Hazard assessment of chemicals, and especially of those likely to impact on aquatic environments, is a relatively highly developed area of toxicological science. None the less, its use as a predictive tool for wider ecosystem protection is somewhat uncertain because, as pointed out earlier, the use of single-species and/or single-chemical tests has little direct relevance to ecosystem protection. Furthermore, the use of single-species tests is more likely to be governed by constraints imposed by the need to handle and culture the species in the laboratory rather than by any key consideration relevant to ecosystem protection. Moreover, such tests do not adequately reflect the diversity of chemicals entering the environment, neither do they cater for the potential interactions between the chemicals or with other environmental compartments and media. Arguably, the most serious limitations arise from the inability of a hazard assessment to address the problem of chemical mixtures and the associated problem of identifying the various chemicals present in complex discharges. In addition, little ecotoxicological data is available for the great majority of chemicals. Thus it follows that, while any output data from a hazard assessment process may generate empirical data, such data may not be adequate for use as input into a risk assessment framework. Given that exposure assessment depends in part on output data from the hazard assessment process, it is clear that failure at this level can compromise the entire process of ecological risk assessment.

RISK-BASED HAZARD ASSESSMENT

Risk-based hazard assessment encompasses statistical methods which have been developed in an attempt to resolve some of the uncertainties inherent in hazard assessment. In particular, these methods are intended to resolve the problems associated with application and safety factors. As risk-based approaches, they stand somewhere between standard hazard assessments and full ecological risk assessment. The risk-based approach is exemplified by the procedures used in the Netherlands (HCN, 1993) for the setting of water quality criteria.

The mathematical basis of this approach is described in detail in OECD (1995) and will not be repeated here. The principal differences between standard hazard assessment and the risk-based approach are that in the latter greater emphasis is placed on NOEC rather than LC50/EC50, and toxicity data are required for a minimum of five species, ideally to be selected at random from the ecosystem under study. A major advantage is considered to be that this allows informed debate in place of safety factors. The system is designed to provide protection for 95 per cent of the species present in an ecosystem with a certainty of 50 per cent. This latter figure can of course be set at a higher level if desired. Several assumptions are made in constructing the basis of the approach, including that:

- the test species are selected randomly from the ecosystem;
- there are no interactions between species in the ecosystem;
- NOEC data are the most appropriate for setting quality criteria;
- NOEC data for five species comprise an adequate dataset;
- the data fit a log–logistic distribution.

While the last assumption can be tested statistically, the others are likely to fuel considerable debate in the future. This is particularly so because of reliance on NOEC data and considering the observations of Chapman *et al.* (1996) outlined above. Given the difficulties of culturing toxicity test organisms, it is unlikely that the first assumption will ever be fully robust. Thus the overall approach cannot be free from the most subjective element of conventional risk analysis, namely that test organisms should ideally be the most sensitive species and should be selected with this in mind. On the whole, the approach can be recommended mainly for its simplicity, notwithstanding the associated computing requirements. Otherwise it is a process which still depends on many of the underlying assumptions governing conventional hazard assessment. Clearly, it is very far removed from an ideal quantitative risk assessment.

EXPOSURE ASSESSMENT

An assessment of chemical exposure is the second element integral to the analysis of the problem which precedes a risk assessment. In this phase an estimation is made of the exposure of organisms to a chemical or an agent following their release, or relating to their presence in the environment, taking into account the transport, fate and uptake of such agents and duration of the exposure (Suter, 1993). It is highly unlikely, however, that the hazard assessment will have generated a comprehensive dataset on the behaviour and fate of a given chemical, thus leaving substantial areas of uncertainty in the estimation of exposure. Exposure assessment is essentially a model-driven process (Suter, 1993) in which the model simulates the environmental fate of the chemical in all media to which the system of interest is exposed. This information is elucidated, at least in theory, as a component of the hazard assessment framework. Such a model needs to simulate not only such parameters as transport, partitioning and transformation but also other relevant parameters such as the properties of the receiving environment. If these parameters are expressed in the form of probability distributions, then this would allow probabilistic fate models to be used to estimate distributions of exposure. The concentrations of a given toxicant can be transformed into exposure, but this would be contingent upon the existence of a knowledge-base which includes the natural history, behaviour and physiology of the target organism. Environmental concentrations are a prerequisite, but by themselves they are not sufficient for the determination of exposure.

Input estimates

Chemical input estimates are a key component of the information required to estimate exposure. In theory these may be based on the analysis of point source discharges, on measurements taken in the wider environment, or both. Very few point sources are, however, monitored intensively enough to enable reliable derivation of mass balance data in this way (see Ellis, 1986). Moreover, the licence conditions, on which monitoring efforts are based, usually only regulate a small proportion of the chemicals that are routinely discharged. In general, the frequency and intensity of monitoring in the wider environment are too low for the data to resolve trends in chemical concentration. Input estimate figures derived in this way are subject to wide uncertainty as evidenced by those published for the North Sea (NSTF, 1993). In some cases the maximum and minimum estimates differ by a factor of 10. This problem has been recognised since 1984 when for the first time such estimates were produced for the North Sea. The reliability of loading figures for rivers, for example, has remained low in the intervening period. Since the North Sea is one of the most intensively monitored marine systems in the world, the quality of chemical input data from other regions is likely to be even less reliable. Usually input estimates are made for a relatively small set of potential chemical inputs, and this further compounds the difficulty of dealing with mixtures of chemicals.

Quality of analytical data

Instrumental methods of analysis are generally employed to determine the concentrations of toxic chemicals of interest, but often measurements made on a given sample by one laboratory bear no resemblance to those made on the same sample by another (Wells, 1993; Wells *et al.*, 1993). Consequently, interpretation of the environmental significance of such datasets may not amount to more than an observation on the spread of errors generated by the scientists concerned. Thus the use of any data, whether generated by current techniques or as part of the historical dataset, requires them to be validated through the use of rigorous quality assurance and quality control. Unfortunately, the numbers of determinands covered by international quality assurance schemes are relatively limited. This is well illustrated by the QUASIMEME scheme (Wells *et al.*, 1993). Even so, the best available data do not allow trends to be calculated with any confidence, and neither is monitoring sufficiently frequent to permit the data to be used to resolve events over short to medium time-scales. A typical example of this is provided by the International Council for Exploration of the Seas (ICES) Co-operative Monitoring Programme (CMP) on levels of contaminants in fish muscle (Fryer and Nicholson, 1993). Thus, even accounting for randomness between yearly variations, it is considered unrealistic to expect to detect any patterns of change over a 2–3 year period. Given that monitoring data are available for an eight-year period, this would give sufficient power to the programme to detect trends in zinc concentrations of 10 per cent per year, and in copper and mercury concentrations of 20 per cent per year, on the assumption that the analytical data are reliable. Furthermore, if large-scale pollution incidents are considered, copper concentrations will have to be changed by 200 per cent and mercury by 400 per cent. Thus it is only by extending the dataset to 20 years that 5 per cent trends can be reliably detected. But the environmental impacts of a chemical over this period, before the trends begin to emerge, could well prove to be unacceptable.

Validation of models

In some cases inputs can be established on a relatively sound basis, and ambient environmental levels of the target chemical can be determined accurately using instrumental analysis on samples taken from the field. Patterns of pesticide usage, for example, can be determined for a particular area where the quantity sold and consumed is known. Also, the quantities of the pesticide entering the aquatic environment can be determined through an analytical campaign. A recent example of this is the extensive risk analysis for the herbicide atrazine carried out for North America (Solomon *et al.*, 1996). Eighty-three per cent of the total consumption of 36,000 tonnes of this herbicide in the USA in 1993 was used in corn cultivation, predominantly in the midwestern states. But, although atrazine entered the environment through diffuse pathways, the regional patterns of its usage, and the quantities used, were well characterised. This contrasts with the case of many other industrial chemicals whose point sources can be traced, even though the patterns of their production, use and disposal are much wider.

In the case of atrazine, this pattern of usage allowed a relatively highly focused sampling and analytical campaign. The analytical results were then used in an attempt to validate the models used for predicting pesticide runoff. It was found that one of these models consistently underestimated the levels of atrazine, while the other consistently overestimated environmental levels due to runoff. This emphasises the need to develop and refine robust models to support the exposure assessment process.

An overview of exposure assessment

Taken together, the very real difficulties of deriving robust figures for input, coupled with uncertainties attached to the analytical data derived from monitoring chemicals in the environment, suggest that valid exposure assessments are likely to be the exception rather than the rule. Calculation of exposure can be simplified in some cases where inputs are regionally concentrated or where exposure is predominantly via a specified pathway. Alternatively, it would be necessary to validate the assumptions through an analytical campaign. On the whole, there appears to be little hope of characterising the exposure of organisms, or trends in exposure, on an ecosystem-wide basis with any great certainty. This represents one aspect of the extreme subjectivity inherent within any risk assessment, and it is difficult to see how this problem can be fully resolved. Intensification of wide field monitoring regimes is likely to be a costly procedure. Cost penalties will also accrue from more intense oversight of point source discharges. In the latter case, however, improvement in data resolution is likely to increase to a greater extent for a given unit cost. This is simply because analysis of concentrated discharges is technically less difficult than analysis involving samples taken from open waters after considerable dilution has taken place. On the other hand, it can be argued that accurate data on contaminants in open waters gives a more reliable indication of ecosystem exposure.

In essence, point source monitoring implies a bias towards the control of chemicals, while the monitoring of ecosystem levels draws environmental protection further towards a strategy for the management of chemicals in the ecosystem. Overall, failure to establish exposure data, coupled with the enormous uncertainty in the measurement of toxicity (particularly in those related to endocrine disruption), means that a risk evaluation can seldom be developed beyond a putative hazard assessment.

Conceivably, it could be argued that uncertainties in exposure data could be compensated for – by manipulating exposure as a variable within the risk assessment framework, for example – but such manipulation would actually violate the critical commitment in risk assessment to establish empirical exposure values (USEPA, 1989). An alternative approach would be to look for both generalised and specific endpoints of response to toxicological stress and to monitor them in the environment. This may be a useful approach in cases where there is a specific response to specific exposure, such as the formation of DNA adducts on exposure to benzo[a]pyrene or the presence of imposex in a population of marine molluscs exposed to tributyl tin (TBT). However, such direct relationships are rare in practice, even in workforces that are occupationally exposed and certainly in natural ecosystems. While it is possible that future research will bring to light more agent-specific biochemical or molecular indicators of dose, the difficulties in assessing exposure are likely to remain a major obstacle to risk assessment, as defined, for some time to come.

ENDOCRINE-DISRUPTING CHEMICALS

Recently a number of reports have appeared on the effects of a number of chemicals which interfere with the endocrine system (Colborn, 1994; DEPA, 1995; MRC-IEH, 1995). Some chemicals with this ability have been identified, including various organochlorine pesticides, the PCBs, the chlorinated dioxins, some phthalate plasticisers and some phenolic compounds. In addition, many biocides are designed to disrupt the hormonal pathways of target organisms. The endocrine system produces hormones which mediate a wide range of homeostatic functions of organisms. They are also involved in the development of the organism and in its reproduction. Generally, they act at very low concentrations within the body and may be produced only for a limited time in the life of an organism. Alternatively, they may mediate different processes at different points in the life-span of the organism. Thus, if a chemical can either mimic a hormone, or interfere in some way with the metabolic and catabolic processes controlling body levels, then it will have the potential to affect a wide range of functional processes of the organism. The best known example of this in the aquatic environment is probably the feminisation of fish exposed to sewage effluents (White *et al.*, 1994). In humans such chemicals are suspected of being involved in the observed decline in male fertility and in the aetiology of some cancers and reproductive disorders. It is thought that some chemicals, dioxins for example, may already be present in the environment at sufficiently high concentrations to affect humans.

Potential effects of these chemicals range from immune suppression to the impairment of cognitive development through *in utero* exposure. Many of the potential endpoints could, therefore, constitute a severe impact on the individual, but they would only be detectable at the population level, possibly over several generations. These possible impacts on human populations also bring to the fore the ethical dimensions arising from variable individual responses mentioned earlier (see pp. 170–173).

Chemicals capable of disrupting the endocrine system serve to illustrate the practical weaknesses of the risk assessment paradigm. The endpoint of the effect is at the sub-organismal level through interference with hormonal activity. Organismal responses, which may affect entire populations (e.g. effects on behaviour and learning ability), have been identified, but they may only be detected at population levels. Ecosystem effects are yet to be described.

These effects may occur at extremely low concentrations and their dose–response relationship is not clear. Moreover, the classical assumptions underpinning dose–response relationships may not be valid in such cases. Finally, exposures still remain to be quantified and are not amenable to assessment. Most importantly, there appears to be no way in which this impact can be predicted on the basis of chemical structure alone, thus ruling out the use of surrogate measures such as QSARs. There is little doubt, therefore, that chemicals with endocrine-disrupting properties will prove to be a substantial and long-term challenge to ecotoxicologists and risk assessors alike.

CONCLUSIONS: RISK ASSESSMENT AND REALITY

What has been said in this chapter may seem at first sight to be somewhat negative with regard to the view it portrays on the sciences of toxicology and ecotoxicology. But it should be seen in the context of the fact that toxicological and ecotoxicological studies are the only basis on which to evaluate the impacts of a chemical on the environment. Provided that the limitations are understood and acknowledged, such studies will continue to increase our understanding and contribute to environmental protection. However, risk assessment procedures are at present poorly served by ecotoxicological and toxicological data derived through chemical hazard assessment and exposure assessment. The major deficiencies include:

- inadequate data for all but a small subset of chemicals entering the environment and a small subset of potential effects;
- lack of precision in chemical input estimates, and lack of direct measurements which compromise exposure assessment;
- test endpoints do not correspond with assessment endpoints;
- doubtful relevance of test organisms to ecosystem conditions and status;
- assessments cannot accommodate newly discovered effects (e.g. endocrine disruption);
- the process does not address the issue of chemical mixtures or chemicals present in them;
- risk assessment cannot be applied to unidentified chemicals;
- in many cases models for exposure assessment are not robust.

In view of the above, risk assessments should be regarded with deep scepticism unless all the areas of uncertainty are defined and the process is carried out with periodic and penetrating reality checks. These concerns are also articulated by Power and McCarty (1996). They argue that the critics of probabilistic risk analysis point to the fact that what the assessors are actually engaged in is a complex process of applying subjective judgements to the determination of risk. But what should be remembered in this context is that regulation, arrived at by whatever means, is at best a compromise between what is achievable and what is desirable. The best way to resolve the manifold uncertainties is to accept that regulations are merely resting places on the road to a goal of zero-discharge (Sprague, 1991).

REFERENCES

Agg, A.R. and Zabel, T.F. (1990) 'Red List substances: selection and monitoring', *J. Inst. Water Environ. Management* 4: 44–50.

ASTM (1991) *Annual Book of ASTM Standards, Section 11: Water and Environmental Technology*, Philadelphia, Pa.: American Society for Testing and Materials.

Bergman, H.L., Kimerle, R.A. and Maki, A.W. (eds) (1986) *Environmental Hazard Assessment of Effluents*, Oxford: Pergamon Press.

Bol, J., Verhaar, H.J.M., van Leeuwen, C.J. and Hermens, J.L.M. (1993) *Predictions of the Aquatic Toxicity of High Production Volume Chemicals. Part A: Introduction and Methodology*, The Hague, The Netherlands: Ministry of Housing, Physical Planning and Environment.

Bro-Rasmussen, F., Lokke, H., Kristensen, P., Laursen, E., Boyd, H.E., Jorgensen, C.E. and Nielsen, K.M. (1996) *The Non-assessed Chemicals in the EU. Report and recommendations from an interdisciplinary group of Danish experts*, Danish Board of Technology 1996/5, ISBN 87-90221-10-9.

Cairns, J. (1989) 'Applied ecotoxicology and methodology', in A. Boudou and F. Ribeyre (eds) *Aquatic Toxicology: Fundamental Concepts and Methodologies*, Boca Raton, Fla.: CRC Press Inc., pp. 275–290.

Cairns, J. (1995) 'The genesis of ecotoxicology', in J. Cairns and B.R. Niederlehner (eds) *Ecological Toxicity Testing: Scale, Complexity and Relevance*, Boca Raton, Fla.: Lewis Publishers, pp. 1–10.

Cairns, J. and Pratt, J.R. (1989) 'The scientific basis for bioassays', *Hydrobiologia* 188/189: 5–20.

Chapman, P.M., Caldwell, R.S. and Chapman, P.F. (1996) 'A warning: NOECs are inappropriate for regulatory use', *Environ. Toxicol. Chem.* 15: 77–79.

Colborn, T. (1994) 'The wildlife/human connection: modernising risk decisions', *Environ. Health Perspect.* 102 (Suppl. 12): 55–59.

Crouch, E.A.C. (1996) 'Uncertainty distributions for cancer potency factors: laboratory animal carcinogenicity bioassays and interspecies extrapolation', *Hum. Ecol. Risk Assess.* 2: 103–129.

DEPA (1995) *Male Reproductive Health and Environmental Chemicals with Estrogenic Effects*, Copenhagen: Danish Environmental Protection Agency.

Edwards, P. (1992) *Dangerous Substances in Water: A Practical Guide*, London: Environmental Data Services Ltd.

Ellis, J.C. (1986) *Determination of Pollutants in Effluents: Part B: Alternative Forms of Effluent Contents: Some Statistical Considerations*, WRc Environment Report No.: TR235, Medmenham, Bucks., UK: WRc Environment.

Finney, D.J. (1962) *Probit Analysis*, Cambridge: Cambridge University Press.

Fryer, R.J. and Nicholson, M.D. (1993) 'The power of a contaminant monitoring programme to detect linear trends and incidents ICES', *J. Mar. Sci.* 50: 161–168.

HCN (Health Council of the Netherlands) (1993) 'Ecotoxicological risk assessment and policy making in the Netherlands – dealing with uncertainties', *Network* 6/7: 8–11.

Johnston, P.A. (1987) 'Acute toxicity of inorganic selenium to *Daphnia magna* (Straus) and the effect of subacute exposure upon growth and reproduction', *Aquatic Toxicology* 10: 335–352.

Johnston, P.A., Santillo, D. and Stringer, R.L. (1996) 'Risk assessment and reality: recognising the limitations', in M.D. Quint, D. Taylor and R. Purchase (eds) *Environmental Impact of Chemicals: Assessment and Control*, Cambridge: Royal Society of Chemistry, pp. 223–239.

Kendall, R.J., Lacher, T.E., Bunck, C., Daniel, B., Driver, C., Grue, C.E., Leighton, F., Stansley, W., Watanabe, P.G. and Whitworth, M. (1996) 'An ecological risk assessment of lead shot exposure in non-waterfowl avian species: upland game birds and raptors', *Environ. Toxicol. Chem.* 15: 4–20.

Matthiessen, P., Thain, J.E., Law, R.J. and Fileman, T.W. (1993) 'Attempts to assess the environmental hazard posed by complex mixtures of organic chemicals in UK estuaries', *Mar. Poll. Bull.* 26: 90–95.

Mount, D.I. (1995) 'Development and use of single species aquatic toxicity tests', in J. Cairns and B.R. Niederlehner (eds) *Ecological Toxicity Testing: Scale, Complexity and Relevance*, Boca Raton, Fla.: Lewis Publishers, pp. 97–104.

MRC-IEH (1995) *IEH Assessment on Environmental Oestrogens: Consequences to Human Health and Wildlife*, Medical Research Council, Institute for Environment and Health, University of Leicester, Assessment A1.

NSTF (1993) *North Sea Quality Status Report 1993*, North Sea Task Force, London, Oslo and Paris Commissions.

NZMoE (1996) *A Proposed Methodology for Deriving Aquatic Guideline Values for Toxic Contaminants*, Wellington, New Zealand: Pollution and Risk Management Directorate, New Zealand Ministry for the Environment.

OECD (1995) *Guidance Document for Aquatic Effects Assessment*, Monograph Series No. 92, Paris, France: Organisation for Economic Co-operation and Development.

Power, M. and McCarty, L.S. (1996) 'Probabilistic risk assessment: betting on its future', *Hum. Ecol. Risk Assess.* 2: 30–34.

Shane, B.S. (1994) 'Introduction to ecotoxicology', in L.G. Cockerham and B.S. Shane (eds) *Basic Environmental Toxicology*, Boca Raton, Fla.: CRC Press, pp. 3–10.

Smith, E.P. (1995) 'Design and analysis of multispecies experiments', in J. Cairns and B.R. Niederlehner (eds) *Ecological Toxicity Testing: Scale, Complexity and Relevance*, Boca Raton, Fla.: Lewis Publishers, pp. 73–95.

Solomon, K.R., Baker, D.B., Richards, R.P., Dixon, K.R., Klaine, S.J., La Point, T.W., Kendall, R.J., Weisskopf, C.P., Giddings, J.M., Giesy, J.P., Hall, L.W. and Williams, W.M. (1996) 'Ecological risk assessment of atrazine in North American waters', *Environ. Toxicol. Chem.* 15: 31–76.

Sprague, J.B. (1969) 'Measurement of pollutant toxicity to fish, I: Bioassay methods for acute toxicity' (Review Paper), *Water Research* 3: 793–821.

Sprague, J.B. (1990) 'Aquatic toxicology', in C.B. Schreck and P.B. Moyle (eds) *Methods for Fish Biology*, Bethesda, Md.: American Fisheries Society, pp. 491–528.

Sprague, J.B. (1991) 'Environmentally desirable approaches for regulating effluents from pulp mills', *Wat. Sci. Technol.* 24: 361–371.

Suter, G.W. (1993) *Ecological Risk Assessment*, Boca Raton, Fla.: Lewis Publishers.

Suter, G.W. (1994) 'Endpoints of interest at different levels of biological organisation', in J. Cairns and B.R. Niederlehner (eds) *Ecological Toxicity Testing: Scale, Complexity and Relevance*, Boca Raton, Fla.: Lewis Publishers, pp. 35–48.

Swackhamer, D.L. and Eisenreich, S.J. (1991) 'Processing of organic contaminants in lakes', in K.C. Jones (ed.) *Organic Contaminants in the Environment*, London: Elsevier, pp. 33–86.

Taub, F. (1989) 'Standardised aquatic microcosms', *Environmental Science and Technology* 23 (9): 1064–1066.

Taubes, G. (1995) 'Epidemiology faces its limits: special news report', *Science* 239: 164–169.

Train, R.E. (1979) *Quality Criteria for Water*, London: Castle House Publications Ltd.

UKEA (1996) *The Application of Toxicity Based Criteria for the Regulatory Control of Wastewater Discharges*, A consultation document issued by the Environment Agency on the proposed introduction of toxicity-based conditions to discharge licences, Worthing, UK: Environment Agency.

USEPA (1989) *Assessing Human Health Risks from Chemically Contaminated Fish and Shellfish*, Report No. EPA-503/8-89-002, Washington, DC: United States Environmental Protection Agency.

Wells, D.E. (1993) 'QUASIMEME – an introduction', *Quasimeme Bulletin* No. 1 (March), Aberdeen: Scottish Office Agriculture and Fisheries Department.

Wells, D.E., Cofino, W.P., Quevauviller, P. and Griepink, B. (1993) 'Quality assurance of information in marine monitoring: a holistic approach', *Mar. Poll. Bull.* 26 (7): 368–375.

Wharfe, J.R. and Tinsley, D. (1995) 'The toxicity-based content and the wider application of direct toxicity assessment to protect aquatic life', *J. IWEM* 9: 526–530.

White, R., Jobling, S., Hoare, S.A., Sumpter, J.P. and Parker, M.G. (1994) 'Environmentally persistent alkylphenolic compounds are estrogenic', *Endocrinology* 135 (1): 175–182.

SUGGESTED READING

Cairns, J. (1995) 'The genesis of ecotoxicology', in J. Cairns and B.R. Niederlehner (eds) *Ecological Toxicity Testing: Scale, Complexity and Relevance*, Boca Raton, Fla.: Lewis Publishers, pp. 1–10.

Suter, G.W. (1994) 'Endpoints of interest at different levels of biological organisation', in J. Cairns and B.R. Niederlehner (eds) *Ecological Toxicity Testing: Scale, Complexity and Relevance*, Boca Raton, Fla.: Lewis Publishers, pp. 35–48.

SELF-ASSESSMENT QUESTIONS

1 Standard laboratory-based toxicity tests often fall short of the defined requirements for hazard assessment because:
   (a) the test organisms most frequently chosen do not survive under laboratory conditions;
   (b) the concentrations of toxicants employed are frequently too low to induce measurable effects;
   (c) it is generally not possible to employ species which are likely to be the most sensitive to a particular toxicant;
   (d) only lethal effects can be recorded;
   (e) it is not possible to carry out tests involving long-term exposure.
2 Hazard assessment is fundamentally based on a comparison of:
   (a) concentrations of a toxicant resulting in lethal versus sublethal effects;
   (b) predicted concentrations of a toxicant released to the environment versus background concentrations;
   (c) relative concentrations of different toxicants in chemical mixtures;
   (d) measured or predicted environmental concentrations versus concentrations known to induce an observed effect;
   (e) potential risks to organisms versus the value of such species as a human food source.
3 Which of the following statements regarding the UK Red List evaluation protocol is incorrect?
   (a) Bio-accumulative potential may be expressed as an octanol/water partition coefficient.
   (b) Provision is made for the consideration of effects of toxicants on higher trophic levels.
   (c) Persistence is expressed in terms of the time taken for the toxicant to decline in concentration to background levels.
   (d) An arbitrary safety factor of 0.01 may be applied to the results of acute toxicity tests in the absence of chronic test results.
   (e) Provision is made for the consideration of the production volumes of chemicals.
4 Which of the following statements concerning endpoints is incorrect?
   (a) The problems of determining cause–effect relationships are greater at the population level than at the organismal level.

(b) Epidemiological studies attempt to relate cause and effect at the population level.
(c) The term 'biomarker' refers to endpoints that can be detected at the sub-organismal level.
(d) Endpoints involving organismal or sub-organismal effects are most commonly employed in hazard assessment.
(e) Endpoints that can be detected at the sub-organismal level are generally preferred by decision-makers.

5 Hazard and risk assessment protocols are relatively well defined for aquatic systems because:
(a) aquatic organisms are more readily cultured in the laboratory than terrestrial organisms;
(b) hazard assessment has largely been driven by the need to regulate chemical discharges to the aquatic environment;
(c) chemicals are more toxic to aquatic organisms than to terrestrial organisms;
(d) toxicity tests can only be carried out using aqueous solutions of toxicants;
(e) understanding of the ecology of aquatic systems is much greater than that of terrestrial systems.

6 Which of the following statements relating to hazard and risk assessments is most correct?
(a) Conventional hazard assessment avoids the use of toxicity testing by basing the assessment on the fundamental properties of chemicals.
(b) Exposure assessment provides an alternative to hazard assessment.
(c) Risk assessment attempts to replace arbitrary safety factors with empirical risk factors.
(d) Environmental exposure of organisms to a chemical can be described adequately by its concentration in the environment.
(e) Risk assessment surmounts the problems resulting from lack of data required for hazard assessment.

7 Which of the following statements relating to industrial discharges is incorrect?
(a) Industrial effluents may contain complex mixtures of chemicals.
(b) Toxicological data are available for a relatively small proportion of chemicals released to the environment.
(c) Hazard assessment depends on the reliable identification of chemicals.
(d) Chemical compositions of industrial effluents are generally well characterised.
(e) The effects of an individual chemical in a discharge on organisms in receiving waters may be modified through interaction with other chemicals in the effluent.

8 Identify the incorrect statement.
The phenomenon of endocrine disruption confounds conventional risk assessment procedures because:
(a) levels of hormones in the body are too low to measure;
(b) dose–response relationships are poorly defined for endocrine-disrupting effects;
(c) endpoints are difficult to define as effects on individuals may only be measurable at the population level;
(d) the chemical structure does not appear to be a useful predictor of endocrine disrupting potential;
(e) effects may occur at exposure levels which are not amenable to assessment.

10
ENVIRONMENTAL IMPACT ASSESSMENT

Dimitri Devuyst

SUMMARY

This chapter focuses on Environmental Impact Assessment (EIA), its various aspects and modern developments. Starting from its birth in the United States of America in the early 1970s as a requirement in that country's National Environmental Policy Act, the evolution of EIA to date in North America, Europe and other countries has been charted. The role of EIA as an essential environmental management tool has been emphasised and presented in a simple but comprehensive manner, with the aid of case studies and illustrations, to facilitate the understanding and application of its fundamental principles and methodology. The basic steps in the preparation of an Environmental Impact Statement have been explained, together with a discussion and scrutiny of various aspects such as matrix methods, networks, the vital role of public participation, monitoring and quality control, as well as the limitations and shortcomings of conventional EIA procedures.

The discussion then moves on to more advanced topics including Cumulative Impact Assessment, Strategic Impact Assessment, socio-economic impacts, integrated dynamic models, area-wide EIA, as well as the current perception of EIA from a strictly scientific perspective. It is argued that much work remains to be done to improve current EIA procedures before they can be relied upon to give better and more acceptable predictions of future impacts.

The current status of EIA in developing countries has also been discussed, together with suggestions of how their EIA procedures could be adapted to local culture, conditions and constraints, and ways in which EIA can be promoted more effectively in those countries.

ACADEMIC OBJECTIVES

The aim of this chapter is to introduce EIA to the reader as an instrument for realising a preventive environmental policy, to give an overview of the basic steps to be followed in preparing an EIA, to discuss the different types of problems inherent in EIA, and to examine possible future developments.

On completion of this chapter the reader should be able to:

- define EIA;
- give an overview of the history of EIA;
- recognise the essential aspects of an EIA procedure;
- determine the use of and need for EIA in decision-making;
- identify problems inherent in EIA;
- discuss future developments in EIA;
- start an examination of EIA systems and procedures in the country of the reader;
- have an idea of the quality of an environmental impact statement.

DEFINITIONS

Environmental Impact Assessment (EIA) is an instrument which is used to aid and improve the decision-making process. The objective of an EIA is to determine the potential environmental, social, and health effects of a proposed development project. It attempts to assess the physical, biological, and socio-economic effects of the project in a way that permits logical and rational decision-making. It also facilitates the development of strategies to mitigate or eliminate any potential adverse impacts of the project (Clark, 1989).

Although there are several definitions of EIA, none is generally or universally accepted. The definition given by Clark (1989) would appear to be the most comprehensive and to the point. According to him

> Environmental Impact Assessment is a procedure for encouraging decision-makers to take account of the possible effects of development investments on environmental quality and natural resource productivity and a tool for collecting and assembling the data planner[']s need to make development projects more sustainable and environmentally sound. EIA is usually applied in support of policies for a more rational and sustainable use of resources in achieving economic development.
>
> (Clark, 1989: 4–5)

The main goal of EIA is to prevent environmental damage or degradation caused by human activities. It provides decision-makers with an account of the implications of a proposed project or action before any decision on it is made. Another goal is to make decision-makers more accountable for their actions by opening up the decision-making process for public scrutiny. The Environmental Impact Statement (EIS), which results from the EIA process, is prepared in a systematic and multidisciplinary manner in order to ensure that planning and decision-making processes are based on relevant inputs from all branches of science.

HISTORICAL ASPECTS OF ENVIRONMENTAL IMPACT ASSESSMENT

The United States of America and the National Environmental Policy Act

Growing pressure from environmental movements, which started in the United States of America in the late 1960s and the early 1970s, led to the National Environmental Policy Act (NEPA) which became law in that country on 1 January 1970. The policy objective of NEPA was to promote and safeguard environmental quality as a national priority. Indeed, it was the introduction of NEPA which resulted in the creation of a new 'action-forcing' mechanism – the EIA. The intended purpose of the EIS, which is the document which results from the EIA process, was to force both public and private organisations and agencies to take the substantive provisions of the NEPA seriously, and to take into account the environmental policy directives of the US Congress in formulating their plans and procedures. This American EIA procedure, and its analytical process established by NEPA, continues to be a model for comparable systems worldwide.

Canada and the Canadian Environmental Assessment Act

On 20 December 1973, the first Canadian EIA process was established by the Federal Government of Canada with the introduction of the Environmental Assessment Review Process (EARP). The purpose of EARP was to ensure that environmental consequences of all federal proposals were assessed for their potential adverse effects early in the planning process before irrevocable decisions were made.

The next notable development was in 1987 when the Minister of the Environment initiated a major reform of the federal Canadian EIA system through nation-wide consultations. The reform reached an important milestone in June 1992 with the passage through Parliament of the Canadian Environmental Assessment Act which was implemented in January 1995. Also, the old Federal Environmental Assessment Review Office (FEARO) was replaced by the Canadian Environmental Assessment Agency (CEAA). Established as a national organisation responsible for putting the Act into practice, the CEAA highlights the following provisions of the Act:

- EIA in Canada now has a statutory basis, reducing legal uncertainties and the need for court interpretations;
- sustainable development is established as a fundamental objective of the federal environmental assessment process;
- public participation is one of the goals of the Act, with increased opportunities for public input to all types of environmental assessment;
- all projects receive an appropriate degree of environmental assessment. The degree depends

largely on the scale and complexity of the likely effects of the project. Consequently, there are four types of environmental assessment: screening, comprehensive study, mediation and panel review.

The European Union and Directive 85/337/EEC

In the European Union, which aspires to political and economic integration and stability, an environmental policy has been in operation since 1973. The EU has established two fundamental principles with which all its environmental policy measures have to comply. According to the first principle, the polluter should pay for the costs of emission control and anti-pollution measures. The second principle, explicitly adopted by the EU, is that 'prevention is better than cure' (Williams, 1988). The implementation of a European Directive on the assessment of environmental effects is a logical step which is in accordance with the second principle.

In March 1985, after many years of deliberation, the Environment Council of the European Communities approved Directive 85/337/EEC which was formally issued on 27 June 1985. It is also known as the 'EIA Directive'. The potential scope of this Directive 'on the assessment of the effects of certain public and private projects on the environment' is very wide. It takes into account both direct and indirect effects of a project on the following factors:

(a) human beings, flora and fauna;
(b) soil, water, air, climate, and landscape;
(c) interactions between the factors mentioned in (a) and (b);
(d) material assets and cultural heritage.

Two strong and conflicting interests influencing the European policy-making process had to be considered in the Directive. They were the need to harmonise environmental legislation throughout the EU on the one hand, and the demand of most of the EU member states to retain their national freedom of action on such matters on the other. The primary purpose of harmonisation was, and still is, to create a 'level playing field' throughout the EU, without which European enterprises in member states with stringent environmental laws would be tempted to re-locate in other member states with less stringent laws for greater profits, thus distorting the market with unfair competition.

Thus, of necessity Directive 85/337/EEC had to be a compromise. It sets minimum standards that may be exceeded, but cannot be ignored, by governments of the member states; and the discretion on its implementation rests to a large extent with the national authorities. Table 10.1 gives an overview of projects that are subject to EIA as required by the Directive. Annex I of the table contains a restricted list of projects for which the preparation of an EIA is mandatory. Also, the Directive invites member states to specify other types of projects, in the areas summarised in Annex II of Table 10.1, which they would like to be subject to an assessment and for which they themselves may establish the necessary criteria and/or thresholds. This EIA Directive was implemented by the twelve member states in July 1988 (Devuyst and Hens, 1991).

Implementation of the EIA Directive in different EU member states was evaluated by the EIA Centre at the University of Manchester in 1993. This study revealed major variations in the way EIA had been implemented in different member states. For example, the number of EISs prepared annually in the Irish Republic and Portugal was only about eighty, compared with a massive 5,000–8,000 in France. But these numbers did not say anything about the quality of the documents, which was also very variable. When it became clear that a majority of the EISs was of insufficient quality, the European Commission made a proposal in January 1996 to amend Directive 85/337 with proposed changes that are concerned mainly with modifications or additions to Annexes I and II.

Worldwide dissemination of EIA principles

The need to incorporate environmental considerations into the process of development is gradually being accepted in all countries around the world. Most countries have either a formal EIA system, or are in an experimental stage with no systematic or structured procedure for preparing EISs. The question is no longer whether the EIA principle

Table 10.1 Projects subject to EIA, as required by EC Directive 85/337/EEC

Annex I	*Annex II*
1 Crude-oil refineries (excluding undertakings which manufacture only lubricants from crude oil) and installations for the gasification and liquefaction of 500 tons or more of coal or bituminous shale per day.	1 Agriculture.
2 Thermal power stations and other combustion installations with a heat output of 300 megawatts or more, and nuclear power stations and other nuclear reactors (except research installations for the production and conversion of fissionable and fertile materials, whose maximum power does not exceed 1 kilowatt of continuous thermal load).	2 Extractive industry.
3 Installations solely designed for the permanent storage or final disposal of radioactive waste.	3 Energy industry.
4 Integrated works for the initial melting of cast-iron and steel.	4 Processing of metals.
5 Installations for the extraction of asbestos and for the processing and transformation of asbestos and products containing asbestos: for asbestos-cement products with an annual production of more than 20,000 tons of finished products, for friction material, with an annual production of more than 50 tons of finished products, and for other uses of asbestos, utilisation of more than 200 tons per year.	5 Manufacturing of glass.
6 Integrated chemical installations.	6 Chemical industry.
7 Construction of motorways, express roads and lines for long-distance railway traffic and of airports with basic runway length of 2,100 m or more.	7 Food industry.
8 Trading ports and also inland waterways and ports for inland waterway traffic which permit the passage of vessels of over 1,350 tons.	8 Textile, leather, wood and paper industries.
9 Waste-disposal installations for the incineration, chemical treatment or land fill of toxic and dangerous wastes.	9 Rubber industry.
	10 Infrastructure projects.
	11 Other projects (such as holiday villages, hotel complexes, waste-water treatment plants, sludge-deposition sites and test benches for engines, turbines or reactors).
	12 Modifications to development projects included in Annex I, and projects in Annex I undertaken exclusively or mainly for the development and testing of new methods or products and not used for more than one year.

is valid and applicable, but rather how it can be operationally incorporated into the planning and management process (Biswas and Qu Geping, 1987).

International organisations, such as the World Bank and the Asian Development Bank, have played an important role in introducing EIA to developing countries. Historically, development projects in those countries have been aimed at removing strategic economic bottlenecks and for creating the basis for economic growth; but little consideration was given to the negative social and environmental impacts that are often associated with such projects. However, with a better understanding of these impacts, increasingly greater emphasis is now being given to formulating and implementing measures to mitigate them (Goodland *et al.*, 1996).

As a result, today the preparation of EISs following certain guidelines has become mandatory for all major development projects seeking the support of institutions such as the World Bank and the Asian Development Bank. For example, since 1989 EIA has been a requirement for the funding of major projects by the World Bank. As a result, during 1989 and 1995 more than a thousand World Bank-funded projects of different type were screened for their potential environmental impacts on regions and sectors. This requirement has to a large extent motivated both decision-makers and institutions in many countries to acquire the art of EIA.

Given that the concept of EIA is rooted in North American culture, it is clear that certain difficulties will arise when attempts are made to implement it in its original form in countries with different cultural norms and values. For example, public participation and transparency of decision-making, both of which are fundamental tenets of EIA, can be a serious problem in countries in which such openness is a strange concept. It would follow, therefore, that since deep-rooted cultural traditions or systems of government cannot be changed, the modalities of implementing EIA may have to be adjusted to suit local conditions and constraints without compromising on its fundamental principles. It is for this reason that, although many of the countries are struggling to put EIA into practice, most are organising their own environmental assessment programmes in one form or another. In Africa, for example, the following countries have acquired expertise in preparing EISs: Botswana, Burkina Faso, Ethiopia, Ghana, Kenya, Mauritius, Namibia, Nigeria, South Africa, Tanzania and Zimbabwe. Countries in Asia such as India, Nepal, Thailand, Vietnam, the Philippines, Indonesia, Malaysia, South Korea and Japan have been actively developing or executing EIA principles. Australia and New Zealand are strongholds of EIA for the South Pacific region, while Latin American countries such as Brazil, Bolivia, Chile, Costa Rica, Honduras and Mexico have acquired varying levels of experience of EIA.

Increasingly greater efforts are also being made in the transition countries of Eastern and Central Europe, and in the newly independent states (NIS) of what used to be the USSR, to incorporate EIA into their decision-making and management process. Hungary and Poland have already experimented extensively with it. In Hungary, for example, the importance of and need for EIA became evident during discussions on the consequences of the Gabcikovo-Nagymaros dam project. Since the countries in this region of the world are in economic transition, societal changes would in time also bring about changes in their environmental legislation to which EIA would be integral.

CONTENTS OF AN ENVIRONMENTAL IMPACT STATEMENT

The Environmental Statement Review Package, developed by Lee and Colley (1992), gives the contents of what should be described and analysed in an EIS. With regard to the summary of the contents given in Table 10.2, the following are needed for an EIS:

- *Description of the development*: the purpose(s) of the development should be described together with its physical characteristics, scale, and design. Quantities of materials needed during construction and operation should be included and, where appropriate, a description of the production processes should be given.
- *Site description*: on-site land requirements of the development should be described, and the duration of each land use should be given.
- *Residuals*: the type and quantities of residual and/or waste matter and energy created should be estimated, the expected rate of production given, and the proposed disposal routes to the environment should be described.
- *Environmental description*: the likely geographical extent of the affected environment should be described.
- *Baseline conditions*: a description of the affected environment as it is currently, and as it could be expected to develop if the project were not to proceed, should be presented.
- *Definitions of impacts*: potential impacts of the development on the environment should be investigated and described. Impacts should be broadly defined to cover all potential effects on the environment, and they should be determined as predicted deviation from the baseline state.
- *Identification of impacts*: methods capable of identifying all significant impacts should be used.
- *Prediction of impact magnitude*: the likely impacts of the development on the environment should be described in exact terms wherever possible.
- *Assessment of impact significance*: the expected significance that the projected impacts will have for society should be estimated. The sources of quality standards plus the rationale, assumptions and value judgements used in assessing significance should be fully described.

Table 10.2 Contents of an Environmental Impact Statement

1	Summary
2	Project description
3	Description of the actual state of the environment
4	Discussion of the environmental impacts of the project
5	Presentation of alternatives and analysis of their impacts
6	Possible mitigation measures
7	Monitoring plan
8	Annex

- *Alternatives*: alternatives to the proposed project should be considered. The alternatives should be outlined, and the environmental implications of each should be presented.
- *Mitigation measures*: all significant adverse impacts should be considered for mitigation. Evidence should be presented to show that proposed mitigation measures will be effective (Lee and Colley, 1992).
- *Monitoring*: the predictions included in the EIS and the effect of the proposed mitigation measures could be evaluated when a permit for the project has been issued. A plan which describes the best way to monitor the project could already be proposed in the EIS.
- *Annex*: interesting additional information, external sources, etc. should be presented at the end of the document in an annex.

METHODOLOGY OF ENVIRONMENTAL IMPACT ASSESSMENT

The basic phases of EIA

Although there are many different regulations for the preparation of an EIA, we can, in the light of experience, identify some basic phases in its preparation. These phases, shown schematically in Figure 10.1, are 'basic' in the sense that the process will not work if they are not attended to.

The process begins with the 'screening' phase in which whether or not the project under consideration is subject to EIA is ascertained. If it is, a decision is made on its significant impacts which should be considered in the EIA. This is called the 'scoping' phase. A draft EIS is then prepared, followed by its review and scrutiny by experts and the public. A final EIS is then prepared, taking into account the comments of experts and the public. After a final review and quality check, the EIS is sent to the decision-makers to arrive at a decision. Evaluation and monitoring of the EIS, and of the project itself, is the last phase of the process.

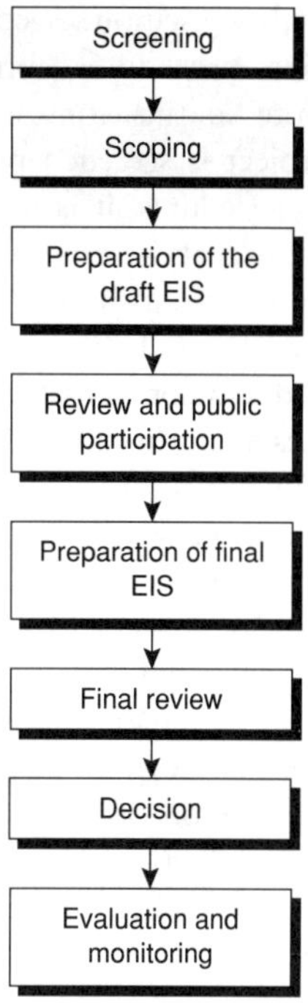

Figure 10.1 The basic phases of EIA

Project selection, significance of impacts and scoping

In an ideal situation EIAs would be expected to be carried out for all human actions with implications for the environment, but clearly it would be very demanding and expensive to do so. It is for this reason that the legal requirement for carrying out EIA has to be limited to only those actions that are likely to have the most detrimental impact on the environment. Only a limited number of selected projects are thus 'screened' positively to require EIA in compliance with legislation. A comparison of the different screening procedures currently in use is interesting to note in this context; it shows that, while a given project is screened positively for EIA by some of the procedures, it is not by the others.

Some EIA systems make use of lists with thresholds. If the proposed project is included in the list, an EIA has to be prepared for it. In other systems a limited initial study (called 'environmental assessment') of the proposed project is made, from which decision-makers can judge the significance of its likely impacts. There are yet other systems that are based on a 'list of exclusions', meaning that projects not included in the list are exempt from EIA. Certain EIA systems have lists with thresholds which apply only to certain places, or lists of exclusion with variable thresholds depending on the area where the project is planned.

A complete description of all the possible impacts of a proposed project would be an enormous task, and very inefficient too when there is a significant or large number of impacts of minor importance. It is for this reason that EIA is concerned primarily with significant environmental impacts. Since only the most serious impacts can be considered, the concept of 'impact significance' deserves to be defined and applied rigorously. It is necessary also to do so because in many cases, as in the NEPA for example, the only threshold for EIA requirement is the term 'significantly', which is defined in US regulations in terms of both 'context' and 'intensity'. 'Context' recognises that significance varies with the setting of the proposed action; it also indicates that an action should be viewed from several different perspectives. 'Intensity', on the other hand, refers to the severity of impacts, to be evaluated according to the various listed factors such as beneficial and adverse impacts, effects on public health and safety, unique characteristics of a geographical area, whether the effects are highly controversial, whether there are highly uncertain effects or unique or unknown risks, etc. (Yost and Rubin, 1989).

'Scoping' is the process by which the significant impacts of a proposed project are identified and selected, after having analysed and examined all its possible impacts. US regulations define scoping as 'an early and open process for determining the scope of issues to be addressed and for identifying the significant issues related to a proposed action'. One of the purposes of scoping is to notify and involve all agencies and individuals who may be concerned about or affected by the proposed project. Another is to identify the issues which should be analysed in depth, and those which should be ignored as not being sufficiently significant (Yost and Rubin, 1989).

The extent of public involvement in the scoping phase varies from country to country. The reasons for involving the public are:

- an acknowledgement of the right of the public to be consulted on major projects;
- a realisation that involving the public in the early stages, and obtaining a measure of their agreement on the impacts/topics to be investigated, would reduce the likelihood of criticism of the EIA at a later date for omitting matters of public concern; and
- an acknowledgement that local people can have greater knowledge of the local environment than the EIA team.

Reasons for involving other agencies in scoping are similar to those given above, but include the following in addition:

- obtaining consensus on the most appropriate methods/techniques to be used in EIA work; that is, those that would give the most accurate predictions (Bisset, 1989a).

Impact identification: checklists, matrix methods and networks

At the outset of an impact analysis the researcher faces the difficult task of identifying and organising potential impacts in a systematic way. The three main classes of quantitative impact identification techniques are: (1) checklists, (2) matrices, and (3) networks.

Checklists

Checklists are standard lists which give an overview of the range of impacts associated with a particular type of project. The checklist of Table 10.3, which is a typical example, is concerned with potential

Table 10.3 Checklist of potential ecological impacts during the construction phase of a transportation project

Transportation project	*Construction Phase*
I Noise impacts	X
A. Public Health	
B. Land use	
II Air quality impacts	X
A. Public Health	
B. Land use	
III Water quality impacts	X
A. Groundwater	
1 Flow and water-table alteration	
2 Interaction with surface drainage	
B. Surface water	
1 Shoreline and bottom alteration	
2 Effects of filling and dredging	
3 Drainage and flood characteristics	
C. Quality aspects	
1 Effect of effluent loadings	
2 Implication of other actions, such as	
(a) Disturbance of benthic layers	
(b) Alteration of currents	
(c) Changes in flow regime	
(d) Saline intrusion in groundwater	
3 Land use	
4 Public Health	
IV Soil erosion impacts	X
A. Economic and land use	
B. Pollution and siltation	
V Ecological impacts	X
A. Flora	
B. Fauna (other than humans)	

Source: Adapted from Westman, 1985
Note: X indicates a possible impact which may be beneficial or adverse, depending on circumstances

ecological impacts during the construction phase of a transportation project.

Although checklists promote thinking about the possible range of impacts in a systematic way and allow a concise summary of the effects to be made, they are often general and incomplete. Moreover, they do not show the interactions between or among the effects, because the same effects may be registered in several places under headings which overlap in content, and the number of categories to be reviewed can be immense, thus distracting attention from the most significant impacts. Furthermore, no statement is made on the likelihood of the occurrence of impacts. Also, because of the subjective nature of the estimates, checklists are often not filled out identically by different investigators (Westman, 1985).

Matrices

Checklists are one-dimensional lists of potential impacts of a proposed project. They are readily expanded to two-dimensional matrices by listing a range of actions associated with the project along the second axis (Westman, 1985).

The best-known matrix of this kind is probably that developed for the US Geological Survey by Leopold *et al.* (1971), which is also known as the 'Leopold Matrix'. Figure 10.2 gives an example of a section of the Leopold Matrix. This matrix consists of a list of development actions displayed horizontally, while a list of environmental components is displayed vertically. The result is that items in one list can be related systematically to all other items in the other list to ascertain whether an impact is likely. Once completed, that is when all likely interactions between a development activity and environmental components have been identified, the matrix becomes a useful visual summary which is easy to understand. The Leopold Matrix can be used to measure and interpret impacts by rating them in terms of their magnitude and importance (e.g. on a common 1–10 scale in which 1 is for the least magnitude of importance and 10 for the greatest). The scores can be preceded by a plus or minus sign to indicate whether an impact is beneficial or adverse (Bisset, 1989a).

The basic Leopold Matrix has been adapted and amended to suit the needs of individual EIA studies. These matrices have serious theoretical and practical limitations, however. They focus on direct impacts between two items, namely the impact-causing factor and a target environmental component affected by that factor. Also, impacts are always identified in Leopold-type matrices by a series of discrete, two-way linkages between development activities and components, but in reality specific environmental features can be affected in a cumulative fashion by a number of different impacts occurring through different pathways. For the reasons given above, Leopold-type matrices cannot deal with this cumulative aspect of impacts. What are called 'networks' (Bisset, 1989a) can, on the other hand, help deal

Instructions

1. Identify all actions (located across the top of the matrix) that are part of the proposed project.

2. Under each of the proposed actions, place a slash at the intersection with each item on the side of the matrix if an impact is possible.

3. Having completed the matrix, in the upper left-hand corner of each box with a slash, place a number from 1 to 10 which indicates the magnitude of the possible impact: 10 represents the greatest magnitude of impact and 1 the least (no zeroes). Before each number place + if the impact would be beneficial. In the lower right-hand corner of the box place a number from 1 to 10 which indicates the importance of the possible impact (e.g. regional vs. local); 10 represents the greatest importance and 1 the least (no zeroes).

4. The text which accompanies the matrix should be a discussion of the significant impacts, those columns and rows with large numbers of boxes marked and individual boxes with the larger numbers.

	a	b	c	d	e
a	2/1				
b	8/8		3/1		

PROPOSED ACTIONS			A. MODIFICATION OF REGIME												
			a. Exotic flora or fauna introduction	b. Biological controls	c. Modification of habitats	d. Alteration of ground cover	e. Alteration of groundwater hydrology	f. Alteration of drainage	g. River control and flow modification	h. Canalisation	i. Irrigation	j. Weather modification	k. Burning	l. Surface or paving	m. Noise and vibration
A. PHYSICAL AND CHEMICAL CHARACTERISTICS	1. Earth	a. Mineral resources													
		b. Construction material													
		c. Soils													
		d. Land form													
		e. Force fields and background radiation													
		f. Unique physical features													
	2. Water	a. Surface													
		b. Ocean													
		c. Underground													
		d. Quality													
		e. Temperature													
		f. Recharge													
		g. Snow, ice and permafrost													
	3. Atmos	a. Quality (gases, particulates)													
		b. Climate (micro, macro)													
		c. Temperature													
	4. Processes	a. Floods													
		b. Erosion													
		c. Deposition (sedimentation, precipitation)													
		d. Solution													
		e. Sorption (ion exchange, complexing)													
		f. Compaction and settling													
		g. Stability (slides, slumps)													
		h. Stress-strain (earthquake)													
		i. Air movements													

Figure 10.2 Section of a Leopold Matrix

Source: Bisset, 1989a (from US Geological Survey)

with impact pathways and their implications for environmental features.

Networks

The method of networks was developed to deal explicitly with the secondary, tertiary, and higher-order impacts which can arise from an initial impact. Sorensen (1971) developed a network for one of several land uses in coastal zones, making it possible to trace secondary and subsequent effects of actions on environmental elements (Westman, 1985). Networks of this kind are known as Sorensen networks. A typical example is given in Figure 10.3.

Figure 10.3 shows, for example, that the construction of high-density apartments and parking areas can result in the removal of trees, and this can have two possible impacts. First, increased surface runoff which can cause flooding and can be addressed by placing flower beds. And second, removal of topsoil which can cause reduced fertility and can be mitigated by planting shrubs.

Impact measurement, prediction of impacts and assessment of impact significance

Once the impacts have been identified, the next step is to measure and predict their extent and nature. When describing and discussing impacts in an EIS, it is useful to note the following characteristics which they exhibit (Bisset, 1989a):

- spatial dimension;
- time dimension;
- reversibility;
- probability of occurrence;
- beneficial or adverse character.

Next, it is necessary to discuss what is called the 'social distribution' of impacts which may also relate to the spatial dimension. As a typical example of this, we may consider certain species, ecosystems or social groups which may be affected by more than one impact. In fact, they may be affected by multiple adverse impacts such as noise, air pollution, odour, etc., or they may be subject to a mixture of adverse and beneficial impacts, as is often the case with social groups. For example, a project may bring beneficial impacts such as jobs and economic prosperity to some in the community, while others endure only its adverse impacts. An assessment of the overall distribution of impacts, and of their cumulative consequences, is therefore an important aspect of EIA studies (Bisset, 1989a).

A study of the various methods currently available for predicting different impacts was made in the Netherlands (VROM, 1985). The following main 'types' of methods were identified in that study:

(a) *physical models* in which illustrative or working-scale models are constructed to represent the environment; they may include:
  - visual representations of the environment by picture, photograph, film or three-dimensional models;
  - working models of the environment using, for example, wind tunnels or wave chambers;

(b) *experimental methods* involving practical field or laboratory work; such methods may include:
  - field experiments in which tests are carried out at the proposed site;
  - laboratory experiments in which tests are carried out in the laboratory by simulating the conditions of the affected environment;

(c) *mathematical models* in which cause–effect relationships are represented by mathematical equations. These models may be:
  - either empirical (black box) models in which the input–output relationship is established from statistical analysis of observations in the environment;
  - or 'internally descriptive', meaning that the mathematical relationships within the model are based on an explicit representation of the mechanisms of the processes occurring within the environment.

  These models range from simple formulations which can be applied manually, to complex dynamic and stochastic models which require computer application.

Other 'types' of methods were also identified in the same study (VROM, 1985), even though they cannot actually predict future changes in the environment. Classified as 'valuation methods', various approaches are used in these methods to calculate

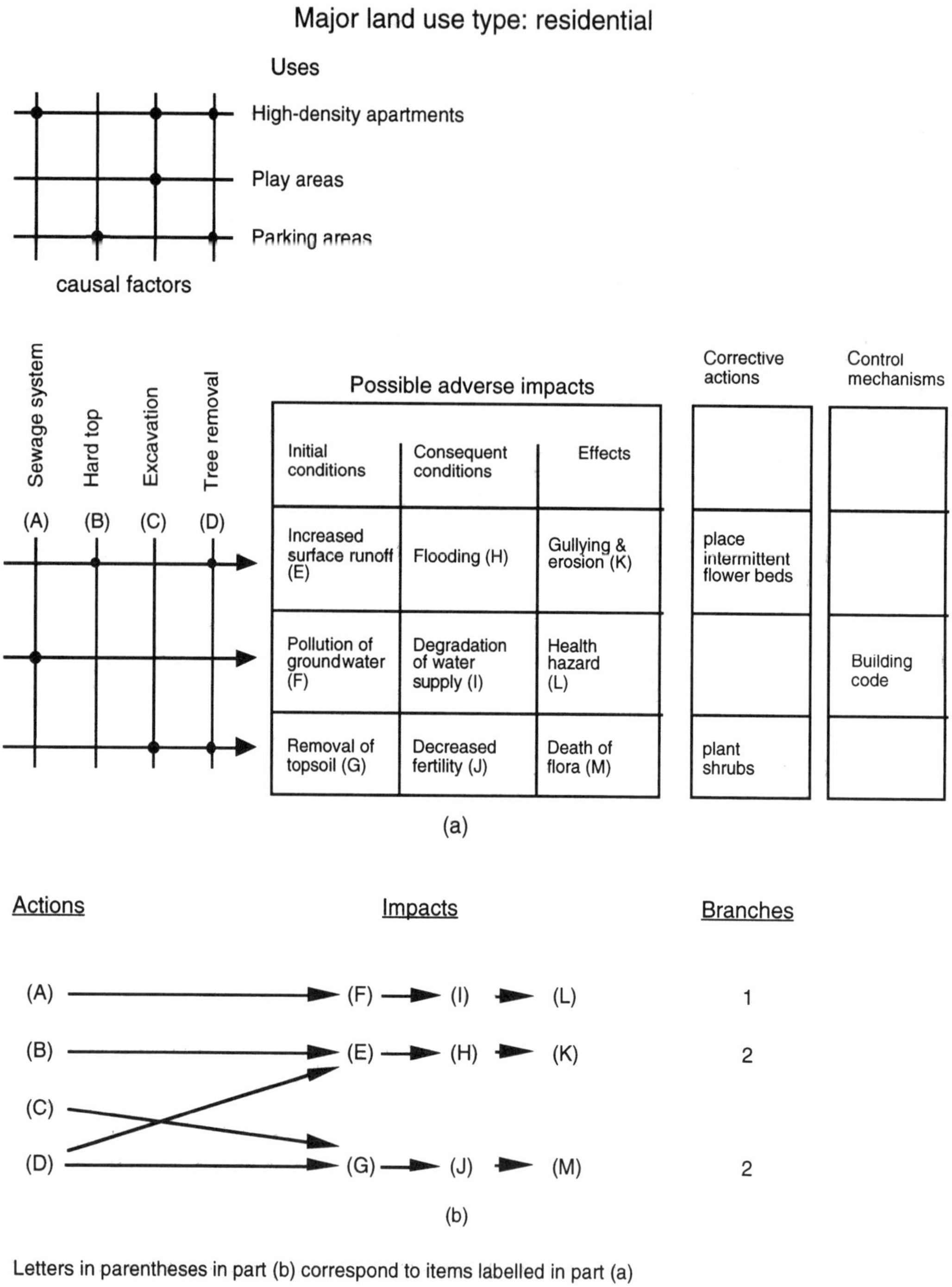

Figure 10.3 Example of a Sorensen network

Source: Westman, 1985

the 'value' of an environmental aspect that will be lost or disturbed as a result of the proposed project or action. Valuation methods are frequently used for describing habitat and landscape qualities. Although these methods are not actually predictive, they are often used for comparing alternative locations in an EIA, based on the value of the biological or visual environment that will be lost or damaged by the proposed activity.

Finally, many of the environmental impacts can be described simply in terms of the number of receptors, or areas of the environment, that will be affected by the proposed activity. This simple 'inventory approach' is widely used to provide a simple basis for comparing alternatives in an EIA. A similar approach involves the use of some of the characteristics of the receiving environment as the basis for comparing alternatives (VROM, 1985).

Once the environmental impacts of a proposed project or activity have been predicted, experts are asked to assess their significance. In doing so they may compare the predicted impacts with local, national or international standards, as well as with the goals of environmental policies.

Alternatives and mitigating measures

Alternatives are essential to an effective EIA system, because it is very important and necessary to be able to compare the impacts of different alternatives. Furthermore, alternatives make it possible for decision-makers to choose the most appropriate option. Alternatives can be of the following types:

- *the no-action alternative*: this is the situation in which the proposed activity does not take place;
- *the most environmentally friendly alternative*: in this case no effort is spared to protect the environment. All the best available techniques are employed to protect it.

Other types of alternatives may also be considered, depending on the aims of individual EISs; such as the 'location alternative', which is concerned with finding the most appropriate site for the proposed project or activity; the 'realisation alternative', which is a study of the different ways in which the project could be set up; and the 'process alternative', which tries to find different ways of making a product in terms of choice of resources, capacity, site layout, management, transportation, etc.

As an instrument, the purpose of EIA is to prevent environmental damage or degradation caused by human activities. Therefore, in order to derive maximum benefits from an EIA, priority should be given to impact prevention in the earliest possible stages of project design and planning. Sound decisions made during the design and planning stages of a project, on such matters as site-selection and selection of appropriate technology, can eliminate a variety of potential impacts and minimise the need for mitigation responses. But, since not all impacts of a given human activity can be prevented through such an approach, mitigation measures would thus be needed only to minimise the major adverse impacts (CEARC, 1988).

Mitigation measures to be employed may include for example the installation of pollution control equipment to reduce water, soil, and air pollution; or landscaping of the site to reduce visual intrusion or noise. Another example is the promotion of alternative modes of transportation for the employees of a new office building to reduce traffic congestion, or changing the shape of the proposed building in order to increase the amount of sunlight falling on neighbouring streets.

Public participation

Public participation is one of the basic tenets of EIA, since one of its main aims is to open up the 'black box' of decision-making in order to make decision-makers accountable to the public for the decisions they make on its behalf.

The extent of public participation in EIA varies from country to country, depending on the social organisation and culture of the country in question. There are, however, other good reasons for public participation. First, people who live locally know far more about the problems and possibilities of their area than outsiders, and therefore they can provide better information on local needs than what is perceived by outsiders from data. And second, certain EIA systems also give the public an important review function, by making all documentation and information available to them and inviting them and local interest groups to scrutinise the accuracy of information to be used in decision-making. This is very important, because withholding of information, or secrecy, only makes people suspicious about the motives of decision-makers. It is emphasised that 'public' in this context not only means the local citizenry; it also includes local administrators, public and private institutions, expert groups, public authorities at different levels, and other social actors (such as unions, consumer groups, environmental groups, etc.) with views on what is needed locally.

Public participation can be organised in different ways and with different means. At its simplest, EISs can be made available for public scrutiny and comment in local authority offices and public libraries, and sent to interested individuals. For certain projects it would be more effective to use newsletters and/or videos for the wider dissemination of information. Public meetings and hearings can be organised for an in-depth scrutiny of the proposed project by individuals, interest groups and other interested parties. Other creative methods, such as opinion polls and questionnaire surveys, may also be considered. One or more of the above methods can be used depending on the type, importance and anticipated impact of the proposed project.

Quality control, evaluation and monitoring

It is important to review the quality of EISs, not least in order to convince those individuals, groups, organisations and others who may have doubts about the need for carrying out an EIA. Also, although the EISs may be in order legally, they may not necessarily be useful for decision-making. Moreover, it is important for decision-makers to have a complete understanding of the EISs so that the decisions they make are clear and deliver the planned objectives. In certain countries independent panels of experts are given the task of reviewing the quality of EISs.

Prediction is a major element of EIA. Since prediction implies a certain level of uncertainty, it is very important to ensure that predictions are made most sensibly using best available methods. Accumulated experience of the past and the present can be helpful in making predictions about the future.

'Evaluation' is aimed primarily at questions of effectiveness. It asks whether the procedures have achieved the objectives set by the policy-makers. And it looks at all the results of a programme or project and compares them with the previously defined policy goals (Munro *et al.*, 1986). It is interesting to note that both evaluation and monitoring in EIA have been neglected in the past.

Many different aspects of an EIA can be evaluated, such as an examination of whether EIS recommendations and mitigation measures have been implemented; a comparison of the actual and predicted impacts; an evaluation of the content and clarity of the reports prepared; an examination of the effectiveness of the different steps in the EIA procedure, etc. (Devuyst, 1994).

MAJOR LIMITATIONS OF ENVIRONMENTAL IMPACT ASSESSMENT, TOPICS OF DISCUSSION AND FUTURE DEVELOPMENTS

EIA and the problems of scale, time, scope and methodology

Experience with EIAs has revealed a number of problems concerned with their scale, time, scope and methodology as described below.

Scale

Cumulative impacts of several small projects, all executed within a relatively short time or concentrated geographically, are almost never considered in an EIS. The limited application of international EIA procedures in projects that are funded by western governments or international agencies but executed in developing countries can also be seen as a problem of scale.

Time

EIA, as it is applied today on a reactive, short-term and project-by-project basis, is not a panacea for long-term global environmental problems. This is mainly because while the time perspective of EIA is short, adverse environmental changes occur imperceptibly over many years and decades to threaten our civilisation. Moreover, prevailing political systems are not geared to solving long-term problems – the time horizon of politicians seldom extends beyond the next election (Ornstein and Ehrlich, 1989).

Scope

Today most of the EISs are concerned with specific projects for which only a limited number of alternatives can be considered. But, clearly, project-specific EISs deny a proper discussion of environmental problems and aspects in the wider context as required, nor do they permit a comprehensive articulation of policy options. This calls for an extension of project-specific EIAs to 'higher-order' EIAs. Such a higher-order EIA is often called a 'strategic environmental assessment'.

A new addition to the list of activities, for which adapted EIA systems need to be designed, is product development. Today EIAs for manufacturing industries inquire into the possible effects of plants on the surroundings, and alternative location or mitigating measures are proposed when there is a problem. But the environmental soundness of the product being produced is almost never in question. It is important here to make a link between EIA and Life-Cycle Assessment (LCA) for developing a holistic, product-oriented analysis.

Methodology

In many countries of the world today there are still major problems with public participation and public access to information. Socio-economic aspects of this kind are often not considered in EIAs, even though their inclusion can reduce major adverse impacts in certain projects. Moreover, because there is no perfect or universal method of developing an EIA, country-specific or region-specific methodologies and guidelines need to be developed with local needs, conditions and constraints in mind: a region in a developing country has different priorities, needs, conditions and constraints from those of a urban region of western Europe, as has a rural area in Asia compared with its counterpart in North America.

At the beginning of the 1990s an international effort was made to evaluate EIA practice with a view to improving its performance in the future. This initiative, called the 'International Study of the Effectiveness of Environmental Assessment', was sponsored jointly by the Canadian Environmental Assessment Agency and the International Association for Impact Assessment (Sadler, 1996).

The problems discussed above thwart the achievement of the major goals of EIA systems. Rees (1988) suggests four initial steps, described below, for recreating EIA so that it can play an important role in achieving sustainable development:

- extend the scope of activities subject to EIA to cover the full range of ecologically and socially relevant public and private sector proposals and actions;
- create a variety of institutional frameworks for environmental assessment adapted to the increased diversity of initiatives and activities to be assessed;
- develop methods for environmental assessment which reflect the discontinuous temporal and spatial dynamics, and the resilience properties of ecosystems;
- implement the foregoing as part of a broader planning and decision-making framework which effectively recognises ecological functions as limiting factors.

Although today's EIA systems are not attuned to solving complex global environmental problems, scientists are now able to offer several suggestions for procedural reform which could have an immediate and profound impact on the role of EIA in our society. Different suggestions to solve these problems and to improve the EIA systems are discussed below.

Cumulative impact assessment

In spite of the recognition of the problem, cumulative effects of human activities are rarely incorporated in traditional project-level EIAs, or discussed in separate documents. Attempts to include what is called Cumulative Impact Assessment (CIA) in today's EIA systems encounter a host of methodological problems. This is not unexpected, given the complexity of ecological systems and the often intricate nature of human–environment linkages.

An analysis of several impact assessments points to the four types of effects in cumulative impact assessment described below:

1 *Linear additive effects*: incremental, small additions are made to, or deleted from, a fixed large store. Each addition has the same effect, whether it is made early or late in a sequence of regional developments.
2 *Amplified or exponential effects*: incremental additions are made to an apparently limitless store. Unlike the previous category, each addition has a greater effect than the one preceding it.
3 *Discontinuous effects*: in this case incremental additions have no obvious consequences until a threshold or a stability boundary is crossed.
4 *Structural surprises*: they have the following characteristics:
   - they are caused by multiple developments within regions and affect a number of ecosystems through terrestrial, aquatic, and atmospheric connections;
   - two features of space and time behaviour can be identified. First, the effects appear locally and abruptly. Second, slow changes gradually exaggerate the intensity of the sudden episodes and spread their effects over larger regions. Gradually, they define a coherent syndrome of structural effects which result in decreasing resilience of the ecosystems;
   - the surprises can be described ecologically, but they emerge from a growing interdependence and interaction among ecological, economic, social, and government regulatory forces.

The last category is the least understood, and it represents the greatest challenge to the understanding and management of cumulative effects (Sonntag *et al.*, 1987). A typical example of cumulative effects is given in Box 10.1.

Development of international EIA procedures

Many countries, and especially international organisations and private companies, develop and pursue activities in many different parts of the world. But the environmental impacts of their activities are not always taken into consideration. In order to address this problem, it is important for organisations such as the United Nations, non-governmental organisations and bilateral agencies to take EIA seriously. The World Bank has made an important contribution to this by introducing EIA in its decision-making process and by developing source-books on EIA. Moreover, national EIA systems should permit collaboration with neighbouring countries to study the environmental impacts of transboundary projects, or projects with transboundary impacts. A framework for this is provided by the United Nations Economic Commission for Europe through the 'Espoo Convention on Transboundary Impacts' of February 1991.

The benefits accruing from what we have said above would enhance international environmental awareness, serve to improve relations between developed and developing countries, and enable the developing countries to identify their own problems more efficiently and cost-effectively than at present. The following points are relevant in this context:

- the participation of developing countries in the EIA process will improve the availability of data on local environmental conditions. Also, local involvement in environmental assessment will give developing countries less or no reason to complain about international assistance imposing on them in a top-down fashion;
- another benefit would derive from a deeper study of local social conditions, leading to a better planning of the social services;
- if properly done, EIA will actually reduce or even eliminate the long-term costs of an otherwise environmentally harmful project;
- an objective oversight and quality control system needs to be created. International organisations must be held responsible for the environmental effects their projects may create (Klein-Chesivoir, 1990).

Strategic environmental impact assessment

Strategic EIA (SEA), sometimes also called 'Programmatic EIA' or 'EIA for Policies, Plans and Programmes', is not a new idea. The NEPA in 1969 required that all major federal actions significantly

BOX 10.1 AN EXAMPLE OF CUMULATIVE IMPACT ASSESSMENT

The Fraser River Estuary study (Sonntag *et al.*, 1987) is a typical example of a research project aimed at detecting and reducing cumulative effects. The Fraser River flows through the Canadian province of British Columbia in the western part of the country. Human settlement has radically changed the Lower Fraser Valley in the last 50 years: dykes have channelled the river, two-thirds of the original wetlands have been drained, and forests on the valley floor have been replaced by agriculture and urbanisation.

The cumulative effects of development on the Fraser Estuary have been numerous and major. A better understanding of the levels at which these effects occur is facilitated by applying a four-level typology for the classification of activities:

1 *Single activity – linear additive:* for example, the Iona Island sewage treatment plant at the western edge of the estuary, which gives only primary treatment to the combined storm and sanitary sewage from the city of Vancouver, continues to discharge effluent onto the banks of the estuary. This has degraded the environment, and a variety of toxic substances have accumulated in both sediments and biota.
2 *Multi-component activity – amplifying:* for example, development of the Roberts Bank super-port involved the construction of a causeway over Roberts Bank to a nearby island for coal storage and loading berths. The island was later expanded. As a result, the wetland habitat has been lost because of infilling, while in the adjacent areas a continuing process of erosion and accretion has resulted in the loss of fish habitat.
3 *Areal activity – discontinuous:* for example, the accumulation of pollutant discharges and loss of shoreline habitats have been caused by diverse developments, ranging from small marinas in the Ladner Slough area to large industrial sites on the North Arm.
4 *Global change – structural surprise:* for example, if scenarios of climate change caused by increasing concentration of carbon dioxide from industrial development worldwide come to pass, sea levels are likely to be raised by the melting of polar ice caps. Raised sea levels will, in turn, increase the risk of failure of the dykes in the estuary to cause massive damage to the extensive urban development on the floodplains.

Four major cumulative effects in the Fraser River Estuary were identified, based on the framework established above:

1 *Agriculture:* the cumulative loss of highly productive agricultural land has become a matter of growing concern because of the province's increasing dependence on imported products which may not be as economically available in the longer term as at present.
2 *Floods:* even though dyke protection is now adequate for the one-in-200-year flood, the estimated value of damage caused by larger floods, which will inevitably occur in the future, continues to rise because of continuing urban expansion into the floodplain.
3 *Toxics:* beginning in the early 1970s, a general concern for the effects of toxic discharges led to an extensive search for toxic substances in the estuary. Evidence of accumulation of toxic substances has been found, particularly in the sediments and biota adjacent to the outfalls where flushing is poor.
4 *Wetlands:* in the early 1970s, a growing appreciation of wetland environments was reinforced by the discovery that juvenile salmon reside in these areas during a critical period of their growth and adaptation prior to migrating out to sea. A study also showed that about 70 per cent of the original wetlands of the estuary and the floodplain had been lost before the turn of the century, mainly as a result of dyking.

Although both planning and impact assessment have often been controversial, it may be argued that the early application and subsequent refinement of several of the methodologies used have contributed to the substantial success achieved in controlling the cumulative effects of development in the Fraser River Estuary, as the following would show:

1 The early introduction of regionally based planning by the Lower Mainland Regional Planning Board began questioning the cumulative effects of development in the region, and this led to the adoption of broad goals and general strategies for achieving them.
2 Planning has become more detailed and specific over time in response to growing pressure of development: the Agricultural Land Reserves programme dealt with agricultural land loss, while the Fraser River Estuary Study specifically responded to environmental degradation of the estuary.
3 There has been a major shift in planning, from emphasising the final 'plan' or product to emphasising the ongoing 'planning' process. This shift has been accompanied by greater involvement of bureaucrats, politicians and the public, and it has facilitated a greater consideration of science and essential values.
4 A more focused consideration of the environmental issues in the estuary began with the development of impact assessment procedures in the early 1970s. Assessments for both waste management permits and foreshore leases have become progressively more comprehensive and detailed (Sonntag *et al.*, 1987).

affecting the quality of the human environment of all agencies of the Federal Government should prepare a detailed statement on environmental impacts (Sigal and Webb, 1989). However, following Sigal and Webb (1989), the NEPA environmental review has seldom, if at all, been applied to the development of broad national policy by any federal agency of the USA. It is clear, however, that if relatively small projects are subjected to EIA, then other actions such as policies, plans, and programmes, which have a much wider scope, should also be screened for adverse environmental impacts.

The term 'tiering' is important in this context. Referring to broader environmental documents prepared at the programme or policy level, it reduces paperwork, explains the relations between several environmental reports, and provides a general framework for project-level EIAs. EIA at the project level is most detailed and quantitative. SEA, on the other hand, is qualitative and indicative, but usually capable only of providing general information. The EIA and the SEA documents supplement each other.

The USA, Canada, Australia and the Netherlands, among others, were the first to experiment with SEA. In order to advance the state-of-the-art of EIA, the US Department for Housing and Urban Development developed the 'Area-wide EIS' during the 1970s. Developed specifically for predicting future impacts of human activities over a larger geographical area, the Area-wide EIS can be considered an EIA of both cumulative and strategic type. It is designed to:

- reduce delays, caused by the preparation of EISs on a project-by-project basis, in the approval and construction of housing developments;
- assess more adequately the aggregate impact of development (McElligott, 1978).

The Commission of the European Communities is also considering the introduction of SEA. A proposal for a Directive on the assessment of certain plans and programmes on the environment was introduced in 1995. Discussions are now taking place.

In 1996 major research was undertaken on the practice of SEA. Key issues to emerge from that research are summarised below (Sadler and Verheem, 1996):

- SEA is useful for strengthening project-level EIA to address cumulative effects and to incorporate sustainability considerations into the decision-making process;
- SEA should be regarded as a 'temporary instrument' whose long-term goal is to establish integrated or 'green' planning;
- results of studies on the environmental and socio-economic effects of proposed policies, plans and programmes should be integrated;
- the scope of application of SEA throughout the world is very wide, ranging from the development of new legislation, through Cabinet decisions, to sectoral policies, programme development, and regional and communal physical planning.

An example of SEA is given in Box 10.2. It shows how SEA can be useful in decision-making for the planning of forest management.

Improving public participation

External review of the environmental process is a cornerstone of EIA systems. Although most of the EIA systems used worldwide originate from the NEPA, external review and public participation, which are its vital parts, are often omitted in those systems, because both governments and private organisations are often afraid of opening-up the decision-making process for public scrutiny, and try to resist doing so with every available means. Agencies with exclusive access to environmental information do not like to undermine their powerful position (as they see it) by disseminating it widely. They fear that hitherto confidential information might be misused by the public. But, given that a major goal of EIA is transparency of decision-making with the widest public articulation of issues, it would follow that EIA can only work in countries with democratic systems of government where the public has easy access to information.

It should be clear by now that an EIS is not a dispassionate and objective collection of unambiguous, hard facts about environmental effects. Even high-quality EISs contain a substantial number of extrapolations and inferences based on simplifying assumptions, debatable theses, and even second-hand data from parties with vested interest (Weiss, 1989).

BOX 10.2 EIA OF THE BARA FOREST MANAGEMENT PLAN, NEPAL

The Nepalese Forestry Sector EIA Guidelines, prepared by the Ministry of Forest and Soil Conservation and IUCN Nepal, clearly mentions that an EIA is mandatory for the preparation of forest management plans for Nepal. The Forest Management and Utilisation Development Project has proposed a plan for production-oriented forest management in the Bara province of Nepal. Because of the particular nature of the proposed protection-oriented forest management plan, it was deemed essential to include an EIA in it for a careful evaluation of the potential environmental impacts.

The assessment work was co-ordinated by the IUCN Nepal. Two main alternatives were examined in its report. The 'do-nothing' alternative reflects the current situation in the province. In this scenario not much would change and all the current positive and negative impacts would be expected to continue. The 'production forestry' alternative, on the other hand, would transform the current passive protection regime into one in which forest production would be actively managed.

More than 150 possible impacts of varying size and significance were identified. They were synthesised into 19 main issues of concern – namely, involvement of people, uncontrolled burning, wood poaching, wildlife poaching, fuel-wood collection, clearing of forest land, employment, transportation, grazing, health, erosion, biodiversity, tenure, institutional matters, harvesting methods, economic activity, awareness and education, silvicultural practices and marketing strategies. Each issue was then analysed by a team of professionals, and a 'best judgement' was attached to the magnitude, extent and duration of each positive and negative impact associated with the two alternatives mentioned above. The severity of impacts was ranked using a numerical ranking system following the format given in the 1993 Nepalese National Environmental Impact Assessment Guidelines and the guidelines for the Forestry Sector.

The results of this 'issue analysis' form the basis of recommendations on how the proposed plan could be improved to enhance its potential for successful implementation in the Bara province.

Some of the most important conclusions to emerge from the EIA were:

1 The second alternative (the 'production forestry' regime) has considerable merit and should be implemented, provided that the recommended changes are incorporated into the management plan. Many of the current negative impacts could be mitigated, or even eliminated, by a careful implementation of the Operational Forestry Management Plan.
2 Some of the impacts identified in the second alternative would require high-level government decisions; the decision-making process should start immediately.
3 The level of success of the second alternative would be determined by the availability of technical resources, time, money, manpower and political commitment.
4 An effective monitoring and evaluation strategy is essential for ensuring the long-term success of the proposed plan.
5 As currently written, the management plan focuses almost exclusively on timber management. Because so much is involved beyond timber, the focus must be broadened. Also, there are several ways in which the plan can be improved.

Some of the mitigation measures needed are these:

1 Careful design of policies and activities, and a fair involvement of people affected by the proposed plan before, during and after its implementation.
2 A careful strategy should be designed to control unregulated burning, grazing, fuel-wood gathering and wood harvesting, while at the same time ensuring that the local needs, which these activities satisfy, are met.
3 Design and careful implementation of an inventory of the environmentally sensitive areas.
4 Incorporation of the principles of biodiversity and conservation into all harvesting and silvicultural activities.
5 Design and implementation of a more equitable marketing strategy, as well as the design and implementation of a scheme for the equitable distribution of economic and material benefits brought by the new plan.
6 Design of a well thought-out educational strategy to promote the new plan and to ensure that the people most directly affected by it have some say in its operational details.

Source: IUCN, 1995.

A broad societal discussion is therefore essential for defining and weighing the hidden subjective values in EIAs.

Mechanisms should be developed to improve the availability of information required in the EIA process. To this end access to basic data, and to results of analysis and evaluations, should be promoted. Poor access to information hinders the development of consistent review documents and diminishes the ability of practitioners to learn from experience.

Socio-economic impact assessment

Social impacts can be defined as changes in relations which occur among members of an institution, community and society as a result of an external change (Bisset, 1989b). Social impacts include changes in such features of social life as:

- relations and obligations to kin;
- marriage patterns;
- relationships between generations;
- visiting patterns;
- political and dispute resolution institutions and processes;
- values, the importance and significance of the various features of the environment and social life of a group of people;
- cultural life, including such aspects as language, rituals and general lifestyle. It is this component which distinguishes a social group from the other groups.

Social impacts affect individuals, institutions, communities and larger social systems. Impacts on psychological attitudes, health and safety, cultural life and social, economic and political structures and processes are all included in the definition of social impact.

EIA procedures in certain countries only consider traditional environmental aspects such as water, air and noise. But the trend today is to opt for a much wider definition of 'environment', including the human environment as well as socio-economic aspects. Typically, social impact assessment would be beneficial in these two cases: (a) the native Aboriginal groups in the remote parts of Australia who are being put under mounting pressure of rapidly growing development plans and projects in those areas, and (b) large groups of people in India displaced by the flooding of their towns, villages and farm land caused by the construction of hydroelectric dams. In both cases large groups of people have been subjected to serious social impacts resulting from developmental activities.

Development of specific methodologies and guidelines

There is now a growing need to develop EIA systems for rural, urban and specific geographical areas. In this context we will outline here some aspects of such systems for developing countries in particular. The main problem in incorporating EIA in project planning, often pointed out by the decision-makers themselves, is concerned with the additional costs it entails, delays in project implementation, and lack of skilled manpower to assess impacts. One of the reasons why EIA is expensive in developing countries is their limited access to technical databases on which project impacts are to be based. It is for this reason that a large amount of baseline data has to be collected, and this is probably the most expensive element in the development of an EIA. Also, given their lack of expertise, limited funds and time constraints, it is essential for developing countries at least to make use of the rapid assessment technique of checklists and/or matrices in conjunction with a set of established guidelines. This would enable the developer, environmental protection authorities and the public to do some orderly thinking with a view to avoiding costly and irreversible mistakes (Lohani and Halim, 1987).

Post-project analysis and scientific research

Both environmental management and EIA are based on a weak scientific foundation as literature on these subjects would confirm. Indeed, there is a strongly held view that the discipline of ecology is in its infancy at present. Admittedly, there is now a large and growing body of knowledge of the factors which affect the operation of ecosystems, but current knowledge is far from being sufficiently complete to permit the construction of accurate causal models. And it is difficult to see how future environmental impacts could be forecast with an acceptable degree of accuracy without such models. To put it simply and bluntly, we do not know many of the implications of the multitude of complex cause-and-effect relationships (Munro *et al.*, 1986). This underlines the need for what is called 'Post-Project Analysis' (PPA).

The United Nations' Economic Commission for Europe summarises the benefits of PPA as follows:

> Efficiency is a key word in this regard. Project proponents may become more supportive of EIA processes when PPAs help to focus mitigation measures on impacts that are real problems rather than on those which are unlikely to materialise. The use of PPAs partly for environmental management of projects and partly for EIA process development gives a double-edged advantage. With regard to project management, some mitigation measures can be dismissed in the face of more certain information about project impacts; such information is obtained through PPA. As for process-development aspects, a better understanding of likely impacts, gained through the study of PPAs done on previous projects, will save a lot of unnecessary trial and error. Governments likewise benefit from the application of PPAs. Because PPAs can often forecast environmental surprises, these can be identified and dealt with before any damage is done. Hence, PPAs inspire confidence that approved projects will operate in an environmentally satisfactory manner. This factor, combined with the environmental management role of PPAs, allows decision-makers to operate even in the face of some uncertainty about environmental consequences of particular projects. Through learning from project implementation, everyone concerned gains a better understanding of the issues and how best to deal with them. The public also benefits from PPA in that useful information about project impacts can be applied to achieve the environmental protection needed and people can see that the impacts are dealt with appropriately. In addition, public involvement during the EIA process review is extended through PPA into the implementation phase of the project. These benefits yield legitimately increased satisfaction with and acceptability of projects.
>
> (Economic Commission for Europe, 1990: 41)

LIST OF EIA RESOURCE CENTRES

Canada
The Canadian Environment Assessment Agency (CEAA)
200 Sacré-Coeur Boulevard
Hull, Quebec K1A 0H3
Canada

United States of America
Council of Environmental Quality
Executive Office of the President
Washington, DC 20500
USA

The United Kingdom
EIA Centre
Department of Planning and Landscape
University of Manchester
Oxford Road
Manchester M13 9PL
UK

The Netherlands
The EIA Commission
Arthur van Schendelstraat 800
PB 2345
3500 GH Utrecht
The Netherlands

Belgium
The EIA Centre
Human Ecology Department
Vrije Universiteit Brussel
Laarbeeklaan 103
1090 Brussels
Belgium

CONCLUSIONS

EIA is essentially a management tool whose purpose is to improve the decision-making process by incorporating environmental matters into it. EIAs should be prepared efficiently – meaning that they should be prepared in a simple and succinct way – so that decision-makers have no difficulty in understanding them. A critical element of an EIA is that the issues and problems under consideration should be subjected to the widest possible articulation by the public and all other groups and actors likely to be affected by the proposed activity or project. It is only then that transparency of decision-making would be assured, together with wider public approval of what is being proposed on their behalf. Easy access of the public and other concerned actors to information on what is being proposed is thus essential, and every effort must be made to facilitate the widest possible dissemination of information.

EIAs have limitations and disadvantages too, however, and by themselves they are certainly no panacea for the increasingly serious environmental problems. These deficiencies would be significantly addressed by developing SEAs for different kinds of policies, plans and programmes. Although there are serious problems in this approach, the fact remains that further development of the art of EIA in this way, and its effective implementation, holds much promise for arresting the mounting environmental damage and destruction being inflicted by human activities.

REFERENCES

Bisset, R. (1989a) 'Introduction to EIA methods', in *Proceedings of 10th International Seminar on Environmental Impact Assessment and Management*, Scotland, UK: University of Aberdeen, 9–22 July 1989.

Bisset, R. (1989b) 'Social impact assessment', in *Proceedings of 10th International Seminar on Environmental Impact Assessment and Management*, Scotland, UK: University of Aberdeen, 9–22 July 1989.

Biswas, A.K. and Qu Geping (eds) (1987) *Environmental Impact Assessment for Developing Countries*, London: Tycooly International.

CEARC (1988) *Mitigation and Compensation Issues in the Environmental Assessment Process: A Research Prospectus*, Canada: Ministry of Supply and Services, Cat. No. En 107–3/14–1988.

Clark, B.D. (1989) 'Environmental assessment and environmental management', in *Proceedings of 10th International Seminar on Environmental Impact Assessment and Management*, Scotland, UK: University of Aberdeen, 9–22 July 1989.

Devuyst, D. (1994) 'Instruments for the evaluation of environmental impact assessment', Brussels, Belgium: Ph.D. thesis, Human Ecology Department, Vrije Universiteit Brussel.

Devuyst, D. and Hens, L. (1991) 'Environmental impact assessment in Belgium', *Environmental Impact Assessment Review* 11 (2): 157–169.

Economic Commission for Europe (1990) *Post-project Analysis in Environmental Impact Assessment*, Geneva: United Nations Publication.

Goodland, R., Mercier, J.-R. and Muntemba, S. (1996) 'Environmental assessment in Africa. A World Bank commitment', *Proceedings of the Durban World Bank Workshop*, Durban, South Africa, 25 June 1995.

IUCN (1995) *EIA of the Bara Forest Management Plan*, Nepal: IUCN, The World Conservation Union.

Klein-Chesivoir, C. (1990) 'Avoiding environmental injury: the case for widespread use of environmental impact assessment in international development projects', *Virginia Journal of International Law* 30 (2): 519–551.

Lee, N. and Colley, R. (1992) *Reviewing the Quality of Environmental Statements. Occasional Paper 24*, Manchester, UK: EIA Centre, Department of Planning and Landscape, University of Manchester.

Leopold, L.B., Clarke, F.E., Hanshaw, B.B. and Balsley, J.R. (1971) 'A procedure for evaluating environmental impact', *Geological Survey Circular* 645, Washington, DC: US Dept. Interior.

Lohani, B.N. and Halim, N. (1987) 'Recommended methodologies for rapid environmental impact assessment in developing countries: experiences derived from case studies in Thailand', in A.K. Biswas and Qu Geping (eds) *Environmental Impact Assessment for Developing Countries*, London, UK: Tycooly International.

McElligott, C.R. (1978) *Area-wide Approach to Environmental Assessment*, Report by the Office of Environmental Quality of the US Dept. of Housing and Urban Development.

Munro, D.A., Thomas, J.B. and Matte-Baker, A. (1986) *Learning from Experience: A State-of-the-Art Review and Evaluation of Environmental Impact Assessment Audits*, CEARC, Hull: Canadian Government Publication.

Ornstein, R. and Ehrlich, P. (1989) *New World New Mind. Moving Toward Conscious Evolution*, New York: Doubleday.

Rees, W.E. (1988) 'A role for environmental assessment in achieving sustainable development', *Environmental Impact Assessment Review* 8: 273–291.

Sadler, B. (1996) 'Environmental assessment in a changing world: evaluating practice to improve performance', *International Study of the Effectiveness of Environmental Assessment*, Canada: Canadian Environmental Assessment Agency and International Association for Impact Assessment.

Sadler, B. and Verheem, R. (1996) *Strategic Environmental Assessment. Status, Challenges and Future Directions*, The Netherlands: Dutch Ministry of Housing, Spatial Planning and the Environment.

Sigal, L.L. and Webb, W. (1989) 'The programmatic environmental impact statement: its purpose and use', *The Environmental Professional* 11: 14–24.

Sonntag, N.C., Everitt, R.R., Rattie, L.P., Colnett, D.L., Wolf, C.P., Truett, J.C., Dorcey, A.H.J. and Holling, C.S. (1987) *Cumulative Effects Assessment: A Context for Further Research and Development*, A background paper prepared for the Canadian Environmental Assessment Research Council, Hull: Canadian Government Publication.

Sorensen, J.C. (1971) 'A framework for identification and control of resource degradation and conflict in the multiple use of coastal zone', MA thesis, Berkeley, California: Dept. Landscape Architecture, University of California.

VROM (1985) *Prediction in Environmental Impact Assessment*, Environmental Resources Limited and Ministerie van Volkshuisvesting, Ruimtelijke Ordening en Milieubeheer and Ministerie van Landbouw en Visserij, The Netherlands: Publications by the Dutch Government.

Weiss, E.H. (1989) 'An unreadable EIS is an environmental hazard', *The Environmental Professional* 11: 236–240.

Westman, W.E. (1985) *Ecology, Impact Assessment, and Environmental Planning*, New York: John Wiley & Sons.

Williams, R.H. (1988) 'The environmental impact assessment Directive of the European Communities', in M. Clark and J. Herington (eds) *The Role of Environmental Impact Assessment in the Planning Process*, New York, London: Mansell Publishing Limited, pp. 74–87.

Yost, N.C. and Rubin, J.W. (1989) 'The National Environmental Policy Act', *NEPA Deskbook*, Washington, DC: Environmental Law Institute, pp. 1–24.

SUGGESTED READING

Periodicals

Environmental Impact Assessment Review is a quarterly publication highlighting approaches to impact assessment, environmental decision-making, and the resolution of environmental disputes. It is published by Elsevier Science Publishing Co., The Netherlands.

Impact Assessment, a publication of the International Association for Impact Assessment.

EIA Newsletter, news and information on EIA by the EIA Centre, University of Manchester.

IAIA Newsletter, news from the International Association for Impact Assessment, Fargo, ND.

Environmental Management which has as field 'applied ecology' in the widest sense, without regard to the disciplinary boundaries created by modern academic study. Published by Springer-Verlag, New York.

Books

Caldwell, L.K. (1982) *Science and the National Environmental Policy Act. Redirecting Policy through Procedural Reform*, University, Ala.: The University of Alabama Press.

Gilpin, A. (1995) *Environmental Impact Assessment. Cutting Edge for the Twenty-first Century*, Cambridge: Cambridge University Press.

Glasson, J., Therivel, R. and Chadwick, A. (1994) 'Introduction to environmental impact assessment', *The Natural and Built Environment Series 1*, London: UCL Press.

Rau, J.G. and Wooten, D.C. (eds) (1980) *Environmental Impact Analysis Handbook*, New York: McGraw-Hill Publishing Company.

Roe, D., Dalal-Clayton, B. and Hughes, R. (1995) *Directory of Impact Assessment Guidelines*, London: International Institute for Environment and Development.

Sadler, B. (1996) *Environmental Assessment in a Changing World: Evaluating Practice to Improve Performance*, Final Report, International Study of the Effectiveness of Environmental Assessment, Canada: Canadian Environmental Assessment Agency and International Association for Impact Assessment.

Sheate, W. (1994) *Making an Impact. A Guide to EIA Law & Policy*, London: Cameron May.

Therivel, R. and Rosario Partidario, M. (1996) *The Practice of Strategic Environmental Assessment*, London: Earthscan Publications Ltd.

Vanclay, F. and Bronstein, D. (1995) *Environment and Social Impact Assessments*, London: Wiley.

SELF-ASSESSMENT QUESTIONS

1 The term 'scoping' has a very specific meaning and it plays an important role in EIA procedures. In what stage of the EIA process is scoping used?
   (a) during the impact prediction phase;
   (b) in post-project analysis for the identification of EIA limitations;
   (c) at the beginning of the EIA process when important impacts are identified.

2 Matrices are used for identifying and organising potential impacts in a systematic way. Which of the following statements refers to one of the major limitations of matrices?
   (a) When matrices are used, impacts are always identified as a series of discrete, two-way linkages between development activities and components; and secondary, tertiary, and higher-order impacts are not considered.
   (b) Matrices are one-dimensional and do not represent interactions between effects.
   (c) Matrices are extremely ineffective in identifying impacts on water and air quality; they are therefore never used for assessments of projects with major impacts on water and air.

3 An analysis of several impact assessments led to the identification of four types of effects in cumulative impact assessment. The discontinuous effects can be characterised as follows:
   (a) incremental additions have no obvious consequences until a threshold or a stability boundary is crossed;
   (b) incremental, small additions are made to, or deleted from, a fixed large store. Each addition has the same effect whether it is made early or late in a regional development sequence;
   (c) incremental additions are made to an apparently limitless store, and each addition has a greater effect than the one preceding it.

4 What is an area-wide EIA?
   (a) An area-wide EIA is applied for developing broad national policy.
   (b) An area-wide EIA predicts future impacts of human activities over a larger geographical area.
   (c) An area-wide EIA is an EIA procedure developed for international organisations such as the World Bank and the United Nations to predict the impacts of their actions.
5 Care should be taken in presenting the information in an EIA. Which of the following statements is correct?
   (a) An EIA is prepared as a detailed document with inventories of all the possible effects of a proposed action using the high-level technical and scientific language of experts.
   (b) An EIA is a brief document in which the potentially severe adverse impacts are mainly discussed. The information is presented in such a way that both non-specialists and decision-makers can understand the meaning.
   (c) An EIA consists of several volumes in which important adverse impacts are discussed at the highest possible level of scientific and technical language, while important beneficial impacts are presented in the style of popular journalism.

11
LIFE-CYCLE ASSESSMENT

Eddy Nierynck

SUMMARY

Life-Cycle Assessment (LCA) has been called 'the environmental management tool of the 1990s'. In theory, LCA has the potential for incorporating environmental quality aspects into the decision-making process. Also, as an instrument for preventive environmental policy, it provides a means for presenting environmental conflicts in an objective way through quantitative analyses which permit subsequent assessment of the environmental consequences associated with products, processes and activities on a 'cradle-to-grave' basis.

But the authority of LCA continues to be undermined by the lack of scientific consensus on several of its major aspects. Although the need for a generally accepted methodology is widely acknowledged, and numerous initiatives are being pursued to that end, there is at present no consensus on a common methodology.

Starting with the historical perspective on the LCA, all major issues relevant to it have been discussed in this chapter, including the main bottlenecks, limitations and prospects for the future. The two boxes in the final section of the chapter, one on a classic milk packaging study and the other on the use of LCA as a tool in the eco-design in electronics, are meant to help bridge the gap between theory and practice.

ACADEMIC OBJECTIVES

On completion of this chapter the reader should be able to:

- define LCA;
- determine the key aspects of LCA methodology;
- identify limitations of the LCA procedure;
- apply the LCA methodology;
- comment on future developments in LCA;
- design a technical framework for case studies.

INTRODUCTION

LCA in the historical perspective

The need to remove the emotional load from environmental discussions on the provision of goods and services to society requires public authorities, trade and industry, and public interest groups to apply objective methods to assess environmental impacts in an integrated fashion. Life-Cycle Assessment (LCA) is a tool with which the environmental consequences of both products and activities can be analysed and evaluated qualitatively.

Studies on life-cycle characteristics began in the 1970s, but the focus then was only on the inventory of energy consumption and emissions. Even so, those studies constituted the blueprint for the subsequent development of modern LCA. Different LCAs, conducted on comparable products, frequently led to different and even opposite outcomes, however. This, together with the recognition of the potential value of LCA as a tool for incorporating environmental quality aspects into the decision-making process, has led to a proliferation of initiatives aimed at developing a generally acceptable and unequivocal methodological framework for LCA. Furthermore, the shift in policy emphasis, from pollution control to source reduction, reflects a growing recognition that laws and regulations as such do not adequately address risks to public health and the environment at large.

When the objective is to improve the overall environmental performance of societal activities, 'products' become a specific target of environmental policy. Accordingly, policies on products are aimed at preventing and reducing the environmental consequences associated with the provision of goods and services. An LCA can support product policies by:

- identifying the phase or phases in the life-cycle of a product which induce major environmental consequences;
- improving the product in a way that enhances its environmental benefits significantly;
- providing information on alternative products, within a given product group, which generate least adverse environmental consequences.

Given that systematic collection and analysis of information on the life-cycle of products is fundamentally important to devising suitable strategies to deal with a wide range of environmental problems, there is an urgent need to establish a scientific foundation on which product policies can be based. To that end the aim of the life-cycle approach is to support the overall goal of sustainability (LCANET, 1996). It is pointed out however that, as a policy instrument, LCA is but one of several environmental management tools currently available. Other decision support tools aiming at sustainability include Substance Flow Analysis (SFA), Material Intensity Per Unit of Service (MIUS), Environmental Impact Assessment (EIA), Risk Assessment (RA), and Environmental Auditing. Some of these tools, such as EIA and RA, focus on processes and/or activities, while others, such as the LCA, are based on the life-cycle concept.

There is now a growing interest in combining the life-cycle concept with tools for specific processes, and to this end efforts are being made to position LCA in relation to other decision support tools. Furthermore, there is a need to specify the relationship between the different life-cycle approaches, especially with regard to the major differences and similarities between the LCA, MIPS and SFA (LCANET, 1996).

Prominent actors in the development of LCA

The Society for Environmental Toxicology and Chemistry (SETAC) is probably the most important international 'think-tank' in the development of the LCA. The workshops of the SETAC support both consensus-building on the methodology of LCA and its acceptable professional practice. SETAC's public forums, and its bi-monthly publication entitled *SETAC LCA Newsletter*, provide ample opportunity for the user community to make inputs to the evolution of the LCA. Although SETAC's *Code of Practice* (Consoli *et al.*, 1993) is not a detailed methodological reference, it nevertheless presents the general principles and a framework for the conduct, review, presentation and use of LCA findings. Its aim is to enhance the quality, transparency and credibility of LCA studies. Five SETAC working groups are involved in addressing the methodological and application-related issues of LCA in a three-year plan, which should culminate in a new 'Code of Practice'.

Within the International Standards Organisation (ISO), efforts to standardise LCA are being made by its Technical Committee 207 as well as by its Subcommittees 5 and 6. The SETAC enjoys official liaison status at the ISO (Victory, 1996). The outcome of the ongoing ISO process should be seen as a crucial effort towards standardisation.

In 1990 the United States Environmental Protection Agency (USEPA) initiated a series of projects for developing the LCA methodology; and the *EPA–LCA Project Update*, which is a bi-annual newsletter, gives information on those projects. The USEPA publication, entitled *Life-Cycle Assessment: Inventory Guidelines and Principles* (Vigon *et al.*, 1993), is an important reference document.

The document, *Life-Cycle Assessment* (CSA, 1994), published by the Canadian Standards Association (CSA), provides technical guidance on an acceptable method for conducting LCAs and on the reporting of the assessed results. Based on the SETAC/USEPA efforts, this guideline is an attempt at harmonising with similar initiatives in the European Union.

What is called the 'Groupe des Sages' was established by the European Commission to advise on the role of LCA in its eco-labelling programme, and the efforts of this body have crystallised into guidelines for the application of LCA in this programme (Groupe des Sages, 1994). The 'European Network for Strategic Life-Cycle Assessment Research and Development' (LCANET), subsidised by Directorate General XII of the European Commission, is essentially a platform for the discussion and exchange of information on the state-of-the-art of LCA. It seeks to identify research needs and to draw up a research programme for the application of LCA (Anon., 1996b). At the European level, LCANET attempts to ensure that developments in LCA take place in a coherent way, while at the same time contributing to the SETAC and ISO efforts. The roles of these leading organisations in the evolution of the LCA can be summarised as follows (LCANET, 1996):

- SETAC focuses on research topics (free research);
- LCANET plans research by defining a common research framework;
- ISO continues to standardise methods and reach agreement on principles.

Both the Netherlands and the Scandinavian countries have also made important inputs to the development of the LCA (Heijungs, 1992; Nordic Council, 1995).

The private sector has been playing a remarkably active role in the development of the LCA. Notably, the Society for the Promotion of LCA Development (SPOLD), whose membership includes large industrial companies, has been funding research projects on LCA. One of the specific aims of SPOLD is to design a common format for lifecycle inventory data. To this end sectoral LCAs have been compiled, from which LCA-averaged inventory data can be produced for the plastics, paper, steel, aluminium and zinc industries among others (Vancolen, 1996).

DEFINITIONS

> Life-cycle assessment is an objective process to evaluate the environmental burdens associated with a product, process or activity by identifying and quantifying energy and materials used and wastes released to the environment, to assess the impact of those energy and materials uses and releases on the environment, and to evaluate and implement opportunities to effect environmental improvements. The assessment includes the entire life-cycle of the product, process, or activity, encompassing extraction and processing of raw materials, manufacturing, transportation and distribution, use/re-use/maintenance, recycling, and final disposal.
>
> (Consoli *et al.*, 1993)

The purpose of LCA may be described as the identification, quantification and evaluation of the environmental consequences associated with a product, process or activity from 'cradle-to-grave' on a scientific and systematic basis, aimed at improving the environmental performance of the product and the process of its production. In principle, LCA addresses products, processes and/or activities. The main focus of the practice of LCA, which was initially on finished consumer products, is now shifting towards improving tools such as those used in design (Curran, 1996).

Although there is now a consensus on the 'cradle-to-grave' approach to environmental evaluation, there is still a fundamental controversy about whether the economic and social aspects should also be considered. Indeed, it is a question which marks the dividing line between LCA and product-line analysis (PLA) (Enquete Commission of the German Bundestag, 1994). Rubik points out the conceptual differences between the two assessments as follows (Rubik and Baumgartner, 1992): while 'ecological assessment' is a more general concept describing an approach to assessment which includes and combines environmental impacts with social and economic impacts, 'environmental assessment' is a narrower concept by comparison, limited only to the impacts on the environment. But, even when economic and social aspects are not formally included, there is consensus that weighting should be given to environmental, economic and social impacts in both assessment and final decision-making, and that the life-cycle approach may be useful in considering economic and social issues too. As such, LCA may be seen as part of a more extensive assessment, as shown in Figure 11.1 (Assies, 1992).

The general approach in an LCA is to include the functioning of 'capital goods', but not their

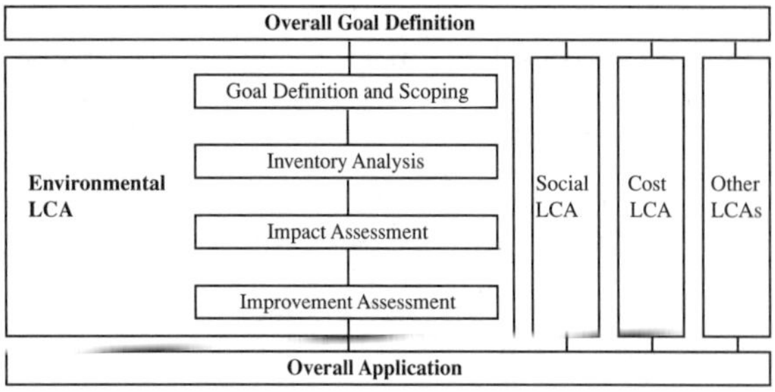

Figure 11.1 The relation between different types of LCA
Source: After Assies, 1992

production; that is, to consider only the consequences of the actual functioning of the systems. Thus the environmental consequences of infrastructure and installations are not taken into account, and all construction activities are excluded from the inventory. Maintenance and repair activities for transport are also excluded. Finally, by definition LCA does not address the societal relevance of the product, process or activity under consideration. The adjective 'objective', if used, can therefore create false expectations (Mazijn, 1994).

The focus of this chapter will be on environmental LCA.

THE METHODOLOGICAL FRAMEWORK

Objectives

The *prime objectives* for conducting an LCA are (Consoli *et al.*, 1993):

- to provide as complete a picture as possible of the interactions of an activity with the environment;
- to contribute to the understanding of the overall and interdependent nature of the environmental consequences of human activities;
- to provide decision-makers with information which identifies opportunities for environmental improvements.

In addition, systematic LCA procedures can facilitate a constructive dialogue among the various societal actors concerned with environmental quality.

The central principles

Udo de Haes (1990) argues that LCA should satisfy four central principles:

- LCA should be developed for comparing products with respect to their environmental impact;
- the entire life-cycle of the product should be taken into consideration;
- all relevant forms of environmental intervention should be taken into account;
- environmental interventions should be quantified as much as possible.

These principles are described below, as elaborated by Udo de Haes (1990).

Comparison of products

The environmental profile of a given product becomes significant only when it is compared either across product lines, or within the product itself (i.e. when the product is modified in some way). However, the environmental profile of a product does not constitute a basis for assessing whether or not it is environmentally sound; it merely indicates whether the product in question is more or less environmentally sound than other products that are functionally similar. In fact, what is important here is not the products that are compared, but their 'units of use'. Products are not compared in terms of their physical attributes such as volume or weight, but in terms of the task they perform. In this context the 'functional units of product' refer to the amount of the product needed to perform a specific function. In the case of packaging of milk, for example, it is not valid to compare a refillable milk bottle with a milk carton because the former can be used many times unlike the latter which can be used only once. A valid 'functional unit of product' is, for example, the 'packaging of 1,000 litres of milk'. As such, the functional unit enables comparisons of different products on the basis of a common function (Groupe des Sages, 1994).

The life-cycle approach

The entire life-cycle of a product should be considered in the LCA, starting from the acquisition of the raw materials, through the production of intermediate products, components and the product itself, to the use of the product and the processing of all solid waste flows generated in its life-cycle including its final disposal. Thus, in the LCA of a plastic bottle, for example, the source of the raw material should be considered to begin with. This would entail a consideration of the extraction of the fossil oil, refining and cracking it, and then converting it into plastic pellets from which the bottles are subsequently blown and their caps injection-moulded. Once the bottles are filled, the modalities of their distribution and use would need to be addressed. Bottles can be refilled or 'reincarnated' into new bottles or alternative products through recycling. Finally, they can be disposed of in various ways such as in landfills or by incineration. The added value of the life-cycle approach is that it extends the debate on environmental performance beyond single issues such as 'solid waste' or 'recycling'.

Environmental interventions

All relevant forms of environmental intervention should be taken into consideration. SETAC defines environmental interventions as '(any) exchange between the anthroposphere (the "economy") and the environment' (SETAC, 1994). If only a limited number of aspects are considered, improvements of these aspects may be accompanied by a deterioration of the other aspects. Therefore, at least the interventions in the following should be considered:

- extraction of primary resources (including the energy carriers);
- emission of harmful substances into water, air and soil;
- aspects of land use.

Other types of intervention can be added to the above list, such as radiation, sound emissions, and other forms of environmental harm.

Quantification

The various forms of environmental intervention should be quantified as much as possible, because it is only then that products intended to perform the same or similar functions can be compared with regard to their environmental credentials. But this does not detract from the importance of qualitative guidelines. Quantification of interventions is usually difficult, however. There is also the need for supplementary qualitative criteria which give an idea of the potential environmental interventions, including the nature of resources and emissions, the re-usability of the product, the extent to which the material of the product can be recycled, and the natural degradability of the product, or its material, following its disposal. Clearly, qualitative and quantitative approaches supplement each other. At the start of the design process, qualitative criteria can to a certain extent make up for the lack of quantitative information. But quantification is essential for the comparison of the actual products, not least because to some degree it constitutes a check on the qualitative criteria used.

The technical framework: the main structure of an LCA

Although there are differences between the various approaches to LCA, a broad consensus on its overall structure appears to be emerging. The methodological framework, recommended in SETAC's *Code of Practice*, comprises four main, interrelated components: 'goal definition and scoping', 'inventory analysis', 'impact assessment' and 'improvement assessment' (Figure 11.2) (Consoli *et al.*, 1993). Furthermore, impact assessment is subdivided into three phases, namely 'classification', 'characterisation' and 'valuation'.

Table 11.1 shows the different components and phases of LCA and their respective stages of development. As will be seen from this table, only the first two components of the methodology mentioned above are well defined.

The term 'Life-Cycle Inventory (LCI)' was introduced in order to identify studies which comprise only the first two components of a full LCA. A major problem is that at present there is no methodology to guide the interpretation of LCI results. In particular, there is no agreement on an impact assessment methodology that would translate the results of the inventory into measures of environmental impact. And, although improvement assessments based on

Table 11.1 State of development of LCA methodology

LCA components or phases	*State-of-the-art of written documents*
1 Goal definition and scoping	Defined
2 Inventory analysis	Defined and understood; needs some further work
3 Impact assessment	
3.1 Classification	Defined and understood: requires further work
3.2 Characterisation	Conceptually defined and partly developed
3.3 Valuation	Conceptually defined; different methods and approaches are currently being used
4 Improvement assessment	Not yet documented

Source: Consoli *et al.*, 1993

LCI information are being conducted, formal or systematic procedures for improvement assessment are yet to be established (Consoli *et al.*, 1993). It is to be noted, however, that policy-makers are mostly interested in the two final components of an LCA.

There are significant differences between the SETAC framework and that which is now being developed by the ISO shown in Figure 11.3. The major difference lies in the inclusion of 'interpretation' and the omission of 'improvement assessment'. 'Interpretation' feeds back into 'goal and scope definition', while 'valuation' may become a part of 'interpretation' rather than a part of 'impact assessment' (LCANET, 1996).

There are a number of changes in the terminology, too; apart from the introduction of the term 'goal and scope definition', 'component' has been replaced by 'phase' and 'step' by 'element' (LCANET, 1996).

THE PROCEDURAL FRAMEWORK: THE MAIN COMPONENTS OF AN LCA

Goal definition and scoping

Goal definition

When initiating an LCA, it is most important to define the objective of the study, in particular its purpose and the intended audience since this would determine the type of analysis to be conducted, the way the results would need to be presented, and the use to which the results would be put. Key questions in goal definition, and possible answers to them, are (Hindle, 1991):

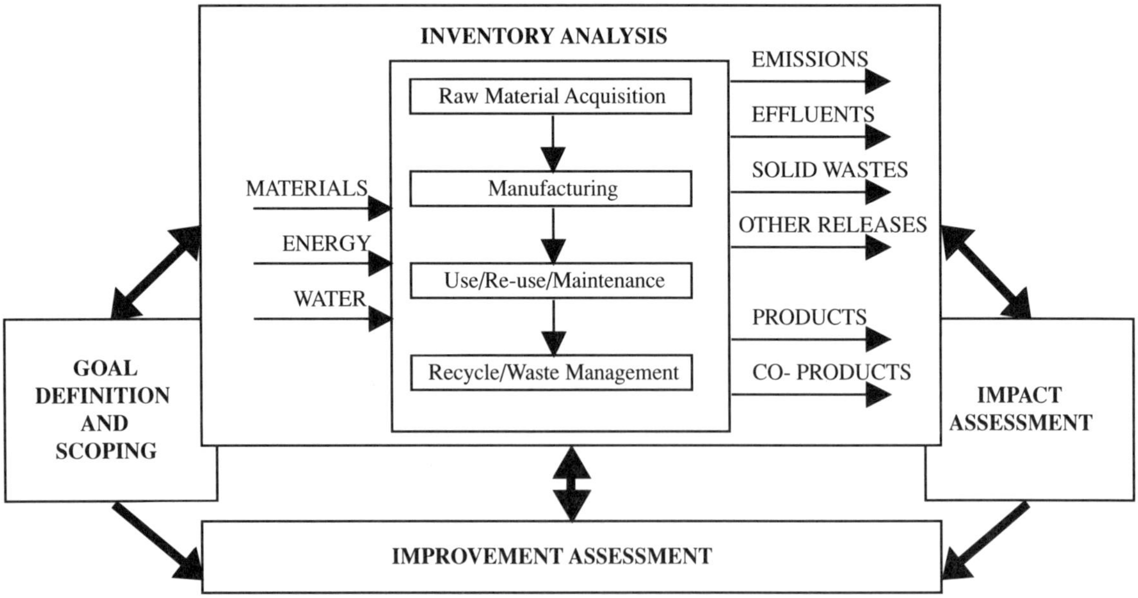

Figure 11.2 A generic life-cycle model
Source: After CSA, 1994

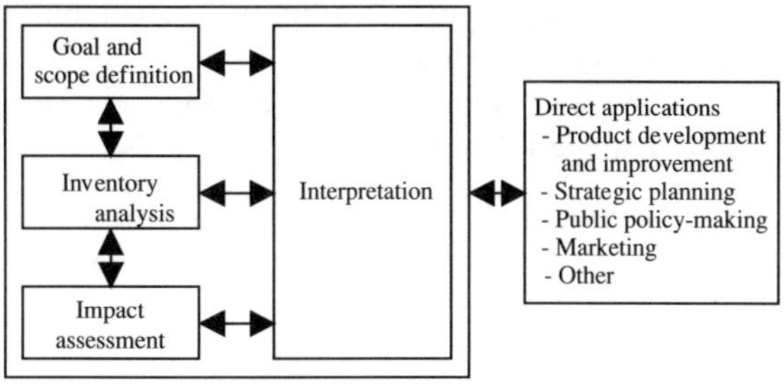

Figure 11.3 Phases of an LCA
Source: LCANET, 1996 (source ISO DIS/14040)

(a) Why is the study being made?
  1 to define the *status quo* of environmental burden;
  2 to determine ways of reducing the environmental burden on a 'cradle-to-grave' basis;
  3 to guide development work; or
  4 to help support advertising claims or stated environmental information.

(b) How will the study be disseminated and to whom?

  The answer to this question depends on whether the target audience is a body of experts, decision-makers, or the general public. Experts require more technical information than a lay audience. The intended use of the study is determined by the extent to which it would be subject to public scrutiny and debate.

A key consideration here is whether the LCA results will be used by a company internally to improve its environmental performance, or whether the results will be used externally to influence public policy for example (Consoli *et al.*, 1993).

Scope of the assessment

The scope determines the system, its boundaries, data requirements, assumptions, and limitations, and the scoping process links the goal with the extent or scope of the study (SETAC, 1994). 'Boundaries of the system' is a term borrowed from thermodynamics – the system is considered as a 'black box' for which the LCA produces the sum of all inputs and outputs. Defining the system boundaries is a crucial step because it determines exactly what is to be included in the LCA and what is to be left out. It is useful to consider the scope in terms of both 'length' and 'breadth'. 'Length' defines where the 'cradle' is in relation to the 'grave' and identifies precisely the network of steps to be considered as constituting the life-cycle. 'Breadth', on the other hand, defines the level of detail of the analysis, specifying the variables considered in each step (White *et al.*, 1995).

A number of critical parameters should be considered in the scope, including the geographical extent or space (local, national, regional, continental or global), and time (product life, time horizon of processes and impacts). In general, criteria for defining the scope should be the obvious ones, such as those items that are relevant to the intended use of the LCA and their quantitative significance.

The functional unit

A critical issue in an LCA is the definition of the functional unit. It is said to be 'the essential ingredient in an LCA to avoid nonsensical results' (Heijungs and Guinée, 1995). The definition of the functional unit establishes the basis for the selection of the alternatives to be investigated, and it is closely related to the objectives of the LCA (Assies, 1992). The functional unit, which should include the re-use processes and an estimate of the lifetime of the product in question, constitutes the basis for comparison and is the measure of the performance which the system under study is intended to deliver. For some products, consumer durables for example, it may prove difficult to define the functional equivalence, and so other factors such as service life, frequency of use, ease of repair, etc., which can have a major impact, have to be taken into consideration. Furthermore, it often turns out that, when alternatives are compared, there is no full equivalence of the benefits; there is only a rough equivalence at best. Both user and consumer behaviour also play an important role. Obvious examples of this are the actual amount of detergent used per wash and the compliance of motor vehicle drivers with speed limits (Enquete Commission of the German Bundestag, 1994).

Examples of functional units are 'units of surface area covered by paint for a defined period of time', 'packaging used for delivering a given volume of beverage', 'the amount of detergent needed for a standard household wash', etc. (Consoli *et al.*, 1993).

Assessment of quality of data

SETAC defines 'data quality' as the degree of confidence in individual input and output data, in the dataset as a whole, and ultimately in decisions based on the data (Consoli *et al.*, 1993). Assessment of the quality of data is an integral and essential part of an LCA, if the results are to be properly interpreted and communicated. Key indicators of data quality are age, frequency and method of collection, geographical scope, time period, completeness, representativeness, accuracy, uncertainty, and estimates of variability (Consoli *et al.*, 1993).

Inventory analysis

Introduction

Inventory analysis is defined as 'a technical, data-based process of identification and quantification of energy and raw material requirements, atmospheric emissions, waterborne emissions, solid wastes and other releases of a product, package, process, material, or activity' (Consoli *et al.*, 1993). The quality of a life-cycle inventory depends on an accurate description of the system to be analysed, while the collection and interpretation of essential data is contingent upon a proper understanding of where each stage of the life-cycle begins and ends (Vigon *et al.*, 1993). The main result of an inventory analysis is the *inventory table*, which is a list of environmental interventions caused by processes within a product system (SETAC, 1994).

A full life-cycle inventory comprises four major stages: 'raw materials acquisition', 'manufacture', 'use/re-use/maintenance', and 'recycling/waste management' (Vigon *et al.*, 1993; CSA, 1994). Although this inventory model is in accordance with SETAC's 'technical framework' (Fava *et al.*, 1991), it redefines three aspects (Curran, 1996):

- transportation activities are disaggregated and not presented as a separate phase;
- materials manufacture, product fabrication and filling, packaging and distribution are combined into a single life-phase;
- recycling and waste management are combined into a single life-phase.

The description of the major stages of the LCI is based on the 'inventory guidelines and principles' published by the USEPA (Vigon *et al.*, 1993).

Raw material acquisition begins with all the activities needed for acquiring necessary raw materials and energy, and ends at the first stage of refinement. Transportation to the point of processing is part of this stage. Outputs from the acquisition of raw materials, such as oil spills, agricultural run-off, and leaching of mine tailings, may have far more serious impacts on the environment than outputs from the subsequent life-cycle stages. Some of these outputs are difficult to quantify; for example, the aesthetic degradation and destruction of a habitat. Nevertheless, it is desirable to characterise these interventions in some way so that they can be accounted for in a subsequent impact assessment.

Manufacturing encompasses three sub-stages, namely taking feed-stocks of raw materials, converting them into final products, and delivering them to the consumers. *Materials manufacture* starts with the receipt of raw materials and includes on-site storage and handling. It involves converting raw materials into a form that can be used to fabricate a finished product. Normally, the process also involves the production and transportation of many intermediate materials. *Product fabrication* involves the processing of the manufactured material to create a product ready to be filled or packaged. The *filling/packaging/distribution* stage includes all manufacturing processes and transportation required to fill, package and distribute a finished product.

The *use/re-use/maintenance* of a product starts after its distribution and ends when it enters a waste stream after being discarded. *Use* includes activities such as consumption, operation of equipment, storage for later use (e.g. refrigeration) and preparation for use (e.g. cooking). *Maintenance* involves repair activities and preventive maintenance (e.g. changing oil). *Re-use* may involve the re-use of a product for its original purpose, or its re-use for a different purpose (e.g. storage of plant seeds in a jam jar). It also includes donations of used items to charities for re-use, as well as the return of materials to a retailer or manufacturer to be re-used for their original purpose (e.g. refillable containers).

Recycling encompasses all activities necessary for taking a material out of the waste stream and delivering it to the manufacturing stage. Recycling starts when a material or product is delivered to a collection system (e.g. when a bottle is placed in a roadside bin, or taken to a recycling centre). *Composting* is the controlled, biological decomposition of organic materials into a relatively stable, humus-like material, while *waste management* systems refer to incineration and land-filling.

It is essential for an LCA to be as comprehensive as possible, but in practice it seldom is. A 'limited' LCA may be conducted to compare a product before and after change(s) are made in the production process(es) to improve its environmental performance. Great caution must be exercised, however, in conducting limited LCAs and in interpreting their results. Studies are being made to determine if such limited LCAs can yield accurate and reliable results (Curran, 1996)

Construction of a life-cycle inventory model

This section is based on the major findings of the SETAC technical framework for life-cycle assessment (Fava *et al.*, 1991; Consoli *et al.*, 1993). Although industry and agriculture are primarily concerned with products, it is the production systems of these activities that consume raw materials and fuels and generate solid, liquid and gaseous wastes. In this context any group of operations that perform some well-defined function can be regarded as a system, and the boundary enclosing the group of operations is referred to as the *system boundary*. The region surrounding the boundary is the *system environment*, which acts as the source of all inputs to the system and the sink for all outputs from it.

Once the boundaries of each system have been determined, a *system flow diagram* is prepared to describe the system graphically. Figure 11.4 shows the life-cycle of a polyvinylchloride (PVC) container as a typical example. PVC is a chlorinated hydrocarbon polymer. The raw materials for the production of PVC are crude oil or natural gas and sodium chloride. The hydrocarbon raw materials are converted to ethylene, following which sodium chloride is electrolysed to produce chlorine (Cl_2), caustic soda (NaOH), and hydrogen (H_2). Ethylene and chlorine are then reacted to produce ethylene dichloride (EDC), which is decomposed by heating to yield vinylchloride monomer (VCM). Finally VCM is polymerised into PVC, the basic material for the production of the container. As will be seen from Figure 11.4, the integral life-cycles constitute complex networks.

The next step is to translate the flow diagram into a 'material balance'. The environmental impact of the system is assessed by measuring or calculating the flows of energy and materials across the system boundary. Systems are defined by specifying the nature of the input and output, and, where necessary, by indicating the processing route to be followed when there is more than one route. In the inventory analysis, the system is defined such that the inputs are all raw materials taken from the environment and the outputs are waste materials released back into it. At some stage the materials exist in the

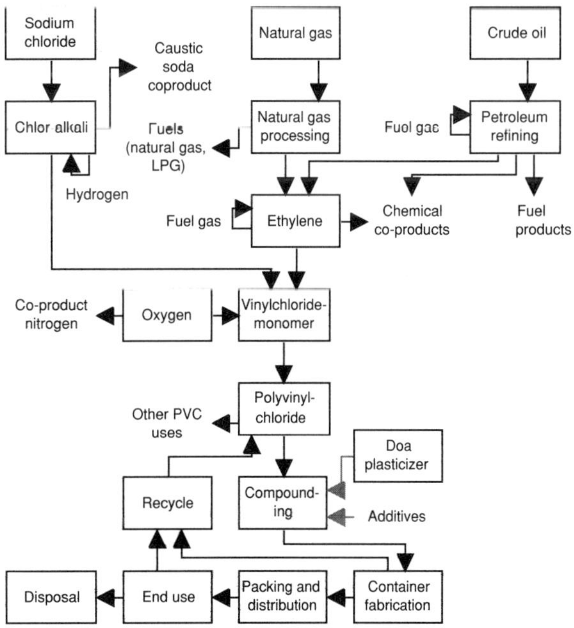

Figure 11.4 The life-cycle of polyvinylchloride (PVC)
Source: Holloway, 1991

system in a useful form as products or services delivered to the consumer.

Three main groups of operations can be identified within a given system (Fava *et al.*, 1991):

1 *the main production sequence:* these are operations concerned with the production, use, transportation, and disposal of the product;
2 *production of ancillary materials*, such as packaging and machinery needed to process the raw materials that feed into the main processing or production sequence;
3 *the fuel production industries* that supply the energy needed to drive the system.

In order to describe the performance of a system, the overall system is divided into a series of subsystems linked to each other through balanced material flows. The subsystems are broken down to a level of detail such that each corresponds to some physical operation. Once all the component subsystems have been identified, each can be regarded as a system in its own right. For each subsystem, data are obtained on the materials and energy inputs and on the products and waste generated. Once this information is assembled for all the subsystems, the performance of the overall system is calculated by establishing mass balances among all the subsystems. The total demand for raw materials and fuels, and the total output of solid, liquid and gaseous wastes from the overall system, is simply the sum of the inputs and outputs of all the component subsystems.

Inputs to the life-cycle inventory analysis

Raw Materials

Raw materials are defined as materials that have been extracted from the earth but have not yet been refined or processed (Vigon *et al.*, 1993).

An important issue associated with raw materials is whether to distinguish between renewable and non-renewable resources (Vigon *et al.*, 1993). A *renewable resource* can be replaced within a socially relevant time frame, certain species of wood for example. Minerals, metals and hydrocarbon fuels (coal, petroleum and natural gas) are deemed *non-renewable*. The intrinsic nature of a resource does not determine its renewability. However, the potential for the replacement of a given material is not by itself sufficient to qualify it as being 'renewable' – replacement or renewal should actually occur for this to be true. *Sustainable resource* usage may be said to occur if or when the replacement rate exceeds the rate of consumption.

Energy

'Energy' represents the total amount of energy consumed by the system. Energy is obtained from a variety of sources, including coal, natural gas, petroleum, hydropower, nuclear power and wood.

Energy requirements are quantified in three different categories:

1 *Process energy* is the energy required to operate and run the subsystem processes.
2 *Transportation energy* is that which is required to power the diverse modes of transportation.
3 *Energy content of material resources* accounts for those products/materials that are made or obtained from energy resources.

Energy released by the combustion of fuels is only part of the total energy associated with the consumption of fuels. Also, the amount of energy expended in the acquisition of fuels consumed is significant in comparison to other energy expenditures. Pre-combustion energy is defined as the energy needed to deliver a usable fuel to the point of usage, and it is the sum of the total amount of energy needed to acquire the raw fuel materials, processing these materials into usable fuels, and transportation.

When energy recovered from the burning of wastes is credited against the energy required by the system for primary production, net energy requirement becomes less than the total energy required by the system as a whole, as in the plastics industry for example.

Water

It is recommended that the 'water volume requirements' should be included in the inventory analysis (Vigon *et al.*, 1993; CSA, 1994). In practice, however, the quantity of water consumed is estimated in terms of the net 'consumptive usage' which, as a life-cycle inventory input, equals the fraction of the total water withdrawn from surface or groundwater sources that is either incorporated into the product, co-products (if any) or wastes, or is evaporated.

Outputs from the life-cycle inventory analysis

Three major categories of environmental releases are quantified in a 'traditional' inventory analysis and reported in terms of 'weight per unit of weight' of product output:

1 *Atmospheric emissions* including all substances classified as air pollutants.
2 *Waterborne emissions* which include all substances classified as waterborne pollutants.
3 *Solid wastes* including all materials that are disposed of from all sources within the system.

Products, defined by their subsystems, as well as co-products are also quantified. Each subsystem has a resulting product with respect to the entire system. The subsystem product may be considered either as a raw or intermediate material with respect to another system, or as the finished product of the system itself (Vigon *et al.*, 1993).

Data gathering

Collecting data for every step of the life-cycle model is a laborious task. And complete accuracy demands measurement of energy inputs for every individual part of the manufacturing plant involved in the process, the efficiency of every transport vehicle used, etc. Plant-specific databases of this type take considerable effort to assemble and, even then, they are not broadly applicable. Instead, a sample may be taken and average data used. Some loss of accuracy may occur in this approach, but this is compensated for by ease of use and broad applicability. The danger with generic databases is that conditions may vary with geography, so that the average data may not be applicable. Typically, most publicly available life-cycle documents present industrial average data, while many private company studies using plant-specific data may still have to use aggregated data for stages that are external to the plant for example. In the end, any model is only as good as the data it uses. Therefore, access to up-to-date, broadly sourced and accurate data is crucially important.

Limitations of the inventory analysis

Inventories are by no means simple tabulations of numbers. The preparation of an inventory according to the listed principles calls for additional choices to be made, and the crucial choices relate to the technology considered and the allocation of the environmental interventions.

Technology

Data on processes cover a considerable range of technologies. The first important choice to be made, namely the choice of the type of process technology, is critically important for calculating environmental interventions. The choice depends primarily on the purpose of the study. Type of technology covers a wide range of data, such as on actual (real average), worst, modern (best real), and future (best available technology) technology. The significance of the level of technology is elaborated by Ekvall (Ekvall *et al.*, 1992). Table 11.2 shows, typically, selected emissions caused by electrolysis used in aluminium production. As will be seen, emissions levels are much lower in furnaces with pre-baked anodes.

Allocation

One or more of the three main situations, described below, may occur where allocation has to be performed (Udo de Haes, 1992).

Table 11.2 Emissions of different furnaces, with pre-baked and with Söderberg anodes

Emissions (g/Kg aluminium)	*Pre-baked anodes*	*Söderberg anodes*
Sulphur dioxide	1.1	2.9
Dust	0.4	4.4
Total fluorides to air	0.4	1.1
Total fluorides to water	0.4	5.3

Source: Ekvall *et al.*, 1992

1 *A multi-output situation* in which a given process yields different valuable co-products. The distillation and cracking of crude oil, for example, generates at least ten different oil products, while the electrolysis of sodium chloride produces chlorine gas, hydrogen gas and sodium hydroxide.
2 *A multi-input situation* in which different products are fed into a single process; an example of this is the incineration of municipal solid waste.
3 *An open-loop recycling situation* in which recycling occurs outside of the given product system.

These three allocation situations are shown graphically in Figure 11.5 (Mazijn, 1994). In all three the impacts of the processes have to be partitioned between the product under consideration and other product systems.

However, allocation can actually be a bottleneck in the inventory phase. The main question here is concerned with the criteria on which allocation should be based. In principle, physical (mass, energy content, volume) or economic (values, prices) criteria can be used. Also, the choice of the partitioning method is important, because it can significantly affect the final result. Boustead states that there is no overwhelming scientific reason for choosing any specific (allocation) procedure and, consequently, the choice is almost arbitrary (Boustead, 1994). But, given the critical importance of allocation, clear documentation of the method(s) used in an LCA is strongly advocated.

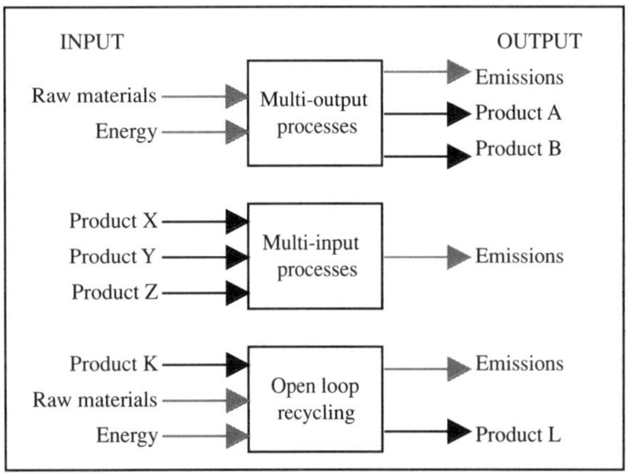

Figure 11.5 The main allocation situations
Source: After Mazijn, 1994

The hierarchy of allocation principles, advocated by the ISO, is as follows (LCANET, 1996):

1 Avoid allocation by detailing into more sub-processes.
2 Minimise allocation by enlarging the system.
3 Physical causality.
4 Economic/market causality.

Impact assessment: interpretation of the results of an inventory

Introduction

The translation of inventory results into environmental impacts is one of the most contentious areas of LCA. The crucial question here is, what do the results mean in terms of the environment? One of the major stumbling blocks in the development of LCA has been the lack of consensus on how to convert the inventory table into environmental impacts. At present there is still no consensus on how this should be done.

Evaluation of inventories involves two discrete but interlinked aspects; namely, aggregation and interpretation. An inventory analysis leads to the quantification of a large number of input and output streams, involving 200 or more parameters. But, in order to be useful to decision-makers, inventory information needs to be simplified through aggregation. This means that a large number of parameters should be combined and condensed into a few but meaningful indicators. But the disadvantage of this is that information is lost through aggregation, not least because the weighting factors used are often subjective and arbitrary.

Interpretation of LCI results: current methods

Excellent reviews are available on current methods of interpreting LCI results (White *et al.*, 1992; Finnveden *et al.*, 1992; Grisel *et al.*, 1994). This section draws heavily on them.

(A) Categories of aggregation

Energy consumption data can be aggregated using the common currency of megajoules. Combining all forms of energy may, however, obscure significant information, because energy may be consumed as the latent energy of raw materials, or that used in processing or transportation. In addition, different energy sources (e.g. oil, nuclear, wind) have different environmental impacts.

Solid waste can be aggregated by weight. There are also models in which quantities of solid waste are expressed in terms of volume using conversion factors; but the main criticisms are that (Finnveden, 1992):

- aggregation of different types of waste, regardless of their environmental characteristics (i.e. without weighting), is scientifically unjustified;
- emissions from landfills (landfill gas and leachate) contribute to almost all environmental problems with which LCAs are concerned. The theoretical aspects of waste disposal are not yet sufficiently developed and their effects are underestimated.

Also, major problems arise when attempts are made to aggregate *emissions.*

The critical volume concept

In this method a theoretical 'critical volume' is calculated by dividing the amount of each pollutant by its maximum legally permissible concentration limit (BUWAL, 1991). These regulatory limits refer to maximum exposure concentrations in the workplace, or maximum emission levels, reflecting tolerable human toxicity levels. Emissions are then aggregated by summing up the critical volumes for air and water respectively. But this concept is criticised on several grounds (White *et al.*, 1992):

- while regulatory standards are set on a national basis, emissions may have global effects;
- regulatory standards often reflect feasibility or cost rather than good science;
- critical volumes cannot be calculated for non-regulated emissions;
- it inappropriately combines different materials with different effects;
- it takes into account only one aspect of emissions, namely human toxicity.

This last criticism is fundamental, because it presumes that human health considerations are both predictive and protective of the environment (White *et al.*, 1992). A weighting system, based on a single effect such as human toxicity, also ignores the fact that different compounds have different effects on both target organisms and ecosystems, and that a single chemical may also produce different effects.

Aggregation according to environmental effects

A sensible approach to aggregation would be to assign an appropriate weighting to an emission reflecting the environmental consequences caused by that emission. In this approach the greenhouse gases would be aggregated, for example, by applying to each gas a weighting which reflects its 'global warming potential'. The same exercise can be repeated for ozone depleting substances, for gases which cause acid rain, and so on. The problem of combining different effects is avoided in this approach, and it ensures that each component has a meaningful environmental currency. An early model based on this approach, developed at the Centre for Environmental Science (CML) University of Leiden, The Netherlands, recommends a number of headings (impact scores) classified into three main environmental impacts, namely 'resource depletion', 'pollution' and 'disturbances' (Udo de Haes, 1992). The environmental problems considered are global warming, ozone depletion, toxic effects (human and ecological), acidification, formation of photo-oxidants; nitrification (aquatic and terrestrial), dispersion of heat, occupational health, and solid waste (hazardous and others).

(B) Aggregation across categories

The ecopoints approach

In this approach, developed in Switzerland, aggregation is carried out by weighting individual emissions or energy consumption using 'ecofactors' to give what are called 'ecopoints' (units of environmental pollution) (BUWAL, 1990). The sum of the ecopoints gives a measure of the total environmental burden and, clearly, the more the ecopoints, the greater the burden. The ecofactor, to be used for calculating the weighting of a substance, is derived from the relationship between the present emission level of that substance (or energy consumption level) and its total loading capacity (or energy use) in the Swiss environment.

Although in this model an inventory table is reduced to a single ecopoint total, the methodology is based on several assumptions. A critical assumption is that the maximum loading capacity of the environment, before irreversible damage occurs, is known for all emissions. The critical loading levels, however, reflect Swiss government targets rather than known threshold damage levels. To summarise, the ecopoints approach may give an indication of the environmental burden, but it says little about the overall environmental impact.

Multi-factorial indices

The inability to deal with the complex relationships between LCA inventories and environmental impact using single-factor weighting systems has provided the impetus for developing more complex, multifactorial indices, such as those used in the Environmental Priority Strategies (EPS) system (Ryding and Steen, 1991). In this separate environmental indices are defined for natural resource use, effects of substances (emissions), materials and processes. Indices for natural resource use and substance effects each consist of several 'ecological scores' that are multiplied together to give a numerical value of the index. Natural resource indices are based on known resources and their 'uniqueness', while environmental indices for substance effects are the products of six separate scores: type and extent of the problem; intensity and frequency of occurrence of the problem; spatial distribution of the problem; duration of the problem; contribution to the problem by the emission of 1 kg of the polluting substance; and feasibility and cost of remedying the problem. Indices are expressed as environmental load units (ELU) per unit of mass for substances, per m^2 for land, and per MJ for energy. It is possible to aggregate the ELUs across environmental compartments.

An advantage of this approach is that it concentrates on the effects of emissions without relying on regulatory limits, and also that the consumption of both raw materials and energy is evaluated. But the calculation of the indices is complex and subject to several assumptions. Assessment of the effects of substances by considering 'frequency of occurrence of the problem' and 'possibility of remediation' introduces elements of risk into the interpretation of environmental impact, even though it cannot be considered as a full risk analysis.

The SETAC approach: impact assessment

SETAC defines 'impact assessment' as 'a technical, quantitative, and/or qualitative process to characterise and assess the effects of the environmental burdens identified in the inventory analysis' (Consoli *et al.*, 1993). *But there is no fixed methodology yet.* Indeed, the lack of agreement on what to include in an impact assessment reflects a wider divergence of views on the objective of the overall LCA process as well as on the relationship between LCA and other environmental management tools (De Smet *et al.*, 1994).

SETAC recommends a conceptual framework for impact assessment comprising three separate steps – namely, classification, characterisation, and valuation (SETAC, 1994).

The purpose of *classification* is to organise data from the inventory analysis into a manageable number of impact categories. Three major categories of impact should be considered: resource depletion, human health impacts, and ecological impacts. These categories can be further divided into specific impact categories which, in turn, can be subdivided into subcategories (Finnveden, 1994). An important issue here is to decide which of the environmental impacts should be included in the impact assessment. But, since at present there is no consensus on a standard list of impacts or on a methodology to deal with the different types of impact, the tentative list of Table 11.3, proposed by Finnveden (1994), may be used.

Clearly, both the spatial and time dimensions of the impact categories differ (Henshaw, 1994). For example, while CO_2 emissions contribute to global

Table 11.3 List of impact categories

Impact category
Resource depletion
Energy and materials
Water
Land (including wetlands)
Human health
Toxicological impacts (excluding the work environment)
Non-toxicological impacts (excluding the work environment)
Impacts in the work environment
Ecological impacts
Global warming
Depletion of stratospheric ozone
Acidification
Eutrophication
Formation of photo-oxidants
Ecotoxicological impacts
Habitat alteration and impacts on biological diversity
*Inflows which are not traced back to the system boundary between the technical system and nature
*Outflows which are not traced back to the system boundary between the technical system and nature

Source: Finnveden, 1994

Note: * Not impact categories, but should be included

warming (global impact), SO_2 may contribute to acidification which is a problem on a regional or local scale.

Characterisation, which should be based on scientific knowledge about environmental processes, is concerned with the analysis, quantification and, where possible, aggregation of impacts within the given impact categories. Impacts can be characterised at different levels of formalisation and detail. There are approaches in which data from the inventory table are related in a generic way to the NOECs or to environmental standards for example. There are yet other approaches in which both exposure and effects are modelled in a generic fashion and applied in a site-specific manner. Presently much attention is being given to the development of equivalence factors for different impact categories such as Global Warming Potential (GWP) and Ozone Depletion Potential (ODP). The process of characterisation may be concluded by normalising the aggregated data per impact category in relation to the actual magnitude of the impacts within that category for a given area. The outcome of *characterisation* provides what is called an 'impact profile' or 'environmental profile'.

Valuation is the process in which data in the different impact categories are weighted in a quantitative and/or qualitative manner so that they can be compared among themselves. When valuation is done with quantitative weighting, it can result in an *environmental index* which is a parameter indicating the perceived harmfulness of a product to the environment (SETAC, 1994). The aim of valuation is to facilitate further interpretation and aggregation of impact assessment data. Although the assignment of weighting is essentially a subjective process, there are tools that have the potential to make valuation both rational and explicit. These include expert judgement and input from interested and/or affected parties.

Improvement assessment

SETAC defines 'improvement assessment' as the component of LCA in which options for reducing environmental impacts or burdens of the system(s) under study are identified and evaluated. But improvement assessment has not yet undergone the consensus examination (Consoli *et al.*, 1993). The CSA guidelines stipulate a set of core principles, listed below, for making improvements that are embodied in the concept of sustainable development (CSA, 1994).

Whenever possible, a product should:

(a) minimise the use of non-renewable resources;
(b) maximise the use of sustainable resources in terms of both raw material and energy;
(c) minimise the total energy consumption;
(d) minimise the use of toxic materials and/or processes;
(e) minimise the release of emissions;
(f) reinforce source reduction, re-use, recycling and recovery activities.

Strategic goals for making specific improvements can be defined from these core principles.

PRACTICAL ISSUES: RUNNING THE PROJECT

The architecture of LCA models

Given the complexity of most LCAs and the multiplicity of their data sources, the use of computer models as an effective analytical tool becomes inevitable. But it is important that such models are structured in a way that allows the processing of the data and permits the information to be stored for use in other projects (Holloway, 1991).

The example of Figure 11.6 shows a comparison of different LCAs for beverage containers made from a variety of materials, including polyvinylchloride (PVC), polyethylene terephtalate (PET), high density polyethylene (HDPE), and from glass. Here the data sources are recorded for future use. Also, the data are incorporated in the process modules which feed the matrices of each material; and this, in turn, leads to the outputs for each of the containers being compared (Holloway, 1991).

Sources and types of data

The use of well-characterised and up-to-date data relating to specific processes is highly recommended,

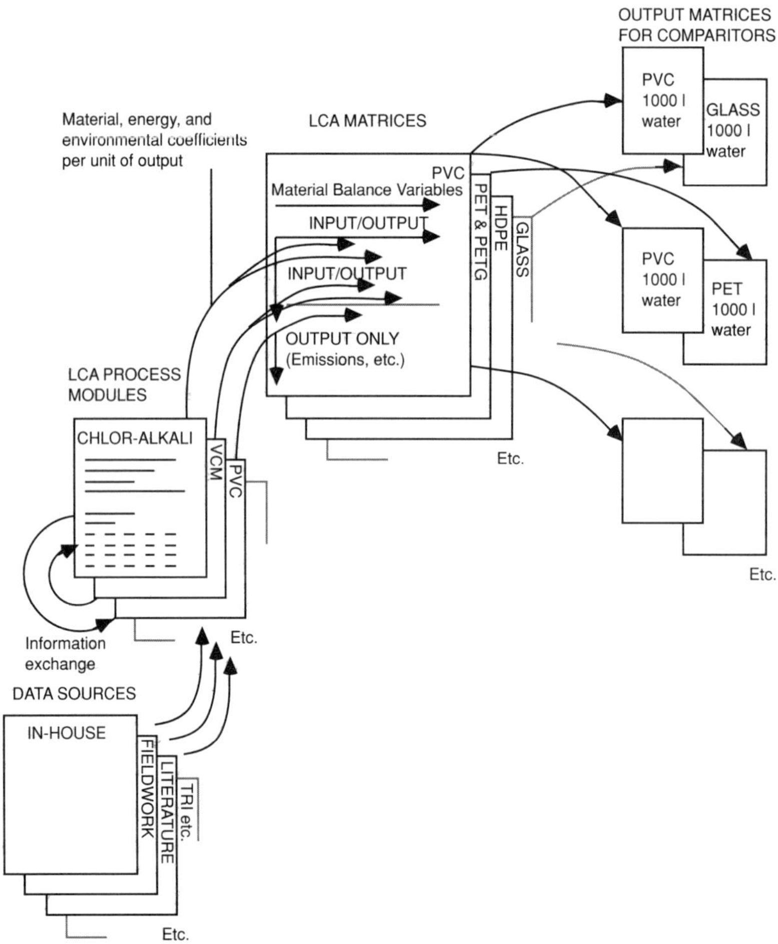

Figure 11.6 LCA model architecture
Source: Holloway, 1991

because processes may change or become more efficient over time. The sources of data can be diverse, however; facility-specific data at the enterprise level for example, or sector-specific data at a more general level. Sources of data can be divided into two main classes (CSA, 1994); namely, *primary* sources and *secondary* sources. The primary sources include facility-specific data which may be accessed from public or other sources, and specific data from other LCA studies. CSA lists examples of primary industrial data sources, including independent or internal reports, periodic measurements, accounting or engineering reports or datasets, specific measurements, and machine specifications. Governmental and industrial electronic databases are an important source of secondary data. These sources can be non-bibliographic and may contain averaged industrial data and/or product specifications. Government documents on data on broad categories of processes are publicly available. In addition, publications in the form of technical books, reports, conference reports and articles often contain much useful information.

The USEPA gives an overview of the different types of data that are used in practice (Vigon *et al.*, 1993). Starting with the most disaggregated, they comprise the following:

(a) *individual process- and facility-specific*: data from a particular operation within a given facility that are not combined in any way;
(b) *composite*: data from the same operation or activity combined across locations;
(c) *aggregated*: data combining more than one process operation;
(d) *industry-average*: data derived from a representative sample of locations and believed to describe statistically a typical operation across technologies;
(e) *generic*: data whose representativeness may be unknown but which describe quantitatively a process or technology.

The 'goal definition and scoping process' determines the level or type of information required.

Software

A wide range of computer tools have been developed to assist in LCA modelling, and there are several publications on the subject. In particular, four programs, namely 'EPS 2.0', 'LCA Inventory Tool', 'PLA Educational Tool' and 'SimaPro 2', are described in detail by Miettinen (1993). An independent review of current LCA software was carried out by the Centre for Environmental Strategy, University of Surrey, UK (Anon., 1996a). The report examines most of the main software packages on the market, including 'The Boustead Model', 'EcoPro 1.3', 'GaBi 2.0', 'KCL-ECO' (with ECODATA database), 'LCAiT 2.0', 'LMS Eco-Inventory Tool', 'Oeko-Base für Windows', 'PEMS 3.0', 'PIA', 'Simapro 3.1', 'SimaTool' and 'The Ecobilan Group's TEAM™ (with DEAM™ database)'. Also, a study entitled *Evaluation of Life-Cycle Assessment Tools*, commissioned by Environment Canada (Menke *et al.*, 1996), profiles 37 software tools for LCA of which five are reviewed in depth; the University of Tennessee has compiled a list of software tools currently available for LCA, including reviews, documentation and demonstration versions, development of criteria for in-depth evaluation, and the evaluation of five software tools, namely KCL-ECO, LCAiT, PEMS, SimaPro and TEAM™. The criteria cover computer requirements, system definition, data and data management, flexibility, calculations and comparisons, and outputs and exports.

Checklist

A checklist, covering most of the decision areas in the execution of an LCA, is a valuable tool for improving communication. A completed checklist, when included, gives the reader knowledge of the following: purpose of the study; the system and its boundaries; the geographical scope of the study; types of data used; data collection and procedures for synthesis; data quality; construction of the computer model; and presentation of the results – and all these enable a better understanding of the results by recognising the differences among the various reports on the same topic. Moreover, the analyst performing the study can use the checklist to help ensure that all the important stages and information categories are included. Tables 11.4 and 11.5 reproduce a typical inventory checklist (Vigon *et al.*, 1993) which consists of two major components: a summary section

Table 11.4 Life-cycle inventory checklist, Part I

LIFE-CYCLE INVENTORY CHECKLIST, PART I – SCOPE AND PROCEDURES
INVENTORY OF: ______

Purpose of the Inventory: (check all that apply)

Private Sector Use
Internal Evaluation and Decision-Making
☐ Comparison of Materials, Products, or Activities
☐ Resource Use and Release; Comparison with Other Manufacturers' Data
☐ Personnel Training for Product and Process Design
☐ Baseline Information for Full LCA

External Evaluation and Decision Making
☐ Provide Information on Resource Use and Releases
☐ Substantiate Statements of Reductions in Resource Use and Releases

Public Sector Use
Evaluation and Policy-making
☐ Support Information for Policy and Regulatory Evaluation
☐ Information Gap Identification
☐ Help Evaluate Statements of Reductions in Resource Use and Releases

Public Education
☐ Develop Support Materials for Public Education
☐ Assist in Curriculum Design

Systems Analysed
List the product/process systems analysed in this inventory: ______

Key Assumptions: (list and describe) ______

Define the Boundaries
For each system analysed, define the boundaries by life-cycle stage, geographical scope, primary process, and ancillary inputs included in the system boundaries.

Post-consumer Solid Waste Management Options: Mark and describe the options analysed for each system.
☐ Landfill ______
☐ Combustion ______
☐ Composting ______
☐ Open-loop Recycling ______
☐ Closed-loop Recycling ______
☐ Other ______

Basis for Comparison
☐ This is not a comparative study ☐ This is a comparative study
State basis for comparison between systems: (Example: 1000 units, 1,000 uses) ______

If products or processes are not normally used on a one-to-one basis, state how equivalent function was established.

Computational Model Construction
☐ System calculations are made using computer spreadsheets that relate each system component to the total system.
☐ System calculations are made using another technique. Describe: ______

Describe how inputs to an outputs from postconsumer solid waste management are handled. ______

Quality Assurance: (state specific activities and initials or reviewer)
Review performed on: ☐ Data Gathering Techniques ______ ☐ Co-product Allocation ______ ☐ Input Data ______ ☐ Model Calculation and Formulas ______ ☐ Results and Reporting ______

Peer Review: (state specific activities and initials of reviewer)
Review performed on: ☐ Scope and Boundary ______ ☐ Data Gathering Techniques ______ ☐ Co-product Allocation ______ ☐ Input Data ______ ☐ Model Calculation and Formulas ______ ☐ Results and Reporting ______

Results Presentation
☐ Methodology is fully described.
☐ Individual pollutants are reported.
Emissions are reported as aggregated totals only.
Explain why: ______
☐ Report is sufficiently detailed for its defined purpose.
☐ Report may need more detail for additional use beyond defined purpose.
☐ Sensitivity analyses are included in the report.
List: ______
☐ Sensitivity analyses have been performed but are not included in the report. List: ______

Table 11.5 Life-cycle inventory checklist, Part II

LIFE-CYCLE INVENTORY CHECKLIST, PART II – MODULE WORKSHEET

Inventory of: ________________ Preparer: ________________

Life-Cycle Stage Description: ________________

Date: ________________ Quality Assurance Approval: ________________

MODULE DESCRIPTION: ________________

	Data Value [(a)]	Type[(b)]	Data[(c)] Age/Scope	Quality Measures[(d)]
MODULE INPUTS				
Materials				
Process				
Other[(e)]				
Energy				
Process				
Precombustion				
Water Usage				
Process				
Fuel-Related				
MODULE OUTPUTS				
Product				
Co-products[(f)]				
Air Emissions				
Process				
Fuel-Related				
Water Effluents				
Process				
Fuel-related				
Solid Waste				
Process				
Fuel-related				
Capital Repl.				
Transportation				
Personnel				

(a) Include units.
(b) Indicate whether data are actual measurements, engineering estimates, or theoretical or published values and whether the numbers are from a specific manufacturer or facility, or whether they represent industry-average values. List a specific source if pertinent, e.g., 'obtained from Atlanta facility wastewater permit monitoring data'.
(c) Indicate whether emissions are all available, regulated only, or selected. Designate data as to geographic specificity, e.g., North America, and indicate the period covered, e.g., average of monthly for 1991.
(d) List measures of data quality available for the data item, e.g., accuracy, precision, representativeness, consistency-checked, other, or none.
(e) Include non-traditional inputs, e.g., land use, when appropriate and necessary.
(f) If co-product allocation method was applied, indicate basis in quality measures column, e.g., weight

describing the procedures and systems used in the study, and a set of worksheets listing and qualifying the data collected.

Peer review

If an LCA is intended to play a part in public debate, then the formation of a peer group to review the process is recommended. Among the benefits of a peer review are that it enhances the scientific and technical quality of the LCA, that the process helps to focus on study goals and data collection, and that it enables a critical screening of the conclusions of the study, thus enhancing its credibility.

APPLICATIONS OF LCA

Major themes of LCA studies

The themes of the published LCA studies have been investigated by Rubik and Baumgartner (1992) and Rubik (1994), and an overview of their findings is given in Table 11.6. Forty per cent of the studies were made in Germany, 18 per cent in Switzerland and 11 per cent in the USA (Rubik, 1994). Most of the studies were on packaging, and this is explained by the vigorous environmental debate on packaging in Germany and Switzerland. However, notwithstanding the decrease in the relative importance of packaging in favour of polymers (intermediate materials), the focus of LCA continues to be on finished consumer products.

Table 11.6 Statistics on published LCA studies

Theme	*Total number 1992*	*% 1992*	*Total number 1994*	*% 1994*
Packaging	59	44.7	101	36.9
Chemical products	12	9.1		—
Building materials	11	8.3	25	9.1
Nappies	10	7.6		—
Waste and recycling	5	3.8	10	3.6
Tableware	4	3.0	0	
Maintenance products and detergents			32	11.7
Energy			24	8.8
Plastics			23	8.4
Agriculture			9	3.3
Household articles			9	3.3
Electronics			7	2.6
Paper products			16	5.8
Totals	132	100	274	100

Sources: Rubik and Baumgartner, 1992; Rubik, 1994

Rubik's work also sheds light on 67 per cent of the clients of the published LCA studies on whom necessary information was available. Twenty-three per cent of the commissioning bodies were private companies and 18 per cent public authorities. The remaining 26 per cent of the studies were made mainly within the framework of research initiatives. Consumers and environmental pressure groups made only marginal inputs of 1 and 3 per cent, respectively. However, due caution should be exercised in analysing the data because many of the studies, conducted internally or by consultants, were classified as 'confidential information' (Rubik, 1994).

Potential use of LCA results

Potential use of LCA information, of which a range is shown in Table 11.7, can be categorised on the basis of whether the user is a private or a public decision-maker, although some uses are common to both (WWF, 1991). Opinions on appropriate uses of LCA vary a great deal, however (Assies, 1992), and several authors have addressed this issue by interviewing the users (Finnveden *et al.*, 1992; Marion *et al.*, 1995).

In general, experts agree that information on product life is most appropriate for at least the following:

- internal decision-making for manufacturers;
- general comparative purposes versus absolute determinations;
- specific use in making decisions on product labels.

Furthermore, results obtained from life-cycle inventories are at present used internally by industry in many different ways and for different purposes, such as (Fava *et al.*, 1992, Vigon *et al.*, 1993):

- for establishing a comprehensive baseline of information on a system's overall resource requirement, energy consumption, and environmental loading for further analysis;
- for identifying the points within the life-cycle as a whole, or within a given process, where the

Table 11.7 Potential uses of product-life information

Manufacturers may use LCA information in their overall product design and manufacture to:

- compare generic materials;
- evaluate effects of resources associated with particular products, including new products;
- compare functionally equivalent products;
- compare different options in a process with the objective of minimising environmental impacts;
- identify processes, ingredients, and systems that are major contributors to environmental impacts;
- supply information for product and procurement audits;
- provide guidance for long-term strategic planning concerning trends in product design and materials;
- help train product designers in the use of environmentally preferred products and materials;
- evaluate claims made by other manufacturers;
- enhance market competitiveness;
- provide information to consumers on the resource characteristics of their products or materials.

Public decision-makers may use LCA information to:

- supply information needed for legislative or regulatory policies that restrict the use of product materials (e.g. bans, taxes, etc.);
- supply information to set standards for product advertising (e.g. defining the recycled content, etc.)
- gather environmental information;
- identify gaps in knowledge and research priorities;
- help evaluate and differentiate among products for labelling programmes;
- provide information to the public about the resource characteristics of products or materials;
- help develop long-term policy regarding overall material use, resource conservation, and reduction of environmental impacts and risks posed by materials and processes;
- evaluate claims by manufacturers;
- evaluate resource effects of source reduction and alternative waste management techniques.

Source: WWF, 1991

Note: This list does not differentiate between potential and appropriate uses of information

greatest reduction in resource requirement and emissions might be achieved;

- for comparing system inputs and outputs associated with alternative products, processes or activities;
- for helping to guide the development of new products, processes, or activities aiming at a net reduction of resource requirements and emissions;
- for helping to identify the areas to be addressed in an impact assessment.

LCA IN CONTEXT

Benefits of the LCA approach

Hindle (1990) summarises the benefits of the LCA approach as follows:

(A) LCA is a *comparative* tool which can be used to compare the overall environmental performance, even when there are many differences between products and/or processes.

(B) LCA is an *inclusive* tool. It is essentially an accounting process with which all necessary inputs to and emissions from a defined system are listed and then totalled.

It is also an inclusive approach because the LCA:

- includes all the components of the system;
- includes all inputs to, and all emissions from, the system;
- integrates over time – that is, it integrates all inputs and emissions over the whole life-cycle;
- integrates over space – that is, all the sites are involved;
- integrates the effects of all the constituent processes in the life-cycle;
- integrates all the issues – for example, different methods of reducing environmental impacts, such as source reduction, re-use or recycling can be included, and their effects evaluated.

Limitations of the LCA approach: future developments

Today the scientific basis of LCA remains inadequate, however. The key issues in the development of LCA, for combining accuracy and objectivity with transparency, are discussed below.

Standardisation of the LCI approach

Standardisation of the LCI framework is of paramount importance. Pertinent questions in this regard are: which aspects should be considered, which data should be used, and how should the data be processed? Comparison will be greatly facilitated if a set of 'model' conventions can be agreed upon to ensure the use of the same system boundaries and the same method of accounting across all models.

Standardisation and availability of LCA databases

Lack of data for the execution of LCI is a serious problem. The need for systematically available data is acknowledged by all parties concerned. But agreements are essential for the creation and maintenance of databases that are up to date, accurate, and based on as many sources as are available. Also, data should not contain interpretations, but should quantify explicitly all relevant environmental aspects, and mean values should be given together with ranges and standard deviations.

Agreed methods for converting LCA inventories into environmental impacts

Given the need to use a standard model, which also implies the use of an agreed database, it is essential that common methods are agreed upon for converting LCA inventories into environmental impacts. Indeed, this is the most contentious area in which a very great deal of work still remains to be done.

Valuation

There is an urgent need for developing conventions and procedures for unequivocally comparing or weighting different impacts, as well as for developing quantitative or science-based approaches to valuation (Consoli *et al.*, 1993).

Improvement assessment

A proper definition of 'improvement assessment' is needed, as is the development of an appropriate methodological framework for it.

LCA and other decision support tools

There is a need first to clarify the status of LCA in relation to other decision support tools; and second, to agree upon a common basis on which LCA impact assessment could be compared with impact assessments obtained with other environmental management tools. In particular, the relationships between the different life-cycle approaches (LCA, MIPS and SFA) should be specified.

CONCLUSIONS

LCA has the potential to become an important environmental management tool incorporating environmental quality aspects in the decision-making process for developing policies on products. Clearly, it is important that any measure of the environmental performance of a product should be determined quantitatively on a 'cradle-to-grave' basis. Unfortunately, the current scientific basis of LCA is far from adequate. However, if consensus could be reached on general principles, system boundaries and data collection, and the outcome of a given LCA study could be reproduced by others, then LCA would be more objective than at present. Given that the development of sustainable strategies aimed at improving the overall environmental performance of societal activities is a fundamental objective now and will be in the future, it is more likely than not that LCA will prove to be a crucial design tool in this regard.

It is important to note, however, that LCA is but one environmental management tool among many. It should, therefore, be considered as being complementary to the other tools, such as those of Substance Flow Analysis (SFA), Material Intensity Per Unit of Service (MIPS), Environmental Impact Assessment (EIA), Hazard and Risk Assessment, Technology Assessment (TA) and Environmental Auditing. The following chapters in this volume are relevant in this context: Chapter 8 (Buekens *et al.*), Chapter 10 (Devuyst), Chapter 12 (Pento) and Chapter 13 (De Weerdt).

BOX 11.1 A 'CLASSIC' CASE STUDY ON MILK PACKAGING SYSTEMS

Aim

This study is on the analysis and comparison of the environmental effects of different packaging systems for pasteurised milk. One-litre packages are considered, based on production and distribution conditions in the Netherlands. Specifications of the packages are given in the following table. Different trip rates are assumed for the refillable systems, and variable system elements (e.g. cap and labelling systems and transit packaging) are considered. Roll-in containers are used for polycarbonate bottles and carton packages, and polyethylene boxes for their distribution. Glass bottles are distributed in polyethylene crates.

Specification of the elements of fresh milk packaging systems

Milk packaging system *Specification*	*Coated carton*	*Glass bottle*	*Polycarbonate bottle*
PACKAGE			
board (g)	25.3	–	–
coating (g)	3.2	–	–
glass (g)	–	480	–
polycarbonate (g)	–	–	70
Total weight (g)	28.5	480	70
Number of trips	1	20/30	50/75
CAP (one-way)			
aluminium (g)	–	0.25–0.3	0.25–0.3
polyethylene (g)	–	4.0	4.0
twist-off (g)	–	4.35	4.35
LABEL (one-way)			
paper (g)	–	1.72	1.72
polyethylene (g)	–	1.5–2.0	1.5–2.0
TRANSIT PACKAGING			
roll-in container (kg) (160 L; 750 trips)	20	–	20
polyethylene box (kg) (20 L; 500 trips)	2	–	2
polyethylene crate (kg) (12 L; 500 trips)	–	1.98	–

Method

The entire product life-cycle of the packaging systems is considered in the analysis, and the 'functional unit' is defined as 'the packaging of 1,000 litres of milk'. The environmental profile of the systems is based on the following considerations:

- raw material used is obtained only from fossil fuels (expressed in megajoules (MJ));
- emissions of potentially hazardous substances (including emissions from the production of energy consumed);
- generation of solid waste (final waste is expressed in units of mass).

Emissions are related to their toxicity and potential for acidification, while the emissions are divided by media-related and substance-oriented standards, totalled, and then expressed in Units of Polluted Air (UPA), Acidification Equivalents (AE) and UPW (Units of Polluted Water).

Methodological assumptions on recycling and incineration of waste are discussed.

Data

Origin and quality of data on the production and manufacture of polycarbonate, glass and cartons are well documented, but filling of the packages is not considered. Distribution of milk and transportation of empty bottles is analysed by a model developed for the Netherlands. Processing of household waste is assumed at 40 per cent by incineration and 60 per cent in landfills. Incineration emissions are not considered, and gaps in datasets are documented.

→

Environmental profiles

The environmental profiles are summarised in the following table.

Environmental profiles of five different milk packaging systems

Functional units (1,000 litres packed milk)		*Polycarbonate 50 trips*	*Polycarbonate 75 trips*	*Glass 20 trips*	*Glass 30 trips*	*Carton one way*
Fossil energy fuels	MJ	366	353	552	494	534
UPW	dm^3	4.90	4.46	3.97	3.70	33.1
UPA	m^3	11.2	11.3	24.9	21.3	62.0
AE	ha	0.304	0.319	1.0	0.806	3.78
Waste	kg	1.37	1.26	6.68	4.63	18.6

Evaluation and conclusions

The overall assessment shows that the polycarbonate system is superior to the carton gable-top system in all the quantified environmental aspects. However, data on emissions from the production of board and paper would appear to be outdated. The glass bottle system is superior to the carton pack system in nearly all respects.

A comparison of the glass system with the polycarbonate system shows the latter to be more attractive in four environmental aspects, with only water pollution being slightly higher. Missing data appear to have a slight bias against the glass alternative. The assumed number of trips does not influence the environmental effects significantly. Finally, significant differences in effects can be expected if more household waste is incinerated as planned and the efficiency of electricity production by incineration is improved. The scores of the carton system will then improve substantially, scores of the polycarbonate system would improve moderately, but those of the glass system not at all.

Source: Mekel and Huppes, 1990

BOX 11.2 LCA AND ITS USE AS A TOOL IN THE ECODESIGN OF ELECTRONICS: 'THE A1000 MODEM' CASE STUDY

Aim

The main goal of the project, as determined by Alcatel Telecom Belgium, was to assess the possibilities and limitations of using LCA as a tool in the development of environmentally less damaging products (ecodesign). The study focused on the A1000 Modem which provides high bit-rate services for Video On Demand, Internet Access, Tele-Learning, Tele-Shopping, etc. to households using conventional telephone wires. The A1000 Modem contains all the basic blocks found in typical electronic consumer products, namely a printed circuit board, a number of electronic components, and an enclosure in which the printed circuit board is placed.

Method

The study began with a 'screening LCA' for the prototype design (Release 1). The results of this preliminary effort were used to pinpoint key environmental issues in the life-cycle of the prototype design. This information provided the input for a brainstorming session in which various product improvement options were identified and then assessed by Alcatel Telecom for their technical and economic feasibility. The improvement options which passed this assessment were implemented in the actual design (Release 2). Finally, a new LCA was made for the improved design which allowed the quantification of the overall environmental improvement achieved by moving from Release 1 to Release 2. As both alternatives fulfilled the same function and had the same expected life-span, it was possible to make a straightforward comparison.

The screening LCAs were performed within a six-month period. They were based on ISO 14040 guidelines and included these four methodological steps: goal and scope definition, inventory analysis, impact assessment and interpretation.

→

Data

The following were identified and quantified using the inventory analysis: emissions, material flows, energy use, and waste streams for different components and processes of Alcatel's production chain including the printed circuit board, the enclosure (metal or plastic), the assembly process (soldering of the electronic components to the naked printed circuit board) and the power unit. But the screening LCA did not give a detailed analysis of the individual electronic components contained in the printed circuit board. Data from the literature on those electronic components were used for that purpose. This effort resulted in an extensive inventory table for both prototype and improved designs.

The impact assessment consisted of the three steps of *classification, characterisation* and *normalisation*. During classification, the relevant impact categories were identified and the loadings of the inventory table were assigned to the following problem themes: energy consumption, consumption of non-renewable materials, water consumption, greenhouse effect, acidification and eutrophication. This classification is based mainly on the Dutch CML methodology, although other classification factors were also included.

In the characterisation stage, the effects of environmental loadings were quantified and aggregated within the defined impact categories. Equivalency factors, such as those of Greenhouse Warming Potential (GWP) and Ozone Depletion Potential (ODP), were used to convert loadings into potential impacts. The results were normalised so that a better interpretation could be made. Normalisation was done by dividing the impact scores by the total magnitude of the effect score for Belgium, taking 1995 as the reference year. In the interpretation stage, the relative importance of the different environmental impacts were weighted against each other.

Comparison of environmental profiles

The environmental profiles of the prototype (Release 1) and the improved design (Release 2) are compared in the following figure. This figure shows that a reduction in environmental impact of 50 per cent and more has been achieved in Release 2 for most of the impact categories compared to Release 1. A further reduction at the level of the category entitled 'feedstock energy' can possibly be made by substituting recycled plastics for virgin material in the production of the housing.

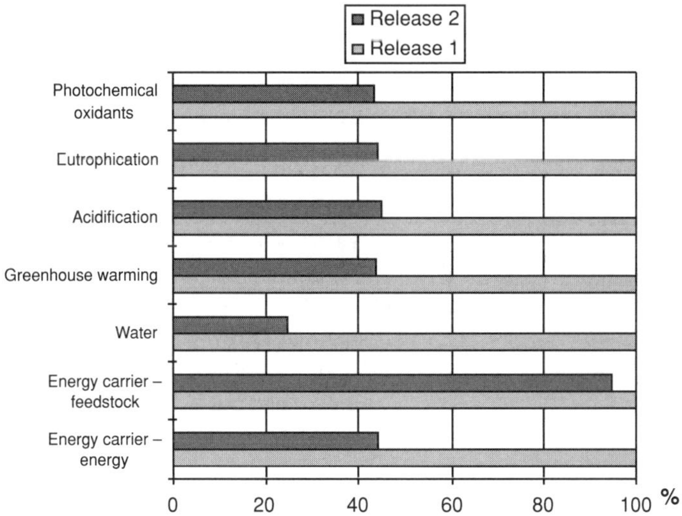

Conclusions

This case study shows that life-cycle assessments can be used to pinpoint critical environmental issues in the design of a product. When applied to the A1000 Modem, the assessment highlighted that the product's use was responsible for the major part of the overall environmental impact, and that the printed circuit board seems to have been the most important component causing that impact. Possible improvements to the design of this particular component may be made by reducing the size and/or thickness of the printed circuit board, and by optimising the processes involved in its production.

→

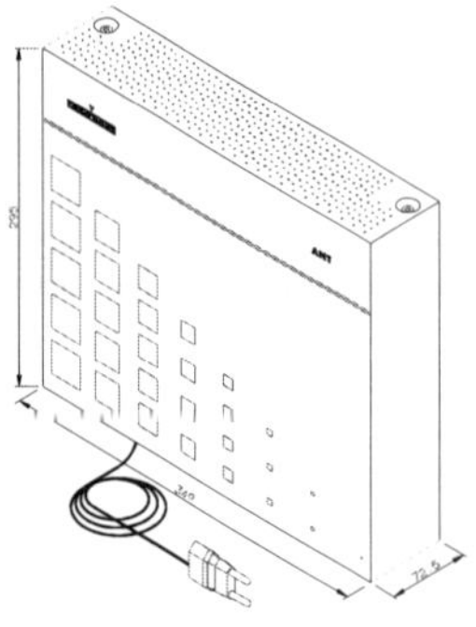

Prototype

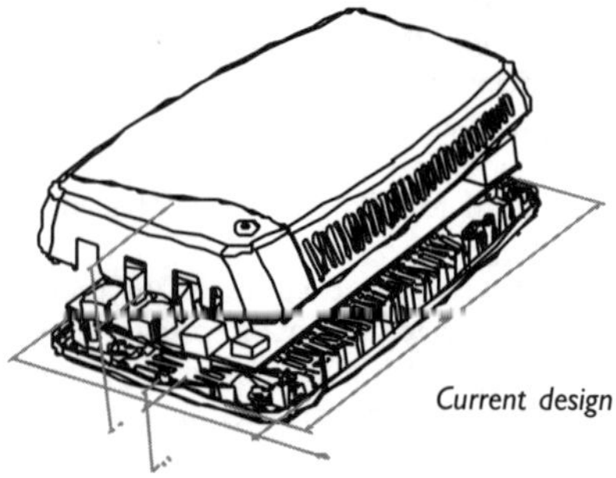

Current design

The current design of the Modem differs remarkably from its prototype design, as will be seen from the above two figures.

The enclosure for the prototype was made of metal and contained a large printed circuit board with an integrated power unit. The environmentally improved design is by comparison more compact and lighter, because it has a plastic housing and a smaller printed circuit board. The power unit of the current release is external, and this permits rapid removal for re-use or recycling.

Source: Ceuterick *et al.*, 1996.

LIST OF LCA RESOURCE CENTRES

Belgium

Department of Human Ecology
Free University of Brussels
Laarbeeklaan 103
1090 Belgium

Flemish Institute for Technological Research (VITO)
Boeretang 200
2400 Mol

Procter & Gamble European Technical Centre
Temselaan 100
1853 Strombeek-Bever
Society for the Promotion of Life-Cycle Assessment Development

SPOLD-Europe
Avenue E. Mounier 3, Box 1
1200 Brussels

Society of Environmental Toxicology and Chemistry
SETAC-Europe
Avenue E. Mounier 3, Box 1
1200 Brussels

Canada

Environment Canada
Solid Waste Division
Place Vincent Massey
351 St Joseph Blvd.
12th Floor
Quebec K1A 0H3

Denmark

Interdisciplinary Centre
Technical University of Denmark
DK 2800 Lyngby

France

Ecobilan SARL, Research and Development
13–15 rue Buffon
F-75005 Paris

Germany

Fraunhofer Institut für Systemtechnik und Innovationsforschung
Breslauer Strasse 48
B-76139 Karlsruhe

The Netherlands

Centre of Environmental Science (CML)
University of Leiden
PO Box 9518
2300 RA Leiden

Institute for Environment and Systems Analysis (IMSA)
Emmastraat 16
1075 HT Amsterdam

Delft University of Technology
Faculty of Industrial Design Engineering
Section of Environmental Product Development
Jaffalaan 9
2628 Bx Delft

Spain

APRODACV
Associación Española para La Promoción de Análisis del Ciclo de Vida
Cardinbal Vives I Totó 41, Entl. 12
08034 Barcelona

Sweden

Chalmers Industrietechnik
S-412 88 Göteborg

IVL
Swedish Environmental Research Institute
Box 21060
S-100 31
Stockholm

Switzerland

Institut für Energietechnik
ETH Zürich
CH-8902
Zürich

Eidgenössische Materialprüfungs- und Forschungsanstalt (EMPA)
Unterstrasse 11
CH-9001 St Gallen

UK

Centre for Environmental Strategy
University of Surrey
Guildford
GU2 5XH

USA

Franklin Associates Ltd.
4121 West 83rd Street, Suite 108
Prairie Village
Kansas KS 66208

Environmental Protection Agency
Office of Research and Development
LCA Research Programme
26 West Martin Luther King Drive
Cincinnati, Ohio
OH 45268

Society of Environmental Toxicology and Chemistry
SETAC-USA
1010 North 12th Avenue
Pensacola
FL 32501–3307

LIST OF LCA-RELATED HOMEPAGES

There are numerous LCA-related websites on the INTERNET. A tentative and incomplete list of LCA-oriented homepages is given below.

CIT Ekologik	http://www.ekologik.cit.chalmers.se/
CPM	http://www.cpm.chalmers.se/
ECDM lab	http://ie.uwindsor.ca/ecdm/journal/
ECOBALANCE, INC.	http://ourworld.compuserve.com/homepages/ecobalance/
Ecocycle	http://www.doe.ca/ecocycle/
Env. Product Development	http://duto02.tudelft.nl/research/mpo/index.html
EPA	http://www.epa.gov/
ETH-ESU	http://www.esu.iet.mavt.ethz.ch/esu/
Franklin Ass. site	http://www.websmart.net/-franklin/falhome.html
ISO 14000	http://www.iso14000.com/scs/ISO14000i
IVAM LCA Database	http://www.ivambv.uva.nl/ivam/product/lcadata.html
LCA pages at RMIT	http://www.cfd.rmit.edu.au/Research/LCA.html
LCA Talk	http://www-personal.umich.edu/~robbtrek/LCA.html
LCANET	http://www.leidenuniv.nl/interfac/cml/lcanet/hp22.htm
Life Cycle Assessment	http://challenge.tiac.net/users/tgloria/LCA/lca.html
Life Cycle Assessment	http://www.websmart.net/~franklin/lca.html

Nortel	http://www.nortel.com/habitat.html
Pre Consultants:	http://www.pre.nl/
SimaPro 3.1 LCA Software	http://www.pre.nl/sp31.html
Tellus LCA Program	http://www.channel1.com/users/tellus/r-lca.html#USPack
The LCA Website	http://www.io.org/~lca
VITO and LCA	http://www.vito.be/ene/lca_en.htm

The Home Page of the United Nations Environment Programme (UNEP) Working Group on Sustainable Product Development is http://unep.frw.uva.nl/. It provides links with LCA sites on the web.

REFERENCES

Anon. (1996a) 'Independent LCA software review' in *LCA News*, a SETAC Europe Publication, Volume 6, Issue 4, Brussels, Belgium, pp. 3–4.

Anon. (1996b) 'New LCA initiative establishes network' in *LCA News*, a SETAC Europe Publication, Volume 6, Issue 5, Brussels, Belgium, p. 5.

Assies, J.A. (1992) 'State-of-the-art', in *Life-Cycle Assessment*, Report of the SETAC Workshop, 2–3 December 1991, Leiden, Netherlands, pp. 1–20.

Boustead, I. (1994) *Ecoprofiles of the European Polymer Industry. Report 5: Co-product Allocation in Chlorine Plants*, Brussels, Belgium: APME's Technical and Environmental Centre.

BUWAL (1990) 'Methodik für Oekobilanzen auf der Basis Ökologischer Optimierung', *Schriftenreihe Umwelt* Nr. 133, Bern, Switzerland.

BUWAL (1991) 'Oekobilanz von Packstoffen; stand 1990', *Schriftenreihe Umwelt* Nr. 132, Bern, Switzerland.

Ceuterick, D., De Langhe, P. and Criel, S. (1996) 'Introducing environmental issues into the design of complex electronic products: a case study', in *Proceedings of the Second International Conference on EcoBalance*, 18–20 November 1996, Tsukuba, Japan, pp. 327–332.

Consoli, F., Allen, D., Boustead, I., de Oude, N., Fava, J., Franklin, W., Quay, B., Parrish, R., Perriman, R., Postlewaite, D., Seguin, J. and B. Vigon (eds) (1993) *Guidelines for LCA: A Code of Practice*, Report of the SETAC workshop, Sesimbra, Portugal.

CSA (Canadian Standards Association) (1994) *Life-Cycle Assessment*, Environmental Technology CSA Guideline Z760–94, Toronto, Canada.

Curran, M.A. (1996) Personal communication.

De Smet, B., White, P.R., Franke, M., Lentz, R. and Owens, J.W. (1994) 'Principles for the development of environmental life-cycle impact assessment', in *First Working Document on Life-Cycle Impact Assesment Methodology*, Report of the SETAC workshop at ETH Zurich, Switzerland, pp. 68–72.

Ekvall, T., Baumann, H., Svensson, G., Rydberg, T. and Tillman, A.-M. (1992) 'Life-cycle assessment: pilot study on inventory methodology and databases', in *Product Life-Cycle Assessments. Principles and Methodology*, Nordic Council of Ministers, Nord 9: pp. 132–171.

Enquete Commission of the German Bundestag on the 'Protection of Humanity and the Environment' (ed.) (1994) *Responsibility for the Future: Options for Sustainable Management of Substance Chains and Material Flows*, Interim report, Bonn, Germany: Economica Verlag GmbH.

Fava, J.A., Consoli, F. and Denison, R.A. (1992) 'Analysis of product life-cycle assessment applications', in *Life-Cycle Assessment*, Report of the SETAC Workshop in Leiden, pp. 125–131.

Fava, J.A., Denison, R., Jones, B., Curran, M.-A., Vigon, B., Selke, S. and Barnum, J. (eds) (1991) *A Technical Framework for Life-Cycle Assessments*, Report of the SETAC Workshop at Smugglers Notch, Vermont, USA.

Finnveden, G. (1992) 'Landfilling – a forgotten part of life-cycle assessments', in *Product Life-Cycle Assessments Principles and Methodology*, Nordic Council of Ministers, Nord 9: pp. 263–277.

Finnveden, G. (1994) 'Comments and suggestions for the classification', in *First Working Document on Life-Cycle Impact Assessment Methodology*, Report of the SETAC Workshop at ETH Zurich, Switzerland.

Finnveden, G., Andersson-Sköld, Y., Samuelsson, M.-O., Zetterberg, L. and Lindfors, L.-G. (1992) 'Classification (impact analysis) in connection with life-cycle assessment: a preliminary study', in *Product Life-Cycle Assessments Principles and Methodology*, Nordic Council of Ministers, Nord 9: pp. 172–231.

Grisel, L., Jensen, A.A. and Klopffer, W. (1994) *Impact Assessment Within LCA*, Brussels, Belgium: SPOLD.

Groupe des Sages (1994) *Guidelines for the Application of Life-Cycle Assessment in the European Union Labelling Programme*, Brussels, Belgium: SPOLD.

Heijungs, R. (ed.) (1992) 'Environmental life-cycle assessment of products: guide and backgrounds', Leiden, Netherlands: Centre of Environmental Science.

Heijungs, R. and Guinée, J. (1995) 'On the usefulness of LCA of packaging', *Environmental Management*, Vol. 19, No. 5, pp. 665–668.

Henshaw, C.L. (1994) 'Spatial detail in LCA', in *First Working Document on Life-Cycle Impact Assesment Methodology*, Report of the SETAC Workshop at ETH Zurich, Switzerland, pp. 41–42.

Hindle, P. (1991) 'The practical application of product

life-cycle analysis. Case study 2: the methodology of a life-cycle analysis', in *Proceedings of IIR Conference*, London, UK.

Holloway, T. (1991) 'Life-cycle analysis: getting started', in *Proceedings of IIR Conference*, London, UK.

LCANET (1996) *Definition Document: State-of-the-art of LCA Methodology and Present and Future Research Needs*, First draft, LCANET homepage.

Marion, J.-Y., Valenduc, G., Vendramin, P., Huybrechts, D., Berloznik, R., Wouters, G., Hellenbrandt, S. and Teichert, V. (1995) *Milieubalansen: Nut en Beperkingen van een Overleg- en beslissingondersteunend Middel*, Brussel, Belgium: Koning Boudewijsstichting – Programma Leefomgeving.

Mazijn, B. (1994) *Van tekentafel tot afvalberg: LCA, een instrument voor eco-design, eco-label en eco-taks*, Monografieen Stichting Leefmilieu nr 32, Kapellen, Belgium: Uitgeverij Pelckmans.

Mekel, O.C.L. and Huppes, G. (1990) *Environmental Effects of Different Package Systems for Fresh Milk*, CML Report 70, Leiden, Netherlands.

Menke, D., Davis, G.A. and Vignon, B.W. (1996) *Evaluation of Life-Cycle Assessment Tools*, Hazardous Waste Branch, Environment Canada.

Miettinen, P. (1993) 'Software tools in life-cycle assessment', in B. Pedersen Weidema (ed.) *Environmental Assessment of Products: A Textbook on Life Cycle Assessment* (2nd edition), Helsinki, Finland: UETP-EEE, pp. 93–107.

Nordic Council of Environmental Ministers (1995) *Nordic Guidelines on Life Cycle Assessment*, Report Nord 20, Copenhagen, Denmark: Nordic Council of Ministers.

Rubik, F. (1994) *Themen der Ökobilanzforschung: eine Übersicht über aktuelle Ökobilanzergebnisse*, Heidelberg: IÖW.

Rubik, F. and Baumgartner, T. (1992) *Evaluation of Eco-balances*, Brussels–Luxembourg: the Commission of the European Communities, CD-NA-14737-EN-C, ECSC-EEC-EAEC.

Ryding, S. and Steen, B. (1991) *The EPS system. A PC-based System for Development and Application of Environmental Priority Strategies in Product Design from Cradle-to-grave*, Swedish Environmental Research Institute (IVL), Report L91–85.

SETAC (Society of Environmental Toxicology and Chemistry) (1994) *First Working Document on Life-Cycle Impact Assessment Methodology*, Report of the SETAC Workshop at ETH Zurich, Switzerland.

Udo de Haes, H.A. (1990) 'Environmental analysis and evaluation of products', in *Proceedings of the Specialised Workshop Life-Cycle Analysis for Packaging Environmental Assessment*, Leuven, Belgium.

Udo de Haes, H.A. (1992) 'General framework for environmental life-cycle assessment of products', in *Life-Cycle Assessment*, Report of the SETAC Workshop in Leiden, pp. 21–28.

Vancolen, D. (1996). 'Life-cycle assessment: a tool towards sustainable development?', in *Conference Proceedings of the 1996 International Sustainable Development Research Conference*, Manchester, United Kingdom: ERP Environment.

Victory, K. (ed.) (1996) *Business and the Environment, ISO 14000 Update*, Arlington, Mass., USA: Cutter Information Corp., March 1996, pp. 1–6.

Vigon, B.W., Tolle, D.A., Cornaby, B.W., Latham, H.C., Harrison, C.L., Boguski, T.L., Hunt, R.G. and Sellers, J.D. (1993) *Life-Cycle Assessment: Inventory Guidelines and Principles*, Cincinnati, Oh.: US Environmental Protection Agency, EPA/600/R-92/245.

White, P., Franke, M. and Hindle, P. (1995) *Integrated Solid Waste Management: A Life-Cycle Inventory*, London, UK: Blackie Academic and Professional.

White P., Hindle, P. and Drager, K. (1992) 'Life-cycle analysis of packaging', in G. Levy (ed.) *Packaging and the Environment*, London, UK: Blackie, pp. 118–146.

WWF (World Wildlife Fund) and The Conservation Foundation (1991) *Getting at the Source: Strategies for Reducing Municipal Solid Waste*, Final Report of the Strategies for Source Reduction Steering Committee, Baltimore, Md.: WWF publications.

SUGGESTED READING

Canadian Standards Association (1994) *Life-cycle assessment*, Environmental Technology, CSA Guideline Z760–94, Toronto, Canada.

Consoli, F., Allen, D., Boustead, I., de Oude, N., Fava, J., Franklin, W., Quay, B., Parrish, R., Perriman, R., Postlewaite, D., Seguin, J. and Vigon, B. (eds) (1993) *Guidelines for LCA: A Code of Practice*, Report of the SETAC Workshop, Sesimbra, Portugal.

Curran, M.A. (ed.) (1996) *Environmental Life-Cycle Assessment*, New York: McGraw-Hill Book Co.

Fava, J.A., Denison, R., Jones, B., Curran, M.-A., Vigon, B., Selke, S. and Barnum, J. (eds) (1991) *A Technical Framework for Life-Cycle Assessments*, Report of the SETAC Workshop at Smugglers Notch, Vermont, USA.

Nordic Council of Environmental Ministers (1995) *Nordic Guidelines on Life-Cycle Assessment*, Report Nord 20, Copenhagen, Denmark: Nordic Council of Ministers.

Society of Environmental Toxicology and Chemistry (SETAC) (1994) *First Working Document on Life-Cycle Impact Assessment Methodology*, Report of the SETAC Workshop at ETH Zurich, Switzerland.

Vigon, B.W., Tolle, D.A., Cornaby, B.W., Latham, H.C., Harrison, C.L., Boguski, T.L., Hunt, R.G. and Sellers, J.D. (1993) *Life Cycle Assessment: Inventory Guidelines and Principles*, Cincinnati, Oh.: US Environmental Protection Agency, EPA/600/R-92/245.

SELF-ASSESSMENT QUESTIONS

Indicate the correct statement(s) in the following questions

1 Life-cycle assessment measures:
(a) environmental safety;
(b) environmental quality;
(c) environmental safety and quality.

2 The output of a life-cycle inventory is an:
(a) inventory table;
(b) environmental improvement;
(c) environmental burden.

3 Life-cycle assessment is a potentially valuable tool for assessing the total environmental load associated with a product. It considers:
(a) local effects;
(b) global effects;
(c) local and global effects.

4 The methodology of constructing a life-cycle inventory is based on the definitions and general principles of:
(a) chemistry;
(b) thermodynamics;
(c) economics.

5 Functional units, which are essential elements of a life-cycle assessment, refer to:
(a) the quantity of the product needed to perform a specific function;
(b) the determination of relevant parameters;
(c) the definition of the scores which constitute the environmental profile.

6 Inventories are by no means a simple tabulation of numbers. The elaboration of an LCA method according to the central principles requires further choices to be made. An important choice to be made is concerned with the technology to be used, because:
(a) data on production processes can vary a great deal both geographically and depending on the technology used;
(b) the choice of process technology is independent of the purpose of the study;
(c) the determination of a functional unit implies quality criteria for the data to be used.

7 A life-cycle inventory analysis leads to the quantification of a large number of input and output streams, and the various parameters are usually combined to produce a convenient number of scores. But the main disadvantage of aggregation is that:
(a) the method is not 'user-friendly';
(b) the effects of all the constituent processes in the life-cycle are integrated;
(c) information is lost.

8 The 'Critical Volume Concept' is a method of aggregation for:
(a) energy and raw material consumption data;
(b) emission data;
(c) solid waste data.

9 The 'ecopoint approach' translates an initiative into:
(a) aggregate data according to effects;
(b) aggregate data across categories;
(c) aggregate data into categories.

12

INTEGRATING ENVIRONMENTAL MANAGEMENT INSTRUMENTS

Material flow analysis

Tapio Pento

SUMMARY

Analysis of material flows resulting from various human activities, using what are called Material Flow Models (MFMs), is the subject-matter of this chapter. Starting from the early years of their development, the evolution to date of these models is presented to give the reader a thorough understanding of how they are constructed and applied, and their limitations. Attention is also drawn to the continuing research and development work aimed at making them more reliable and attuning them to the real problems and situations of environmental planning and management.

The chapter begins with a general introduction and definitions, following which the different types of material flows have been discussed in detail together with the principles for the construction of models for them. Various types of static models have been discussed, including their limitations and the assumptions on which they are based. The discussion then moves on to a detailed consideration of the dynamic models with emphasis on recent developments and future possibilities. The concepts have been presented, and the arguments developed, with particular reference to examples taken from the paper and pulp industry of Nordic countries in which a variety of such models have been used effectively.

ACADEMIC OBJECTIVES

On completion of this chapter the reader should have a good understanding of:

- the reasons why it is necessary to build and apply comprehensive and reliable Material Flow Models (MFMs) to environmental planning and management;
- the different types of MFMs currently available, how they are applied in practice, and their strengths and limitations;
- how the current problems of scoping, aggregation and allocation compromise the reliability of these models, and possible ways in which those problems could be overcome;
- continuing research and development work aimed at improving the reliability of MFMs.

MATERIAL FLOW MODELS AND THEIR PURPOSE

Innumerable materials and substances of great variety flow in the natural environment in which they are constantly being transferred and transformed within and between different ecosystems. In the course of history man has developed his own material flow systems which take inputs from the natural environment in the form of materials and energy. These inputs are transformed by human use or activities into outputs such as effluents and wastes which need to be treated before they are disposed of into the natural environment so that the pollution caused by them is reduced or eliminated if possible. In terms of a chemical analogy, we may think of the anthroposphere as the 'system' in which the inputs are processed to produce the outputs. The 'system', the inputs and the outputs thus constitute the 'human materials flow system'. The 'system' in fact represents the interface between the natural environment and the anthroposphere (Figure 12.1) within which all human activities take place. A necessary condition for

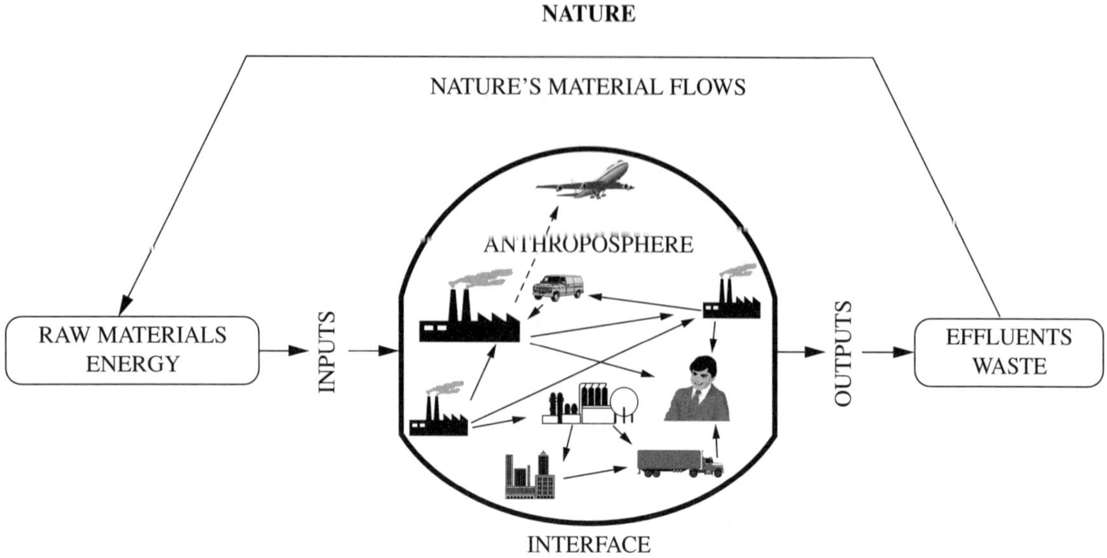

Figure 12.1 Material flows in the natural environment and the anthroposphere

achieving strong sustainability is that both the contents and quantities of the input and output vectors should be part of, and compatible with, the flows occurring in the natural environment.

Different kinds of models are used to analyse and study the flows of different materials in the natural environment and in the anthroposphere. For example, ecological models have been used to analyse the functioning of the environment, as well as to study how different materials flow within and between ecosystems. Models are used in social sciences, on the other hand, to analyse material flows created by selected human activities. The equilibria models of economics, national or regional input–output models, and the mathematical programming models of economic or industrial sectors are typical examples of such models, while yet other models, the engineering and logistic models for example, are often used for the analysis of flows of materials and products in response to the planning and control needs of corporate management.

The purpose of Material Flow Models, in the sense in which they are to be understood here, is to analyse the directions, contents and volumes of material flows within the anthroposphere, and to measure at least some of the inputs from, and outputs to, the natural environment caused by the flows. This definition covers many types of models. Some focus only on materials, such as the input of one or more of the elements into the anthroposphere. Others analyse products, or groups of products, and their materials in different ways, beginning with the elements taken from the natural environment to make them, through components and parts to the final product and even to material recovery and re-use. In the absence of a better name, the name of 'Material Flow Models' (MFMs) will be used throughout in the generic sense, even though it does not give a good or comprehensive description of product-based models.

The German Enquete Commission on 'Protection of People and the Environment' sets out the goals of MFMs as: 'In order to develop a concrete model of sustainable development it is necessary to analyse material flows at corporate, regional, national and global levels (material flow analysis). The corresponding information systems shall . . . provide a basis for operating directions for the actors concerned (material flow management)' (Enquete-Kommission, 1993: 18). A material flow analysis uses the results of MFMs as a basis for re-directing man-induced flows in an attempt to change the inputs and outputs to a desired direction. MFMs may also produce or require value analyses of the flows across the anthroposphere–environment interface, in terms of weightings of different materials taken and effluents released for example. Furthermore, they may facilitate experimental trials and simulations of alternative or planned material flow systems. Such normative MFMs are increasingly becoming a preferred tool of environmental planning and policy-making.

The ultimate objective of MFMs, regardless of how they are used or developed, is to provide information on how the magnitudes and/or directions and/or combinations of the flows within the anthroposphere could be directed towards a state in which the input and output of flows across the anthroposphere–environment interface are minimised and are of such quality that they can be accommodated by the natural environment as a part of its own flows.

TYPES OF MFMS AND THEIR CONSTRUCTION

Flows and processes are the building blocks of MFMs, as shown typically in Figure 12.2. Flows refer to the transport of materials, components, products and energy from one process to the next in the flow system, while processes refer to the recovery, transformation and sorting of materials or products at a particular point or area of the system. Typical processes include the extraction and transformation of materials from the natural environment, assembly of components into a product, degradation of the product by consumption, and its recovery for re-use or re-processing or disposal by incineration or in a landfill. Each MFM includes a limited set of flows and processes within its scope.

As most of the flow systems in the anthroposphere are much too complex to be analysed in complete detail, they are usually aggregated into macro flows and processes. For example, flows of newsprint from paper machines to printers and the associated processes are well documented. But it is still not possible or practicable to try to include in a single model all the processes of hundreds of paper mills and thousands of printing machines, let alone the hundreds of millions of newspaper readers some of whom also act as paper collectors for recycling. And so the simplest method of combining the different units into a single process (or flow aggregate), to be used in a model, would be to include an 'average' paper or printing machine and an 'average' reader and paper collector. Although this approach reduces the model to a manageable size, it may introduce significant deviations from reality. For example, when there is a change in the elements or aspects on which the average is based, or if the volume or type of the activities is changed, the calculated average may be a very poor approximation to that of the changed situation. Thus, the average SO_x production, calculated on the basis of some old data from a number of pulp mills, may show a large bias when the amounts produced by different mills vary over business cycles, or when production systems or the products themselves change. Indeed, this is very much the reason why the SO_x produced per ton of pulp

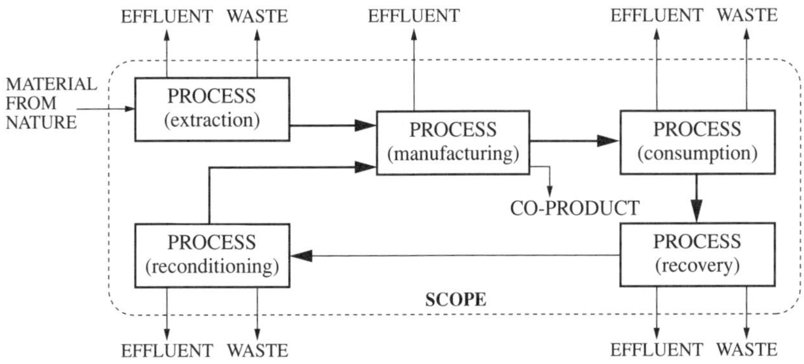

Figure 12.2 Building blocks and the scope of material flow models

varies widely from mill to mill. It is possible to make a more detailed representation of the processes and flows by including greater details in the rule of aggregation; for example by incorporating the largest volume producers independently into the model, or by including the input and output distributions of processes in the models. Such a strategy would clearly allow a more faithful representation of the real flow system, but it would complicate the models to such an extent that their cost-effectiveness would fall below those in which a rougher aggregation is used.

In general, flows are easier to aggregate than processes, because the materials in a flow do not change physically, and the measurement of inputs and outputs involves only a limited number of factors such as distance, frequency, loading and type of transportation used. Aggregation of processes is much more difficult by comparison, because processes often combine different materials and products at different rates to make other different products, at the same time producing widely varying amounts and types of effluents and waste as by-products. Moreover, processes are often not fixed because their yields of products and effluents change from season to season depending on capacity utilisation rate and the like. For example, the effluent produced per ton of pulp made in a digester 'kraft' pulp mill always fluctuates, sometimes widely. Such fluctuations are caused mainly by varying capacity utilisation rates and by seasonal factors. A typical seasonal factor, common in the Nordic countries, arises from the feeding of frozen wood containing snow into the digester. Another problem is that industrial processes are often closely guarded trade secrets on which little reliable information is available.

The aggregation of processes for modelling is easier in some industries than in others. The incorporation of hundreds of newsprint machines into a single process in a model is actually straightforward when compared with the aggregation of, say, one hundred engineering job shops. This is because, while newsprint is a rather homogeneous product which is made everywhere using essentially the same process, the practices and work processes of different engineering shops can differ fundamentally. For example, a product may be made in one shop from a piece of raw metal with a considerable amount of manual machining, while other shops may employ raw material-saving automated casting processes to make the same product. The aggregation of such processes, in terms of an 'average shop' or an 'average process', is thus extremely difficult and impracticable.

Troublesome problems may arise in aggregation, even in the case of newsprint, from one or more of the following: widely differing materials employed by different mills (sometimes varying from 100 per cent virgin pulp to 100 per cent de-inked pulp); type of energy used; range and amount of effluents generated per ton of newsprint produced; and wide fluctuations in processes caused by seasonal factors, business cycles and even the geographical location of mills. At present aggregation rules to be employed in the MFMs are in the early stages of their evolution and development, and much work remains to be done in this potentially fruitful area of research.

Once the transport of materials in the flows and processes has been modelled, the next step is to calculate the associated inputs to, and outputs from, the natural environment. The most difficult modelling problem at this stage is concerned with how to allocate inputs and outputs of materials or products in the model. One approach involves the allocation of inputs and outputs to the co-products, defined as products that are made simultaneously in a production process. As an example, take the 'tall' oil which is produced as a co-product in soft wood 'kraft' pulp manufacture. The question is, what proportion of the energy generated in the cooking process, and which amounts of the effluents and wastes produced, should be assigned to 'tall' oil and to pulp? Should all be assigned to pulp, or a part to 'tall' oil which has only a small fraction of the commercial value of pulp?

Another allocation problem occurs in flow systems in which a large part of the materials and products are recycled outside the scope of the model. The question which now arises is this: if virgin paper and recycled paper are to be analysed in two different models, how should one divide the materials, energy and effluents, attributed to the virgin paper and its recovery, between the two models? One approach would be to allocate zero per cent to recycled paper on the assumption that, in so far as recycling is concerned, virgin waste paper 'falls from heaven' with no environmental implication. But this is likely to make the results of both models meaningless. One principle here is that the scope of the model should be sufficiently wide to embrace all large flows of co-products. Otherwise both allocations and the final results would tend to become subjective decisions or constructs of the model-builder.

There are several time-based methods of allocating inputs and outputs in dynamic MFMs. In some of these methods materials may be produced in a different time period than when they are re-processed or consumed. If a model is based on temporal allocation, it would assign all inputs and outputs of the flows and processes to the same period, without the carry-over of materials and their environmental impacts to the next period. Materially based allocation, on the other hand, carries over the impacts and disregards the time of their production. In other methods the inputs and outputs of capital equipment and some durable materials could either be allocated fully to the period when they occur, or amortised over the time period during which they are useful economically, technically and environmentally. The method of 'life-span production-based overhead' is often considered practicable when amortisation is opted for. In this method the total environmental impact, created during the construction of the capital good, is divided by the quantity which it will produce during its life-cycle. The resulting ratio, called the 'overhead ratio', is then multiplied by the quantity of products made with the capital good during the time period in question.

The flows and processes of materials and products within the anthroposphere are often so convoluted that sometimes it is not possible to include even the flow system of just one material in the anthroposphere within the scope of a single model. Many high-volume materials, which also have high inputs and outputs, tend to have flow systems that are too complex to be included in a single model, unless aggregated at a very high level. This necessitates the narrowing-down of MFMs by representing the flows and processes within a tighter and workable scope. Such workable 'scopings' have been developed by restricting a model to cover only those flows and processes that are directly connected to an input element, or to a material or product, a region, an industrial sector, or to all or some of the output materials. In many cases 'scoping', developed as above, may be reinforced and made more effective by incorporating the evolving standards of Life-Cycle Assessment (LCA) models.

In some MFMs only *one* time period is considered in a static framework, while other MFMs are dynamic in character in the sense that they incorporate *several* time periods. The results obtained from static models show absolute values of this type: X kilograms of effluent A are produced in this flow system in *one* given time period. Dynamic models, on the other hand, produce such absolute measures for *each of the several* time periods under consideration. And this feature of dynamic models makes it possible to study both directions and magnitudes of intertemporal changes, which is valuable in the planning of material flow policies.

Table 12.1 shows some MFMs which have been selected as typical examples of the main directions of recent model development. Each type of model shown in this table is discussed in some detail below, while only an outline is given of the Life-Cycle Assessment models which have been presented in full elsewhere in this volume. The latest dynamic models have been emphasised in the following discussion, because they are considered by the author to be more promising, especially as a tool for managing material flows.

Table 12.1 Classification of material flow models

	Time element	
Scope limited via:	*Static*	*Dynamic*
Input	Substance Flow Analysis (SFA) Material Intensity Per Unit Service (MIUS)	
Region	Regional flow models	
Product	Life-Cycle Assessment (LCA)	
Industry	Industry flow graphs	Sectoral LP models
	Industrial metabolism / materials balance models	Life-Cycle Assessment with Delta analysis (LCA-D)
	Average Life-Cycle Assessment (A-LCA)	Joined time projection models (JTP)

STATIC MATERIAL FLOW MODELS

The static MFMs calculate the input and output vectors of a flow system within a defined scope for some specified time period – say one year. They are applied in environmental planning mainly to identify areas in which improvements could be made. For example, they can identify hot-spots at which a large amount of a critical material is being used, or a large amount of a particular effluent, or effluents, is being produced or discharged, and so on.

Substance Flow Analysis (SFA)

The research group at the Environmental Centre of the University of Leiden in the Netherlands has developed an 'economy–ecology' model which takes one input element, identifies all its flows, and calculates the accumulated inventory across the entire anthroposphere (i.e. economy). The flows are then aggregated to determine the environmental effects caused by the contents of the output (i.e. ecology). This method, which is applied to a specific region, is called 'Substance Flow Analysis' (SFA). It has been applied, for example, to study the flows, inputs and outputs of cadmium and nitrates (Udo de Haes *et al.*, 1988; van der Voet *et al.*, 1993). Aggregation may be problematic in a SFA model, because the inclusion of all the flows of an input material requires a great deal of detail which is contrary to the principle of aggregation. Such modelling thus becomes especially tedious when the material being studied has a complex flow structure, as in the case of iron, paper and many plastics. It is for this reason that the SFA method has been applied mostly to toxic materials and to those that have relatively few branches in their flow charts. In the case of toxic materials, the mere identification of the flow pattern and inventory is valuable information for managing material flows.

Material Intensity Per Unit Service (MIPS)

The concept of the 'Material Intensity per Unit Service' (MIPS) (Schmidt-Bleeck 1992; Liedtke and Manstein, 1994; Liedtke *et al.*, 1994) is now being made operational by several researchers at the Wuppertal Institute in Germany. The MIPS is essentially a measure of resource intensity, for it shows the amount of materials needed to satisfy a particular human need. The objective of MIPS modelling is to determine ways in which human needs could be satisfied with the minimal input of materials and energy from the environment. In principle, the use of MIPS entails first the identification and specification of a social need. It then calculates how that need could be satisfied with a minimal amount of environmental disturbance, defined by the amount of materials to be removed or disrupted. For example, the material intensity of a newspaper would be the sum of all the inputs from nature needed to make and consume it. This sum would include the amount of wood needed to make the paper; the amount of earth disturbed and ore removed in making the necessary machines; the amount of energy consumed (e.g. in energy equivalents of coal, petroleum, etc.) in its manufacture, printing and transport; and the water used in the pulp and paper-making processes. Weights, developed by expert panels, are then assigned to the contents of the inputs to derive the MIPS of the newspaper.

If the service provided by the newspaper could be supplied by another product, such as the Internet or another newspaper produced at a lower material intensity, then the use of the latter should be preferred. The MIPS values of alternative flow systems can be used in this way to direct regional, national or global material flows towards those with the lowest resource intensity.

Resource intensity calculations, such as in MIPS, are hampered by the fact that resource inputs in different flow systems are not identical in terms of sustainability, neither are the effects of their outputs on the environment. Expert panels are set up to address this problem by assigning weights to the different components of the flow system. But this sometimes leads to another problem: the experts often disagree among themselves, since there is no standard method in use. A typical disagreement could be the following: a ton of water should be given a lower weight than a ton of wood, and a ton of wood should be given a lower weight than a ton of tin (used in Internet switches for example) because, while water can be renewed by purification almost immediately and wood can be re-grown over a decade or two, known reserves of tin are expected to run out by the next century and tin is not a renewable resource. Would it therefore be reasonable to assign the weights as follows: 1 for water, 10 for wood and 100 for tin? Or would it be more reasonable to assign a weight of, say, 10,000 to tin, considering that its exhaustion would compromise the ability of future generations to meet their own needs?

The concept of MIPS modelling can be compared to that of neo-classical 'economic choice' models that were pioneered by Lancaster (1966). These models are based on the hypothesis that the consumers have to satisfy their needs for products that are seen as combinations of different (weighted) attributes available at different prices. However, as with Lancaster's models, the extent to which MIPS modelling can be applied to environmental decision-making is not self-evident. It is well known in microeconomics that the determination of the needs of individuals, and their aggregation, presents formidable theoretical problems. A typical illustration of this is that, while a car satisfies transportation needs, it may also offer the consumer (car owner) other kudos (such as satisfaction of owning a car, status symbol) which a bicycle or public transport cannot. Such multidimensional scaling and aggregation of all possible utility values derived by the members of the public from each 'service' is still a subjective art, not an objective science. Nevertheless, application of the MIPS to major material flows, such as those of steel, will enhance current knowledge of what could and might be done to reduce the environmental impacts of current practices.

Regional flow models

The scope of regional flow models is obtained by limiting the study to the anthroposphere of a small region, and by restricting the analysis of the flows therein to the most important goods and processes, each of which is an aggregate of similar entities in the region (Brunner *et al.*, 1994: 171). An aggregate could represent all private households, or the production sector or agriculture within the region. Data is collected on the flow patterns of the aggregates, together with essential data for the processing of both the flows and the wastes generated. For example, the 'processing' of phosphorus by private households would be determined by calculating the total consumption of domestic phosphorus-containing products (mostly cleaning agents) in the region, and by tallying the quantities which end up in different waste sinks.

The regional models analyse types, volumes, and sources of all flows. They have been employed to determine sources of unwanted waste and to find reasons for and ways of reducing waste by improving the process or by recycling. These models vary in their level and type of aggregation. Some, intended for a detailed regional management of material flow outputs, can deal with very elaborate flow structures with little aggregation, while a higher level of aggregation is used in others in order to measure more general effects of wastes. The regional models offer a comprehensive view of all the flows within a given region, but the magnitude of the different types of flows tends to increase the size of the model very substantially. And this makes such models inefficient even when only limited aggregation is used, especially if the region to be studied has many diverse types of consumers, producers, flows and products.

Life-Cycle Assessment (LCA)

Life-Cycle Assessment is presented in greater detail elsewhere in this volume (Chapter 11); we will therefore give only an outline here.

The classical LCA is a material flow model in which the scope of the system is restricted to the flows associated with a particular product. No aggregation is used, and the data is site-specific; that is, drawn directly from those flows and processes through which the materials of the product, and the product itself, go through. While the site-specific nature of the model improves its accuracy and makes it more practical for its intended use in manufacturing enterprises, it also makes the results less applicable to regional or national studies, because the flows in the model may not be representative of larger aggregates.

Industry-based static MFMs

Industry-based MFMs consider several products of an industry in a single flow system. For example, a model of the forest industry could include in its flows and processes different types of energy, raw materials, pulps, minerals, papers, boards, recovered waste papers and other materials.

Industry flow graphs

Material and product flow graphs within an industry have been developed by environmentalists and industrial economists (Bilitewski, 1993; Enquete-Kommission, 1993). Typical graphs of this kind show both magnitude and direction of flows and sometimes also the economic incentives influencing the flow patterns. Many of the graphs, such as that for cadmium flows in Germany prepared for the Enquete Kommission (1993), are similar to the economic part of the SFA models described earlier in the sense that often they only identify the flows without actually analysing the input and output vectors associated with them.

Industrial metabolism: materials balance models

The second law of thermodynamics is the guiding principle of MFMs based on industrial metabolism. According to this principle, the amounts of materials and energy remain constant in any flow or process in a given system. This principle has been applied by aggregating industrial processes and flows to sets of chemical equations (Ayres *et al.*, 1989). In the manufacturing process of chlorine and caustic soda, for example, the amounts of the different chemicals are calculated from the reaction equation: $2NaCl + 2H_2O = Cl_2 + 2NaOH$. The objective here is to develop what is called 'industrial metabolism' (meaning the way in which industry takes, processes and delivers materials and energy) through chemically aggregated metabolism to calculate the industrial flows and processes and their effluents (Ayres and Ayres, 1994).

Many of the material flow models are based on this approach in which the processes and sub-processes are represented by chemical equations. Such equations, with which engineers are familiar, make a comprehensive representation of the mechanics of the larger models, and this permits insight into the flow patterns of elementary materials. However, these models have been criticised on the ground that they are essentially theoretical models which do not represent practical flows. This, together with the lack of exact theory for many of the processes, limits the applicability of industrial metabolism models only to particular areas of chemical and other process industries, and makes it somewhat questionable in many other fields. For example, the process equations of making 'kraft' pulp are well known in principle, but variations in raw materials, cooking parameters, and other factors make the process significantly stochastic in nature. Consequently, the input, output, and effluent quantities calculated from deterministic chemical equations would in many cases be far from the actual quantities, and this would cause large errors in environmental planning.

Average Life-Cycle Assessment (A-LCA)

Most of the Life-Cycle Assessments made today do not show the flow of a single product from its cradle to its grave. Instead they show the flow system of a characteristic average product. Several homogeneous or similar products are aggregated by taking representative averages of flows and processes; these averages are then used to build a life-cycle. The main reason why the Average LCAs (A-LCAs) are made is because average data, which they use, is more readily available. It is also believed that results obtained from A-LCAs are more general, and thus more usable in regional or national material flow management.

Simple A-LCAs are made either in the form of tables showing the materials, energy, effluents and wastes associated with a particular flow or process, or in the form of a more complex flow structure. However, such calculations do not necessarily show the functional relationships between the flows and their magnitudes so that they could be changed for sensitivity testing or simulation. The model developed by Virtanen and Nilsson (1993) shows flows of printing paper from the Nordic countries to Germany. Their continuing work is concerned with developing three different models to analyse three distinct scenarios, with different rates of used paper recovery and end-treatment of the recovered materials. Modern LCA software makes it relatively easy to develop systems comprising defined flows and processes through which both material and products flow according to a set of equations. A good example of such a system is the massive A-LCA model of pulp and paper industry in the Nordic countries and Germany, developed by Kärnä and his associates (1994, 1996) using the modular KCL-ECO software. In this model about 140 flows and processes are linked together with some 420 flow equations.

DYNAMIC MATERIAL FLOW MODELS

Static MFMs are not an optimal tool for managing material flows, mainly because they do not correspond to the basic idea of material flow management which is to improve the flows of a system from their current state. Indeed, the application of a static model to improve an identified hot-spot may actually exacerbate the problem or create more acute problems elsewhere in the flow system. Second, the inputs and outputs associated with a static solution do not necessarily give guidance on how the flows should be improved. One solution for this has been to use what is called an 'endpoint' analysis. In this analysis one model is prepared for the existing flow system (or a part of it), and a second model for another 'alternative' part which may represent the future goal. For example, the first model could represent the inputs and outputs associated with the use of diesel fuel, and the second those of rapeseed oil fuel. The two sets of inputs and outputs can then be compared to see whether or how improvements could be made by changing over from diesel fuel to rapeseed oil. Unfortunately, an 'endpoint' analysis does not give any indication of the dynamic management plan with regard to the development path leading from the present to the future. Also, often there are many alternative paths, each with different flow structures and necessary intermediate steps leading to the final outcome. Given the current status of modelling standards and techniques, and the absolute input and output vectors of the two endpoint models with their different scopes, allocation and aggregation are at best only barely comparable and may give results that are totally misleading. And this undermines efforts at determining the direction along which material flows should be managed for making improvements.

Dynamic models, on the other hand, are a much more powerful tool for material flow management. Capable of indicating changes towards a targeted goal, they show explicitly the intermediate steps and flow structures that are necessary for achieving that goal. Giving relatively less emphasis to the absolute inputs and outputs that are subject to modelling errors, these models analyse the dynamic changes which are much less prone to bias. Finally, many comprehensive dynamic models, such as those described below, are capable of simulating alternative futures for decision-makers.

Sectoral models

In the last few decades economists have developed 'sectoral' models of material flows. In these models the input–output matrices of the earlier models have been elevated to Linear Programming (LP) models in an attempt to make both flow structures and model details more like those of industry. The first models of this kind for the paper and pulp industry were developed in the late 1970s (Adams and Haynes, 1980), and very extensive structures have been developed since then in both Europe and North America (Kallio *et al.*, 1987; Zhang *et al.*, 1993). The newer versions of these models typically analyse many geographical markets of products, as well as new and recycled raw materials that might be used in their production. They also provide economic equilibria solutions over a series of time periods. These models, which have already been applied to analyse recycling rates, are being expanded to include material, effluent and waste considerations in a straightforward manner.

The price mechanism is used in LP models to determine the flows of both products and materials within and between markets in different time periods. This is both an advantage and a weakness, however, because the price structure obtained from each solution of such a model enables a straightforward analysis of the economic tools to be used for redirecting material flows. But the cost of this price information is that the supply and demand functions of each product and raw material, in each of the markets and in each of the future time periods, have to be estimated before the model can be used. These exacting data requirements of LP models, together with their relatively high sensitivity to errors in data, may have been an obstacle to their wider application in sectoral material flow analysis.

Life-Cycle Assessment with Delta analysis (LCA-D)

LCA with Delta analysis attempts to answer the basic question of material flow management: What will happen to the inputs and outputs from and to the environment within a flow system if a particular change is made in the material flows? A typical example of an LCA-D model is that used by Kärnä and Pajula (1996), based on an aggregated average LCA (Box 12.1). The LCA-D method is essentially concerned with analysing the changes which occur in the inputs and outputs when a small change is made to the material flows. More precisely, a *small* change, 'delta', is made in one or more of the material flows of a model, followed by an analysis of the resulting deltas (changes) in input and output. The directions and magnitudes of the deltas are used as a decision-making aid in material flow management.

The LCA-D method has one significant advantage over static LCA models. It is this: since the methodology of making LCA inventories is still being developed, the magnitudes of material and energy inputs and effluent and waste outputs in an LCA inventory are subjective, to a certain degree at least, and open to interpretation. The results of LCA-D models, on the other hand, present absolute data on the different inputs and outputs in their respective scales. It is known that the methods and principles used in scoping, aggregation and allocation may cause even larger variations in the absolute results. The endpoint method, based on the use of two different static models, is particularly prone to variations. The approach in the LCA-D is to combine both of the models of the endpoint method within the scope of a single model, and to study the effect of a small change from one to the other. The results of the LCA-D are thus less prone to vary with scope, allocation and aggregation, because they analyse changes in the system rather than the absolute magnitudes of inputs and outputs produced by different systems.

However, the LCA-D method is not suitable for analysing situations in which the changes under consideration are larger than 'delta' (which is small by definition). The structure of the system being studied will not change if the delta changes are sufficiently small, meaning that the scope, aggregation and allocation used in the model will remain intact. However, if the changes are large, the structure of the flow system may change sufficiently to invalidate the scope, aggregation and allocation, thus invalidating the delta approach itself.

Kärnä and Pajula (1996) have analysed the changes in inputs and outputs caused by a small rise in paper recovery rate, together with the corresponding increase in the use of de-inked pulp in newsprint manufacture. They found that a small increase in the use of recycled fibre did not affect waste and effluent production at the average de-inking plant or newsprint mill considered in the model; but, clearly, a large change could invalidate many of the principles and assumptions on which the model is based. Higher production of de-inked pulp would require a larger de-inking facility which, in turn, would affect the yield, energy used and the amount of de-inking sludge produced per ton of pulp made. Similarly, the increased proportion of de-inked pulp at the newsprint mill will change both inputs and outputs per ton of newsprint produced. Also, when the changes in the flow to be studied are sufficiently large to affect the model's construction, it is advisable to use the construction methods of Joined Time Projection (JTP) models described below.

Joined Time Projection models (JTP)

Joined Time Projection models can be viewed as a set of average LCA inventories for several time periods, made under the same rules concerning scoping, aggregation, allocation, etc. Model periods are constructed with scenarios obtained by combining estimated future developments in various parts of the basic system and its inputs (Pento, 1994a, 1994b, 1996; Huuhtanen and Pento, 1995). For example, the functional unit of a national JTP model may be the predicted annual consumption of newsprint in future years. But this predicted demand is only one of

BOX 12.1 ANALYSIS OF DIRECTION INSTEAD OF MAGNITUDES WITH LCA-D

Kärnä and Pajula (1996) have employed the KCL-ECO inventory software to build an average LCA model of pulp and paper flows in Finland and Germany. Material flows in the model are represented by a set of 423 linear equations within a 2723 × 2723 flow matrix, which the software solves by a sequential method to determine the flow quantities for a set of functional units. The functional units are specified in terms of tons of paper of a certain grade (newsprint) consumed. The material flows given by the solution are in balance at each process.

When increased post-consumption recovery of newsprint is considered as a policy option, it is necessary to calculate the environmental effects of the policy. The inventory figures of energy, and more than sixty effluents of the static base-model shown in the figure below, are not very informative in this regard because they are magnitudes which cannot be easily compared to anything. A 'delta' analysis is thus necessary, and it is done by constructing another model which increases both recovery and re-use of papers by a small amount and modifies all flows sufficiently to achieve material balance at each process. An inventory is calculated for this model, from which the base-model inventory is deduced to show the direction. The interpretation of the resulting 'delta' inventory is both straightforward and informative. It can show, for example, if a policy would lead to reduced energy consumption if certain changes are made to the effluents.

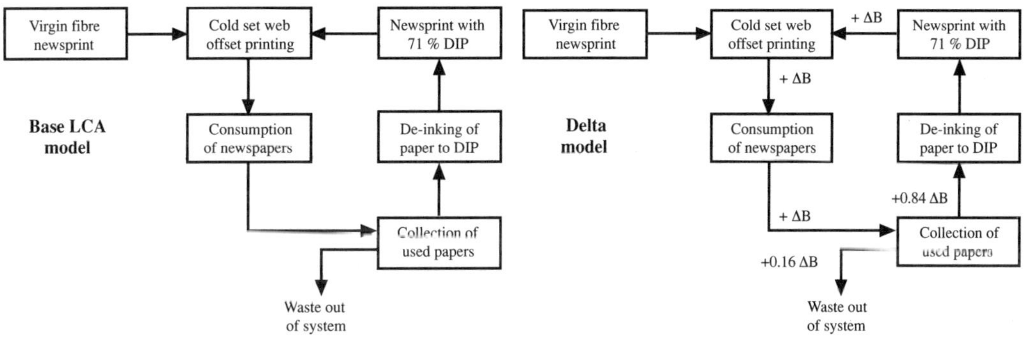

If a large value of 'delta' is used, the resulting large flow changes may invalidate the model. For example, a large increase in recovery may mean that a current 71 per cent de-inked pulp (DIP) content of a newsprint could rise to 100 per cent. But, because the energy consumption and effluent production of a paper machine, which makes newsprint from 71 per cent DIP, are different from those of a machine which uses 100 per cent DIP, it is clear that appropriate changes to the LCA-D model would have to be made. One solution for this would be to make the base-model, as well as two delta models with positive and negative 'deltas', so that the two could be compared against each other. Also, larger changes may require corresponding changes to be made in many parts of the model. Such problems make LCA-D modelling laborious.

the determinants of the flows in the MFMs in future periods, as will be seen from Figure 12.3. Similarly, lack of de-inking capacity, or a delay in its construction or in the construction of other necessary processes and facilities, may prevent a rapid rise in the use of recovered papers (Box 12.2). This is taken into account in determining material flows through the input of projected capacities. Technology and its predicted development may sometimes alter the flows substantially, as has been the case with improved de-inking technology and the resulting higher yields.

In addition to the forecasts, used as inputs to the models for the different time periods under consideration, the JTP method facilitates the study and simulation of alternative policies. In fact, it is possible to include in the construction of such a model a policy whose effects are to be studied. Thus, if a policy of mandatory higher recovery rates of paper is considered for example, then an increasing paper recovery rate (say from the current 40 per cent to a planned 55 per cent over a five-year period) can be fed into the model.

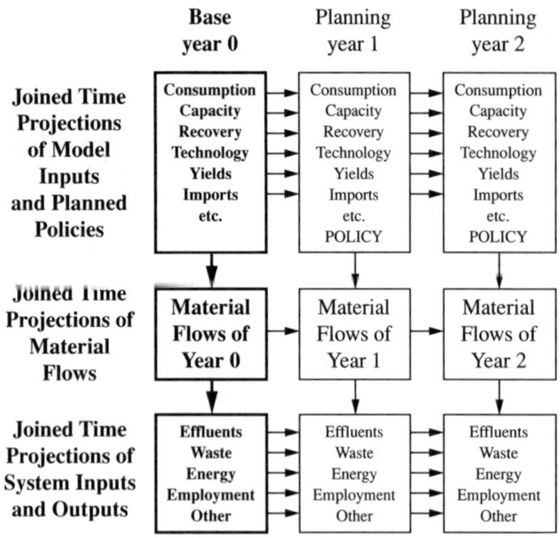

Figure 12.3 The construction of a Joined Time Period model

Source: Gronow and Pento, 1995

The basic flow structure of the base-year model is used without change to make models of future years, each of which represents a future scenario obtained by synthesising the predicted changes in all the factors used as inputs. Thus the model of each future year is essentially a material flow model whose flows and processes generate its inputs and outputs. When changes in directions, as well as changes in material flows, produced by the inputs are sufficiently large, it is necessary to change some of the flows and processes of the model. For example, if the paper recovery rate increases rapidly from one year to the

BOX 12.2 AVOIDANCE OF SUB-OPTIMISATION IN DYNAMIC MATERIAL FLOW MODELS

The lack of landfill capacity in Germany has been one of the driving forces behind the development of German recycling laws and ordinances. Paper and board constitute 20–40 per cent of materials disposed of in German landfills, and the objective of the packaging ordinance of 1991, and the proposed waste paper ordinance, was to reduce this amount by redirecting such solid wastes for collection and re-use. This, together with the fact that recycled paper and board, made from recovered paper, consume less virgin fibre and energy than when they are made from new pulp, clearly shows that future policies should favour collection and recycling.

After the enactment of the packaging ordinance in Germany paper and board have been collected, in compliance with the ordinance, rather than being dumped in landfills. But, since there was no process in place at the time capable of dealing with what turned out to be a huge and growing amount of recovered paper and board, they had to be kept in temporary storage at a considerable cost to the taxpayer. Subsequent changes to the ordinance, together with incineration, export and new treatment methods have since brought the situation under control.

Although the collection of printed paper was not mandated by law, it was accepted by the paper industry voluntarily, and de-inking capacity was increased to match requirement. As a result, the amount of waste paper disposed of in German landfills decreased during 1993–1996. But then, an MFM analysis of the resulting material flows indicated that the policy may have been counter-productive with regard to the total amount of solid waste generated. Because, as will be seen from the figure below, although the amount of waste paper going to landfills decreased, it is most likely that this decrease was more than offset by the increased amount of de-inking sludge and reject, and by the ash from the incineration of a part of the sludge of lower quality waste. An MFM analysis thus points out, as in this case, the fallacy of such apparently 'logical' solutions.

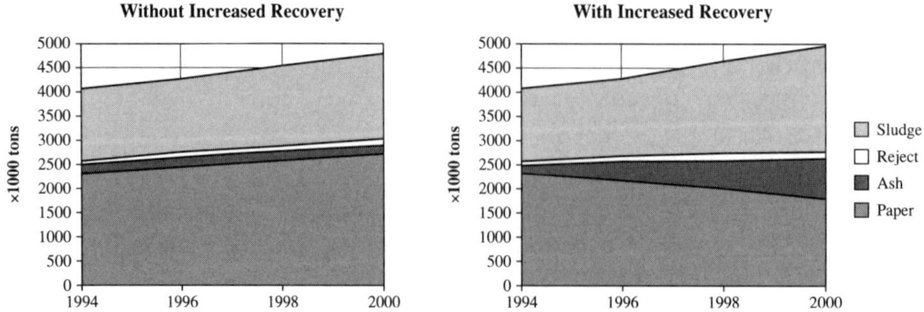

next, the inputs and outputs generated by the processes of de-inking, newsprint manufacture, and paper recovery all have to be changed to reflect the altered composition of aggregated units in each average process.

An MFM, made with the JTP method, thus makes possible the analysis and simulation of forecasts of material flow developments in the future as a function of factors which influence the flows. It also allows the study and simulation of alternative policies with regard to future changes, and produces the inputs and outputs associated with the future scenarios.

CONCLUSIONS

Material Flow Models (MFMs) are potentially capable of analysing accurately the flows of materials in a given system and related processes. The purpose of the exercise is to identify areas of the system in which improvements could or should be made in order to make it more environment-friendly. Starting with the basic Life-Cycle Analysis (LCA) models, in which no aggregation is used and data is site-specific, increasingly sophisticated dynamic models are now being developed. In some of these models even relevant economic or political policy instruments can be incorporated to predict their effectiveness or otherwise.

In their present state of development, however, the reliability of such models is constrained by the lack of accurate and objective methods of aggregation, weighting and allocation, among others. The rapid increase in the size of these models, which occurs with the increasing scope and complexity of the system being studied, is also a problem. A great deal of work thus remains to be done to address these and other problems in this very interesting and fertile area of research and development.

REFERENCES

Adams, D.M. and Haynes, R.W. (1980) *The Softwood Timber Assessment Market Model: Structure, Projections and Policy Simulation*, Forest Science Monograph No. 22, Supplement to *Forest Science* 26 (3).

Ayres, R.U. and Ayres, L.W. (1994) *Chemical Industry Wastes: A Materials Balance Analysis*, Working Paper, INSEAD, Fontainebleau.

Ayres, R.U., Norberg-Bohm, V., Prince, J., Stigliani, W.M. and Yanowitz, J. (1989) *Industrial Metabolism, the Environment, and Application of Materials-Balance Principles for Selected Chemicals*, Laxenburg, Austria: International Institute for Applied Systems Analysis.

Bilitewski, B. (1993) 'Mengenströme in der Papierindustrie 1992', *Das Papier* 47 (10A), Oktober: V88–V92.

Brunner, P.H., Daxbeck, H. and Baccini, P. (1994) 'Industrial metabolism at the regional and local level: a case study on a Swiss region', in R.U. Ayres and U.E. Simonides (eds) *Industrial Metabolism – Restructuring for Sustainable Development*, Tokyo: United Nations University Press.

Enquete-Kommission 'Schutz des Menschen und der Umwelt' des Deutschen Bundestages (hrsg.) (1993) *Verantwortung für die Zukunft – Wege zum nachhaltigen Umgang mit Stoff- und Materialströmen*, Bonn: Economica Verlag.

Gronow, T. and Pento, T. (1995) 'Life cycle inventories and joined material projections in national environmental planning', in A. Frühwald and B. Solberg (eds) *Life-Cycle Analysis – A Challenge for Forestry and Forest Industry*, EFI Proceedings No. 8, Joensuu: European Forest Institute.

Huuhtanen, J. and Pento, T. (1995) 'Effects of printing paper recycling on the age of fibres – an analysis of material flows in Germany', in *Proceedings of the First Ecopapertech Conference*, Helsinki, Finland, 6–9 June 1995, 267–278.

Kallio, M., Dykstra, D.P. and Binkley, C.S. (1987) *The Global Forest Sector: An Analytical Perspective*, Chichester: Wiley.

Kärnä, A., Engström, J., Kutinlahti, T. and Pajula, T. (1994) 'Life cycle analysis of newsprint: European scenarios', *Paperi ja Puu* 76 (4): 232–237.

Kärnä, A. and Pajula, T. (1996) 'An application of the LCA methodology in paper system analysis', *Preprints of the International Conference on Application of Life Cycle Assessment in Agriculture, Food and Non-Food Agro-industry and Forestry: Achievements and Prospects*, 4–5 April 1996, Brussels, Belgium.

Lancaster, K.J. (1966) 'A new approach to consumer theory', *Journal of Political Economy* LXXIV (2), April: 132–157.

Liedtke, C. and Manstein, C. (1994) 'MIPS – Ein neues Mass für ökologisches Wirtschaften', *Lüneburger Umweltsymposium* 26–28 October 1994.

Liedtke, C., Manstein, C. and Merten, T. (1994) 'MIPS, resource management and sustainable development', Paper presented at the Conference on the Recycling of Metals', Amsterdam, 19–21 October 1994.

Pento, T. (1994a) *Modelling National Material Flows for Paper Recycling Policy*, Working Paper 150/94, Jyväskylän yliopisto, Taloustieteen laitos, Jyvzäskylä, Finland.

Pento, T. (1994b) *Painopaperien kierrätys Saksassa*, Jyväskylä, Finland: University of Jyväskylä, Department of Economics and Business Administration, Publication 95/1994 (recycling of printing papers in Germany).

Pento, T. (1996) 'Dynamic Life-Cycle Assessment with the Joined Time Projection technique', Paper presented at the 17th Annual Meeting of the Society of Environmental Toxicology and Chemistry, Washington, DC, 17–22 November 1996.

Schmidt-Bleeck, F. (1992) 'Toward Universal Ecology Disturbance Measures', Wuppertal Institute Position Paper, Wuppertal.

Udo de Haes, H.A., Huppes, G. and Guinée, J.B. (1988) 'Materials balances and flow analysis of hazardous substances. Accumulation of substances in economy and environment', *Milieu* 3 (2): 51–55.

Virtanen, Y. and Nilsson, S. (1993) *Environmental Impacts of Waste Paper Recycling*, London: Earthscan Publications.

Voet, van der E., Egmond, van L., Klein, R. and Huppes, G. (1993) *Cadmium in the European Community: A Policy Oriented Analysis*, Leiden: Centre of Environmental Science.

Zhang, D., Buongiorno, J. and Ince, P.J. (1993) *PELPS III – A Microcomputer Price Endogenous Linear Programming System for Economic Modeling*, Madison, Wis.: United States Department of Agriculture, Forest Service, Forest Products Laboratory, Research Paper FPL-RP-526.

SUGGESTED READING

Ayres, R.U. and Simonides, U.E. (eds) (1994) *Industrial Metabolism – Restructuring for Sustainable Development*, Tokyo: United Nations University Press.

Enquete-Kommission 'Schutz des Menschen und der Umwelt' des Deutschen Bundestages (hrsg.) (1995) *Umweltverträgliches Stoffstrommanagement. Band I: Konzepte*, Bonn: Economica Verlag.

Schmidt, M. and Schorb, A. (hrsg.) (1995) *Stoffstromanalysen in Ökobilanzen und Öko-Audits*, Berlin: Springer-Verlag.

SELF-ASSESSMENT QUESTIONS

1 It is necessary to limit the scope of a material flow model to a subsection of the flows within the anthroposphere because:
   (a) the model becomes too complex and unmanageable otherwise;
   (b) usually data is available only on subsections of real material flow systems;
   (c) subsections are closer to geographical regions.

2 What is meant by aggregation in material flow models?
   (a) aggregation of model results for presentation;
   (b) adding together of the same and similar processes;
   (c) addition of the quantities of material flows for determining process volumes.

3 What types of allocation are needed in material flow models?
   (a) allocation between main products and by-products;
   (b) allocation between inputs of a production process;
   (c) allocation of the effluents of the products of virgin materials to recycled raw materials.

4 Why is a dynamic material flow model less sensitive to errors in data and model construction than a static model?
   (a) the dynamic nature of the model makes small variations insignificant;
   (b) a dynamic model facilitates the study of directions instead of magnitudes;
   (c) dynamic models are inherently more stable and error-free.

5 A static material flow model, which analyses the inputs and outputs from and to the environment caused by a product's flows from cradle to grave, has a special name. What is it?
   (a) a product flow model;
   (b) a life-cycle assessment model;
   (c) a static sectoral model.

6 An LCA-D model may not be used for analysing changes which modify present material flows substantially. Why is this?
   (a) the flow paths of the model change and invalidate the base-model;
   (b) the flows of the models do not balance;
   (c) the processes may change so much that the original model may become invalid.

13

ENVIRONMENTAL AUDITING AND ENVIRONMENTAL MANAGEMENT SYSTEMS

Hildegard De Weerdt

SUMMARY

The concept of environmental auditing of industrial enterprises has been much discussed in recent years. Its general purpose is 'to minimise unconscious environmental risks'. The 'position paper' of the International Chamber of Commerce (ICC) is a good basis for the development of a national industrial programme for environmental auditing. In all countries where environmental auditing is in practice (the USA, Canada, the Netherlands, Great Britain, Scandinavian countries, South Africa, Australia, Japan) experience shows that the full utility of this management tool can be best achieved if it is voluntary, and its findings are confined to the exclusive use of company management. Environmental organisations, however, tend to form more structured audits of a less voluntary nature which give public access to the data obtained.

Checklists, preferably automated, are considered important components in setting up audits.

Recently, environmental auditing has been encouraged as a part of environmental management systems. Beside this trend, we see a push towards the development of standards and, possibly, the registration/certification of these environmental management systems.

It seems that at last the principle of environmental auditing is no longer restricted to industrial activities. More recent proposals involve applying the principle of environmental auditing to environmental programmes in communities, offices, hospitals and schools.

ACADEMIC OBJECTIVES

Upon completion of this chapter students should be able to:

- discuss the definitions of environmental auditing;
- understand why an audit can be an important management medium;
- list the elements necessary for an environmental audit to be effective and yield maximum benefits;
- understand the meaning of EMAS;
- see environmental auditing as the initiating step for the introduction of EMS;
- know all the steps of the ICC flow chart of an audit;
- recognise different types of audit reports;
- understand that auditing is not limited to industrial activities.

INTRODUCTION

Historical context

'Environmental auditing' is a relatively new term but one which is coming into wider use. However, it means very different things to different people. Much of what is now being done under the heading of 'environmental auditing' has been done for a long time under different names (environmental review, screening, investigation, quality control, survey assessment) (UNEP, 1988).

Environmental auditing originally came into existence at the end of the 1970s in the United States, in response to the introduction of stringent environmental legislation carrying heavy penalties. The original emphasis at that stage was on ensuring that a company was in compliance with legal requirements (Tomlinson, 1987).

In reviewing a facility's expansion potential, for example, it will be important to note whether or not the plant is located in an area that is in compliance with ambient air quality standards. In discussing the air pollution situation of the facility, the audit report in this case will outline current emissions and their respective permitted values.

Although the concept of environmental auditing of industries has been much discussed in recent years, no clear international position has yet been put forward on the meaning of this concept, the responsibility for audits and the methodology that should be followed (Buckley, 1991).

The Commission of the European Communities has agreed on a voluntary Environmental Management and Audit Scheme by companies. Companies taking part and conforming to the European Community's agreed green auditing norms would be entitled to use a logo (EMAS Statement of Participation) in their annual reports, letterheads and advertising, so that the system would become a marketing as well as a management tool (CEC, 1993).

Definitions

In a broad context, environmental auditing means checking on certain aspects of environmental management; it implies testing and verification.

The definition that the International Chamber of Commerce presents in its position paper on environmental auditing is generally accepted:

> A management tool comprising a systematic, documented, periodic and objective evaluation of how well environmental organisation, management and equipment are performing with the aim of helping to safeguard the environment by:
> 1 facilitating management control of environmental practices;
> 2 assessing compliance with company policies, which would include meeting regulatory requirements.
>
> (ICC, 1989)

Although the definition of the Commission of the European Communities (CEC, 1992a) is almost identical, its range of application is broader. (e.g.it is also applicable to local authorities).

Functions

Companies may carry out an audit (Adriaenssens *et al.*, 1991):

(a) as a basis for management (e.g. for the early identification of environmental problems);
(b) as a planning instrument (e.g. environmental auditing can anticipate changes in legislation);
(c) as an information source (e.g. for real-estate deals and company take-overs);
(d) as a 'basis' document for insurance (e.g. to control risk);
(e) as an instrument for training personnel (e.g. environmental aspect in training staff);
(f) as a medium of external communication (e.g. to the public, to authorities).

The broad purpose is to provide an indication to company management of how well environmental organisations, systems and equipment are performing (Sadler, 1985).

Essential elements

The practice of environmental auditing involves critical examination of the operations on site and, if necessary, identification of areas for improvement which can assist the management in meeting requirements. The essential steps are: collection of information, evaluation of that information and the formulation of conclusions, including identification of aspects in need of improvement. Suggested procedures are described later (see pp. 257–260).

As is shown in Figure 13.1, environmental auditing can be applied at any stage of a company's production process.

For environmental auditing to be effective and yield maximum benefit, the following elements are necessary (ICC, 1989):

1 *Full commitment of management.* Management, from the highest levels down, must openly support a useful and systematic environmental auditing programme. Such commitment is demonstrated

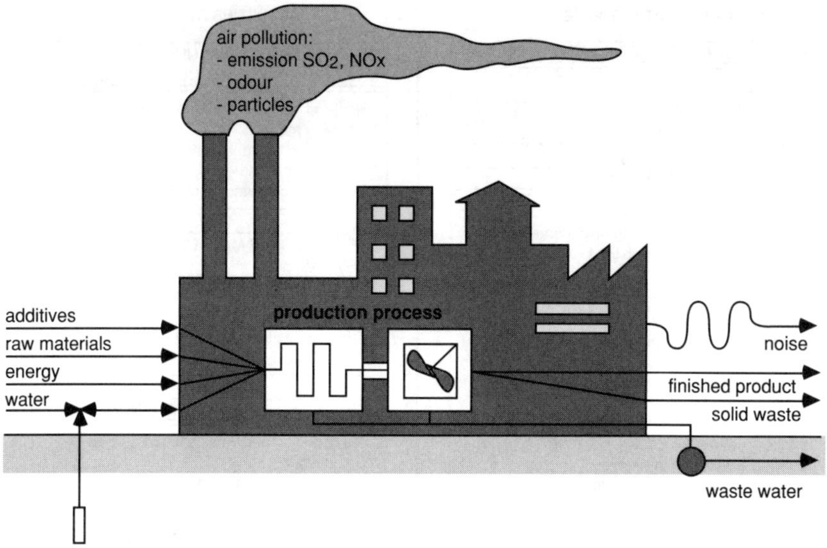

Figure 13.1 A production process showing input and output

by (for example) personal interest and concern, implementation in formal company policy, adoption of high standards, allocation of appropriate human resources and active follow-up of recommendations.

2 *Audit team objectivity.* The principal members of the audit team should be sufficiently detached to ensure objectivity. How this is best arranged will depend upon the size and structure of the company concerned and the nature of the specific audit.
3 *Professional competence.* Team members should be appropriately qualified to provide a technically sound and realistic appraisal and to command respect. The skills required fall under the headings of general environmental affairs and policy, specific environmental expertise, operational experience and knowledge of environmental auditing.
4 *Well-defined and systematic procedures.* To ensure comprehensive and efficient coverage of relevant matters, procedures such as those outlined in the flow chart in Figure 13.2 should be adopted (for details, see 'Methods of Environmental Auditing', this chapter).
5 *Written reports.* It is obvious that the process should be properly documented and that a clear report should be submitted to the appropriate company management. This report should contain factual and objective observations.
6 *Quality assurance.* Some quality assurance mechanism is desirable to maintain the quality of the auditing system itself and to provide assurance to the company of the consistency and reliability of the process.
7 *Follow-up.* Clearly, the full value of audit can only be realised if there is active implementation and follow-up of matters identified.

METHODS OF ENVIRONMENTAL AUDITING

As shown in Figure 13.2, environmental auditing is characterised by a well-defined and organised structure in three steps (ICC, 1989).

Pre-audit activities

Preparation for each audit covers a number of activities, including selecting the review site and audit team, developing an audit plan which defines the technical, geographic and time scope, obtaining background information on the plant (for example, by

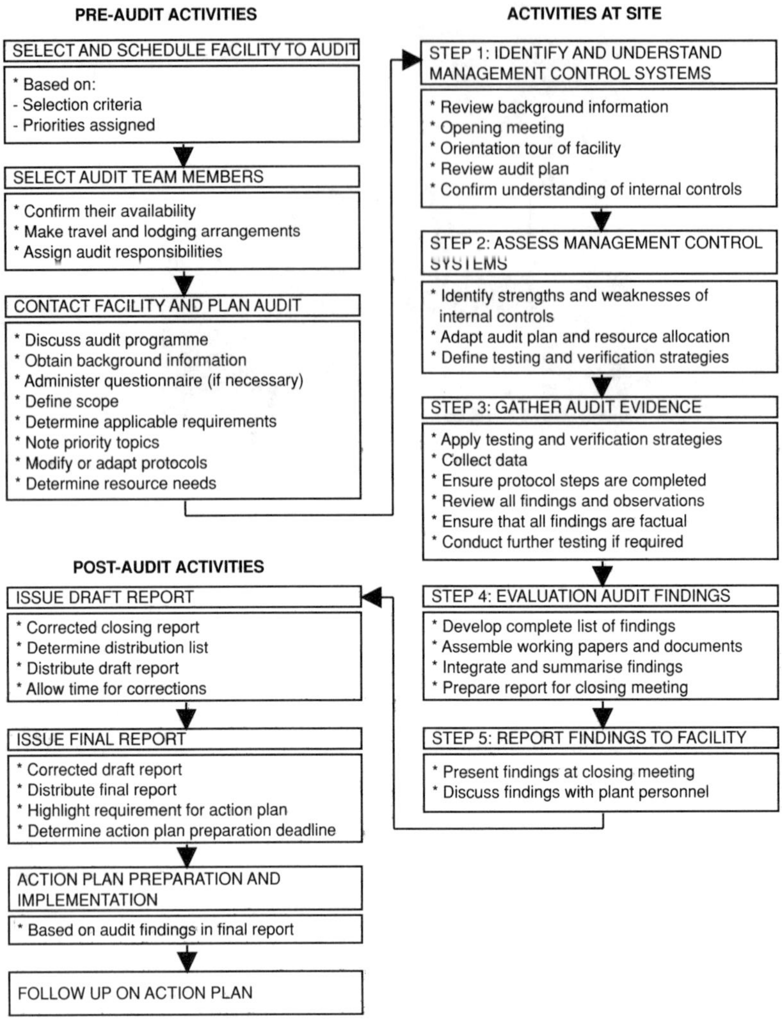

Figure 13.2 Typical environmental work flow
Source: ICC 1989

means of a questionnaire) and the criteria to be used in evaluation programmes. The purpose of these activities is to minimise time spent at the site, and to prepare the audit team to operate at maximum productivity throughout the on-site portion of the audit.

With respect to the composition of the audit team, there are both advantages and disadvantages of including a member from the site being audited. Advantages include:

1 the insider's knowledge of the specifics of the plant regarding both physical installations and organisational patterns;
2 associating a local employee with the audit report may make it appear more credible to the plant's workforce.

The main disadvantage is that the insider may have difficulties in taking or expressing an objective viewpoint, especially if this might be seen as criticism of his or her superior or immediate colleagues. Independent consultants may provide assistance, especially to smaller companies, in the event of a lack of in-house expertise.

Activities at the site

The audit activities at the site typically consist of five basic steps:

Step 1 *Identify and understand management control systems.* Internal controls are incorporated in the facility's environmental management system. They include the organisational monitoring and record-keeping procedures, formal planning documents such as plans for prevention and control of accidental release, internal inspection programmes, physical controls such as containment of released material, and a variety of other control system elements. The audit team seeks information on all significant control system elements from numerous sources through the use of formal questionnaires, observations and interviews.

Step 2 *Assess management control systems.* The second step involves evaluating the effectiveness of management control systems in achieving their objectives. In some cases, regulations specify the design of the control system. For example, regulations may list specific elements to be included in the plan for responding to accidental releases. More commonly, team members must rely on their own professional judgement to assess adequate control.

Step 3 *Gather audit evidence.* In this step the team gathers evidence required to verify that the controls do in practice provide the intended result. Team members follow testing sequences outlined in the audit protocol which have been modified to consider special conditions at the site. Examples of typical tests may include review of effluent monitoring data in order to confirm compliance with limits, training records in order to confirm that appropriate staff have been trained, or purchasing department records in order to verify that only approved waste disposal contractors have been used. All the information gathered is recorded for analysis, and as a record of conditions at the time of the audit. If a control element is found to be in some way deficient, the condition is recorded as a 'finding'.

Step 4 *Evaluate audit findings.* After the individual controls have been tested and team members have reached conclusions concerning individual elements of the control system, the team meets to integrate and evaluate the findings and to assess the significance of each deficiency or pattern of deficiencies in the overall functioning of the control system. In evaluating the audit findings, the team confirms that there is sufficient evidence to support the findings and summarises related findings in a way that most clearly communicates their significance.

Step 5 *Report findings to facility.* Findings are normally discussed individually with facility personnel in the course of the audit. At the conclusion of the audit, a formal final meeting is held with facility management to report all findings and their significance in the operation of the control system. The team may provide a written summary to management which serves as an interim report prior to the preparation of the final report.

Once involved in the auditing process, one will be confronted with an enormous pool of data. To manage the analysis of large volumes of data and identify trends and patterns, the use of specially developed software is a welcome help (Speeding *et al.*, 1994).

Post-audit activities

At the end of the on-site audit, two important activities remain: preparation of the final report and development of a corrective action programme. The final audit report is generally prepared by the team leader and, after review by those in a position to evaluate its accuracy, it is presented to the appropriate management.

Action plan preparation and implementation

Facility personnel, sometimes assisted by the audit team or outside experts, develop a plan to address all findings. This action plan serves as a mechanism for obtaining management approval and for tracking progress towards its completion. It is imperative that this activity takes place as soon as possible so that management can be assured that appropriate corrective action is planned. A primary benefit of the audit is lost, of course, if corrective action is not taken promptly.

Follow-up on action plan

Follow-up to ensure that the corrective action plan is carried out and all necessary corrective action is taken care of is a very important step. This may be done by an audit team, by internal environmental experts, or by management.

Although not all audit programmes necessarily contain each step, the design of each programme generally makes provision for each of the activities described.

SETTING UP ENVIRONMENTAL AUDITS

Types

The first step in carrying out any environmental audit is to establish its purpose, which in turn determines the type of audit required (Bedrijfsinterne Milieuzorg, 1989).

The three main types of audits are juridical, technical, and organisational.

Juridical audit

A juridical audit involves a review of (1) the country's environmental policy goals; (2) how well existing legislation is achieving those goals; and (3) how legislation could best be modified (Buckley, 1991). Some areas to consider include the country's policies on the ownership, use and management of natural sources, and what laws and regulations the country has for the control of pollution and maintenance of environmental quality.

Technical audit

A technical audit reports the results from measurements of air and water pollution, solid and hazardous waste, radioactive materials, PCBs and asbestos (Tomlinson, 1987). For example, the air emission source forms may include the type of source, type and date of installation, capacity of the control device, and the location, height and emission rate of the emission point.

Organisational audit

This type of audit includes information about the management structure in a company, the means of internal and external communication, and education and training programmes. It reveals details, for instance, of the history of the plant, and the names of the plant manager, environmental co-ordinator, purchasing agent, maintenance supervisor and laboratory manager (Blakeslee and Grabowski, 1985).

Methodological approaches

Industrial auditing reports have an explicitly confidential character which makes it difficult to collect them.

There are three basic styles for making an environmental audit-report. In all three, the interview is probably the method most widely used for gathering information (Bins-Hoefnagels, 1991).

Descriptive analysis

This type of audit is frequently used. It can be carried out either by an internal or an external auditor. It simply describes the actual situation and compares this with existing and foreseeable regulations. From this comparison, a number of regulations and a management plan can be devised.

In *Environmental Audit and Business Strategy*, Ledgerwood *et al.* (1992) illustrate the audit of a waste transfer station. Box 13.1 provides the circumstances in which the audit took place, the organisation of the audit, the main results and recommendations.

BOX 13.1 AUDIT CASE OF A WASTE TRANSFER STATION

M.J. Carter Associated were commissioned by the consultancy office Shanks and McEwan (Southern) Ltd to undertake an independent comprehensive environmental audit of the waste transfer station at Hendon, London. The site occupies approximately 2.6 hectares and processes up to 1,200 tonnes per day of solid household, commercial and industrial wastes. The aim of the audit was to assess the aspects of control and monitoring of the facility which relate to the environmental quality at the site and the surrounding area. These aspects were compared with regulatory standards and the company code of practice. Two members of M.J. Carter Associated's technical staff undertook the audit. A telephone call was made to the site manager explaining the reasons for the audit and a letter was sent detailing the information needed to undertake the audit, namely:

- a detailed plan of the facility;
- copies of the planning permission and site licence;
- an operational working plan;
- details of monitoring procedures and monitoring results;
- waste disposal arrangements;
- details of the final disposal points of all wastes;
- maintenance records;
- schedules of environmental control and landscape works;
- waste disposal authority site inspection reports;
- any other reports *re* environmental investigations;
- any correspondence with the public regarding environmental issues.

Based on the collected information, it was also apparent that one working day would be sufficient to undertake the audit visits.

The inspection of the facility, written down in a detailed report, revealed that considerable efforts had been made to design a structure and yard layout which would cause minimum environmental disturbance, especially the sensitive locations to the east of the site. The most significant deficiency was the lack of an oil interceptor in the surface drainage system in the site yard. There was no procedure for the regular inspection of the oil interceptors in the foul water drainage system from the waste transfer hall. No arrangements were made to monitor noise at the sensitive locations adjacent to the site. The facility is designed so that vehicles discharge wastes directly into the compactor hoppers, so there was no suitable opportunity adequately to inspect the majority of the waste. The audit report contained recommendations for improvements to the surface water drainage system and the instigation of monitoring procedures for the surface water gullies, foul water system oil interceptors and noise levels at sensitive locations.

Source: Ledgerwood *et al.*, 1992

Analysis by checklists

Checklists are handy and help to ensure that information is not lost. The United States Environmental Protection Agency (USEPA), which works with this system of checklists, publishes a comprehensive *General Protocol for Environmental Audits* (USEPA, 1989). The auditor's checklist provides a step-by-step worksheet to be used by audit team members conducting the audit. Each worksheet is divided into four columns. The first column is the citation of a federal regulatory requirement; the second column provides instructions to the auditor; the third column is for the auditor's comments. In the fourth column the auditor gives each finding a unique notation that is to be determined by the agency. This is to encourage early classification of findings. For example, Figure 13.3 shows part of the USEPA checklist on 'Drinking Water'.

Analysis by checklists and quantitative methods

This method is basically the same as that described in the previous subsection, but the processing is more advanced. For each module, questionnaires are prepared. These questionnaires deal with the relevant subjects of the different modules. Most questions need to be answered with yes (y) or no (n) or 'Not Applicable' for the organisation. A few questions require the personal opinion of the auditor.

Facility: ________ Auditor: ________ Date: ____

Regulatory citation	Auditors' Check-list	Comments
	Community Water Supply Systems	
40 CFR 141	• Water supplied by the facility complies with the following primary maximum contaminant levels - Inorganic chemicals - Organic chemicals - Turbidity - Microbiological contaminants - Natural radioactivity - Man-made radionuclides	
40 CFR 141.31-36	• The facility is operating in accordance with reporting, public notification, and record-keeping requirements.	
40 CFR 141.40-42	• The water supply system is monitored for organic chemicals, sodium, and corrosivity characteristics.	

CFR refers to the Code of Federal Regulations.

Figure 13.3 Part of EPA's drinking water checklist
Source: USEPA, 1989

The 'yes/no' questions have a corresponding numbered scale for situations in which the truth lies somewhere in between. If the scale is y/n = 15/0, for instance, a score of 13 or 14 would indicate that the auditor agrees almost completely with the statement, while a lower score would indicate less agreement.

The individual scores are added up, multiplied by 100, and divided by the total number of points possible (50 minus the points 'not applicable') to express the overall performance level on a scale of 0–100 per cent.

Both the individual and the overall scores may be used to point out the strengths and weaknesses of the organisation being assessed. An example of such a questionnaire is given in Figure 13.4.

As indicated previously (see pp. 257–260), one can distinguish different modules in an audit. Not all of them will be relevant in all cases, nor will they all have equal significance or require equal emphasis. Usually it is not necessary for the auditor to go through all the steps of an audit.

Although environmental audits vary in type and in some methodological approaches, they have certain generic characteristics (see pp. 257–260). The approach of the ICC (1989) is characterised by a well-defined and organised structure, careful methodical investigations and strong emphasis on reporting to all appropriate management. These characteristics form the basis for the reputation environmental auditing has earned for providing reliable and useful information to management in all settings where it is practised around the world.

IMPLEMENTATION OF AUDIT POLICY IN SOME COUNTRIES

United States of America

The Environmental Protection Agency (EPA) operates at the federal level and encourages states and companies to use environmental audits as a management instrument. The lines of action of the EPA's policy have been fixed since 1985 (USEPA, 1989).

(#) Contacts with Government, Agencies and Third Parties	50 points
	score
1. (#) The Industry	**(15 points)**
(#) Has a person been appointed for the communication with the Industry concerning Environment and are his responsibilities and authorities clearly defined? (y/n=15/0)	yes/no ___
2. (#) Third Parties	**(20 points)**
(#) Have people in the organisation been appointed to liaise concerning Environment with third parties like consultants and are their responsibilities and authorities clearly defined? (y/n=20/0)	yes/no ___
3. (#) Neighbourhood/Neighbours	**(15 points)**
(#) Have people in the organisation been appointed to liaise concerning Environmental issues with neighbours and social groups in the neighbourhood and are their responsibilities and authorities clearly defined? (y/n=15/0)	yes/no ___
(External) Contacts x 100 / (50 -) = ______%	

Figure 13.4 Part of a quantitative checklist about 'contacts with government, agencies and third parties'

Source: BP Nutrition, 1991

The EPA favours voluntary environmental audits and encourages the active involvement of the company's board of directors in the audit.

EPA's recent 'Environmental Audit Privilege Act' (USEPA, 1995) promotes disclosure of self-audits. The new Act should make a company immune from administrative, civil, or criminal penalties if the following conditions are complied with: the company promptly and voluntarily discloses the violation to the regulatory agency; takes prompt steps to correct the problem; and fully co-operates with the regulatory agency (Hatfield, 1995).

South America

A major positive change can be observed in the environmental management field, for example in Mexico and Brazil: although rather hesitatingly, international (mostly chemical) companies are disclosing their auditing experiences. Nevertheless, the balance between socio-economic considerations and environment remains a big challenge (Foster *et al.*, 1995; Lèbre La Rovere and d'Avignon, 1995).

Europe

Commission of the European Communities (CEC)

From April 1995, the so-called Eco-Management and Audit Scheme (EMAS) came into force (CEC, 1993). The Eco-Management and Audit Scheme is set within the context of the EC's 5th Action Programme on the environment, *Towards Sustainability* (CEC, 1992b). The new range of instruments, market-based instruments offer probably the most innovative development in the EC's 5th Action Programme: EMAS is a market-based tool developed to introduce market forces in the environmental field by promoting competition between industrial activities on environmental grounds. Eco-labelling for products is another example of a market-based regulation.

Although a regulation, participation under EMAS is voluntary. Manufacturers registering under the Eco-Audit scheme must commit themselves to carrying out an environmental audit and an

EC ECO MANAGEMENT AND AUDIT SCHEME	This site has an environmental management system and reports to the public on its environmental performance in accordance with the Community Eco-management and audit scheme. **(Registration No ...)**

Figure 13.5 One of the EMAS Statements of Participation
Source: CEC, 1993

environmental statement by an independent external 'agreed auditor'. Therefore, newly 'competent authorities' are set up to which companies apply for registration. EC member states must in turn promote the establishment of national 'environmental auditing professional bodies' which are responsible for choosing the auditors, firms and individuals who are entitled to perform audit verifications.

Companies registered under EMAS will be entitled to use the 'Eco-Audit Statements of Participation', as illustrated in Figure 13.5. This logo indicates (all) the production site(s) in a company which participate in the EMAS scheme.

In October 1996 (CEC, 1996) 381 companies were registered, of which 293 were German.

Member states may take measures to encourage participation of companies in the Eco-Audit scheme. These could include economic incentives and technical assistance. Non-profit companies are encouraged to participate as well.

International organisations involved in auditing

The sole purpose of this section is to give a range of organisations involved in environmental auditing as a principal or as a side activity.

Organisations with auditing as a principal activity

Among professional organisations dedicated to furthering the development and professional practice of environmental auditing are the Institute for Internal Auditing (IIA), the Environmental Auditing Roundtable (EAR), the Environmental Auditing Forum (EAF) and the Institute for Environmental Auditing (IEA). These are all US initiatives (Cahill, 1995).

The British Environmental Auditors Registration Association (EARA) is a professional organisation which defends the rights of external environmental consultants (Thomson *et al.*, 1995).

The Network on Environmental Management and Auditing (NEMA Network) is a commendable British initiative that gathers together researchers, consultants and companies to discuss the promotion of EMAS (Hillary, 1992).

Organisations which include environmental auditing as a side activity

Recently, internationally recognised organisations have added environmental auditing to the list of environmental management instruments they prefer to work with. Some of these organisations are primary environmentally oriented like EPA, UNEP, ICLEI and Friends of the Earth. Others, like the already discussed CEC (1993), the ISO organisation (ISO, 1994) or the World Bank (1995) recently produced position papers on environmental auditing and environmental management systems.

ENVIRONMENTAL AUDITING: AN INTEGRAL COMPONENT OF ENVIRONMENTAL MANAGEMENT SYSTEMS

Auditing for compliance is only the beginning

Most internal environmental audit programmes begin with compliance audits. In the electrical utility industry in the UK, most companies now conduct compliance audits (Iwanski and Thurman, 1995). Many are beginning to consider expanding their programmes to include the examination of the adequacy of environmental management controls. This kind of auditing tells management the extent of risk and potential liability – valuable information for any well-run business.

It is important for the understanding of the auditing approach that one appreciates the difference between compliance and management audits. Willig (1995) clarifies this difference with an interesting example.

Typically, environmental compliance audits identify specific areas of non-conforming, e.g. improperly labelled printed circuit board transformers, and report the problem to management with a simple recommendation to corrective action. The recommendation is then acted on to bring the facility into compliance, perhaps by marking printed circuit board transformers with labels specified in the regulations. The root cause of the non-conforming is not examined.

In a management audit, the auditor probes for the cause of the problem, which could be a lack of procedures, insufficient training of personnel, or something else.

Examples of questions that the auditors would be asked in a management audit include the following:

- Has the importance of compliance with printed circuit board regulations been communicated to employees by the facility manager?
- How was responsibility for labelling assigned?
- Did the person responsible for training have adequate training?
- Was there a self-assessment programme at the facility to ensure that labelling responsibilities were carried out?
- Does the facility have a system for correcting deficiencies noted in the self-assessments?

This type of questioning gets at root causes and allows auditors to recommend actions that not only correct the existing conditions but enhance the capability of the audited facility to avoid future problems.

This approach lets management know that the facility has a programme in place to manage for compliance over the long term. It helps management assess the quality of the compliance programme and what actions are needed to strengthen it.

Audits of management controls will need more than basic audit experience and knowledge of regulatory requirements. It is imperative that the team has a high level of collective experience in understanding facility operations, how organisations and procedures work, as well as management control systems.

Management audits require that much more time and auditors' experience be invested in audit planning, performing and reporting. Crafting good interview questions, audit plans, and checklists is time consuming. Correspondingly, more time is spent interviewing management and staff in the facility.

Many organisations, having written their environmental policy and carried out environmental (management) audits, find difficulty in translating recommendations into action, but one tool which they are using to facilitate the implementation of environmental policy is a so-called Environmental Management System (EMS).

What is an Environmental Management System?

Although research of the EMS *per se* is not part of this chapter, it is essential to clarify the common stages within an EMS that organisations use to try to ensure that the environment is considered in policy and processes.

An environmental management system is defined by the British Standards Institution (BSI, 1994) as 'the organisational structure, responsibilities, practices, procedures, processes and resources for determining and implementing environmental policy'.

Gilbert (1993) summarises the basic stages of an organisation's EMS:

- a policy statement indicating commitment to environmental improvement and conservation and protection of natural resources;
- a set of plans and programmes to implement policy within and outside the organisational culture;
- the measurement, audit and review of the environmental management performance of the organisation against the policy, plans and programmes;
- the provision of education and training to increase understanding of environmental issues within the organisation;
- the publication of information on the environmental performance of the organisation.

These stages of audit, measurement and review closely resemble elements of quality management systems which have been developed to ensure that the activities, policies and processes of organisations conform to the specified requirements.

The best-known environmental management systems are the European Environmental Management and Audit Scheme (CEC, 1993), the Dutch 'Notitie Bedrijfsinterne Milieuzorg' (1989), and recently the ISO 14001 (1994) environmental management system in preparation. Figure 13.6 depicts a part of this EMAS, showing the different elements of an environmental management system.

The auditing process has a central place in all current described EMSs, with the environmental audit being at its very core (Bhargava and Welford, 1996)

Evaluating management systems as part of environmental audits

To conduct an audit of a management system, an auditor needs to select those programmes that are to be evaluated and to audit components of each of the selected programmes:

- organisation;
- administrative procedures;
- staff assignments, documentation, reports and records;
- implementation.

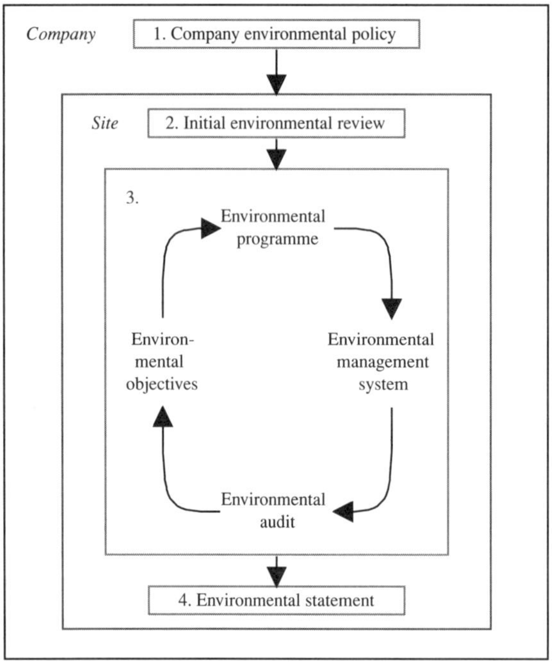

Figure 13.6 Partial overview of the Eco-Management and Audit Scheme
Source: Hillary, 1993

Much of the organisation and procedural review of management systems can be conducted through interviews and evaluations of the programme's documentation. And if the systems can be identified ahead of time, protocols or checklists can be developed to guide the auditors in investigation.

One of the most important aspects of the audit, however, is to ensure that written procedures are, in fact, carried out effectively. The auditor can determine that the system is effective only if it is tested or evaluated on a routine basis. In other words, any audit of management systems should include an evaluation of actual practices as well as documented procedures.

Cahill (1995) points out possible reasons why applying environmental management systems audits can cause more difficulties:

- Performance appraisal. An audit of management systems is truly an indicator of personal performance, and therefore it will always have a performance appraisal flavour to it. This means that interviews will have an additional tension that must be dealt with by the auditor.
- Identifying the root cause of a problem requires extra digging and investigation by the auditor. This will take even a very experienced auditor additional time.
- Lack of standards. There will be many requirements placed on an organisation (e.g. increased environmental awareness) that will have no standards against which they can be evaluated.

Standardisation of Environmental Management Systems

Environmental Management Systems standards have been developed to provide organisations with a framework to implement an EMS within their organisation.

The introduction of voluntary standards means that environmental audits undertaken today are substantially different from those undertaken eight or so years ago. Techniques have been influenced by training in ISO 9000 or BS7750, whose disciplines are applicable to environmental auditing. Much can be learned, for example, from the application of ISO guidance on quality auditing (laid down in ISO 10011–1: Quality Systems Auditing) to environmental auditing.

Environmental auditing is central to the effective implementation of an environmental management system and the EMS standards put much emphasis on the role of audit to monitor continuous improvement processes.

The following Environmental Management Systems standards are common (Starkey, 1996):

1 *National Standards*:
   - The British Standard BS7750 (1994).
   - The Irish Standard IS310 (1994).
   - The Canadian Standard CSA Z750–94-A (1994).
2 *European Standards*:
   - Council Regulation No. 1836/93 of 29 June 1993 allowing voluntary participation by companies in the industrial sector in a Community eco-management and audit scheme. This is commonly known as the Eco-Management and Audit Scheme (EMAS) (CEC, 1993). Although it is a regulation rather than a standard, EMAS has the same aim as the national standards, i.e. standardisation in the field of environmental management systems.
3 *International Standards*:
   - The International Standard ISO 14001 Environmental Management Systems – Specification with guidance for use.

All of the currently available national and international standards for EMS have an audit component. The importance of environmental audits has increased during the last three years, with the publication of the British Standard in 1992 and 1994 and the launch of the Eco-Management and Audit Scheme in 1993.

The Technical Committee 207/SC 2 of the International Standards Organisation published the ISO standards for environmental auditing in September 1996 (Stans, 1995).

It must also be pointed out that there are problems associated with the growing number of EMS standards. The proliferation of standards is creating great uncertainty in organisations wishing to implement a quality environmental management system. Many wish to receive early certification. Others are using the proliferation of initiatives as an excuse not to subscribe to any, and many organisations are undoubtedly waiting to see the full range of standards and regulations on offer (Shimell, 1994).

These standards and the EMS approach in general have been criticised in that they do not go far enough in environmental terms; that they are defensive, bureaucratic and do not provide an adequate framework to tackle urgent environmental issues and promote sustainable management practices (Netherwood, 1996).

FUTURE DEVELOPMENT

In the past few years, a number of interesting trends have surfaced within the discipline of environmental auditing. As described earlier in this chapter, we see a push towards the development of standards and possibly national registration/certification.

Computers (especially portables) are now increasingly being used in a variety of ways in the support of audit programmes: they help to maintain on-line regulatory databases; they store or update audit reports and action plans.

It also seems that the principle of environmental auditing is no longer restricted to industrial activities. More recent proposals involve applying the principle of environmental auditing to environmental programmes in communities, offices, hospitals and schools.

Box 13.2 gives background information on a pilot project, still in progress, concerning the development of an environmental auditing and management system in Flemish secondary schools (Borgo *et al.*, 1994).

CONCLUSIONS

Environmental auditing is not conducted under formal procedures, pursuant to rules and standards that allow comparison and judgement of compliance. To be an effective management tool it should be complemented by publicity campaigns aimed at

BOX 13.2 THE GREEN SCHOOL PROJECT AS A CASE FOR IMPLEMENTATION OF EMS IN THE NON-PROFIT SECTOR

The Green School Project (GSP) aims to stimulate positive feelings towards the environment in secondary level pupils by applying environmental management systems in their own school. The management system was designed to include seven focus areas: energy, waste, water, traffic, materials, green spaces and the kitchen canteen. The EMS is introduced in the school by means of a system based on 15 logical steps. In summary this involves: drafting an environmental policy, choosing one or more focus areas after a brief audit, setting up project groups and working groups, making an inventory of the environmental situation at school, refining the policy, introducing new measures and recording the results. These steps are defined to provide sufficient guidelines whilst leaving enough flexibility for the individuality and creativity of the schools.

A working package including a original computer program is available for each focus area. The working package contains:

- a manual which guides the school through the implementation of the EMS;
- a package of measures to improve the environmental situation at the school;
- well-organised documentation, with background information outlining the environmental problems connected with the focus area.

During the school years 1993–1994 and 1994–1995 the Green School Project was tested out in eight pilot schools. Since September 1995, all secondary schools in Flanders (the northern part of Belgium) can join in. At the end of the school year 1995–1996 more than 230 out of 1,000 secondary schools were registered as participants.

Schools participate on a voluntary basis, with teachers trying to integrate the project into their subject/lessons, or dealing with it during breaks, outside school hours, or during project days, etc.

assuring external audiences of the company's responsible behaviour in the environmental area.

Environmental management systems are recent instruments and experience with them is somewhat limited, especially in non-industrial organisations. The recommendation of the worldwide agreed recommendations of Agenda-21 (UNCED, 1992) to deploy such systems more widely is based on theoretical judgements and not on solid facts and figures relating to environmental effectiveness and efficiency.

In Europe, the Fifth Environmental Action programme formed the basic idea behind EMAS as an innovative piece of EC legislation. It is one of the instruments which puts the principles of prevention of pollution and continuous improvement into practice. The implicit assumption is that the market will reward companies for participating in the audit scheme and, as a result, market pressure will encourage an increasing number of firms to join. Thus, company energies will be tapped and environmental performance improved. Besides, the scheme provides a means for businesses to manage their environmental performance and satisfy the public desire for more environmental information. Ultimately it allows management to be in control (Hillary, 1993).

The concept of environmental auditing in companies has been much discussed in recent years, but since no clear international position has yet been put forward on the meaning of this concept, or the responsibility for audits and the methodology that should be followed, we must keep in mind that auditing can provide only limited control and awareness of potential issues. Without other management systems the audit is a very limited tool in modern environmental management (Friedman, 1988). This is not to say that audits are not beneficial.

An environmental auditing programme constitutes an integral part of the general management system, assisting a company in organising and managing effective environmental programmes. Environmental auditing clearly covers the operational stage of a facility's existence. Environmental impact assessments, undertaken before the facility is constructed and without well-defined decommissioning procedures when its operational life is terminated, are also advisable. Experience shows that the full utility of this management tool can be best achieved if its use is voluntary and its findings are for the exclusive use of company management in carrying out their responsibility to correct all deficiencies promptly (Almgren, 1990).

It remains difficult to make use of industrial audit reports for the implementation and improvement of auditing methodologies because of the confidential nature of those reports.

REFERENCES

Adriaenssens, P., Artois, J., Houthuys, F., Luts, Ph., Nelis, M., Onraedt, E., Can San, E. and Vyvey, Q. (1991) 'Milieu-audit als Element van Bedrijfsinterne Milieuzorg. Model van Vragenlijst', Master of Science thesis, Antwerp, Belgium: Universitaire Instelling Antwerpen.

Almgren, R. (1990) *Miljörevision*, Stockholm, Sweden: Industriförbundets Förlag AB.

Bedrijfsinterne Milieuzorg (1989) *Notitie aan de Tweede Kamer der Staten Generaal, vergaderjaar 1989–1990, 20633 n° 3*, Den Haag, The Netherlands: VROM.

Bhargava, S. and Welford, R. (1996) 'Environmental auditing', in R. Welford (ed.) *Corporate Environmental Management. Systems and Strategies*, London: Earthscan Publications Ltd, pp. 118–139.

Bins-Hoefnagels, I.M.J. (1991) 'Milieudoorlichting', in *Milieumanagement in Stappen*, Deventer, The Netherlands. Kluwer, Bedrijfswetenschappen.

Blakeslee, W. and Grabowski, T.M. (1985) *A Practical Guide to Plant Environmental Audits*, New York: Van Nostrand Reinhold.

Borgo, E., Galle, I., De Weerdt, H. and Hens, L. (1994) 'The Green School Project', *Integrated Environmental Management* 94 (3): 18–20.

BP Nutrition (1991) 'Quantified EHS Auditing Questionnaire', Internal document.

British Standards Institution (BSI) (1994) *British Standard for Environmental Management Systems: BS7750BSI*, London.

Buckley, R. (1991) 'Guidelines for environmental audit', in *Perspectives in Environmental Management*, Berlin, Heidelberg, New York: Springer-Verlag, pp. 120–163.

Cahill, L.B. (1995) 'Evaluating management systems as part of environmental audits', in J. Willig (ed.) *Auditing for Environmental Quality Leadership*, New York: J. Wiley & Sons, Inc.

Canadian Standards Association (CSA) (1994) *CSA Standard Z750–94. A Voluntary Environmental Management System*, Toronto: Canadian Standard Association (CSA).

Commission of the European Communities (CEC) (1992a) 'Proposal for a council regulation allowing voluntary participation by companies in the industrial sector in a community eco-audit scheme', *COM* (91) 459 final, CEC, Luxembourg.

Commission of the European Communities (CEC) (1992b) *Towards Sustainability.* A European Community Programme of Policy and Action in Relation to the Environment and Sustainable Development, Luxembourg: CEC.

Commission of the European Communities (CEC) (1993) 'Council Regulation No. 1836/93 of 29 June 1993 allowing participation by companies in the industrial sector in a Community eco-management and audit scheme', *Official Journal of the European Communities*, L168/1–18, 10 July 1993, Luxembourg: CEC.

Commission of the European Communities (CEC) (1996) 'List of EMAS registered sites', CEC, DGXI, Brussels.

Foster Knight, C., Wells, R.P. and Pratt, L. (1995) 'After the crisis', *The Environmental Forum* 12 (3), June 1995, Washington, DC: Environmental Law Institute, p. 21.

Friedman, F.B. (1988) 'Environmental auditing and environmental management', in *Practical Guide to Environmental Management*, Washington, DC: Environmental Law Institute, pp. 57–84.

Gilbert, M. (1993) *Achieving Environmental Management Standards. A Step-by-step Guide to Meeting BS7750*, London: Pitman Publishing.

Hatfield, M.O. (1995) 'The Environmental Audit Privilege Act', *The Environmental Forum* 12 (3), June 1995, Washington, DC: Environmental Law Institute, p. 21.

Hillary, R. (1992) *The Network for Environmental Management and Auditing (NEMA)*, London: Imperial College for Science, Technology and Medicine, University of London.

Hillary, R. (1993) 'The Eco-Management and Audit Scheme: a practical guide', *Business and the Environment Practitioner Series*, Oxford: Technical Communications Ltd.

International Chamber of Commerce (ICC) (1989) *Environmental Auditing. Report No. 468*, Paris: ICC.

International Standards Organisation (ISO) (1994) *Committee Draft ISO/CD 14000. Environmental Management Systems – Specification with Guidance for Use*, Geneva: International Standardisation Organisation.

Iwanski, M.I. and Thurman, J.R. (1995) 'Management audits versus compliance audits', in J.T. Willig (ed.) *Auditing for Environmental Quality Leadership*, New York: J.Wiley & Sons, Inc., pp. 31–37.

Lèbre La Rovere, E. and d'Avignon, A. (1995) 'Emerging environmental auditing regulations in Brazil and the prospects for implementation', *Industry and Environment* 18 (2–3), September: 11–14.

Ledgerwood, G., Street, E. and Therivel, R. (1992) *Environmental Audit and Business Strategy*, London: Pitman Publishing, pp. 76–78.

National Standards Authority of Ireland (1994) *IS310: Environmental Management Systems – Guiding Principles and Requirements*, Dublin: National Standards Authority of Ireland.

Netherwood, A. (1996) 'Environmental management systems', in R. Welford (ed.) *Corporate Environmental Management. Systems and Strategies*, London: Earthscan Publications Ltd, pp. 35–59.

Sadler, B. (ed.) (1985) 'Audit and evaluation in environmental assessment and management: Canadian and international experience', in *Proceedings of the Conference on Follow-up/Audit of E.I.A. Results*, International Association for Impact Assessment, Toronto, Canada, October 13–16.

Schimell, P. (1994) 'The relationship between eco-audits and environmental management systems', *Eco Management and Auditing* 1 (2): 17–21.

Stans, J. (1995) 'Development of ISO standards for environmental auditing', *Industry and Environment* 18 (2–3), September: 11–14.

Starkey, R. (1996) 'The standardisation of environmental management systems', in R. Welford (ed.) *Corporate Environmental Management. Systems and Strategies*, London: Earthscan Publications Ltd, pp. 59–91.

Thomson, D.R., Bacon, R.A., Tarling, J.P. and Baverstock, S.J. (eds) (1995) *The Environmental Auditors Registration Association (EARA) Register of Environmental Auditors*, London: Earthscan Publications Ltd.

Tomlinson, P. (1987) 'Audits in environmental assessment', *Environmental Monitoring and Assessments* 8 (3): 187–198.

United Nations Conference on Environment and Development (UNCED) (1992) *Agenda 21*, Rio de Janeiro, June 1992.

United Nations Environment Programme (UNEP) (1988) 'Environmental auditing', *Industry and Environment* 11 (4).

United States Environmental Protection Agency (USEPA) (1989) *General Protocol for Environmental Audits at Federal Facilities*, Washington, DC.: EPA.

United States Environmental Protection Agency (USEPA) (1995) *The Environmental Audit Privilege Act. S. 582*, Washington, DC.: EPA.

Willig, J.T. (ed.) (1995) 'Auditing for environmental quality leadership', *Beyond Compliance to Environmental Excellence*, New York: J. Wiley & Sons, Inc., pp. 31–37.

The World Bank (1995) 'Environmental auditing', *Sourcebook Update*, August 1995, no. 11, Washington DC.: The World Bank.

SUGGESTED READING

Nath, B., Hens, L. and Devuyst, D. (eds) (1996) *Sustainable Development*, Brussels, Belgium: VUB Press.

Pearce, B. (1995) *Urban Eco-Auditing and Local Authorities in Europe. The Sustainable City. A European Tetralogy. Part I*, Luxembourg: Office for Official Publications of the European Communities.

Tibor, T. and Feldman, I. (1996) *ISO 14000. A Guide to the New Environmental Management Standards*, Chicago: Irwin Professional Publishing.

Welford, R. (ed.) (1996) *Corporate Environmental Management. Systems and Strategies*, London: Earthscan Publications Ltd.

Willig, J.T. (ed.) (1995) *Environmental TQM*, New York: Executive Enterprises Publications Co., Inc.

SELF-ASSESSMENT QUESTIONS

1 Mark the terms which belong to the auditing vocabulary:
   (a) environmental assessment panel;
   (b) internal auditor;
   (c) checklist;
   (d) council regulation;
   (e) utilitarian tradition.

2 Which elements of environmental auditing are not necessarily effective or do not yield maximum benefit?
   (a) full management commitment;
   (b) audit team objectivity;
   (c) professional competence;
   (d) well-defined and systematic procedures;
   (e) written reports;
   (f) quality assurance;
   (g) follow-up;
   (h) all of the above;
   (i) none of the above.

3 What is the purpose of the EPA's 'Audit Privilege Act'?
   (a) promotion of disclosure of self-audit results;
   (b) penalising environmental violations.

4 In the general ICC audit work flow, the 'follow-up action plan' is:

(a) a pre-audit activity;
(b) an activity at site;
(c) a post-audit activity.

5 The disposition of an audit in a plant can be carried out at the following levels:
(a) policy, management systems, legal, technical;
(b) standards, equipment, legal, technical;
(c) education, communication, legal, technical.

6 Companies registered under the Eco-Audit Scheme will be entitled to use the logo Statements of Participation. This idea comes from:
(a) CFR;
(b) CEC;
(c) BATNEEC.

14

ACCOUNTING FOR ECO-EFFICIENCY

Stefan Schaltegger

SUMMARY

In this chapter the framework of environmental accounting is discussed, focusing on companies and the impact of their activities on the environment. First, *environmentally differentiated traditional accounting*, which is a part of traditional accounting, is discussed as a tool for measuring environmentally induced financial impacts in monetary terms. It is explained how, as the basis of most internal management decisions, it attempts to satisfy the needs of the external stakeholders of companies for financial information. The system of *ecological accounting* is then discussed as a tool for measuring the ecological impact of a company on the environment in terms of non-monetary physical units. This is followed by a discussion of *eco-efficiency*, which measures the economic performance of a company relative to the environmental impacts caused by its activities, together with an explanation of how the systems of traditional accounting and ecological accounting are integrated in eco-efficiency through the *Eco-rational Path Method* (EPM). All essential aspects of the above concepts are explained with the help of diagrams, tables and case studies, along with the concepts of *value added*, *environmental impact added,* etc.

ACADEMIC OBJECTIVES

After completing the chapter the reader should have a good understanding and working knowledge of the following:

- the concepts of value added, environmental impact added, environmentally differentiated accounting and ecological accounting and their respective sub-areas, and many other related topics and aspects;
- how the systems of traditional accounting and ecological accounting are integrated through the Eco-rational Path Method (EPM) to derive measures of eco-efficiency;
- the concept of eco-efficiency, its role and how it can be applied at the corporate level to assess the status of a company in terms of its economic performance relative to the environmental impacts of its activities.

ECO-EFFICIENCY AS AN OPERATIONAL MEASURE

Although the concept of *eco-efficiency* was first introduced and discussed in the literature by Schaltegger and Sturm (1990), it gained popularity only after the publication of the book entitled *Changing Course* by Schmidheiny (1992). Some other aspects of the concept and its limitations have been elaborated by Gladwin *et al.* (1995). In practice the term 'eco-efficiency' is used to convey different meanings with little precision. To begin with, therefore, it is important to clarify the concept of the term as it applies to the corporate level.

In general, efficiency is a measure of the output compared to the input. The higher the output for a given input, or the lower the input for a given output, the more efficient is an activity, product, company, etc. Since the purpose of any economic activity is to manage scarcity in the best possible manner, efficiency is a characteristic of economically rational management.

Efficiency is a dimensionless notion in its most general sense. While the units in which input and output are measured can differ depending on the subject matter, the dimensions of efficiency become concrete when the context is specified. In the business context, for example, if input and output are measured in financial terms, *profitability* becomes the measure of efficiency, commonly referred to as financial efficiency. Some of the measures of economic or financial efficiency are contribution margins, return on sales, return on assets, etc. Economic efficiency indicates whether, and for how long, social activities can be economically sustained. If, on the other hand, efficiency is measured in technological terms, then usually the focus is more on physical units such as kilograms. Technological efficiency is also called *productivity*. The ratio of the 'best possible' efficiency to the 'actually achieved' efficiency is called *X-efficiency* (Leibenstein, 1966). Efficiency is not confined to financial or technological dimensions, and different dimensions can be combined to calculate *cross-efficiency* figures such as the profit generated by each employee.

Using the definition of efficiency given above and appropriate units, the ecological efficiency of an activity can generally be expressed as the ratio of the desired output to what we will call the *environmental impact added* of that activity (Schaltegger and Sturm, 1990). Environmental impact added (EA) is defined as the sum of all the assessed environmental impacts caused by the activity. Thus,

$$\text{Ecological Efficiency} = \text{Desired Output}/\text{Environmental Impact Added} \quad (1)$$

This definition of ecological efficiency does not cover all aspects of sustainable development, however. The same is also true for value added (or added value). For example, the calculation of value added does not take into account an assessment of whether it was achieved by increasing the economic opportunities for the poor, if the products and services were oriented towards satisfying the basic needs of the workforce whose efforts led to the added value, and so on. It is clearly important to correlate environmental impact added with value added because no economic activity is without environmental impacts.

The management of a company is said to be ecologically efficient if the ratio of its products sold, or functions accomplished, to the resulting environmental impact added is high. Two different measures of ecological efficiency have been proposed in this context (Schaltegger and Sturm, [1992] 1994):

- ecological product efficiency;
- ecological function efficiency.

Ecological product efficiency is the ratio of a unit of the product (product unit) to the environmental impact added it would create over its whole life-cycle or a part thereof. That is,

$$\text{Ecological Product Efficiency} = \text{Product Unit}/\text{Environmental Impact Added} \quad (2)$$

Environmental improvements made by companies are often reported in communications on their total product efficiency (e.g. number of cars produced per unit of environmental impact). Product efficiency can be improved with a whole range of pollution prevention methods and strategies, including end-of-pipe devices, reduced input, substitution of resources, etc. Although in principle the improvement of product efficiency is desirable and to be recommended, some products will never be as ecologically efficient as others in meeting a certain demand. In meeting the demand for transportation, for example, a car will always be less ecologically efficient than a bicycle.

Ecological function efficiency takes a broader view compared with ecological product efficiency. It measures the environmental impact that will be caused, or has been caused, per unit of time in serving a specific function. Typically, a function could be defined as the transportation of a person over a certain distance. The alternative, which causes the minimum environmental impact in serving a given function, is said to have the highest ecological function efficiency, defined as

$$\text{Ecological Function Efficiency} = \text{Served Function}/\text{Environmental Impact Added} \quad (3)$$

Ecological function efficiency can be improved by substituting high eco-efficiency products for low eco-efficiency products (e.g. a bicycle instead of a car), by reducing the demand for the function to be served (e.g. use of carpools which lead to reduced demand for the use of private cars), by prolonging the life-span of products (e.g. longer corrosion guarantee for cars), and by improving product efficiency. Environmental interest groups often prefer to measure the environmental record of a product according to its overall ecological function efficiency (for example, the ecological function efficiency of a car in transporting a person over a given distance compared to that of a bicycle or public transport in performing the same function).

Both these measures of ecological efficiency – ecological product efficiency and ecological function efficiency – are useful, but their adequacy depends on their intended purpose. The ratios, given by equations (2) and (3), can be applied to different levels of aggregation, such as to a unit product, a strategic business unit, or to the total sales of a company. In this context it is important to consider total sales and the absolute environmental impact, because a large number of ecologically efficient products can be more harmful than a small number of ecologically inefficient products.

The cross-efficiency between the economic and ecological dimensions, or in other words the 'economic-ecological efficiency', is the ratio of the value added, created by a company, to the resulting environmental impact added (Schaltegger and Sturm, 1990). In most cases it makes sense to define it in terms of a relationship between some kind of economic performance and environmental impact as

$$\text{Economic-Ecological Efficiency (eco-efficiency)} = \text{Value Added} / \text{Environmental Impact Added} \quad (4)$$

Economic-ecological efficiency is often referred to as 'eco-efficiency'. Clearly, the eco-efficiency of a product or activity will increase when its added value increases and/or its overall environmental impact decreases. Note that in equation (4) the measurement of neither the economic aspect nor the ecological aspect is fixed. The ratio of equation (4) should therefore be calculated by using such measures which provide the best information consistent with the aim of the analysis. If economic and ecological concerns are considered to have the same importance, then the numerator and the denominator should be given equal weighting as in equation (4). But, depending on their values, different groups are likely to assign different weightings to economic and ecological performance. Alternatively, as in the appraisal of investment on pollution prevention devices for example, the marginal product of eco-efficiency may be used representing the ratio of environmental improvement made per unit of extra cost (thus measuring cost-effectiveness).

Environmental indicators should always be set in the context of economic performance, because it would then be possible to assess the progress made by companies, products, nations, etc. with regard to their '*eco-efficiency portfolio*', shown in Figure 14.1. Environmental performance is plotted along the vertical axis of this figure, and the economic performance along the horizontal axis. The eco-efficiency portfolio shown in Figure 14.1 has four distinct areas:

- *Green stars* (upper right-hand corner) are sites, products, etc. with low environmental impact added and high economic performance. While low costs are achieved through integrated clean technologies, their environmental impact will already have been optimised during their development. Because they incur relatively low environmental impacts, there are markets where consumers are willing to pay a price premium for such environmentally friendly products. Costs savings as well as higher prices result in higher economic performance.
- '*Dirty cash cows*' (lower right-hand corner) result from strategies of quantitative growth. They are characterised by relatively high profitability and high environmental impact added. 'Dirty cash cows' are in danger of losing their profitability as soon as external environmental costs (economic externalities) are internalised, through the imposition of environmental taxes for example.
- *Green question marks* (upper left-hand corner) are environmentally friendly, but they achieve relatively low economic performance. In the long term they cannot prevail because of their economic weakness.
- '*Dirty dogs*' (lower left-hand corner) have high environmental impact added and a low or negative economic performance. They are economically uninteresting and cause enormous environmental damage. They should be eliminated, or improved both economically and ecologically.

Portfolios which bring environmental and economic data together are a very powerful analytical tool for assessing the relative eco-efficiency status of different production sites, products, companies, etc. They can also serve to benchmark environmentally 'good' competitors in relation to the company in

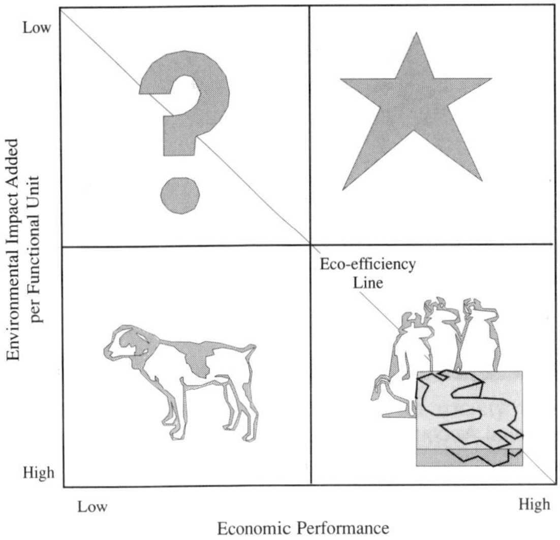

Figure 14.1 The eco-efficiency portfolio
Source: According to Schaltegger and Sturm, 1992

question (Ilinitch and Schaltegger, 1995). Also, once decisions have been taken on the basis of recorded and assessed data, appropriate strategies can be developed for making improvements. All movements above the eco-efficiency line, shown in Figure 14.1, represent improved eco-efficiency.

However, the definition of eco-efficiency is fraught with two major difficulties. First, how can environmentally induced financial impacts be considered and assessed when measuring economic performance? And second, how can environmental impact added be measured accurately?

ENVIRONMENTAL ACCOUNTING

Modern accounting, which began in the fifteenth century in northern Italy, has been the most important corporate system of information gathering and analysis for more than a hundred years. Because of continuing environmental degradation and the problems associated with it, increasingly greater importance is now being given to the gathering and analysis of information on environmental issues and problems. There is thus a growing need to adapt traditional accounting systems to this new situation. In order to assess the eco-efficiency of a company and its operations, the stakeholders of the company would be interested in two main categories of information with regard to environmental issues and problems:

- environmentally induced financial impacts;
- environmental impacts of the company.

Given the different characteristics of these two categories, it is not surprising that many different perceptions and examples of 'environmental accounting' are now to be found (Schaltegger and Stinson, 1994). As a result, there is a correspondingly large, varied and growing body of literature on different aspects of what is generically called 'green accounting' (Adams, 1992; CICA, 1992; Epstein, 1995; Gray *et al.*, 1993; Gray, 1994; Gray and Laughlin, 1991; Price Waterhouse, 1991; Quirke, 1991; Rubenstein, 1994; Schaltegger *et al.*, 1996; Spitzer and Elwood, 1995).

According to the 'Tinbergen rule' though, which applies generally in economics and public policy, an instrument becomes inefficient and ineffective when it is used to pursue different aims that are not absolutely complementary (Tinbergen, 1968). None of the aims can be achieved effectively and efficiently under such circumstances. Therefore, different instruments should be applied to deal with non-complementary issues. With regard to accounting, this is why different accounting systems should be used to deal with different sets of issues. Every accounting system provides specific information for different groups of stakeholders. A stakeholder of a company is by definition an individual, or a group, who has an interest (stake) in the company and can influence, or is influenced by, the activities of the company and its policies. Stakeholders can be managers, employees, shareholders, suppliers and customers of the company, or they can be tax agencies, environmental pressure groups, neighbours, etc. Figure 14.2 shows important stakeholders and their interest in different *categories* and *systems* of accounting. In this figure the three important groups of stakeholders (management, shareholders and regulators) are shown vertically, and the traditional and ecological accounting systems horizontally.

As will be seen from this figure, both systems of accounting deal with environmental issues and are, therefore, part of 'environmental accounting'. The three areas, shaded in dark grey, represent the environmentally differentiated traditional accounting systems. As part of traditional accounting, they measure

technologies and the consequences of various environmental taxes (e.g. CO_2 tax). Environmental issues in other traditional accounting systems include insurance, product liability, mortgage, bank credits, etc.

To summarise, it is important to consider environmentally induced financial aspects in traditional accounting, because it is only by doing so that it becomes possible to make a correct calculation of the profitability and eco-efficiency of a company.

Ecological accounting

As we have pointed out earlier, traditional accounting systems should not only incorporate the financial impacts of environmental issues, but should also be enlarged by including *ecological accounting systems*, or at least the essential features thereof.

The systems of ecological accounting, shown in light grey in Figure 14.2, measure the ecological impact of a company on the environment. Also, as we have already pointed out, measurements to be made for them are in physical units unlike those for environmentally differentiated traditional accounting. Furthermore, like the traditional accounting systems, ecological accounting systems can also be divided into three sub-systems:

- internal ecological accounting;
- external ecological accounting;
- other ecological accounting.

Internal ecological accounting is designed to gather ecological information only for the purposes of internal management. It is complementary to traditional managerial accounting. Methods of measuring the impact of a company's products and processes on the natural environment are necessary for making good management decisions, as they are for monitoring pollution discharges and for calculating eco-efficiency. Account tracking material flows, energy use, and emissions are a core part of internal ecological accounting. Also, a system of internal ecological accounting, preferably a sophisticated one, is a necessary precondition for any environmental management system.

In *external ecological accounting*, which is the counterpart of traditional financial accounting, data are gathered and made available to the external stakeholders interested in environmental matters, including the general public, the media, shareholders, environmental funds and pressure groups. In the last ten years hundreds of companies have published such external environmental reports, thus giving the public at large an opportunity to scrutinise the environmental impacts of their activities. Many of these are extensive annual reports containing detailed information on pollutant discharges. A core question of external ecological accounting is the consolidation of environmental impact added of the different production sites of a company in its group report.

Other ecological accounting systems, which measure data in physical units, are a means for regulators to control compliance. They are necessary for the correct assessment of environmental taxes to be levied on CO_2, volatile organic compounds (VOC), etc. Clearly, such taxes to be paid by a company cannot be assessed without correct information on relevant pollutant discharges and emissions released by that company. Apart from the tax agency and the environmental protection agencies that are primarily interested in specific information on the discharges of specific pollutants, an increasing number and variety of stakeholders, such as banks (acting as money-lending agencies) and insurance companies, are now also asking for reliable information on the environmental impacts of companies. Clearly, ecological accounting is necessary for the calculation of eco-efficiency, because it permits the determination of the environmental impact added of companies, products, etc.

CALCULATION OF ECO-EFFICIENCY

Eco-efficiency may be calculated by internal as well as external stakeholders, but whereas the management can obtain necessary information for this from managerial and internal ecological accounts, external stakeholders have to rely on financial and external ecological accounts, or on other traditional and ecological accounting systems, for such information. For reasons of clarity, the calculation of eco-efficiency is explained in the following sections from an internal managerial perspective only. Further discussion on environmental accounting will be found in Schaltegger *et al.* (1996).

Environmentally corrected managerial accounting information

Managerial accounting is the central tool and basis of most internal management decisions (Cowe, 1988; Horngren and Foster, 1987; Kaplan and Atkinson, 1989; Polimeni *et al.*, 1986). It is defined as 'the identification, measurement, accumulation, analysis, preparation, interpretation, and communication of information that assists executives in fulfilling organisational objectives' (Horngren and Foster, 1987: 2).

Correct financial information, obtained or derived from managerial accounting, is a prerequisite for determining eco-efficiency. Information obtained from environmentally differentiated managerial accounting is often used in financial accounting for communication with external stakeholders. Other environmentally differentiated accounting systems also derive most of their information from managerial accounting. Given this central function of managerial accounting, it must provide correct information to support the most economic way of managing a company. As environmental issues continue to exercise increasingly greater influence on economic performance, it is clear that they must be considered with due rigour in managerial accounting. In order to calculate the profitability of products and the performance of production devices accurately, all environmentally induced costs and revenues, and other costs or revenues, must be tracked, traced and allocated correctly. In the past limited attention has been given to the calculation and allocation of environmentally induced costs. And pollution prevention and clean-up costs have often been considered as overhead costs and allocated to all products – even to the 'clean' products which do not incur such costs. As a result, the 'dirty' products and product groups have been internally cross-subsidised, while the profitability of clean products has been underestimated. Furthermore, management costs relating to machine occupation, personnel and raw materials have often not been considered in calculating the costs of waste generation.

Environmentally induced costs can be increased or reduced through environmental protection efforts. Additional costs can include fines, fees, clean-up costs, increased production of unwanted output, etc. On the other hand, savings may be made through the optimal use of resources, reduced unwanted output, less fines and fees, etc.

Environmentally induced revenues can be direct or indirect. Direct revenues include, for example, gains made from the sale of recyclables, increased sale of products, and higher prices of the products sold. Indirect effects are more intangible by comparison and can include, for example, enhanced corporate image, greater customer and employee satisfaction, transfer of know-how, and the development of new markets for environmentally benign products.

The main issues of managerial accounting are concerned with the tracking, tracing and allocation of environmental costs, as well as with investment appraisal. These issues are relevant to environmentally differentiated accounting, as they are to ecological accounting. Once the environmentally related financial impacts have been considered, managerial accounting enables a correct calculation of the financial performance indicators such as the contribution margin of products, return on net assets for different production sites, net present value of investment projects, etc.

Internal ecological accounting

Ecological accounting is that sub-area of environmental accounting which deals with the activities, methods and systems for recording, analysing and reporting the impacts of a defined economic system on the natural environment. It attempts to answer this question: how can environmental impact added be accurately measured?

Conceptually, the approach of internal ecological accounting corresponds to that of traditional managerial accounting. To this extent internal ecological accounting serves as:

- an analytical tool for detecting ecological weaknesses and strengths;
- a method which supports decisions on relative environmental quality and provides a basis for environmental measures and eco-efficiency;
- a tool to control environmental effects caused directly or indirectly;
- a neutral and transparent vehicle for internal communication (and also external in an indirect way).

Internal ecological accounting represents what is called a *satellite accounting system* whose purpose is to calculate the impacts of businesses on the environment. Measurements to be made for ecological accounting are expressed in physical terms, unlike in environmentally differentiated traditional accounting. The physical units to be used for this can be of a quantitative or qualitative nature, although preference is usually for the quantitative if at all possible. All environmental impacts originate from activities that are deemed to create value or utility. But these economic activities also cause *environmental intervention* such as emissions, use of resource, etc. which can be recorded in an inventory (e.g. an emissions inventory). However, given that almost all economic activities have impacts of varying intensity on nature, and that they inevitably cause complex and many hitherto unknown interactions, it is not possible to record all environmental interventions in an inventory.

Ecological accounting activities, which usually start with the investigation of individual stages of production, can be broadened to the analysis of products, sites and businesses. The *subject matter* of ecological accounting comprises the resources used, environmental interventions, and ecological assets. The resources used, to be considered in this context, include non-renewable material resources and energy carriers (e.g. ores, oil), as well as renewable material resources and energy carriers (e.g. wood, water).

Environmental interventions include material emissions and pollution in terms of substances, or compounds of substances such as CO_2, NO_X, VOC, as well as energy emissions in the form of heat, radiation, noise, etc. Ecological assets (meaning ecological capital goods) include land, forests, biodiversity, etc.

Internal ecological accounting deals with the recording of environmental interventions, and allocation and assessment of environmental impacts. In the context of *recording, tracking and tracing* of environmental interventions, the purpose of recording is to prepare an inventory in which all the environmental interventions are registered in physical terms. *Allocation* refers to the allocation of environmental impacts to products, processes and activities. It is actually a procedure for assigning environmental impacts to the individual stages of production, sites and the products which cause them. Impact *assessment* is concerned with the determination of the relative severity of environmental interventions, and with the calculation of the environmental impact added. Impact assessment is a technical, quantitative, and/or qualitative process of classifying, characterising and assessing the effects of both resources consumed and pollutants released into the environment.

Recording

Quantified data are favoured in ecological accounting to measure the extent of environmental interventions (in terms of resources used, and material and energy emissions) and ecological assets (e.g. land, forests, water reserves, etc.). All internal and external accounting reports should contain quantitative data, together with necessary comments to facilitate their correct interpretation by the reader.

Ecological assets (e.g. forests, land) can be registered and described in what is called an 'eco-asset sheet'. Natural assets are seldom accounted for at the corporate level since only a few of the manufacturing companies own sizeable land assets or other natural resources. The *eco-asset sheet* is thus not a true counterpart of the financial balance sheet of a company. Rather, to put it simply, it is a 'photograph' of a company's natural assets on a given date, listing all the ecological assets the company owns or occupies on that date. The eco-asset sheet contains all the ecological assets of the company, including an inventory of the various species of flora and fauna inhabiting its land, forests, etc. Recently some companies (e.g. Dow Chemicals) have started to assign explicit monetary values to their ecological assets by, for example, establishing harvesting rights in rain forests. Though still 'raw', the eco-asset sheet is nevertheless a valuable tool which enables management to recognise both potential and manifest financial implications of their ecological assets. In the past many companies have underestimated the importance of the financial liability of their ecological assets.

The basic idea of the *environmental impact added statement* corresponds to the income statement in traditional accounting. Flows of material and energy inputs, as well as the resulting outputs into the natural environment, are recorded in these statements and assessed for each accounting period.

Environmental interventions can be accounted for only if they can be recorded, tracked and traced. The use of accounts to record and present financial flows is widely accepted. Therefore, given the need to find practical ways of dealing with data, and because efficient procedures for recording material and energy flows have already been established in traditional accounting systems, it would be sensible to use a similar procedure for ecological accounting, too. Clearly, data on environmental interventions will not be recorded in monetary terms in such a system (which is adapted from traditional accounting), but in physical terms in an 'input–output' account. In such a system of ecological accounting, first the material and energy flow diagrams must be prepared for the manufacturing processes. Information on material and energy flows needed for this can only be obtained from detailed observations of actual production. Table 14.1 shows a typical example of an 'input–output' account.

In the left-hand column of Table 14.1 all the material and energy inputs into production (natural resources, semi-manufactured goods, raw materials, etc.) are listed with an identification number. All desirable and undesirable material and energy outputs of production (desired output, emissions, waste, etc.) are shown on the right also with an identification number. The material inputs are divided into mineral resources, biomass, water, energy carriers and other materials. Energy carriers (e.g. oil) are registered as material resources.

Table 14.1 Typical example of an input–output account

Group 10 Material inputs	*Group 20 Material outputs*
100 Mineral resources	200 Products (environmental impact added carrier)
101 Biomass	201 Re- and downcyclats
102 Water	203 Emissions
103 ...	2050 Landfill
104 Fossil energy carriers	2051 Water emissions
1040 Crude Oil	20510 TOC
1041 Coal	20511 Sulphur
1042 Gas	20512 Water
105 Regenerative energy carriers	20513 ...
106 Materials	2052 Air emissions
1060 Polystyrol	20520 CO_2
1065 Re- and downcyclats	20521 NO_x
1066 ...	20522 VOC
	20523 ...

Source: Schaltegger and Sturm, [1992] 1994

The recording of energies should be separated from the registration of material flows in order to avoid double counting and confusion. According to the law of conservation of mass and energy, total mass and energy on the input side of the account must be the same as that on the output side. In other words, the left and the right sides of the ecological account must be in balance since neither mass nor energy can be destroyed. When there are too many items or activities to be accounted for, separate accounts can be opened for one or more of the items listed in the 'aggregate' account. That is, there could be a separate account for the oil used, another for CO_2 emissions, and so on.

Effluents can be transferred from one account to a subsequent account, or to the aggregate account, as in basic book-keeping, except that the transactions will be entered in kilograms. Following this procedure, the incoming and outgoing mass flows can be aggregated and defined according to their origin. This is analogous to a profit-and-loss account in traditional accounting.

Information on the quality of data is extremely important for the interpretation of results as well as for the search for better alternatives. Quality of data should be measured, calculated and estimated, and, if necessary, secondary data from the literature (e.g. an industry average) or from suppliers should be used. And data should be classified according to quality. Measured data usually reflect the specific situation in the best possible way. Nevertheless, it is helpful to scrutinise the methods used for data generation and collection in order to be able to assess the reliability and quality of data, and to determine how they compare with other data.

It makes no economic sense to aim for a full inventory of all mass and energy flows and, in any case, this goal can rarely be achieved. Methods of data collection usually take several years to develop, improving gradually until the marginal benefit of more detailed information matches the marginal costs of collection. Efficient collection and management of data certainly needs a computerised system, especially considering that the collection of high-quality data in sufficient quantity and detail is a burdensome and demanding task which has to be performed patiently and with great care.

Allocation

In managerial accounting, the purpose of *activity based accounting* is to calculate the total induced costs of an activity and of a cost carrier (product, product group, etc.). Activity based accounting is undertaken in order to determine exactly where a company generates its revenue from and where it loses money. There is a similar feature in internal ecological accounting, too. 'Ecological activity based cost accounting' aims to show where environmental interventions are caused and which products, product groups, etc. contribute most to the overall environmental impact added.

There is, however, a major difference between managerial cost accounting and ecological accounting. It is that, in the latter, environmental interventions are allocated before they are assessed. Usually they are assessed in a second stage, according to their relative environmental impact, in order to calculate the environmental impact added. In traditional cost accounting, on the other hand, no assessment is necessary because the monetary values (prices and costs) already constitute an assessment. Also, in analogy to managerial cost accounting, internal ecological accounting distinguishes between what are called environmental impact added centres, carriers, drivers, allocation rules and allocation keys, as described below.

Environmental impact added centres describe 'places' where material and energy are processed, or where the respective flows enter the natural environment. Examples of such centres are stages of production, production sites, incinerators, sewage plants, etc. 'Environmental impact added centre accounting' helps to identify the places where 'overhead environmental interventions' of jointly used clean-up facilities are actually created, or where they actually occur.

Environmental impact added carriers are analogous to cost carriers in traditional cost accounting. They describe a product, a product group, division, etc. that is seen to be responsible for creating value added as well as environmental impact added. Ideally, such carriers should correspond to the cost carriers, because it is only then that eco-efficiency can be calculated. 'Environmental impact added carrier accounting' identifies the products, processes or activities which cause the environmental interventions allocated to the environmental impact added centres. An analogue of this, to be found in managerial accounting, is the allocation of costs to the cost centres and to the cost carriers.

Environmental impact added drivers initiate environmental impacts. CO_2 emissions which cause the greenhouse effect, and VOC emissions which cause photochemical smog, are typical examples of these drivers.

Environmental impact added allocation rules formulate general procedures for ensuring the correct allocation of environmental interventions, while *environmental impact added allocation keys* describe the relationship between an environmental impact added carrier and the environmental interventions caused. If environmental interventions are caused jointly by the elements (e.g. scrubbers, sewage plants, etc.) of a multi-purpose clean-up facility, for example, then the individual intervention of each element must be traced and allocated to the carrier responsible (a typical example is given in Box 14.1). But, in order to do so as accurately as possible, representative environmental impact added drivers and allocation rules must be defined, and this calls for basically the same allocation procedure as in cost accounting. The importance of the chosen allocation rule becomes especially clear when its influence on the contribution margins of products, as well as on directing capital investments, is considered.

Impact assessment

Sometimes a complete inventory provides enough information in physical units to reveal what the main environmental problems are and where they are created. In such a case priorities for environmental protection, pollution prevention, etc. can be defined on the basis of the inventory. However, in most cases the inventory contains an enormous amount of unassessed and detailed information which cannot be accurately interpreted by management. An impact assessment of the inventory data is then necessary. Impact assessment is a technical, quantitative, and/or qualitative process of classifying, characterising and assessing the effects of the resources required and the recorded environmental loading. The ecological assessment of environmental interventions, which

BOX 14.1 A TYPICAL EXAMPLE OF ALLOCATION

With reference to the schematic representation of allocation given below, consider this problem: the waste waters of environmental impact added centre A (1 kg TOC/litre of water) and environmental impact added centre B (2 kg TOC/litre of water) are treated in a sewage plant. The water released into the river contains 0.090 kg of TOC/litre of water. How should the TOC discharged into the river be allocated to A and B, and to the environmental impact added carriers?

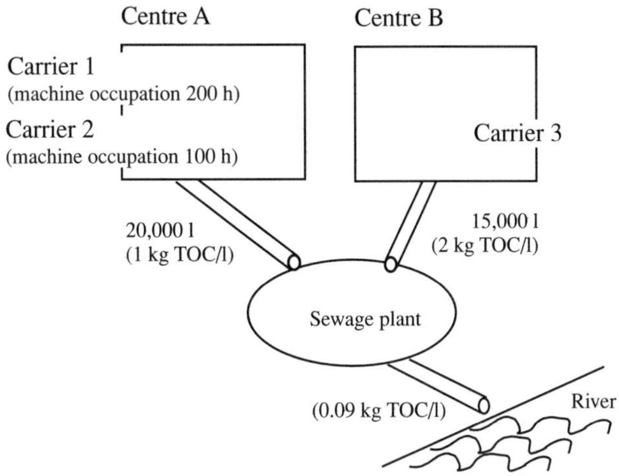

TOC released by environmental impact added centre A = 20,000 l × 1 kg/l = 20,000 kg
TOC released by environmental impact added centre B = 15,000 l × 2 kg/l = 30,000 kg
TOC discharged into the river = 35,000 l × 0.09 kg/l = 3,150 kg

Allocating with respect to the water load into the sewage plant,
TOC discharged by A into river = 3,150 kg × 20,000/35,000 = 1,800 kg
TOC discharged by B into river = 3,150 kg × 15,000/35,000 = 1,350 kg

Then, allocating on the basis of machine occupation time,
Allocation to environmental impact added carrier 1 = 1,800 kg TOC × 200/300 = 1,200 kg TOC
Allocation to environmental impact added carrier 2 = 1,800 kg TOC × 100/300 = 600 kg TOC
Allocation to environmental impact added carrier 3 = 1,350 kg TOC (the same as that for environmental impact added centre B)

often calls for the reduction of a multitude of physical measures to a few units (or even just one unit or index) of measurement, should be undertaken only after the aggregation of an environmental impact added centre (or an environmental impact added carrier) account has been made. The advantage of this approach is that different methods of assessment can be based on the same inventory data and compared with one another. For this reason ecological accounting is not restricted to the current level of knowledge about the harm caused by environmental interventions, but it also allows for the application of new weightings at any time in the future.

In the literature there are many different approaches to *impact assessment*, as well as a host of variants that are adopted in practice.

Mathematically, environmental impact added is calculated by first multiplying the physical quantity of each emission released (e.g. CO_2, CH_4, etc.) by an appropriate weighting assigned to it (e.g. 1 to CO_2, 11 to CH_4). The results of all such multiplications, for all the relevant emissions, are then

BOX 14.2 IMPLEMENTATION OF ECOLOGICAL ACCOUNTING AT CIBA'S PIGMENTS DIVISION

Ciba's Pigments Division implemented the concept of ecological accounting in 1994. An interdisciplinary working group of six to eight staff members was established for the purpose. Its task was to introduce an environmental information system to meet the specific needs of the Division. The goal of the working group was to assess the relative positions of Ciba's different pigments and colours in the eco-efficiency portfolio as a basis for decision-making. After an intensive initial phase, the working group held monthly progress meetings.

The first step for the group was to identify the existing sources of information (production information system, information files prepared for the regulating agencies, etc.). One staff member was given the task of evaluating software with the object of improving the management of information. In order to establish a systematic and continuing environmental information system, the concept of ecological accounting was implemented for the key products of the division. The basic system, established after about one year, has since been continually upgraded. At present one part-time environmental manager is responsible for information collection, allocation and impact assessment. The results are discussed in annual management meetings, and used in making decisions on both environmental management and strategic management.

added to give the environmental impact added. The weighting of a given emission reflects its relative contribution to the overall environmental impact of the problem in question. When necessary, such individual contributions can be expressed in CO_2 equivalents, as in the case of greenhouse gases for example. Detailed discussion on this topic will be found in the literature on Environmental Impact Assessment (EIA) and in Schaltegger and Sturm ([1992] 1994).

The Eco-rational Path Method (EPM): integration of traditional accounting with ecological accounting

As will be seen from equation (4), eco-efficiency is the ratio of value added to environmental impact added, or the ratio of an economic performance indicator (e.g. contribution margin) to an ecological performance indicator. The calculation of eco-efficiency, therefore, requires the integration of economic information (flow of funds such as income, expense, revenues, costs, etc.), derived from traditional accounting, with environmental information (environmental interventions such as emissions, resource use, etc.) derived from ecological accounting. Figure 14.3 shows the process by which economic and environmental information can be integrated, and how the results of the process can be used in decision-making.

The Eco-rational Path Method (EPM), shown schematically in Figure 14.3, has been developed to meet this requirement for integration. It describes a structured decision process to pilot the businesses of a company to eco-efficiency. EPM is a straightforward and practical procedure for integrating the economic dimension with the ecological dimension, shown respectively on the left and right of Figure 14.3. This process of integration distinguishes the three aspects; namely, *accounting*, *judgement* and *decision*.

In module I (Figure 14.3) the monetary results of managerial accounting, including environmental

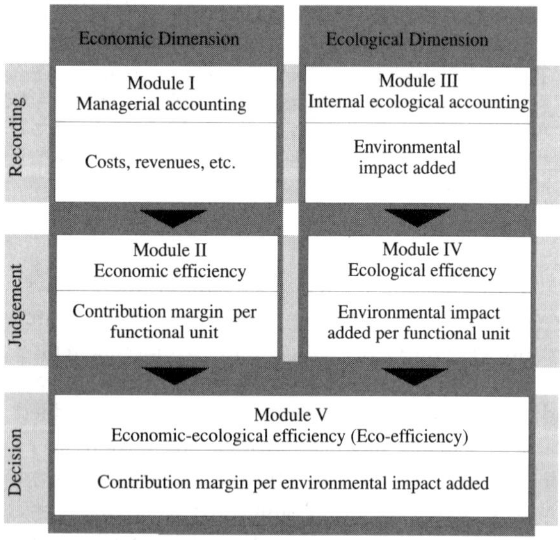

Figure 14.3 The Eco-rational Path Method
Source: Schaltegger and Sturm, 1994

compliance costs and earnings, are evaluated. Module III represents the evaluation of the ecological harm caused in units of environmental impact added, calculated from ecological accounting. An economic performance indicator is calculated in Module II as a measure of economic efficiency, such as Net Present Value (NPV), Return On Net Assets (RONA), or the Contribution Margin (CM). The key numbers needed for this will be found in any well-functioning managerial accounting system. For the allocation of environmentally induced costs, the environmental impact added caused by each cost centre must be known. Module IV calculates ecological efficiency in terms of environmental impact added per product, product group or some other functional unit. Modules II and IV are both steps in which data needed for making separate judgements on ecological or economic efficiency are supplied. All these four steps are integrated in Module V by combining the measures of economic and ecological efficiency, and by calculating the quotient of an economic performance indicator per environmental impact added (e.g. the monetary contribution margin created per environmental impact added of a product).

The 'factual' goal of integrating economic indicators with ecological performance indicators is to measure economic-ecological efficiency (eco-efficiency). Integration of environmental issues into the decision-making process in this way enables both the internal and external stakeholders to make decisions that would move matters towards the 'green cash cow' in Figure 14.1, so to speak. The 'political' purpose of integrating and weighting economic and ecological effects is to consider and meet the requests of different stakeholders. Depending on the need for information and the goals of stakeholders, economic performance or growth can be measured in terms of sales, revenues, contribution margin, net present value (NPV), etc. There is also a need for *project-related* financial and environmental information to support rational eco-integrated investment decisions.

CONCLUSIONS

Eco-efficiency is defined as the ratio of value added to environmental impact added, or, more generally, economic performance achieved per environmental impact caused. To calculate eco-efficiency, environmental impact added, which is the sum of all environmental interventions caused and assessed according to their relative environmental impact, must be known. It is then correlated with value added, since no economic activity is without environmental impacts. Environmental accounting provides the information needed for determining both environmental impact added and eco-efficiency, while the Eco-rational Path Method (EPM) enables the integration of economic information with environmental information. Improvements achieved in eco-efficiency can be judged and compared visually from the eco-efficiency portfolios of different products, production sites, etc. Finally, the goal of eco-efficiency is to achieve high economic performance with low environmental impact added.

REFERENCES

Adams, J. (1992) *Accounting for Environmental Costs: A Discussion of the Issues Facing Today's Businesses*, Norwalk: FASB.

CICA (Canadian Institute of Chartered Accountants) (1992) *Environmental Accounting and the Role of the Accounting Profession*, Toronto: CICA.

Cowe, R. (ed.) (1988) *Handbook of Management Accounting* (2nd edition), Aldershot: Gower.

Epstein, M. (1995) *Measuring Corporate Environmental Performance*, Chicago: Irwin.

Gladwin, T., Krause, T. and Kennelly, J. (1995) 'Beyond Eco-Efficiency: Towards Socially Sustainable Business', *Sustainable Development* 3: 35–43.

Gray, R. (1994) 'Environmental Accounting and Auditing: Survey of Current Activities and Developments', *Accounting and Business Research* 24 (95): 285–286.

Gray, R. and Laughlin, R. (eds) (1991) 'Green Accounting, Special Issue of Accounting', *Auditing and Accountability Journal* (AAAJ) 4: 3.

Gray, R., Bebbington, J. and Walters, D. (1993) *Accounting for the Environment*, London: Chapman Publishing.

Horngren, C. and Foster, G. (1987) *Cost Accounting. A Managerial Emphasis* (6th edition), Englewood Cliffs, Calif.: Prentice-Hall.

Ilinitch, A. and Schaltegger, S. (1995) 'Developing a Green Business Portfolio', *Long Range Planning*, No. 3.

Kaplan, R. and Atkinson, A. (1989) *Advanced Management Accounting*, Englewood Cliffs, Calif.: Prentice-Hall.

Leibenstein, H. (1966) 'Allocative Efficiency versus X-Efficiency', *American Economic Review*, 56: 392–415.

Polimeni, R., Fabozzi, F. and Adelberg, A. (1986) *Cost Accounting, Concepts and Applications for Managerial Decision Making*, New York: McGraw-Hill.

Price Waterhouse (1991) *Environmental Accounting: The Issues, the Developing Solutions*, New York: Price Waterhouse.

Quirke, B. (1991) 'Accounting for the Environment', *European Environment* 1 (5): 19–22.

Rubenstein, D. (1994) *Environmental Accounting for the Sustainable Corporation*, Westport, Conn.: Quorum.

Schaltegger, S. and Stinson, C. (1994) 'Issues and Research Opportunities in Environmental Accounting', Discussion Paper No. 9124, Basel: WWZ.

Schaltegger, S. and Sturm, A. (1990) 'Eco-Rationality' (in German: Ökologische Rationalität), *Die Unternehmung* 4: 273–290.

Schaltegger, S. and Sturm, A. ([1992] 1994) *Environmentally Oriented Decisions in Firms. Ecological Accounting Instead of LCA* (in German: Ökologieorientierte Entscheidungen in Unternehmen. Ökologisches Rechnungswesen statt Ökobilanzierung), Bern: Haupt (2nd edition).

Schaltegger, S. and Sturm, A. ([1995] 1996) *Eco-Efficiency through Eco-Controlling. For the Implementation of EMAS and ISO 14000*, Zürich: Vdf (German version: Zürich/Stuttgart: Vdf/Schäffer–Poeschel).

Schaltegger, S., Müller, K. and Hindrichsen, H. (1996) *Corporate Environmental Accounting*, Chichester: John Wiley & Sons.

Schmidheiny, S. (1992) *Changing Course*, Frankfurt: Artemis & Winkler.

Spitzer, M. and Elwood, H. (1995) *An Introduction to Environmental Accounting as a Business Management Tool: Key Concepts and Terms*, Washington, DC: USEPA.

Tinbergen, J. (1968) *Economic Policy*, Freiburg: Rombach.

Vedsø, L. (ed.) (1993) *Green Management*, Copenhagen: Systime.

SUGGESTED READING

Schaltegger, S., Müller, K. and Hindrichsen, H. (1996) *Corporate Environmental Accounting*, London: John Wiley & Sons.

SELF-ASSESSMENT QUESTIONS

1 What is meant by 'eco-efficiency'? (Choose one or more.)
   (a) Environmental impact caused in producing one unit of a product.
   (b) The fastest way to achieve sustainable development.
   (c) The economically cheapest approach to environmental protection.
   (d) Value added created per unit of environmental impact added.
   (e) Economic performance in relation to environmental impact added.

2 What is 'corporate environmental accounting'? (Choose one or more.)
   (a) A system to account for the environmental impacts of floods, earthquakes, pollution, deforestation, etc.
   (b) The basis of environmental auditing, because it provides all the information needed to complete the check-list of an environmental audit report.
   (c) A sub-area of accounting which deals with the activities, methods and systems for the recording, analysis and reporting of the environmentally induced financial impacts as well as the ecological impacts of a defined economic system (e.g. a company, plant, region, nation, etc.)
   (d) An attempt to account for all the aspects of sustainable development including environmental impacts, social development, and economic performance.

3 What are the main issues in ecological accounting? (Choose one or more.)
   (a) To count the number of species in a geographical area and to draw a map showing their distribution.
   (b) To calculate the monetary value of environmental degradation.
   (c) To record, allocate and assess the environmental impacts of material and energy flows.

4 What is the idea of the 'Eco-rational Path Method (EPM)'? (Choose one or more.)
   (a) The EPM is a strategic concept for more rational management of resource use.
   (b) The EPM serves to integrate traditional accounting with ecological accounting in order to measure eco-efficiency.
   (c) The EPM evaluates the financial impacts of environmental protection measures, thus supporting economically rational management.

5 Environmentally differentiated accounting (choose one or more)
   (a) is not relevant for corporate environmental protection because it deals only with the financial aspects;
   (b) is very important for corporate environmental protection because it improves corporate economic efficiency;
   (c) is increasingly important for the economic performance of companies.

6 Calculation of eco-efficiency. Answer the following questions using the data given in the table below.

Environmental impact added carrier	*Contribution margin (£) per unit*	*Environmental impact added of sales*	*Units (g) sold*	*Number of grams (g) needed to colour 1 m²*
A	5	7,500	2,500	2.5
B	2	800	200	0.5
C	3	9,000	3,000	3.0

(a) What is the ecological product efficiency of the environmental impact added carriers A, B and C?
(b) What is the ecological function efficiency of the environmental impact added carriers A, B and C?
(c) What is the eco-efficiency of the environmental impact added carriers A, B and C (functional efficiency, product efficiency and eco-efficiency of total sales)?
(d) How can eco-efficiency be calculated if the environmental impact added carrier does not correspond with the cost carrier?

15
ENVIRONMENTAL ACCOUNTING

Edeltraud Günther

SUMMARY

This chapter is concerned with environmental accounting in the particular context of companies and business organisations. It is explained how, faced with increasing degradation of the natural environment and mounting public concern about environmental problems and issues, companies are now having to take account of environmental aspects in their corporate decision-making and strategic planning. Given that it is the natural environment which provides all the material and energy resources for industrial production, it is argued that it is in the long-term interest of companies to integrate all relevant aspects of sound environmental management into their general management regime. To this end a number of monetary and non-monetary management instruments have been proposed, and it is explained how industry at large can benefit much in the long term by adopting eco-oriented management systems that are both sympathetic to the needs of environmental protection and compatible with the demands and aspirations of the stakeholders.

ACADEMIC OBJECTIVES

On completing this chapter the reader should be able to understand and acquire working knowledge of the following:

- how the various stakeholders can influence the decision-making process of companies;
- the eco-orientation of the primary and secondary activities within what is called the 'value circle';
- the status of environmental management as a management goal;
- the effect of eco-orientation on the process of investment appraisal;
- eco-oriented cost accounting;
- the use of non-monetary instruments, especially eco-balance for information gathering and decision-making;
- the role of environmental management accounting within the EMAS system.

THE NEED FOR ENVIRONMENTAL MANAGEMENT ACCOUNTING

The need for integrating ecological aspects into the general management of companies and business organisations arises from the increasing danger and damage to the environment being caused by human activities, as well as from the rapidly growing environmental awareness of all social groups. The aim of eco-oriented control – in the form of 'environmental management accounting' – is to safeguard the vital interests of companies and organisations against potential or manifest threats arising from the continuing degradation of the natural environment on which they depend for material and energy resources. The need to develop environmental management accounting is also an acknowledgement of the essential unity of the entire natural system in which the single components are connected to, and interact with, each other and are vitally important to the long-term future of companies and organisations. To this extent environmental management accounting, or simply 'environmental accounting', can be defined as an internal set of management instruments which enable companies to make their operational and strategic decisions through the gathering of external, eco-oriented information, as well as through the quantitative and/or qualitative analysis of that information.

THE STAKEHOLDER APPROACH

The stakeholders, defined as stockholders, suppliers, creditors, the public and others, provide the impetus for the eco-orientation of companies and businesses. Since companies and their economic activities impact on both the natural environment and also the wider societal environment, it is obviously necessary for any eco-oriented system of environmental accounting to take into consideration all these aspects and their implications. Activities with environmental impact involve stakeholders such as stockholders, suppliers, customers, as well as competitors in their role as specific actors in an industrial or business setting. They, as well as the government and the public, can be counted on to participate in the wider environmental aspects acting as the so-called 'regulatory' groups. The societal environment, on the other hand, embodies the general conditions under which companies have to operate. These conditions can be technological, economic, social, ecological or legal in nature, as shown schematically in Figure 15.1. In conventional systems of control these aspects are catered for through the consideration of the traditional flows of goods and services.

Clearly, if there are any changes in the organisational or operational structure of a company, or its activities, with implications for the environment, they should be considered and incorporated into the design and application of the relevant instruments of environmental accounting. It is necessary, therefore, to examine and monitor changes in the corporate structure of companies, as well as changes in their activities, that may have a bearing on the environment.

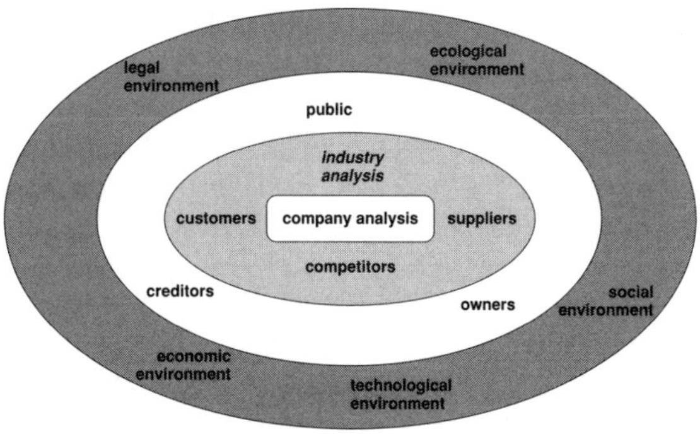

Figure 15.1 The 'societal' and 'task' environments

The reasons for integrating ecological aspects into environmental accounting do not directly arise from any ecological deficiency as perceived by the company itself. Rather, the impetus for integration comes from the stakeholders of the company, who also determine the degree of integration necessary (Freeman, 1984). In this respect they can influence management decisions through one or more of the following:

- legal obligations;
- economic considerations;
- moral and ethical arguments.

Legal obligations: The need for governmental control through legislation arises from the failure of the market to manage resources that are ecologically scarce, but economically available, in an environmentally sound manner. Relevant agencies of the government can influence the decisions of companies and businesses through the enforcement of environmental laws and acts, as well as by granting financial aid to them.

Economic considerations: Ecological demands can be articulated either by the customers, suppliers and the competitors, or by the stockholders, creditors and the public at large. For a company, these demands can have an effect at three different levels: on the potential for profit, on the profit itself, and on liquidity.

Moral and ethical arguments: A third reason for integrating ecological aspects into management decisions can be, at least to a certain degree, normative considerations for limiting the exploitation of nature by humans.

The societal environment

Technological conditions

Part of the societal environment comprises both technical developments, which can be important to the future of the company because of the new possibilities they offer, and constraints on future plans. In this regard technological development is characterised by three distinct elements:

- the state of science;
- the state of technology;
- the generally acknowledged rules of technology (meaning that the majority of the operators of certain systems already use a specific process).

Economic conditions

Economic conditions, as measured by macroeconomic indicators such as the economic growth rate, comprise the basic information which companies and businesses need for planning and assessing their strategic development. They can adopt the following devices to improve both their competitiveness and strategic advantage in the marketplace (Porter, 1986):

- reducing prices of standardised products in the low-price sector of the market;
- charging higher prices when they can be realised through the specific advantages of a product in terms of its quality, service or image;
- the strategy of niche-marketing when there is demand in the market for a specific good or product and when it is profitable to do so.

Analysis of specific conditions in industrial companies shows that there is a correlation between the ecological demands made on them and their specific production parameters. In other words, the production characteristics of an industry determine the amount of environmental damage it is likely to cause. The greater volume of waste water produced by a 'dirty' paper-making factory, compared to a smaller volume produced by a 'clean' factory, is a typical example of this. Thus, the degree to which an industry will be exposed to public demands for cleaning up its act will be determined by its pollution-abating and environmental credentials. Clearly, some industries would be more exposed to public criticism than others.

Social conditions

Often public demands are also made on companies arising from social conditions that are not directly related to their primary activity of production (Dyllick, 1989). For this and other reasons, in general it is essential for companies to be able to deal with new stakeholders in a satisfactory manner, through public discussions for example. This may necessitate an analysis of public attitude and opinion in the wider context of public relations. Results of such an analysis make it possible to anticipate any potential demands the stakeholders may make for cleaner production. Also, when such analyses are carried out over time, they reveal shifts in people's opinions and attitudes. Such information can be of great value to management for developing strategic plans for the future.

Ecological conditions

Given that the natural environment is the source of all materials and energy needed for production, it goes without saying that all relevant environmental aspects, such as resources consumed and pollution caused to soil, water and air, must be included in the environmental accounts of companies. Furthermore, it is essential for companies to be able to respond quickly to changes in the environment, and to develop products and services which will prove to be environmentally compatible in the future.

Legal conditions

The need for legislation arises primarily from the fact that the free market has largely proved incapable of managing resources that are scarce ecologically, but available at a price reflecting their demand. The banning of trade in ivory is a typical example of this. In such a situation governments may adopt the 'punish and reward' approach whereby considerate use of resources, or compliance with law, is rewarded with incentives, while non-compliance is punished with fines, levies or criminal sanctions. In this way, as well as through the granting of financial aid, governments and their agencies enforce environmental laws and acts. In many cases the focus of environmental legislation is on the 'precautionary principle', the 'polluter pays principle', or the 'principle of co-operation'. The following regulatory instruments are used, as necessary, for implementing and enforcing these principles (Wicke, 1993):

- environmental reporting;
- moral persuasion;
- financial aids and incentives;
- taxes;
- environmental certificates and environmental burdens.

Influence of the societal environment on eco-orientation

Since eco-orientation is influenced by the societal environment, all aspects of such influence should be meticulously examined for their potential effects on the eco-orientation of a company or organisation. While the prevailing technological conditions can prevent or promote the integration of ecological aspects, economic conditions such as national growth rate, or the competitiveness of an industry, can create or limit the possibility for the implementation of eco-oriented measures. Moreover, apart from the social and ecological aspects, environmental legislation and its various modifications in the past have had a tremendous effect on the eco-orientation of companies. Therefore, these aspects must also be taken into consideration in formulating environmental accounting systems for companies and organisations.

The 'task environment'

The stakeholders

While the societal environment comprises the general conditions for economic activity, what is called the 'task environment' includes stakeholders who make concrete demands on companies. Freeman (1984: 84) defines a stakeholder 'as any group or individual who can affect, or is affected by, the achievement of the firm's objectives'. Thus the task environment of a company includes all the stakeholders (customers, suppliers, competitors and stockholders) who have a direct influence on its decision-making process. The public and the government also have an influence on the task environment, which can be more or less strong by comparison depending on circumstances.

The consumers

The products, or range of products, which a company makes are in response to a certain consumer demand. However, with growing public awareness of environmental issues and problems, consumers are now insisting more and more on the environmental credentials of both companies and their products by asking for information on whether the products were made in an environmentally friendly way, whether the raw materials were grown or procured without damaging the environment, and so on. But, while business must respond to such demands and follow this new and changing orientation of the market, a 'bottleneck' in the supply–demand equation can occur, in terms of both quality and quantity, if attempts are made to re-orient production too hastily in order to satisfy the increasingly discriminating customer base. In practice, eco-orientation is often most difficult to realise in the retail sector, at least in so far as strategic business planning is concerned, mainly because public endorsement of the eco-orientation of a company or its products cannot be guaranteed. Studies show that when consumers are highly conscious economically, it is not certain if, or the degree to which, their environmental awareness could influence their actual consumption pattern or behaviour. It is clear, none the less, that as these new environmentally conscious consumers begin to dominate the market in time, the public will become even more aware of the environment, and this is certain to have a profound impact on their purchasing behaviour. The environmental 'quality' of a product will then become a prerequisite for its marketability.

The suppliers

The suppliers can be persuaded to supply goods and products, that are more environmentally compatible than hitherto, through the implementation of an eco-oriented demand strategy. In order to guarantee the environmental friendliness of a product, both raw materials and other supplies needed for its production will have to be tested for their environmental friendliness. Secondary raw materials or supplies, if used, will also have to be checked in order to save costs and energy.

The competitors

Since a company must promote and maintain its competitiveness and strategic advantage in the market, a comprehensive analysis of the strategies adopted, or being adopted, by both its old and new competitors is obviously very important to it. Such an analysis usually comprises the following elements:

- identification of potential competitors, both existing and emerging;
- assessment of their strengths and weaknesses;
- analysis of their present and perhaps future strategies for eco-orientation;
- preparation of a reaction profile of the competitors.

An analysis of this type focuses on factors such as price, quality, image and product range. With regard to eco-orientation in this context, the central question is whether the competitiveness of a given company will be improved or diminished when ecological aspects are integrated into its strategic development plan. For most industries there is no clear-cut answer to this question at present, in spite of growing public awareness of environmental issues and problems.

The creditors

As for the creditors, their demands arise from risks which implicitly pervade every economic activity of a company and, naturally, both the creditors and the credit institutions do all they can to minimise their risk. They may, and usually do, scrutinise whether or not the company they do business with is creditworthy, and increasingly this means an examination of the effect of the company's eco-orientation on its profitability. One way of managing such risks is through a questionnaire in which each borrower is required to give detailed information on any current or future environmental damage to which it has financial liability.

The stockholders

There is a growing realisation among stockholders that eco-orientation is essential for companies. This is reflected by the increasing rate of financial investment now being made, especially by the socially responsible and progressive companies, for maximising their efficient use of natural resources. Moreover, companies are now being required to provide more information, and in greater detail, on their activities to potential stockholders. Companies adopting such eco-friendly measures prominently display them in their promotional literature, and thereby they are able to recruit new stockholders without much difficulty. By contrast, companies which do not take eco-orientation seriously are at risk of being shunned by new or even existing stockholders. Financial investment in 'green projects' and environmental stocks is becoming increasingly popular with investors.

Influence of the 'task environment' on eco-orientation

The decisions of a company are influenced by the components of the task environment. For not only do the company and its management have to serve the interests of their shareholders, but they also have to meet the demands of various other groups, collectively called the stakeholders. In fact, the stakeholders also determine the degree to which a company is affected by their own ecological questions and demands.

ENVIRONMENTAL MANAGEMENT AS AN ENVIRONMENTAL GOAL

An empirical study shows that, in general, companies are directly concerned with aspects of ecology, and that they are sensitive to the views of their stakeholders. The question now is whether they could integrate this concern and sensitivity into corporate decisions that are largely about the economic aspects, and, if so, how? An analysis of the results of a theoretical survey and some empirical studies shows that, as far as management is concerned, profit is not the only goal, and that there is a set of management goals rather than a single goal (Günther, 1994; Baum *et al.*, 1994). This means that all the goals should be examined, together with the relationships that may exist among them. According to an empirical study of 483 companies in Germany (Figure 15.2), 53 per cent saw a positive relationship between environmental management and competitiveness, while 61 per cent perceived a negative relationship between environmental management and profit. This finding has a temporal perspective too, because while in the long term the companies saw the possibility for making greater profits with environmental management than without it, in the short term they were only concerned with the cost implications of environmental management. As for costs and profits, the study showed that while 16 per cent of the companies saw a potential for greater receipts, only 5 per cent believed that there was potential for reduced costs. The reason for this could be that savings made from lower inputs of materials and energy, and from less waste generated, are perceived by the companies as results of greater production efficiency rather than accruing from the application of environmental management.

THE 'VALUE CIRCLE' AS A STRUCTURE FOR ENVIRONMENTAL MANAGEMENT ACCOUNTING

In order to be able to analyse environmental impacts on management, and especially on accounting, environmental management should not only be seen as a management goal but also in terms of its consequences on the different activities to be analysed. What is called the 'value chain', proposed by Porter (1986), is an instrument for such an analysis of the different activities of a company. However, for its application to environmental management, it has to be translated into a 'value circle' shown schematically in Figure 15.3 and elaborated in Table 15.1. The 'value circle' enables companies to analyse the strengths, weaknesses, opportunities and threats (SWOT) to the activities that are strategically important to them. More specifically, it is an instrument for the SWOT analysis of companies whose strategic activities generally fall into two categories: primary activities concerned with procurement, production, sales and disposal, and secondary or supporting activities such as marketing and accounting.

MONETARY INSTRUMENTS FOR INFORMATION GATHERING AND DECISION-MAKING

Investment appraisal

The purpose of investment appraisal is to obtain information on current activities of companies and alternatives to them, if any. Such information serves as

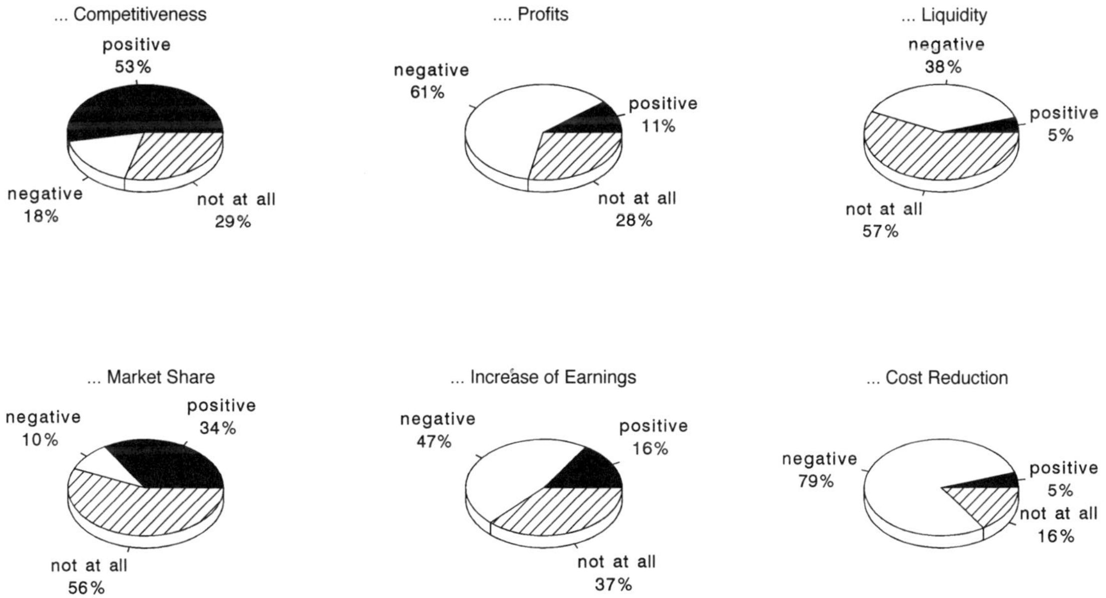

Figure 15.2 Corporate policy and environmental management

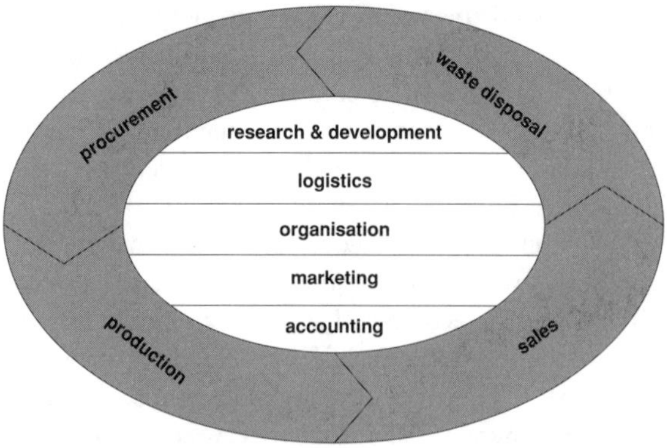

Figure 15.3 The 'value circle'

a basis for making sound decisions on investment (Rückle, 1992). In general, companies invest in environmental protection measures voluntarily, or because they are obliged to do so in order to comply with environmental legislation. Such investments can either be for specific environmental protection measures, or they can be for the enlargement, replacement or rationalisation of plants, processes and equipment into which environmental protection measures are integrated. The process of making corporate decisions on investment comprises a number of phases as shown in Figure 15.4 and described below.

Phase 1: This phase is concerned with the concept of the proposed investment. All relevant problems and issues, including those relating to environmental protection, are systematically examined in this phase

Table 15.1 The different activities of the 'value circle', their importance and especially the environmental aspects relevant to them

Activity	*Importance*	*Environmental aspects*
Procurement	50 per cent of manufacturing costs are material costs	• Ecologically favourable substitute products • Product evaluation for material and energy
Production	Transformation of material and energy Formation of undesirable outputs	• Choice of location: soil, water, air and other resources • Planning of the production programme • Production process: limits for emissions, state-of-the-art
Sales	Customer orientation Increased competition	• Ecologically favourable distribution • Retro-distribution • Service orientation
Disposal	Handling of undesirable outputs (conducts)	• Reduced water, soil and air pollution • Increase of secondary raw material through recycling
R & D	60 per cent of the costs are for R & D	• Decision: 'end-of-the-pipe' technology is integrated technology • Product innovation • Process innovation • Technical innovation • Organisational innovation • Life-cycle analysis
Logistics	Goods, resources, people and information have to be at the right place at the right time	• Logistics in ecology for retro-distribution • Ecology in logistics (e.g. transportation, stock)
Organisation	Legal requirements	• Regulators of waste, water and air • Management responsibility • Ecology circles • Project management
Marketing	Product policy Distribution policy Communication policy Conditions policy	• Ecologically favourable products • Ecologically favourable distribution • Communication of environmental management • Eco-oriented calculation of prices
Accounting	Instruments for decision-making	• Eco-oriented monetary instruments (e.g. cost accounting, investment appraisal) • Eco-oriented monetary instruments (e.g. eco-balance)

idea of investment	search for investment	appraisal of investment	decision for investment
➧ requirements of the stakeholders ➧ analysis of the former investment structure	➧ entrepreneurial alternatives: - avoidance - reduction - substitution - recycling - disposal ➧ possibilities for passing on - to customers - to suppliers - to assurance companies - to the government ➧ possible consequences of passivity - cost of opportunity - legal sanctions - pollution rights - solution of negotiation	➧ selection of alternatives that do not fit minimum requirements ➧ investment appraisal if monetary valuation is possible - cash inflows and cash outflows - discount rate - planning horizon method ➧ non-monetary instruments (in any case)	➧ comparison of the results ➧ decision about internalisation of external effects ➧ selection of the optimal investment

Figure 15.4 The process of making decisions on investment

to ensure that the proposed investment will lead to better corporate performance and greater profitability through an improvement, replacement or enlargement of the activities of the company. Both the views and demands of the stakeholders are taken into consideration in the process.

Phase 2: This phase is characterised by a search for alternatives which can be relied upon to improve the investment structure. Accordingly, due attention is given to measures for the prevention, reduction and disposal of pollution, as well as for the substitution, re-use and recycling of resources. These measures can be aimed merely at complying with statutory regulations for environmental protection, or, depending on the circumstances of the company, at deriving additional environmental benefits. The question of whether or not to pass on the environmental costs to the public is also addressed in this phase. Since an investment decision is based on the comparison of all possible options, the option of not implementing eco-oriented measures is also examined.

Phase 3: In this phase, which is concerned with investment appraisal, the alternatives which do not meet the minimum requirements laid down by the company itself, or by its stakeholders, are rejected. The remaining options are then assessed for their financial viability with respect to projected revenue and expenditure that can refer to both internal factors and external effects (e.g. third-party liability). All such data are evaluated and subsequently used in appraising the proposed investment. Since the issue of environmental pollution is an important element of the process, it is clear that the method to be used for investment appraisal must also be tested for its suitability for the relevant issues of environmental protection.

Phase 4: In this phase a decision is made on the option that is most attractive to the company, having considered all relevant facts and issues. As for the ecological aspects, they can be considered either with reference only to the internal effects, or to the external effects as well. However, if it proves impossible to come to a definite decision, then the non-monetary aspects can be considered in choosing the optimal alternative.

Eco-oriented cost accounting

Internal and external ecological costs

Ecological costs can be defined as all the costs of ecologically scarce resources and energy, plus the costs of avoidance, reduction, substitution, recycling and disposal (Roth, 1992; Kloock, 1992). Two different kinds of costs have to be considered in this context: costs of the internalised effects that are already part of the company's cost accounting, and the costs of the external effects that may or may not have been considered (Bonus, 1988; Heinen and Picot, 1974). External effects, which arise from the use of resources, have a direct influence on the benefits accruing to other people. These effects are not considered in the price mechanism, and those affected by them have no power to interfere with them. They can be caused by production or consumption activities, and they can influence the production function of the company or the benefit function of other people. External effects are characterised by non-compliance with the 'polluter pays principle', the passing on of costs to third parties, and exclusion from the market mechanism. External ecological costs are caused by the use of ecologically scarce raw materials and energy, and by the external impacts of production such as emissions and pollution discharges. Clearly, internalisation of external costs has important implications for the competitiveness of companies and should, therefore, be given careful consideration. Figure 15.5 shows how a decision on this issue can be arrived at.

Differentiation and extension of cost accounting

Ecologically relevant costs and revenues can be defined for each activity of a company. Statements of such differentiated costs and revenues are needed for the eco-oriented financial control of companies in the interests of making all relevant decisions transparent. Differentiated cost accounting thus means that ecological costs have already been internalised. If they have not, the scope of cost accounting used should be enlarged to include ecological costs.

For each activity of a company, the relevant costs and revenues can be recorded as shown in Table 15.2. An illustrative case study, dealing with the concept of what is called 'target costing' in a developing country, is given in Box 15.1.

Evaluation of the ecological aspects in monetary terms is a very difficult part of an eco-oriented cost accounting system, and the difficulty is compounded by inconsistencies between economic scarcity and

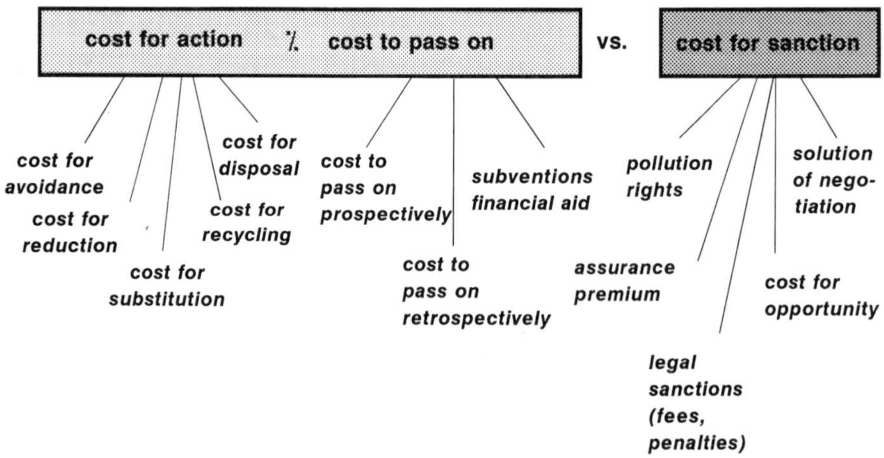

Figure 15.5 The ecological 'net effect'

Table 15.2 Costs, revenues and ecological impact

Costs/revenues	*Type of impact (+/–)*	*Ecological impact*
Turnover	–	Reduced turnover
Other revenues	+	Sale of licences
Cost of materials	–	Rationalisation
Personnel costs	+	Training
Depreciation	+	Ecological investment
Taxes/fees	+	Fees to be paid for waste water discharged
Other costs	–	Contribution to a recycling system

ecological scarcity. Internalised costs can be evaluated on the basis of the market price of actions needed for avoidance, reduction, substitution, recycling and disposal. On the other hand, the target of evaluating the costs of external effects is to determine the consequences of alternative possibilities for the allocation of resources. These costs can be defined by companies with respect to different options for environmental protection, including measures for avoidance, reduction, substitution, recycling and disposal. In addition, costs of gathering and processing information, to be presented to the decision makers, must also be taken into consideration.

The ecological 'net effect': When there are internalised effects, the differentiation of costs has an information effect, too, in the sense that the company has to decide whether or not to take the external effects into consideration. In this context the ecological 'net effect' in environmental management entails comparing an action, whose costs can be properly passed on to a third party, with the response of that party to those costs. Three different categories of costs and revenues may be compared:

- costs of action (c_a);
- costs that can be passed on to a third party (c_p);
- costs of sanction (c_s).

Costs of action: These costs can be differentiated in terms of costs of avoidance, reduction, substitution, recycling and disposal, as well as costs of information gathering and decision-making (e.g. decisions on investment appraisal).

Passing on of the costs: The next step is to ask whether a third party can and will pay the costs for environmental management. The feasibility of passing on these costs to a third party can be explored through consultation with the various stakeholders to determine the level of their willingness to pay and the willingness of the suppliers to act. When the government offers financial aid or subsidy for specific projects, this may amount to an indirect passing on of the costs to the government. Thus there are three possible ways of passing on the costs of environmental management actions: to the potential customers, retrospectively to the suppliers, and through financial aid or subsidy from the government.

Costs of sanctions: The difference between the costs of action and the costs passed on to third parties has to be compared with the costs of sanction (i.e. costs of non-compliance arising from not implementing mandatory environmental protection measures). The following are relevant in this context: pollution rights, insurance premium, legal sanctions (fees, penalties), opportunity costs, and negotiated settlement.

Possibilities for integrating ecological aspects: There are different possibilities for integrating ecological aspects into management decisions. For example, a company can, if it wishes, only take into consideration the internalised costs, or it can evaluate the external effects to serve as information goals, or for use in decision-making.

The process of cost accounting: After evaluating and recording the ecological costs and revenues (cost category accounting), they can be used for planning and controlling the different divisions of the company (cost centre accounting) and its products (cost unit accounting) (Schreiner, 1992).

Cost category accounting: The purpose of 'cost category accounting' is to record and evaluate all the costs of a company, and this involves a careful examination of all its activities. The primary costs can be directly passed on to cost centre accounting. The secondary costs can also be passed on to cost centre accounting through the internal transfer system. Unit costs and overheads are separated in cost category accounting, and the overheads mostly comprise ecological costs.

Cost centre accounting: Cost centres can be fully or partly eco-oriented, or not eco-oriented at all; most often they are eco-oriented in part, and, when so, methods of cost separation have to be used in accounting.

BOX 15.1 CASE STUDY OF THE 'TARGET COSTING' OF SOLAR ENERGY COMPARED WITH TRADITIONAL FORMS OF ENERGY IN A DEVELOPING COUNTRY

Description of the case

Mr Somalis, who is the owner of an electricity generating company in a developing country in central Africa, is also the largest producer of solar energy systems there. Because of the current predominance of traditional forms of energy such as oil, coal and gas, and the high production costs of solar energy equipment, his solar systems are not widely used in that country. In some communities, however, they are being used for watering and cooling the fields and for providing electricity to family homes. This situation is referred to as the 'island solution' because, even though the level of solar radiation is high throughout the year in central Africa, only a limited number of communities benefit from solar energy.

Currently the country is facing massive environmental problems arising from coal burning, among many others, as well as from rising costs of traditional forms of energy, especially oil which has to be imported. Mr Somalis sees this as an opportunity to promote his environmentally friendly and promising solar systems. He asks his Deputy, Mr Asap, to look into the production of the solar system using the concept of what is called 'target costing'. Mr Asap knows from experience that target costing is an instrument for the cost-oriented realisation of customers' needs and wishes which can increase the product's acceptance in the target market. In other words, both price and costs of the product depend very much on the relevant or target market.

For an assessment of the current competitiveness of solar systems, Mr Asap gets in touch with the production and marketing department of Mr Somalis's company, whereupon the Production Manager sends him the following data on the three system components of the 'island solution':

System components	*Investment costs ($/Wp)*
Solar module	12.0
Static converter	1.5
Installation	2.0
Total	15.5

The costs quoted above are in the US dollar equivalent of the local currency, while Wp is a measure of the maximum rated power of a solar module. In the opinion of the production department, the solar module is the only component subject to depreciation. At present the maximum rated capacity of solar systems, which can be produced by Mr Somalis's company, is 1,000 Wp.

The marketing department informs Mr Asap that the current price of electricity is 0.4 $/kWh (kWh is a measure of electric power), whereas the cost of generating electricity from solar energy is about 1.2 $/kWh. The annual average amount of electricity consumed in the country is 60 kWh per family. Moreover, the marketing department could also provide a target price of 0.8 $/kWh for solar energy at a capacity rating of 1,000 Wp (the target price was calculated by the marketing department on the basis of an estimated price–sales function). The marketing department calculated that, in general, the profit margin would be 7.5 per cent for a trade margin of 20 per cent.

On the basis of the above information, Mr Asap can now calculate the target production cost of the solar system.

Solution

Since the solar module is the only component of the system subject to depreciation, Mr Asap calculates its target price and costs as follows:

Current cost of generating solar electricity is 1.2 $/kWh, and the investment costs are 15.5 $/Wp. Therefore, a target price of 0.8 $/kWh corresponds to investment costs of 10.33 $/Wp (= $15.5 \times 0.8/1.2$) for the system as a whole.

When costs of the static converter and installation are excluded, the target price of the solar module reduces to ($10.33 - 3.5 =$) 6.83 $/Wp; and, when the trade margin of 20 per cent and profit margin of 7.5 per cent are deducted, the target price of the module reduces further to 5.05 $/Wp. In order to achieve this target price, though, Mr Asap has to examine ways in which production costs could be reduced within the company with higher production output; that is, through economies of scale.

Unit cost accounting: The purpose of 'unit cost accounting' is to calculate the ecological cost of a product by directly allocating the costs of both primary and secondary raw materials used in the production of each unit. Overheads for primary and secondary processes are allocated to the different cost centres.

Treatment of ecological costs: Ecological costs can be treated within traditional cost accounting, and no other cost accounting system is needed for this if there is a suitable method in place for the purpose. The integration of ecological aspects into traditional cost accounting makes it possible to control and optimise the production process.

NON-MONETARY INSTRUMENTS FOR INFORMATION GATHERING AND DECISION-MAKING

Apart from the differentiation and application of monetary procedures, non-monetary instruments for information and decision-making are often used specifically to assess ecological requirements and to provide eco-oriented information. Such instruments must be capable of identifying ecological weaknesses and risks to the company, in addition to revealing its competitive advantages and gaps in the market. The objective of the exercise is to gather as much relevant information as possible in quantitative terms.

The eco-oriented instrument for information and decision-making which is most widely used at present is that of 'eco-balance' (Hallay and Pfriem, 1992). This non-monetary instrument may be seen as an inventory of the environmental impacts of the activities of a company. Because of its wide and varied application in companies, the concept of eco-balance is continually being upgraded in the light of practical experience in order to make it more effective.

An eco-balance is an eco-oriented instrument for information gathering and decision-making which includes all flows of materials and energy in the activities of a company. The aim is to record, describe and assess all ecologically relevant resources used and output produced, thus revealing both ecological opportunities and risks. An eco-balance can be prepared for the company as a whole, or for its individual production processes or products, and the data thus obtained refer to flows of material and energy in the entire company, or its individual parts, as the case may be. On the input side a differentiation is made between materials and energy, while on the output side both products and their impacts are distinguished. The prerequisite for drawing up a useful eco-balance is information on both materials and energy, as well as a suitable method for assessing them. The assessment can take the form of a purely physical valuation, or it can be done through what is called an 'ABC analysis'. While the internal functions of an eco-balance are aimed at eco-oriented control, from an external point of view the focus is on the communication between the company and its stakeholders. A typical example of an eco-balance is given in Table 15.3.

In preparing the eco-balance of a company as a whole, the starting point is the company itself, and its input and output in particular. To begin with, the company is considered in its entirety as a 'black box', the input into which are materials and energy and whose output are products as well as wastes generated during production. According to the suggestion of the Institute for Ecological and Economic Research (Hallay and Pfriem, 1992), the drawing up of an eco-balance is characterised by three levels: at the first level the relevant quantities of the input and output are determined, together with the assignment of a balance number, a marking, and a number for the category of the material. At levels two and three all input factors are subdivided, as far as possible and necessary, and their composition analysed. The importance of an eco-balance to a company lies in the fact that it contains a complete record of all the physical factors relating to production. The aim is to obtain information on the ecological impacts of the company as a whole. What is called 'process balance' is concerned, on the other hand, with the examination of the processes whereby input is transformed into output. The prerequisite for a process balance is the breaking down of the entire production process into individual or partial processes for examination. Such a detailed examination is also undertaken for both input and output. The purpose of 'product balance' is to consider the ecological effects of each product during its whole life-cycle (cradle-to-grave). For this it is necessary to record systematically and

Table 15.3 Example of an eco-balance

	Input	Stock		Output	
Stock					
B.1. Land (m^2)					**B.1. Land (m^2)**
1.1. Sealed	7,425	54,885		10,748	1.1. Sealed
1.2. Green	2,658	358,927		43,458	1.2. Green
1.3. Build-over	4,348	105,502		30,126	1.3. Build-over
B.2. Buildings (m^2)					**B.2. Buildings (m^2)**
2.1. Production		58,778		1,255	2.1. Production
2.2. Distribution and storage	2,956	70,074			2.2. Distribution and storage
2.3. Administration		13,764			2.3. Administration
B.3. Equipment (piece)					**B.3. Equipment (piece)**
3.1. Production machines	273	5,047		443	3.1. Production machines
3.2. Office equipment	376	5,924		167	3.2. Office equipment
3.3. Office and communication machines	337	2,245		142	3.3. Office and communication machines
3.4. Vehicle fleet	34	131		45	3.4. Vehicle fleet
3.5. Technical facilities	38	191		32	3.5. Technical facilities
Flow					
I.1. Circulating goods (kg)					**O.1. Products (kg)**
1.1. Raw materials	2,981,144			4,093,129	1.1. Hosiery
1.2. Semi-finished and finished goods	2,080,104			169,406	1.2. Outer wear
1.3. Auxiliary materials	146,622	3,220,210		791,420	1.3. Transport packing
1.4. Ancillary	1,295,679	7,164		2,094,244	1.4. Product packing
					O.2. Waste (kg)
			803	20,678	2.1. Hazardous waste
			73,731	1,111,223	2.2. Recyclables
			216	886,400	2.3. Residual waste
				54,960	2.4. Building rubble
I.2. Energy (kwh)				116,042,437	**O.3. Waste heat (kwh)**
2.1. Gas	15,913,843				
2.2. Electricity	38,303,027				
2.3. Fuel oil	47,532,996	252,363			
2.4. District heating	4,476,544				
2.5. Fuel	10,068,390				
I.3. Water (m^3)					**O.4. Waste water (m^3)**
3.1. Tap water	243,082			301,031	4.1. Tap water
3.2. Raw water	152,953				4.2. Raw water
					O.5. Air emission (m^3)
				247,010	5.1.1. NO_X
				197,315	5.1.2. SO_2
				39,028,882	5.1.3. CO_2

assess all materials and energy used in production, as well as all emissions caused during production. Ideally, this exercise should be carried out for all the phases in the life-cycle of a product and, clearly, this is contingent upon receiving all relevant information from both suppliers and consumers.

The term 'balance' is used here to mean that the sum of materials and energy entering the system as input is equal to the sum of materials and energy leaving it as output. Input must be equal to output because, according to the first law of thermodynamics, neither material nor energy can be destroyed or created, they can only be transformed from one state into another.

After recording all relevant data with regard to the different balances which must prevail, they are then assessed in physical units at first. An 'ABC analysis' is carried out in the next step; that is, all relevant materials and energy are assessed and classified with regard to their risk potential. As far as a company is concerned, risks can arise from non-compliance with legal requirements, social demands, internalised costs, as well as from the production stages themselves. Within the process balance it is possible to make an assessment of the main processes, and possibly also for each individual or partial process. In practice, the entire company may be subdivided into individual or partial processes for identifying and controlling the extreme points (hot-spots) where problems are most likely to occur. At the product level, on the other hand, the ecological relevance of single material lines is assessed. When the eco-balances of a company have been drawn up, it is expected to take the necessary measures. If several alternatives are available, the 'demand for action' has to be compared against 'chances for ecological improvement'. Decision grids can be useful in making such a comparison. Certainly, demands for action should be given very careful consideration, because inaction may have potential for ecological damage that would tarnish the public image of the company and lower the motivation of its employees. As for the 'chances for ecological improvement', it may include such items as personnel factors, process and product changes, as well as the existence of alternatives.

If by following the steps outlined above the management of a company successfully establishes an environmental management system within that company, and provided that the system complies with the EC eco-audit regulation (described in Chapter 13 of this volume), then it will qualify to participate in the Eco-Management and Audit Scheme (EMAS).

CONCLUSIONS

Driven by the demands of stakeholders, there is now an increasing requirement to integrate ecological aspects into the decision-making process of companies at the management level. Accordingly, in the interests of long-term competitiveness and the profitability of companies, it is now becoming increasingly necessary to incorporate in their management decisions the demands of the societal environment (technological, economic, social, ecological and legal aspects) as well as of the task environment and the stakeholders (customers, suppliers, competitors, creditors and other stockholders). In this regard the 'value circle' is an important instrument for analysing the eco-orientation of all the activities of a company, mainly because it can identify the strengths, weaknesses, opportunities and threats (SWOT) implicit in or arising from those activities. In particular, two categories of activities need to be examined: the primary activities such as procurement, sales and disposal, and the secondary activities such as marketing and accounting. In this regard accounting plays an important part, especially because the monetary and non-monetary instruments of accounting make it possible to integrate the ecological aspects into management decisions. The monetary instruments can also be used for distinguishing investment appraisal from cost accounting.

Eco-balance is an important non-monetary instrument for calculating all the material and energy flows associated with the activities of a company. The aim is to record, describe and assess all ecologically relevant resources and output. After recording all such data, an ABC analysis can be carried out to classify all material and energy flows with respect to their risk potential.

By following the procedures described in this chapter, a company can introduce into its management regime an efficient and effective mechanism for dealing with environmental problems created by its production activities.

REFERENCES

Baum, H.-G., Coenenberg, A.G., Günther, E. and Wittmann, R. (1994) 'Unternehmenspolitik und Umweltschutz', *Zeitschrift für betriebswirtschaftliche Forschung* 46 (1): 81–100.

Bonus, H. (1988) 'Möglichkeiten der Internalisierung externer Effekte als Instrument der Koordination von Unternehmenszielen und gesellschaftlichen Zielen', *Schriften des Vereins für Socialpolitiks*, 207–226.

Dyllick, T. (1989) *Ökologisch bewußte Unternehmensführung: Der Beitrag der Managementlehre*, St Gallen: Schriftenreihe ÖBU/ASIEGE.

Freeman, R.E. (1984) *Strategic Management: A Stakeholder Approach*, Marsfield, Mass.

Günther, E. (1994) *Ökologieorientiertes Controlling*, München: Vahlen.

Hallay, H. and Pfriem, R. (1992) *Öko-Controlling – Umweltschutz in mittelständischen Unternehmen*, Frankfurt am Main/New York: Campus.

Heinen, E. and Picot, A. (1974) 'Können in betriebswirtschaftlichen Kostenauffassungen soziale Kosten berücksichtigt werden?', *Betriebswirtschaftliche Forschung und Praxis* 26 (4): 345–366.

Kloock, J. (1992) 'Kostenrechnung mit integrierter Umweltschutzpolitik als Umweltkostenrechnung', in W. Männel (ed.) *Handbuch Kostenrechnung*, Wiesbaden: Gabler, Betriebswirtschaftlicher Verlag, pp. 929–940.

Porter, M.E. (1986) *Wettbewerbsvorteile: Die Spitzenleistungen erreichen und behaupten*, Frankfurt/New York: Campus.

Roth, U. (1992) 'Umweltkostenrechnung', Masters thesis, Wiesbaden.

Rückle, D. (1992) 'Investition und Finanzierung, München', in U. Steger (ed.) *Handbuch des Umweltmanagements: Anforderungs- und Leistungsprofile von Unternehmen und Gesellschaft*, Beck, Juristischer Verlag, pp. 451–467.

Schreiner, M. (1992) 'Auswirkungen einer umweltorientierten Unternehmensführung auf die Kosten- und Leistungsrechnung', in W. Männel (ed.) *Handbuch Kostenrechnung*, Wiesbaden: Gabler, Betriebswirtschaftlicher Verlag, pp. 941–952.

Wicke, L. (1993) *Umweltökonomie*, München: Vahlen.

SUGGESTED READING

Bundesumweltministerium/Umweltbundesamt (ed.) (1995) *Handbuch Umweltcontrolling*, München.

Gray, R. (1990) *The Greening of Accountancy: the Profession after Pearce,* Chartered Association of Certified Accountants, Research Report No. 17, London.

Gray, R. (1992) 'Accounting and Environmentalism: An Exploration of the Challenge of Gently Accounting for Accountability, Transparency and Sustainability', *Accounting, Organisation and Society* 17 (5): 399–425.

Günther, E. (1994) *Ökologieorientiertes Controlling*, München: Vahlen.

Steger, U. (ed.) (1988) *Handbuch des Umweltmanagements: Anforderungs- und Leistungsprofile von Unternehmen und Gesellschaft*, München: Gabler, Betriebswirtschaftlicher Verlag.

SELF-ASSESSMENT QUESTIONS

1 Why is environmental management accounting necessary for companies?
   (a) political pressure;
   (b) risk of damage to the environment;
   (c) increasing environmental demands of all stakeholders;
   (d) demands of the customers.
2 How can customers influence the eco-orientation of a company?
   (a) through their boycott of the company's products;
   (b) through the possibility that a company can increase its market share by offering environmentally friendly products;
   (c) there is no way in which the customers can influence a company's decisions.
3 Describe the relationship between environmental management as a management goal and other management goals in the short and the long term.
   (a) in the short term environmental management is the main goal, whereas in the long term maximisation of profit is necessary for the sustainable development of the company;
   (b) in the short term profit maximisation is the main goal, whereas in the long term environmental management is necessary for the sustainable development of the company.
4 Why should companies include environmental aspects in their R & D decisions?
   (a) because the R & D department is the most important department and therefore environmental aspects should be considered there;
   (b) because the main part of the costs, and therefore the opportunity for influencing the production process, is determined in the R & D department;

(c) because environmental aspects can only be included in R & D decisions.

5 Describe the different phases of an eco-oriented investment process. Which of the alternatives mentioned below is correct?
(a) phase 1: search for investment; phase 2: concept of the investment; phase 3: appraisal of the investment; phase 4: decision on the investment;
(b) phase 1: search for investment; phase 2: concept of the investment; phase 3: decision on the investment; phase 4: appraisal of the investment;
(c) phase 1: concept of the investment; phase 2: search for investment; phase 3: appraisal of the investment; phase 4: decision on the investment.

6 Which kind of costs are compared in ecological net effect?
(a) costs of action minus costs passed on to third parties against costs of sanction;
(b) costs passed on to third parties minus costs of action against costs of sanction;
(c) costs of sanction minus costs passed on to third parties versus costs of action.

7 Explain the meaning of 'balance' in 'eco-balance'. Which of the following statements is correct?
(a) the term 'balance' is used because, with an eco-balance, a company achieves a balance between its environmental goals and results;
(b) the term 'balance' is used because the input side (material and energy) equals the output side (products and conducts).

PART II.2

ECONOMIC INSTRUMENTS

16

ECONOMIC INSTRUMENTS FOR ENVIRONMENTAL MANAGEMENT

Aviel Verbruggen

SUMMARY

A variety of economic instruments have been described in this chapter for reducing and controlling pollution created by human activities, together with an examination of their advantages, disadvantages and the optimal ways in which they could be applied to achieve best results.

The chapter begins with a description of how the philosophy of dealing with environmental problems has shifted from the implementation of curative measures at the end of pollution chains to structural measures aimed at redirecting human activities in a less-polluting and sustainable way. It is argued that since the free market economic system is a major organising force in modern industrial societies, citizens have much to learn about economics, especially because the price mechanism is a major source of information and it guides our decision-making in all matters large and small. To this end the basic concepts of economics are then discussed, with particular reference to the environment. However, considering that the basic economic model is too simplistic to deal effectively with the environmental problems of the real world, important adjustments to it have been discussed, including public goods and externalities. This is followed by a discussion of the most favoured economic instruments, their application and the criteria for selecting them, all in the context of public policy-making on the environment.

Finally it is pointed out that while economic instruments are potentially the best vehicle for moving towards sustainability, they are but one 'tool' among many needed for reorienting both the behaviour and attitude of people towards global sustainability.

ACADEMIC OBJECTIVES

On completion of the chapter the reader should have a good understanding of the following:

- the pollution chain, its different elements, and how it facilitates policy-making on environmental problems and issues;
- the basic concepts of the free market economic system, such as supply and demand, consumers and producers, markets, economic utility, etc.;
- economic instruments used in environmental management, such as taxes, levies, marketable permits, deposit refund systems, etc., the criteria for their selection and why such instruments should be preferred to those of command-and-control for moving towards sustainability;
- the shortcomings of economic instruments, and of the free market system, in dealing with environmental management problems, and how they could be addressed, at least in part, through the mechanisms of public goods and externalities.

THE POLLUTION CHAIN

A flow chart of the pollution chain with policy feedback is shown in Figure 16.1. The major policy problem in dealing with environmental degradation is the variety of pollution sources. Practically all human activities in an industrial society consume energy and natural resources and, as a result, these activities harm the environment to an extent determined by their nature and scope. Pesticides and fertilisers used in modern agriculture, tourists visiting natural sites and discharging wastes into them, aircraft and motor cars transforming fuels and air into carbon dioxide, nitrogen oxides, etc. are typical examples of such activities. All these gaseous, liquid and solid waste flows are called 'emissions' caused by human activities. The amount and quality of emissions depend on the extent and type of human activity from which they arise, and appropriate scenarios are used to model the processes of such activities for developing policies to protect the environment.

Although emissions are concentrated at their sources and disperse into the atmosphere, often they settle down on soil and water bodies through dry or wet deposition. Ambient conditions, the presence of other substances and their concentration, and a host of other factors determine the levels of concentration and exposure of pollutants which, in turn, determine the intensity of their impact on human health and well-being, on the natural and cultural environment, and on the economy. It is for this reason that transmission and impact models have received much attention in the environmental sciences.

However, costing of the various impacts of environmental pollution remains a major unsolved problem in the analysis of the pollution chain. How can one weigh the conservation of an animal species against the improvement of the economic conditions of a group of people? Is the electricity from nuclear power plants worth the increased risks of weapons proliferation and cancer deaths? Answers to these difficult questions are needed when preparing effective and efficient policy measures. By definition, a comprehensive policy sets the correct goals, selects the appropriate instruments, proposes the correct measures and employs the right means. But all this requires specification of type, extent and timing.

Policy response to a given pollution problem can be divided into five approaches that have developed over time, as shown in Figure 16.1. The first response to harmful impacts is curative reaction: sick people are sent to hospitals, endangered species are bred in artificial environments, etc. In the second stage attempts are made to reduce exposure by taking effect-oriented measures such as for the collection of sewage and wastes. These measures concentrate harmful substances in designated sites so that they cause less harm to people and nature than when they are disposed of haphazardly. An environmental activity sector has now emerged in the economy whose purpose is to make profits and pay salaries, aiming at survival if not growth. But soon attention will focus on emissions themselves, raising questions such as, would it be possible to confine the harmful substances at the place where they are produced by installing gas cleaners, filters, catalysts, etc.? If so, this would significantly improve efficiency in separating waste flows from human activities. Measures of this type, called 'source-oriented measures', are still considered by many as the highest level of policy-making.

It is becoming increasingly clear, however, that the laws of thermodynamics cannot be avoided with regard to the harmful substances produced. According to these laws, one waste stream can always be transformed into another, but the total mass will remain constant while the quality of the whole will deteriorate. It is for this reason that greater attention is now

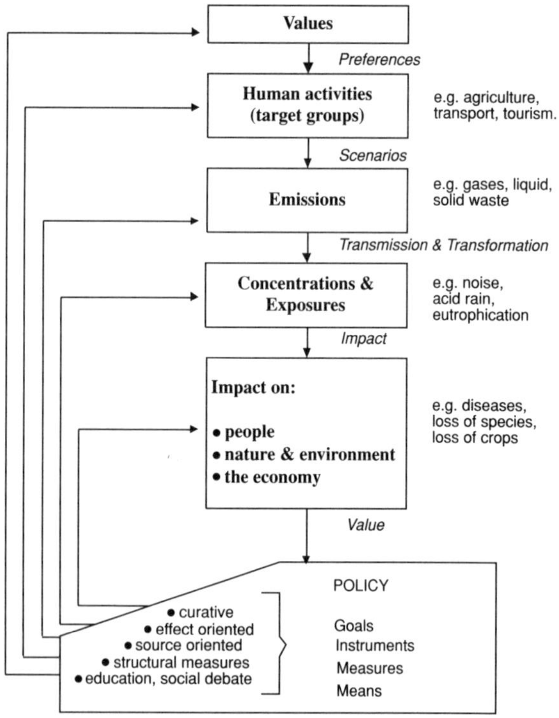

Figure 16.1 The pollution chain

being given to structural measures in an attempt to redirect human activities themselves, including the adoption of clean technologies to improve production processes, the design of consumer products to facilitate their easy recycling cheaply, and urban planning requiring fewer car trips. At the same time, both the quantity and quality of energy and natural resources processed by our industrial societies should be decreased significantly. But how can we accomplish all this? Clearly, it will require changes in preferences, attitudes, customs, technologies, processes, procedures, industrial sectors, urban settlements, land uses, international trade, etc. But no government can bring about these changes without the full commitment of its citizens and captains of industry. No doubt such an exercise will require skilful manipulation of all the policy instruments in a comprehensive and concerted manner.

Values may be changed at the highest level whereby people would prefer less harmful activities to those which create greater emissions. Even so, societies must change fundamentally before a majority of the population would prefer, for example, bicycles to cars or domestic holidays to those in distant and exotic places. But, to begin with, it is necessary to define the target groups (e.g. home-owners, commuters, electricity generators) for policy purposes. Changing attitudes and behaviour, together with a modification of the technologies used by the various target groups to foster and promote sustainability, will then be the major challenge for environmental policy.

Economic theories contribute in good measure to the resolution of the problems posed by the pollution chain in three different areas. First, the depiction of human activities in macroeconomic models has been an area of economic theory and practice for several decades. Indeed, it may be said to be the firmest ground from which to start improving our knowledge of human activities when mapping out alternative scenarios of development.

The second area is concerned with the 'money value' standard used by economists. Money allows the economist to convert very different things into a common general equivalent, thus making comparisons possible. A simple example of this is the following: lemons and peaches are different physical entities which usually have different values. Let the price of lemons and peaches be $5/kg and $2/kg respectively. Then, according to the economist, 1 kg of lemons would be equal to 2.5 kg of peaches in value. This pricing system thus makes it possible to compare things that are physically very different. And so it was that economists were called upon to place values on the various impacts of pollution and to balance the impacts. But this role of economists has been challenged by ecologists who contend that it is impossible to assign monetary values to items such as the existence of a species, the conservation of natural beauty, the life of human beings, etc. While economists (Pearce and Turner, 1990) refer to these special items as 'existence values', most economists agree that it is of little use in practice to assign numerical (monetary) values to them. However, notwithstanding the limitations of the yardstick of 'money', it is the only one available at present capable of comparing a large number of items of very different nature on a simple numerical scale. It is also helpful to policy-makers who have to make comparisons between very different items every day.

In formulating a comprehensive environmental policy, as in daily life, it is necessary to make decisions which entail balancing means and outcomes that are usually of very different nature and some of which are extremely difficult to quantify; but, when a decision is taken, it implicitly means that certain values have already been assigned to all the factors pertinent to that decision. The economists then argue that it is better to specify, or try to specify, explicitly the values assigned to the various factors. However, since this topic is tangential to our discussion we will not dwell on it any further; neither will we delve into the methods developed by economists to calculate the values to be assigned to various goods and services outside the price system (Braden and Kolstad, 1991), except to point out that most of these methods try to assign some 'economic use value' to, for example, environmental amenities by analysing the behaviour of people when they buy houses or go on holiday, etc. Although trade-offs are necessary, it may be true that economic valuation on its own is not sufficient to direct decision-making, and that there must be scope for the political process to play its part, too.

The third area of the pollution chain, in which economics can make a contribution, is in preparing the comprehensive environmental policy plan for the authorities responsible. Strategic planning and management are needed for this, together with the full use of the insights and experience accumulated in corporate business practice (Winsemius, 1986). Because the policy must address all relevant target groups at the proper time and with appropriate incentives, it is necessary to make full use of the market mechanism. As the predominant carrier of information and incentive, it is only the price mechanism which can organise entire populations towards a common goal in a sustainable way (Arrow, 1974). Although the price system has a central role in formulating efficient environmental policies, it cannot by itself cure our environmental problems. On the contrary, a strong public policy is essential, and the economic instruments are but one (albeit particularly powerful) collection of instruments at hand. The performance of economic instruments is greatly enhanced when they are set in a clear and enforceable regulatory framework and accompanied by correct information and technologies adapted for problem-solving. Furthermore, many corrections to the spontaneous price system need to be made before it is able to convey the correct information for environmental protection to the various actors in the marketplace.

SOME ECONOMIC CONCEPTS

Microeconomic theory is concerned with the study of the behaviour and activities of economic agents, both as individuals and actors in the marketplace. Since a great number of textbooks is available on the subject, we will only discuss here a few of the concepts which will be used later to develop our arguments.

The consumer

The first agent of interest to microeconomics is the consumer, who is said to maximise utility by choosing particular quantities of the goods and services available in the market at particular prices. It is assumed that the utility of the consumer increases with increasing quantities he can buy, and so he will increase his consumption until his expenses reach his budget limit. But the problem facing the consumer is how much of his budget he should spend on the goods and services offered in the marketplace in order to maximise his own utility. When this problem is stated formally, it will be found that the best a consumer can do is to spread spending on goods and services in such a way that the marginal (extra) utility received from a marginal (additional) amount of the good, divided by the price of the good, is the same for all the goods purchased. When he obeys this rule, he cannot reach a higher utility level by reshuffling his budget, because the marginal utility per marginal dollar spent is the same for all purchases.

It is possible to derive from this 'optimising behaviour' of the consumer a demand curve for goods and services with regard to the particular prices established in the market and the budget of the consumer. Figure 16.2 shows the demand curve for one particular good (or service), along with the relationship between the quantity demanded (horizontal axis) and the price of the good or service (vertical axis).

Demand curves of normal goods and services have a negative slope because the higher the price, the lower will be the demand (and vice versa). The demand curve represents the consumer's willingness-to-pay for the goods, reflecting his marginal utility. The area under the demand curve represents total utility. For example, when consumption is q^* units, the consumer's total utility is represented by area $OASq^*$ in Figure 16.2. When price p^* is charged for each unit of the good, the consumer will fix his

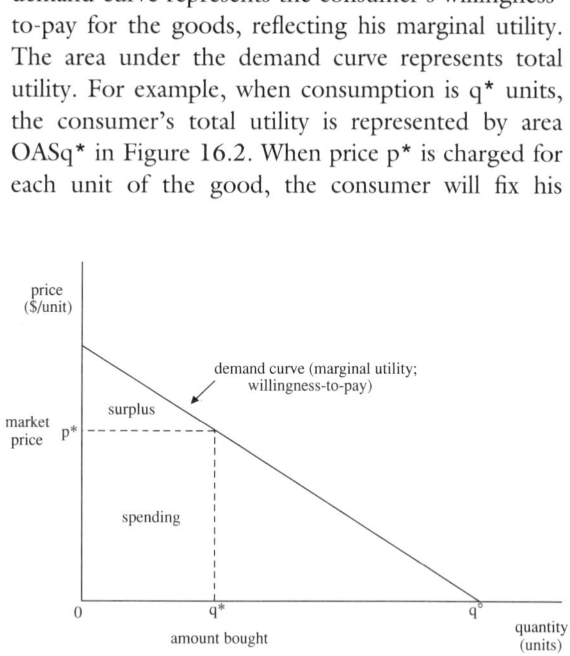

Figure 16.2 Demand for a good or service

consumption at amount q*. Consuming less (i.e. a quantity between 0 and q*) would be unwise, because the marginal utility of an extra unit (the ordinate of the curve) would then be higher than price p* which he must pay. Consuming more than q* would bring the marginal utility of extra consumption to below price p*, and so it would cost more than what he would get. When consumption is q*, the consumer receives the utility represented by the area OASq*, but he must trade the amount Op*Sq* (= p* times q*) to the seller. The remainder, that is the area of the triangle p*AS, is called 'consumer surplus'. The consumer surplus is the real value to the consumer of his consumption q*, and not the sale (Op*Sq*) made in the marketplace which is used for calculating the gross national product (GNP). Measurement of willingness-to-pay, and estimation of the extent of (or changes in) consumer surplus, lies at the heart of all methods for assessing the benefits of goods and services. This is even the case when the goods or services are not traded in a market, as in the case of environmental amenities for example.

It is important at this stage to point out that prices influence people in deciding their levels of consumption. When a good carries a zero price (also called a 'free good') there will be a large demand for it, up to the amount q° in the example of Figure 16.2. At the positive price of p*, consumers will refrain from consumption to level q*, and they will refrain further when the price rises. Clearly therefore, it would be necessary to put a (high) price on resources, energy, nature and the like in order to conserve them.

The impact of price on consumption depends on the slope of the demand curve. This impact is represented by a function which economists call the 'price elasticity of demand':

$$\varepsilon = \frac{dq/q}{dp/p}$$

It is in fact the relative change in the quantity demanded caused by a relative change in price. If the absolute value of ε (|ε|) approaches zero, price changes do not affect demand very much. When (|ε|) is large, on the other hand, a small change in price has a large impact on demand. Price elasticity of many goods and services is small in the short term, because people have insufficient time to change their consumption pattern. In the long term, however, higher prices convey the message of the relative scarcity of the good or service in question, and people react accordingly. In other words, long-term price elasticity tends to be significant or large.

The producer

Demand in the market is met by the supply of goods and services provided by the producers. The producers maximise their profits, which are by definition the difference between sales, $(p.q)$, and costs of production, $C(q)$, of the goods or services.

Formally

$$\underset{q}{Max}\ p{\cdot}q - c(q)$$

in which $p = \delta C/\delta q$ or, price = marginal cost.

In addition, the following inequality must hold in order for profits to be non-negative

$$p \geq \frac{C(q)}{q}$$

meaning that price should be greater than or equal to average costs. The supply curve of a producer is therefore the segment of its marginal cost curve above average cost

Another way of approaching optimality by the producer is by minimising costs, C(q), subject to the constraint that a particular output target q* is met. The cost function is then written as the sum of the expenditures on the various production factors which a producer must buy in order to pursue his business. Production factors are labour, capital, land, energy, and natural resources used in particular technologies to produce the desired outputs.

The producer also faces an optimisation problem similar to that faced by the consumer, which is: how much of each production factor (e.g. labour) should he buy or hire in the market in order to minimise the overall costs in producing the desired output q*? The solution to this problem is also very similar to that of the consumer's. The producer should acquire those amounts of production factors that equate the marginal productivity in monetary terms (= the marginal product of each production factor divided by its price) for all production factors.

Again the price system proves to be a powerful instrument, because raising the price of a factor that has to be conserved (e.g. natural resources) would lead to higher productivity in its use, and also to the adoption of its cheaper substitutes (because lower-cost factors would be substituted for higher-cost factors). A typical example of this phenomenon is to be found in the energy market. When energy prices were high in the early 1970s and especially in the 1980s, energy was used more efficiently (conserved) and lower-cost energy sources were substituted for those with higher costs. The resulting energy conservation was greatest in the 1980s, because by that time better know-how and technologies to conserve energy became available. As with consumers' reaction to higher prices, producers also have greater freedom of action in the long term compared to the short term because, in the long term, new technologies are developed, new or improved equipment is installed and new people are trained, and thus it is possible to achieve conservation much more easily than with old equipment, old technologies and old attitudes.

The market

In economic theory a workable market is populated by many consumers and producers. All of them listen to the unique signal of the market price, according to which they optimise the purchases of their products, production factors, and of consumption or production technologies. The individual demand curves of consumers are added along the quantity axis, and the same can also be done with the supply curves of producers. This creates market supply and demand (Figure 16.3). The equilibrium price, p*, and quantity, q*, occur at the intersection of these curves. At this point the sum of consumers' and producers' surplus is at a maximum. The latter surplus is the difference between sales (p*.q* or area Op*Sq* in Figure 16.3) and costs, OBSq*, which is the area under the supply or marginal cost curve of producers. Efficient equilibria are said to be attained when all markets in the economy are free and function well on the whole; and it is only then that the sum of producers' and consumers' surpluses are maximised. This is why the price system is advocated so strongly by economists.

Unfortunately, reality is much more complex than the world created by the assumptions of economic theory. Most of the theory assumes static conditions (i.e. all changes occur overnight) or comparatively static conditions (i.e. two equilibrium states occur at different points in time). Also, although such arguments are advanced in a deterministic way, most environmental problems are dynamic in nature and subject to uncertainty (e.g. the exploitation of life-support systems with unknown carrying capacity). It is very important, therefore, to exercise due care when applying economic analyses and results to environmental problems and issues.

An understanding of the principles and functioning of markets is, however, valuable to all policy-makers who must address their respective constituencies that are characterised by the diversity of their activities. For example, the working of the energy market in the last 20 years can be analysed in market terms. Prices escalated in the first oil crisis of the early 1970s because demand overwhelmed supply. While most of the policy-makers reacted to this situation by investing capital to increase supply, many of the users (households and companies) reacted by restraining demand through conservation efforts. The net result of this was that increased supply and tempered demand led to over-capacity, which forced prices downwards; but low prices frustrate conservation efforts, and so consumption started to take off again. Also, policy measures, which deny the power of market mechanisms, create black markets in

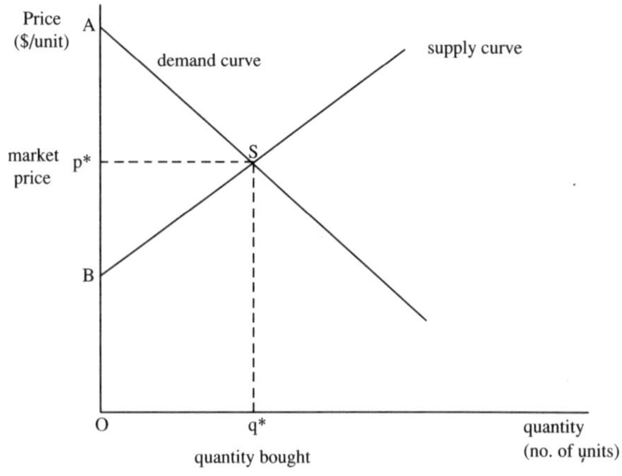

Figure 16.3 Market equilibrium

unsold stocks and are responsible for wasting valuable resources.

ADJUSTMENTS TO THE BASIC ECONOMIC MODEL

The basic economic model can be heavily criticised for its shortcomings in representing the complexities of real life. Aware of this, economists have been continually proposing adjustments to the model in order to make the theory operational. In this section we deal with two major and inter-related problems which turn up in environmental economics – 'public goods' and 'externalities'.

Public goods

Markets are created spontaneously when private goods are produced and exchanged. The quantities which the consumers are willing to buy at certain prices are added, and so are the quantities which the producers are willing to supply. The addition of quantities in this way is normal, because goods consumed by one cannot be consumed by another (e.g. food, clothing, housing, etc.). Other goods, however, can be used by many people at once with little or no impact on their use by others. There is no rivalry in consumption, and usually consumption by an additional person does not create extra costs. Everyone should therefore be allowed to use these goods as much as he or she wants and at zero price. For most of these goods it would be rather difficult to exclude some persons from using them, because the policing of such an exclusion would entail high costs.

Although the consumers benefit from the use of public goods, they do not have to translate their willingness-to-pay into real purchasing power. There is therefore no market demand for public goods, and the producers are not ready to react by offering such goods. No public goods would be offered in a purely private economy in which infrastructure, education, culture, public health and safety would be available in very low and sub-optimal amounts. It is thus necessary for public authorities to look after public goods and to offer them in a sufficient quantity so that the difference between consumers' surplus and supply costs is maximised (see Figure 16.3).

A special type of public good in the environmental area is the 'commons' (e.g. fisheries, rivers, natural scenery, clean air, a stable climate, etc.). A distinguishing feature of a 'commons' is that everyone has access to it. No one can be excluded from its use, or from entering it, because private property rights *vis-à-vis* a commons are too difficult to define and enforce, and also because commons are managed as a joint property by a larger group of people. In the latter case, if some members of the group abuse the commons, they may be denied access to it and/or to its use. Hardin (1968) has clearly pointed out the danger of the over-exploitation of such a public good in terms of the 'Tragedy of the Commons'. We will briefly discuss this 'tragedy' with the help of Table 16.1 which represents the pay-offs for an individual decision-maker. Imagine that Greenpeace, which is an NGO, has raised money to save some animal species – a project which may or may not be successful. An individual can choose to do nothing for this project, represented by zero pay-off shown at the left of Table 16.1. If the species does become extinct, the decision-maker is harmed (equal to the amount of –1); but, if the species is preserved, the individual experiences happiness (pay-off equal to +1). The numbers in Table 16.1 are obtained by adding the pay-off for the individual's personal attitude to the pay-off which derives from the results of the action. Judging by these joint pay-offs, the obvious strategy for the individual would be to do nothing because, regardless of whether the commons are destroyed or saved, the pay-off would be greater when the individual does nothing (the bottom line of Table 16.1 contains larger numbers than the top line). But, if every single individual follows the strategy of doing nothing, the likelihood that the commons would be destroyed becomes very great and all pay-offs to the individual become negative. On the other hand, society and all its individual members would be better off if the

Table 16.1 The 'Tragedy of the Commons', pay-off table of the individual decision-maker

	Commons are:	
The individual:	Destroyed (–1)	Saved (+1)
Contributes (–1)	–2	0
Does nothing (0)	–1	+1

commons are saved through the contribution of all individuals. Then, every individual would enjoy the 0 pay-off, shown in the upper right-hand corner of Table 16.1.

It is a major task for all societies in the world, and especially public authorities, to organise their members or constituencies in a way that enables them to contribute in full measure to the preservation of such common goods (Bromley, 1992). This is urgently needed for preserving the fragile life-support systems of both humans and natural ecosystems. Also, this unique but challenging responsibility of public authorities calls for the full-scale deployment of all possible means to gain control over the various pollution chains created by human activities (see pp. 307–310).

Externalities

Earlier (pp. 310–313), it was shown how the consumers maximise their utility by selecting the optimal quantities of goods, and how the producers maximise their profit by engaging the optimal production factors. Both parties act in full freedom, unhindered by the actions of others, and the only relationship between them is through market transactions that are freely entered into and which bear no special costs. However, there are many variables which appear in the utility functions of real consumers that are beyond their control. For example, if we go for a quiet walk in a nature park and motorcyclists cross our path, their noise lowers our utility significantly. We certainly did not want this. It was imposed on us by third parties that are far out of any market transaction. In order to avoid such unpleasant interruption, we must go to more distant nature parks where motorcycles are not allowed, or call upon the regulatory authorities to limit the freedom of motorcyclists in nature parks. A producer can also run into similar problems when, for example, the river water he or she uses as a (free) resource is polluted by discharges from a factory located upstream.

Such effects are called 'externalities'. Baumol and Oates (1988) have stipulated two conditions which, when satisfied, constitute an externality by definition:

> Condition 1: an externality is present whenever some individual's (say A's) utility or production relationships include real (that is, non-monetary) variables, whose values are chosen by others (persons, corporations, governments), without particular attention to the effects on A's welfare.
>
> Condition 2: the decision-maker, whose activity affects others' utility levels or enters their production functions, does not pay (receive) in compensation for this activity an amount equal in value to the resulting costs (or benefits) to others.
>
> (Baumol and Oates, 1988: 17)

This definition rules out cases in which someone deliberately does something to affect someone else's welfare (e.g. by crime). If no transaction costs are involved, all externalities can be avoided by imposing the right prices on all activities carried out or experienced by people. In reality, however, utility and production functions are distorted by the decisions of others, and this makes the market equilibria fall far short of social optima (Bromley, 1986). Such a situation is often illustrated by the following example. Imagine a power company with coal-fired generators, but without flue-gas cleaning equipment. The company offers electricity along a particular supply curve S°, leading to the market equilibrium at (p°, q°) shown in Figure 16.4. But, while the company imposes large external costs on others (acid rain, particles, greenhouse gases), it does not consider these costs as part of its own cost function. Therefore, the government should levy a tax on the company's harmful activities that will shift the supply curve to the left, to its socially correct position S*, leading to equilibrium at (p*, q*). But this would mean smaller quantities (q*, < q°) at higher prices (p* > p°).

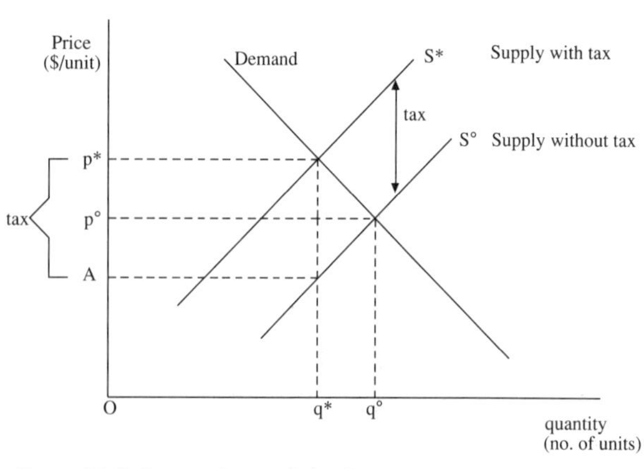

Figure 16.4 Externality and the Pigouvian tax

The share of this tax, taken up by the consumers or the producers, depends on the slope of the demand curve (in Figure 16.4 p°p* is the price rise for consumers, and Ap° the share of the tax rate on the producer not covered by the price rise). In other words, when the demand for the good is price-inelastic, consumers will bear a larger share of the tax burden.

Taxes of this type, called Pigouvian taxes, are very popular in economics as a remedy for externalities (Field, 1994), but they are difficult to quantify and realise in practice.

COST TRADE-OFFS FROM A PRIVATE AND SOCIAL PERSPECTIVE

Even though one may feel uneasy about many hypotheses and theoretical constructs of economists, the fact remains that both people and institutions always try to maximise their benefits commensurate with minimum costs. In this section we will describe this type of behaviour graphically – first for a private decision-maker and then from a social perspective.

Private decisions

In the context of the private use of natural goods, specifically energy, let us assume that a consumer wants to keep his house at a comfortable ambient temperature. To do so, he can spend a part of his heating budget on conservation measures (e.g. an efficient boiler, measuring and monitoring equipment, wall insulation, etc.), and the remainder on fuel or electricity for heating. Two types of cost, shown in Figure 16.5, are relevant to this hypothetical case. The cost curve, AA′, slopes upwards when energy conservation measures are not implemented, but with little conservation effort avoiding costs would be lower (point A′) although energy consumption would be higher. Thus, energy conservation first requires lower avoiding costs; but, to reach really low levels of energy use, avoiding costs may be high (point A). If there is no charge on the use of energy, people will consume more and more of it as long as curve AA′ slopes downwards in the direction of A′.

By imposing a tariff structure, the cost of energy consumed would be billed to the user along the cost curve BB′. The intercept, OB, represents a fixed amount which may be an entry fee or a standing charge, while the slope of BB′ represents the price per unit of energy. Clearly, the greater the consumption of energy, the higher would be the bill. A rational user will add the cost curves AA′ and BB′ to obtain the combined curve CC′ and aim for consumption q* which corresponds to the minimum point on CC′. It is clear that q* will be high if the slope of BB′ is small, and/or when AA′ is already high in the neighbourhood of A′ or runs up steeply. An energy conservation policy should, therefore, be oriented towards high user prices (i.e. a steep BB′ curve). This may be achieved either by lowering the fixed term OB and increasing the variable term, or by imposing taxes on energy use whereby BB′ would be rotated upwards around point B. On the other hand, greater energy conservation can be achieved by lowering the conservation costs AA′. This can be accomplished through improved technology, better information, subsidies, removal of barriers, etc.

Two points need to be emphasised with regard to Figure 16.5. First, the positive slope of the total cost curve CC′ at higher usage (i.e. at or near C′) is due mainly to the price of energy (i.e. slope of BB′). Lower prices will shift q* to the right and higher prices will shift q* to the left. Thus the price mechanism is a very powerful tool in setting particular usage levels of energy (and of other natural resources).

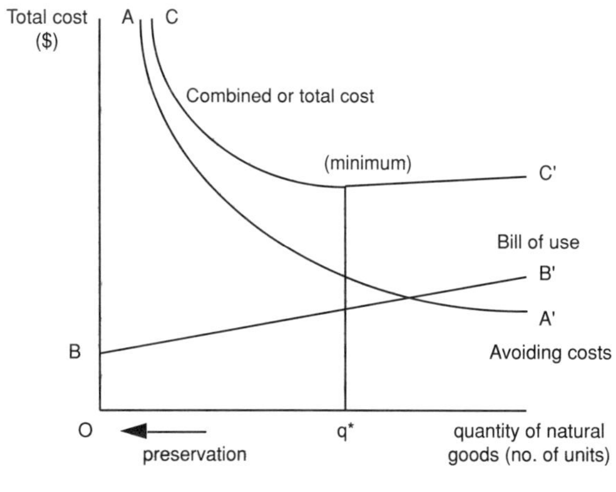

Figure 16.5 Private minimisation of total costs of an energy service

BOX 16.1 LOW ENERGY PRICES

While in most developed countries energy prices are set with the objective of covering at least the full financial costs of supply, in many developing countries they do not cover the economic or full financial costs of supply except in the case of some petroleum products. This means that energy consumers, many of whom produce other goods, lack the incentive to use energy efficiently, to select the most economic fuel, or to use the technology best suited to their needs. Fuels are used inefficiently in the transport sector when fuel prices are below border prices or when there are price differentials among alternative fuels (petrol, diesel, kerosene) distorting the market. In the case of electricity, a recent survey of electricity tariffs in 60 developing countries showed that, on average, between 1979 and 1988 tariffs declined from 5.2 cents/kWh to 3.8 cents/kWh (with respect to the constant value of the US dollar in 1986) – or less than half the level of those in the OECD countries. During the same period, power supply costs in those countries tended to be higher.

The survey found that average tariffs, in the case of nearly 80 per cent of the utilities in those countries, did not cover the long-term marginal costs of supply. In addition to encouraging waste in the end-use of energy and making energy-efficiency, conservation, and technology initiatives financially unattractive, these tariff policies were also found to put considerable financial strain on the energy sector. As a result, there was widespread inefficiency as well as underfunding of operations and maintenance which, in turn, caused further wastage of energy on the supply side.

Source: World Bank, 1993

Second, in general AA′ will increase sharply at low levels of usage (i.e. towards A), requiring large price increases (i.e. slope of BB′) in order to push q* further to the left. In other words, the impact of the price mechanism reaches its limits in the absence of new technical development or opportunity (that would lead to a lowering of AA′).

Social decision-making

While it is relatively straightforward to define and enforce property rights in the case of energy use, it is much more difficult to do so for most environmental goods. As a typical illustration of this, consider a power company which emits noxious substances through its stacks. Two types of cost are relevant here. First, the emissions will cause damage to people, property, nature, etc. (Craincross, 1993). Although extremely difficult to measure, it is reasonable to assume that the costs of damage would increase with higher levels of emissions (curve DD′ in Figure 16.6). Second, when the company lowers its emission level, it will incur abatement costs (e.g. of flue gas washers, catalysts, cleaner fuels, etc.). The more the emissions are reduced, the more the abatement costs will grow (curve AA′ in Figure 16.6). Adding costs DD′ and AA′ to give the combined cost CC′, it is possible to look for the minimum occurring at q* in Figure 16.6. There is, however, something special about curve CC′ in the sense that it is derived from two cost curves that are not incurred by the same decision-maker. Abatement costs AA′ are clearly shouldered by the polluter who has changed his technology, while damage costs DD′ are externalities imposed on innocent neighbours or other people. These people bear the harmful effects of the emissions, yet they have no control on emission levels. If both parties, polluters and victims, do not come to an agreement to address the problem, the optimum for the polluters would be around A′ and that for the victims around D. When the polluters are free to do what they want, there will be a very high level of pollution (A′) causing a correspondingly great damage (D′). Sometimes the victims can obtain a court ruling forbidding all further emissions. This would lead to a lower damage (D) but higher abatement costs (A).

It is here that governments can intervene to weigh both curves AA′ and DD′ on an equal footing in order to find a value of q* which is equal to the social optimal level of emissions. For ease of analysis, the total cost curves (top diagram of Figure 16.6) have been converted into marginal cost curves (shown in the bottom diagram of Figure 16.6). The minimisation of (AA′ + DD′) over q requires MAC to be equal to MDC (point S at level q*). We stress

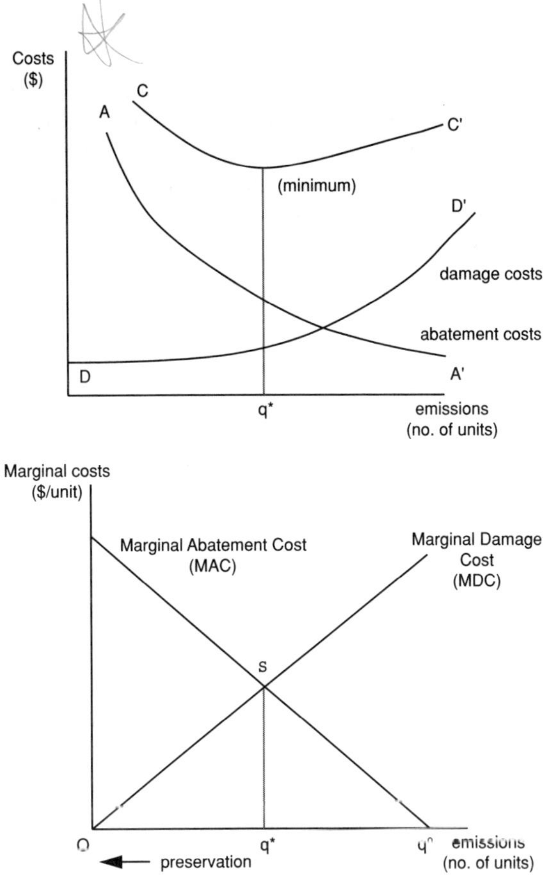

Figure 16.6 Cost minimisation from a social point of view

that the polluter only knows, or wants to know, about the MAC curve, while the victims only take the MDC into account. The best social position occurs at point S. The polluter must reduce emissions from level q° to q* (i.e. a reduction of q*q° units of emission) at a cost equal to the triangular area of q°Sq*. The remaining emission level q* (i.e. Oq* units) would cause damage whose costs would be equal to the triangular area of OSq*. This means that externalities are not equal to zero in the social optimum. The public policy-maker accepts this externality because, at any point left of the vertical line q*S, abating a unit of pollution will cost more than the cost of the damage it would cause (i.e. MAC lies above MDC). Polluting more than q* would not be optimal either, because it would be cheaper to abate than to carry the damage (i.e. MAC lies below MDC). However, in practice policies are generally not derived from a clear and extensive analysis of the cost curves of damage and abatement. On many occasions standards and permits to pollute are issued on the basis of maximum supportable abatement costs for polluters, i.e. they are required to implement Best Available Technologies Not Entailing Excessive Costs (BATNEEC) without a full consideration of how much damage will be left over after the implementation of BATNEEC. When the damage is too high, the case may be brought before the court and, in general, the court will either allow the polluter to continue his activities, or order a complete stop to the polluting activities.

In order fully to understand the prescriptions of economists for curing environmental problems, it is essential for the reader to be acquainted with the marginal cost curves shown in Figure 16.6. In the following section the reader will be assumed to have gained a full understanding of these curves.

INSTRUMENTS FOR REACHING THE SOCIAL OPTIMUM

The instruments favoured by economists to regulate environmental pollution are briefly described in this section. Unfortunately, there is insufficient space here fully to discuss the advantages and disadvantages of the various instruments. We will, therefore, limit our discussion mostly to their mechanics. Later (pp. 320–323), a list of criteria, used for judging policy instruments, will be given with some reference to the instruments to be described here. For our discussion we will make ample use of the lower diagram of Figure 16.6, redrawn as Figure 16.7.

Negotiation

Some economists advocate that governments should refrain from intervening in pollution affairs and leave the problem to the free market. In all cases, they argue, the market actors will enter into negotiations whereby pollution levels would settle down to the optimum q* (Figure 16.7). When the polluter is liable for the damage caused, the victim will claim compensation that would be at least as high as the marginal costs of damage caused by pollution. Starting at point q°, the polluter would be reluctant to pay the high damage cost represented by q°T. Instead, he would prefer to reduce emissions along the MAC function. He will reduce emissions in this way as long as MDC is higher than MAC (i.e. as long as the claim per unit of pollution is higher than the unit cost of emission reduction). However, once point q* is reached, the polluter would prefer to pay the claims of the victim rather than paying the additional cost of abatement.

Even if the polluter is not liable for the emissions, negotiations will bring both parties to the optimum q*. At the emission level of q° damage is so high (q°T) that victims will actually pay the polluter to reduce emissions. As long as MDC > MAC, an agreement between both parties can be found, because the polluter would be interested in further abating the emissions if he is paid by the victim to do so. But when MDC < MAC, the victim's willingness-to-pay falls below the abatement costs, finally reaching point q*.

This elegant solution to the pollution problem is of course contingent upon smooth and successful negotiation between polluters and victims, but transaction costs will prove to be excessively high, especially if the victim is a member of the general public. Negotiations work well and successfully only when the contending parties, polluters and victims, are almost equally strong. A typical example is that of two companies on the same river, both almost equally strong in terms of power and financial resources, one polluting the process water of the other. Most environmental problems, though, are caused by the activities of thousands of polluters (see Figure 16.1) producing negative externalities on the commons and on a multitude of third parties. In such cases negotiation can only take place through the political process.

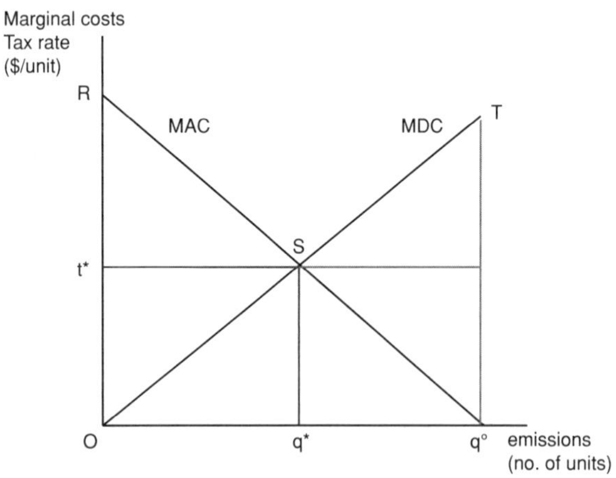

Figure 16.7 Economic policy instruments for controlling pollution

It is sometimes argued that level of income often influences the allocation of liability, but clearly it would be more equitable and appropriate to assign liability to the polluter based on the provisions of the 'Polluter Pays Principle'.

Negotiation is certainly interesting as an instrument, but it is actually used only in a relatively small number of cases. Relevant and responsible public authorities have to intervene in by far the majority of pollution problems and disputes which occur, equipped with the other instruments described below (Bromley, 1986).

Standards

A standard is the most commonly used instrument of pollution control. For a given problem, a standard comprises a set of target levels of emissions (or emission reduction) for different pollutants which may not be exceeded, together with procedures for measuring and controlling these levels and the enforcement rules to be applied when the targets are not met. From the economist's point of view, the standard in our reference problem (Figure 16.7) would be a permit to pollute up to the amount Oq*; or, in other words, an obligation to reduce emissions by the amount q*q°. Imposition of this standard on the polluter would mean changing abatement costs by an amount equal to the area of q*Sq°.

Setting q* as a standard implies knowledge of the MDC curve. But, since there is a great deal of uncertainty about MDC, in practice standards are generally set rather arbitrarily. Often the 'economic carrying capacity' of the polluter, relative to the weight of the abatement costs (area q*Sq°), prevails in setting the standard as q*. When the polluters know that standards will be used as an instrument they tend to exaggerate their abatement costs, pretending that

MAC is higher than it really is. Thus the polluters overestimate their costs (q*Sq°), in an attempt to force the regulators to set a lower standard at Oq*.

Taxes (charges and levies)

This instrument is widely advocated by economists. The level of the Pigouvian tax to be imposed on polluters is represented by point S in Figure 16.7 which is the intersection of the MAC and MDC curves. With a tax rate of t* per unit of emission, at point q° the polluter will have to pay a tax of t*.q° (i.e. the rectangular area Oq°.Ot* in Figure 16.7). At this point the polluter will find that it is better to reduce emissions, and he will do so as long as MAC is lower than the tax rate t*; but at point q*, and beyond towards R, further reduction is no longer profitable (MAC > t*). So the polluter's total costs consist of a tax t*.q* (the rectangular area Ot*Sq*) plus abatement costs incurred in reducing emissions from q° to q* (triangular area q*Sq°). For a given tax rate of t*, the polluter will incur lower costs at no pollution level other than q*. As for the victims, their damage burden will remain equal to the triangular area OSq*.

The optimal Pigouvian tax represents the marginal cost of pollution damage at the optimum level of emissions, q*. It is pointed out that lack of sufficient and reliable information is a major problem in setting the optimal tax level, t*, as it is in finding the optimal standard, q*.

There are many other items and activities, apart from emissions, which can also serve as the basis for taxation. The flow chart of the economic throughput process, given in Figure 16.8, shows the wide range of tax incidence possibilities in this regard. However, economists generally prefer taxes on input factors and on emissions to taxes which intervene directly in the economic choices of production and consumption (European Communities, 1994).

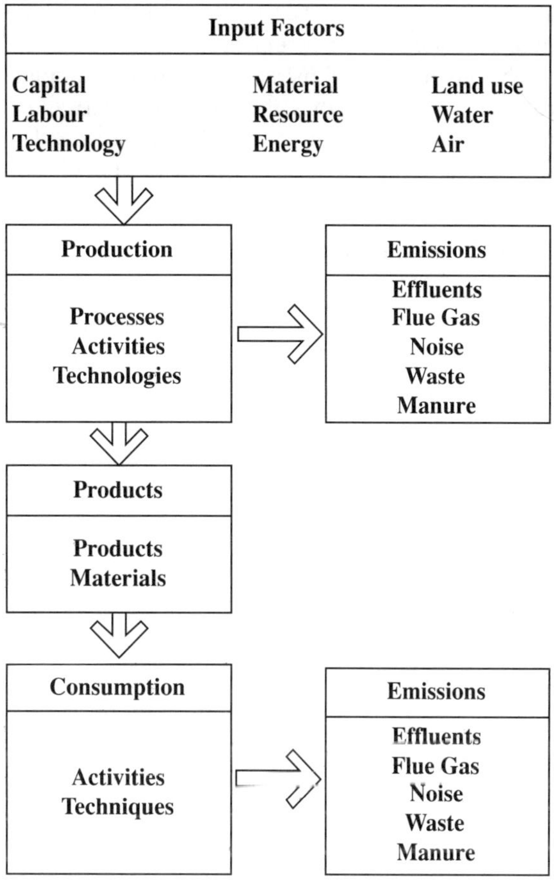

Figure 16.8 Tax incidence

Marketable permits

Marketable permits are a very attractive instrument from an economic point of view. The government issues an amount of emission permits equal to Oq* (Figure 16.7) in a simple format (e.g. one permit allows the polluter to emit 1 ton of SO_2 annually). This amounts to setting a vertical supply curve of emission permits at the optimal point q*. The polluters, each with its own MAC curve, bid for the permits and this leads to a market in MAC (shown in Figure 16.7) representing its willingness-to-pay (or demand curve) for permits. Demand for and supply of permits reach equilibrium at point S, with a market permit price equal to Ot*. Again polluters will be motivated to abate pollution from the original level q° up to the optimal level q*, incurring costs represented by triangle q*Sq°. Depending on the way the permits are issued by the public authorities, polluters will either pay for the first acquisition (e.g. when an auction takes place) or receive them without charge (e.g. permits are distributed to present polluters according to a set of rules). The pollution permits may be traded in the permit market.

The instrument of marketable permits combines the main advantage of standards, which guarantees that the pollution limit of q* will not be exceeded, with the acknowledged efficiency of the tax system.

Deposit refund systems

Application of standards, taxes or marketable permits requires the monitoring of emissions which may not always be feasible when the sources are numerous (e.g. littering, disposal of lubrication oil by car owners). Deposit refund systems can be useful instruments in such situations. In these systems the potential polluter pays a deposit when buying a good, and receives a refund if certain conditions are met; for example, proof that he has returned the good to a specified place, or that a certain type of damage has not been done by the good or to the good. The instrument of deposit refund can also be applied to dangerous substances such as mercury. When applied to such substances, these schemes motivate the user to take greater care in using them than he would otherwise, especially in minimising spillage, as well as to minimise the quantity of such substances he holds in stock. Although there are other instruments such as subsidies, liability rules, etc., the most popular instruments of the economist are those which we have discussed above.

CRITERIA FOR SELECTING SUITABLE POLICY INSTRUMENTS

This section draws primarily on the work of Bohm and Russell (1985). They have proposed six criteria that are listed below and briefly discussed. In selecting a particular instrument for application, it is obviously important to ensure that it is practical, effective, suitable for the circumstances under which it is to be used, and, above all, that it will achieve its objectives.

Static efficiency

The word 'static' here means that pollution abatement is performed by each polluter using a given technology and at a given location. Efficiency is measured by the total expenditure a polluter has to make in order to meet specified reduction targets. And, clearly, the sum of the abatement costs of all the polluters should be considered in evaluating the efficiency of the particular policy instrument implemented.

Taxes and marketable permits are generally superior to standards when efficiency is considered. For example, if a particular tax per kg of SO_2 emission is levied, every polluter will react to it by reducing emissions up to the point where the MAC becomes equal to the tax rate (see Figure 16.7). Through a uniform tax, or through the price of permits when marketable permits are in use, the marginal costs of the last unit of emission abatement by the several polluters will become equal to each other. This will guarantee the realisation of the overall abatement effort at least cost. In order for standards to attain the same efficiency (minimum cost), it is necessary to specify a different standard for every different abatement cost curve, thus ensuring that the MAC curves at the permitted standards are equal for all the sources. This is clearly more tedious than when a uniform tax rate is levied. The latter becomes especially appealing when the damage cost function may be assumed to be linear, resulting in a constant MDC (i.e. a horizontal line in Figure 16.7) which is exactly the optimal tax rate.

Dynamic efficiency

This involves actions, encouraged or promoted by the instrument in the longer term, offering incentives to the polluters to alter their technologies for production and/or abatement, or to relocate in other areas where emissions are less damaging. An instrument should also give continuing encouragement to polluters to seek better technologies or other measures with which to minimise the pollution damage they cause.

Taxes and marketable permits perform well in this respect because of the financial incentives they offer. Tax motivates the polluter to save money by polluting less, while marketable permits generate revenue for companies who sell their permits to other companies that are less efficient in pollution abatement.

BOX 16.2 AN EMPIRICAL ANALYSIS OF ALTERNATIVE REGULATORY APPROACHES IN THE USA

On the subject of efficient environmental policies, there is at present a major gap between a vast body of theoretical literature on one hand, and a surprisingly small amount of empirical analysis on the other. Recent research into alternative regulatory approaches in environmental policy in the US paper-and-pulp industry contributes to bridging this gap. This research focused on three alternative regulatory approaches affecting the industry: technology standards, ambient standards, and benefits-based standards. The first approach is dominant in water quality policy, mandated by the Clean Water Act of 1972. It requires polluting sources to adopt the 'best available' or 'best practicable' pollution control technologies. The second approach is prevalent in policy on air quality, mandated by the Clean Air Act of 1970. It requires polluting sources to install pollution control equipment to the extent necessary to meet the ambient standards. The third approach also stems from the Clean Air Act. For pollutants that are not hazardous to human health, standards may be based on a comparison of environmental benefits and control costs.

In an attempt to evaluate the efficiency of the three regulatory approaches, the environmental benefits and control costs of 60 to 68 paper-and-pulp mills were analysed during the period from 1973 to 1984. Reductions in the amounts of the following pollutants were analysed: organic waste (i.e. biological oxygen demand, BOD), particulates (i.e. total suspended particulates, TSP), and total reduced sulphur (TRS). The relevant costs and benefits, obtained from this analysis, are summarised in the following table.

Costs and benefits of three alternative regulatory approaches (in million ECU, 1984)

Approach/ pollutant	*Mills with B>C/mills analysed*	*Total benefits*	*Total costs*	*Net benefits*
'Technology'/BOD	07/68	30.4	80.2	– 49.8
'Ambient'/TSP	22/60	20.9	19.8	+ 1.1
'Benefits'/TRS	29/60	72.1	46.3	+ 25.8

BOD = biological oxygen demand; TSP = total suspended particulates; TRS = total reduced sulphur

The environmental benefits of the technology-standard approach exceeded the costs in only 7 out of the 68 mills investigated, and the overall net benefits of this water quality policy turned out to be negative. The approach used in air quality policy performed better – the environmental benefits exceeding the costs in 22 out of the 60 mills investigated. The overall net benefits of the policy were slightly positive. The benefits-based approach was found to perform best in that net benefits occurred in 29 out of the 60 mills investigated, with significantly positive overall net benefits.

Two major conclusions may be drawn from this work. First, the efficiency of environmental policies is enhanced when they take environmental consequences into account. Second, efficiency is further increased when local conditions (e.g. the number of people benefiting from improved environmental quality) are taken into account.

Source: Luken and Clark, 1991

Standards usually do not stimulate innovation. Also, because sometimes tight standards only apply to new investments, they actually encourage established companies to pollute more by retaining and operating their existing (old) plants, processes and equipment.

Furthermore, once a tax demand is issued, a polluter will pursue the optimal level of abatement as soon as possible because he would minimise his costs by doing so. Also, compliance with standards will be delayed, or may even be ignored, unless non-compliance is prosecuted by law.

Intensity of information

This is a measure of the data, and of the level of predictive modelling skill, needed by the pollution control agency to implement a given instrument. Its importance lies in the desire for achieving an efficiency target when many different polluters impact on the environment differently (see the request for individually tailored standards when overall least abatement costs are pursued, discussed on p. 320).

In fact, all economic instruments assume a thorough knowledge of abatement and damage cost functions in order to be able to pinpoint the optimal emission level q^* and the corresponding marginal cost (distance q^*S or Ot^* in Figure 16.7). With taxes or marketable permits, however, efficiency in abatement is achieved with less information than with standards.

Ease of monitoring and enforcement

Emissions need to be measured for assessing compliance, preparing bills, and to audit self-reporting. However, many types of emission are invisible and/or fugitive, and vary with production levels, equipment malfunction and methods of operation. Also, measurements are subject to errors and mistakes, for example when discrete sampling is used or when the polluter obstructs entry to the premises. Enforcement actions to force violators back into line may include administrative fines, civil or criminal court proceedings, or indirect actions such as blacklisting.

The problems and costs of monitoring and enforcement are related to the technical status of the polluting processes, as they are with the legislative and regulatory regime of the country in question. For example, marketable permits are much easier to organise in the USA, where all parties are accustomed to the functioning of the free market, than in most other industrialised countries. It would appear, therefore, that monitoring and enforcement tasks do not vary a great deal in a given setting between the various policy instruments.

Flexibility in the face of economic change

'Flexibility' here means the ease with which a policy instrument, designed to maintain certain emission targets, can be adjusted to changes in the economy, when such changes occur in terms of inflation, industrial growth and other economic indicators.

High flexibility for adjustment allows the regulated polluter to take proper decentralised action. Low flexibility, on the other hand, necessitates a recalculation and re-issuing of measures by the controlling agency. This involves costly information gathering and administrative effort, and may leave the door open for political interference and lobbying.

Marketable permits have much to recommend themselves in this respect, because decentralised market forces take over once the public authority has set the overall target at q^*. New polluters can enter the market by offsetting emissions caused by existing polluters. The problem of inflation is addressed by the fact that permits respond to supply and demand. It is also possible to reduce the overall pollution level q^* by the agency by withdrawing some of the permits from the market. All that is necessary in the permit scheme is to find a good mechanism for issuing and administering permits and for monitoring the performance of the market actors.

Taxes, on the other hand, are less flexible than marketable permits, even though they are easier to adjust than standards which, by their nature, involve complex legislation and/or agreements with individual polluters.

Political considerations

Political considerations affect society's choice of policy instruments, at least as much as cogent arguments about the relative merits of the above criteria. Three distinct political considerations are especially important. The first is distributional, the second ethical, and the third concerned with broader aspects of economic stabilisation.

The distribution of the costs of implementing a policy instrument often determines its viability. A policy instrument may be voted down if a majority of the voters, or a powerful or vocal group of voters, incurs significant financial costs from it. Taxes are unpopular with polluters mainly because tax bills, which represent 'mere transfers' from the polluters' pocket to the public treasury, are a part of the cost to be borne by the individual polluter. Marketable permits also imply significant transfers when they are reauctioned frequently.

For many people pollution is a violation of natural harmony, and they regard its creation as being little short of a crime. To this extent regulatory orders work well when backed up by criminal sanctions. Tax systems, on the other hand, allow for the immoral 'buying' of the right to pollute. However, a large and increasing number of people in the environmental movement now acknowledge the positive results which can be achieved through the application of economic instruments.

It is argued, especially by the major polluters, that stringent policy measures would ruin economic welfare and that closure of factories, and resulting mass unemployment, would be part of our future if such measures were to persist. There is empirical evidence of job loss when polluting activities are closed down, as there is of job creation in sectors and countries playing a maverick role in environmental protection. Also, a prima facie danger of bias arises when policy-makers offer lucrative markets to large capital groups, by authorising them to build waste incineration plants for example, instead of actively promoting waste prevention strategies.

The performance of the various economic policy instruments under the above criteria depends to a large extent on specific circumstances. For example, is there a well-established and operational environmental protection agency? Do free markets work? Is the overall regulatory and legislative machinery capable of administering environmental regulation and enforcement effectively? A scrutiny of the proposed instruments against the criteria listed above will at least improve the choice of instruments to be used.

CONCLUSIONS

Environmental problems are numerous, complex and difficult to solve. They arise from a large variety of human activities and are difficult to monitor. It is clear, however, that for sustainability it is necessary to widen the scope of policy from curative to structural. The latter requires the involvement of millions of people acting as members of society or as representatives of the various target groups (car drivers, entrepreneurs, farmers, family heads, etc.). It is not possible to try to direct the activities of all these people, or to modify their behaviour, with instruments that may be perceived by them as 'big sticks'. Instead, the approach should involve influencing the mechanics of decentralised decision-making processes. The price mechanism, which is a very efficient carrier of information as well as a guide, pervades the daily lives of millions of people. Also, next to education and legislation, the economic system is the major and effective organising force in modern industrial societies. Not surprisingly, therefore, economists propose to exploit this mechanism to the full in an attempt to set up a comprehensive and effective environmental policy.

Several economic policy instruments offer insight into better policy-making, especially taxes (charges) and marketable permits. However, for an optimal use of these instruments, it is necessary to understand the principles of decision-making followed by the economic agents (consumers and producers) and pursued by public authorities (environmental protection agencies). The choice among the various economic policy instruments can be facilitated by checking them against a list of criteria. From an economist's point of view, efficiency should be ranked highly in that list, although the exigencies of practical and political realities will determine the final choice. Even so, attention to the opportunities offered by economic policy instruments may significantly improve policy-making, thus opening better perspectives for sustainable development.

REFERENCES

Arrow, K.J. (1974) *The Limits of Organizations*, New York: W.W. Norton & Company.

Baumol, W.J. and Oates, W.E. (1988) *The Theory of Environmental Policy* (2nd edition), Cambridge: Cambridge University Press.

Bohm, P. and Russell, C.S. (1985) 'Comparative Analysis of Alternative Policy Instruments', in A.V. Kneese and J.L. Sweeney (eds) *Handbook of Natural Resource and Energy Economics, Vol. I*, Amsterdam: Elsevier Science Publishers, pp. 395–460.

Braden, J.B. and Kolstad, C.D. (1991) *Measuring the Demand for Environmental Quality*, no. 198 in the series *Contributions to Economic Analysis*, Amsterdam: North-Holland.

Bromley, D.W. (ed.) (1986) *Natural Resource Economics. Policy Problems and Contemporary Analysis*, Boston, Mass.: Kluwer Nijhoff Publishing.

Bromley, D.W. (ed.) (1992) *Making The Commons Work*, San Francisco: Institute for Contemporary Studies.

Cairncross, F. (1993) *Costing the Earth*, Boston, Mass.: Harvard Business School Press.

European Communities (1994) Potential Benefits of Integration of Environmental and Economic Policies: An Incentive-based Approach to Policy Integration, *European Communities Environmental Policy Series*, Brussels, pp. 336–337.

Field, B.C. (1994) *Environmental Economics. An Introduction*, New York: McGraw-Hill.

Fisher, A.C. (1981) *Resource and Environmental Economics*, Cambridge: Cambridge University Press.

Hardin, G. (1968) 'The Tragedy of the Commons', *Science*, 162, December: 1243–1248.
Luken, R.A. and Clark, L. (1991) 'How Efficient are National Environmental Standards? A Benefit–Cost Analysis of the United States Experience', *Environmental and Resource Economics, vol. 1*, Dordrecht: Kluwer Academic Publishers, pp. 385–413.
Pearce, D.W. and Turner, R.K. (1990) *Economics of Natural Resources and the Environment*, London: Harvester Wheatsheaf.
Winsemius, P. (1986) *Gast in eigen huis. Beschouwingen over milieumanagement*, Alphen a/d Rijn: Samson H.D. Tjeenk Willink.
World Bank (1993) *Energy Efficiency and Conservation in the Developing World*, Washington, DC: World Bank.

SUGGESTED READING

Baumol, W.J. and Oates, W.E. (1988) *The Theory of Environmental Policy* (2nd edition), Cambridge: Cambridge University Press.
Braden, J.B. and Kolstad, C.D. (1991) *Measuring the Demand for Environmental Quality*, no. 198 in the series *Contributions to Economic Analysis*, Amsterdam: North-Holland.
Field, B.C. (1994) *Environmental Economics. An Introduction*, New York: McGraw-Hill.
Fisher, A.C. (1981) *Resource and Environmental Economics*, Cambridge: Cambridge University Press.
Kneese, A.V. and Schultze, W.D. (1985) 'Ethics and Environmental Economics', in A.V. Kneese and J.L. Sweeney (eds) *Handbook of Natural Resource and Energy Economics, Vol. 1*, Amsterdam: Elsevier Science Publishers.

SELF-ASSESSMENT QUESTIONS

Since the economic instruments have been explained in the text with the aid of figures, six questions are set below with the object of giving students an opportunity to practise using such figures. The following apply to all the six questions:

q = mercury emitted by the polluter in kg/year;
t = tax rate imposed on mercury emission in $/kg/year by the environmental agency;
MAC (marginal abatement cost) = 12–q;
MDC (marginal damage cost to the victim) = 2q.

The policy is to be directed towards minimising overall costs from a societal point of view. (*Hint:* draw a graph showing MAC and MDC.)

1 What level of mercury emissions is optimal in kg/year?
(a) 4 kg/year;
(b) 10 kg/year;
(c) 11 kg/year.
2 What reduction in mercury emissions will be required from the polluter in kg/year?
(a) 2 kg/year;
(b) 8 kg/year;
(c) 1 kg/year.
3 What is the optimal tax rate t* in $/kg/year?
(a) $2/kg/year;
(b) $6/kg/year;
(c) $8/kg/year.
4 What abatement costs will the polluter incur?
(a) $11;
(b) $32;
(c) $52.
5 What would be the overall expense of the polluter?
(a) $64;
(b) $59;
(c) $33.
6 What damage (in $) would be inflicted on the victims?
(a) $7;
(b) $16;
(c) $25.

17

APPLICATION OF ECONOMIC INSTRUMENTS FOR ENVIRONMENTAL MANAGEMENT

James Winpenny

SUMMARY

Prices and market mechanisms are used in Economic Instruments (EIs) to induce the desired change in the behaviour of producers and consumers. EIs have several potential advantages over pure regulation, notably that they enable polluters to comply with standards at a lower cost to themselves and to society. But there is less certainty of their impact. In practice EIs tend to complement, rather than substitute for, regulations.

This chapter focuses on a group of EIs commonly used to control pollution and waste disposal; namely, charges and taxes, deposit refund schemes and marketable permits. For air pollution, the use of marketable permits is gaining ground, mainly in the USA, while emission charges are less common and yet to be proven. The situation is reversed for controlling water pollution in which effluent charges are a more common and effective method compared to marketable permits that are seldom used. A variety of EIs are in use for solid waste; of these the deposit refund schemes have been the most effective and successful to date.

The EIs mentioned above are not the only ways of using prices, markets and self-interest to serve environmental policy. For instance, in developing countries a wider range of measures is evident, such as the use of customary rights in regulating the use of common property resources, reforms in the pricing of natural resources, capturing economic rents for society, etc. Although EIs are often discussed with the needs and experience of the developed OECD countries in mind, it is important to acknowledge this wider perspective.

ACADEMIC OBJECTIVES

The chapter should be read in conjunction with Chapter 16 of this volume. After completing this chapter the reader should have a good understanding of the following:

- the concept of EIs, their advantages and disadvantages, and the extent to which the advantages are realised in practice;
- how the EIs relate to other policy measures for pollution reduction;
- how EIs are used for controlling air and water pollution, and solid waste disposal, and the advantages and disadvantages of marketable permits and emission charges in controlling such pollution;
- how, and the extent to which, EIs can generate and promote incentives to producers and consumers to create less pollution;
- how EIs could be used effectively in developing countries and in the countries in economic transition.

SCOPE OF THE CHAPTER

Concepts and definitions

In most general terms, an economic instrument (EI) is a policy measure which uses prices and market mechanisms to induce the desired change in the behaviour of producers and consumers. An EI appeals to the financial self-interest of agents, rather than to their observance of laws and regulations or their response to exhortation or altruism. EIs give agents the freedom to respond in their own ways to incentives. As the previous chapter showed, under certain conditions this produces the 'optimum' response from the point of view of public interest.

EIs are often used in combination with other measures, such as regulations, information campaigns, and exhortation. Although EIs are often presented as alternatives to regulation – frequently referred to as 'command and control' – in practice they usually depend on an effective system of regulation and monitoring.

EIs should be differentiated from two other types of measure, both involving financial payments. The first is the 'user charge', which is a payment levied on a person, household or a firm in return for certain environmental services. The treatment of waste water through sewage collection and treatment is one example. Others are the collection and disposal of solid waste and garbage, and charges for entry into areas of natural beauty or wildlife which merely defray the cost of upkeep and administration. However, in none of these instances is it the intention to discourage the use of the services in question.

The second example of a financial measure, which is not an EI, is a fine or penalty imposed to enforce a system of regulations. Thus, a person found disposing of waste illegally, a firm discovered releasing toxic substances into a river, or a hunter who kills a protected animal, may be fined for misdemeanour. The purpose of fines is to enforce an absolute standard or prohibition, not giving the agent an option of whether or not to comply. However, enforcement is often lax in practice, and the potential environmental offender may be able to form a judgement about the probability of being caught, of being convicted, and of having to pay the fine. If the benefit of offending exceeds the weighted probability of paying the fine, the rational criminal may persevere and vice versa if the benefit is less. In this case the fine is more akin to an EI.

Many, perhaps the majority, of the EIs in operation today are set too low to have a strong incentive effect (OECD, 1994). Although they may be intended to be an incentive and designed accordingly, the payments involved may not induce the required change in behaviour. A common reason may be that these schemes are introduced with the limited ambition of raising enough funds to cover the cost of administering the environmental policy concerned.

This situation should be distinguished from another common case in which an EI generates no net revenue for the public purse because its proceeds are redistributed to users as subsidies, or applied to a specific environmental project. In these 'carrot and stick' cases, which are commonly resorted to for mobilising public support for EIs, there may be an incentive for users to reduce pollution and other forms of environmental harm, and to invest in more environmentally benign practices.

Context and scope

There are three main services that the natural environment provides for mankind. First, natural resources such as trees, fertile soil, coal and water which are used in production or for direct consumption. In many cases it is both feasible and desirable to charge for these goods and services, but in other cases, public goods such as sunlight and fresh air for example, pricing is not feasible. One important class of EIs applies to these resources when they are used directly in production and consumption.

Second, the environment provides a 'sink' for the receipt and assimilation of wastes resulting from consumption or production. If the sink is overloaded by excessive emissions or waste disposal it becomes polluted. A second major class of EIs is concerned with pollution and waste charges.

A third environmental function is the provision of life-support services such as climatic regulation, photosynthesis, pollination, protection from ultra-violet rays, genetic resources, etc. It is not usually possible, necessary or desirable to charge for these services.

This chapter will be principally concerned with the second of these types of EI mentioned above, namely those addressing pollution and waste disposal. Much of the debate on EI has focused on this area, reflecting the concerns of developed and industrialised countries. However, it is essential for countries, at all stages of development, to get their resource prices 'right' as well by removing distortions caused by so-called market and policy failures (Panayotou, 1993; OECD, 1995). A brief discussion of EIs in this broader context can be found on pp. 337–338.

Most of this chapter will be concerned with the use of EIs for controlling air and water pollution and solid waste disposal. According to an authoritative recent survey (OECD, 1994), this coverage reflects the actual prevalence of EIs. There are other major environmental problems, such as aircraft noise, liquid animal manure and ozone depletion that are also susceptible to the use of EIs; but experience with them is more limited and the options less clear-cut.

Relation of EIs to other policy measures

Much of the debate on EIs creates the impression that EIs and regulation – the 'command and control' methods – are the only alternatives. There are, however, other measures, such as the use of exhortation, persuasion, demonstration, appeal to the national good, etc. which do not fall into these two categories. Yet other actions involve creating or clarifying people's use of, or access to, environmental resources ('reforming property rights') and facilitating agreements on the use of common resources ('reducing transactions costs') that are basic to the success of specific environmental policies. They can be regarded as indirect supporting actions, but are none the less important for that.

Projects and spending programmes are also part of the armoury of environmental policy: in some cases projects can substitute for policy measures, traffic management schemes for road pricing for example; while in others they can complement each other, as in investments in cleaner public power stations alongside increases in electricity prices and pollution taxes.

There is a much wider group of economic policy actions, which have environmental implications and which can be manipulated for environmental gain even though this is not their prime purpose (components of economic reform and structural adjustment programmes for example). The various ways of classifying environmental policy measures are illustrated in Table 17.1.

As Table 17.1 shows, governments have an array of policies at their disposal to encourage their citizens to alter their behaviour towards the environment, and often there is synergy between different measures. For instance, the four kinds of targeted measure often work best in combination and, in any case, they need to be fully backed up by indirect and supportive measures. In addition, the effect of targeted policies can easily be negated by general economic measures which force incentives in contrary directions.

Table 17.1 Types of environmental policy measure

Type	*Examples*
Targeted	
Command and control	Air and water quality laws, waste disposal regulations
Economic instruments	Charges, taxes, subsidies, deposit refund systems, tradable permits, resource pricing
Exhortation	Publicity, propaganda, 'good practice' demonstrations
Projects and programmes	Afforestation, installation of cleaner technology
Indirect and supportive	
Property rights reforms	Land, leasehold tenure, rights of access, community control
Reduced transactions costs	Better information, simpler legal procedures, titling, legal reforms
General economic measures	Structural adjustment and economic reform programmes, exchange rate, budget, monetary policy, foreign investment code

Source: OECD, 1995

ADVANTAGES OF ECONOMIC INSTRUMENTS: A SUMMARY

Regulations and EIs

A regulation aims to influence environmental behaviour through control and prohibition. Certain practices, such as tipping radioactive wastes into public drains and exporting ivory, may be banned by law, while other activities may be subject to ceilings or discretionary permits (for example release of waste water into rivers or lakes, hunting whales and game animals). Physical planning, land-use zoning and building permits are also attempts at preserving local amenity and environmental quality. Some regulations are based on the impact of pollution on a receptor (e.g. control of the release of particulate matter according to the preservation of minimum air quality in a specific air quality 'basin'), while others attempt to control emissions by banning selected production processes and types of equipment.

Some types of regulatory measures have financial penalties built into them, such as the use of fines to ensure compliance and the use of licence fees typically to cover the costs of administration. However, as we have pointed out in earlier (pp. 325–326), fines are not intended to be an EI; but, where enforcement is weak, the probability of paying a fine may act like an economic incentive rather than a categorical measure.

The dichotomy between regulations and EIs should not be exaggerated. Regulations can provide 'fail safe' limits within which economic instruments can work. For example, a pollution tax, proportional to the amount and/or concentration of the pollutant, could be backstopped by an absolute prohibition on certain kinds of pollution, such as heavy metal pollution or pollutants in excess of a stipulated concentration. Moreover, the apparatus of regulation (environmental quality standards, performance monitoring, and enforcement systems) is a prerequisite of certain market-based solutions. The concept of 'tradable permits' for air pollution control (see below) depends on a highly developed system for setting local air quality standards, monitoring air quality, inspecting factories and premises, and enforcement through fines and other penalties.

The principal categories of EIs are the following (OECD, 1994): *charges and taxes* – intended to penalise anti-social practices and the use of certain products – are often justified by the Polluter Pays Principle, according to which the party causing pollution should bear the costs of that pollution falling on others. They directly raise the costs of the offending party, thus producing an incentive for that party to reduce or cease the practice concerned. In most cases charges and taxes will raise revenue, unless they discourage entirely the polluting practice or the use of the polluting matter in question. Charges refer to the *quid pro quo* for the provision of a service (public treatment of effluent for example), whereas taxes are an unrequited levy, although the distinction is not always possible to uphold. There are many examples of charges and taxes, such as effluent charges and product charges, on harmful substances for example; and differential taxes, as on leaded and unleaded petrol, or according to the size of vehicle engine. Outside the realm of pollution, charges and taxes could apply to licence fees for hunting or fishing, entry to nature reserves, levies on the growth of crops considered to be erosive, taxes on pesticide, etc.

Subsidies, which are another type of EIs, are the opposite of taxes and charges in the sense that they are aimed at promoting the consumption of environmentally friendly products and services. Examples include subsidies on commercial heating and cooking fuels instead of alternatives such as charcoal and wood-fuel; subsidy to non-conventional power generation such as from wind, waves, photovoltaics, etc.; subsidy to fertilisers aimed at discouraging extensive modes of cultivation, etc. However, the disadvantage of subsidies is that they absorb public revenues and are notoriously prone to be diverted and abused. They also offend the Polluter Pays Principle.

Deposit refund schemes encourage users or producers of a potentially polluting good to dispose of it safely. In such a scheme a surcharge is imposed on the price of the item, which is refunded when it is returned or safely disposed of. Deposit refund schemes are widespread for metal cans, glass bottles and plastic containers; there are also a few examples for fluorescent light bulbs, vehicle batteries and motor vehicle hulks (OECD, 1994). Some of these schemes were introduced by producers for their own commercial reasons, while others have been promoted by governments for environmental reasons.

Marketable permits (MPs) can be created in situations where pollution of air and water is subject to an upper limit. Individual polluters may be allowed to buy and sell emissions 'quotas'. Where MP schemes exist, firms who introduce more efficient or less polluting processes economise on their original 'quotas' and are free to sell them to other firms who cannot manage on their own quotas – for example, by selling them to those trying to increase the output of an old plant, or to new firms entering the area for the first time.

In a wider sense, MPs work wherever environmental obligations or needs can be exchanged through market processes. There are cases of US electric power companies fulfilling their obligations to reduce greenhouse gas emissions, not by modifying their own processes but by investing in afforestation in Central America. Urban authorities in the western states of the USA are buying water rights from farmers as a cheaper alternative to investing in their own supply sources. Under the US Wetland Mitigation Banking Programme investors receive credit for creating and enhancing wetlands, which can be sold to developers who modify other wetland areas.

Advantages and disadvantages of EIs

It is explained in the previous chapter how in theory EIs are superior to regulations in terms of economic efficiency. EIs, both MPs and pollution taxes, have the following advantages:

1 They penalise polluting firms but reward firms and utilities which make their processes less polluting.
2 They provide continuing incentive to firms to search for pollution-saving materials and production processes, unlike the 'command and control' methods which usually provide a once-for-all motive to change production methods or equipment. Under MPs, firms in need of quotas are forced into the market all the time, and even 'clean' producers have an incentive to reduce their pollution to the point where the opportunity cost of further pollution (that is, what they could earn by selling the quota) equals the cost of abating that pollution. Under pollution taxes, the same effect is also produced by the continuing spur to minimise taxes. Both methods are 'technology forcing' and both provide continuing incentive for pollution reduction.
3 They are more flexible than regulations; a polluting firm can continue its operations at a cost penalty, or can choose the means with which to reduce pollution if it so wishes.
4 Under MPs and pollution taxes, pollution abatement is likely to be carried out by firms that can do it most efficiently. Regulation, on the other hand, is enforced to ensure that all firms comply with the set standards regardless of their age, size or technological endowment. Some firms may only be able to comply at high cost while others might be forced to close down. In theory, EIs enable pollution abatement to be achieved at least total cost. Under the MP variant, the total amount of permissible pollution, which is fixed and divided up into quotas, is distributed to the firms concerned. If the system works as intended, the necessary abatement is carried out by firms which can do it most efficiently. Under certain conditions, the price of quotas, which is determined by the market for quotas, will be equal to the optimal pollution tax (see Chapter 16).
   In the case of a pollution tax, if each polluter is taxed at the same rate per unit of pollution, each firm will adjust its abatement effort until there is no difference between paying the tax and carrying out more abatement. At this point the marginal cost of abatement is equalised across the industry, and it becomes impossible to reduce aggregate abatement costs any further.
5 Both methods can be used to generate public revenue. In the case of MPs the initial allocation of pollution quotas can be auctioned or sold to bidders. The alternative is to award quotas on the basis of present or historical emissions, known as 'grandfathering'.

On the other hand, EIs have certain shortcomings compared to regulation:

1 Pollution taxes and charges are an uncertain method of achieving a given environmental quality standard. In theory, it is possible to attain either the desired amount of pollution abatement, or the total cost of doing so, but not both simultaneously unless the pollution tax is fixed with exceptional accuracy (see Chapter 16). Taxes will only achieve the desired environmental standard if the authorities have accurate information about polluters' abatement costs, and if the polluters respond in the way they are expected to.
   Formally regulations will be preferred to EIs if the marginal damage curve is steeper than the marginal abatement cost curve in the relevant range. This corresponds to the real-life situation where, on the one hand, extra pollution would have a high social cost, but, on the other, the marginal abatement cost curve is relatively flat so that firms have no strong incentive to abate any further in preference to paying the tax. In this situation a great deal would depend on the authorities setting the pollution tax at the right level. If it is set too low, pollution will be much greater and its social costs more onerous than what would

be acceptable or desirable; if, on the other hand, the tax is set too high, 'too much' abatement will occur at an 'excessive' cost.

The advantage of MPs over pollution taxes is that they work within a system of regulations which place a ceiling on total emissions. Lowering of emission ceilings over time would redouble the incentive for participants to curb their emissions. It is also possible for environmental groups to buy quotas which they themselves do not intend to use, thus tightening the pressure on polluters.

2 Taxes and charges raise the total fiscal burden on the household and company sector unless taxes are reduced elsewhere, and this makes them potentially unpopular. Regulation raises private costs too, but in a less transparent way. Moreover, under a flat rate system of emissions charges a polluting firm would pay tax on every unit of pollutant, whereas emissions up to the permitted amount are free under a system of regulation. This is one reason why many firms currently prefer regulation to EIs. Governments' sensitivity to public reaction to higher taxes is undoubtedly one reason why many EIs are set at levels that only recoup administrative costs – levels that are too low to have the desired incentive effect.

3 An accusation, which is often made, is that MPs give firms a 'licence to pollute'; but it is a false accusation because MPs can only work in a situation where pollution standards are in place, emissions from individual firms can be monitored and the standards enforced. 'Emissions trading' is a way of making the enforcement of standards more efficient, and it is compatible with any level of pollution desired by society. A similar argument could be made to refute the allegation that pollution taxes enable firms to 'pay their way out of their environmental obligations'. In fact, the sole purpose of such taxes is to penalise polluters, and, if set judiciously, they can help to achieve the level of pollution desired by or acceptable to society.

A comparison of marketable permits and emission charges

If EIs are chosen in preference to a purely regulatory approach, the choice for pollution control would be between MPs and emission charges, but in order to derive the full benefit of MPs there must be significant differences in the marginal pollution abatement costs of different polluters, and firms must have necessary technological opportunities for carrying out more abatement. Also, there should be a sufficient number of buyers and sellers in the market. The location of sources and receptors should not be important, otherwise a permit for a given amount of emissions would have a different environmental cost depending on who held it. Finally, the rules governing the working of the MP market should be simple and transparent; otherwise high transaction costs would limit the number of MPs traded (OECD, 1994).

If or when these conditions are met, MPs would have the following advantages over emission charges:

- They would give the authorities direct control over the quantity of emissions, without having to guess the response of polluters. This is important when the cost of pollution damage rises rapidly with scale, for example where there are threshold levels of pollution. In these circumstances the regulating authorities need to be certain about the maximum level of emissions.
- The market both sets and continually adjusts the price of permits. This is important in inflationary conditions when the real level of emission charges will decline unless adjusted frequently.
- Regulators are familiar with permits, and the step of making them transferable may be more acceptable to them than that of introducing a new system based on effluent fees (Oates and Cropper, 1992).

The disadvantages of MPs compared to emission charges are:

- Existing permit holders may use them as anti-competitive devices, for example to restrict new entrants, or to prevent competitors from expanding.
- Firms may be shy of using the market for MPs because of their reluctance to give (implicit) information about their costs to their competitors.
- The pre-conditions for deriving the full benefits of MPs are quite stringent.

AIR POLLUTION

Survey of EIs in use

As a generalisation, the USA has favoured the use of MPs, while the European countries have preferred to use emission charges. A number of schemes for *marketable permits*, in use in the USA to control air pollution, are given in Box 17.1.

Although Canada does not have a formal MP scheme, the province of Ontario allows its electricity utility to transfer permissible emissions between its various stations under the Acid Rain Control Programme. CFC producers are also allowed to trade production rights internally. In Germany, a form of 'netting' is allowed between new and existing plants under the TA Luft Programme.

There are relatively few examples of *emission charges* to control air pollution. Since 1992 the province of British Columbia in Canada has levied a fee on the level of pollutant emission allowed in permits. The fee is reduced if it can be shown that actual emissions are below those for which the permit is issued, but if actual emissions turn out to be higher than the permitted level a new and more expensive permit is required. In France certain large installations are subject to a charge on acidifying pollutants, the proceeds of which are largely returned to the same firms to subsidise their efforts at abatement.

In Japan charges are made for emissions of sulphur dioxide, based partly on current and partly on historical levels. Since 1993 the revised Clean Air Act of the USA has authorised states to levy emission charges as a condition of issuing permits. Sweden levies a charge on NO_x emissions by large power producers, the charge being applied alongside the use of permits. All proceeds are rebated to the installations subject to the charge in proportion to the energy produced. Poland requires stationary source polluters to obtain a point source emission permit, the fee for which is based on the type and quantity

BOX 17.1 MARKETABLE PERMITS FOR AIR POLLUTION IN THE USA

In order to meet its obligations under the Montreal Protocol, the USA aims to reduce, and shortly to phase out entirely, the production and use of ozone-depleting chemicals (CFCs). In the meantime, rights to produce and use CFCs have been allocated to existing producers, who may transfer them to producers in other countries that have signed the Protocol.

The Acid Rain Programme aims to reduce emissions of sulphur dioxide (SO_2) with a target ceiling for the year 2012. Utilities are given sulphur emission allowances; if they emit less than the allowance, they can exchange or sell the unused quota to other utilities, or save them for future years. A small part of the allowances is reserved for an annual auction.

The Emissions Trading Policy, which started in 1982, is the best-known scheme of MP. It allows states or localities latitude in how the Clean Air Act is implemented. Emission Reduction Credits (ERCs) – emission reductions achieved beyond baseline levels – can be used in various ways by firms earning them. The ERC, which is the currency used in emissions trading, can be stored or spent according to the offset, bubble, banking and netting policies (Tietenberg, 1990). 'Offsetting' requires major new or expanding sources in areas of poor air quality to secure offsetting emission reductions from existing firms. The 'bubble' concept treats multiple emissions within one air quality zone as if it were from a single source, and allows ERCs to be exchanged within the bubble. 'Netting' allows firms modifying or expanding their plant to avoid stringent new source review processes, provided their net increase in emissions (allowing ERCs earned in other parts of the firm) is within an established threshold. Banking allows emission sources to store ERCs for future use.

Two other MP programmes are mainly aimed at vehicle emissions. The Oxygenated Gasoline Programme allows refiners, blenders and importers credit for dealing with or using gasoline (petrol) with a greater than required oxygen content. They may transfer or sell these credits to refiners or importers whose gasoline falls short of required oxygen levels. Similarly, under the Low Emission Vehicle Credit Programme, producers who sell more clean-fuel vehicles than required obtain credits which can be traded with others. Between 1982 and 1987 there was also a programme allowing refineries to trade the rights to add lead to gasoline.

Finally, 'Emissions Averaging' is a feature of several programmes. It allows a plant or firm to average its emissions over different parts of its operations from the point of view of meeting air quality regulations. It applies, for instance, to heavy-duty engines in a fleet of trucks, and in programmes for the reformulation of gasoline.

of the air pollutant in question. China also has an emission fee system, the proceeds from which are made available for loans to enterprises making investments in pollution control.

In this connection *product taxes* on petrol (gasoline) and other mineral oils are widely used. They have a potentially major impact on air pollution, although this is not currently their major purpose in most cases. Many countries levy different rates of tax according to the product's polluting potential – for example, on leaded and unleaded petrol, diesel oils of different quality, etc. This has led to a sizeable switching between categories of fuel in countries such as Finland, Sweden and the UK.

Assessment of experience

Marketable permits

One of the principal motives of MPs is to reduce the cost of compliance with regulatory standards. A number of empirical studies, comparing abatement costs imposed by the regulatory ('command and control') method with the least-cost techniques, show that potential cost savings can be very large (Tietenberg, 1990). The performance of MPs is improving with time, as firms become more familiar with the various schemes and as the regulations governing the programmes are modified in the light of experience. The volume of trade has been substantial, especially under the programmes for Emissions Trading and lead additives in petrol. Several other programmes are of very recent origin, and much is expected of them.

Nevertheless, the interim verdict on MP in the USA is highly qualified. Under the Emissions Trading Programme, for instance, there has been active trading whereby the parties involved have benefited from substantial cost savings. But much of this trade has been internal to firms. External trades have been discouraged by extensive and complicated procedures which have ruled out some trades and raised the transaction costs of others (Oates and Cropper, 1992). For example, every potential trade has to be approved by the regulator, and trades involving non-uniformly mixed pollutants will only be approved after the applicants have carried out dispersion modelling, because of the fear of creating localised 'hot-spots'. The majority of trades have been internal as a result, involving uniformly mixed pollutants, especially volatile organic compounds, and/or contiguous facilities. The comparative rarity of ERC banking schemes, which enable holders to save credits for future use, has limited their incentive effect (Tietenberg, 1990).

The brief trading rights programme to add lead to petrol was much more successful by comparison, and it indicated the conditions necessary for MPs to flourish. Towards the end of the programme, half of all lead added to petrol was obtained through trading, with cost savings to buyers roughly estimated at hundreds of millions of dollars. Trading was very lightly regulated, and the refiners were familiar with markets for products and additives (Oates and Cropper, 1992).

The experience of trading in sulphur dioxide under the US Acid Rain Programme is very recent, and so far its achievement has been modest. Although the programme, which was mandated by the 1990 amendments to the Clean Air Act, only started in 1993, it allowed firms and utilities much more freedom to meet the emissions standards laid down by the Act than had been allowed in the previous command-and-control approach. In practice relatively little trading has taken place so far, but abatement costs have fallen drastically with the opening up of new compliance options. In particular, firms have switched to low-sulphur coal, the cost of which has fallen sharply due to market trends and the deregulation of road transport. The choice between different compliance options has also been distorted by different states' incentive regimes for the use of local coal and investment in abatement, rather than trading in allowances.

It is anticipated that, if the cost of low-sulphur coal rises in future and utilities make more use of scrubbers, differences in marginal abatement costs will increase between firms, and allowance trading will become more attractive as a result. In the meantime, the potential cost reductions offered by allowance trading remain sizeable. It is estimated that the annual cost of compliance under a command-and-control regime for SO_2 emissions would be \$7 billion if every utility had to install scrubbers, and \$4.3 billion if uniform emission rates were applied to individual power plants. These figures should be compared with costs of $2.5 billion for constrained allowance trading (with internal transfers) and $1.4 billion for completely flexible inter-utility trading (Burtraw, 1996).

The German offset and netting provisions have rarely been used. The reasons are that the geographical areas over which emissions can be substituted are small, substitution between substances is not allowed, and because different firms cannot be involved since emissions savings are not formally tradable. It is also thought that many of the firms that might be interested face similar abatement options and abatement cost curves, limiting the incentive to trade (OECD, 1994).

Emission charges

In general, the effectiveness of existing air pollution emission charges is yet to be proved, either because they have been introduced very recently or because the charge levels have been too low to have any incentive effect. In several cases the charge is used in conjunction with a permit. The Swedish scheme is one of the few with a clear incentive impact.

The incentive level of the Canadian and Japanese schemes is unclear. However, in Sweden emission charges have reduced emissions by 30–40 per cent compared to the base period, much more than expected in fact. The US charge scheme is also expected to have an impact in cases where control costs are low (OECD, 1994).

The level of fees in the Polish emission charging system is considered too low to have any incentive effect. Moreover the fees bear no relation either to the costs of abatement or the costs of pollution damage (Wilczynski, 1990). In the Chinese scheme the charge level is considered too low to induce any extra abatement. In any case, firms can offer financial difficulties as a reason to avoid paying the charge (Bernstein, 1991).

WATER POLLUTION

Survey of EIs in use

Unlike in the case of air pollution, EIs applied in the water sector are based mainly on emission charges, while MPs have been sparingly used with little success.

The only cases of tradable water pollution rights are from the USA – in Colorado (Dillon and Cherry Creek Reservoirs), Wisconsin (Fox River) and North Carolina (Tar-Pimlico Basin). The Fox River scheme is the best known of these. The lower Fox River in Wisconsin, which is polluted by organic discharges from ten pulp and paper mills and four sewage works, has unacceptably low dissolved oxygen levels in certain places and under certain seasonal conditions. Firms were issued with five-year permits defining their waste-load allocation which they could trade with each other, subject to certain conditions to be discussed below.

Effluent charges, on the release of waste water from point sources to surface water bodies, are more common; examples are found in Australia, Belgium, Canada, France, Germany, the Netherlands, Portugal, Spain and the USA (OECD, 1994). In each case the basis for charging is the pollution load. In most cases the charge is based on actual measurement, the remainder being flat-rate fees. In every case the revenue is earmarked for improvements to water quality and/or subsidies for pollution abatement.

The best-known schemes of this type are those in France, Germany and the Netherlands. Their primary motive has been to raise funds to finance water quality improvement, and any incentive effect has been secondary. In France river basin agencies impose water pollution charges on firms based on a standardised pollution parameter which takes into account the daily COD content, suspended solids, nitrogen, phosphates, toxic materials and dissolved salts. The revenues are devoted to subsidies to communes and industrial firms for traditional treatment processes as well as for cleaner technologies.

The German water pollution charge is intended to give polluters an incentive to clean up their discharges. The basis of charges is a menu of polluting substances, COD and heavy metals. The charge rate is reduced by 75 per cent if a firm can demonstrate its compliance with new standards before it is obliged to do so. The charge, administered closely with the permit policies, is combined with the offer of subsidies for investments in pollution abatement.

The Netherlands has one of the oldest and most effective systems of water pollution charges. Its purpose is to finance a major programme of investment in treatment plants by industry and municipalities. The rate has been set at a sufficiently high level to produce a strong secondary incentive effect as well. Levied on the BOD and heavy metal content, it is administered in combination with a permit system. The level of charge has risen regularly in parallel with pollution criteria which have been steadily tightened.

Effluent charges are not to be confused with sewerage charges; the latter are charges on users of sewerage and waste water treatment services in return for specific services rendered.

Assessment of experience

Marketable permits

The Fox River scheme in the USA has been extensively analysed. Results of analysis show that so far firms have made little use of their opportunities to trade discharge permits. For example, prior to 1989, only one trade had taken place – a paper mill, which shifted its treatment activities to a municipal waste water treatment plant, transferred its rights to the latter. This was despite the predictions of a sophisticated model of discharges and marginal abatement costs which indicated that trading could produce sizeable potential cost savings, estimated at $7 million per annum compared to the costs of uniform regulations. This disappointing outcome was attributed to several factors. Many of the firms involved were in the product market competing with each other, and it was not in their competitive interests to sell permits to rivals. The discharging municipalities (eg. sewage works) were, on the other hand, under less pressure to minimise costs. Also, firms were not certain about whether or how the selling of their current permits could prejudice the future allocation of permits. Various rules were imposed to constrain trading, and this in effect increased the transaction costs of permits. In any case, pollution control costs were very low, and there was thus little motivation to reduce them further through trading (Hanley, 1993).

A crucial reason why MP schemes are rarely used to control water pollution is undoubtedly the importance of the location of discharges with regard to the quality of the receptor water body. The regulating authority cannot be indifferent to the swapping of discharge permits if this leads to a concentration of effluents in particular sections of a water body, thus creating pollution 'hot-spots'. The polluting potential of a given discharge varies depending on its location. An analogous problem is concerned with the timing of discharges which have a larger polluting impact at times of low flow, or when the ambient temperature is high. Regulators may seek to tackle this problem by dividing the water body into different zones, with trading only allowed within, rather than between, them; or other restrictions may be placed on trading, such as by making them dependent on season or on river flow characteristics. But such solutions add to the transaction costs of MPs and make trading of permits less attractive.

Another general constraint on MPs is the nature of the market itself in which the agents operate. MPs are predicated on well-functioning markets populated by cost-minimising firms. But competitive factors may dictate firms not to buy or sell permits in order to deprive competitors of a temporary advantage (Hanley *et al.*, 1990). Some firms may also think that their active use of the market for permits could divulge information about their costs and general trading position to rivals. In theory, large firms can also buy up the supply of MPs to create entry barriers to new firms. The geography of point-source water pollution is such that the market for MPs will often be very thin.

Effluent charges

As a general rule, most effluent charges have been implemented as a means for generating revenue for improving water quality, and not usually for creating or promoting abatement incentive effects. Also, fees have tended to be low and related to average or expected discharges. However, there have been national variations on this theme, as will be seen from what follows.

There is general agreement on the lack of the incentive effect in the French system. In most cases the level of charge, which is tied to a cost of living index, does not offer any incentive to a firm to abate its discharges. Industrial sectors, such as chemicals and pulp and paper in which polluting discharges have been greatly reduced, are also those that have benefited most from the subsidies administered as a part of the scheme. Thus it is the subsidy, rather than the tax, which explains the reduction in pollution (Andersen, 1994).

The German water pollution charge (higher than that in France) is generally thought to have contributed to reduced industrial pollution, and has increased over time. Initially the costs of administration were very high, due in part to close monitoring and partly to the large number of pollutants covered by the tax. Increases in the tax are announced well in advance in order to give polluters time to plan further abatement. Observed improvement in water quality is due in part to the provision of subsidies funded by the pollution tax.

The Dutch waste water charge has been studied on various occasions, and there is a consensus that it had a major impact on the observed reduction in industrial organic pollution which fell by two-thirds between 1970 and 1980. In one study 60 per cent of the companies taking abatement measures attributed this to the charge levied (Sunman, 1992). The incentive effect is considered to have been particularly strong in the areas of chemicals, food, beverages and tobacco. The charge for heavy metals has, however, been less successful. In practice, a combination of regulation, negotiation and charges has been necessary to deal with heavy metal pollution (Bernstein, 1991).

Special considerations apply to the use of effluent charges in socialist economic regimes. For example, water pollution charges have generally been ineffective in former Yugoslavia. Their rates, set well below marginal abatement costs without adjustment for inflation, have not always been collected from public enterprises in financial difficulty. Moreover, the charges were based on concentration levels rather than pollution loading, and this encouraged firms to dilute their effluent rather than clean it (Kosmo, 1989). In China, however, industrial effluent fees have helped to ensure compliance with state regulations. The environment protection agency and the polluting firm connive at payment of the local pollution charge, which is then recovered from the polluter's parent firm or sponsor. A large part of the charge is rebated to the polluter for spending on environmental protection work (Bernstein, 1991).

SOLID WASTE

Survey of EIs in use

The accumulation and disposal of solid wastes has a number of environmental impacts, such as on local amenities, public health, use of energy, spread of landfill sites, etc. It is an example of non-point-source pollution in the sense that the sources of solid waste are widely spread throughout the population. And this complicates the task of measuring the volume of waste generated and of volumetric pricing. Another important consideration is that charges should be set at a level that does not encourage illegal or antisocial disposal of rubbish.

Hazardous waste is a special case. Although charges are made in some countries to cover the cost of their collection and safe disposal, regulation is the preferred policy in most cases.

Various EIs have been applied to control this problem. *User charges* apply to the collection and disposal of waste. Most of these are levied on firms and households to recover the cost of the service. Waste disposal charges apply to material received by landfills and incinerators, while *product charges* apply to anti-social, non-returnable items such as plastic bags and lubricating oil. Finally, deposit refund schemes can be implemented to encourage the return and re-use, or safe disposal, of items such as bottles, cans and old cars. Some examples of EIs for solid wastes are described below.

User charges

Practically every OECD country has a system of municipal waste user charges. They are usually levied on households at a flat rate, and on firms either at a flat rate or as a volumetric charge which varies with the amount (volume) of waste taken away. The volumetric charge is often intended to have an incentive effect, giving the waste generator a motive for reducing the volume of waste. For instance, some US cities have a 'pay-per-bag' system and variable garbage can rates. In one New Jersey town, for example, only garbage containers bearing a pre-paid sticker are collected, while Seattle has a complex garbage collection rate giving households a clear incentive to generate less waste (Bernstein, 1991).

Waste disposal charges

A relatively small number of countries operate schemes in which charges, based on weight, are made on wastes received at disposal sites such as landfills and incinerators. The highest unit charge of this kind is levied in Denmark. Some states of the USA also impose disposal charges on anti-social items such as tyres and used oil.

Product charges

In several countries a tax is levied on batteries and lubricating oil to defray the cost of the safe disposal of these troublesome items. No incentive effect is intended. In a larger group of countries a tax is imposed on packaging items, and the incentive effect is implicit in some of these schemes. The most common items subject to this charge are non-returnable containers, plastic bags, fertiliser, pesticide and car tyres. Eight countries, including Denmark, impose a charge on packaging, mainly of glass bottles and metal cans, but also of cardboard boxes.

Deposit refund schemes (DRSs)

DRSs are becoming a popular way of encouraging the recycling, re-use or safe disposal of anti-social items. Some of the DRSs have been introduced by private companies for their own commercial reasons. Several countries, such as Greece, Norway and Sweden, have deposit schemes for old car hulks, and in two of these countries the refund is greater than the deposit. Many countries have such schemes for glass bottles, plastic beverage containers and metal cans. Other items covered include fluorescent light bulbs (Austria) and vehicle batteries (USA). All of these schemes have an incentive purpose; namely, to encourage users to return the item rather than carelessly throwing it away.

Assessment of experience

Although *user charges* are operated mainly as a means of covering the cost of the collection and disposal service, several schemes using variable rates claim some incentive effect too. For instance, in the first year of the implementation of Seattle's variable garbage rate structure, waste collection in that city fell by 30 per cent compared to the previous year. In towns in New Jersey and Pennsylvania, in which pay-per-bag garbage systems have been introduced, the volume of collection has fallen by 25 per cent in one case and by 50 per cent in the other. On the other hand, some increase in illegal and anti-social disposal was also noted, probably as a side-effect of these systems (Bernstein, 1991).

Several *waste disposal charge* schemes are claimed to generate and promote incentive effects. The Danish Waste Charge was introduced in 1987, based on the weight of waste supplied to landfills and waste incinerators net of waste abducted from these sites. In the first two years of its operation this scheme led to a reduction of 12 per cent in waste supplied and increased waste abduction. But, as a side-effect, there was a minor increase in illegal dumping, together with much greater use of building wastes as filling material for other projects. However, as increasingly greater amounts of re-usable materials are separated, the residual waste becomes correspondingly more noxious (OECD, 1994).

There is little evidence to suggest that *product charges* have any incentive effect. Most of them are introduced to raise money for environmental improvement, and charge levels are too low to affect users' behaviour. The convenience and cost of non-returnable packaging and containers stands them in good stead, despite the tax. Exceptions to this are to be found in Finland and Norway where returnable containers have an overwhelming market share, and in Italy where the use of plastic shopping bags fell by 20–30 per cent soon after the imposition of a tax.

Deposit refund schemes are the most effective and successful EI for solid waste management. In countries in which such schemes are used for car hulks, 80–90 per cent of car bodies are normally returned to authorised disposal sites. For metal cans the return rate in Australia and Sweden is 80–90 per cent. The following statistics from a number of countries, in which DRSs are applied, are interesting to note: return rates are seldom less than 60 per cent for plastic bottles and containers, 90–100 per cent for beer and soft drink containers, and 40–80 per cent for containers of wine and spirits (liquor). DRSs are popular with users too: they reward good behaviour, rely on voluntary compliance, and have clear environmental goals. Administered by private operators in certain cases, DRSs require little or no involvement of regulatory authorities.

A BROADER VIEW: EIs IN DEVELOPING COUNTRIES AND IN THE COUNTRIES IN ECONOMIC TRANSITION

Our discussion has so far focused on various EIs used in developed industrialised countries to control pollution. But, as we have pointed out earlier (p. 327), there is a much wider array of measures to influence behaviour towards the environment, all acting on people's economic incentives. Countries in economic transition (meaning the recently democratised countries of Eastern and Central Europe) and developing countries may well find it advantageous to draw on this wider spectrum of measures. In the first place, there is greater market distortion in those countries, and therefore wider scope for correcting market and policy failures (though they are present in practically all countries at all stages of development). Second, there are often well-functioning systems of traditional and communal management of resources that can be refined and built upon. Third, these countries often lack the sophisticated regulatory and fiscal systems necessary for the performance of EIs (Barde, 1994). In short, both developing countries and the countries in transition are thus likely to make greater use of the definition and allocation of property rights, together with necessary reforms of resource prices, than adopting classical EIs that are primarily or exclusively aimed at pollution control.

The canvas is wide so to speak, and we will give a few examples here to illustrate the point. What are called *customary rights* are allocated as a means of controlling over-fishing in a number of areas in countries such as Brazil, the Ivory Coast and Sri Lanka. Also, communal tenure and stewardship rights have been awarded to forest-dwellers in Papua New Guinea and the Philippines to discourage over-exploitation of forests (Panayotou, 1994). There are many documented examples of local environmental degradation being controlled by collective action taken by the users and other affected parties who devise their own systems of rewards and sanctions (McNeeley, 1988; Ostrom, 1990).

Reforms in the *pricing of natural resources* have taken many forms. It should be recalled that the economists' ideal prescription is for resource prices to reflect fully the cost of production, the user cost (depletion), and the environmental costs resulting from production, use and disposal (OECD, 1995). While the pricing of farm irrigation water usually falls far short of this ideal, there are other cases in which prices can be raised closer to their full cost in the above sense. Active markets for groundwater in India and Bangladesh imply that water is being exchanged at its full opportunity cost (Winpenny, 1994). The same is true of the tradable water rights schemes in Chile (Hearns, 1995). Raising the prices of electric power, petroleum products, and petrol closer to their respective economic cost, which are staple ingredients of Structural Adjustment Programmes, is equivalent to the imposition of a product tax on these potentially polluting items.

Capturing the economic rents of natural resources for society is an element in some reform programmes, especially in the forest sector. Various attempts have been made in Western and Central Africa to reform the system of charges for logging concessions to reflect the full economic rent (or surplus) from exploitation, as in Congo, Ghana, Zaïre and the Ivory Coast (Panayotou, 1994). In a general sense, the same principle applies to urban street space, a scarce resource, the users of which enjoy an economic benefit for which they do not normally pay.

Turning to more conventional EIs, Malaysia has an *effluent charge* on discharges from palm oil mills (Box 17.2) which goes back to the 1970s. A similar scheme was later applied to discharges from rubber factories. Both schemes were successful in dramatically reducing polluting discharges and forcing the pace of technological development in the treatment of waste.

Korea also has a long-standing emission charging system which applies to both air and water pollution. Facilities which exceed mandated emission levels are penalised by a levy based on pollutant concentration. The emission charge is thus more akin to a fine to reinforce a regulatory standard than a pure incentive. Its incentive effect has been diminished by the low

BOX 17.2 THE MALAYSIAN EFFLUENT CHARGE SYSTEM

Legislation was passed in 1974 to allow economic incentives in the form of effluent charges to complement regulatory controls on discharges. All dischargers were required to pay a fee to obtain a licence to discharge waste into public water bodies. The licence fee varied with the level of waste discharged. In effect it was therefore a discharge fee which was assessed on the class and location of premises, quantity of waste discharged, nature of the pollutant, and the existing level of pollution.

In 1977 the discharge fees and standards were combined in a regime applying to palm oil mills, and the first fees were collected in 1978. The licence fee was initially set at US$3 per ton of biochemical oxygen demand (BOD) up to the permitted standard, with a surcharge of $100 per ton for amounts in excess of this. These rates were fixed at levels such that the annual fee for untreated discharges exceeded the capital cost of building treatment facilities, thus giving polluters a clear incentive to abate their effluent.

Standards have become more stringent over time. In spite of a 50 per cent increase in the number of palm oil mills between 1978 and 1982, and a steady increase in palm oil production, the total BOD load released into public water bodies fell from 222 tons per day in 1978 to just 5 tpd in 1984. The discharge fee was combined with a subsidy to research into effluent treatment technology which has achieved some success.

The fee was not economically optimal in that its level was related to the capital cost of abatement rather than the marginal costs of environmental damage. Nevertheless, it was effective in achieving a dramatic reduction in pollution loads from this sector.

Sources: Vincent, 1993; Panayotou, 1994; O'Connor, 1995

level of the charge; also, the basis of the charge, namely pollutant concentration, has encouraged dilution of discharges rather than abatement (O'Connor, 1995). China's pollution charges are in effect also non-compliance fees, the revenues from which are used to finance pollution control investments.

Product taxes, in the form of a differential tax on leaded and unleaded petrol, are used in Taiwan and Thailand. Korea and Taiwan also operate *deposit refund schemes*; Taiwan's is considered to be the more effective of the two, applying as it does to bottles, cans, waste paper, tyres, lubricating oil, batteries and pesticide containers. By the fourth year of its operation, the scheme had attained a recycling rate of 80 per cent (O'Connor, 1995).

There are examples of *tradable permits* in Singapore and Poland. Singapore has operated a permit auction scheme for the right to import and use ozone-depleting substances. Half the amount of the quota is allocated on the basis of historic shares; the remainder is auctioned, with the proceeds being retained by the government. This has led to a rapid and sizeable reduction in the demand for CFCs (O'Connor, 1995). In Poland, a tradable industrial emissions scheme has recently been introduced on a pilot basis. The first phase of the project involves two large enterprises, one in the steel sector and the other in the power sector. The exercise is intended as a demonstration, whose scope will be widened in due course (Panayotou, 1994).

CONCLUSIONS

It is important to place the advantages of EIs in the proper perspective. Although EIs are often presented as alternatives to regulation, in practice they usually depend on an effective complementary system of monitoring and control. EIs should be distinguished from user charges and fines, although in practice there are areas of overlap. EIs may be designed with the intention of having an effect on incentives, but in practice they tend to be set at too low a level to have the desired impact. In any case, there are many situations in which EIs are best used in combination with other measures such as regulation, publicity and persuasion.

The main focus of the chapter has been on EIs to control air and water pollution and solid waste disposal, reflecting their actual prevalence in the OECD countries. For the control of air pollution, marketable permits (MPs) may be more promising than emission charges, and there is growing experience in the USA of the use of MPs for various kinds of air pollutants. The situation is reversed in the case of water pollution where effluent charges, which are a common practice in Europe, appear to be more effective than MPs; experience to date in the USA in this regard has been very limited and unconvincing. There are various options for the control of solid wastes, the most successful of which has been the deposit refund scheme.

A broader perspective on the scope for EIs would include reforms to property rights and the pricing of natural resources. EIs work best when users of the environment have firm tenure over their resources. Governments can influence the use of natural resources by extracting economic rents for public benefit. With regard to this broader view, a number of other non-OECD countries have experience of having successfully applied price and market mechanisms in the cause of environmental policy.

REFERENCES

Andersen, M.S. (1994) *Governance by green taxes: making pollution prevention pay.*

Barde, J.-Ph. (1994) 'Economic instruments in environmental policy: lessons from the OECD experience and their relevance to developing countries', *OECD Development Centre Technical Papers* No. 92, Paris.

Bernstein, J.D. (1991) *Alternative approaches to pollution control and waste management: regulatory and economic instruments*, Discussion Paper, World Bank/UNDP, April.

Burtraw, D. (1996) 'Trading emissions to clean the air: exchanges few but savings many. Resources', *Resources for the Future*, Washington, DC (Winter).

Hanley, N. (1993) 'Controlling water pollution using market mechanisms: results from empirical studies', in R.K. Turner (ed.) *Sustainable environmental economics and management: principles and practice*, Belhaven Press.

Hanley, N., Hallet, S. and Moffat, I. (1990) 'Why is more notice not taken of economists' prescriptions for the control of pollution?', *Environment and Planning* A, 22: 1421–1439.

Hearne, R.R. and Easter, W.K. (1995) 'Water allocation and water markets: an analysis of the gains-from-trade in Chile', *World Bank Technical Paper*, 315.

Kosmo, M. (1989) *Economic incentives and industrial pollution in developing countries*, Environment Department Division Working Paper No. 1989–2, Washington, DC: World Bank.

McNeely, J.A. (1988) *Economics and biological diversity: developing and using economic incentives to conserve biological resources*, Gland, Switzerland: IUCN.

Oates, W.W. and Cropper, M.L. (1992) 'Environmental economics: a survey', *Journal of Economic Literature* XXX (June): 675–740.

O'Connor, D. (1995) *Applying economic instruments to developing countries: from theory to implementation*, Singapore: Economy and Environment Program for South East-Asia.

Organisation for Economic Co-operation and Development (OECD) (1994) *Managing the environment: the role of economic instruments*, Paris: OECD.

Organisation for Economic Co-operation and Development (OECD) (1995) *The economic appraisal of environmental projects and policies: a practical guide*, Paris: OECD.

Ostrom, E. (1990) *Governing the Commons: the evolution of institutions of collective action*, Cambridge: Cambridge University Press.

Panayotou, T. (1993) *Green markets: the economics of sustainable development*, International Centre for Economic Growth/Harvard Institute for International Development, San Francisco, Calif.: ICS Press.

Panayotou, T. (1994) 'Economic instruments for environmental management and sustainable development', Paper prepared for UN Environment Programme, Nairobi, seen in draft.

Sunman, H. (1992) *The application of charging schemes for the management of water pollution: experience and prospects*, London: Environmental Resources Ltd.

Tietenberg, T.H. (1990) 'Economic instruments for environmental regulation', *Oxford Review of Economic Policy* 6 (1):

Vincent, J.R. (1993) *Reducing effluent while raising affluence: water pollution abatement in Malaysia*, Cambridge, Mass.: Harvard Institute for International Development.

Wilczynski, P. (1990) *Environmental management in centrally-planned non-market economies of Eastern Europe*, Environmental Working Paper No. 35, Washington, DC: World Bank.

Winpenny, J. (1994) *Managing water as an economic resource*, London: Routledge/Overseas Development Institute.

SUGGESTED READING

Bernstein, J.D. (1991) *Alternative approaches to pollution control and waste management: regulatory and economic instruments*, Discussion Paper, World Bank/UNDP, April.

Oates, W.W. and Cropper, M.L. (1992) 'Environmental economics: a survey', *Journal of Economic Literature* XXX (June): 675–740.

O'Connor, D. (1995) *Applying economic instruments to developing countries: from theory to implementation*, Singapore: Economy and Environment Program for South-East Asia.

Organisation for Economic Co-operation and Development (OECD) (1994) *Managing the environment: the role of economic instruments*, Paris: OECD.

Organisation for Economic Co-operation and Development (OECD) (1995) *The economic appraisal of environmental projects and policies: a practical guide*, Paris: OECD.

Panayotou, T. (1993) *Green markets: the economics of sustainable development*, International Centre for Economic Growth/Harvard Institute for International Development, San Francisco, Calif.: ICS Press.

Panayotou, T. (1994) 'Economic instruments for environmental management and sustainable development', Paper prepared for UN Environment Programme, Nairobi, seen in draft.

Tietenberg, T.H. (1990) 'Economic instruments for environmental regulation', *Oxford Review of Economic Policy* 6 (1):

Turner, R.K. (ed.) (1993) *Sustainable environmental economics and management: principles and practice*, Belhaven Press.

Winpenny, J. (1994) *Managing water as an economic resource*, London: Routledge/Overseas Development Institute.

SELF-ASSESSMENT QUESTIONS

1. How do economic instruments (EIs) differ from user charges and fines?
2. Apart from EIs and regulations, what other environmental policy measures does the government have? Suggest combinations of measures to produce the maximum impact.
3. What are the principal kinds of EIs? What advantages are claimed for them?
4. When will regulation be preferred to EIs?
5. What are the advantages and disadvantages of marketable permits (MPs) compared with emission charges?
6. What factors explain why US experience with MPs for controlling air pollution has been more successful than the use of MPs for water pollution?
7. In what ways do emission charges for air and water pollution depart from the economic optimum? Which schemes come closest to the economic ideal?
8. Which EIs have been most successful in tackling the problem of solid waste?
9. What are the special problems and opportunities for applying EIs (a) in developing countries, and (b) in countries in economic transition?
10. How would you seek to convince (a) an environmental activist and (b) an engineer of the benefits from using EIs?

18

INTERNATIONAL TRADE AND ITS ENVIRONMENTAL INTEGRITY

Aaron Cosbey

SUMMARY

This chapter contains a discussion of the challenges of environmental management in today's world of expanding international economic activity and increasing recognition of the international nature of many of the environmental problems. Environmental regulators in the WTO-member countries would like to create a level 'playing field' for both domestic and foreign producers, and to address environmental problems that are international in nature, but they are constrained by international trade rules which demand non-discrimination and 'most-favoured nation' treatment, and which may not permit discrimination at the border on the basis of how a product is produced. There are some exceptions, however, to which reference has been made. Case studies are presented to illustrate these elements of the international trading system. However, in spite of these challenges and constraints, there are still a number of effective tools at the disposal of the environmental manager, multilateral environmental agreements being an example. The strengths and weaknesses of these agreements have been discussed, together with a range of purely domestic instruments including product standards and eco-labelling. Even so, the task of the environmental manager in a global economy will remain difficult until there is better integration of trade, environment and development in policy-making.

ACADEMIC OBJECTIVES

This chapter is designed to give the reader a basic knowledge of the multilateral trading system operated by the World Trade Organisation (WTO), together with an understanding of how this system of rules might affect the environmental manager's task in developing policies which affect internationally traded goods.

On completion of this chapter the reader should be able to:

- understand and discuss the basic principles of the multilateral trading system;
- understand and discuss the environmental exceptions to those principles;
- apply the legal instruments of the GATT to address environmental problems concerned with imports, specifically those with or without international impacts, as well as those based on either production-related or consumption-related problems;
- apply the legal instruments of the GATT to address environmental problems concerned with exports;
- discuss the North–South issues relevant to various alternative trade measures.

INTRODUCTION

The continuing debate between environmentalists and advocates of free trade intensified in recent battles over the passage of the North American Free Trade Agreement (NAFTA). Also, the conclusion of the Uruguay Round of multilateral negotiations has been as fractious and emotional as any public policy debate in recent memory. Battle lines, drawn like a patchwork, separated not only developed and developing nations, 'greens' and free-traders, but also ran within the communities, as with the North American environmental movement which was painfully divided over the passage of the NAFTA.

Does this 'new' set of issues have any special implication for environmental management? After all, trade is simply economic production which crosses borders, and environmental managers have had years of practice in regulating the environmental effects of economic production. In fact, the task of environmental management is seriously complicated by borders which are open to trade, and by the existence of other jurisdictions. Although many of the environmental problems now facing the world are international in nature, each jurisdiction has its own ecological conditions, priorities and methods of dealing with environmental issues and problems.

The goals of environmental management remain the same in the context of international trade, but the means by which they are attained may change. The greatest change stems from the increasing desire to change the behaviour of actors in other jurisdictions; that is, we may wish to change our neighbours' behaviour and attitude to the environment if only because their behaviour and attitude will affect our own quality of life, directly or indirectly, through environmental spill-over effects. But the obvious problem here is that the environmental manager only has limited, if any, authority outside his or her own jurisdiction; moreover, options for exerting such influence are constrained by established rules governing how nations must treat one another. The General Agreement on Tariffs and Trade (GATT), which is the most important set of rules governing international trade, is the focus of this chapter.

THE MULTILATERAL TRADING SYSTEM

A brief history of the GATT

Established in 1947 as an agreement between 23 countries, the GATT was originally meant to be the third post-war pillar of international co-operation, along with the International Monetary Fund and the World Bank. Unfortunately, the full agreement, constituted as the International Trade Organisation (ITO), was never ratified by the USA because of Congressional opposition. Consequently GATT, which was originally just one section of the agreement detailing trade rules and tariff concessions, was the only piece salvaged from the full agreement.

The international community used that piece to the greatest extent possible. Though hamstrung by the strange circumstances of its institutional birth, the GATT nevertheless proved to be a sufficiently flexible instrument. Indeed, the fact that it has been the subject of seven subsequent 'rounds' of multilateral negotiation bears testimony to its flexibility. The focus of initial negotiations was the lowering of tariff barriers between nations. But, more recently (beginning in earnest in the Tokyo Round of the 1970s) emphasis has also been on drafting codes of agreement for dealing with non-tariff barriers such as subsidies, technical standards, rules for government procurement, and so on.

The latest round – the Uruguay Round – lasted a gruelling seven years and produced agreements on a range of new issues, including trade in services, trade-related investment measures, and trade-related intellectual property. It also produced an amended version of the GATT, referred to as GATT 1994. The most noteworthy result of the Uruguay Round was undoubtedly the establishment of the World Trade Organisation (WTO) – an overall structure containing all the various agreements (see Figure 18.1) – which in some ways finally established what was lost with the still birth of the ITO. The WTO, created with 125 member states, is a much stronger institution than the GATT which it replaced. With minor exceptions, all members of the WTO must now accede to all its agreements, whereas previously codes on such things as subsidies and technical barriers had been plurilateral. The WTO also contains a strengthened and integrated dispute settlement mechanism spanning all its agreements. A good summary of the organisation, its history and structure is given in WTO (1998).

Several aspects of the multilateral trading system are of special relevance to the environmental manager. First, there are two fundamental principles on which the system has been based from the beginning: the Most-Favoured Nation (MFN) Principle and the Principle of National Treatment. Second, there is the GATT Article dealing with quantitative restrictions (Article XI), and the two Agreements covering application of standards and other regulations at the border: the Agreement on Technical Barriers to Trade (TBT) and its annexe, and the Agreement on Sanitary and Phytosanitary Measures. Finally, there is the GATT Article outlining when nations may exercise general exceptions to the rules (Article XX).

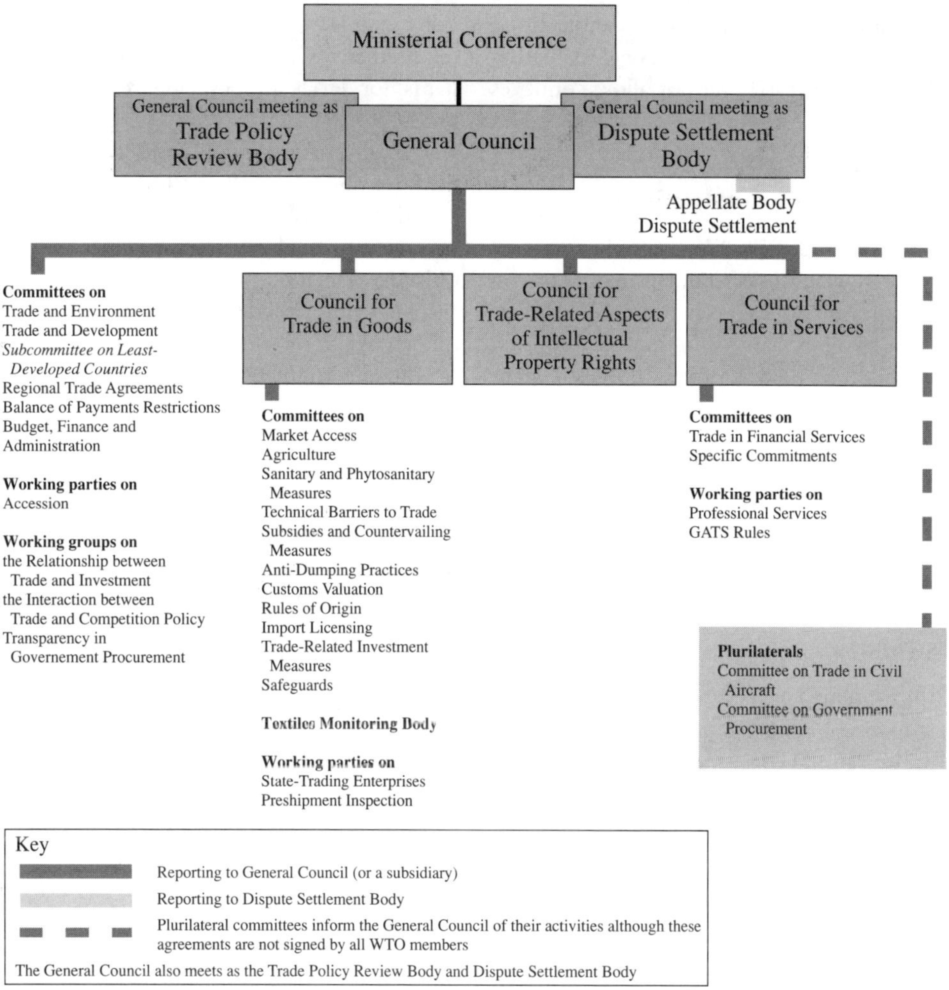

Figure 18.1 WTO structure

Note: All WTO members may participate in all councils, committees, etc., except Appellate Body, Dispute Settlement panels, Textiles Monitoring Body and plurilateral committees.

The most-favoured nation principle

The MFN is contained in GATT Article I. It states that WTO members must accord to any product, originating in or destined for any other country, the same treatment to all other third members. In other words, all members are 'most-favoured' and none may benefit from concessions not offered to the rest.

The principle of national treatment

Contained in GATT Article III, this principle addresses how a country may treat imports. It requires that any treatment applied to an imported product, such as taxes, regulations or requirements, should be no less favourable than that applied to like domestic products. But, as we will see later, the definition of a 'like' product becomes very important in this context. For example, is a circuit board manufactured using CFCs 'like' one manufactured domestically without using ozone depleting substances? If so, Article III will not allow countries to discriminate between them at the border.

Article XI: quantitative restrictions

According to GATT Article XI, the only forms of prohibitions or restrictions a country may place on imports or exports are duties, taxes and other charges. The aim is not to allow bans or quotas on imported goods which might be used to protect domestic manufacturers, or on exports, particularly raw materials, where restrictions might be used to subsidise domestic processing industries unfairly by blessing them with a cheap supply of inputs.

The agreements on technical barriers to trade (TBT) and sanitary and phytosanitary measures (SPS)

The TBT Agreement covers any kind of restriction placed on imports or exports based on product standards and technical regulations, and it describes how such restrictions must be prepared, adopted and applied. Technical regulations mandate that the characteristics of a product (or of the process and production methods, if they affect the final product's characteristics) must meet some requirement. A technical regulation might require, for example, that all imported cars must be equipped with catalytic converters, or that a product must be labelled in a certain way or must undergo certain tests before entering the country. Standards are non-mandatory descriptions of the characteristics of a product (or of the process and production methods) specified by a recognised body, such as a national standard-setting body or an international agreement. Standards may form the basis of technical regulations at the national level. Eco-labels, discussed below, are considered to be standards.

The Agreement on Sanitary and Phytosanitary Measures (SPS) covers a subset of such measures: those which apply specifically to protection of human or animal life from risks arising from additives, contaminants or disease-causing organisms in their food, or from the spread of pests from imported products. For example, a country may have certain quarantine requirements for imported livestock, or may set standards for levels of pesticide residues on imported produce.

The TBT Agreement basically requires that any measure must meet the national treatment principle, and that it must not be 'more trade-restrictive than necessary to fulfil a legitimate objective'. Legitimate objectives include 'national security requirements; the prevention of deceptive practices; protection of human health or safety, animal or plant life or health, or the environment'. The key word in the TBT phrasing is 'necessary'; Wirth (1994) points out that there is as yet no written jurisprudence clarifying how one would judge a measure necessary or unnecessary in the context of the TBT. This is somewhat unsettling to regulators because, in other contexts discussed below, 'necessary' has been interpreted to mean 'least trade-restrictive'. In other words, if there were a less trade-restrictive way in which an objective might be fulfilled, the measure under examination might be deemed unnecessary.

The TBT encourages, but does not force, countries to implement technical regulations that are based on international standards where such standards exist, and sets up a notification system by which countries should inform trading partners if or when they decide to adopt standards that are more stringent than those regarded as norms. A country adopting stricter standards must be able to argue that existing norms are ineffective or inappropriate for the fulfilment of the objectives pursued.

The SPS Agreement, whose drafting seems to have been motivated at least in part by a long-standing dispute between the USA and the EU over trade in hormone-treated beef, goes slightly further than the TBT in this regard. It too encourages countries to implement measures in line with established international norms. Specific bodies such as the Codex Alimentarius, which is a subsidiary of the UN's Food and Agriculture Organisation and the World Health Organisation, are recognised as legitimate bodies for establishing international norms.

Unlike the TBT, however, the SPS Agreement requires that any standards which exceed such norms should be 'based on scientific principles and . . . not maintained without sufficient scientific evidence'. But the meaning of this passage is not clear, in spite of an explanatory text. The USA, in its Statement of Administrative Action (which is the implementing legislation for the Uruguay Round), interprets this requirement as 'a country must follow the mandated scientific procedures for risk assessment, but the process of setting a standard of appropriate risk for any country is purely a public policy choice'. According to this particular interpretation, which is meant to reassure US citizens that their environmental laws would not be subject to international scrutiny, any country may set SPS standards higher than international norms provided that it has followed scientific principles for assessing risks; but, when applying those risks to determine standards, it can do anything it feels appropriate for its constituents. This untested interpretation, of course, has no international legal force.

The general exceptions: Article XX

What we have said above explains how countries are expected to behave according to the WTO rules of conduct: they may not arbitrarily discriminate against imports from a particular country (MFN), or do so to the benefit of domestic producers (national treatment), and so on. Since the GATT of 1947, however, countries have recognised that in certain circumstances there should be exceptions to these rules of conduct. Article XX spells out those circumstances that are general in character (as distinct from those related to security, covered in Article XXI). There are three sections of this article that are of particular interest to the environmental manager: the *chapeau*, or the heading section, and subsections (b) and (g).

The *chapeau* states that

> Subject to the requirement that such measures are not applied in a manner which would constitute a means of arbitrary or unjustifiable discrimination between countries where the same conditions prevail, or a disguised restriction on international trade, nothing in this Agreement shall be construed to prevent the adoption or enforcement by any contracting party of measures:

And the relevant subsections are

(b) necessary to protect human, animal or plant life or health;

(g) relating to the conservation of exhaustible natural resources if such measures are made effective in conjunction with restrictions on domestic production or consumption.

The text of the *chapeau* is fairly standard; it simply seeks to ensure that a measure is actually used to fulfil the objectives stated in the subsections, rather than to protect domestic producers unfairly. A part of the US Clean Air Act, for example, was recently found to contravene this requirement (see WTO, 1996). It specified that reformulated gasoline imports would be subject to cleanliness requirements that are quite different (and, in some cases, more stringent) from those applied to US producers. The ostensible reasoning was that importers' data on production was not of good enough quality to warrant equivalent treatment. But at least one US government official is on record in Congressional testimony as admitting that the measures were designed with the protection of the US refiners in mind. A WTO dispute settlement appellate body ruled that the measure in question was, in fact, a disguised restriction on international trade, and therefore did not qualify for an Article XX exception.

Article XX (b) seems to give broad leeway to regulators, allowing any measure for the protection of human, animal or plant life or health to contravene GATT rules. However, the key here is the interpretation of the word 'necessary'. GATT dispute panels have interpreted this to mean 'least trade-restrictive' (see Box 18.1). In practice this means that a dispute panel might decide that, although a measure does meet the *chapeau* requirements – and the case is not that of protectionism masquerading as environmentalism – its objectives could nevertheless be achieved in another way that would be less disruptive of international trade. A multilateral agreement, which is an example of such a less trade-restrictive alternative, is given in Box 18.2. This interpretation considerably narrows the scope for the application of the XX (b) exception.

Article XX (g) deals with the conservation of exhaustible natural resources. Again, this appears to be an important exception for environmental management. A recent WTO decision ruled that clean air, for example, is such a resource (WTO, 1996); and another WTO dispute panel heard similar arguments about the conservation of wildlife (see Box 18.1).

BOX 18.1 THE US–MEXICO TUNA/DOLPHIN DISPUTE

The most famous trade versus environment clash is possibly the GATT US–Mexico Tuna–Dolphin dispute. It stemmed from the US Marine Mammal Protection Act (MMPA) which bans the importation of tuna fish from the eastern tropical Pacific Ocean caught using purse seine nets, unless the fishing country has implemented policies to protect dolphins from being killed during fishing with nets. Dolphins normally swim above shoals of tuna and become trapped in the nets during fishing. The acceptable rate of mortality was to be set retroactively at 1.5 times the US rate for the season in question.

In 1990 Mexico complained to the GATT that the MMPA was a quantitative restriction, prohibited under Article XI. The USA defended the law, arguing first that it was acceptable under Article III in that it applied equal treatment to imported and domestic tuna production, and that therefore it was not a quantitative restriction as prohibited by Article XI. It also argued that, if the panel rejected its first argument, it should grant an exception under Article XX (b) which is on the protection of animal health, or under XX (g) which is on the conservation of exhaustible natural resources.

The dispute panel (whose report was never adopted) ruled against the USA on a number of grounds. First, it held that Article III did not apply, because the MMPA did not apply equally to 'like' products from foreign and domestic production. The panel ruled that tuna caught with or without a dolphin protection regime are essentially alike, thus denying the USA the ability to discriminate at the border on the basis of how a good is produced, if the process or method of production does not ultimately affect the good's characteristics. This meant that the MMPA was in fact a quantitative restriction, and that it would need to be granted an exception in order to be found legal.

The panel ruled out the granting of exceptions under Article XX, because Article XX (b) only allows exceptions where the regulation in question is 'necessary', and the panel interpreted this to mean 'least trade-restrictive'. It ruled that international agreements would have been less restrictive, and that the USA had failed to try hard enough to achieve this. It also ruled that neither XX (b) nor XX (g) was intended to allow for the protection of animals or resources outside the borders of the regulating country, thus denying the legality of extra-territorial regulation to protect the environment (this ruling was later contradicted by the so-called Tuna–Dolphin II panel ruling which was also not adopted. It ruled on the EEC's objection to the secondary embargo imposed under the same legislation).

The panel also found the retroactive nature of the mortality limit to be an arbitrary restriction, since Mexican fishermen could not know until the end of the season the rate they should aim at.

The panel did, however, uphold the right of the USA to label tuna imports according to the manner in which the tuna was produced (caught). The eco-label in question informs consumers about whether or not the tuna is 'dolphin-safe'.

There are three requirements in this subsection which limit its application. The first is that any such measure should be undertaken in conjunction with domestic restrictions on consumption or production. For example, a country cannot ban the import of ivory, ostensibly to conserve elephant herds, while allowing domestic producers to continue unrestricted harvesting or processing. Nor could it ration oil exports on conservation grounds without enacting domestic plans for restricting consumption.

The second limitation arises from the interpretation of the words 'relating to', interpreted in various GATT dispute panels to mean 'primarily aimed at'. In other words, the measure in question cannot just incidentally relate to conservation; it must be *primarily aimed at* conserving an exhaustible resource. A typical example of this was Canada's requirement that all salmon and herring fish caught in the waters of the western coast of Canada be landed in Canada. The argument advanced was that such landings were necessary for a conservation objective – the collection of accurate catch data for setting catch limits. But the panel ruled that a 100 per cent landing requirement was not necessary for effectively achieving this objective, and that the measure in question could not therefore be primarily aimed at conservation. The implication was that the measure was in fact designed to create jobs in Canada's fish processing industry.

The third limit on the use of XX (g) as an exception is illustrated in Box 18.1. According to one GATT dispute panel interpretation (Tuna–Dolphin) of XX (g), the exhaustible resources under consideration cannot be outside the jurisdiction of the country enacting the measure in question. Thus, a country could not, according to this interpretation, enact trade restrictions designed to protect global resources such as the ozone layer, planetary biodiversity or migratory fish stocks. Some (see Charnovitz, 1991) have argued that this interpretation is not justified by precedent or negotiating history, and, as noted in Box 18.1, the most recent panel to rule on this question (Tuna–Dolphin II) found no grounds for such a restrictive interpretation. Neither of the two Tuna–Dolphin panel reports was officially adopted by the GATT, however. And, because even adopted reports do not constitute jurisprudence in any case, the GATT legality of extra-territorial regulations remains unresolved. A succinct summary of the implications of the two Tuna–Dolphin panel rulings is given in Goldberg (1994).

ENVIRONMENTAL MANAGEMENT IN THE GLOBAL CONTEXT: MANAGING IMPORTED GOODS

The obligations to which most of the countries of the world have committed themselves in the WTO will obviously affect the types of regulations the environmental manager can enact. We have already seen some examples of the ways in which domestic regulations may fall foul of international rules of conduct. To make clear what instruments may or may not be used under current WTO law, we will now systematically examine the various types of environmental problems which may occur, and search for tools which the regulator can use for addressing them. To do this, we will first examine the problems specific to imported goods, and then turn to export-related problems.

For the purposes of discussing imports, it is useful first to divide environmental problems into those which occur in the consumption (use and disposal) of a product, and those which occur in its production, as shown in Table 18.1. Examples of the former would be the pollution created by a car, or landfill problems associated with the disposal of nappies. Examples of the latter would be the effluent from the factories manufacturing those cars and nappies. The importance of this distinction will become clear in the course of our discussion.

The second distinction, used in Table 18.1, is between environmental problems whose effects are limited to the country of production or consumption, and those with 'spill-over' effects creating international problems. An example of the former is the unsustainable draw-down of aquifers, or bodies of surface water, caused by agricultural irrigation. While the final product may be exported, the environmental effects in such a case are confined to the country of origin (provided, of course, that the water in question does not straddle national boundaries). An example of the latter is the burning of coal to generate electricity, whereby sulphur and nitrogen emissions from the plant cause acid precipitation in the neighbouring countries. For various reasons environmental managers may be concerned about importing goods manufactured with energy generated in this way.

In what follows, the issues discussed graduate from the least to the most controversial, starting at the top left of Table 18.1 and ending at the bottom right.

Table 18.1 Types of import-related environmental problems, and the various trade-related instruments available for addressing them

Type of problem	*Purely domestic environmental impact*	*International environmental impact*
Problems arising from the *consumption* of goods and services	• Product standards	• Product standards • Multilateral environmental agreements
Problems arising from the *production* of goods and services	• Eco-labels	• Multilateral environmental agreements • Eco-labels • Trade bans or restrictions* • Trade sanctions* • Border tax adjustments* • Countervailing duties*

Note: *Of questionable legality under WTO rules, see discussion, pp. 353–355.

Consumption-related problems with purely domestic impacts

These are problems arising from the use and disposal of goods whose impacts occur within the borders of the consuming country. Of the four types of problems listed in Table 18.1, these are the simplest to address within the rules of international trade. Regulators in the importing country are free to draft product standards, specifying for example that cars must attain a certain level of fuel efficiency, as long as such standards apply equally to both domestic and imported cars in line with the principle of national treatment.

However, such regulations still need to respect the TBT requirement that they do not create unnecessary obstacles to international trade. The USA recently imposed what is called the Corporate Average Fuel Efficiency (CAFE) standards on cars, specifying average fuel efficiency levels to be achieved by a manufacturer's entire fleet marketed in the USA, but not including trucks or vans. At first glance these standards seem to comply with the TBT specifications, since they apply equally to imported and domestically produced vehicles. The EU complained, however, stating that by setting up the system in the way it was (based on fleet mileage), the USEPA in effect discriminated against foreign manufacturers whose exports to the US consisted of only a few large gas-guzzling models, such as Mercedes-Benz (Germany) and Rolls-Royce (UK). The dispute panel eventually ruled in favour of the USA.

Consumption-related problems with international impacts

Problems of this type are not usually of special concern to regulators; they treat most of the consumption-related problems in the same way, regardless of the scope of their impact. For example, if carbon emissions from the use of cars are a cause for concern because of their potential for climate change (an international problem), regulators might respond by applying product-related measures, such as those discussed on pp. 348–349, to both imported and domestically produced vehicles.

However, if the effects of a country's consumption-related problems create environmental damage in other countries, the latter are likely to complain loudly. Typically, international disputes over shared water resources, such as rivers and seas, provide strong evidence of this. Thus, pressure may be brought to bear on regulators to enact stronger measures than they would contemplate if only domestic environmental damage was the issue. Moreover, the citizens of many countries demand environmental protection measures which take international impacts into account.

This does not present a problem from the perspective of trade rules, however, because in general product standards can be as strong or weak as a country deems appropriate, provided that they are based on scientific principles as we have already pointed out. The wording of the SPS Agreement with regard to science leaves open the question of whether a country should take risks to non-citizens into account when deciding on the appropriate level of protection; but this is a hypothetical point which is not likely to be tested.

This is due in part to the fact that, when faced with issues of international concern, many countries choose to create Multilateral Environmental Agreements (MEAs). In order to reduce the use of ozone-depleting substances, for example, the international community drafted an agreement limiting their consumption, production and import. The MEAs, and their relationship with the multilateral system of trade rules, are discussed in greater detail on pp. 351–353.

Production-related problems with purely domestic impacts

There are two main types of production-related problems:

1. Type I: The good itself is harmless, but there are problems inherent in the process, or the production method (PPM), used for its manufacture, as in the case of a factory emitting pollution.
2. Type II: The production of the good itself is the problem, rather than the manner of its production. This is the case, for example, in the harvesting (which can be thought of as production in the economic sense) of endangered species such as rhinoceros and tiger.

In this section we look specifically at the subset of production-related problems which create purely domestic environmental impacts. Accordingly, we will focus on Type I (PPM-related) problems; Type II problems will be discussed on pp. 350–353, where we will consider international environmental impacts.

Trade measures to address these problems are understandably controversial. What rights do regulators have to address environmental problems that are confined to the exporting/producing country? For example, country 'A' may decide to ban the import of leather shoes from country 'B' on the ground that the tanning process in B pollutes its own domestic waterways.

Such measures are called 'extra-territorial' because they aim to change producer behaviour in a country which is outside the territory of the regulator. They are almost always based on PPM standards, as opposed to product standards; that is, they are not concerned with the characteristics of the product itself but with the processes and methods used in its production.

Why should the regulators be concerned if the environmental damage in question does not occur within the regulating country? The answer to this question has two parts: the first part is about economic competitiveness. If domestic producers are required to clean up their acts while the foreign producers do not, then foreign competitors will presumably have lower production costs. The second part, which flows from the first, refers to the environmental status of the country in question. Domestic regulators may be struck with what is called the 'regulatory chill', meaning that they are unable to set appropriate standards for fear of businesses relocating to less-green pastures. Most empirical work suggests that pollution control costs are in fact too low to be major factors in relocation decisions – of the order of 2–3 per cent (Low and Yeats, 1992) – although the threat in specific cases can be very real. The International Institute for Sustainable Development (1994) has argued that differing environmental conditions and different policy priorities across jurisdictions make it necessary to tolerate different production-related standards, provided that the environmental damage involved is purely domestic.

The argument over this principle, called the 'PPM issue', is still to be settled, and the issue remains controversial. A key question, which is still to be answered definitively, is whether or not the WTO rules permit PPM-related trade measures. In other words, could a country ban, for example, the import of fur from animals caught in leg-hold traps, but at the same time allow imports of farmed fur?

The response of most in the trading community to the above would be 'no'. It is worth recalling that GATT Article III prohibits less favourable treatment being accorded to imported goods than to like domestically produced goods. The key here is the interpretation of 'like'. Chakarian (1994) explains that in GATT jurisprudence 'like' goods are those whose product characteristics are similar, but that how they were produced is of no consequence. Thus, a shrimp, caught in a net which uses turtle excluder devices, is 'like' a shrimp caught in a net which does not (and which therefore kills endangered turtles), because it is impossible to tell the two apart physically when they arrive at the border. According to this interpretation of 'like', countries which use trade measures to restrict imports of the latter are in fact treating imported goods less favourably than like domestic goods; they are thus in breach of Article III. Where a domestic industry in the product does not exist, such measures would in any case probably breach Article I since they would discriminate arbitrarily among 'like' products from different countries, thus violating the MFN principle.

What, then, is left in the regulatory 'toolbox' to deal with PPM-related problems whose impacts are confined to the exporting countries? The only tool that has been used in such cases to date is eco-labelling; but it is a relatively 'soft' instrument.

Eco-labels

An eco-label is a mark of distinction on a product's label which is awarded on the basis of its environmental merit. There are two basic types: 'single-issue' and 'multiple-criteria'. Single-issue eco-labels certify that a product has met one particular standard. For example, a product may be certified as recycled, dolphin-safe, or manufactured without the use of CFCs. Only one criterion is specified, and the product either meets that criterion or does not. A multiple-criteria label certifies the product as being environmentally friendly, based on a number of different criteria. One Dutch eco-label for cut flowers, for example, considers weighted scores in four different categories: energy used in transport, application of fertiliser, wastes produced, and crop protection remedies. A variant of this type of label assigns grades, like the five-star rating system for hotels, to certify different levels of environmental friendliness.

Single-issue labels are more likely to be used to address the kinds of problems we are considering here; namely, production-related damage which is confined to the country of production. For example, there is work under way internationally to determine how forest products could be certified on the basis of whether or not they came from 'sustainably harvested' stocks (of course defining sustainable harvesting is the difficult part of this work.). A traded product bearing such a certification would give consumers information about the process by which it was produced, even if the effects of its production method – forest eco-system integrity in this case – were felt only in the exporting country. The dolphin-safe label, which applies to US tuna imports, is an example of such a system currently in use.

It is not enough for regulators simply to set up the labelling system, because eco-labels are only effective if they are implemented in conjunction with a comprehensive and continuing campaign of public education. The public needs to know about the environmental problem being addressed; otherwise they will not be motivated to buy eco-labelled products. They need to know that the labelling system exists, so that they can look out for it; and they need to understand what the label means, so that they can make an informed choice between labelled and non-labelled goods. The daunting work involved in public education is the most common stumbling block for eco-labelling systems world-wide.

Another common problem with eco-labelling systems is the lack of openness and accessibility for foreign producers. Openness, particularly during the design of the system, is important for ensuring that eco-labels do not become protectionist in their effect. The Dutch systems of eco-labelling for cut flowers have been criticised for including environmental costs of transportation in their criteria, because they automatically penalise foreign producers. Verbruggen *et al.* (1995) suggest that the inclusion of transportation costs may have been intended as protection for Dutch producers. Had it been possible for foreign producers to participate in discussions on the choice and weighting of criteria, this issue might not have been so problematic.

Accessibility is also important. Most eco-labelling schemes involve complex certification procedures which can be particularly expensive and difficult for foreign producers, unless their special needs are considered in designing the system.

Discussions now taking place at the WTO may result in voluntary and mandatory eco-labelling schemes under the rules of the TBT 'Code of Good Practice', which is essentially a set of guidelines addressing issues such as openness and accessibility of the standard-setting process. It should be noted, however, that there is as yet no consensus in the WTO on the question of whether or not such voluntary standards should even be allowed to apply to PPMs.

Production-related problems with international impacts

There is a stronger argument for trade measures to address production-related problems which cause international impacts, because such impacts do in fact involve the regulator's jurisdiction. For example, an imported computer, whose manufacturer used CFCs as a cleaning solvent, can cause a tangible effect on the citizens of the importing country by depleting the stratospheric ozone layer.

That said, it remains a good question as to which environmental effects are in fact international and which are purely domestic. Is the destruction of bio-diversity by rain-forest clearing an increasingly serious loss of a global resource – which some argue is the patrimony of mankind – or is it a loss only to the citizens of the country in which the clearing takes place? How should we regard the depletion of pelagic fish stocks which may travel through the waters of dozens of countries? Would it be an environmental loss only to the country in whose waters they happen to be when they are caught? What if they are taken on the high seas? There are no easy answers to such questions, and they remain to be addressed at the international policy level.

For the purposes of GATT law, however, the difference is unimportant. Whether the environmental effect in question is international or purely domestic in scope, the WTO rules as currently interpreted still do not allow trade measures to address production-related environmental damage on the basis of the definition of 'like' products discussed earlier. This interpretation leaves only two types of instruments available to regulators: MEAs and eco-labelling.

Multilateral Environmental Agreements (MEAs)

The MEAs are an important instrument available to environmental managers to address production-related problems which have international environmental effects. For example, in order to deal with the problem of drift-net fishing on the high seas – a production method which is devastating to marine ecosystems – the international community adopted a UN resolution (44/225) in 1989 calling for a moratorium on drift-net fishing, and negotiated the Wellington Convention for the Prohibition of Fishing with Long Drift-nets in the South Pacific.

Such agreements are notoriously difficult, however, and take many years to conclude. And what is finally achieved is always less than what the most concerned countries had hoped for. Such agreements are therefore not an attractive option if the environmental problem at hand is urgent and/or severe.

Facilitating the process of drafting MEAs involves an understanding of why their progress is so slow. Consider the example of country 'A', which has laws in place, wants similar enforcement in other countries through an MEA, but country 'B' is reluctant to sign on. Since country B has not adopted 'green' measures in the first place, we can assume that the issue is not as high on its list of priorities as it is on country A's. This may be due to different ecological conditions which cause lower environmental impacts in country B for the same damaging practice. Or, it may be because country B's level of development is such that it has more pressing issues to address. For example, the top environmental priority for many developing countries is the supply of clean water, the lack of which kills more than 25,000 people daily. In view of this, and given their meagre financial and human resources for dealing with environmental problems, many of the developing countries have less stringent measures to address environmental issues such as air quality.

There is a clear need for agreements with appropriate mechanisms, such as the Multilateral Fund (Box 18.2) of the Montreal Protocol, to finance ozone-saving initiatives in countries that are unable to give high priority to the issue. Money, technical training and new technologies need to be offered to such countries as part of the MEAs. This means that more than just the Ministries of Environment need to be involved; but such cross-departmental arrangements are difficult to make at the best of times, and ever more so now because of shrinking foreign assistance budgets of the OECD countries.

In addition to such 'carrots' (i.e. incentives and assistance) for encouraging green behaviour, a small number of MEAs also use 'sticks' in the form of trade measures to help enforce their intent and to help bring more participants on board. A few of these are listed in Table 18.2.

Table 18.2 shows that trade measures in MEAs can serve several purposes. The measures in the Wellington Convention, for example, address Type I production-related problems – PPMs which are environmentally damaging – by banning the import of fish caught with drift-nets. Where an MEA addresses such problems, trade measures are often contemplated to discourage what is called 'free-riding'. A 'free-rider' in this context is a country which does not participate in an agreement, but enjoys the environmental benefits which that agreement brings. In this particular case, it was felt that such a strategy would be less attractive if any goods produced in violation of the terms of the agreement were banned from the markets of the signatories.

CITES is typical of wildlife-related MEAs; it treats production itself (i.e. the taking of species of flora and fauna from the wild) as problematic – a Type II production-related problem. In agreements dealing with such problems, trade measures are often contemplated for restricting the commercial flow of the products in question, the ultimate goal being to reduce production in the first place. In theory this goal can be achieved with controls on production and consumption alone, without the need for trade measures. For

BOX 18.2 THE MONTREAL PROTOCOL

The Montreal Protocol on Substances which Deplete the Ozone Layer, concluded in 1987, set production and consumption limits on the principal ozone-depleting substances (ODSs) – namely, chlorofluorocarbons and certain halons. It was a protocol to the earlier Vienna Convention for the Protection of the Ozone Layer, and was drafted in response to increasing evidence that those substances were depleting stratospheric ozone which protects the earth's surface from damaging ultraviolet radiation.

The Protocol was amended in 1990 to prohibit several other compounds and to accelerate the schedules for phasing them out. Recognising that developing countries would have difficulties in making the necessary transition, the parties also established a Multilateral Fund to provide financial and technical assistance to countries adopting ozone-friendly technologies.

The parties were worried about the challenges posed by non-parties to the Convention. There were two worries in particular: the first was that non-parties would simply increase their production of ODSs and export them to the parties; and second that the non-parties would also export to the parties goods whose manufacturing processes release ODSs into the environment. The first worry was addressed by banning trade in ODSs with non-parties. Transfer of technology for producing controlled ODSs from parties to non-parties was also discouraged.

As for the second worry, the parties contemplated addressing it through measures restricting import of goods made with such substances from non-parties. However, this option, which constitutes a trade restriction imposed on the basis of PPM criteria, has since been set aside, in part because of the conflict it would engender with the rules of international trade. The issue is still to be addressed.

The Protocol has a number of features that are of interest to the environmental manager. It is an instrument which addresses production-related environmental problems using trade measures to increase its effectiveness. But these measures are in potential conflict with WTO rules because they are quantitative restrictions which may be challenged by a non-party to the protocol arguing that such measures impair or nullify its WTO rights. PPM-based trade measures were contemplated to increase the effectiveness of the Protocol further, but the idea was rejected. The Protocol incorporates a mechanism for bringing less developed parties on board, as well as for helping them financially and technically for phasing-out the prohibited substances.

Table 18.2 Three MEAs which use (or contemplate using) trade measures to address production-related issues

Agreement	*Issue*	*Trade Measure*
The Montreal Protocol	Substances which deplete the ozone layer	Parties have banned the import of ODSs from non-parties. Parties also considered, but rejected for the time being, banning imports of goods manufactured using ODSs from non-parties
The Wellington Convention	Fishing with long drift-nets in the South Pacific	Parties may take measures *consistent with international law** to prohibit the import of fish or fish products caught using drift-nets
CITES	Trade in endangered species of flora and fauna	Parties restrict or prohibit trade in particular species, depending on the degree of endangerment of the species in question

Note: *Since international law is unclear on the issue of PPM-based regulations, it is not clear which kinds of measures this qualification leaves available

example, both the exporting and the importing countries could restrict the harvest or sale of endangered species (in fact they do so under CITES). In practice, however, Type II problems rely on the enforcement of relevant measures, and, in the absence of effective enforcement as at present, a good deal of trade in endangered species continues on an illegal basis. Trade controls are a particularly valuable tool, because there is more rigorous control at the border crossing point than at the point of harvest or at the final point of sale.

In the Montreal Protocol trade measures have been designed to address both types of problems. Type I problems involve the production of goods, such as integrated circuit boards, using ozone-depleting substances (ODSs). The goods thus produced are themselves not harmful to the environment, but the PPM used is harmful to the ozone layer. The measures contemplated (and rejected for the time being), such as banning the import of such goods, were aimed at discouraging free-riding by restricting the markets to which they could be exported.

The Protocol also addresses Type II problems, out of a desire to stop the production of ODSs in the first place. Seen from this perspective, the production of ODSs is itself a problem. Again, in theory this could be addressed simply by controlling the production and consumption of ODSs, but the signatories recognised the practical value of measures for limiting imports of ODSs from non-signatory countries.

It is not yet clear what the relationship will be between such trade measures in MEAs and WTO rules. The measures could be perceived as contravening WTO rules because trade restrictions or bans are prohibited under Article XI. Furthermore, Article XX exceptions may not be interpreted in such a way as to allow regulations which discriminate on the basis of PPMs. Most of the WTO members are also parties to the major MEAs, and so they are unlikely to challenge the restrictions to which they have committed themselves. The real issue here is concerned with trade restrictions applied under an MEA to a non-party country which is a WTO member. In such a case the non-party country might contend that its rights, ensured by the WTO, were being violated. Discussions are taking place at the WTO on the most appropriate way in which to deal with this political bombshell before it explodes. The NAFTA broke new ground in 1995 when it specifically named three MEAs, and two bilateral agreements, to which it would defer in the event of conflicting obligations (Article 104).

Eco-labelling

This instrument was discussed earlier as the only one in current use to address production-related problems with purely domestic impacts. It is also used, of course, to address issues of international concern. The EU eco-label for paper products is a typical example. It is a multiple-criteria label which incorporates, among other things, carbon emissions from the production process. Such labels thus inform consumers about the impact of products on the environment: global climate change for example.

Instruments of questionable WTO legality

A number of instruments (Stevens, 1994) for addressing production-related environmental problems are listed in the bottom right-hand quadrant of Table 18.1. Although their legality with regard to the WTO rules is questionable, they are discussed here for two reasons: first, they have all been proposed at various times by the environmental community and enjoy strong support in spite of the dismal view of them taken by international trade rules. It is therefore necessary to know how they work and in what ways they might contravene the WTO rules. Second, even though current interpretation rules against their use, it is impossible to predict how the trade–environment debate will evolve in the future. The policy interface between trade and environment is still very much an 'unsculptured marble' whose final image is being shaped even now. The official position of the USA, for example, is that unilateral extra-territorial trade measures are a WTO-legal way in which to address production-related problems, despite rulings such as by the Tuna–Dolphin panel. Clearly, until this situation is clarified with greater certainty, environmental managers will only have ill-defined guidelines to follow in applying trade measures.

Trade bans or restrictions

Trade bans or restrictions prevent goods from entering a country. For example, recently the USA implemented a ban on shrimp imported from countries which do not mandate turtle-excluder devices in their shrimp nets. These devices are intended to reduce the mortality of endangered turtles which drown in conventional shrimp nets.

The affected exporters, mostly South-East Asian countries, took the matter to the WTO for a ruling. What is at issue here is whether this ban violates

Article XI of the GATT, or whether it qualifies for an Article XX exception. In fact, the WTO dispute panel ruled against the US in this case.

Trade sanctions

Trade sanctions operate in the same way as trade bans or restrictions by preventing goods from entering a country. In the case of sanctions, however, restriction on import is not aimed at the goods in question but at other goods with greater economic significance and greater potential for influencing policy in the exporting countries. For example, when the USA wanted to put pressure on Norway to stop whaling in compliance with the rules of the International Whaling Commission, it banned the import of Norwegian fish products. But, as Norway does not export whale products to the USA, bans or restrictions would not be effective in this case. Like trade bans or trade restrictions, such measures are generally seen as contravening Article XI, and therefore illegal under WTO rules.

Border tax adjustments

Environmental managers are increasingly turning to market-based instruments for environmental protection. One such instrument is a PPM-based charge, or tax, assessed on domestically produced goods. A typical example of this is the 'carbon tax' which has been the subject of much public discussion recently. This tax is assessed on energy sources in relation to the amount of carbon emitted by their use.

Imposing such a tax is problematic, however, because domestic products, manufactured with energy on which the tax has been paid, will have to compete with imports that would be cheaper since no such tax will have been paid on the energy used in their production. Thus the carbon tax merely transfers the pollution to another jurisdiction. Also, massive opposition to the tax is generated by its adverse implications for both jobs and economic competitiveness. Indeed, such opposition was a key factor responsible for the delay in implementing the EU's proposed fossil fuel tax.

Some hold the view that these problems can be overcome with a PPM-based 'border tax adjustment'. This adjustment is a WTO-legal mechanism which rebates exporters at the border for taxes paid domestically. The tax to be paid on imports at the border is assessed in relation to an equivalent tax which will have been paid by domestic firms. So, if a domestic firm pays value-added tax, for example, it can have that tax rebated on goods it ships abroad. This will assure domestic producers that imports will be subject to the same tax, thus creating an even playing field.

However, current interpretations of WTO rules only allow such adjustments for taxes levied *directly* on products, such as value-added taxes, sales taxes and excise duties. *Indirect* taxes, such as social security charges, payroll taxes, corporate income tax and so on, are not eligible for border adjustment. The logic for this is that it would make little sense for foreign manufacturers to pay such taxes, and so each country should decide for itself the level at which it wishes to assess them. The same logic may apply to PPM-based environmental taxes, although there is some uncertainty on this score (Arden-Clarke, 1994). For example, taxes on the energy used to produce a good would probably be ruled as being indirect.

Countervailing duties

In the same way that many environmentalists would like to use the existing border tax adjustment mechanism to address environmental problems, it is often suggested that the existing countervailing mechanisms should be 'greened' as well. The existing mechanism is a duty imposed on imports, based on an assessment made by the importing country that the goods are being traded below their true cost; that is, they are being 'dumped'. Dumping may be due to government subsidies for a firm's exports, or it may be a deliberate pricing strategy by a firm aiming to gain or increase its market share. In either case, countervailing duties may be assessed in order to bring the prices up to their 'proper' level.

The environmental analogy should be obvious. Countries with high environmental standards may regard goods produced in countries with lower standards as being subsidised by lax regulation. They may feel that, because the exporting government allows dirtier (and therefore cheaper) production, the resulting goods are being 'subsidised'; that is, the prices of such goods do not reflect the true costs of production when environmental considerations are taken into account. Such manufacturers are said to be engaging in 'eco-dumping', and the call is for countervailing duties as a remedy.

Measures such as these would almost certainly be ruled illegal by the WTO because the WTO rules have a very specific definition of subsidies and they do not encompass implicit subsidisation by lack of regulation. Rather, they define subsidies as financial contributions to industry, or income or price support.

In practice, eco-countervailing duties would be difficult to administer, however, because they would require the calculation of subsidies implicit in particular levels of regulation in different countries. Apart from the methodological difficulties involved, there would also be ample scope for vested domestic interests to subvert such a process to their own ends.

MANAGING EXPORTED GOODS

So far we have discussed ways in which the rules of the multilateral trading system impact on measures applied to imported goods; but exported goods have an environmental dimension too, albeit more limited by comparison. In this context there are at least two areas of WTO rules with which environmental managers should be familiar, namely export bans and subsidies.

Export bans

Export bans or restrictions are often considered and imposed for conservation purposes. For example, if a country is in danger of losing significant amounts of its wildlife habitat because of logging it may impose a ban on the export of logs, as Indonesia did in 1985 (followed by a ban on the export of rattan in 1988); or if a country is worried about the prospect of dwindling reserves of non-renewable (minerals, oil) or potentially exhaustible (fish, forests) reserves, it may do likewise.

Such bans are, however, of questionable legality under WTO rules, and the aims they espouse can often be easily met with measures which do not violate WTO rules. Typically, Article XI is contravened by import or export restrictions. When this happens, the imposing country is likely to appeal for an exception under Article XX (g) which may be invoked in order to protect exhaustible natural resources. The catch here is that such measures must be taken in conjunction with domestic restrictions on production and consumption. In the case of an export ban on logs, for example, there would need to be laws restricting the domestic harvest of logs.

In the case of the Indonesian ban, for example, the charge made by foreign manufacturers was that it was 'protectionism dressed up in a green cloak'. Their contention was that Indonesia was trying to ensure a cheap supply of raw materials to foster a domestic processing industry for forest products. Under pressure from foreign governments, who threatened a GATT dispute proceeding, Indonesia lifted the ban.

In this particular case Indonesia's aims of forest conservation could probably have been met by domestic regulation of logging in sensitive areas, although the government argued that such regulations could not be enforced and that 'pirate' logging would still continue. In the case of non-renewable resources such as oil, restrictions may be placed on drilling. Even if a country felt the need to impose restrictions above and beyond such domestic conservation laws, it would not need to do so by imposing a ban. Instead, it could simply impose export taxes which are entirely legal under WTO rules. Furthermore, such taxes can be set at levels which in effect amount to an export restriction.

Subsidies

In the same way that many of the measures discussed above aim to penalise environmental 'bads' using taxes, levies, bans, etc., governments may occasionally want to reward or assist environmental 'goods' through use of subsidies. According to WTO rules, limited subsidies are allowed for improving the environment, specifically:

> assistance to promote adaptation of existing facilities to new environmental requirements imposed by law and/or regulations which result in greater constraints and financial burden on firms, provided that the assistance:
> (1) is a one-time non-recurring measure; and

(2) is limited to 20 per cent of the cost of adaptation; and
(3) does not cover the cost of replacing and operating the assisted investment, which must be fully borne by firms; and
(4) is directly linked to and proportionate to a firm's planned reduction of nuisances and pollution, and does not cover any manufacturing cost savings which may be achieved; and
(5) is available to all firms which can adopt the new equipment and/or production processes.

While this set of criteria appears to be rather limiting, there are a number of other WTO-legal ways in which governments can subsidise environmental improvements. The most important among them is the funding of research and development for improving or protecting the environment, as well as for improving the efficiency of technologies.

CONCLUSIONS

Two important and overarching conclusions emerge from what has been said in this chapter. The first is that the rules defining the trade–environment interface lack precision, certainty and often clarity. They are evolving through a process of change in response to on-going debate. It is impossible to say whether or not today's constraints on environmental managers, imposed by current WTO interpretations, will cease tomorrow, because since the dispute panel process does not rely strictly on precedent new interpretations are possible at any time. Indeed, the important concept of a 'like' product has been variously interpreted by GATT dispute panels on different occasions. Moreover, on-going discussions in the WTO's Committee on Trade and Environment may yet produce amendments to, or new interpretations of, the rules governing trade and environment.

The second conclusion to emerge is that the task of 'greening' international trade is not the sole purview of the environmental manager. If trade and environmental concerns are indeed to be reconciled, it will take co-operative efforts by policy-makers with diverse backgrounds in trade, finance, environment, and international development. Until and unless policy-makers with such diverse and essential talents can work together, especially at the institutional level (for example, in drafting MEAs that will satisfy the needs of all, or in working to dismantle escalating tariffs and other trade barriers that have negative environmental and developmental impacts in practically all countries, North and South), the debate will continue to produce much heat and little light.

REFERENCES

Arden-Clarke, C. (1994) *Environmental Taxes and Charges and Border Tax Adjustment – GATT Rules and Energy Taxes: A Critique for the World Wide Fund for Nature*, Gland, Switzerland: World Wide Fund for Nature.

Chakarian, J. (1994) 'PPMs and the GATT', in OECD *Trade and Environment Processes and Production Methods*, Paris: OECD, pp. 113–120.

Charnovitz, S. (1991) 'Exploring the Environmental Exceptions in GATT Article XX', *Journal of World Trade* 25 (5): 37–55.

Goldberg, D.M. (1994) *GATT Tuna–Dolphin II: Environmental Protection Continues to Clash with Free Trade (CIEL Brief No 2)*, Washington, DC: Center for International Environmental Law.

International Institute for Sustainable Development (1994) *Trade and Sustainable Development Principles*, Winnipeg, Canada: IISD.

Low, P. and Yeats, A. (1992) 'Do "Dirty" Industries Migrate?', in P. Low (ed.) *International Trade and Environment*, Washington, DC: World Bank, pp. 89–103.

Stevens, C. (1994) 'Synthesis Report: Trade and Environment: PPM Issues', in OECD *Trade and Environment Processes and Production Methods*, Paris: OECD, pp. 7–22.

Verbruggen, H., Kuik, O. and Bennis, M. (1995) *Environmental Regulations as Trade Barriers for Developing Countries: Ecolabelling and the Dutch Cut Flower Industry* (CREED Working Paper Series No. 2), London: International Institute for Environment and Development.

Wirth, D.A. (1994) *The Role of Science in the Uruguay Round and NAFTA Trade Disciplines* (UNEP Environment and Trade Series No. 8), Geneva: UNEP.

WTO (1998) *WTO Trading Into the Future*, 2nd ed. Geneva: WTO.

WTO (1996) *United States – Standards for Reformulated and Conventional Gasoline: Appellate Body Report, Action by Dispute Settlement Body* (WT/DS2/9), Geneva: WTO.

SUGGESTED READING

Bhagwati, J. and Daly, H. (1993) 'Debate: Does Free Trade Harm the Environment?', *Scientific American* 269 (5): 41–57.

Esty, D.C. (1994) *Greening the GATT: Trade, Environment and the Future*, Washington, DC: Institute for International Economics.

Ewing, K.P. and Tarasofsky, R. (1996) *The Trade and*

Environment Agenda: Summary of Major Issues and Proposals, Bonn: IUCN Law Centre.

Jackson, J.H. (1992) 'World Trade Rules and Environmental Policies: Congruence or Conflict?', *Washington and Lee Law Review* 49: 1227–1284.

OECD (1994) *Trade and Environment: Process and Production Methods*, Paris: OECD.

OECD (1994) *The Environmental Effects of Trade*, Paris: OECD.

Runnalls, D. and Cosbey, A. (1992) *Trade and Sustainable Development: A Survey of the Issues and a New Research Agenda*, Winnipeg, Canada: International Institute for Sustainable Development.

Twum-Barina, R. and Campbell, L. (1994) *Protecting the Ozone Layer through Trade Measures: Reconciling the Trade Provisions of the Montreal Protocol and the rules of the GATT* (UNEP Environment and Trade Series No. 6), Geneva: UNEP.

SELF-ASSESSMENT QUESTIONS

1 What did the GATT dispute panel (Tuna–Dolphin I) find objectionable about the US decision to ban tuna imports from Mexico?
(a) tuna and dolphins are not considered 'like' goods by multilateral trade rules;
(b) it was extra-territorial, unnecessary and discriminated between 'like' goods on the basis of PPMs;
(c) it violated the GATT's most-favoured nation principle as laid out in Article I.

2 Why is it important to learn about trade measures which are considered illegal by trade rules, such as green countervailing duties and border tax adjustments?
(a) because such measures should never be used by environmental managers;
(b) because such knowledge will be useful in WTO dispute panels, should the measures ever be challenged;
(c) because such knowledge will be useful if the rules are eventually interpreted to allow them to be used.

3 PPM-based environmental problems in a traded product are
(a) related to the physical characteristics of the product;
(b) those which have 'spill-over' environmental impacts beyond the country of production;
(c) not obvious on examining the product at the border.

4 Multilateral environmental agreements
(a) are a potentially WTO-legal way to address production-related environmental problems;
(b) are typically in violation of WTO rules;
(c) are impossible to formulate in such a way as to address the needs of developing countries.

5 Differing environmental standards between countries
(a) are a sign of policy failure by the various countries' environmental managers;
(b) are a result of different environmental conditions and different priorities;
(c) are considered illegal by current interpretations of WTO rules.

PART II.3

LEGAL INSTRUMENTS

19

INTERNATIONAL ENVIRONMENTAL POLICY

Elizabeth R. DeSombre

SUMMARY

International environmental policy is difficult to make, due to the lack of a higher authority within the international system to impose or enforce regulations. Consequently, individual nation-states have incentives to 'free ride' on environmental actions take by others, thus undermining the ability of others to protect the international environment. Nevertheless, some of the nation-states have been able to take collective action to address many global environmental problems. They have been persuaded to do so by increased scientific knowledge, supportive domestic public, international organisations to assist in overcoming problems of collective action, and by tools such as financial and technical assistance, economic sanctions, and improved monitoring and compliance mechanisms. Although international co-operation to address global environmental problems remains difficult, individual nation-states have learned how to negotiate more effective international environmental agreements, and to participate and comply with regulations that have been collectively adopted. Improvements can still be made in addressing equitably the disparities between developed and developing countries when making international environmental policy, and in increasing the willingness of nation-states to agree to a more intrusive monitoring of their compliance.

ACADEMIC OBJECTIVES

Aims

- to gain familiarity with the constraints that the international system imposes on the ability of individual nation-states to make environmental policy at the international level, and with ways in which environmental problems challenge the traditional focus of international relations;
- to learn about how the difficulties of policy-making within the international system can be overcome to allow international co-operation for managing environmental problems.

Objectives

On completion of this chapter the reader should be able to recognise the factors that make environmental policy more difficult to develop and implement at the international level than within a country or a city. The reader should be able to identify the main actors within the international environmental system, understand how the 'tragedy of the commons' operates at the international level, and appreciate how commons and transboundary issues create different sets of incentives and difficulties for addressing global environmental problems. The reader should also gain knowledge of the instruments of international environmental policy, ways in which individual nation-states can be persuaded to take part in managing environmental problems at the international level, and how to select instruments that may be most appropriate under different circumstances.

THE INTERNATIONAL SYSTEM AND ITS IMPLICATIONS FOR INTERNATIONAL ENVIRONMENTAL POLICY

First and foremost, a student of international environmental policy should understand that no sovereign state can ever be required to join an international agreement, or to enforce a particular regulation. The international system is thus anarchic in that there is no overarching authority (in this case a world government) that can dictate to the constituent parts what they must do. In national politics, on the other hand, there is generally a legislature which makes laws that bind the actors who might prefer not to be bound, and a judicial system that can hold these actors to account for their actions and sanction (and enforce) punishment on those who do not comply. At the international level, although there are bodies such as the United Nations and international conferences which act like legislatures, they are not empowered to make laws to bind the states which do not wish to be bound. And, although there are international courts and tribunals, no sovereign nation-state can ever be forced to appear before them, or to accept punishment from them.

This prevailing structure means that individual nation-states wishing to make environmental policy at the international level can do so only if they are willing to comply with the policies they have made, and submit to dispute resolution procedures to address problems that occur as a result of their action or inaction. The question is, under what conditions would they be willing to do so?

ENVIRONMENTAL PROBLEMS AND THEIR IMPACT ON THE INTERNATIONAL SYSTEM

Many environmental problems occur across national boundaries, and cannot therefore be addressed by any nation-state acting alone. Typical examples are: ozone-depleting substances harm the ozone layer regardless of which country they are emitted from; polluting chemicals, discharged intentionally or accidentally into the Rhine in Germany or Switzerland, flow downstream to impact those who depend on that river in the Netherlands; and emission of sulphur dioxide in central USA causes acid rain in New England and Atlantic Canada, and that emitted in the UK reaches Scandinavia. Even if it were once possible for individual nation-states to protect their own populations from environmental problems, they can no longer ensure that human activities elsewhere will not harm their territory or populations. Likewise, actions that once might have seemed to be purely national – cutting down a forest on one's own territory – can have global consequences; in this instance by contributing to global climate change or by decreasing the world's biodiversity.

A cost is often attached to environmental protection, however, and individual nation-states are often unwilling to bear it alone, especially when any positive action they take can be undone by the lack of (or inadequate) environmental protection measures taken by other states. This is what is known as the 'tragedy of the commons' on the international scene. Nation-states that would benefit collectively from taking actions to protect the environment may not be willing to take these actions if they cannot be assured that the other concerned states will also take necessary actions, and, more importantly, comply with the regulations accepted by all. For, while all the states stand to gain collectively by taking action to protect the environment, each state would clearly be better off if it did not have to take costly action for environmental protection while all others did. The state that did not participate would then benefit from the cleaner environment created by the group, but the consequences of its lack of environmental protection would be shared collectively by the group. In other words, the 'free-riding' state would benefit from the positive utility (i.e. lack of cost) from its refusal to protect the environment, while the negative utility (i.e. actual environmental damage) would be shared by all (Hardin, 1968). States being rational actors, each knows that the others have an incentive to free-ride, and that co-operation cannot be ensured. No one wants to undertake the full cost of environmental protection only to have its benefits shared by others, and, at the same time, bear the costs of environmental damage caused by the free-riders. Clearly, co-operation is not likely under these circumstances, even though everyone stands to benefit from it. This dynamic, noted by Hardin and others, is certainly not limited to the international system; it does, however, create additional difficulties when the actors involved are nation-states rather than individuals.

At the international level nation-states are generally sensitive to, and often preoccupied with, their position relative to other states (Waltz, 1979; Grieco, 1988), especially in terms of power and influence. This is understandable in so far as issues of national security are concerned, considering that national defence, pride, rivalry, promotion of national self-interest and other such traits are well embedded in the international body politic and have been so from the time when the concept of nation-states became accepted. It is difficult to see, however, why these traits should hinder, or be allowed to hinder, the formulation and implementation of effective international environmental programmes that would benefit all nation-states. For example, if all the nation-states benefit from cleaner air resulting from an international programme in which they all co-operate, then why should it matter to State A if its air is less clean than that of State B, as long as the air quality of both states continues to improve? To some degree it does *not* matter, and this is one of the factors that may make international co-operation easier on environmental issues than on other issues; but to the extent that one state may have to pay more for cleaner air than another, or that the amount State A pays for cleaner air benefits B more than itself, issues of competition may none the less come into play.

Other aspects may also add to the difficulty of environmental management, especially at the international level. For instance, in general the uncertainty inherent in addressing environmental problems is magnified when addressing global environmental problems, and may involve a number of factors that may become more severe across borders than they would have been locally. Determining the cause of environmental damage is difficult even when it is close by, but when it could be due to factors anywhere in the world it may be much more difficult to find the source. For example, when lakes in Scandinavia began to acidify, no one realised that air pollution could travel long distances, and the initial contention by Norway and Sweden that damage to fish stocks in their lakes was due to industrial activities in other European countries was met with scepticism (Levy, 1993). Thus the mere identification of a trans-boundary environmental problem can be difficult. The difficulty is compounded if it is costly to address the cause of the problem, or when it is not clear that there is actually an environmental problem, or, if there is, what its cause is.

It is even harder when, as with acid rain, the environmental damage takes on a 'directional' character rather than acting as a true 'commons'. In the commons of Hardin's 'cow pasture', each herder gains from individual actions (adding an additional cow to the pasture and reaping the benefits of owning an additional cow), but also loses (due to the additional stress on the pasture) from these actions, even if the herder does not lose as much as is gained. Clearly, all individual actors would benefit from a collective solution to their resource problem, since they would all be better off by regulating their behaviour than by not doing so. However, when some of the nation-states are net exporters of pollution (or cause resource degradation) while some others are net importers, the group loses its collective incentive for regulation. Consequently, some of the actors bear costs without benefits, while some others receive benefits without costs. Early efforts to protect the Rhine from chloride pollution took on some of the characteristics of this problem; nation-states which contributed most of the pollution were not harmed by it, while the down-stream states (particularly the Netherlands) had little early success in convincing the upstream states to stop polluting (Bernauer, 1996). In general, since no sovereign nation-state can be forced to agree to something it does not want to accept, it is particularly difficult to convince the net exporters of environmental damage of the reasons why they should change their behaviour.

Even when an international environmental policy *can* be made, it is difficult to ensure that individual nation-states actually comply with the rules they have set for themselves. Indeed, even when there are agreed rules, the actors are subject to the same individual incentives to free-ride, and there is no judiciary with the authority to punish them for violating the rules. Non-compliance may be due to a number of reasons. Some nation-states may join international agreements with no real intention of changing their behaviour to comply, while some developing countries may simply be unable to meet the requirements they have agreed to because of lack of resources or inadequate administrative structure. More importantly, individual nation-states may not be able to keep track of the actors and their activities within their own borders. As a result they may not be able to comply despite best intentions.

Although it may be difficult for developing countries to ensure compliance by their citizens, especially if they lack the necessary regulatory apparatus, non-compliance is by no means unique to the developing world. In fact, most of the black-market environmental activity is probably occurring in developed countries, simply because much of the industrial activity takes place in those countries. Individual actors have the same incentives to free-ride as nation-states, and their non-compliance is often harder to trace. One of the problems with early regulations on intentional dumping of oil at sea, for instance, was that when an oil slick was discovered, it was almost impossible to trace it to the tanker responsible (Mitchell, 1994). Other typical examples are detection of those fishing above their quota, and the smuggling of the horns of rhinoceroses, despite international agreements to protect these endangered animals, which is very difficult to deal with because the horns can be ground up quickly and easily. Even black-market chlorofluorocarbons (CFCs), banned under the international agreements to protect the ozone layer, have found their way into the American market (Wald, 1995). Clearly, it is not possible to check every fisherman and every traveller to ensure compliance. Even if it were, nation-states would resist this intrusion into their sovereignty. Thus it is difficult and often impossible for any national government to monitor all the activities taking place within its own borders, let alone within the borders of other nation-states.

This situation is made more difficult by the fact that, at the international level, most of the monitoring is done through a 'self-reporting' system whereby individual nation-states are required to report information on relevant activities under international agreements. Reporting in a timely and consistent manner, though, is rare. Non-compliant nation-states would clearly be reluctant to report. And yet, lack of reporting, which is rampant, does not necessarily mean non-compliance, because the nation-state in question may not know what is happening within its own borders, or may not have the capacity or incentive to report it (USGAO, 1992).

The potential for non-compliance can undermine the ability to make agreements in the first place. Just as individual nation-states hesitate to undertake costly programmes to address global environmental problems unless others agree to do the same, they will also be wary of agreeing to measures for which the compliance of others cannot be assured; and clearly, international co-operation on environmental matters cannot take place without adequate assurance of compliance.

The potential for non-compliance, or non-participation, can also make it difficult to assess the effectiveness of existing rules and regulations. For example, if the environment does not recover as predicted after the implementation of a set of rules and regulations, it becomes difficult to find the reasons why things went wrong. Was it because the rules and regulations fell short of what was actually needed, as was the case in the early attempts to prevent ozone depletion for instance? Was it because the problem is caused by other factors that had not been discovered or regulated, as was the case with transboundary acid rain? Or was it because of non-compliance by some of the actors so that the effectiveness of the rules and regulations could not be tested, as was the case with some of the fisheries treaties including the International Convention for the Regulation of Whaling? The difficulties of successful fishery regulation may be due to undetected non-compliance. There is certainly evidence now of past non-compliance (Brown, 1994) in treaties on the regulation of whaling. Clearly, if it is difficult to detect non-compliance, it is also correspondingly difficult to identify problem areas in which rules and regulations need to be tightened.

EVOLUTION OF INTERNATIONAL ENVIRONMENTAL POLICY

Despite the numerous obstacles to international co-operation on environmental issues, it is interesting to note that international action to address environmental problems has developed over time into a robust activity. Historically, early environmental co-operation was concerned almost exclusively with the protection of threatened migratory wildlife. One of the most effective early treaties of this kind was the 1911 Fur Seal Convention in which biological indicators were employed to manage fur seals sustainably.

Although nation-states have been co-operating to manage natural resources for more than a century, the United Nations Conference on the Human Environment, held at Stockholm in 1972, is generally accepted as the beginning of the modern international environmental movement. Attended by more than a hundred nation-states and numerous non-governmental organisations, collective human responsibility for environmental protection on a global scale was addressed in that conference, together with the promotion of the idea that environmental protection was important for human social and economic development. The Declaration on Human Environment, and a number of resolutions that emerged from that conference, brought a new focus on air and water pollution on a global scale. Perhaps most importantly, the conference led to the creation of the United Nations Environment Programme (UNEP) which has taken on an important co-ordinating and leadership role in international environmental policy since its inception.

The publicity and enthusiasm generated by the Stockholm conference led to the negotiation of a number of international environmental treaties in the early 1970s. These treaties addressed such issues as the protection of wetlands, prevention of ocean dumping, regulation of trade in endangered species, and the preservation of world cultural and natural heritage. Although broadly inclusive, these treaties generally reflected the concerns of the developed countries which initiated their negotiations.

By the 1980s the focus of international environmental policy shifted to include the concerns of developing countries, and the idea of 'sustainable development' took hold. The World Commission on Environment and Development (known as the Brundtland Commission, after the Prime Minister of Norway who headed it), which is an international commission created by the United Nations, defined sustainable development as 'development that meets the needs of the present without compromising the ability of future generations to meet their own needs' (WCED, 1987: 43). International co-operation began to focus on the needs of developing countries, while at the same time attempting to protect the environment. Subsequent treaties, such as the 1987 Montreal Protocol on Substances that Deplete the Ozone Layer, thus included measures to help address the developmental concerns of non-industrialised countries (see Box 19.1). At the same time, as evidenced by the Montreal Protocol, environmental issues began to be seen even more clearly as global in scope but elusive in character. It became clear that potential impacts of ozone depletion and global climate change could lead to environmental disaster, and that action to prevent them would have to be taken before the potential harm was manifest. It was also clear that all countries, rich and poor, would have to be involved in the prevention of these impacts.

Twenty years after Stockholm, the United Nations Conference on Environment and Development, held in Rio de Janeiro in 1992, addressed the interaction between environment and development more explicitly. The declaration, negotiated at that conference, promoted the idea that the needs of the least developed countries should be given special priority and that developed countries should bear a special responsibility for working towards sustainable development. Treaties signed in the wake of the Rio conference addressed issues such as biodiversity, global climate change, and desertification, all of which have significant North–South equity implications. Recent treaties generally consider differential obligations for developed and developing countries, and include funding mechanisms and provisions for technology transfer.

THE INSTRUMENTS OF INTERNATIONAL ENVIRONMENTAL POLICY

The evolution of environmental policy shows that there are various reasons why nation-states agree to work collectively at the international level to address environmental problems, and that there are ways in which even recalcitrant nation-states can be persuaded to join, and live up to, international environmental

BOX 19.1 OZONE DEPLETION

The ozone layer lies in the stratosphere, 20–50 km above the earth's surface. It shields the earth from 99 per cent of the sun's ultraviolet radiation which could otherwise cause, among others, skin cancer, immune system deficiency and agricultural damage. The first ozone-destroying substances were found to be chlorofluorocarbons (CFCs). These chemicals, invented in the 1930s, have been widely used because they are non-toxic and stable. Their chemical stability, however, prevents them from degrading until they reach the stratosphere where they are broken down by sunlight, and this triggers a chain reaction which destroys many molecules of ozone at the same time. In the early 1970s it was theorized that chlorine in the stratosphere could set off a chemical chain reaction to destroy ozone. This was followed by the discovery that CFCs can migrate into the stratosphere and stay there for decades, eventually breaking down to emit large amounts of chlorine. The combination of these two findings provided evidence for the possibility of anthropogenic destruction of the ozone layer, even before any manifestation of such destruction was observed.

This evidence was sufficient for some environmental NGOs, particularly in the USA, to take action by boycotting CFC products such as spray cans. The USA took unilateral action in 1977 to ban the use of CFCs in all non-essential aerosols. This move immediately reduced the world consumption of CFCs by 50 per cent, although benefit to the ozone layer was temporary because other nation-states increased their use of CFCs, and new substances, such as halons, were found to harm the ozone layer as well.

International action was called for, and the first step was taken with the signing of the Vienna Convention in 1985. This agreement included a general obligation 'to take appropriate measures to protect the ozone layer' (United Nations, 1985, Article 2) and established mechanisms for co-operation in research, monitoring, and exchange of data. Although the Convention did not actually stipulate a reduction in ozone-depleting substances, the scientific research undertaken as a follow-up, as well as a resolution directing the United Nations Environment Programme (UNEP) to open negotiations for a protocol with binding controls, contributed to the international call for the protection of the ozone layer. This convention/protocol approach resulted in the Montreal Protocol in 1987, which required a 50 per cent cut in the use of CFCs. It also banned trade in controlled substances with non-parties as a way of inducing others to join, and allowed developing countries an extra ten years to meet the deadlines set. Although this agreement was signed by the European Union and 27 other nation-states, most of the developing countries stayed out of it, preferring not to preclude their use of an important class of chemicals used in the process of development unless they were compensated for their sacrifice. The fact that these countries could rapidly increase their use of ozone depleting substances, and thereby wipe out any environmental gains made by the sacrifices of the nation-states already regulating these chemicals, gave the developing countries a high degree of negotiating power. And, under the London Amendments, the parties to the agreement set up the Multilateral Fund to provide financial help to developing countries which joined the agreement. Most did. The London Amendments also increased the number of controlled substances, and for the first time called for a full phasing-out of CFCs by the year 2000. Amendments in 1992 in Copenhagen again brought forward the phasing-out dates and brought under control additional ozone-depleting substances. Although the problem has not been completely solved, significant international steps have been taken to reduce human impact on the ozone layer.

agreements. It is therefore important to examine the tools with which international environmental policy is made and enforced.

Policies to address international environmental issues are formulated by the actions and negotiations of a group of like-minded countries which decide to co-operate internationally to address environmental problems. Instruments such as international law and non-binding agreements and declarations are the most visible manifestations of this co-operation, but scientific research and individual or state leadership can also be important in bringing about international action to address scientific problems. Nation-states that may be reluctant to participate, or live up to these agreements, can be induced to join or comply with these international efforts either by promises or threats, and disputes that may arise can be settled by an appropriate process set up in the course of international co-operation.

Although international environmental co-operation can be difficult, there are nevertheless numerous incentives for achieving it; perhaps the most important incentive is that many of the global environmental problems cannot be managed otherwise. Another important incentive is that, through the services of international organisations, created to address environmental problems, the participating states can gather and share information, monitor both the environment and state actions, and reassure each other that they are taking the collective steps needed to help avert the international 'tragedy of the commons'.

International Law

As indicated implicitly above, the main instrument of international environmental policy is international treaties that are sometimes bilateral but most often multilateral. In legal terms, they are often called 'conventions'. These treaties can be regional or global, depending on the scope of the environmental problem to be addressed and the expectations of the actors involved. There are two main approaches to the making of international environmental treaties: in the first, authority for regulation is delegated to a committee or an organisation, while in the second, a 'convention' or a 'protocol' constitutes the regulatory instrument. In practice, however, a treaty may contain elements of both.

A treaty of the delegatory type involves empowering a treaty organisation (composed of representatives of the participating states) to make decisions on regulation, generally through majority voting (such as a two-thirds or three-quarters majority). These decisions are binding on the member states of the treaty and do not require formal ratification. This arrangement thus allows the policy to respond much more quickly to available environmental information. Most fisheries treaties, as well as the International Convention for the Regulation of Whaling, are of the delegatory type in which there is an obligation to make annual 'schedules' of catch. The same procedure is also adopted in the Convention on International Trade in Endangered Species of Fauna and Flora for changing the status of, and therefore also the regulations pertaining to, endangered species.

But, because nation-states cannot be bound by international law without their consent, and also because this process allows for regulations to be made with majority voting, it is essential to include an appropriate 'opt-out' mechanism for those nation-states which do not agree with the majority decision. Usually there is a specific procedure for this, called an 'objection' or 'reservation' procedure, whereby a state can indicate its intention not to be bound by a regulation passed in this manner. Unfortunately, nation-states which benefit most from such a regulation may also be most likely to opt out of it, thereby free-riding on actions taken by the others. The other problem is that, if and when a nation-state opts out, others may follow, either because they wish to free-ride on those remaining, or because they do not wish those opting out to free-ride on them if they stayed in. Indeed, the opt-out mechanism may sometimes lead to a situation in which none of the relevant actors is bound by the regulation in question. For example, there was one whaling season in which all the whaling states in the Antarctic opted out of the quotas for Antarctic whaling (Birnie, 1979).

Although there are many complaints about opt-out provisions in delegatory treaties, the fact that no nation-state can be bound by international law against its will makes them a necessary evil in a treaty of this type. Without them few nation-states would agree to regulation passed by less than unanimous voting. However, when a nation-state does opt out, it is possible that scientific evidence, domestic pressure or diplomatic pressure can persuade it to withdraw its objection to the regulation in question.

The most common approach to recent environmental treaty-making is the convention/protocol format in which an initial convention is first negotiated, followed by the negotiation of protocols and amendments. Often, the initial convention is what is called a 'framework' convention because it includes a general approach to the issue without requiring specific obligations to reduce environmental damage. It may call for co-operation on research to increase the understanding of the problem, and may include a plan for negotiating binding controls. These protocols, to be negotiated later, must be fully agreed to by the representatives of all the participating nation-states and subsequently ratified by them, thus making the process more time-consuming than the delegatory approach discussed above; but, because these conventions and protocols are fully negotiated, they are generally not subject to opt-out procedures. On the other hand, nation-states may sign or ratify protocols and amendments at different speeds, and so a situation can still exist in which different states are held to different standards. Nevertheless, both the recently popular convention/protocol approach and the delegatory process have an advantage over earlier forms of treaties in that a completely new treaty can be negotiated at any time if circumstances change. This flexibility is essential to international environmental policy-making because both knowledge and understanding of environmental problems can change rapidly, thus making existing treaties, based on inadequate knowledge or understanding, ineffective and outdated.

Another aspect that needs to be considered in deciding how to negotiate a treaty is whether to involve all possible nation-states, limit the number to those specifically interested in the issue, or to include only those willing to take strong regulatory action. Since the negotiation of international treaties requires all parties involved to agree to the provisions before signing, such treaties often fall prey to the 'least common denominator' level of regulation that is acceptable to those involved. Some therefore advocate including only those nation-states willing to undertake serious commitments, or those that contribute to the problem. Unfortunately, two potential problems may arise from limiting the number of states involved in a treaty negotiation. First, if a nation-state, which is a major contributor to the environmental problem in question, is unwilling to participate, then, clearly, any regulation to which that state is not a party would be of little consequence. Second, if the states not involved are developing countries, or those that may contribute little to the problem or care little about it, the potential for those states to contribute to the problem in the future is sufficient reason for including them in the initial negotiations because, from their perspective, it is only fair for them to have a say in the rules they might eventually be expected to abide by. More importantly, because they cannot be bound without their consent, the populous developing countries excluded from the negotiating process of an international convention may decide to stay outside that process and refuse to reduce the impact of their activities on the global environment.

Limited participation works best when the issue in question is what may be called 'geographically bounded' rather than one of global commons. The Mediterranean Action Plan is a typical example of a geographically bounded regional agreement to protect a variety of aspects of the Mediterranean Sea involving all the major states in the region. In this case there was no particular reason to involve those outside of the region, and doing so would probably have hampered negotiations.

An important aspect of making environmental treaties is whether to negotiate a multi-issue treaty, or to focus fairly narrowly on to a specific issue. A typical example of a multi-issue treaty is the United Nations Convention on the Law of the Sea which addresses almost every issue relating to oceans. Multi-issue treaties allow for direct issue-linkage within the negotiating process, so that nation-states with different priorities can trade off their positions and agree to something they might not want in return for a different regulation that is important to them. In addition, many environmental problems *are* linked to each other, and addressing one aspect without including all can lead to proliferating regulations that do not take into consideration the effect of the rules pertaining to a nation-state outside its narrowly defined issue area. For example, regulations on forestry will have an impact on biodiversity and global climate change and vice versa. These links may not be taken into consideration if the issues of forestry, biodiversity and global climate change are addressed separately. On the other hand, the Law of the Sea Convention took eight years to negotiate, and it was so politicised that some of the most important nation-states in the process did not ratify it. It is for this reason, at least in part, that narrowly defined environmental treaties are now the most common type.

Customary international law – the set of binding principles that come into being through state practice and sense of obligation – also plays a part in shaping international environmental policy, although it is less important than treaty law. Evidence supporting that a nation-state has traditionally acted upon, or recognised, a certain set of obligations can be used in international court cases or in dispute settlement procedures. An example of this type of law is the obligation that a state should not 'use or permit the use of its territory in such a manner as to cause injury . . . in or to the territory of another' (Sands, 1995: 244). This principle was deemed customary international law and was invoked to settle a transboundary air pollution dispute between the United States and Canada in 1935. It is frequently cited as a principle of international law, even though it has not been negotiated into a formal treaty.

Soft Law

Also important in international environmental policy is what is known as 'soft law' (Birnie and Boyle, 1992: 26); so called because, although not technically binding legally, it nevertheless takes on the character of international law. For example, 'framework conventions' that are in the convention/protocol form of international treaty often contain no legally binding obligations but are nevertheless declarations by nation-states of what they intend to do with regard to regulation. Conventions of this type often set into motion the process by which legally binding rules are made. International organisations sometimes also adopt 'codes', such as the International Maritime Dangerous Goods Code which lists goods that are considered hazardous for shipping and sets up processes for labelling and documenting them (Birnie and Boyle, 1992). Although such codes are not legally binding, they are often taken seriously, amended regularly, and they bring about a change in the behaviour of the nation-states involved. The use of soft law allows nation-states to indicate their collective intentions to address international environmental problems (and sometimes sets out principles about the way in which they intend to do so) when they may not yet be ready to accept binding obligations.

Similar to, and sometimes considered a version of, soft law are declarations made by international organisations such as the United Nations General Assembly. For example, UN General Assembly Resolution 44/225 recommended a moratorium on 'all large-scale pelagic driftnet fishing on the high seas by 30 June 1992'. Although this recommendation was not legally binding, many nation-states took action to change their own behaviour in order to comply with this moratorium and encouraged or persuaded others to do the same. Large international conferences, such as the 1972 United Nations Conference on the Human Environment, or the 1992 United Nations Conference on Environment and Development, and many others that received less international attention, also lead to declarations that are legally non-binding but nevertheless indicate the intentions of nation-states. The Rio Declaration on Environment and Development, for example, urges all nation-states to 'co-operate in a spirit of global partnership to conserve, protect, and restore the health and integrity of the Earth's ecosystem', but it does not stipulate specific modalities with which to do so. More importantly, it declares that 'the developed countries acknowledge the responsibility that they bear in the international pursuit of sustainable development in view of the pressures their societies place on the global environment and of the technologies and financial resources they command' (UNCED, 1992, Principle 7). Evidence that such resolutions are taken seriously can be found in the fact that the United States released 'interpretative statements' on several of the Rio Declaration principles, including the 7th quoted above, indicating in this case that it did not accept any international obligations or liabilities (Grubb *et al.*, 1993). Thus, declarations by international organisations are often a precursor to more serious environmental policy.

Scientific co-operation

Declarations, resolutions and framework conventions mentioned in the previous section ('Soft law') often also lead to joint scientific undertakings; indeed, sometimes that is their purpose. Nation-states are reluctant to take action when there is uncertainty about the causes and effects of environmental damage. It is then incumbent upon the international scientific community to try to remove that uncertainty by providing scientific evidence showing how anthropogenic causes lead to environmental damage. A typical example of this is the Intergovernmental Panel on Climate Change (IPCC), which emerged in 1988 from an international conference on the possible causes and effects of global warming at a time when there was little consensus that the earth's temperature was rising, or that, if it was, such a change was due to human activities. But, by the time of the Rio conference in 1992, there was sufficient evidence to show that anthropogenic activities were causing global warming, and, consequently, all the nation-states represented at that conference were willing to negotiate a Framework Convention on Climate Change. This particular scientific collaboration was international in character, pursued at the instigation of two international organisations – the World Meteorological Organisation (WMO) and the United Nations Environment Programme (UNEP). Once the scientists involved were able to reach consensus and present their evidence, their conclusions were persuasive even to those reluctant to participate in creating the Framework Convention on Climate Change.

This convention was actually unusual in the sense that the collaborative scientific effort was in place even before the framework convention was formulated (although it did grow out of an international meeting on the subject). In most cases it is the framework convention itself, or the process which creates it, that sets up official international scientific co-operation in research and monitoring. Most international environmental treaties include some elements of scientific collaboration, either through the setting up of independent scientific bodies associated with the treaty organisation, or by calling for increased state efforts in research, or by sharing the co-ordination of research efforts. Clearly, it is through such a process that nation-states can be persuaded to accept an increasingly strict regulatory regime, especially as the problem becomes increasingly more serious. As evidence of increasing environmental damage emerges, or new causes of damage are discovered, additional protocols may be negotiated within existing conventions to deal with the problem. This may be done through a system of yearly 'schedules' of the annual regulatory system which is based, at least in part, on the latest scientific evidence on the state of the environment.

Leadership

Leadership by one or more nation-states on an environmental policy can be persuasive in getting other nation-states to take action as well. A nation-state, which takes unilateral action to protect an international environmental resource, is indicating by its action that it believes the issue to be so important that it will take action even though it may not be able to solve the problem alone. Often this initial action, and subsequent international leadership, is in response to domestic pressure led by environmental or scientific organisations. If a nation-state contributes sufficiently to the problem, as the United States did initially in the depletion of the ozone layer, unilateral action *can* sometimes have a significant positive effect on the global environment. Nation-states can provide moral leadership in this way, as did those which participated in the Long-Range Transboundary Air Pollution agreement in Europe and agreed to go further than what was required by that agreement. In this case, even before measures to reduce sulphur dioxide emissions were in place, a group of nation-states formed what came to be known as the '30 Percent Club', so called because the stated objective of every member of the Club was to reduce its sulphur dioxide emissions by that amount (Levy, 1993). Although such declarations are not legally binding, they put pressure on the other nation-states either to adopt similar policies, or run the risk of appearing to be 'guilty' if they do not. Unilateral or leadership actions of nation-states can also serve as examples or case studies demonstrating that it is possible to take action to protect the environment, and that it need not be prohibitively costly to do so.

The 'carrots'

For some developing countries, however, environmental action can be prohibitively expensive, especially if global environmental protection merits lower priority in their national agenda than feeding a starving population, paying back an external debt, or addressing purely local environmental problems. In many cases they may feel, with justification, that they are being asked to forgo opportunities for development already enjoyed by the richer countries. This argument is strengthened by the fact that much of the global environmental damage and destruction has been caused, and is being caused, by the richer countries in pursuit of their industrial development. In this situation funding mechanisms and technology-transfer programmes can be helpful to the poor developing countries in meeting the obligations they might be persuaded to undertake for protecting the global environment.

The Montreal Protocol Multilateral Fund, which was one of the first of such global mechanisms, provides funding to help developing countries to meet the obligations they undertook to protect the ozone layer. Indeed, without this help only a few, if any, of the poor developing countries would have joined the international agreements to protect the ozone layer. A similar but broader international funding mechanism is the Global Environmental Facility (GEF) which provides funding to address issues of global warming, biodiversity and international waters, as well as ozone depletion. Both these mechanisms work to some extent on the principle of 'incremental costs' in that they provide funding only for those aspects of projects that cost the recipient state more than it gains in local benefits, thus essentially paying for the global environmental benefit of the project. In both cases implementing agencies of the United Nations and the World Bank carry out projects in the recipient developing countries. Bilateral and multilateral foreign aid can also be used to help developing countries meet their global environmental obligations.

Technology transfer is a more controversial aspect of assistance from the developed to the developing countries. The importance of technology transfer is generally expressed through declarations and even treaties, and developing countries are particularly eager to receive this type of assistance. However, the developed countries – the donors – may be reluctant either to give up intellectual property rights to technologies invented by them, or to forgo the profits they could make by selling their technologies to the recipient developing countries.

The 'sticks'

A device which is used frequently is the threat of economic sanctions against nation-states which do not participate in, or comply with, international environmental agreements. Such action is often a type of issue-linkage. Sometimes sanctions are part of an international agreement itself, as with the Montreal Protocol banning trade in ozone-depleting substances by any nation-state which is not party to the agreement. More frequent but less legitimate is the threat or imposition of trade sanctions by one nation-state on other states which refuse to undertake environmental obligations. Frequently employed by the United States, this strategy has also been used by European countries – for example in the case of an Austrian policy to import only tropical timber that had been harvested in a sustainable manner. Refusal of the United States to import tuna fish from those countries not using 'dolphin-safe' methods was a controversial manifestation of this type of sanction. The United States has also used the threat of economic sanctions in an attempt to persuade reluctant nation-states to participate in the International Convention for the Regulation of Whaling and the Convention on International Trade in Endangered Species of Flora and Fauna, to withdraw the objections of those states to these treaties, and to increase the level of their compliance with them (DeSombre, 1995). Threats or imposition of economic sanctions, though effective, are often seen as internationally illegitimate, especially if they are imposed outside of the framework of the international treaty which the imposing party supports. Often these sanctions are themselves in violation of international trade law, and this means that a nation-state breaks one type of international law in support of another. Furthermore, sanctions may provoke the threatened state to retaliate, and this may lead to a trade war.

The reason why economic sanctions are controversial is because of the interaction between free trade and environmental protection. While there is no inherent conflict between the two, the system of international policy for each of the two issue areas has evolved separately and can therefore come into conflict. The agreements governing international trade generally do not allow the use of economic instruments for political purposes. Those who see trade sanctions as an effective tool of international environmental policy generally argue for a reform of international trade in a way that would allow economic tools to be used for environmental protection.

Dispute settlement procedures

The final instrument of international environmental policy is concerned with dispute resolution. In general, there are clauses in a treaty stipulating how disputes pertaining to the specific agreement are to be dealt with, either through a specific process set up by the treaty itself, or through referral to an existing international judicial body such as the International Court of Justice. But, although the participating nation-states may have agreed upon a certain way in which to resolve disputes that may arise, they still cannot be forced to participate in the process or to submit to any punishment that is meted out. Some do participate none the less, and even accept punishment. They may do so because some actors within the state believe that the system is impartial and that they should actually take the prescribed action; or they may do so in order to perpetuate the international legal system in general so that others, who are later found to be in the wrong, will also accept the decisions of the system. Few disputes are actually brought to any judicial system, however. The setting up of a system, which everyone deems fair and effective, serves an important function in convincing the parties that the agreement as a whole will work. Otherwise the parties would be reluctant to join.

There are thus a number of ways in which nation-states can adopt environmental obligations, as there are ways in which international environmental policy can be made. Scientific evidence can be used to convince nation-states that an environmental problem exists; they can be spurred on by the leadership of some nation-states within the system, and can be persuaded by means of the 'carrots' and 'sticks' that it is in their best interest to make environmental policy collectively.

NON-STATE ACTORS AND INTERNATIONAL ENVIRONMENTAL POLICY

Although it is nation-states which actually agree to undertake environmental commitments, there are a number of ways in which non-state actors can also influence the process. Scientists, environmental non-governmental organisations (NGOs) and industry all contribute to the making of international environmental policy. Indeed, without their concern, often expressed first at the domestic level, international environmental policy would rarely be made. This is because one or more nation-states must be sufficiently interested in the environmental issue in question in the first instance. It is only then that they would seek international co-operation involving all interested actors in an attempt to address it.

The role of scientists is important throughout the process. Often they are the first to discover environmental damage, or to theorise that certain substances may be harmful even before damage becomes evident. Thus the process of addressing both domestic and international environmental issues often begins with scientists. Their role becomes even more important in international issues, however, because nation-states must be convinced collectively that there is an environmental problem, resulting from anthropogenic causes, before they would be willing to take costly action. Also, transboundary scientific investigation contributes to enhancing public awareness of environmental issues and problems.

The work of environmental NGOs also raises public awareness of environmental issues and problems, both domestically and across national boundaries through the sharing of information. Nationally they alert the public to environmental problems, lobby government actors to take action on international environmental policy issues, and lead consumer boycotts of products deemed harmful to the environment. In the case of whaling, for instance, scientists helped alert the public to declining populations of whales, while some of the environmental NGOs successfully convinced the international community that whales were worthy of concern as particularly important species (see Box 19.2). Often, relevant NGOs are invited to participate as observers in international conferences on environmental policy, and allowed to speak at the discretion of the chair.

Industrial actors are also important players in environmental regulation at the international level, not least because industrial activities are responsible for creating environmental problems in the first place. Their motivation is different, however, and can even be contrary, because environmental protection policies often have cost implications for industry. Consequently, more often than not industrial actors actively resist any regulation that would increase their costs or reduce profits. Often they do so by calling into question the scientific justification for taking preventive or remedial action. On the other hand, industry can also play an important role in promoting international environmental regulation in different ways. First, industry may prefer international regulation to domestic regulation for a variety of reasons; for example, future domestic regulation could be more stringent than international regulation. More importantly, since environmental

BOX 19.2 WHALING

Commercial whaling has been taking place since at least 1100, and whale products have been used as a source of oil and protein. Because whales live in what is literally a global commons – the open oceans – they are vulnerable to over-capture. As technology advanced and human population grew, whales, like fish, became easier to catch, and their population began to decline. Once such depletion begins, no individual actor or state wishes to regulate its own behaviour if it cannot be certain that others will do the same, even though all would collectively benefit from regulation.

The main treaty regulating whaling is the International Convention for the Regulation of Whaling (1946), designed to ensure collective regulation of whaling and thereby to overcome the 'tragedy of the commons'. This agreement set up an International Whaling Commission (IWC), which meets annually to determine a 'schedule' of how many whales can be caught. Nation-states are allowed to 'opt out' of regulations within the annual schedule, by stating that they do not wish to be bound by specific provisions of the treaty. Unfortunately, at times this procedure has meant that none of the major whaling nations has been bound by the regulations passed by the organisation (Birnie, 1979).

Regulation of whaling is difficult in part because of the uncertainty in regulating a resource that cannot be measured directly or accurately. When scientists themselves disagreed about how depleted a particular whale stock was, what incentive could the whalers have to make sacrifices in order to protect a stock whose depletion was in question? More importantly, because of concern that the whaling nations might pull out of the IWC rather than adhere to a quota they considered to be too restrictive, the quotas have been set consistently higher than that recommended by the organisation's own Scientific Committee (Birnie, 1979; Andresen, 1989). During most of the 1960s and the early 1970s, the whalers could not catch as many whales as they were allowed to despite their best efforts – a clear indication that the quotas were set too high.

It is also extremely difficult to monitor compliance because the oceans are very large, and because only a few of the whaling nations are willing to submit to searches or agree to an international observer programme (although the latter was finally implemented in a limited way for whaling in the Antarctic). Searches and monitoring programmes are costly, and so it was left to individual whalers and whaling nations to report their catches. Because of this system of self-reporting, and since the IWC only has the power to ask individual whaling nations to investigate and punish infractions, there is not much data on this type of non-compliance. More serious is unreported non-compliance which is harder to detect. For example, post-Soviet Russia reported to the IWC that the whaling statistics reported by the former Soviet Union were incorrect; the USSR had taken humpbacks, right, and blue whales in large numbers, and reported their catches as other non-protected species (Brown, 1994).

The USA used threats of economic sanction to persuade the whaling nations to join the agreement and to submit to the regulations passed by the organisation. The most serious regulatory step taken by the IWC was a moratorium on commercial whaling which began in 1986. By that time it was clear that all stocks had been seriously depleted, and both scientists and the NGOs lobbied hard for the protection of whales. The moratorium was initially for a ten-year period, but it has not yet been lifted, despite the fact that some stocks have recovered substantially. Some of the whaling nations have since withdrawn from the organisation and/or recommenced commercial whaling, which they believe can be done sustainably. The issue for many at this point seems to be a possible conflict between sustainable harvest and whether (as some environmentalists believe) there are some species which should not be harvested at all for moral reasons.

regulation usually bears a cost, industry in a highly regulated country can lose its competitive edge over its counterparts in other countries where regulation is lax or not enforced. In this situation industry would prefer to be under international regulation because it would provide a level 'playing field' for all with regard to competition. Second, if a certain industry has developed substitute products or processes to address a given environmental problem, it would obviously prefer that as many actors as possible be bound by the same regulations on those substitutes so that their market share is maximised. Industrial actors within a nation-state thus have ample motivation to push for the creation of international environmental policy, or at least not to hinder it.

Once a policy has been created at the international level, industrial actors often remain involved as advisers and consultants. They also participate in international meetings to ensure that their interests are considered and safeguarded. It is in their interest to do so, because, after all, those who have developed a particular manufacturing process, or a product, know best how to retrofit it with new technology. Similarly, the manufacturer of substitute chemicals knows best the circumstances in which they can or should be used. The presence of industrial actors within the international system can lead to a situation in which chemical solutions are recommended for solving chemical problems so to speak, rather than undertaking a complete review of the way industrial activities are carried out. Even so, it is hard to imagine, given the influence of industry within nation-states, a realistic system at the international level that would not include industrial actors as important players.

Finally, international organisations, such as the United Nations and its subsidiary bodies, financial organisations, and organisations created by international environmental treaties themselves, have a role of their own in the creation and maintenance of international environmental policy. Most international environmental agreements create some type of international organisation to co-ordinate efforts under the agreement, and to hold regular meetings which inevitably result from the treaty process. In addition to collecting and disseminating information and funding, these organisations may monitor compliance with agreements, and they generally manage the process to ensure that the implementation of regulations leads to environmental improvement. Other international organisations, such as funding agencies or international scientific organisations, may also be involved in the process. This often leads to a set of 'nested' institutions, all of which have a role in implementing international environmental policy.

Non-state actors thus play a central role in the creation and maintenance of international environmental policy, even though it is the nation-states which actually commit themselves to changing their behaviour. But, while the nation-states are held to account for the environmental policy they have made, it is through the actions of individuals and organisations that behaviour is actually changed.

CONCLUSIONS

Despite the difficulties inherent in the prevailing international system, many of the nation-states have been persuaded to make and participate in international environmental policies to address a number of global environmental issues. Currently there are at least 150 multilateral agreements devoted specifically to environmental issues (UNEP, 1991). In addition there are nearly a thousand multilateral and bilateral treaties that have at least one provision addressing environmental protection issues (Weiss *et al.*, 1992). These numbers indicate in part the increasing discovery of damage which humans have been causing to the natural environment; they also demonstrate a recognition by nation-states that some environmental problems must be addressed internationally, and that they are increasingly willing to do so.

Not only are nation-states agreeing to an increasing number of international environmental agreements, they are also taking steps to change their behaviour. Typically, the whaling nations agreed to give up the practice of whale harvesting for ten years so that whale stocks could recover; and the developed countries have completely phased out their use of CFCs, a chemical that was previously considered indispensable, and have provided funding to the developing countries to help them do the same.

In order to continue this progress, those concerned about environmental protection must be willing to work within the constraints of the prevailing international system and accept the fact that, for the present at least, nation-states are sovereign entities that will not often submit to supranational authority. Those responsible for making international environmental policy must also be willing to accept that pollution of the natural environment does not respect national boundaries and that some of the environmental issues can only be addressed through international co-operation. In many instances nation-states would be willing to co-operate to that end, if they can be reassured that their collective sacrifice now will be rewarded later by a cleaner environment for all.

The remaining major challenge is to move towards better monitoring of compliance, even if it means more intrusive practices, in order to ensure that all parties to a given agreement comply with its requirements. This is likely to encourage greater participation of nation-states in international agreements. And in situations of unequal power, due either to the level of development, or the geography of environmental damage or destruction, special consideration must be given to the weak who might be unwilling or unable to undertake necessary environmental protection measures.

Even the most recalcitrant nation-states have been showing their increasing willingness to make international environmental policies to address issues about which they are convinced, often by participating in international policy-making processes, and this certainly augurs well for the health of the global environment.

REFERENCES

Andresen, S. (1989) 'Science and Politics in the International Management of Whales', *Marine Policy* 13: 99–117.

Bernauer, T. (1996) 'Protecting the Rhine River against Chloride Pollution', in R.O. Keohane and M.A. Levy (eds) *Institutions for Environmental Aid: Pitfalls and Promise*, Cambridge, Mass. and London: MIT Press, pp. 201–232.

Birnie, P. (1979) *International Regulation of Whaling: From Conservation of Whaling to Conservation of Whales and Regulation of Whale Watching*, Volumes I and II, New York, London and Rome: Oceana Publications.

Birnie, P.W. and Boyle, A.E. (1992) *International Law and the Environment*, Oxford: Clarendon Press.

Brown, P. (1994) 'Soviet Union Illegally Killed Great Whales', *Guardian*, 12 February, p. 13.

DeSombre, E.R. (1995) 'Baptists and Bootleggers for the Environment: The Origins of United States Unilateral Sanctions', *Journal of Environment and Development* 4 (1): 53–75.

Grieco, J.M. (1988) 'Anarchy and Co-operation', *International Organisation* 42 (3): 485–507.

Grubb, M., Koch, M., Thomson, K., Musons, A. and Sullivan, F. (1993) *The 'Earth Summit' Agreements: A Guide and Assessment*, London: Earthscan Publications, Ltd.

Hardin, G. (1968) 'The Tragedy of the Commons', *Science* 162: 1243–1248.

Levy, M.A. (1993) 'European Acid Rain: The Power of Tote-Board Diplomacy', in P.M. Haas, R.O. Keohane and M.A. Levy (eds) *Institutions for the Earth: Sources of Effective International Environmental Protection*, Cambridge, Mass. and London: MIT Press, pp. 75–132.

Mitchell, R.B. (1994) *Intentional Oil Pollution at Sea: Environmental Policy and Treaty Compliance*, Cambridge, Mass. and London: MIT Press.

Sands, P. (1995) *Principles of International Environmental Law*, Manchester and New York: Manchester University Press.

UNCED (1992) 'Rio Declaration on Environment and Development', in S.P. Johnson (1993) *The Earth Summit*, The United Nations Conference on Environment and Development (UNCED), London: Graham & Trotman.

UNEP (1991) *Register of International Treaties and other Agreements in the Field of the Environment*, Nairobi: UNEP.

United Nations (1985) *Vienna Convention for the Protection of the Ozone Layer*, New York: United Nations.

United States General Accounting Office (USAGO) (1992) 'International Environment: International Agreements Are Not Well Monitored', *Report to Congressional Requesters* GAO/RCED-92-43, Washington, DC: United States General Accounting Office.

Wald, M.L. (1995) 'Smuggling of Polluting Chemicals is Described', *New York Times*, 17 September, p. 30.

Waltz, K.N. (1979) *Theory of International Politics*, Reading, Mass.: Addison-Wesley Publishing Company.

Weiss, E.B., Magraw, D.B. and Szasz, P.C. (1992) *International Environmental Law: Basic Instruments and References*, New York: Transnational Publishers.

World Commission on Environment and Development (WCED) (1987) *Our Common Future*, Oxford and New York: Oxford University Press.

SUGGESTED READING

Conca, K., Alberty, M. and Dabelko, G.D. (eds) (1995) *Green Planet Blues: Environmental Politics from Stockholm to Rio*, Boulder, Colo.: Westview Press.

Haas, P.M., Keohane, R.O. and Levy, M.A. (eds) (1993) *Institutions for the Earth: Sources of Effective International Environmental Protection*, Cambridge, Mass. and London: MIT Press.

Keohane, R.O. and Levy, M.A. (eds) (1996) *Institutions for Environmental Aid: Pitfalls and Promise*, Cambridge, Mass. and London: MIT Press.

Matthews, J.T. (ed.) (1991) *Preserving the Global Environment: The Challenge of Shared Leadership*, New York and London: W.W. Norton & Company.

Porter, G. and Brown, J.W. (1996) *Global Environmental Politics* (2nd edition), Boulder, Colo.: Westview Press.

Susskind, L. (1994) *Environmental Diplomacy: Negotiating More Effective Global Agreements*, Oxford and New York: Oxford University Press.

SELF-ASSESSMENT QUESTIONS

1 In the context of international environmental policy, the implications of anarchy are that (choose those which apply):
   (a) there is no higher authority that can enforce international regulations without the consent of the participants;
   (b) there is no order within international environmental relations;
   (c) nation-states need to be persuaded to join international agreements since they cannot be forced to do so;
   (d) nation-states never feel obliged to live up to international commitments they have made.

2 The advantages of a convention/protocol approach to international environmental policy include all but which one of the following?
   (a) the ability to update regulations as new scientific evidence emerges;
   (b) the ability to involve nation-states in a scientific and regulatory process without necessarily requiring binding obligations initially;
   (c) the possibility for nation-states to opt out of specific provisions made within the agreements;
   (d) the opportunity to involve nation-states in differing levels of obligation, if they adopt subsequent protocols or amendments at different paces.

3 Describe the various ways in which a nation-state can be persuaded to join international environmental agreements.

4 What incentives do nation-states have to address commons problems as opposed to transboundary problems which do not have commons aspects? (choose all that apply):
   (a) in a true commons problem nation-states gain only positive utility from actions that harm the international environment, and therefore they have little incentive to address these problems;
   (b) in a true commons problem the state actors who gain positive utility from actions that harm the international environment also suffer some degree of negative utility from these actions, and therefore they will gain from a collective response;
   (c) in a directional transboundary environmental problem the 'upstream' state has very little incentive to protect the resource in question, because it is not harmed by its own actions;
   (d) in a directional transboundary environmental problem the 'upstream' state will feel responsible for its actions, and therefore has an incentive to take action to prevent environmental damage.

5 Describe the roles that non-governmental organisations (NGOs) can play in creating and maintaining international environmental policy (choose all that apply):
   (a) they can put pressure on their governments to sign, make, and uphold international agreements;
   (b) they can make international agreements to protect the environment;
   (c) they can help raise public awareness of environmental problems;
   (d) they can expose non-compliance with international agreements.

6. Which of the following best expresses the legal status of declarations by international organisations, or by international conferences, at the international level?
   (a) they have no legal status and are essentially meaningless;
   (b) they have the full force of international law;
   (c) they are legally binding but not enforceable;
   (d) they can be seen as 'soft law' and can be used to convince nation-states to take legally binding actions.

7 Although international environmental policy can be made in a number of different ways, in what order are the following steps towards international co-operation for environmental protection most likely to take place?
   (A) International scientific investigation.
   (B) A formal treaty with binding obligations.
   (C) A framework convention which sets out general but non-binding obligations.
   (D) Domestic concern in one or more nation-states about the possibility of an environmental problem.
      (a) A, C, D, B
      (b) D, C, A, B
      (c) A, C, B, D
      (d) C, D, A, B

8 If a developing country with a large and growing population refuses to join an international agreement to address a global commons problem relating to industrial processes, what strategy is likely to be most useful for the other nation-states to adopt?
   (a) Do nothing, because it is not important for this country to take part in international environmental policy since it is at a low level of development.

(b) Threaten economic sanctions if it does not join the agreement.
(c) Offer funding and technology transfer to help the country meet the obligations it would incur under the agreement if it joined.
(d) Offer scientific proof that the environmental problem exists and that it has an effect on the population of the developing country in question.

9 If a developed country refuses to join an international agreement addressing a global commons problem relating to industrial processes, what strategy is likely to be most useful for the other nation-states to adopt?
(a) Do nothing, because it is not important for this state to take part in international environmental policy.
(b) Threaten economic sanctions if it does not join the agreement.
(c) Offer funding and technology transfer to help the country meet the obligations it would incur under the agreement if it joined.
(d) Offer scientific proof that the environmental problem exists and that it has an effect on the population of the country in question.

10. For which of the following reasons may industry wish to support international environmental regulation? (choose all that apply):
(a) If it is already regulated domestically, it needs assurance that its international competitors are similarly regulated as well.
(b) If it makes products that can serve as substitutes for regulated substances, it wants to develop new markets for these products by helping to create a situation in which as many actors as possible are bound by the regulations in question.
(c) Industrial actors want a cleaner environment as much as anyone else.
(d) It may fear that domestic regulations will be stricter than international regulations.

20

ENVIRONMENTAL LAW IN THE USA

William Luneburg[1]

SUMMARY

Environmental law in the United States of America is the subject matter of this chapter. Starting with an historical perspective, an overview is given of how US environmental law has evolved from largely judge-made law to predominantly statutory and administrative law – and from state law to law whose basic parameters are established at the national level – in an attempt to address environmental impacts caused by human activities as industrial and technological development gathered pace.

The philosophy underpinning US environmental law, as well as the methods used for its implementation including subsidies, taxes, public information, regulation of market access, technology standards and quality standards, is then discussed, together with an explanation of how regulatory decisions often reflect the results of cost–benefit analyses, risk assessment and other germane aspects.

Attention then focuses on recent experiments with market-based strategies to reduce pollution, based on the theory that they can produce equivalent or better environmental performance at less cost than traditional command-and-control regulation. Discussion then moves on to other developments at the national and state levels, suggesting that US law is moving away from detailed, nationally prescribed environmental policy towards increased flexibility, voluntarism, and respect for local conditions and constraints in setting and implementing strategies for environmental protection. Finally, it is argued that although US law is largely an indigenous creation, shaped by distinctive factors such as history, political culture and the free market system, it has been influenced to some degree in recent years by international environmental law, and that this trend is expected to continue in the future.

ACADEMIC OBJECTIVES

After reading this chapter the reader should have a good understanding and working knowledge of the following:

- the three principal sources of law in the United States – the common law, statutes, and administrative regulation – along with the processes of their creation;
- the allocation of authority between the national and state governments as it affects the area of environmental regulation;
- the historical development of US environmental law, from a common law base to detailed statutory and regulatory prescription;
- the underlying factors and considerations which influence the scope and nature of modern US environmental law;
- the types of regulatory approaches now employed in US environmental law, including those based on economic incentives and the free market system;
- recent trends in the development of legal approaches in the USA to problems of environmental contamination, including the impact of recent international environmental law.

INTRODUCTION

Today environmental law in the USA comprises a vast assemblage of statutes, administrative regulations and guidelines, and judicial doctrines, but, since it has evolved in large part over the past twenty-five years, often in response to specific actual or perceived 'crises', the law is largely devoid of any systematic integration or sense of priority. For example, an important factor in the original enactment of the Comprehensive Environmental Response, Compensation, and Liability Act of 1980 (CERCLA, also known as the 'Superfund' law) was the discovery of high levels of Mirex in fish in Lake Ontario, which were ultimately traced to the Love Canal in New York.

In fact, some type of regulatory or public management scheme is to be found in practically all areas of environmental concern such as air and water pollution; the generation, transportation, and disposal of hazardous wastes; clean-up of toxic waste sites; endangered species and ecosystem protection; and conservation of natural resources. The huge investment of public and private resources – whether scientific, technical, legal, or otherwise – in an effort to maintain or enhance environmental quality which underlies this massive legal structure has clearly enjoyed some substantial successes. For instance, the air and surface water bodies are cleaner today than they were in 1970; toxic waste contamination of groundwater has been limited in some areas; and various animal species have been rescued from extinction. But significant environmental problems remain, and the public debate is increasingly focusing on ways in which to set priorities for future action and to achieve further environmental improvement at a socially acceptable cost. Inevitably, this debate will result in some, perhaps significant, changes in the legal framework for environmental protection.

Meanwhile, despite the fact that US environmental law has developed in the past largely in response to domestic concerns, it appears that international law will be an increasingly important factor in reshaping at least some aspects of national and state responses to environmental issues.

At the outset of this chapter, which gives an overview of environmental regulation in the United States of America, it is important to have a clear sense of the general trends now occurring in the evolution of the law in this area.

First, at both the national and state levels, there has been a consistent and increasing displacement of judge-made law ('common law') by statutory and administrative regulation because of the complex nature of the scientific, technical, and social problems created by environmental issues.

Second, this development has been accompanied by a growing and ultimately predominant role of national law in either displacing state and local regulation or, more commonly, establishing minimum standards which must be met in all states. Today, however, there appears to be widespread support for increasing the flexibility of individual states to craft regulatory structures that are more respectful of the unique conditions confronting each of them. In fact, there is substantial evidence to show that in a variety of areas, including ecosystem protection and environmental permitting, individual states have adopted, or are in the process of doing so, more innovative methods and approaches than the national government has prescribed in the past.

Finally, whichever level of government is vested with the authority to plan and implement environmental protection measures, it is clear that market-based strategies, meaning those regulatory methods (such as that of emissions trading) which attempt to utilise the purported potential of a free market economy to produce better environmental quality at a lower cost, are viewed by many as a method of choice in dealing with some, though perhaps not all, environmental problems. There has been limited 'real world' experience with these new approaches, although experience to date suggests some basis for at least qualified optimism that they would bring significant progress at a lower cost.

Decision-making in both the public and private sectors may result in significant adverse consequences for the environment. For example, the national government of the United States owns vast tracts of land which can be utilised for a variety of purposes, including resource extraction, recreation, and ecosystem preservation. It also owns, operates or controls various facilities, such as weapons manufacturing plants, storage sites and hydroelectric projects, whose impact on the surrounding environment may be catastrophic. Moreover, the potential for harm posed by private facilities which discharge wastes into the ambient air, or into surface or groundwaters – which are essentially considered public resources – has been used to justify elaborate pre-construction and operating permit systems created by the national and state governments. Regardless of whether the government is faced with a decision to sell or permit the use of public land for certain purposes, to build or decommission a public project, or to issue a permit to a private individual or a company allowing discharges into the surrounding environment, state or national law generally requires that public decision-makers give some special consideration to and, in some cases, institute appropriate mitigating measures to combat the potential adverse environmental consequences flowing from their decisions. The following discussion reflects the distinctive approaches taken by US law-makers in regulating both governmental and private decision-making when public health or the environment is potentially at risk.

It is emphasised that the corpus of national and state law in the USA dealing with environmental protection is so massive that in this chapter we cannot aspire to offer anything approaching a detailed discussion of the specific laws and regulations. Rather, we will attempt to portray the broad outlines of the various legal approaches now in use, or proposed, with specific examples of regulatory mechanisms, methods of assessment, and decision-making models taken from particular statutory schemes. Moreover, since most existing US law has largely been shaped by indigenous factors, and since international environmental law has assumed its current prominence as a body of law only in recent years, our discussion of the domestic impact of international law will be brief of necessity, although that impact will no doubt grow substantially in the not-too-distant future.

THE LEGAL BACKGROUND

No meaningful discussion of US environmental law is possible without a preliminary and general exploration of the nature and evolution of the legal system in the United States, including both the sources of law and the allocation of authority among governmental institutions. We will take environmental law as the focus of our discussion, although many of the descriptions and observations would also apply outside of that particular area of the law.

The 'common law' as a protection for the environment

Until the early decades of the twentieth century, law in the United States was largely what is called 'common law'; that is to say, rules developed by judges in controversies involving private parties – rules which purported to be rooted in more than the personal predilections of individual judges. The results of prior cases controlled similar later cases under the doctrine of *stare decisis*, although departure from precedent was permitted under appropriate circumstances. This system had existed for hundreds of years in England and was transplanted to the American Colonies. Even after the American Revolution, English precedents continued to influence decision on cases in American courts. The common law tradition, thus 'received' in the United States, included an embedded hostility to the products of legislatures, most aptly captured in the rule of 'statutory construction' to the effect that statutes in derogation of the common law should be 'strictly construed'.

Accordingly, during this period, if any body of law in the USA concerned itself with issues of environmental protection, it was usually the common law and, in particular, the doctrines of the common law defining rights in private property. Under the common law, the ownership of a real property included the right of the owner to its exclusive use, possession and enjoyment and, correlatively, the right to exclude its occupation and use by others. Inevitably, however, some uses of one person's property interfered with another person's use or enjoyment of his or her property. A factory belching smoke, which renders a nearby house almost uninhabitable, is a typical example of this. To address this, the common law courts developed various doctrines designed to mediate conflicting uses, including the doctrines of trespass and nuisance. For example, under the doctrine of private nuisance, a landowner could obtain an order from a court ('injunction') to prevent conduct on another's property where that conduct was considered to be a 'substantial' infringement of the use or enjoyment of the plaintiff's property. However, as the pace of industrial development accelerated during the latter part of the nineteenth century, there was a tendency of courts, at least in some states, to consider the perceived social and economic benefits of the conduct allegedly constituting the nuisance in determining what, if any, remedy to give to the injured party.

The common law also recognised that the general public had certain rights with regard to outdoor air, lakes and rivers, and that the government, as its representative, should be able to obtain a judicial remedy against private conduct which interfered substantially with those rights. In fact the government was seen as the trustee of such resources with legal obligations to protect them. The public nuisance and public trust doctrines were elaborated by the common law courts as a matter of state and federal law in recognition of the public rights inherent in natural resources.

Even today in the USA, the common law doctrines of trespass, public and private nuisance, and public trust, among others, remain vital sources of protection against environmental degradation, including the public health effects thereof. The courts have been willing in some cases to relax or modify the allocation of the burdens of proof required to establish what alleged damages are attributable to defendants' conduct; to recognise a broader range of public health and other injuries deserving either compensation or other remedial relief; and to impose more wide-ranging and stringent duties on public authorities to act as stewards of natural resources.

However, the common law now operates largely to give relief to injured private parties, while the implications of broader public interests fall within the purview of state and federal environmental statutes and administrative regulations. Moreover, those same statutes have increasingly created remedies for private parties to seek relief which would otherwise have been denied them under the prevailing doctrine of common law.

A variety of factors have been responsible for this change from a legal system, dominated by the common law, to one which is largely the creation of legislatures and administrative agencies; but basically it was the scientific and industrial revolutions which created social, economic and environmental problems whose development, scope, and complexity:

1 overtook the ability of courts to create and adjust the legal doctrine in a timely fashion;
2 called for social planning on a scale and of a type for which the courts were entirely inadequate both by instinct and nature;
3 required information-gathering methods and expertise unfamiliar to the judicial process;
4 demanded more broadly based public accountability for the results of the necessary decisions to be made.

The development of statutory protection for the environment

As the nineteenth century drew to a close, some of the most glaring environmental consequences of the Industrial Revolution were those which raised health concerns; typically, the air in some of the cities was polluted by soot from the burning of coal, and public water supplies were at risk of serious contamination from water-borne bacteria and other pollutants. Many cities were thus forced to enact smoke abatement ordinances and to adopt sanitary codes to protect drinking water.

During this period there was little legislation at the national level dealing with what are now considered to be environmental issues. As it turned out, however, what did exist at the time had great significance for the future. First of all, areas of public land were reserved for what ultimately became a vast system of national forests, parks, and wildlife sanctuaries. In addition, the Congress enacted the Rivers and Harbours Act of 1889 which prohibited the disposal of 'refuse' in navigable waters without a permit from the Secretary of the Army. This Act, which was originally meant as a mechanism for avoiding hazards to navigation, was resurrected in the 1960s to limit pollutant discharges. And ultimately in the early 1970s it constituted the basis for the first national pollutant discharge permit system which was replaced by the permit programme mandated by the Federal Water Pollution Control Act of 1972.

Early federal and state legislation focused separately on problems of public health and the maintenance of certain environmental amenities, including the spiritual, aesthetic and recreational values of wild and scenic areas. But in the early 1970s these separate strands converged into modern American environmentalism as a consolidated concern for the protection and enhancement of public health and welfare, with welfare being defined broadly to include aesthetic appreciation and other non-economic values. Stated briefly, environmental degradation was perceived as a threat to the health of both the body and the spirit.

This consolidated emphasis is to be seen in the resulting statutory instruments. For example, the Clean Air Act of 1970, as amended in 1977 and 1990, requires the improvement of air quality both to reduce identified public health hazards and to protect and improve the visibility of those vast treasured natural vistas to the east and west of the Mississippi River.

Until the 1960s, most of the legislative activity in the environmental area was at the state and local levels. But since then the law has slowly broadened its focus to embrace problems of air pollution and water pollution, generally employing, among other things, regulatory instruments such as permit programmes for new and existing sources of pollution. However, the continued decline in the quality of public natural resources, as evidenced by increasing ozone pollution, super-tanker oil spills, and rivers so dirty that they could catch fire, along with warnings by some scientists of potentially more catastrophic environmental disasters to come, has created a perceived need for a more energetic, expansive and demanding legal regime to combat the identified problems. Given the 'New Deal' legacy of the 1930s and the Second World War, which optimistically suggested that the national government could summon the resources, wisdom and dynamism to combat major social and other serious threats to the public, it was predictable that increasingly greater attention would be focused on the United States Congress as the institution best suited for determining the manner and pace of environmental protection. Accordingly, in a little over a decade – from the enactment of the National Environmental Policy Act of 1969 to the enactment of the Comprehensive Environmental Response, Compensation and Liability Act of 1980 – the content of environmental law in the USA has come to be determined largely by congressional mandate and elaboration by federal agencies. The law has also expanded in scope to encompass most of the areas of identified environmental degradation. The following list graphically illustrates this development, including as it does some of the most significant federal environmental statutes with the dates of their enactment or amendment:

1969 National Environmental Policy Act
1970 Clean Air Act Amendments
1972 Federal Environmental Pesticide Control Act
Federal Water Pollution Control Act Amendments
1973 Endangered Species Act
1974 Safe Drinking Water Act
1976 Resource Conservation and Recovery Act
Federal Land Policy and Management Act
National Forest Management Act
Toxic Substances Control Act
1977 Clean Air Act Amendments of 1977
Clean Water Act of 1977
Surface Mining Control and Reclamation Act
1980 Comprehensive Environmental Response, Compensation and Liability Act
Low-Level Radioactive Waste Policy Act
1986 Emergency Planning and Community Right-to-Know Act
Superfund Amendment and Reauthorization Act
1987 Water Quality Act
1990 Clean Air Act Amendments of 1990
Oil Pollution Act
Pollution Prevention Act
Pollution Prosecution Act

Implementation of statutory directives: legislative delegation, executive supervision, and policy-making by agencies

The nature of the statute-making process and the product

Under Article I of the United States Constitution, the Congress is the pre-eminent policy-maker within those limits of law-making that are reserved for the national government. Statutes are enacted with the required concurrence of majorities in both Houses: the Senate and the House of Representatives, each constituted differently. The former is composed of two senators from each state (100 in all), and the 435 members of the latter are apportioned on the basis of the population of the states. The President's role in the law-making process is limited. Other than having power to propose legislation for enactment, his or her principal legislative function is to approve, either affirmatively or through inaction, bills which have passed both Houses, or to disapprove those which he or she dislikes (in which case two-thirds of both Houses can pass the measure over the 'veto'). This pattern of law-making is largely followed in each of the 50 states of the USA.

The product of the legislative process is significantly different from that of the common law process in terms of the rules created. A common law 'rule' is identified by an examination of the major and minor premises necessary for making a decision on the facts presented to the court in a controversy between the parties. In some instances it may take a series of cases to clarify exactly what is the governing rule for decision-making on a particular type of controversy.

Statutes, on the other hand, present their 'rules' in a transparent form; each word enacted by the legislature has the force and effect of law and conveys – in the context of the other related statutory language – the governing requirements with more (or less) clarity. Rules of legislative origin tend to be more detailed, often including elaborate exceptions or waiver provisions, than those rules which derive from the common law method. In fact, of all the statutes currently in force in the United States, environmental legislation is clearly among the most voluminous and detailed of any. For example, the Clean Air Act, as amended in 1990, is more than 500 single-spaced pages in one of its published forms. Indeed, since 1970 the pattern of statutory development in the environmental area has been from less to more detail in response to a variety of factors, including a greater appreciation of the complexity and scope of environmental issues, and distrust of the willingness of those persons charged with the implementation of the statutory schemes (i.e. the administrative bureaucracy) to do what the Congress believes should be done to deal adequately with the problems of environmental degradation.

Statutory implementation through delegation by the legislature

Traditionally, in grappling with the complex issues posed by modern social regulation, especially on environmental matters, both state and federal legislatures have relied on administrative agencies to fill in the details of the pertinent public policies created by statutes. In many cases, though, legislatures have lacked the time, energy, interest and knowledge necessary to provide answers to the multitude of questions which arise in establishing and implementing their own policies. It may thus come as a surprise to non-lawyers, and even to some lawyers for that matter, that a 500-page statute such as the Clean Air Act requires additional elaboration to make it work. Yet the complexity of the issues presented is such that very much greater elaboration, running into thousands of pages of regulations and guidance documents, is indeed essential.

In the USA, and particularly at the national level, agencies are the creations solely of the legislature; that is to say, neither state constitutions nor the national constitution mandate that they exist. Accordingly, they are limited in their powers and duties to those bestowed upon them by the legislature. Action in excess of those limitations (i.e. *ultra vires*) is generally subject to judicial reversal at the suit of affected persons.

While most of the lower and mid-level positions in agencies are filled by civil servants with some type of tenure protection, high-ranking agency officials are usually the appointees of the chief executive of the relevant governmental unit (e.g. the President in the case of the national government and the Governors in the case of the states). It is the bestowal of the appointment power – when vested in the chief executive who is himself or herself popularly elected – which is seen as one of the important ways in which policy-making by an agency pursuant to the legislative mandate (i.e. delegation) can be squared with public accountability for decision-making in a democratic society. Executive power of removal of high agency officials, when it exists – which is not always the case – is another means of ensuring that accountability. Authority of the legislature to amend the statute, which creates the agency or gives it powers, and to authorise or refuse to authorise the expenditure of public money for agency programmes, are other important control mechanisms which are frequently used.

Presidential supervision and co-ordination of the national bureaucracy

At the national level in the USA, Presidents have over the years struggled mightily to supervise, control, and co-ordinate the actions of agencies created by the

Congress whose impact on the economy and other aspects of people's lives has expanded exponentially. But their success in this regard has been limited for a variety of reasons, including the fact that most of the bureaucracy, on which the power and responsibility for day-to-day administration rests, has a tenure which outlasts the term(s) of any President. It is also to be noted that, even though Article II of the United States Constitution vests the 'executive power' in the President alone and directs that he or she 'take care that the laws be faithfully executed', the President has no legal power to direct an agency or its officials to take any action, or a particular action, under a statute which vests decision-making power in the agency or its head. Moreover, most statutes delegate decision-making power not to the President but to other administrative agency officials. In other words, the President's power over the federal bureaucracy rests to a large extent on his powers of 'persuasion', although he does have certain constitutional and statutory prerogatives and political leverage whereby agency officials can ignore his wishes only at their peril.

There is a variety of national agencies, charged with the responsibility for implementing the statutes, which concern themselves primarily with matters of environmental protection. Chief among them is the United States Environmental Protection Agency (USEPA). The Council on Environmental Quality, located in the Executive Office of the President, has some general responsibility for overseeing the activities of agencies, particularly with regard to the implementation of the National Environmental Policy Act (which requires environmental impact assessment by all federal agencies), but it lacks regulatory authority. The Depart-ments of the Interior and Agriculture not only administer certain regulatory schemes of environmental significance, such as the Endangered Species Act, but they also have responsibility for the management and use of important federal lands such as National Parks and Forests. This listing is not comprehensive by any means, neither does it include the many other national agencies whose primary statutory missions do not encompass the protection and enhancement of environmental quality, but whose actions can nevertheless have direct and deleterious effects on it.

The co-ordination of this diverse collection of bureaucracies is essential in order for the national government to avoid acting at cross-purposes with itself so to speak, even though in some cases the statutes enacted by the Congress may not in fact permit such co-ordination. Recent Presidents have attempted this co-ordination by issuing various Executive Orders, including a series requiring the preparation of a cost–benefit analysis for each proposed and final administrative regulation of major significance.

One of the most recent of these Executive Orders (12866) was issued by President Clinton in 1993. Like its predecessor in the Reagan and Bush Administrations (12291), this Order establishes the Office of Management and Budget (OMB), located in the Executive Office of the President, as a primary institution for co-ordinating national domestic policy, including that relating to the environment, through its power to review and comment on agency rules and related cost–benefit analyses; but prior to 1993, OMB's role in this process had proved to be very controversial for a variety of reasons. It was alleged, for example, that the OMB had deliberately delayed the promulgation of important rules and secretly introduced legally irrelevant considerations into the decision-making processes of agencies. These charges were made in connection with USEPA's rule-making efforts.

In response to these concerns, the Clinton Executive Order opened up the review process to greater public scrutiny and limited the ability of the OMB to delay rule-making by the agencies. Moreover, unlike its predecessor, the new Executive Order seems fully to accept that costs and benefits are difficult to quantify in some cases, but that they are nevertheless significant and deserve equal consideration in the cost–benefit context. This recognition is especially significant in the area of environmental law in which many of the interests (e.g. visibility of pristine areas in the national parks) need to be protected by statute, even though their economic values are difficult to quantify.

In evaluating the significance of this Executive Order, it must be noted that even where a cost–benefit analysis is required by the Order an agency may not use the cost–benefit calculation in making decisions on an issue which the Congress has precluded from cost considerations. A typical example of this is USEPA's establishment and revision of the National Primary Ambient Air Quality Standards which specify the permissible concentration of various widely emitted pollutants in the outdoor (ambient) air. As intended by the Congress, public health protection *alone* should be the guiding factor in the adoption of those standards.

The process of decision-making by agencies and judicial review

The action to be taken by an administrative agency can focus on a class (or classes) of entities or circumstances. Alternatively, it may direct its attention to an individual within a broader class. Legally, the first type of action comes under rule-making, while the second is considered to be adjudication. Unless otherwise stated, the product of rule-making by an agency looks like a statute enacted by the Congress, and often has the same legal effect as that of a statute. The product of agency adjudication, on the other hand, resembles a judicial order issued in a case involving contesting parties.

In the environmental area, a typical example of an agency 'rule' (or regulation) would be a requirement that a particular type of industrial facility (e.g. a copper smelter) emit no more than 10 pounds of sulphur dioxide per hour. An adjudication, on the other hand, would entail a proceeding wherein the agency would find, after hearing the evidence presented, that a particular copper smelter had violated the aforementioned 'rule' by emitting 15 pounds of sulphur dioxide per hour. An appropriate penalty would then be imposed on the guilty party.

Not only do both the rules made by agencies and adjudication concern themselves with the regulation of the behaviour of the private sector, they also control the allocation of rights to use and exploit natural resources owned by the government (e.g. mining, timber harvesting, and grazing on public land).

Generally speaking, the processes of making rules and issuing adjudicatory orders are laid down by statute (including the Administrative Procedure Act) and by an agency's own procedural rules. Often, but not always, the adjudicatory process resembles a trial with agency employees acting as prosecutors, judges and intermediate appellate officials (though these roles are vested in different individuals who function entirely independently of one another). Also, the sanctions may include 'cease and desist' orders and the imposition of so-called civil penalties which, in the case of the Clean Air Act for example, can amount to $25,000 per day for violation of the applicable regulations.

Because the rule-making process sets the standards of acceptable conduct for large classes of persons or entities, and may therefore impose significant costs along with benefits, its design has proved to be a matter of considerable controversy over the years. The basic rule-making process, mandated by the national Administrative Procedure Act, includes a notice published in the Federal Register (the daily 'newspaper' of the national bureaucracy) which contains a draft of the agency's proposed rule along with a detailed explanation of the background to it and its various provisions. Public comments on the rule are solicited, and those that are submitted must be considered by the agency. The final step is the agency's action in adopting, or refusing to adopt, the proposed rule with or without changes, accompanied by the issue of an explanation for its action. This is called 'informal' or 'notice-and-comment' rule making. In the environmental area major rule-making by agencies can result in rules of Byzantine complexity whose published explanatory material can run into hundreds of pages.

Some statutes require a more elaborate rule-making process, at times including the right to a hearing and cross-examination. The Clean Air Act is an example of a statute which elaborates substantially on the basic 'notice-and-comment' model. However, the resource costs of such proceedings can be so great, and their perceived benefits so small, that informal rule-making is normally the preferred option.

Whether an agency action is classified as a rule or an adjudicatory order, it is generally subject to review by the national courts, but only at the suit of those persons who are injured in fact by the agency action and where the injury is to an interest that may be protected or regulated by the statutory provision(s) relevant to the legal challenge. The 'doctrine of standing', which determines who can ask for judicial reversal of agency action, is a complicated speciality of the law which determines the subject matter of the jurisdiction of the national courts. Since the 1960s the national courts have recognised that harm to environmental interests, such as aesthetic enjoyment of forests and streams (though not 'mere interest' in the sense of 'concern'), was sufficient to confer standing on conservation and other groups active in the protection of the natural environment.

In recent years, in the context of several environmental cases, the Supreme Court has restricted the standing and the related 'doctrine of ripeness' governing the timing of judicial review, thereby restricting access to the national courts for the review of certain types of administrative action. Even so, the law of standing to sue in the national courts is still quite liberal by pre-1970 standards. Moreover, the Congress itself can broaden access to some degree through the authorisation of citizen suits in most environmental laws (e.g. the Clean Air Act and the Clean Water Act), and it has in fact done so. Such provisions allow interested persons to sue the following: (1) the USEPA and other agencies charged with duties for the protection of the environment (e.g. the duty to avoid governmental action which may result in the extinction of an endangered species), or to compel agency officials to carry out their responsibilities as required by law; and (2) private entities who are alleged to be in violation of state or national environmental regulations which comprise the framework for national regulatory programmes. Many of the environmental statutes contain clear and unambiguous deadlines for agency action in order to ensure that the administrative process responds quickly to perceived risks to public health and welfare. For instance, the Clean Air Act Amendments of 1990 contain more than fifty specific deadlines for major rule-making. Frequently a basis for citizen suits is the failure of the USEPA to meet those deadlines.

Once a suit for the judicial review of an agency action is brought properly before a reviewing court, the court is required to determine whether the agency has:

1 violated or otherwise disregarded some clear legal constraint on its discretion;
2 reasonably interpreted ambiguous statutory provisions and agency regulations;
3 considered only legally permissible factors in its decision;
4 avoided unreasonableness in resolving policy and factual issues;
5 complied with applicable procedural requirements.

While the reviewing court's role is limited in the sense that it is not generally authorised to substitute its judgement for that of the agency merely because it disagrees with the agency decision (that is, it must give 'deference' to the agency within the realm of its 'expertise' and statutory authority), courts have consistently required agencies to take a 'hard look' at the issues presented for decision, to respond in a well-reasoned fashion to the objections raised, and to explain their decisions in a way that does not leave significant gaps of reasoning. This mode of judicial review, which has developed largely in the context of environmental cases which arose in the 1960s and 1970s, became the mainstream of the law of judicial review of administrative action in the USA at the national level.

It is not possible, in the brief compass of this chapter, to examine the varying administrative processes and modes of judicial review of the states of the USA. However, many of the basic approaches which characterise national administrative law, as described above, are similarly reflected in state practice, although, as in so many other areas, the states are now showing increasingly greater creativity in modelling administrative procedures than they have in the past.

The complications of the federal system

If environmental law in the USA were solely the product of the Congress and the national bureaucracy, the resulting legal system would be complicated enough. However, the Constitution of the United States of America establishes a federal system wherein certain enumerated powers are granted to the national government, while others are reserved for the states comprising the Union. Matters are further complicated by the overlapping of the powers of the national and state governments in a multitude of areas, thus creating intricate inter-relationships between state and federal law. Actual or apparent conflicts between state and federal law are mediated in a variety of ways, although national law displaces inconsistent state law in those areas in which the Congress has the constitutional power to act. In a word, the brand of federalism of the USA exhibits those characteristics of divided powers and overlapping spheres of authority which also characterise the relationships that exist among the three main branches of the national government.

At the national level there are four heads of enumerated power that are of particular relevance to the environmental area. They are the powers of the Congress:

1 'To lay and collect Taxes [including individual income and corporate taxes] . . . to provide for the . . . general welfare of the United States';
2 'To regulate Commerce with foreign nations, and among the several states . . .';
3 'To dispose of and make all needful rules and regulations respecting the territory or other property belonging to the United States';
4 'To make all laws which shall be necessary and proper for carrying into execution the foregoing powers . . .'.

The first head of power authorises the collection of revenue which can be distributed as subsidies to promote, for example, research and development in the area of environmental protection, or to assist the states in establishing pollution control programmes or to build public sewage treatment facilities. It also authorises the imposition of charges to discourage particular types of disfavoured behaviour (e.g. emission fees).

The second head of power is the most common constitutional basis for national environmental legislation in view of the Supreme Court's consistent modern interpretation that this so-called 'Commerce Clause' extends to the regulation of any activity which alone, or in conjunction with other activities, has or is likely to have a significant effect on interstate commerce. As so construed, most economic activities which create pollution fall within the authority of the Congress to regulate.

The third head of power allows the Congress to manage public lands in such a way as either to preserve or destroy their environmental values, including their vital roles as parts of functioning natural ecosystems. This power may even be extended to protect public lands from activities carried out on private property that may impinge on the uses of the land as designated by the Congress. The significance of this power is fully grasped only when it is realised that almost one-third of the nation's territory (particularly to the west of the Mississippi River and in Alaska), or more than 284 million hectares, is owned by the national government.

The fourth head of power authorises, among other things, the elaborate administrative structure created by the Congress to carry out the policies embodied in national environmental statutes.

Where the Congress enacts legislation which falls within one of these heads of power and does not otherwise fall foul of constitutional limits, it displaces inconsistent state law under the Supremacy Clause of the Constitution. In fact, the force of pre-emption arising from a particular national statute may be such as to fence off an entire area (to ensure the safety of commercial nuclear power generating plants for example) from direct state regulation. But whether or not state law is to be displaced is a question for the Congress itself to determine in the first instance and, where the legislation is unclear on the point, for the judiciary to determine as a matter of construction of the legislation.

One of the persistent questions which arises pertains to the ability of a state to resist a particular assertion of national power. The Civil War interred the doctrines of secession and nullification of national law. Even so, the states have not been reduced to mere administrative instruments of national power. In a recent case, involving the problem of ensuring an adequate supply of disposal sites for low-level radioactive wastes, the Supreme Court examined the limits on congressional power to conscript state legislative and administrative processes for national purposes. First of all, the Court noted that the Congress could offer the states various financial incentives to 'encourage' their co-operation in carrying out national policy, and to withdraw those incentives if co-operation was not forthcoming. Alternatively, the Congress could offer the states the choice to make their regulations consistent with national standards, or face pre-emption of state law by national regulation. In fact, this is the most common pattern of national environmental statutes. It is found, for example, in the Clean Air and Water Acts and is known euphemistically as 'co-operative federalism'. Finally, however, the Court indicated that what the Congress could *not* do was to mandate, through the direct imposition of legal sanctions (other than withdrawing federal monies), that the states employ their legislative or administrative processes to regulate in a manner determined by the Congress.

Where the Congress has not pre-empted an area of state law-making for national law, the states are generally free to exercise their reserved powers to legislate for public health, safety and welfare (the 'police power'). This authority is in fact a common basis for state environmental regulation. But even here the Constitution of the USA imposes some significant limits. For example, since one of the important purposes animating the adoption of the United States Constitution in 1787 was the creation of a national market free of internal tariff and other barriers to trade, the Supreme Court has construed the Commerce Clause, when considered in isolation, to invalidate state and local regulations which discriminate against or unduly burden inter-state commerce. This is the so-called 'Dormant Commerce Clause Doctrine', many of whose most recent applications by the Supreme Court have involved attempts by the states to prevent or deter the disposal within their borders of solid or hazardous wastes generated in other states. To this end differential fees imposed on in-state and out-of-state refuse have been a common device employed both by the states and localities. Such schemes have so far been struck down by the Court, however, because it has found that there was no reason for the discrimination, apart from the fact that the refuse originated outside of the state imposing the restriction. Even where a state regulation does not directly and explicitly discriminate against inter-state commerce, the Court scrutinises the burdens imposed on that commerce to determine whether they are indeed justified by a legitimate interest of the state which enacts the restriction. It should be noted, however, that the Congress itself can authorise precisely those state restrictions which the Dormant Commerce Clause Doctrine would invalidate. This fact may give the Congress leverage in 'encouraging' the states to adopt regulations which implement national policy.

Another important restriction on state legislative authority arises from the 'Due Process Clause' of the Fourteenth Amendment to the United States Constitution which has been construed to require compensation from the state when it 'takes' private property for public use. A similar restriction, though in express terms, applies to the national government under the Fifth Amendment. Recently, in the context of a case where a state barred the residential development of a beach-front property in consideration of ecological and other aspects, the Supreme Court imparted some 'bite' to these provisions. It held that, when a regulation on the use of private property destroys all economically beneficial or productive values, compensation must be paid to the landowner by the government.

While most environmental regulations do not destroy all economically valuable usages of property, there are several types of environmental schemes which are now vulnerable to what are called 'takings' challenges. For example, development of wetlands under the Clean Water Act requires national permits, but the denials of such permits have been successfully challenged as 'takings'. The Endangered Species Act imposes criminal penalties for the destruction of any part of the habitat of a listed endangered species where that destruction causes the death of a member of that species. This restriction may mean that some currently undeveloped land may have to remain undeveloped, thus raising the 'takings' issues.

There are various general patterns of state and national authority to be seen in the environmental area. In some areas the Congress has entirely pre-empted state regulation. This has traditionally been the case with regard to emission standards for new passenger vehicles, although special provisions have been made for California and other states in which ozone levels significantly exceed the National Ambient Air Quality Standards. As to other matters, the Congress has left the states to regulate as they wish. This is because it may have no other constitutional choice, given the limitations on the scope of national powers. But this is indeed a rare case today in the area of environmental law where national law generally has something to say on almost every issue of significance. Finally, under the 'co-operative federalism model', the Congress may establish, or authorise an agency of the national government to establish, minimum standards of environmental quality or other aspects of an environmental programme. State programmes may be more, but not less, stringent in their restrictions, but in most cases they must be reviewed and approved by the USEPA or some other national agency. Where a state fails to take the necessary action to establish the required programme, or to implement an approved programme, the USEPA or some other agency is required to adopt, implement and enforce a programme for the state in question.

UNDERLYING POLICIES INFLUENCING LEGAL DOCTRINE AND STATUTORY LAW

Modern environmental law of the USA reflects the complex interplay of a variety of underlying considerations, some of which are not easily harmonised. Some of the most significant include the following.

Economic efficiency

Of overarching significance is the fact that the USA purports to operate a free market economy; that is, an economy whose functioning depends largely on the individual decisions of persons and firms and not on the government. This particular brand of economic theory, which has found a secure home in the USA, postulates that scarcity of human and natural resources means that not all needs can be fulfilled; that those needs should be met in order to bring greatest benefits to the society at least cost; and that, if it functions as theory predicts, the market system of supply and demand would result, through the decisions of the market actors, in a distribution of resources and products that would maximise social welfare based on existing wealth distribution. The government's role in the economy should, according to the theory, be limited to perfecting the market where that is feasible.

For example, a factory's smoke or other pollutants can be viewed as imposing costs of various kinds on its neighbours. These 'external' costs are, in fact, costs of production which should be reflected in the price of the product of the factory. If they are not, consumers will purchase more of that factory's product than they should where there are comparable goods whose resource costs are less than the total costs (including external costs) incurred by the polluting factory's output. In this situation, the government might impose a charge on the factory representing health and other costs caused by the pollution. In this way the costs imposed by the pollution will be 'internalised' and the product priced accordingly. The common law nuisance doctrine results in some cost internalisation, but it is limited for a variety of reasons, including the difficulty in some cases of proving who has caused the damage and of identifying and assessing the full scope of the damage attributable to the pollution. In authorising the collection of clean-up costs, along with damages reflecting injuries to the natural environment, the Comprehensive Environmental Response, Compensation and Liability Act of 1980 seeks to internalise the environmental costs of toxic wastes disposed of improperly by waste generators, transporters and owners of disposal sites. However, its liability rules function in ways that only fitfully assure that each responsible party bears its share, and only its share, of the damage attributable to its acts or omissions.

More broadly, because it is difficult to identify and assess the full scope of damages that can be traced back to the source of pollution, taxes or other charges imposed on manufacturers per unit of emission or effluent released under the authority of a statute, or administrative regulation, have not proved to be a feasible regulatory strategy as a general approach to environmental cost internalisation. There is growing experience with somewhat similar programmes, however. Under the Clean Air Act, for example, each major source of atmospheric emissions must obtain a permit from the state or the USEPA to operate. In order to obtain and keep a required permit, an annual fee is imposed on each permittee wherein the amount of the fee is set in such a manner that, when considered in aggregate, the collected fees cover the costs of the establishment and administration of the permit system. Since the fees are based on the tonnage of emissions, there is some incentive for the permittees to reduce emissions in order to reduce their annual fees.

In the classical world of the economist, other strategies are also available for reducing pollution to the point where any further reduction would impose greater costs than the resulting benefits. According to what is known as the 'Coase Theorem', bargaining by the polluter, and by persons affected by the pollution, will result in an efficient allocation of resources regardless of whether the polluter is given the legal right to pollute, or the victims are given the legal right to be free of pollution. From this perspective, the role of the government would be to reduce or eliminate the 'transaction costs' relating to the lack of information on who the polluter is and how to value pollution damage, because such costs may interfere with the bargaining process. The Coaseian theory has been influential in the design of a variety of statutory approaches to pollution control, including programmes which authorise trade in permits and those which confer rights to others to emit certain air pollutants (e.g. SO_2 as a precursor to acid rain).

When it comes to governmental decisions regarding whether and how to regulate activities perceived to have adverse environmental effects, however, the economists' concern for the maximisation of social welfare, as measured in monetary terms, is seen in many different contexts in the USA. For example, the Clinton Executive Order, which requires cost–benefit analysis for major national rules, includes as an important element of its underlying regulatory philosophy the proposition that, in choosing from among regulatory approaches, agencies should select those which maximise net benefits. Moreover, some statutes, including the Clean Air Act and the Clean Water Act, require the USEPA at least to consider and, in some cases, balance costs and benefits in deciding on the level of stringency of emission and discharge standards. However, the concern for costs of regulatory actions does not pervade all national environmental laws. For example, the Endangered Species Act forbids both government and private action which threatens protected (listed) species, regardless of the costs involved.

Risk assessment

Closely allied to concerns for economic efficiency is the felt need to quantify the magnitude of risks to human health and natural systems posed by man-made products and human activities. This encompasses the areas of risk identification, assessment, management and communication. For example, given a yearly average exposure to a certain concentration of a particular air pollutant in an area where the population is of a certain size and geographical distribution, what is the statistical likelihood of a member of that population contracting cancer attributable to that pollutant (e.g. 1 in 100, 1 in 10,000, or 1 in 1,000,000, etc.)?

Sensible answers to this and similar questions can be used to determine the possible benefits of a proposed regulation designed to limit exposure to the pollutant in question and can, therefore, be factored into the cost–benefit equation. In addition, the results of the risk assessment process can purportedly be used to compare and rank environmental hazards. Public resources can then be targeted on those risks that have the highest probability of occurrence.

In theory, risk assessment appears to be a sensible tool for public policy analysis. However, uncertainties abound in identifying the existence of health and environmental risks, in quantifying the magnitude of those risks, in comparing risks across environmental media, and ultimately in assessing the adverse health and welfare effects that are identified. For instance, the results of human health studies, often conducted in occupational settings following the onset of a disease, may be seriously undermined as reliable data by what are called 'confounding effects'. For instance, did smoking cause the cancer in question, or was it the pollutant identified in the occupational environment? Although the number of such confounding variables can be limited in a laboratory setting, animal studies used for predicting the effects, if any, of toxic chemicals on humans can prove to be equally questionable as the scientific 'proof' of a risk. Can it be assumed, for example, that certain animal organs react to substances in the same way as those of humans? Can the effects of a relatively high intake of chemical X be the basis for predicting adverse health effects of that same substance in humans exposed to much lower doses of chemical X over a longer period of time? In the case of comparative risk assessment, can a death resulting from cancer caused by one substance be compared to developmental abnormalities in children caused by a different material, or by the same material but through a different environmental medium?

When it comes to factoring the results of risk assessment into a cost–benefit analysis, there are the additional economic and moral problems of putting a value on the lives saved, or on the pain and suffering avoided. Furthermore, in comparing risks for regulation, agency decision-makers must take into account the fact that the public often ranks risks in a significantly different order than experts on the basis of a variety of 'non-scientific' criteria, including the degree to which risks are assumed voluntarily by the victims.

However, despite these uncertainties, various national environmental laws require risk assessment in determining the need for and type of regulation. For instance, this is the case with the Federal Insecticide, Fungicide, and Rodenticide Act which requires pre-marketing approval of pesticides to protect against 'unreasonable' adverse effects. The same is true of the Toxic Substances Control Act which authorises the USEPA to test and regulate new and existing chemical substances to avoid 'unreasonable risk of injury to health or the environment'.

Environmental justice and equity

Environmental hazards to public health are not uniformly distributed throughout the population. Some sub-groups, such as asthmatics, may be more susceptible to certain pollutants, or more likely to suffer adverse effects, than other groups. In order to deal with this variability of impact, some environmental quality standards, such as the National Ambient Air Quality Standards, adopted in pursuance of the Clean Air Act, are set at levels that are necessary to protect the most vulnerable groups.

In recent years another type of inequality has become a focus of attention in the United States: allegedly some social and racial groups are disproportionately affected by certain types of pollution. High levels of air pollution in major urban areas may be of particular concern to the residents of inner cities who are often poor and non-white. Research findings also suggest that hazardous waste disposal sites, which generally require government permits, are more commonly to be found in or near areas populated by such minority groups.

Approaches to address this type of problem of environmental 'equity' or 'justice' have included suits brought under the constitutional provisions and statutes which seek to guarantee equal treatment of all citizens by the government. Such legal actions have, however, been largely unsuccessful because of the inability to prove that the alleged disproportionate effects flowed from forbidden discriminatory actions of official decision-makers. However, the Clinton Administration has adopted an Executive Order mandating that all federal agencies consider and attempt to avoid the disproportionate effects of their programmes. This Order has also proposed remedial actions, including the denial of permits authorising activities with potentially adverse environmental effects. There is evidence to show that this Order, along with heightened sensitivity to the problems of environmental equity generally, may be having some effect in modifying governmental decision-making. However, the degree to which such a concern can influence action would vary, depending on the scope and discretion of the agency in question to consider and decide on an equitable distribution of environmental risks among various population groups.

METHODS OF ENVIRONMENTAL PROTECTION ADOPTED BY UNITED STATES LAW

Since a detailed discussion of each significant national and state statute dealing with environmental protection is not possible in a chapter such as this, we will only describe the types of legal strategies that are most commonly encountered. First, the methods employed to encourage or command private sector conduct will be surveyed. The discussion will then turn to the structure of governmental decision-making in so far as it affects environmental quality.

Private sector approaches

Dissemination of information

Dissemination of information to market actors can achieve a variety of beneficial purposes, including the reduction of transaction costs which, when high, may interfere with the ability of private bargaining to produce an economically efficient result.

On another front, eco-labelling programmes can alert consumers to the environmental costs of the products they purchase, thus enabling them to structure their consumption in a way that minimises the demand for resource-intensive or environmentally destructive products. While several states have been implementing modest programmes in this area, legislative and administrative activity at the national level has been limited and directed largely at product claims that are, or may be, false or deceptive.

Certain federal statutes, such as the Federal Insecticide, Fungicide, and Rodenticide Act, require covered products to bear necessary labels warning about health and other hazards that may arise from the use of such products. Also, the Emergency Planning and Community Right-to-Know Act requires companies to file annually toxic chemical release forms and emissions inventories with the USEPA and the states. This information, which is available to the public for use and planning purposes, must include estimates of the annual quantity of toxic chemicals released into each environmental compartment (e.g. air, water, soil).

Where the Congress has mandated the creation of a permit system for discharges to the environment, it has also required elaborate periodic monitoring and reporting by the permittees of the actual amounts discharged by them. By law these reports are available to the public, as is most of the information used in the prosecution of citizen suits against companies violating the terms and conditions of their permits. Also, the Freedom of Information Act, which makes all records held by the national government available to any person on request, unless the particular record requested falls within one of the Act's exemptions, can be used to gain access to information for citizen suits or other purposes. This is the case regardless of whether the information was generated by government employees or obtained from sources outside the government.

Subsidies

The Congress has appropriated funds to subsidise a variety of projects benefiting the environment. These funds have been used for research in the private sector to develop promising new pollution control technologies. Other programmes underwrite the building of public transit systems to reduce passenger car usage, and of municipal sewage treatment facilities to help eliminate significant sources of water pollution.

Imposition of charges on disfavoured activities

Although the Congress has not relied exclusively on its powers to impose taxes and other charges to discourage environmentally destructive behaviour, there are however a few examples of the application of such instruments at the national level. For example, in 1989 it imposed an escalating tax on the production of chlorofluorocarbons (CFCs) to accelerate the phase-out of these ozone-depleting substances and to encourage the development of their substitutes. Under the Clean Air Act, each major source is required to pay an annual fee for an operating permit, the money thus generated being earmarked to support the administration of the permit programme itself.

Regulation of market access

Several of the statutes authorise pre-market review and approval of new products that may present an unreasonable risk to human health or the environment. They also authorise, if necessary, the imposition of restrictions on the production or use of such products. The Federal Insecticide, Fungicide, and Rodenticide Act and the Toxic Substances Control Act are typical examples of this approach.

Liability and insurance schemes

Traditionally, compensation has been paid to persons injured by environmental contamination under common law doctrines such as those of strict liability for the results of ultra-hazardous activities and contractual insurance. However, some statutes go beyond these schemes to broaden liability by, for example, reducing the burden of proof linking a particular defendant with the harm done, and by creating new remedies (such as damages to public resources caused by toxic contamination) not previously afforded by the common law. Perhaps the most famous of these statutes is the Comprehensive Environmental Response, Compensation and Liability Act of 1980 which imposes strict, joint and several liability for the cleaning up of hazardous waste sites on present and former owners of such sites, as well as on the generators and transporters of the waste.

Technology standards

Based in part on the theory that each source of pollution should be required to reduce its discharges as much as is 'feasible', environmental statutes commonly require that new, modified, and existing facilities install 'end of the pipe' controls and/or modify their operating processes to mirror the environmental performance achieved by 'good' emission discharge reduction technology. First, the regulatory authority is required to identify the industrial class (or sub-class) and the manufacturing processes that are sufficiently similar in terms of emissions, equipment and other matters to justify their subjection to a standard of categorical (i.e. uniform) application. This is followed by the identification of the discharge or emission limits which can be achieved in practice as a general matter within that category by the use of: (1) some existing technology; (2) some method which is likely to become operational in the near future; or (3) some means that can be hoped for (these alternative approaches characterise different statutory schemes). The stringency of the limits thus identified may be relaxed on the basis of economic cost, or on other considerations such as the adverse environmental consequences of certain types of technology. Waivers from the standards may be obtained under some schemes in a limited number of cases (e.g. to permit innovative technology, to avoid very significant adverse economic effects). Examples of these standards under the Clean Air Act include MACT (Maximum Achievable Control Technology) for toxic air emissions, LAER (Lowest Achievable Emission Rate) for non-toxic pollutants emitted by new or modified sources in heavily polluted areas, and BACT (Best Available Control Technology) for new or modified sources in 'clean' areas.

Some of the most commonly heard criticisms of technology standards, which may be more or less accurate depending on the context, include the following: they impose unnecessary economic costs because they are uniform nationally, even though the pollution at issue may be a serious problem only in certain areas; they impose impossible burdens of information-gathering on administrative agencies; they discourage research and development of new technologies because the affected industries would not want to comply with more stringent limits that might thereby become attainable; and they discourage entry into the market of new firms because of the higher costs imposed.

Environmental quality standards

These describe either in quantitative terms (e.g. in parts per million) or narratively (e.g. 'no significant deterioration') the quality sought in the environmental medium (e.g. ambient air, surface water, groundwater) to be protected in the interests of public health and welfare. The National Ambient Air Quality Standards for the six commonly emitted pollutants – sulphur oxides, nitrogen dioxide, particulate matter, carbon monoxide, ozone, and lead – constitute perhaps the most significant example of environmental quality standards to be found in national US law. Adopted by the USEPA, they apply uniformly across the country. They are directly enforced through the establishment and implementation by the states of specific emission controls applied to stationary and mobile sources of air pollution, not by the courts. Adoption of water quality standards, on the other hand, is a matter for individual states, subject to the approval of the USEPA. As such they vary from state to state, and even from one stream or river segment to another within the same state. Thus they are subject only to modest national control with regard to their stringency.

Permit programmes

Whether or not a particular environmental scheme is based largely on technology standards (as the Clean Water Act has been until recently) or environmental quality standards (as in the case of the Clean Air Act's provisions dealing with commonly emitted non-toxic pollutants), or both, the Congress has often chosen to ensure the implementation of control measures through the requirement that new and existing sources obtain permits for their construction and/or operation from the state in question or the USEPA. Such permits are issued only after there has been an opportunity for members of the public to comment on the contents of the terms of the proposed permit. The permits are also subject to judicial review by state or federal courts. The permit systems, which conform to the requirements of the USEPA, are generally adopted as a matter of state law and run by state or local agencies. However, if a state fails to establish acceptable permit programmes, the USEPA is mandated to promulgate and administer such programmes and delegate, if appropriate, the power to implement them to a state or local agency.

Permit systems have a variety of advantages, several of which are crucially significant to environmental law in the USA. First, they can be used to present in a single document the diverse array of environmental requirements which apply to a particular facility, thus clarifying what those requirements entail for the permittee in terms of environmental performance. Second, the permit facilitates enforcement by governmental authorities and private citizens to the extent it forecloses arguments at the enforcement stage that the source is (or is not) subject to a particular regulation. Also monitoring reports, prepared pursuant to permits, can be used to prove violations of applicable requirements. Third, the permit applications, the terms of the permit themselves, and the monitoring reports provide reasonably accurate information on the type and amount of actual environmental discharges from a source along with the 'baselines' of allowable discharges. Such information is essential to the operation of many market-based strategies which involve 'trading'. Fourth, the permit process generates information (a 'source inventory') which is often essential for future environmental planning. For example, information submitted to the permit authorities may indicate the most important sources of toxic pollution in a watershed, and it may also assist in identifying the most appropriate allocation of discharge limitation responsibilities. Finally, and related to the fourth point above, is the fact that the very process of gathering information on the discharges of a given source may reveal to that source how its pollution could be prevented, and/or how its current discharges could be reduced cheaply.

Both the USEPA and individual states are increasingly emphasising the value of 'facility-wide permits' which can be used to establish 'caps' on allowable discharges or emissions. This would allow increases or decreases in discharges or emissions of particular pollutants from discrete points within the facility, without the need for prior administrative review and approval. Operational flexibility of firms would thus be enhanced, along with a potential reduction of their compliance costs. Moreover, there has been some development on multi-media permits which encompass in one document applicable regulations pertaining to all environmental media (e.g. air, surface water, groundwater). In constructing such permits, both managers and regulatory authorities would be able to focus on resolving real or apparent conflicts among regulations, in addition to minimising the aggregate environmental risk posed by discharges from a given source, instead of merely shifting risks from one form to another that is equally or more dangerous.

Market-based strategies: trading

Economists have argued that, if the law aims to achieve some general environmental quality objective, such as attaining a certain level of air purity in an air basin, it makes more economic sense to focus control requirements on those sources which can limit their emissions or discharges at least cost. But how is one to choose whether to control Plant A rather (or more stringently) than Plant B? It is argued that it would be futile, and cost-ineffective, for a bureaucrat to collect all the necessary information for making an accurate judgement on the matter because the owners of the sources have, or should have, all the necessary information. Moreover, not only are they in a much better position to know what choices are the 'best' for their individual circumstances, but they also have strong incentives to choose 'correctly'. But how can a system be designed to 'delegate' that choice to the emission sources in question without forfeiting the ability to achieve the desired level of air quality?

Reliance on the 'market' is the answer, and schemes to implement these insights are generally referred to as 'trading'. Given the baseline of 'lawful' emission levels for each plant, each source can choose to reduce its emissions to that level by installing control equipment, changing fuels or processes, shutting down, reducing operating levels, or by implementing other strategies; or by buying emission reductions of the relevant pollutant, not otherwise required by law, which have been achieved, or will be achieved in the future, by other plants capable of reducing their emissions more cheaply.

Such a system has several basic attributes, including:

1 an environmental 'goal' which can be quantified sufficiently to allow monitoring to determine its achievement in practice;
2 a 'baseline', in terms of emissions or discharge levels, from which 'excess' reductions of each source participating in the 'market' can be measured (the baseline must be consistent with the chosen environmental goal);
3 protocols and emission monitoring methods to identify accurately the actual emission levels at the sources participating in the market;
4 the ability of the sources to buy and sell, without substantial restriction, the 'excess' emission reductions achieved;
5 Some kind of system to facilitate the identification of willing buyers and sellers of 'excess' emission reductions;
6 effective systems to identify sources not complying with required emission limits and to impose credible sanctions to discourage 'cheating'.

There are at present two primary types of market-based emission trading systems operating in the USA. The first is the Emission Reduction Credit (ERC) programme in which emission reductions beyond the levels required to attain and maintain the National Ambient Air Quality Standards create 'credits' (ERCs) which can be saved for future use or sale (i.e. banked). They can also be traded among existing sources to comply with applicable emission limits, or sold to new or modified sources to allow their growth or change without accompanying increases in aggregate emission levels. The second type of system is known as the 'cap-and-trade' system (also called the 'allowance' or 'budget' programme) wherein the law:

1 establishes a regulatory limit (cap) on emissions from a discrete group of sources;
2 allocates allowances to emit to each of those sources, in aggregate these allowances being consistent with the attainment or maintenance of the 'cap';
3 authorises the trading of allowances in order to achieve cost-effective compliance with the 'cap'.

In fact, environmental groups can buy these 'allowances' and retire them, thus reducing allowable emission levels.

The case studies given in Boxes 20.1 and 20.2 respectively, are typical examples of trading schemes that are currently operating in the USA. While much of the experience to date with this trading has been in the area of air pollution control, the USEPA has started to encourage its application in water pollution control too. Other than cost saving, one of the projected benefits of market-based strategies is the incentive they may offer for the development of less-polluting products and processes. Since excess emission reductions can earn money for those firms able to achieve them, it is in the interest of both such firms and outsiders to invest in the development of better pollution control methods and technologies.

Sanctioning systems

Environmental regulation can be entirely ineffective without an effective sanctioning system. In fact, experience at both state and national levels prior to 1970 seems to confirm this.

The administrative agencies in the USA not only issue orders to cease and desist from violations of environmental regulations, they can also impose civil penalties (sometimes amounting to as much as $25,000 per day) for those violations. Similarly, the courts can issue injunctions to stop illegal behaviour and assess penalties for the same. In addition, they can impose *criminal* fines and imprisonment which the agencies cannot. In all such matters the government is aided by standards of proof which do not generally require proof that the defendant purposefully violated a particular regulation of which it was aware. In other words, liability for civil and criminal penalties in the environmental area is often not confined to 'intentional', 'wilful' or even 'negligent' behaviour. These aspects of United States law have

BOX 20.1 NEW SOURCE REVIEW (NSR) AND EMISSION TRADING

The Clean Air Act includes a variety of programmes which combine preconstruction permit review, technology standards, and emission trading. One of these programmes is referred to as non-attainment NSR (for New Source Review).

In those areas of the United States which do not currently comply with the National Ambient Air Quality Standards, a 'major' source of air emissions (e.g. one emitting 100 tons or more annually of a particular pollutant) must obtain a permit from the state, or from the USEPA in some cases, to construct and operate the plant or facility in question. Such a permit will be issued only if certain very stringent conditions are met, including the following:

1 The new source must control its emissions down to what is identified as the Lowest Achievable Emission Rate (LAER) for the category of source of which it is a member.
2 The new source's emissions, reduced to LAER levels, must be more than offset by emission reductions achieved no later than the start-up of the new facility by an existing facility or facilities usually located in the area of the new source's location. This may mean that the owner of the new source will have to purchase excess emission reductions (ERCs) from the owner(s) of other facilities.

The first requirement is justified in part by the determination that one of the appropriate costs to be incurred by a new facility choosing to locate in a badly polluted area is the installation of controls, or process design, which minimises emission levels. The second is justified by the requirement of the Clean Air Act that each area of the United States should attain the National Ambient Air Quality Standards by a staged reduction in emissions, with the attainment to occur on a certain specific date (e.g. 31 December 1998). In this context, any new emissions may disrupt the control strategy for attainment, or what is otherwise called the 'reasonable further progress' (RFP) curve.

BOX 20.2 THE ACID RAIN PROGRAMME

The 'Acid Rain Programme' (Title IV of the 1990 Clean Air Act Amendments) is the first 'cap-and-trade' programme for stationary source emissions required by the Clean Air Act. With regard to SO_x, the Act imposes a nationwide cap of approximately 8.9 million tons of SO_2 annually on the electric utility sector. This cap is to be attained in two phases. During Phase I, which began in 1995, the 111 highest emitting utility plants are subject to an intermediate SO_2 emission reduction limit. Phase II begins in 2000 when almost all other utility units with output capacity of more than 25 megawatts, as well as all new utility units, will be required to reduce their SO_2 emissions. Also, Phase I units must make further emissions reductions in Phase II. These required reductions may occur in a variety of ways, including the use of low-sulphur coal or oil, installation of stack-gas scrubbers, and curtailing the operation of plants or shutting them down.

The unique aspect of the SO_2 strategy begins with the transformation of the required intermediate emission reduction goals into the allocation of allowances among the Phase I plants based on their fuel consumption during 1985–1987. The Act itself specifies, for each of the affected 111 utility plants, what its initial annual allowance allocation will be during Phase I. An allowance is an authorisation to emit one ton of SO_2 during a calendar year. Allowances are fully transferable by sale or otherwise. During Phase II, existing units will be allocated their allowances each year (not exceeding 8.9 million tons in total) using formulae based on their utilisation rates, emission rates, and fuel consumption during 1985, 1986, and 1987. Emissions in excess of allowances held by a covered facility are prohibited and subject to various penalties. Affected units must either comply with applicable emission limits under the acid deposition programme, or hold sufficient allowances to emit SO_2 to cover their annual SO_2 emission totals.

New utility sources will not be allocated allowances after 2000 and will receive limited allowances before that. Thus they must purchase their allowances to emit SO_2. Existing utilities may also find it cheaper to purchase allowances to cover emissions in excess of their initial allowance allocations than to install control technology or take other measures necessary to reduce their emissions to match their initial allowance allocations. It is important to note that, while allowances which are traded have a date before which they cannot be used, the utilities may 'bank' their allowances for future use.

→

As of 1 January 1996, market activity in allowances had been limited. The price of each allowance has often been less than $200, much cheaper than originally forecast. In order to entice the utilities to the compliance option, vendors of scrubbers and low-sulphur fuels have apparently reduced their prices. Accordingly, even though allowance trading has been limited, it appears that the Acid Rain Programme as a whole has to date achieved substantial cost savings for the utility sector over a purely command-and-control approach because of the choice afforded by compliance options.

The programme has not, however, been without its critics. One concern of particular significance relates to the protection of unusually sensitive ecosystems from the effects of acid loading. Since the SO_2 'cap' is nationwide, there are no restrictions on trading from one area of the country to another. This means that SO_2 emissions may be concentrated in particular areas.

not, however, escaped criticism on a variety of counts, including the over-deterrence of legitimate behaviour with draconian penalties, and the impropriety of branding as 'criminal' behaviour that which could have been 'thoughtless' and not motivated by evil intent.

On the non-criminal side, there are various contexts in which the assessed amount of the penalty is required to include the economic benefits reaped by violators as a result of their non-compliance with the law (e.g. avoidance of pollution control expenditures over a number of years). For example, under the Clean Air Act, not only did the Congress mandate that civil penalty assessments recoup such benefits, but there is also a programme, administered by the USEPA, specifically designed for that purpose. The thinking here is that, unless such economic benefits are recovered by the government, violators of environmental regulations would retain competitive advantage over those firms which willingly comply with their legal obligations. This arrangement undermines the force of the law and diminishes the practical incentives for compliance.

Natural resource damages

Where damage to the environment has been done, and it severely affects natural ecosystems and their components that are deemed to be in the public domain or otherwise under public management and control, the question arises regarding the types of remedies that should be available. In the case of oil pollution and contamination resulting from the improper disposal of hazardous wastes, the Congress has imposed liability on those deemed responsible for damages to public natural resources in favour of national and state governments as trustees of those resources. The preferred remedy is the restoration or replacement of the damaged resource. If this is not feasible or excessively expensive, then the value of the losses incurred must be assessed and collected. The valuation process itself is subject to considerable controversy, however, since it involves, of necessity, the assigning of a monetary value to things that are often not traded in the 'market' (e.g. the use of a wetland as a nesting ground for migratory birds) and for which, therefore, there is no established price. The Contingent Valuation Method (CVM), which involves surveys of selected groups of the public to elicit their answers to hypothetical questions relating to, for instance, their willingness to pay to clean up or preserve a certain aspect of the natural environment, is one of the accepted (but still very much debated) methodologies used in the process of damage assessment.

Public sector approaches

Almost any action taken or funded by the national or state governments (e.g. highway construction projects and issuing of permits of various kinds) can have a direct or indirect adverse impact on the quality of the natural environment. For years every agency at the national level has pursued its distinctive mission without focusing attention on the environmental consequences of its actions. However, although it has taken almost a quarter of a century, today most agencies do accept – albeit grudgingly – that they cannot cavalierly brush aside objections to projects and programmes they wish to pursue on the excuse that environmental protection is not their concern. At the same time, outside of those few national agencies such as the USEPA and the Department of the Interior, which have been consciously designed to protect environmental resources, the environmental consequences of governmental action may in fact be given little weight in the decision-making process.

Environmental assessment

A variety of specific statutes require that particular national agencies consider certain types of environmental concerns in their decision-making. Beyond this, there are two statutes which apply to all federal agencies, requiring rigorous assessment of the effects of their action on the natural environment. The National Environmental Policy Act of 1969, which is discussed more fully in Box 20.3 and which is a government-wide mandate for general environmental assessment, has been copied by a number of the states. But it does not require that agencies avoid adverse environmental consequences, only that they should be considered as part of an agency's decision-making process. The Endangered Species Act, on the other hand, not only applies to all national agencies, but it also directs them to ensure that, unless an exception is granted – and they rarely are – no action authorised, funded, or carried out by them should jeopardise the continued existence of any species that has been listed by the Secretary of the Interior as being endangered or threatened, or results in the destruction or adverse modification of the critical habitat of those species. Inter-agency consultation and assessment is required both to identify such potential effects and to avoid them.

Management of public natural resources

The national government owns huge tracts of land, much of it to the west of the Mississippi river and in Alaska, some of which has been set aside as national parks and wildlife reserves. In other parts of the public domain, timber harvesting, mining, and/or grazing of livestock is common and permitted at fees which grossly underestimate the market value

BOX 20.3 THE NATIONAL ENVIRONMENTAL POLICY ACT OF 1969

The National Environmental Policy Act (NEPA) makes environmental protection a part of the statutory mission of all national agencies of the federal government of the USA. More precisely, this Act puts a legal obligation on all such agencies to consider the environmental consequences of their policies, programmes or actions in the decision-making process. This was not the case before 1969 when many of the agencies had refused to consider the adverse environmental effects of their actions on the ground that the US Congress had not authorised them to do so.

Both to implement this general mandate and for other purposes, no later than the time an agency makes a proposal for a 'major' federal action that may have a significant impact on the environment, the initiating agency must have prepared a final environmental impact statement (EIS) which canvasses the beneficial and adverse effects of the proposed action, alternatives to it, and the ways to mitigate its identified adverse effects. Prior to the final EIS stage, agencies are required to prepare what are called 'environmental assessments' (EAs) to determine whether the proposed action requires a full-scale EIS. If an agency determines that its proposal falls below the threshold for requiring the preparation of EIS, it issues what is called a FONSI (i.e. Finding Of No Significant Environmental Impact). A draft EIS, if prepared, is then circulated among concerned federal agencies for their comments prior to issuing a final statement, which by law must accompany the proposal through the decision-making process of the agency. Other than helping agencies to carry out their mandated duty to consider the environmental impacts of their actions, NEPA impact statements are meant to provide the President, the Congress, and the public at large with information they may need for evaluating and responding to the action proposed by the agency.

The NEPA, especially in its early years, generated a significant amount of litigation as citizens sued to halt various national actions either because NEPA statements had not in fact been prepared when they should have been, or because the statements were inadequate in their consideration and treatment of adverse environmental effects. Over the years national agencies have more fully embraced their NEPA obligations, although compliance has not always (or even most of the time) resulted in actions very different from those which would have been taken without consideration for environmental consequences.

The President's Council on Environmental Quality adopts and revises regulations directed at all national agencies. These regulations place NEPA obligations on agencies and offer guidance on how they should be met.

of the commodities extracted by private parties. The government has also funded vast irrigation schemes in the western USA and hydroelectric projects across the country.

Over the years the Congress has enacted and amended a variety of statutes dealing with the management of these national lands and resources, including the Federal Land Policy and Management Act (FLPMA) and the National Forest Management Act (NFMA) of 1976. While some of these statutes, such as the Wilderness Act of 1964, embody a dominant 'preservation' philosophy, others such as the FLPMA and the NFMA incorporate the twin goals of 'multiple use' and 'sustained yield'. The former purportedly allows simultaneous production of compatible resources through sound land-use planning, while the latter aspires to long-term sustainability of timber and other resources. In practice, however, public management has more often than not attended primarily to the interests of commodity users and less to the interests of sustainable ecosystems supporting a variety of renewable resources.

THE RELATIONSHIP BETWEEN DOMESTIC AND INTERNATIONAL ENVIRONMENTAL LAW IN THE USA

To date the development of environmental law in the USA has largely been in response to distinctive historical, political, social and economic factors that are mainly domestic. Its content has not been substantially dictated or determined by international influences, except in a limited number of areas. In fact, it is interesting to note that one of the more important precedents for the development of international environmental law (the so-called Trail Smelter arbitration of 1935 which imposed liability on a Canadian smelter for damages caused in the USA by its sulphur dioxide emissions) relied on the principles of common law nuisance developed by the US Supreme Court.

However, in the 1970s, Canada 'turned the tables' on the USA by demanding significant reductions of sulphur and nitrogen oxide emissions from the United States of America which allegedly caused acid rain damaging Canadian property and resources. While in the early 1980s there was some reluctance at the political level in the USA to acknowledge responsibility for this damage, ultimately the Congress responded by enacting Title IV of the Clean Air Act Amendments of 1990 which, through an innovative market-based system, promised a 50 per cent reduction in US emissions of sulphur dioxide by the turn of the century.

This statute also contained Title VI which was intended, in part, to satisfy the obligations of the USA under the 1987 Montreal Protocol on Substances that Deplete the Ozone Layer and the subsequent amendments to that protocol adopted in London in 1990. Prior to 1990, the USEPA had in fact adopted various regulations to reduce the production and consumption of stratospheric ozone-depleting substances in an attempt to begin the implementation of the Protocol as a matter of domestic US law.

One of the most important areas in which international law is beginning to have a significant impact on domestic law is concerned with alleged trade barriers created by various federal statutes. For example, under the authority of the Marine Mammal Protection Act of 1972, the USA imposed a ban on tuna imports from Mexico and four other countries because the method of catching the fish was deemed to result in too large an incidental kill of dolphins. In 1991 a panel, convened pursuant to the General Agreement on Tariffs and Trade, found this embargo to violate provisions of the treaty which defined impermissible trade restrictions. Then in 1994 another GATT panel ruled against US corporate average fuel economy standards for automobiles on the ground that they established separate criteria for determining the overall fuel economy of domestic and imported fleets. Finally, following the successful conclusion of the Uruguay Round of GATT negotiations in 1993 and ratification of the resulting agreement by the Congress in 1994, the newly created World Trade Organisation (WTO) ruled in 1996 that a regulation governing reformulated gasoline under the federal Clean Air Act discriminated unfairly against foreign refiners, because standards for domestic refiners were based on the quality of gasoline they produced in 1990 while foreign refiners were held to a standard based on the overall average quality of gasoline in the USA.

The negotiation and implementation of the North American Free Trade Agreement (NAFTA) of 1992, entered into by the United States, Canada, and Mexico, raised other environmental issues. The signatories are directed to 'avoid arbitrary or unjustifiable distinctions' in levels of environmental quality that would create 'unjustifiable discrimination' against goods from another country or that would 'constitute a disguised restriction on trade'. A Commission on Environmental Co-operation was created under the environmental 'side agreement' of NAFTA with the authority to investigate the status of compliance with environmental restrictions in all three countries to ensure that lax compliance efforts do not create trade advantages for one country over another. Citizens or governments may file complaints with the Commission and, if it finds a persistent pattern of failure to enforce environmental laws effectively, it can impose fines or trade sanctions.

Finally, US courts have occasionally had to confront questions regarding the extra-territorial reach of US environmental law. For example, the National Environmental Policy Act of 1969 may require an agency of the national government to determine the environmental costs of a nationally funded or directed activity taking place abroad. Similarly, the Endangered Species Act may mandate the protection of species living abroad from the potentially destructive actions of a US national agency. Litigation on these issues has not, however, resolved the reach of these statutes definitively.

RECENT TRENDS IN THE DEVELOPMENT OF US ENVIRONMENTAL LAW

While public support in the USA for environmental protection has remained high ever since the 1970s, a consensus among national and state decision-makers seems to have developed. It is that while fundamental goals such as clean air and water, and ecosystem protection, remain sound, there may be better ways to achieve them than have been embodied in the legislation of the last 25 years – ways that are more cost-effective and respectful of local differences. Increasing experimentation with market-based systems, as illustrated by the Acid Rain Programme, is only one manifestation of this new agenda. Focus on pollution prevention and integrated environmental regulation to avoid problems before they arise, and to reduce aggregate environmental risk in a co-ordinated fashion, represents yet another aspect of the evolving philosophy of environmental regulation. In some states steps to implement these new strategies are evidenced by innovative revisions to their environmental permit systems.

Where detailed and prescriptive regulations must be retained, there is an appreciation that unnecessary, duplicative and conflicting agency requirements should be eliminated. It is felt, furthermore, that the burdens imposed on the private sector to maintain records and prepare reports for the USEPA and other agencies should be reduced by streamlining and consolidating requests for information. Where, as is the case increasingly, smaller businesses fall within the scope of environmental laws, it is recognised in various programmes that the government has an obligation to offer technical and other assistance to facilitate and encourage the compliance of this group, and that its interests should be specially considered at the policy-formation stage.

While civil and criminal sanctions remain important backstops to effective regulation, there is growing interest in encouraging environmental auditing as a means for regulated entities to discover and cure violations on their own and earlier than might otherwise be the case. In order to create incentives to conduct regular audits and to improve compliance, many states have already enacted, or are considering, laws to create privileges that would shield the results of environmental audits from environmental agencies and/or limit their use as evidence in enforcement proceedings. While the USEPA has yet to endorse a broad audit privilege, which it fears may interfere with effective enforcement when it is most called for, it has adopted a variety of policies which permit penalty mitigation in instances where a company institutes an auditing programme, cures detected violations, and otherwise co-operates with enforcement officials. It has also extended some degree of protection to small businesses where violations are detected as a result of compliance assistance received from state agencies.

Both the USEPA and the states realise that the existing structure of largely prescriptive, 'command-and-control' regulation can deter companies from searching for and implementing strategies capable of greater and faster pollution reduction, or at least the same degree of environmental protection, at a lower cost. With this realisation, the USEPA is now experimenting with co-operative agreements with various industrial firms which commit themselves to changing their processes or installing controls to reduce their emissions or discharges, by at least as much as or possibly more than that required under existing regulations, in lieu of complying with those regulations. However, while the USEPA has attempted to involve citizen groups as parties to this process, there is the concern that, unless existing statutes and implementing regulations are changed, firms subject to these agreements will still face the risk of citizen-initiated enforcement actions for violations of the applicable regulations.

There appears to be a growing acceptance of the idea that the states should be given more flexibility than at present under national law to design and administer programmes for environmental protection that are better suited to local needs and circumstances. Because of increasing interest in spurring economic development, both the USEPA and various states are starting to experiment with so-called 'brownfields' initiatives. Such initiatives seek to reduce the perceived burdens of 'inflexible' environmental regulation, where they are deemed to be overly protective, on businesses which might establish or expand facilities in areas that have hitherto been heavily industrialised but have lately seen their industrial base shrink dramatically. At the same time, many feel that in all events, state legislatures and agencies must be held to certain national minimum standards in order to prevent local interests from effectively undercutting environmental protection. It is only then that the 'race to the bottom' among states in regulating pollution can be minimised.

Developments at the international level are increasingly focusing attention on the United States. Whether it is the implementation of the UN Framework Convention on Climate Change, actions of the World Trade Organisation that may impact on federal and state law, the effects of ISO 14000 (including its focus on environmental management systems) with regard to compliance obligations and the behaviour of US companies operating in the US and abroad, or a host of other issues, the traditionally insular system of legal regulation for environmental protection in the USA will inevitably reflect themes, requirements, and innovations developed globally.

CONCLUSIONS

It can be said that during the last 25 years the USA has approached environmental problems with an intensely prescriptive type of legal regime centralised at the national level. However, experiments with new approaches which incorporate market incentives, operational flexibility for regulated entities, some degree of voluntary action on the part of the private sector in improving environmental performance, along with locally determined strategies, appear to represent trends for future development. Hopefully, this development will not threaten the substantial progress which has been made to date in protecting public health and welfare, and endangered ecosystems. At the same time, the themes, concerns, and approaches of the international community will no doubt be reflected to an increasing extent in domestic US law, as well as in the implementation of policies designed to protect and improve environmental resources.

NOTE

1 The author would like to acknowledge the editorial and research assistance in the preparation of this chapter of Cynthia D. Driscoll, a third-year law student at the University of Pittsburgh School of Law.

REFERENCES

Menell, P.S. and Stewart, R.B. (1994) *Environmental Law and Policy*, Boston, Mass.: Little, Brown & Co.

Plater, Z.J.B., Abrams, R.H. and Goldfarb, W. (1992) *Environmental Law and Policy: Nature, Law, and Society*, St Paul, Minn.: West Publishing Co.

(Both books are excellent and comprehensive collections of important cases and textual discussions of the areas covered in this chapter.)

SUGGESTED READING

Breyer, S. (1993) *Breaking the Vicious Circle/Toward Effective Risk Regulation*, Cambridge, Mass.: Harvard University Press.

Grad, F.P. (multi-volume set, periodically updated) *Treatise on Environmental Law*, New York: Matthew Bender.

Hays, S.P. (1987) *Beauty, Health & Permanence: Environmental Politics in the US 1955–1985*, Cambridge: Cambridge University Press.

Liroff, R.A. (1986) *Reforming Air Pollution Regulation: the Toil and Trouble of EPA's Bubble*, Washington, DC: The Conservation Foundation.

Mandelker, D.R. (1992 edn, with updates) *NEPA Law and Litigation*, USA: Callaghan & Co.

Rodgers, Jr., W.H. (multi-volume set, periodically updated) *Environmental Law*, St Paul, Minn.: West Publishing Co.

Sax, J. (1980) *Mountains Without Handrails: Reflections on the National Parks*, Ann Arbor, Mich.: University of Michigan Press.

SELF-ASSESSMENT QUESTIONS

1 The common law approach to pollution control proved to be inadequate in part because:
 (a) there were no judge-made rules dealing with pollution problems;
 (b) the common law was a product of English courts and could not be applied in the United States;
 (c) courts were incapable of creating the types of planning mechanisms necessary for effective environmental protection.

2 One of the characteristics of the federal system found in the United States is:
 (a) the existence of a specific provision in the national Constitution, expressly giving the Congress the power to regulate for the protection of the environment;
 (b) the effective power of the Congress to deal with most environmental issues pursuant to its authority to regulate inter-state and foreign commerce;
 (c) the invalidity of national laws which conflict with state laws on the same subject.

3 Most of the environmental law (by volume) in the United States is found in:
 (a) statutes enacted by the Congress;
 (b) regulations and guidance documents adopted by national and state administrative agencies which implement and elaborate on statutory provisions;
 (c) statutes enacted by the state legislatures.

4 Which of the following concerns is not clearly reflected in US environmental law as it currently exits?
 (a) Economic efficiency;
 (b) Risk reduction;
 (c) The moral superiority of claims for environmental protection.

5 The national government of the United States came increasingly to dominate in setting environmental policy after 1970 because:
 (a) the states had failed to use their authority in the past to vigorously combat environmental problems;
 (b) the states had no experience in regulating for environmental protection;
 (c) the states lacked the power to regulate environmentally dangerous conduct.

6 Which of the following statutes is characterised primarily by its requirement that all national governmental decision-makers consider the adverse environmental effects of their decisions?
 (a) The Clean Air Act.
 (b) The Administrative Procedure Act.
 (c) The National Environmental Policy Act of 1969.

7 Which of the following regulatory approaches is criticised for its failure to promote environmental protection in an economically efficient manner?
 (a) Tradable rights to pollute.
 (b) Comparative risk assessment.
 (c) Technology standards for new and existing sources of pollution.

8 The trend in US environmental law for market-based systems of regulation is illustrated by:
 (a) the Endangered Species Act;
 (b) the Acid Rain Programme;
 (c) the Federal Insecticide, Fungicide, and Rodenticide Act.

9 Which of the following does not characterise the sanctioning system that exists in the USA for violations of environmental law?
 (a) A requirement that criminal penalties be imposed only in cases where defendants have acted with actual knowledge that what they were doing was unlawful.

(b) Sizeable civil penalties amounting to as much as $25,000 per day of violation.
(c) Provisions authorising the recovery of damages to natural resources owned, managed or regulated by the government.

10 Which of the following statements does not correctly characterise recent developments in US environmental law?
(a) There is less desire for national prescription of the details of state environmental policy.
(b) There is increasing interest in giving private industry more flexibility in meeting environmental goals.
(c) The cost of environmental regulation is becoming less relevant as a consideration in designing approaches to environmental protection.
(d) Developments at the international level are having increasingly greater impact on the approach to environmental regulation.

21
EU ENVIRONMENTAL LEGISLATION

Katrien Debeuckelaere and Liam Cashman[1]

SUMMARY

This chapter aims to give an overview and understanding of EU environmental legislation, its evolution, the competence of the Community in environmental protection, preparation of legislation, and the control of both implementation and application of legislation.

The reader should be aware of the fact that this chapter only covers EU environmental legislation. It does not cover the radiation protection legislation of the European Union, for example, which has been adopted under the EURATOM Treaty.

The chapter charts how the EC Treaty has evolved to give increasingly greater weight to the protection of the environment. Different types of legislative acts, used for adopting EU environmental legislation (e.g. regulations, directives, decisions and recommendations), and how the legislative process works, including the role of the European Parliament, are explained. An overview is given of the EU's sectoral legislation adopted to date (e.g. for water, air, noise, chemicals, nature conservation, waste and general measures) with the aid of two short case studies. Also explained is how EU environmental legislation is enforced, together with the roles of the European Commission and the European Court of Justice.

ACADEMIC OBJECTIVES

Aim

The aim of this chapter is to give an introduction to EU environmental legislation, the competence of the Community in environmental protection, the legal measures which the Community can take, and the limitations of such measures.

Objectives

After reading it the reader should understand that the competence of the Community, the contents of environmental legislation, the process of adoption and the control of implementation and application, are all different from national legislation; that the Community does not adopt environmental legislation for the member states on every aspect of environmental protection, and that legal measures which the Community can adopt are different from national laws. Furthermore, the reader should be able to find and identify EU environmental legislation, know how to interpret it, and, if in doubt, know where to look for Community interpretation.

DEVELOPMENT OF EU ENVIRONMENTAL LEGISLATION

EU environmental legislation is crucial in the sense that member states of the EU often rely heavily on Community laws to shape their own national legislation.

The Treaty of Rome

The Treaty of Rome, under which the European Economic Community was established in 1957, did not contain specific provisions for environmental legislation. The concepts of the environment, environmental protection, etc. are thus absent in the Treaty of Rome. However, the need for environmental legislation became apparent as an aspect of the harmonisation of standards, and the first piece of environmental legislation – Council Directive 67/548/EEC on the approximation of laws, regulations and administrative provisions relating to the classification, packaging and labelling of dangerous substances (OJ 1967, L 196/1) – was adopted on 27 June 1967.

On 22 July 1971, the Commission made the first communication to the Council on environmental policy. This communication opened a vigorous debate on the question of whether environmental matters were best dealt with at the Community level, or by means of intergovernmental agreements between the member states and by co-ordinating national environmental policies. The Commission, supported by the European Parliament, was in favour of Community action (Krämer, 1990a).

The Paris Summit

A major turning point in Community environmental legislation was the declaration of the heads of state or government at the Paris Summit, held in October 1972, where the suggestion of the Commission, expressed earlier in its communication, was finally accepted. The declaration stated that

> Economic expansion is not an end in itself; its first aim should be to enable disparities in living conditions to be reduced. It must take place with the participation of all the social partners. It should result in an improvement in the quality of life as well as in standards of living. As befits the genius of Europe, particular attention will be given to intangible values and to protecting the environment, so that progress may really be put at the service of mankind.
> (Paragraph 3 of the preamble to the Declaration).
>
> The Head of State or of Government emphasised the importance of a Community environmental policy. To this end they invited the Community institutions to establish before 31 July 1973, a programme of action accompanied by a precise timetable.
> (Point 8 of the Final Declaration)
> (OJ 1973, C 112/5).

The First Environmental Action Programme

The First Environmental Action Programme was adopted by a declaration of the European Community Council, and of the representatives of the governments of the member states, meeting in the Council of 22 November 1973 to discuss the programme of action of the European Communities on the Environment (OJ 1973, C 112/1).

The Paris Summit called for Community action in the environmental field, and the Action Programme was adopted not only by the Council but also by the representatives of the governments of the member states. Some member states, France in particular, maintained that intergovernmental co-operation was preferable to a Community policy on environmental matters (Krämer 1990a).

The Second, Third and Fourth Environmental Action Programmes (OJ 1977, C 139/1; OJ 1983, C 46/1; OJ 1987, C 328/1) were also adopted by the Community and the representatives of the governments of the member states. The Fourth and Fifth Environmental Action Programmes were adopted after the Single European Act came into force.

These Environmental Action Programmes specified priorities and scheduled actions which the Community envisaged taking in different sectors of the environment. The Fifth Environmental Action Programme represented something of a departure from previous action programmes, in as much as it set out an in-depth conceptual analysis of how best to develop environmental policy. For example, it advocated a wider use of non-regulatory approaches to solving problems.

These Environmental Action Programmes were in fact political declarations of intent which did not form the legal basis of the Community's environmental measures. Nevertheless, they represented more than a mere voluntary undertaking by the Community and the member states. Development of the Environmental Action Programmes required profound studies and preparation. And the decision to adopt the Programmes by the Community and the member states was carefully considered because of their potential impacts (Krämer, 1990a, 1990b).

The articles of the EC Treaty relating to the environment (see pp. 406–407) are reinforced by these Programmes, together with suggestions for concrete measures, a timetable and priorities. Their adoption by the Community and the member states meant that they were in agreement with the Programme, and, consequently, with the measures contained therein. This was also reflected in the preparation of annual budgets. Funds needed for implementing the measures contained in the Programmes were allocated in the budget.

In the case of C91/79 and C92/79 (ECR 1980, pp. 1099 and 1115), concerned with the legal force of these two directives – one on detergents and the other on the sulphur content of liquid fuels – adopted under the Community programme to eliminate barriers to both trade and the environmental programme, the Italian government observed that the subject matter of the directives lay at the fringe of Community powers. But the Court ruled that, 'It should be observed that the directive had been adopted not only within the programme of action of the Communities on the environment; it also comes under the general programme for the elimination of technical barriers to trade. In this sense it is validly founded upon article 100.'

It can be concluded from this court ruling that the validity of Community measures could be based on a Community programme. By adopting an environmental action programme, the Community had agreed on matters that were to be the subject of Community legislation, and therefore that such matters were consequently within the competence of the Community (Krämer, 1990a).

Since the coming into force of the Treaty on European Union (see pp. 406–407), there is now an explicit legal basis, enshrined in Article 130, section 3 of that Treaty, for adopting general environmental action programmes. Both the European Parliament and the Economic and Social Committee are formally given a role in the adoption procedure of such programmes.

The Single European Act and the Treaty on European Union

On 17 February 1986, the Single European Act was signed in Luxembourg, and it came into force in 1987. On 7 February 1992 the Treaty on the European Union was signed in Maastricht (it is also known as the Maastricht Treaty), and it came into force in 1993.

The Single European Act (S.E.A.) is very important to the European Community's environmental legislation, because it contains a new and specific provision, called 'Title VII', on the environment (Articles 130r–130t)[2] which is enshrined in the Treaty establishing the European Economic Community, and also because Article 100a is incorporated in the same treaty.

Title VII describes the objectives of Community action on the environment and sets out the basic principles which should be respected, such as the Polluter Pays Principle, the principle that preventive action should be taken, and the principle that environmental protection should be integrated into other Community policies. Also enshrined in Title VII is what is now known as the 'Principle of Subsidiarity' (although this term is not explicitly used in the Title). This principle is expressed in a provision which requires the Community to take action relating to the environment to the extent to which environmental objectives could be better attained at the Community level than at the level of individual member states.

The Treaty on European Union (TEU) resulted in a change of name of the Treaty Establishing the Community – from the 'Treaty Establishing the European Economic Community' to the 'Treaty Establishing the European Community' (the 'EC Treaty'). References in subsequent legislation are therefore made to EC instead of EEC. Also, the environment was mentioned for the first time specifically in Article 2 of the EC Treaty, which defines the basic tasks of the European Community. It is referred to in terms of 'sustainable and non-inflationary growth respecting the environment'. As amended by the TEU, Article 3 of the EC Treaty also explicitly mentions policy-making on the environment as one of the activities of the Community.

After the S.E.A. came into force, the Principle of Subsidiarity emerged as a key concept in Community policy in general, and not just for environmental policy. This is reflected in the TEU, which amended the EC Treaty in order to make an explicit reference to this principle. Article 3b now provides that, in areas where the Community does not have exclusive powers (such as the environment), it shall take action, in accordance with the Principle of Subsidiarity, only if and in so far as the objectives of the proposed action cannot be sufficiently achieved by the member states and can therefore, by reason of the scale or effects of the proposed action, be better achieved by the Community. Thus, since the coming into force of the S.E.A., the Principle of Subsidiarity has graduated from exclusively a principle of Community environmental policy to a general Community principle.

The TEU has also made changes to Title VII of the EC Treaty (which is now known as Title XVI), by introducing the Precautionary Principle and by making it possible to adopt some legislation under the Title by Qualified Majority Voting (QMV) instead of unanimity.

Since the coming into force of the S.E.A., a given set of environmental measures can have either Article 130s or 100a as its legal basis, depending on the content of the measures in question (i.e. whether the measures are principally aimed at the protection of the environment, or primarily concerned with the realisation of the internal market). Article 100a is the appropriate legal basis for environmental measures necessary for the realisation of the internal market, and thus it requires harmonisation of national legislation. Other articles of the Treaty can also form the legal basis of environmental measures in cases where other aspects of Community policy are involved and are considered to be preponderant in importance (e.g. measures covering environmental protection, agriculture, transport or economic policy).

Environmental measures, legally based on Article 130s, leave scope for the member states to adopt more stringent measures (Article 130t) on condition that the latter are compatible with the Treaty and are notified to the Commission. On the other hand, measures that are legally based on Article 100a do not offer this freedom to the member states. This is because the implementation of stricter measures is subject to a notification procedure aimed at facilitating harmonisation of legislation throughout the Community, which is one of the main objectives of Article 100a.

The legislative process for the adoption of measures legally based on Article 130s is different from those based on 100a, as will be explained later (see pp. 408–409).

LEGISLATIVE ACTS

Forms of legislative acts (Article 189 of the Treaty)

The following legislative measures can be adopted to reinforce the Treaty provisions and the Environmental Programmes: regulations, directives, decisions, recommendations and opinions (Article 189).

A *regulation* is meant for general application; it is binding in its entirety and is directly applicable in the member states. When a regulation has been adopted, the member states do not have to take national measures to implement the act. Thus a regulation has to be more detailed of necessity than a *directive*. Once a regulation is published in the Official Journal of the Communities (OJ), it comes into force on the date specified in the text, or on the 20th day following publication.

A *directive*, on the other hand, is binding as to the result to be achieved in each member state to which it is addressed. However, the member states have the choice of both form and method of implementing the directive in their national legislation. In practice this means that a directive is a legal measure which requires action from the member states to translate it in their respective national legislation. At the end of a directive there is always a specific article, or articles, stipulating the date by which the member states should implement the measures to which the directive refers.

Although Article 189 of the Treaty states that the member states have the choice of both form and method of implementing a given directive in their national legislation, this choice is limited in practice. This is because the implementation of a directive into national legislation is subject to certain conditions established by the European Court of Justice (ECJ). And these conditions, summarised below, can and do constrain freedom of choice:

- the measures may not be adopted by way of acts which contain a delegation of power and can be amended unilaterally by local authorities;
- they have to be binding measures;
- they have to be published, and thus they create rights for third parties.

For example, the ECJ has ruled on several occasions that circular letters cannot be considered to implement directives correctly because they do not fulfil the conditions enumerated above (cases C 102/79, C 96/81, C 97/81, C 173/83, C 239/85, C 116/86).

A directive is notified to the member states concerned and is also published in the OJ, although the latter is not obligatory as per the Treaty.

A *decision* is binding in its entirety upon those to whom it is addressed. It is notified to the member states concerned, but its publication in the OJ is not obligatory.

Regulations, directives and decisions contain in their preamble the legal basis of the measure, the reasons for its adoption, the origin of the proposal, and the required opinions. *Recommendations* and *opinions* are not legally binding.

In matters of environmental protection most Community measures are in the form of directives. This means that the member states have to take necessary steps to incorporate them into their national legislation within the time limit stipulated in the text of each directive. The member states are also required to communicate the measures taken to implement the directives. In practice, however, there is often a delay in implementation, and sometimes the member states are not able to communicate adopted measures on time; the Commission, as guardian of the Treaty, has to ensure that the member states do implement the directives. The procedure to control the implementation of directives is explained on pp. 414–417.

The legislative process and the legal basis of environmental legislation

The institutions involved in the Community legislative process are the Commission, the Council, the Economic and Social Committee, and the European Parliament (EP). The procedure to be followed depends on the legal basis of the proposed legislation.

To begin with, the Commission undertakes the preparation of a proposal for legislation, which is then elaborated within the Commission. It is normal practice during preparation to consult with the member states and other parties who may be concerned. Once the proposal is deemed to be in its final version, it has to be formally adopted by the Commission. After adoption the proposal is referred to the Council, as well as to the Parliament when what is called a 'co-decision' procedure is followed.

A proposal based on the first paragraph of Article 130s can be adopted by the Council on Qualified Majority Voting (QMV), following a co-operation procedure with the EP as well as consultation with the Economic and Social Committee. The qualified majority vote involves a system of weighting in which the more populous member states have more votes. The number of votes each member state has is set out in Article 148 of the Treaty.

A proposal based on paragraph 2 of Article 130s requires unanimity in the Council in order for it to be adopted, unless it concerns a matter on which the Council agrees in advance to allow a decision by a QMV. The EP is consulted (i.e. it is given a weaker role than when the proposal is based on the first paragraph of Article 130s), as is the Economic and Social Committee.

A proposal based on Article 100a of the Treaty requires the co-decision of the Council and the European Parliament (i.e. the Parliament has a stronger role), as well as consultation with the Economic and Social Committee. The role of the European Parliament has been strengthened by both the S.E.A. and the TEU, because the S.E.A. introduces into the process a co-operation procedure and the TEU a co-decision procedure.

As explained earlier, environmental legislation can be legally based on either Article 130s or 100a of the Treaty, depending on the purpose and subject of legislation. The proposal submitted by the Commission also includes the legal basis which the Commission deems appropriate. There may however be disagreement between the Commission, Council and Parliament. For example, the Council may change the legal basis of a measure, with the risk that the Commission or the Parliament will refer the case to the European Court of Justice.

Directive 89/428/EEC, which is on procedures for harmonising programmes for the reduction and eventual elimination of pollution caused by waste from the titanium dioxide industry (OJ 1989, L 201/56), provides a typical example of the disagreement mentioned above. In this, Article 100a was given by the Commission as the legal basis of the proposal. But the Council changed it to Article 130s, after consulting with the EP which was in favour of Article 100a. The Commission attacked this change, and, with the EP as intervening party, lodged an appeal to the European Court of Justice (ECJ). The ECJ decided in its judgement of 11 June 1991 (case 300/89) in favour of Article 100a as the correct legal basis on the following grounds, and the directive was annulled:

1 The fact that Article 130r provides for the integration of the environment into other Community policies indicates that Community legislation does not have to be based on Article 130s simply because that legislation refers to the environment.
2 Environmental measures can have an impact on the competitive position of enterprises, consequently a measure aimed at eliminating the distortion of competition can contribute to the realisation of the internal market and falls within the scope of Article 100a.
3 Para. 3 of Article 100a expressly refers to the need to base environmental measures on a high level of protection. This clearly indicates that environmental objectives can be attained by means of legislation based on Article 100a.

While in Case 300/89 the ECJ confirmed that Article 100a was the correct legal basis, in two subsequent cases it ruled in favour of Article 130s (Cases 155/91 and 187/93). In the first of these cases, the Commission, again supported by the EP, contested the Council's use of Article 130s instead of 100a for an amendment to the Framework Waste Directive (Directive 91/156/EEC). This time the ECJ agreed with the Council's choice of legal basis stating that, where the chief objective of the legislation was environmental, Article 100a could not be justified even if the legislation also had a bearing on the functioning of the internal market. The ECJ employed a similar reasoning in the second case, in which the EP challenged the use of Article 130s instead of Article 100a for a Community regulation on the shipment of wastes (Council Regulation (EEC) 259/93).

The problem of the legal basis of Community environmental legislation is not concerned only with Article 100a or 130s. Other articles of the Treaty can also serve as the basis of environmental legislation, and thus they too may be the subject of disputes as to their appropriateness. However, the above examples now provide useful guidance on how to determine the correct legal basis.

An overview of existing environmental legislation

Community environmental legislation is prepared and followed by Directorate General XI (DG XI) of the Commission. The Commission is divided into twenty-four DGs, each with a specific competence and task, although some measures do come within the competence of two or more DGs. A compilation of EC environmental legislation was published by the Commission in 1992 and updated in 1996.

Community environmental legislation is divided into these sectors: water, air, noise, chemicals, nature conservation, waste and general measures. The text of the measures is published in the OJ. Also, the *Celex* database, which is a public Community database, can be consulted for research on legal measures. This database contains complete texts of all legislative measures of the Community.

This brief overview of existing environmental legislation is certainly not exhaustive, but it gives an introduction to the main Community acts in this field.

Water

Community legislation to protect and improve the aquatic environment can be subdivided into legislation to protect the aquatic environment from pollution by certain dangerous substances; legislation to protect water according to its specific use; legislation relating to specific industries; and legislation relating to the source of pollution.

Council Directive 76/464 on pollution caused by certain dangerous substances discharged into the aquatic environment of the Community (OJ 1976, L 129/23) is the Framework Directive on measures to protect the aquatic environment from pollution by substances listed in the annex to the directive. This annex contains List I substances that are particularly dangerous because of their toxicity, persistence and bio-accumulation, and List II substances. Article 6 of the Directive provides that for List I substances the Commission has to establish Community limit values. Member states can also opt for emission standards instead of limit values, provided that these standards are based on quality objectives.

To date Community limit values and quality objectives have been set for:

- mercury (Council Directive 82/176, OJ 1982, L 81/29; Council Directive 84/156, OJ 1984, L 74/79);

- cadmium (Council Directive 83/513, OJ 1983, L 291/1);
- hexachlorocyclohexane (Council Directive 84/491, OJ 1984, L 274/11);
- carbon tetrachloride, DDT, pentachlorophenol (Council Directive 86/280, OJ 1986, L 181/16);
- aldrin, dieldrin, endrin, isodrin, hexachlorobutadiene and chloroform (Council Directive 88/347, OJ 1988, L 158/35);
- dichlorethane, trichlorethylene, perchloroethylene and trichlorobenzene (Council Directive 90/415, OJ 1990, L 219/49).

According to the Annex to the Directive List I, substances for which no Community limit values or quality objectives have so far been established have to be considered as List II substances. Also, member states have to establish programmes (Article 7) to reduce pollution of the aquatic environment caused by List II substances.

Community legislation to protect water according to its specific use covers:

- the quality of surface water intended for the abstraction of drinking water (Council Directive 75/440, OJ 1975, L 194/26) supplemented by a directive on the methods of measurement and frequencies of sampling and analysis of such water (Council Directive 79/869, OJ 1979, L 271/44);
- the quality of bathing water (Council Directive 76/160, OJ 1976, L 31/1);
- the quality of fresh waters needing protection or improvement in order to support fish life (Council Directive 78/659, OJ 1978, L 222/1);
- the quality required of shellfish waters (Council Directive 79/923, OJ 1979, L 281/47);
- the quality of water intended for human consumption (Council Directive 80/778, OJ 1980, L 229/11).

These directives contain provisions relating to the achievement of required water quality. They impose an obligation on member states to take necessary measures to that end, and (in most cases) require them to establish plans or programmes to ensure that the required quality standards are met within the time limit set in each directive.

Community legislation on the reduction of water pollution from specific industries is limited to the titanium dioxide industry (Council Directive 78/176, OJ 1978, L 54/19; Council Directive 82/883, OJ 1982, L 378/1). A directive to harmonise national programmes for the reduction of titanium dioxide pollution (Council Directive 89/428, OJ 1989, L 201/56) was annulled by the ECJ (see pp. 408–409) before it was re-adopted under a new legal basis (Council Directive 92/112/EEC, OJ 1992, L 409/11).

Community legislation on the protection of water, which deals specifically with the source of pollution, in fact comprises a directive on the protection of waters against pollution caused by nitrates from agricultural sources (Council Directive 91/676, OJ 1991, L 375/1), and a directive concerning urban waste water treatment (Council Directive 91/271, OJ 1991, L 135/40).

The Nitrates Directive requires member states to identify nitrate-polluted waters and to take a range of measures to limit agricultural nitrate use in the catchments. The Urban Waste Water Directive requires member states to meet certain standards for urban waste water collecting systems and for the treatment of urban waste water. These obligations are to be met in several stages, depending on various factors such as the sensitivity of receiving waters, and the population equivalent of the urban area under consideration.

The Community is party to several international conventions on the protection of the marine environment and the Rhine. It should be noted that in this capacity the Community has mixed competence along with the member states.

In 1996 the Commission presented a communication to the Council and Parliament in which it discussed the possibility of a major new piece of Community water legislation – a Water Resources Framework Directive – which is a more comprehensive and integrated approach to water management (Com (96) 59 final, 21 February 1996).

Air

Community efforts to protect air quality through legislation can be categorised into legislation on pollution caused by vehicles, product-related legislation,

legislation concerning ambient air quality, and legislation relating to industrial activities.

Community legislation on pollution caused by vehicles comprises the following:

- a directive on the approximation of laws relating to measures to be taken against air pollution by gases from positive-ignition engines of motor vehicles (Council Directive 70/220, OJ 1970, L 220/1; last amended by Council Directive 94/12, OJ 1994, L 100/42);
- a directive relating to pollution caused by diesel engines used in vehicles (Council Directive 88/77, OJ 1988, L 36/33; Council Directive 91/542, OJ 1991, L 295/1).

Product-related legislation, also concerned with transport, is comprised of these directives:

- a directive to approximate laws concerning the lead content of petrol (Council Directive 85/210, OJ 1985, L 96/25; Council Directive 87/416, OJ 1987, L 225/1);
- a directive on the sulphur content of certain liquid fuels (Council Directive 75/716, OJ 1975, L 307/22; Council Directive 87/219, OJ 1987, L 91/19; Council Directive 93/12/EEC, OJ 1993, OJ 74/81);
- a directive to control volatile organic compound (VOC) emissions from petrol storage and distrib ution (Council Directive 94/63/EC, OJ 1994, L 365/24).

The directive on the lead content of petrol provided that unleaded petrol had to become available throughout the Community from October 1989. It also contains provisions enabling the member states to prohibit regular leaded petrol on their territory.

Community legislation on air pollution caused by industrial activities is comprised of the following directives:

- reduction of air pollution from industrial plants (Council Directive 84/360, OJ 1984, L 188/20);
- limitation of emissions of certain pollutants from large combustion plants (Council Directive 88/609, OJ 1988, L 336/1, amended by Council Directive 94/66/EC, OJ 1994, L 337/83);
- prevention of air pollution from new municipal waste incineration plants (Council Directive 89/369, OJ 1989, L 163/32) and existing waste incineration plants (Council Directive 89/429, OJ 1989, L 203/50);
- incineration of hazardous waste (Council Directive 94/67/EC, OJ 1994, L 365/34).

For the first two directives, the member states are required to establish programmes for existing plants to reduce air pollution as specified in the annexes to the directives. New industrial plants must use the best available technology not involving unreasonable cost.

Directives establishing ambient air quality limit values in the territories of the member states cover the following air pollutants:

- sulphur dioxide and suspended particulates (Council Directive 80/779, OJ 1980, L 229/30; last amended by Council Directive 89/427, OJ 1989, L 201/53);
- lead (Council Directive 82/884, OJ 1982, L 378/15);
- nitrogen dioxide (Council Directive 85/203, OJ 1985, L 87/1; amended by Council Directive 85/580, OJ 1985, L 372/35).

These directives set limit values for each of the pollutants which have to be reached within a certain timeframe. The member states are required to elaborate their plans for reducing air pollution and for meeting the limit values. If a member state finds it difficult to reach the stipulated limit value of one or more of the above pollutants, it is required to inform the Commission.

There is also a Community directive dealing with the problem of air pollution by ozone (Council Directive 92/72/EEC, OJ 1992, L 297/1). It provides for ozone monitoring, and for informing and warning the population of high ozone levels.

It is to be noted in this context that the Community is a party to the Vienna Convention on the Protection of the Ozone Layer (1985), and to the Montreal Protocol on Chlorofluorocarbons (CFCs) and Halons (1987).

Noise

Community legislation on noise pollution is concerned with certain products, and in particular with transportation, construction machinery, and household appliances. Noise pollution caused by transportation is covered by directives concerning:

- motor vehicles (Council Directive 70/157, OJ 1970, L 42/16; amended by Council Directive 92/97/EEC, OJ 1992, L 371/1);
- motorcycles (Council Directive 78/1015, OJ 1978, L 349/21; last amended by Council Directive 89/235, OJ 1989, L 98/1);
- subsonic aircraft (Council Directive 80/51, OJ 1980, L 18/26; amended by Council Directive 83/206, OJ 1983, L 117/15; Council Directive 89/629, OJ 1989, L 363/27; and Council Directive 92/14/EEC, OJ 1992, L 76/21).

Community noise legislation on construction machinery and equipment comprises directives on:

- the determination of noise emissions of construction plant and equipment (Council Directive 79/113, OJ 1979, L 33/15, last amended by Commission Directive 85/405, OJ 1985, L 233/9);
- the permissible sound power level of compressors (Council Directive 84/533, OJ 1984, L 300/123; amended by Commission Directive 85/406, OJ 1985, L 200/11);
- the permissible sound power level of tower cranes (Council Directive 84/534, OJ 1984, L 300/130; amended by Council Directive 87/405, OJ 1987, L 220/60);
- the permissible sound power level of welding generators (Council Directive 84/535, OJ 1984, L 300/142; amended by Commission Directive 85/407, OJ 1985, L 233/16);
- the permissible sound power level of power generators (Council Directive 84/536, OJ 1984, L 300/149; amended by Commission Directive 85/408, OJ 1985, L 233/18);
- the permissible sound power level of powered hand-held concrete-breakers and picks (Council Directive 84/537, OJ 1984, L 300/156; amended by Commission Directive 85/409, OJ 1985, L 233/20);
- the limitation of noise emitted by hydraulic excavators, rope-operated excavators, bulldozers, loaders and excavator-loaders (Council Directive 86/662, OJ 1986, L 384/1; amended by Commission Directive 89/514, OJ 1989, L 253/35).

Household appliances are regulated by a Framework Directive (Council Directive 86/594, OJ 1986, L 344/24) and the Directive on Lawnmowers (Council Directive 84/538, OJ 1984, L 300/171; last amended by Council Directive 88/181, OJ 1988, L 81/71).

Chemicals

Legislation in the chemical sector is targeted at regulating products, laboratory practices and biotechnology. Council Directive 67/548 on the approximation of laws, regulations and administrative provisions relating to the classification, packaging and labelling of dangerous substances (OJ 1967, L 196/1) has been amended on numerous occasions. It was last amended by Commission Directive 94/69/EC adapting to technical progress for the twenty-first time (Directive 67/548, OJ 1991, L 338/23). With the adoption of the sixth amendment of the directive (Council Directive 79/831, OJ 1979, L 259/10), a notification system was introduced for new chemicals (i.e. substances not in circulation in the Community on 18 September 1981) before they could be put on the market.

Council Directive 88/379 on the approximation of laws, regulations and administrative provisions of the member states relating to the classification, packaging and labelling of dangerous preparations (OJ 1988, L 155/2; amended by Directive 93/18/EC, OJ 1993, L 104/46), covers dangerous preparations as distinct from dangerous substances.

Council Directive 76/769 on the approximation of laws, regulations and administrative provisions of the member states relating to restrictions on the marketing and use of certain dangerous substances and preparations (OJ 1976, L 262/201) was last amended by Directive 94/60/EC, OJ 1994, L 365/1.

Following the major disaster in Seveso in Italy, a directive was adopted to cover major accident hazards of certain industrial activities (Council Directive 82/501, OJ 1982, L 230/1; last amended by Council Directive 88/610, OJ 1988, L 336/14). The primary objective of the directive is to prevent major accidents. But, since it is not always possible to prevent them, the directive imposes an obligation on member states to take the necessary measures to ensure that, for the industrial plants which fall within the scope of the directive, emergency plans are established, a notification procedure is set up, and the public is adequately informed in an emergency.

Other Community legislation in the chemical sector covers the following:

- protection of animals used for experimental and research purposes (Council Directive 86/609, OJ 1986, L 358/1);
- good laboratory practices for tests on chemical substances (Council Directive 87/18, OJ 1987, L 85/40);
- inspection and verification of good laboratory practice (Council Directive 88/320, OJ 1988, L 145/35; amended by Council Directive 91/442, OJ 1991, L 238/25);
- use of genetically modified organisms (Council Directive 90/219, OJ 1990, L 117/1) and the deliberate release into the environment of such organisms (Council Directive 90/220, OJ 1990, L 117/15);
- asbestos (Council Directive 87/217, OJ 1987, L 85/40; amended by Council Directive 90/656, OJ 1990, L 353/19);
- pesticides (Council Directive 91/414, OJ 1991, L 230/1).

Nature conservation

The two most important Community instruments in this sector are a directive on the conservation of wild birds (Council Directive 79/409, OJ 1979, L 103/1; amended on several occasions), and a directive on the conservation of natural habitats and of wild fauna and flora (Council Directive 92/43/EEC, OJ 1992, L206/7).

The first directive establishes a comprehensive framework for wild bird conservation, involving the protection of the habitats of 182 bird species and sub-species, and the regulation of hunting and commercial exploitation. Special Protection Areas (SPAs) must be designated by the member states in order to ensure habitat protection.

The second directive establishes a similar framework for a range of other wild species (plants and animals), as well as a range of habitat types considered to need conservation action at Community level. The directive aims to create a Community network of protected sites known as NATURA 2000, made up of sites designated under the Directive as well as those designated under Directive 79/409/EEC.

The Community is a party to several international nature conservation conventions such as the Convention on the Conservation of European Wildlife and Natural Habitats, known as the Berne Convention (Council Decision 82/72/EEC, OJ 1982, L38/1).

While not itself a party to the Convention on International Trade in Endangered Species (CITES), the Community implements the Convention through a Community regulation (Council Regulation EEC/3626/82, OJ 1982, L 384/1; amended on several occasions).

Waste

Community legislation in the waste sector comprises Framework Directives on the one hand, and directives on specific types of waste on the other.

The Framework Directive on Waste is Council Directive 75/442 (OJ 1975, L 194/47), amended by Council Directive 91/156 (OJ 1991, L 78/32); and the Framework Directive on Hazardous Waste is Council Directive 78/319 (OJ 1978, L 84/43), replaced by Council Directive 91/689 (OJ 1991, L 377/20), which was subsequently amended by Council Directive 94/31/EC (OJ 1994, L 168/28).

The Framework Directive on Waste gives a definition of waste and imposes an obligation on member states to establish waste management plans for achieving the objective of the directive (i.e. to dispose of waste without endangering human health or damaging the environment). Since the declaration of the Council Resolution on Waste Policy (OJ 1990, C 122/2), prevention of waste has become more important, and its increased importance is reflected in the amended directive.

The Framework Directive on Hazardous Waste contains a list of 40 categories of hazardous wastes. The member states are required to take appropriate measures to regulate their collection, storage and treatment, and to elaborate their management plans. This directive is supplemented by a regulation, which replaces an earlier directive, on the supervision and control of waste shipments within, into and out of the Community (Council Regulation 259/93, OJ 1993, L 30/1).

Community directives on specific types of waste are concerned with:

- disposal of waste oils (Council Directive 75/439, OJ 1975, L 194/23; amended by Council Directive 87/101, OJ 1987, L 42/43);
- disposal of PCB and PCT (Council Directive 76/403, OJ 1976, L 108/41);
- waste from the titanium dioxide industry (see under 'Water' and Council Directive 82/883, OJ 1982, L 378/1);
- containers of liquids for human consumption (Council Directive 85/339, OJ 1985, L 176/18). This has been replaced by a wider directive on packaging waste (Council Directive 94/62/EC, OJ 1994, L 365/10);
- protection of the environment, and in particular of the soil, when sewage sludge is used in agriculture (Council Directive 86/278, OJ 1986, L 181/6);
- batteries and accumulators containing certain dangerous substances (Council Directive 91/157, OJ 1991, L 78/38);
- incineration of hazardous waste (see under 'Air' and Council Directive 94/67/EC, OJ 1994, L 365/34).

General measures

The Directive on Environmental Impact Assessment (Council Directive 85/337, OJ 1985, L 175/40), and the Directive on Access to Information on the Environment (Council Directive 90/313, OJ 1990, L 158/56), are the most important instruments in this field. The second directive requires member states to take appropriate measures to ensure access by the public to information on the environment held by public authorities. Access to Commission and Council documents relating to the environment is covered by a general code of conduct concerning public access to Commission and Council documents (see Commission Decision 94/90/ECSC, EC, Euratom, OJ 1994, L 46/58).

The Community has adopted a regulation to create a Community eco-labelling scheme (Council Regulation 880/92, OJ 1992, L 99/1). It provides for the award of an 'eco-label' to products meeting certain environmental criteria. The objective is to promote products with a reduced impact on the environment, and to give consumers better information about them.

Another Community regulation promotes voluntary participation by industrial companies in a Community Eco-management and Audit Scheme (Council Regulation (EEC) 1836/93, OJ 1993, L 168/1). The aim of this regulation is to promote continuing improvements in the environmental performance of industrial activities.

CONTROL OF THE IMPLEMENTATION OF EU LEGISLATION IN THE MEMBER STATES

The Commission (Article 169 of the Treaty)

The Commission, as guardian of the Treaty, is responsible for ensuring that Community legislation is properly implemented in the member states, and it does so through a process of monitoring and control. The aim is to ensure that directives are incorporated into national legislation correctly and on time, and that the provisions of the directives are properly applied in practice.

Formal incorporation of Community Directives into National Legislation

As already mentioned, each directive has a time limit within which necessary national measures are to be adopted and communicated to the Commission. If not, the Commission can take action under Article 169 of the Treaty. In practice this means that the Commission will start legal proceedings against those member states who fail to adopt, or communicate to the Commission, their national legislation necessary to give effect to a directive. The first step in such a procedure is to send a letter of Formal Notice, which contains the alleged infringements of Community legislation, to the member state concerned about the intent to proceed. The member state is given a set time, specified in the letter, by which it has to reply to the letter of Formal Notice. The case is closed if the Commission is satisfied with the member state's reply. If, on the other hand, the member state does not reply, or does not give a satisfactory reply, the Commission can proceed with the case and decide to send what is called a 'Reasoned Opinion'. If the member state does not reply to the letter of Formal Notice, the Reasoned Opinion can invoke, in addition to the alleged infringements set out in the letter of Formal Notice, Article 5 of the Treaty which requires member states to co-operate with the Commission in order to facilitate the performance of the Community's tasks. The Reasoned Opinion gives the member state a time limit within which to submit its observations.

After this stage of the procedure, the Commission can decide to proceed further with the case by submitting it to the ECJ.

Conformity of national law and practical application

The Commission examines the implementing measures, communicated by the member states, to determine whether they have given full and accurate effect to the directive in question. If the implementing measures are considered to be inadequate (i.e. they do not conform to the requirements of the directive), the Commission can start proceedings against the member state concerned under Article 169 of the Treaty. The brief case study given in Box 21.1 is illustrative.

The Commission also monitors the practical application of directives. For example, it tries to ensure that, where environmental directives require member states to adopt pollution reduction programmes, such programmes are actually put in place. It is relevant in this context to mention a complaints procedure which the Commission operates. Under this procedure, an individual or organisation can lodge a complaint with the Commission against a member state on the basis of evidence that that member state has not respected Community law.

The complaints procedure works as follows: The Commission is informed by letter, and/or by audiovisual material, about an alleged infringement of Community law by a member state. The alleged infringement is then examined and registered as a formal complaint, and the complainant is notified about it. The Commission then investigates the alleged infringement and seeks more information from the complainant if necessary. It is a strict policy of the Commission to keep the identity of the complainant confidential. The member state concerned is informed about both facts and the legal basis of the alleged infringement, and is asked to comment on them. When the necessary information is received, and the facts and the alleged infringement of Community law have been examined, the Commission either decides to close the case if it is deemed to be weak and/or the evidence insufficient, or to start the formal procedure under Article 169 of the Treaty. The complainant is kept informed on developments throughout the case.

BOX 21.1 *COMMISSION* v. *BELGIUM*, CASE C-133/94, JUDGEMENT OF THE ECJ OF 2 MAY 1996

In 1994, using Article 169 of the EC Treaty, the European Commission applied to the ECJ for a ruling alleging that Belgium had failed to incorporate Directive 85/337/EEC (environmental impact assessment) into its national law completely and correctly. In May 1996, the ECJ gave its judgement, agreeing with the Commission that in several respects the Belgian legislation was less complete than the Directive required. For example, whereas the Directive required all 'integrated chemical installations' to be subject to an environmental impact assessment, it was restricted to integrated chemical installations with a processing capacity of 100,000 tonnes in the Flemish region of Belgium. The ECJ ruled that this was too narrow. This case typically illustrates the role of the Commission in checking the conformity of national legislation and in taking legal action against a member state when necessary.

Discussions between the Commission and the member state concerned can take place during different stages of the procedure, aimed at finding a remedy to the alleged breach of Community legislation.

Apart from formal complaints, the Commission also investigates alleged infringements that are the subject of petitions to the EP, or the subject of written and oral questions to the EP.

The European Court of Justice

The last stage in an infringement procedure is the Commission's decision to bring the matter before the Court of Justice. It is then up to the ECJ to consider the matter and decide whether the member state concerned has failed to fulfil its obligations. The observations of the ECJ and the final decision are written in the judgement of the ECJ.

Until a case has been brought before the ECJ, it is in a confidential stage and only the Commission can make a formal decision on whether or not to disclose certain facts of the case to the public. Hearings before the ECJ are open to the public; and the opinion of the Advocate-General, and the judgement, are published *in extenso* in the Reports of Cases Before the Court (ECR). The same information is also to be found in the *Celex* database.

In the first judgement the ECJ cannot impose a sanction on a member state which it finds to be in breach of its Treaty obligations. However, if the member state concerned fails to comply with the first judgement, the Commission can initiate a second legal procedure under Article 171 of the Treaty. If the member state concerned is judged to have failed to comply with the first judgement, then, by virtue of power conferred upon it by the TEU, the ECJ can impose a lump sum fine on the member state.

The rights of individuals in the member states

Direct effect of EC directives

The doctrine of 'Direct Effect' means that an individual, or a private body, can invoke the provisions of a directive against a governmental or other public body, even when that directive is not implemented in national legislation. The aim is to offer individuals the opportunity to invoke the provisions of a directive where a member state has failed to implement that directive. It is based on the following ruling by the ECJ:

> It would be incompatible with the binding effect attributed to a directive by Article 189 to exclude, in principle, the possibility that the obligation which it imposes may be invoked by those concerned. In particular where the Community authorities have by directives imposed on the Member States the obligation to pursue a particular course of conduct, the useful effect of such an act would be weakened if individuals were prevented from relying upon it before their national courts and if the latter were prevented from taking it into consideration as an element of Community law.
>
> (Case C 41/74, ECR 1974: p. 1337)

A directive has no direct effect *per se*, but it may contain certain provisions which have direct effect whilst other provisions of the same directive do not. The conditions for direct effect can be summarised as follows:

- the obligation must be clear and unconditional;
- it can only be invoked by an individual or a private body against a governmental or public body, but never against an individual or a private body. The ECJ has ruled that the doctrine can only be invoked against a party which is 'an emanation of the state'. The concept of 'an emanation of the state' does not only mean governmental departments, but includes all public bodies at whatever level and can include in certain circumstances public corporations which received special powers from the state for providing public services, and the powers are such that they go beyond the normal rules which apply in relations between private parties (Case C 188/89, 12 July 1990).

Article 177 of the EC Treaty – the preliminary reference

Article 177 of the Treaty makes it possible for a national court to refer a legal matter to the ECJ for a ruling, when it deems it necessary so to do in order to enable it to give a judgement on the interpretation of the Treaty, the validity and interpretation of acts of the institutions of the Community, and on

BOX 21.2 *LAPPEL BANK*, CASE C-44/95, JUDGEMENT OF ECJ OF 11 JULY 1996

In December 1993, the British government designated the Medway Estuary and Marshes as an SPA under Directive 79/409/EEC, but in doing so it excluded the *Lappel Bank* which is a small but rich area for birds. This exclusion was made in order to facilitate a port expansion scheme. An expert non-governmental organisation challenged the exclusion in the British Courts, arguing that the Directive required that boundaries of SPAs be fixed according to strict ornithological criteria. The highest British Court, the House of Lords, made an Article 177 reference to the ECJ (see pp. 416–417 of this chapter), which ruled in favour of the non-governmental organisation. Apart from being a good illustration of Article 177, this case shows the practical importance of Community nature conservation legislation for wetlands and other endangered habitats.

the interpretation of the statutes of bodies established by an act of the Council where those statutes so provide. The possibility of such a referral becomes an obligation when the matter, referred to the ECJ, is raised in a court which decides in the last instance.

The ECJ is the highest court to give an authoritative interpretation of the meaning of Community legislation, following a preliminary ruling given by a national court on a matter raised by an individual. There have been a number of such preliminary rulings of much importance on environmental legislation. The case study given in Box 21.2 is instructive.

CONCLUSIONS

EC environmental legislation, which can take the form of regulations, directives or decisions, is currently based on Article 130s or 100a of the Treaty. These articles are reinforced by the Environmental Action Programmes. The Paris Summit was the first major milestone in Community policy-making on the environment. The Single European Act introduced the current legal basis for environmental measures into the Treaty, and the Treaty on European Union has further reinforced it.

The Community legislative process is complex, with different procedures depending on the content of the proposed legislation. In this context and in particular, the role of the European Parliament has been strengthened by both the Single European Act and the Treaty on European Union.

The control of implementation of Community legislation plays an important part in ensuring the effectiveness of legislation. The complaints procedure helps the Commission to exercise its role as guardian of the Treaty. It is impossible, however, for the Commission to be informed on all the problems which arise in the application of Community measures. Individuals can also invoke Community measures in national courts, subject to certain conditions.

NOTES

1 The research on which this chapter was based was concluded in mid-1996. Any views expressed are not necessarily those of the Commission.
2 Articles 100a, 130r, 130s and 100t are described in the Appendix (see pp. 418–419).

REFERENCES

CEC (1973) 'Declaration of the Council of the European Communities and of the Representatives of the Governments of the Member States', Council Meeting of 22 November 1973 on the Programme of Action of the European Communities on the Environment, OJ 1973, Communication 112/1.

CEC (1977) 'Resolution of the Council of the European Communities and of the Representatives of the Governments of the Member States', Council Meeting of 17 May 1977 on the continuation and implementation of a European Community Policy and Action Programme on the Environment, OJ 1977, Communication 139/1.

CEC (1983) 'Resolution of the Council of the European Communities and of the Representatives of the Governments of the Member States', Council Meeting of 7 February 1983 on the continuation and implementation of a European Community Policy and Action Programme on the Environment (1982 to 1986), OJ 1983, Communication 46/1.

CEC (1987) 'Resolution of the Council of the European Communities and of the Representatives of the Governments of the Member States', Council Meeting of 19 October 1987 on the continuation and implementation of a European Community Policy and Action Programme on the Environment (1987–1992), OJ 1987, Communication 328/1.

CEC (1990) Council Resolution of 7 May 1990 on Waste Policy, OJ 1990, Communication 122/2.

CEC (1992) *Toward Sustainability: A European Community Programme of Policy and Action in Relation to the Environment and Sustainable Development*, COM (92) 23 final, vol. II.

Krämer, L. (1990a) *EEC Treaty and Environmental Protection*, London: Sweet & Maxwell.

Krämer, L. (1990b) 'Droit communautaire de l'environnement, Juris Classeurs', *Fascicule 1900*, 11.

Krämer, L. and Kromarek, P. (1991) 'Droit communautaire de l'environnement', *Revue Juridique de l'Environnement* 4: 473–500.

SUGGESTED READING

Annual reports to the European Parliament on the control and application of Community law, Brussels and Luxembourg: The European Commission.

Brinkhorst, L.J. (1993) 'Subsidiarity and European Community Environment Policy, a Panacea or a Pandora's Box', *European Environmental Law Review* 8: 18.

Charpentier, J. (1988) *La Protection de l'environnement par les Communautés européennes*, Paris: Editions A. Pedone.

European Commission (1995) *The Institutions of the European Union*, Brussels and Luxembourg: The European Commission.

European Commission (1996) *Building the European Union*, Brussels and Luxembourg: The European Commission.

General Report on the Activities of the European Communities, Annual Publication of the European Commission, Brussels and Luxembourg: The European Commission.

Hession, M. and MacRory, R. (1994) 'Maastricht and the Environmental Policy of the Community: Legal Issues of a New Environmental Policy', in D. O'Keeffe and P.M. Twomey (eds) *Legal Issues of the Maastricht Treaty*, London: Chancery Law Publishing, pp.151–167.

Kiss, A. and Shelton, D. (1993) *Manual of European Environmental Law*, Cambridge: Grotius publications.

Krämer, L. (1991) 'The Implementation of Community Environmental Directives Within Member States: Some Implications of the Direct Effect Doctrine', *Journal of Environmental Law* 3 (1): 39–56.

Krämer, L. (1993) *European Environmental Law Casebook*, London: Sweet & Maxwell.

Report on Bathing Water Quality, Annual Publication of the European Commission, Brussels and Luxembourg: The European Commission.

Verhoeve, B., Bennett, G. and Wilkinson, D. (1992) *Maastricht and the Environment*, Arnhem: Institute for European Environmental Policy.

Winter, G. (1996) *European Environmental Law: A Comparative Perspective*, Aldershot: Dartmouth Publishing Co.

APPENDIX

EC Treaty, Title XVI, Environment

Article 130r

1 Community policy on the environment shall contribute to the pursuit of the following objectives:
- preserving, protecting and improving the quality of the environment;
- protecting human health;
- prudent and rational utilisation of natural resources;
- promoting measures at the international level to deal with regional or world-wide environmental problems.

2 Community policy on the environment shall aim at a high level of protection taking into account the diversity of situations in the various regions of the Community. It shall be based on the precautionary principle and on the principles that preventive action should be taken, that environmental damage should as a priority be rectified at source and that the polluter should pay. Environmental protection requirements must be integrated into the definition and implementation of other Community policies.

3 In preparing its policy on the environment, the Community shall take account of:
- available scientific and technical data;
- environmental conditions in the various regions of the Community;
- the potential benefits of action and the costs of action or lack of action;
- the economic and social development of the Community as a whole and the balanced development of its regions.

4 Within their respective spheres of competence, the Community and the member states shall co-operate with third countries and with competent international organisations. The arrangements for Community co-operation may be the subject of agreements between the Community and the third parties concerned, which shall be negotiated and concluded in accordance with Article 228.

The previous subparagraph shall be without prejudice to Member States' competence to negotiate in international bodies and to conclude international agreements.

Article 130s

1 The Council, acting in accordance with the procedure referred to in Article 189c and after consulting the Economic and Social Committee, shall decide what action is to be taken by the Community in order to achieve the objectives referred to in Article 130r.

2 By way of derogation from the decision-making procedure provided for in paragraph 1 and without prejudice

to Article 100a, the Council, acting unanimously on a proposal from the Commission and after consulting the European Parliament and the Economic and Social Committee, shall adopt:

- provisions primarily of a fiscal nature;
- measures concerning town and country planning, land use with the exception of waste management and measures of a general nature, and management of water resources;
- measures significantly affecting a member state's choice between different energy sources and the general structure of its energy supply.

The Council may, under the conditions laid down in the preceding subparagraph, define those matters referred to in this paragraph on which decisions are to be taken by a qualified majority.

3 In other areas, general action programmes setting out priority objectives to be attained shall be adopted by the Council, acting in accordance with the procedure referred to in Article 189b and after consulting the Economic and Social Committee.

The Council, acting under the terms of paragraph 1 or paragraph 2 according to the case, shall adopt the measures necessary for the implementation of these programmes.

4 Without prejudice to certain measures of a Community nature, the member states shall finance and implement the environment policy.

5 Without prejudice to the principle that the polluter should pay, if a measure based on the provisions of paragraph 1 involves costs deemed disproportionate for the public authorities of a member state, the Council shall, in the act adopting that measure, lay down appropriate provisions in the form of:

- temporary derogations and/or
- financial support from the Cohesion Fund to be set up no later than 31 December 1993 pursuant to Article 130d.

The Council, acting unanimously on a proposal from the Commission and after consulting the European Parliament and the Economic and Social Committee, shall decide what action is to be taken by the Community.

The Council shall, under the conditions laid down in the preceding subparagraph, define those matters on which decisions are to be taken by a qualified majority.

Article 130t

The protective measures adopted in common pursuant to Article 130s shall not prevent any member state from maintaining or introducing more stringent protective measures. Such measures must be compatible with this Treaty. They shall be notified to the Commission.

EC Treaty, Article 100a

1 By way of derogation from Article 100 and save where otherwise provided in this Treaty, the following provisions shall apply for the achievement of the objectives set out in Article 7a. The Council shall, acting in accordance with the procedure referred to in Article 189b and after consulting the Economic and Social Committee, adopt the measures for the approximation of the provisions laid down by law, regulation or administrative action in member states which have as their object the establishment and functioning of the internal market.

2 Paragraph 1 shall not apply to fiscal provisions, to those relating to the free movement of persons nor to those relating to the rights and interests of employed persons.

3 The Commission, in its proposals envisaged in paragraph 1 concerning health, safety, environmental protection and consumer protection, will take as a base a high level of protection.

4 If, after the adoption of a harmonisation measure by the Council decided by a qualified majority, a member state deems it necessary to apply national provisions on grounds of major needs referred to in Article 36, or relating to the protection of the environment or the working environment, it shall notify the Commission of these provisions.

The Commission shall confirm the provisions involved after having verified that they are not a means of arbitrary discrimination or a disguised restriction on trade between member states.

By way of derogation from the procedure laid down in Articles 169 and 170, the Commission or any member state may bring the matter directly before the Court of Justice if it considers that another member state is making improper use of the powers provided for in this Article.

5 The harmonisation measures referred to above shall, in appropriate cases, include a safeguard clause authorising the member states to take, for one or more of the non-economic reasons referred to in Article 36, provisional measures subject to a Community control procedure.

SELF-ASSESSMENT QUESTIONS

1 Specific provisions for Community action in the field of environmental protection are part of the Treaty. Indicate which of the following statements applies:
  (a) A specific title on environment was introduced in the Treaty of Rome because it was one of the objectives of the Treaty.
  (b) The Single European Act introduced Articles 130r to 130t in the Treaty.
  (c) The member states agreed at the Paris Summit that Community action in the field of environment was better than intergovernmental agreements, and the Treaty was immediately amended accordingly.

2 The legislative process of Community acts involves several Community institutions. Indicate which of the following statements is correct:
  (a) The European Parliament adopts the legislation.
  (b) The Commission prepares the proposals of legislation.
  (c) The legal basis of Community measures is included in the proposal of the Commission and cannot be changed by another Community institution.

3 A directive is a Community measure which is directed to the member states. Indicate which of the following characteristics apply to a directive:
  (a) It must be published in the OJ.
  (b) It has no binding force.
  (c) The Commission has to control whether directives are implemented in the member states within the time provided for in each directive.

4 Indicate the correct alternative(s): Community environmental legislation is mainly taken in the form of
  (a) Environmental Action Programmes;
  (b) Directives;
  (c) Recommendations.

5 As the guardian of the Treaty, control of implementation of Community environmental legislation is one of the tasks of the Commission. Which of the following statements apply in this context?
  (a) An individual has no right to invoke Community environmental legislation before a national court.
  (b) The European Court of Justice is the highest court to give an authoritative interpretation of the meaning of Community legislation, and a national court, deciding in the last instance, has the obligation to refer a question of interpretation to the ECJ when such a question is raised.
  (c) The Commission has a special team of inspectors which operates within different member states to control the implementation and application of Community environmental measures. This team reports to the Commission, and infringement procedures against member states are initiated on this basis.

6 The legal basis of Community environmental legislation is either Article 130s or 100a of the Treaty. Which of the following statements are correct in this context?
  (a) The role of the European Parliament varies according to the legal basis.
  (b) Whether measures should be based on Article 130s or 100a is very clearly defined in the Treaty and does not cause confusion.
  (c) The European Court of Justice has ruled that directives relating to waste must be based on Article 130s.

PART II.4

INSTRUMENTS FOR ENVIRONMENTAL COMMUNICATION AND EDUCATION

22

ENVIRONMENTAL EDUCATION AND COMMUNICATION

Complementary tools for environmental management

Walter Leal Filho

SUMMARY

Environmental education, awareness and training are considered to be important tools in any attempt at achieving sustainable development. A fourth tool may be added to this list; namely, 'communication'. In both developing and developed countries the work of environmental managers can be greatly facilitated, and their capacity enhanced, through the application of appropriate skills in environmental education and communication. Indeed, such skills are indispensable in achieving the desired goals of any environmental management programme, regardless of geo-political considerations.

Against this background, this chapter focuses on the contribution of environmental education and communication to environmental management. To begin with, some background information is given on the meaning of the terms 'environmental education' and 'environmental communication'; an analysis is then made of their relationship with environmental management. Case studies on biodiversity and coastal areas are given to illustrate some of the ways in which the concepts of environmental education and communication may be integrated to provide an essential tool of environmental management.

ACADEMIC OBJECTIVES

Education and communication are essential tools in environmental management, and both their relevance and importance are being increasingly acknowledged. This state of affairs suggests that further information on the compatibility of these terms – education and communication – needs to be elaborated in order to ensure that they are fully integrated in the context of environmental management programmes. On the basis of this, the aims of this chapter are: (1) to provide background information on environmental education and communication; (2) to inform readers of the relevance of education and communication to environmental management; and (3) to show, by means of examples, how education and communication could be integrated with environmental management. On completion of this chapter readers should be able to:

- acknowledge the importance of environmental education and communication in environmental management;
- discuss the approaches and methods used in integrating education and communication with environmental management;
- identify ways in which education and communication may be included as part of environmental management programmes.

INTRODUCTION TO ENVIRONMENTAL EDUCATION AND COMMUNICATION

The search for alternative ways of promoting sustainable economic growth began with the publication of the book entitled *Silent Spring* (Carson, 1962), and the process has continued with the publication of *The Limits to Growth* (Meadows, 1972) and the report entitled *Our Common Future* (WCED, 1987). Driven by the need to promote sound environmental training, the debate has been raised to the highest international level, culminating in a recommendation enshrined in Agenda-21 (UN, 1992) stating that the 'principles of environmental education and sustainable development should be part of the training of specialists in all areas'.

Although the need for environmental education has been recognised since the 1960s (Chiappo, 1978; McCarson, 1978), its relevance as a tool for sustainable living (Leal Filho, 1993), and the need for incorporating environmental education in the framework of training programmes, has only recently been widely acknowledged by international organisations such as the United Nations Environment Programme (UNEP), the United Nations Development Programme (UNDP), UNESCO and the European Union (EU). Nowadays environmental education is seen as an important tool in fostering a sense of environmental awareness amongst professionals in all areas and at all levels (e.g. workers, managers and professionals such as doctors, engineers and lawyers). As acknowledged by Atchia (1995), perhaps the greatest need for environmental education is in the training of in-service professionals as well as of those training to enter the profession of environmental management. This is so because, in order for them to be able to make sound decisions on environmental issues and problems, environmental managers and decision-makers must have a comprehensive and holistic view of the 'environment'.

The most widely accepted definition of environmental education has been proposed by the International Union for the Conservation of Nature (IUCN). In its 'Workshop on Environmental Education', held in Carson City (Nevada, USA) in 1970, environmental education was defined as follows:

> Environmental education is the process of recognising values and clarifying concepts in order to develop skills and attitudes necessary to understand and appreciate the inter-relatedness among man, his culture and his bio-physical surroundings.
>
> (IUCN, 1970: 7)

More recently, Leal Filho (1996a: 6) has defined environmental education as 'a process directed towards increasing the general level of public concern about environmental dynamics as well as fostering awareness of the need for public participation in order to promote environmental conservation'. In addition to the discussions on its definition, the description of the principles and practices of environmental education has been the subject of several works (e.g. Disinger, 1983; Hale, 1993; Leal Filho and Hale, 1992, 1994). According to Leal Filho (1996a), the prime objective of environmental education is to help people to interpret the environment and to stimulate them to:

- see their surroundings with an appreciative but critical eye;
- acquire a range of relevant skills;
- understand something of the processes of the physical world, and to gain a basic knowledge of the ecological principles and relationships between organisms and the environment;
- understand something of the economic, technological, planning and political processes which affect their use of the environment;
- develop some insight into other people's environment, life-styles and problems;
- develop concern both for their own environment and that of others;
- acquire a basis on which to make informed decisions about environmental issues that affect them and society at large, and to develop the willingness to do so.

The following are some traditional areas of concern for environmental education:

- environmental education in schools;
- teacher training and environmental education;
- environmental education at the university level;
- environmental education and communities;
- environmental education projects;
- networking and environmental education;
- adult environmental education.

As can be seen, the use of environmental education in the context of environmental management programmes is not yet a well-established area, although this situation is gradually changing. Environmental education is, however, well-established as both a requirement (WCED, 1987) and a tool for sustainable development (Schneider, 1993). Indeed, it has been consistently mentioned in the international literature (e.g. Bain, 1994; Hardoy, 1992; McCormick, 1995) as an item that needs to be taken into account in promoting sustainable development. In this context it is clear that, as modern environmental education continues to evolve, it will increasingly pervade applied disciplines such as those of environmental management and environmental legislation which are certain to be developed further in the late 1990s and beyond.

One of the means through which the noble goal of promoting people's environmental education can be reached is through the wide dissemination of environmental information; or, in other words, through *environmental communication*, which may be defined as a modality of communication concerned with environmental matters and which takes the particular characteristics of environmental issues into account, including their complexity, with the ultimate aim of providing information which is (a) clearly understood, (b) unambiguous and (c) scientifically sound (i.e. does not contain technical inaccuracies).

Whereas the context here is the 'environment' (other contexts could be, for example, politics or economics), 'environmental communication' refers to the *modality* which is used for the transmission of information. Considering the modality *per se*, this line of thinking suggests that for both communicating environmental issues and for disseminating environmental information, increasingly greater use should be made of the effective vehicles of the mass media such as radio, television and the press (Alaimo and Rooney, 1980) in various geographical regions (e.g. de Bruin, 1991; Leal Filho, 1994; Leal Filho and Padgham, 1994).

Compared with environmental education, whose process of evolution goes back to the 1970s, or even earlier if various other separate developments are taken into account, environmental communication is a fairly new area and as such it has not received the same level of academic attention as environmental education. Some work has, however, been done by authors such as Hansen (1993) who looked at the emphasis given to environmental issues by the mass media, and by de Haan (1995) who focused on environmental awareness and the mass media. Others, such as Axelrod and Lehman (1993), have examined the elements which motivate people to take action on environmental issues and problems. Ray and Guzzo (1993) have analysed the consequences of exaggerating environmental threats by the media; their work has thrown some light onto the various matters concerned with the relationship between the media and the environment – matters that were discussed at a specialist workshop on environmental journalism held in 1995 (Leal Filho and Bynoe, 1995).

Part of the problem in dealing with environmental matters through the media stems from the fact that, even though there are many people with *educational* backgrounds who are interested in environmental education, the management and mass dissemination of environmental information requires *communication* skills as well as knowledge of the *environment*. And it is difficult to find people who are knowledgeable in environmental matters *and* competent in media communication skills, especially in developing countries.

In addition, two specific differences, which can also be problematic, are to be noted in this regard – one relating to geography and the other concerned with style and approach:

1 *Geographical differences*: empirical studies made by Petri (1988), Friedman and Friedman (1989) and Kepplinger (1989), and the more recent surveys by El-Hosseiny (1994) and Bandeira (1996), show that the emphasis given to the coverage of environmental matters in developed and developing countries differs significantly. While in the developed countries environmental issues are systematically approached, giving particular attention to the precision of the contents of what is reported, in developing countries environmental topics are often regarded as issues of secondary importance which, more often than not, are only sporadically dealt with.
2 *Differences in approach*: in both developed and developing countries the mass media sometimes tend to overemphasise the effects or impacts of environmental problems. A classic example of this was the media reporting of the eruption of Mount Pinatubo in the Philippines in 1991. It was described in some reports as 'likely to cause an overheating of the planet', which in fact did not happen. According to Ray and Guzzo (1993), such overemphasis or scare-mongering by the media can be counterproductive, because it creates a false anticipation among the public and generally erodes the credibility of the reporting media in such matters.

Also, the environmental information passed on to the public separately through radio, television and the press is often not evenly dealt with and lacks objectivity (Alvaredo, 1985; Leal Filho, 1990; El-Hosseiny, 1994). Despite their many advantages, both television (which relies on the attractiveness of moving images) and radio (which is based on the transmission of oral reports) lack the long-term perspective one can often find in the printed medium. As observed by Leal Filho (1990, 1991), the press has a number of special characteristics, listed below, which reinforce its potential role in raising environmental awareness and illustrate its potential usefulness in environmental conservation.

1 *Provision of permanent registers*: under normal circumstances, environmental articles published tend to be available in perpetuity. If properly managed, newspaper articles can be kept for re-reading, consultation and further reference for many years. This represents the most cost-effective way of retrieving environmental information, not requiring tapes or electronic equipment for its use.
2 *Comprehensive approach*: printed articles facilitate the establishment of links between the topic under consideration and other issues; it is thus possible to widen the scope of a given problem or issue as much as is deemed necessary or desirable.
3 *Plurality of sources*: printed papers and articles, containing as they do information emanating from a variety of sources (e.g. scientists, economists, governmental authorities), can be gathered for subsequent comparison and cross-checking for consistency.
4 *Cost-effectiveness*: for their price, newspapers and news magazines represent the cheapest information source available.

The concept and definition of 'environmental literacy' is also important in the context of environmental education and communication. It may be defined as:

> a process of increased awareness on environmental issues both on a local and regional scale, complemented by a knowledge of the causes and consequences of environmental problems, including particularities relating to their impact on an individual's life-style.

This definition includes awareness, for example, of pollution within the city or region where an individual lives. An individual who is not aware of the main environmental issues or problems where he or she lives, or of the impact of these problems on his or her livelihood, may be defined as 'environmentally illiterate'. It is to be noted, however, that, thanks to advances in environmental communication, increasingly fewer people now belong to this category.

But the management of environmental literacy is not a simple process. In order properly to inform people about facts and figures on the environment, information needs to be disseminated systematically and in a way in which it is easily understood by those for whom it is intended. The following approaches are usually adopted in communicating environmental issues:

- formal education (schools, kindergartens);
- informal education (outdoor activities, fieldwork);
- mass media (television, radio, newspapers);
- other sources (talks among friends, religious events, etc.).

The ultimate aim of environmental education and communication is to provide the public with information on the environment with two objectives: first, to make them environmentally literate; second, to make them commit themselves to the cause of environmental protection and conservation.

ENVIRONMENTAL EDUCATION AND COMMUNICATION IN THE FRAMING OF DECISIONS INVOLVING THE ENVIRONMENT

Before we begin to examine the linkages between environmental education, communication and awareness, and how they relate to environmental management, it is important to conceptualise their connections and relationships. Although there is no consensus among experts on the stage or level at which the impact of information on environmental awareness is actually felt, all these elements are nevertheless closely related, and ought therefore to be conceptualised as a whole. Figure 22.1 illustrates the relationships between these elements by using a 'step' approach. As can be seen, even though a hierarchical relationship can be built in the 'bottom-up' fashion, the complexity inherent in environmental matters may mean that some of the 'steps' may be 'jumped' over. One could also invert

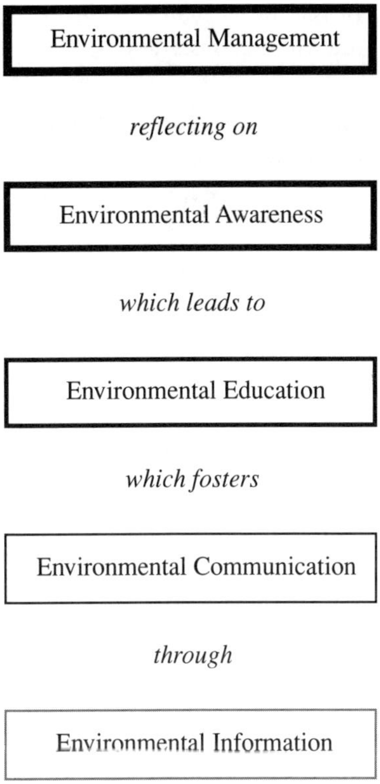

Figure 22.1 Relationship between environmental education, communication, awareness and management

the picture and start with 'environmental management' in what would then be in the 'top-down' fashion. Thus the choice of what would be the best approach under a given set of conditions would depend very much on the context of those conditions.

However, it is widely acknowledged that, in order to derive maximum benefits regardless of the focus in question, decisions concerning the environment need to be made on a sound technical basis, although this is not the only requirement. In order fully to enlist public support, decisions also need to be communicated effectively. Annals of environmental management show that there have often been cases in which lack of communication and co-operation has led to misunderstanding, failed communication and conflicts over values. In such cases the systematic use of educational and communication skills, especially among the main actors, can prove very useful. In general, the main actors, who are often also the main target groups, in education and communication in the context of environmental management are:

- governmental agencies;
- the private sector,
- non-governmental agencies.

The above grouping is based on the recognition that the first group is responsible for making laws for controlling and regulating production processes, the second for the production of goods and services, and the third for the monitoring (as citizens or on behalf of citizens) of the ways in which society behaves.

In addition to the above, there are specific sub-groups which need to be taken into account, especially for communicating matters relating to environmental management to industry and enterprises. These sub-groups are government officials in charge of industrial activities; the producers themselves in various industrial sectors; the consumers (or those who acquire the goods or services through their purchasing power); and the intermediaries such as handlers, transporters and storers. The intermediaries may have a potential impact on the environment, depending on the extent of their involvement in the supply, distribution, handling and disposal of products.

In the past there has been insufficient opportunity for the public to participate in environmental issues, and their involvement in environmental matters, especially in decision-making, has been limited too. Perhaps because of this and other reasons many people did not fully realise until recently the links between economics, politics and the environment (Jacobs, 1991). This state of affairs is gradually changing, however, as more and more people become aware of the need to be actively involved in decision-making processes. There is also a growing public interest in influencing the outcome of initiatives which may lead to structural changes in the physical environment.

The co-ordination of inputs from both the public and the agencies in charge of environmentally related developments, or from other initiatives which could alter the environmental status of a certain site, is the only reliable way of ensuring that a given project or scheme is environmentally sound, morally acceptable and ethically correct. To this end both education and communication have a key role to play, because they facilitate an understanding of the various strategic elements in the short, medium and long term.

In the short term they enable people to access information on the aims of a proposed project and its intended purpose; they facilitate an understanding of the environmental impacts of a proposed project, thus making possible a well-informed discussion on the merits of the project and an assessment of the expected gains against possible losses; and they enable a consideration of the specific characteristics of a proposed project, including the problems it may create and the final decision on its viability. In the medium term they enable compilation of the impacts of the project in terms of both landscape changes and the production of pollutants during its implementation, and in the long term they provide an overview of the possible improvements brought about by the project once it is completed.

In many countries education and communication are also an integral part of environmental impact assessment (EIA) studies that are now mandatory for projects which may have an impact on the environment. Such EIAs thus take into account not only the physical aspects of a project or scheme, but also people's opinions and attitudes to it. Clearly, projects that are strongly resisted by the public at large, or when there is a great deal of resistance by the local people, are less likely to be implemented than those enjoying wider public support. It is also clear that skilled public relations management would be needed when there is a conflict of interest or public opposition (Brown, 1995).

There are situations, however, in which public resistance can be unfounded – for example, when there is sound scientific evidence to show that the actual impacts of a project would be far less than rumoured. Education and communication can be a powerful tool in such situations for dispelling the public's unfounded fear and suspicion, and thus overcoming their resistance.

Box 22.1 gives an overview of the application of environmental education and communication to biodiversity conservation. As a typical application of education and communication, an excerpt, taken from the briefing notes on biodiversity, was circulated among the participants of the Second Conference of the Parties of the Convention (2nd COP), held in Jakarta, Indonesia, in November 1995. It helped to convince those who had reservations.

ENVIRONMENTAL EDUCATION AND COMMUNICATION IN ENVIRONMENTAL MANAGEMENT: EXAMPLES OF APPROACHES

One of the main advantages of developing effective education and communication strategies in the context of environmental management is that, through this approach, goals such as the training of manpower are more easily achieved. The second case study, given in Box 22.2, refers to an example from the Caribbean region (UNEP, 1992) where, in an attempt to promote natural resources management, links were established among the universities of the region.

The example in Box 22.2 shows that, by pooling the efforts of institutions working under different settings but sharing the same interests, it is possible to include environmental management as part of their training programme.

In Europe also there are many examples showing that education and communication are becoming part of the routine of industrial activities. As observed by Elkington (1994), environmental reporting is now a matter to which a growing number of European enterprises are paying attention. There are two reasons for this. The first is that the environmental performance of companies is being included as one of the indicators of their success. The days when pollution did not matter much, and procedures such as recycling were only to be found occasionally, are long gone. Companies are now looking at ways of 'greening' their act, and this inevitably means the implementation of sound environmental management policies. In such a corporate 'culture' trade associations, employer federations and other groups often prefer to disseminate progress relating to their work within their own circles instead of using the mass media for putting the message across. This is partly because they are aware that no wider interest from the public can be expected, except under extraordinary circumstances.

The second reason is that, partly as a result of wide provisions for training in environmental management, there is now a committed and motivated

BOX 22.1 ENVIRONMENTAL EDUCATION AND COMMUNICATION IN BIODIVERSITY CONSERVATION

The conservation of the world's biodiversity is one of the priorities on the environmental agenda for the 1990s. However, biodiversity conservation cannot be seen in isolation from the wider environmental conservation process; neither can it be divorced from the political, social and economic aspects of the environmental debate. On the contrary. In fact, due to the influence of such considerations on efforts to protect the world's plant and animal species, biodiversity conservation is among the most complex issues facing modern society. It is also an issue in which environmental education, environmental communication and environmental management may play an important role for the following reasons:

- the information being provided may be more easily understood by the relevant group(s);
- specific information required by different categories of people (e.g. farmers, NGO members, industrialists) may be met;
- the debate on biodiversity issues may be integrated with those on social and economic matters, as well as with those on biological and ecological issues;
- education and communication would encourage the development of a conservation ethic among decision-makers and other groups;
- education and communication would motivate people to play their role in the conservation process;
- education and communication would command support for environmental conservation policies and regulations in general, and for biodiversity conservation in particular.

Due consideration to education, communication and environmental management, when applied to biodiversity conservation, may represent the difference between short-term changes brought about by laws and decrees, and long-term changes achieved through access to information and the motivation of people which empowers them to be part of the conservation process.

Although the implementation of programmes for raising awareness on biodiversity is a matter for individual countries to decide, it can be best achieved by ensuring that the management aspect is duly covered. In most cases biodiversity conservation turns out to be a question of sound environmental management to ensure that today's resources are not exhausted, and that the fauna and flora of countries, which are among their main assets, are not destroyed in the pursuit of economic goals.

BOX 22.2 DEVELOPMENT OF A CONSORTIUM OF CARIBBEAN UNIVERSITIES FOR NATURAL RESOURCES MANAGEMENT, WITH SPECIAL REFERENCE TO OCEAN AND COASTAL AREA RESOURCES

The objective of this project, run by UNEP's Regional Office for the Caribbean in Jamaica, was to improve university programmes for the management of natural resources of the Caribbean States and Territories through better integration of environmental issues into scientific, technical and degree training programmes.

As a result of the project, the 'Consortium of Caribbean Universities for Natural Resources Management' was established as a programme of the Association of Caribbean Universities (UNICA) in which 15 institutions agreed to participate through a Working Agreement. An Organising and Programme Committee was established, and the Secretariat of the Consortium came into operation in August 1989 at the University of the Virgin Islands (UVI). The Consortium has maintained a strong emphasis on marine and coastal resources.

Other outputs of the project included:

(a) Consortium-wide educational standards at the Bachelor, Diploma (first-year graduate) and master degree levels in resource management, with specific reference to marine and coastal area resources in the Caribbean region. Teaching and training materials were distributed in English, French and Spanish.
(b) Undergraduate courses in resource management were outlined as a co-operative venture involving several Caribbean universities, to be followed up with modules for Diploma and Masters degree programmes.
(c) An outline for a regional one-year Diploma course at the graduate level was developed and made available in English and Spanish.
(d) Sixteen faculty members were trained in curricula development.

workforce trained in environmental management. The expertise of such a workforce can greatly enhance a company's efforts to increase its productivity, and, at the same time, minimise the environmental impact of its activities. Unfortunately, as of now this state of affairs pertains almost exclusively to the industrialised nations. In the majority of developing countries, on the other hand, inadequate public awareness of the link between environmentally sound use of natural resources and sustainable development, coupled with a shortage of managers and technicians equipped with essential skills and knowledge of environmental management, is a major impediment that is still to be addressed.

On both counts, however, education and communication have been playing a vital role for the following reasons:

(a) environmental education and communication initiatives (the latter in particular) relating to environmental management may produce results in the short term, but the medium and the long term are the usual time-frames for results to be seen in the field of environmental education *per se*;
(b) communication approaches and strategies, when integrated with environmental management, may reach small groups as well as larger sections of the population, depending on desired goals;
(c) partly as a result of (a) and (b) above, the importance of integrating education and communication into environmental management is becoming increasingly clear.

INTEGRATION OF ENVIRONMENTAL EDUCATION AND COMMUNICATION WITH ENVIRONMENTAL MANAGEMENT

The United Nations Development Programme (UNDP), which produced the *Handbook and Guidelines for Environmental Management and Sustainable Development* (UNDP, 1992), places 'education, information and the environment' as a theme in the section on 'Managing the Environment in a Sustainable Manner'. According to UNDP, environmental education and information are important tools of environmental management – in addition to community participation – because they foster responsible and effective participation in the framing and monitoring of decisions on the quality of the environment. However, a question to be answered here is: how should environmental education and communication be integrated in the framework of environmental management programmes?

The answer to this question is not simple, however, mainly because the problem to which it refers has various and different aspects. Perhaps one of the most effective approaches, which may be used for integrating education and communication in environmental management, is by establishing genuine partnerships between institutions working in the fields of education and communication, as well as among those working in the field of environmental management. This approach makes sense from an ethical viewpoint, and it may also bring the following advantages:

- the parties may mutually feed on their experiences;
- their pool of expertise may enable them to deal with a wide range of situations and to work in a variety of contexts;
- experience gathered by the partner institutions may enable them to look holistically at problems and to seek reliable ways of addressing them effectively.

An example, which illustrates the integration of education and communication in the context of training initiatives in environmental management, is the European Masters Programme in Environmental Management (EMPEM). This programme offers an opportunity for the training of those who wish to follow a career in this field, regardless of whether they come from industry, government or non-governmental agencies. The course is attended by personnel from industry, while teachers in the course spend some time in industry, often during the preparation of dissertations. Thus the format of the course offers interesting opportunities for establishing or reinforcing university–industry linkages.

As a part of the programme, which is run in various European countries, provisions have been made for a module on environmental education and communication. A module format, which shows a way of integrating environmental education and communication as part of environmental management training courses (both short-term or long-term courses such as a Masters programme), is shown in Figure 22.2.

Contents

a) Environmental Education

Theoretical contents

1. Concept and definition of Environmental Education (EE).
2. Organisations concerned with EE (e.g. UNESCO, UNEP, WWF, IUCN).
3. Historical evolution of EE.
4. EE in Europe.
5. Outline of the rationale for EE.
6. EE in environmental management and in environmental conservation.
7. Nature, scope and structure of EE projects.
8. Assessment of EE projects.

General discussion and conclusion of the first part.

Practical contents

1. Formulation of the aims and objectives of EE.
2. Formulation of the aims and objectives of EE for a given group.
3. Examination and critical comment on a sample of EE projects.
4. Evaluation of the various roles and functions of EE in management.
5. Preparation, execution and monitoring of EE projects in management.

b) Communication

Theoretical contents

1. The media in today's world.
2. Structure of the media – main vehicles of communication.
3. The background against which EE and communication activities have developed.
4. The need for environmental communication in Europe.
5. Communication and public awareness of environmental issues.
6. The didactic use of environmental information.
7. The integration of environmental information into management.

General discussion and conclusion of the second part.

Practical contents

1. Critical analysis of environmental information.
2. Case studies: the media and environmental education in Europe.
3. Integration of communication in environmental management schemes.
4. Techniques for the communication of environmental information.
5. Formulation of strategies for the dissemination of environ-

Figure 22.2 Module on environmental education and communication in environmental management

The academic objectives of the module are:

(a) to inform students on the concept, principles, practices and applicability of environmental education;
(b) to raise awareness of the role of the media in the provision of environmental information;
(c) to make students familiar with the relationships between the topics and the techniques that may be used to communicate environmental issues as part of environmental management initiatives.

The list of themes or topics which can be dealt with by combining education and communication with environmental management is almost endless. Some of the themes, which may be used as a basis for such an integration, are:

- Alternative energy sources.
- Appropriate technologies.
- Case studies on environmental management.
- Climate change.
- Concepts of environmental management.
- Emerging technologies.
- Environmental management applied to sustainable development.
- Environmental management, and energy use and conservation.
- Environmental impact assessment.
- Environmental law, policy and legislation.
- Environmental management education.
- Environmentally sound buildings.
- European policies and directives on environmental management.
- Practices of environmental management.
- Recycling.
- Standards.
- Trends in environmental management in industrialised and developing countries.
- Wastes disposal and management.
- Water conservation.

This list of themes is by no means comprehensive. Some of the topics, such as environmental impact assessment, would appear to have preference (Clark *et al.*, 1984) when compared to others, while issues such as climate change are now becoming increasingly more important (Leal Filho, 1995). The above list none the less outlines some of the areas in which education and communication can help in the implementation of environmental management programmes. Environmental problems may be of particular interest to environmental managers, but so should prudent ways of using natural resources with a minimum of environmental damage.

Regardless of which themes are seen as being interesting, the most efficient way of developing integrated education and communication approaches to environmental management is by formulating, planning and subsequently implementing such strategies and by ensuring that suitable monitoring and evaluation strategies are also in place. The opportunities for education and training being offered in the context of environmental management systems standards, such as the European EMAS, the world ISO 14000, or even systems within individual countries such as the British Standard BS 7750, mean that there is now a wide range of opportunities that ought to be exploited fully.

A word of caution would be appropriate, however, because, as observed by O'Connor and Turnham (1990), a properly devised environmental strategy must place regulatory measures in a broader context, including economic instruments and social measures, along with environmental education. This aspect needs to be taken seriously into account in order to avoid unrealistic expectations of what education and communication can accomplish by themselves.

As stated by Leal Filho (1996b), the following items need to be considered in the integration of environmental education and communication with environmental management:

1 definition of current environmental problems, and of problems that may be solved by implementing a certain environmental management or conservation initiative;
2 identification and selection of suitable measures through which the problem or its consequences may be reversed;
3 selection and adoption of an environmental education and communication strategy, aimed at showing to the public involved the steps that each person might need to take in order to cope with the new situation, thus minimising the inconvenience that might result from a project or other environmentally related initiative to be implemented.

Following item 3 above, the selection of a suitable approach, together with the preparation of information materials explaining important procedures, would increase the chances of success of an environmental management initiative.

CONCLUSIONS

In this chapter we have described the contribution that environmental education and communication can make to environmental management. As we have demonstrated, the ability of environmental management programmes to achieve their goals can be substantially enhanced by implementing effective environmental education and communication strategies. Indeed, this is amply illustrated by typical case studies on initiatives in which education and communication have been integrated into environmental management for the conservation of resources and biodiversity. Provided that due consideration is given to education and communication, and that the methodology used is appropriate, their integration into the process of environmental management can be achieved without much difficulty.

REFERENCES

Alaimo, S.J. and Rooney, L.D. (1980) 'Students' Perceptions of Environmental Problems and Sources of Information', *Environmental Education* 12: 17–21.

Alvaredo, R. (1985) *Media and Environmental Information*, Harmondsworth: Penguin.

Atchia, M. (1995) 'Environmental Education and Training from the 1970s to the 1990s', *Proceedings of International Conference: Reviewing Belgrade*, European Research and Training Centre on Environmental Education, Bradford: University of Bradford.

Axelrod, L. and Lehman, D. (1993) 'Responding to Environmental Concerns: What Factors Guide Individual Action', *Journal of Environmental Psychology* 13: 149–159.

Bain, S. (1994) *Report on the Environmental Education for Sustainability in Small Island Developing States Workshop*, Barbados: Caribbean Conservation Association.

Bandeira, M. (1996) 'Communicating Environmental Issues', Unpublished M.Phil. thesis, University of Bradford.

Brown, V. (1995) 'Who Manages? On Becoming an Environmental Manager', in V. Brown *et al.* (eds) *Risks and Opportunities. Managing Environmental Conflicts and Changes*, London: Earthscan.

Carson, R. (1962) *Silent Spring*, Boston: Houghton Mifflin.

Chiappo, L. (1978) 'Environmental Education and the Third World', *Prospects* 8 (4): 456–465.

Clark, A., Brown, D. and Davidson, R. (1984) *Perspectives on Impact Assessment*, Dordrecht: Reidel Publishing Company.

de Bruin, M. (1991) 'Improving Environmental Media Coverage in the English Speaking Caribbean', *Environment – The Message and the Media – A Caribbean Perspective*, Jamaica: CARIMAC, University of the West Indies and Caribbean Conservation Association, pp. 10–15.

de Haan, G. (ed.) (1995) *Umweltbewusstsein und Massenmedien*, Berlin: Akademie Verlag.

Disinger, J. (1983) 'Environmental Education's Definitional Problem', *ERIC Information Bulletin No. 2*, ERIC Clearinghouse for Environmental Education, Ohio: Ohio State University.

El-Hosseiny, A. (1994) 'Environmental Information in the Egyptian Media', Unpublished M.Phil. thesis, University of Bradford.

Elkington, J. (1994) 'Benchmarking Corporate Environmental Reports – The Challenge to Business', *EEMAA Review* no. 2: 2–3.

Friedman, S.M. and Friedman, P. (1989) 'Environmental Journalism', *Environment* 31 (5): 7–37.

Hale, M. (ed.) (1993) *Ecology in Education*, Cambridge: Cambridge University Press.

Hansen, A. (ed.) (1993) *The Mass Media and Environmental Issues*, Leicester: Leicester University Press.

Hardoy, J. (ed.) (1992) *Environmental Problems in Third World Cities*, London: Earthscan Publications.

International Union for the Conservation of Nature and Nature Resources (IUCN) (1970) *International Working Meeting on Environmental Education in the School Curriculum. Final Report*, September 1970, Morges: IUCN.

Jacobs, M. (1991) *The Green Economy: Environment, Sustainable Development and the Politics of the Future*, London: Pluto Press.

Kepplinger, H.M. (1989) 'Artificial Horizons: How the German Press Presented and the Population Received Technology in Germany from 1965–1988', Paper presented at the Fourth International Congress of Professors, World Peace Academy, London, 25–29 August.

Leal Filho, W.D.S. (1990) 'The Use of the Media as a Source of Information and Environmental Awareness in Formal Teaching', *Environmental Education and Information* 9: 113–118.

Leal Filho, W.D.S. (1991) 'Environmental Education and Environmental Information', Unpublished Post-Doctoral thesis, University of Hamburg.

Leal Filho, W.D.S. (1992) 'Using the Press for Environmental Education', *Nature & Resources* 28: 38–46.

Leal Filho, W.D.S. (1993) 'Environmental Education: The Key to Sustainable Living', *The Bulletin*, Geneva: Centre For Our Common Future.

Leal Filho, W.D.S. (1994) 'Raising Awareness on Environmental Risks', in W. Leal Filho, J. Padgham, J.U. Ahmad, A.H. Khan and S. Safiullah (eds) *Communicating Environmental Risks – Environmental Education as a Tool for Environmental Disaster Management*, Dhaka: Centre for Environmental Studies, Jahangirnagar University, pp. 1–19.

Leal Filho, W.D.S. (1995) 'The Role of Environmental Education in the Training of Scientists on Global Environmental Changes in Developing Countries', in D. Waddington (ed.) *Global Climate Change: Science, Education and Training*, NATO ASI Series, vol I, 29, Berlin: Springer-Verlag.

Leal Filho, W.D.S. (1996a) 'Raising Awareness on Biodiversity', in W.D.S. Leal Filho, C. Dykes and Z. Murphy (eds) *Raising Awareness on Biodiversity: Commonwealth Examples*, Vancouver: The Commonwealth of Learning.

Leal Filho, W.D.S. (1996b) 'Integrating Environmental Education and Environmental Management', *Environmental Management and Health* 2: 80–82.

Leal Filho, W.D.S and Bynoe, P. (eds) (1995) *Proceedings of the International Workshop on Environmental Journalism*, 28–30 April 1995, Bradford: ERTCEE.

Leal Filho, W.D.S. and Hale, M. (1992) *Promoting International Environmental Education*, Proceedings of the International Workshop on Environmental Education held in Rio de Janeiro, Brazil, 4–6 June 1992, London: DES.

Leal Filho, W.D.S. and Hale, M. (eds) (1994) *Trends in Environmental Education World-wide*, London: London Guildhall University.

Leal Filho, W.D.S. and Padgham, J. (eds) (1994) *Communicating Environmental Risks – Environmental Education as a Tool for Environmental Disaster Management*, Ottawa: IDRC.

McCarson, S. (1978) *Environmental Education: Principles and Practices*, London: Edward Arnold.

McCormick, J. (1995) *The Global Environmental Movement*, London: John Wiley & Sons.

Meadows, D. (1972) *The Limits to Growth – A Report for the Club of Rome's Project on the Predicament of Mankind*, New York: Earth Island.

O'Connor, D. and Turnham, D. (1990) *Environmental Management in Developing Countries. A Conference Summary.* Paris: OECD.

Petri, M. (1988) 'Das "Umweltbild" in Medium Tageszeitung. Empirische Untersuchung zum Unfeld des Themesbereichs Umwelterziehung', Unpublished dissertation, Kiel University, Germany.

Ray, D. and Guzzo, L. (1993) *Environmental Overkill*, New York: Harper.

Schneider, H. (1993) *Environmental Education: An Alternative Approach to Sustainable Development*, Paris: OECD.

UN (1992) *The United Nations Conference on Environment and Development: A Guide to Agenda 21*, New York: UN Publications Service.

UNDP (1992) *Handbook and Guidelines for Environmental Management and Sustainable Development*, New York: UNDP.

UNEP (1992) 'Overview of Education, Training and Awareness Projects Implemented Within the Framework of the Caribbean Environment Programme: 1981–1991', *CEC Information Paper*, Jamaica: UNEP.

World Commission on Environment and Development (WCED) (1987) *Our Common Future*, Oxford: Oxford University Press.

SUGGESTED READING

Asian Development Bank (1990) *Economic Policies for Sustainable Development*, Manila: ADB.

Erocal, D. (ed.) (1991) *Environmental Management in Developing Countries*, Paris: OECD.

Interagency Environmental Technologies Office (IETO) (1994) *A Bridge to a Sustainable Future. National Environmental Technology Strategy*, Washington, DC: IETO.

UNEP (1992) *Environmental Training World-wide*, Kenya: UNEP.

SELF-ASSESSMENT QUESTIONS

1 What is environmental education?
(a) an approach to be used in schools;
(b) an approach to be used in communities;
(c) an approach which, due to its nature, may be used in both schools and in the community context.

2 What is environmental communication?
(a) a strategy to inform on issues related to matters such as air and water;
(b) a strategy to inform people in terms of general environmental matters in a didactic way;
(c) a strategy to inform people on the environment, paying due attention to the various factors involved.

3 Community participation in environmental management is:
(a) partly important;
(b) important and indeed essential in some cases;
(c) not important.

4 In which of the following areas will education and communication be useful to environmental management?
(a) only at the local level;
(b) at both local and national levels;
(c) at the local, national and the international levels.

5 The most effective approach to integrating education and communication in environmental management is through:
(a) partnerships between institutions;
(b) individual works of the institutions concerned;
(c) occasional co-operation between institutions.

6 Which of the following groups need to be taken into account in communicating industry-based matters relating to environmental management?
(a) industrial producers only;
(b) consumers only;
(c) producers, consumers and the intermediaries (middle agents).

7 Which of the following could be considered as the main actors and target groups for education and communication in environmental management?
(a) government and non-governmental agencies;
(b) government agencies;
(c) government agencies, non-governmental agencies and the private sector.

8 Which of the following could be considered to be the most efficient in developing education and communication approaches to environmental management?
(a) formulation, planning and implementation;
(b) planning, implementation and formulation;
(c) planning, formulation and implementation.

9 Which of the following is the best possible time to integrate environmental education and communication with an environmental management programme?

(a) once the scheme is completed;
(b) once the scheme is running;
(c) at the planning stage.

10 How can environmental education and communication improve the ability of an environmental management initiative involving a community?
(a) by being used in an integrated fashion;
(b) by being employed separately;
(c) by being used at a later stage.

23
ENVIRONMENTAL CONFLICT MANAGEMENT

Henry van Asselt and Emiel Wubben

SUMMARY

The roots of environmental conflicts are often to be found in the different values which the parties to a conflict attach to the environment. Indeed, the divergence of these values, which often prioritises short-term gains over the long-term survival of mankind, can lead to miscommunication, misunderstanding and conflicts. However, a healthy environment is a precondition for the long-term survival of humans, unlike labour relations and politics that are transitory and temporal in nature. Consequently, the management regime required for resolving environmental conflicts differs from that needed for the resolution of non-environmental conflicts.

Ever since the 1970s there has been a growing interest amongst those involved in environmental affairs to resolve environmental disputes out of court in some form of organised direct contact. The Mutual Gains Approach (MGA) is such an alternative type of environmental conflict management, and has much to recommend it on several counts. It is for this reason, among others, that the MGA will be the focus of discussion in this chapter. The mechanics of how the MGA works in practice are explained with the aid of case studies, together with its basic principles and advantages, especially with regard to the amicable conflict resolution it makes possible and the mutual benefits it brings to all the parties in the dispute. Also, the essential characteristics of the MGA have been compared and contrasted with those of the 'classical' (conventional) approach to environmental conflict resolution which is fundamentally adversarial in nature.

ACADEMIC OBJECTIVES

On completing this chapter the reader should have a good understanding and working knowledge of the following:

- the basic assumptions of conflict resolution that are usually subsumed under the heading of 'alternative environmental dispute resolution' (AEDR);
- the 'classical' approach to environmental dispute resolution, how it works in practice, and its deficiencies;
- the concept of the Mutual Gains Approach to environmental conflict resolution, its principles, how it is applied in practice, the ingredients needed for its successful implementation, and its advantages over the 'classical' approach.

BACKGROUND

Definition of an environmental conflict

The term 'environmental conflict' is to be understood here as being synonymous with a delay in the decision-making process, arising from a controversy amongst the relevant stakeholders concerning the potential negative environmental impact or impacts of current or proposed activities. Five types of conflict may be identified: intra-personal (within the self), inter-personal (between individuals), intra-group (within a group), inter-group (between groups) and international (between nations) (Borisoff and Victor, 1989). Political conflicts are usually managed through the policy process – a process within governments whereby decisions are made on who gets what, when and how. The policy process serves to bring together the conflicting aspirations and demands of the populace, and it functions as a principal mechanism for conflict resolution (Stephenson and Pops, 1990). In this chapter we will not discuss the political conflict at length, but instead address the intergroup type of conflicts that are characterised by the interests of various stakeholders in the local activities of companies.

The question is, when does a conflict become an 'environmental' conflict? In order for a conflict to qualify as an environmental conflict it must relate to an environmental issue such as pollution prevention, saving of raw materials and energy, improved public health, etc. In general, the complexity of a conflict grows rapidly, and it becomes increasingly more difficult to resolve, as the number of stakeholders involved in it increases. Often many parties (stakeholders) may be involved in a conflict, such as:

1 Environmental pressure groups whose focus is on the long-term quality of the environment.
2 Citizens in proximity to the site in question who want to live in a clean environment, or at least subscribe to the Not-In-My-Back-Yard (NIMBY) syndrome.
3 The company or the entrepreneur whose interest is to run the proposed business or project in the most profitable manner.
4 The central government whose principal goal is to serve the long-term interests of society, which means considering the views of all the parties to the conflict.
5 Experts in different disciplines, such as health, energy conservation, economics, etc., whose task is to predict the consequences at both macro and micro levels.

Clearly, the greater the number of stakeholders involved in a conflict, the more difficult it would be to resolve. This is because all the stakeholders will have to be addressed separately in order to arrive at a preferred solution, and also because a common resolution should be reached which accommodates their diverse interests. The mass media play an important role in the process, often by propagating or promoting specific interests, even though by themselves they constitute what may be called a 'secondary' stakeholder and not a real stakeholder in terms of the categories listed above.

The diverse goals and interests of authorities, communities and companies are well-known sources of environmental conflict. For example, the local government, which represents the interests of the community as a whole in a clean-up conflict, will emphasise the direct or indirect health effects, the urgency for taking action, the liability of companies, and the indirect social costs. The industrial company involved may, on the other hand, emphasise lack of knowledge, its autonomy in formulating and implementing the appropriate strategy, and its direct claim on scarce resources. Ultimately, the approach to the management of a given conflict would be determined by whether the issue in question is concerned with conflicting values, ideas, belief systems or resources.

We may thus conclude that, although the issues in an environmental conflict may differ from case to case (legal standards, a new technology, entrepreneurial behaviour, etc.), all conflicts have a common basis none the less; namely, the multifaceted confrontation of diverse environmental values, social values and business values. In this context specific public–private relationships may be seen as reflections of our more open, multi-party orientation. In this chapter we are especially concerned with the common, location-specific conflicts involving a number (or a large number) of stakeholders.

Historical background and sources of environmental conflict

Environmental conflicts are as old as civilisation itself. They have occurred throughout recorded history, growing from small to large, to the truly appalling, then becoming acceptable or at least bearable sooner or later. But the abatement of pollution has usually been contingent upon the formulation and enforcement of necessary legislation. In England, for example, the first law to control air pollution was passed in 1273 whereby the burning of coal in London was forbidden when the monarch was in residence there. It was actually part of a wider conflict about air quality which had become a major issue since coal had replaced wood in lime production, beer brewing, iron forging and domestic heating. Seven centuries ago, in the year 1306, there was still an official ban on the use of coal and, around 1660, the National Smoke

Abatement Society was formed to combat air pollution in London. The issues of primary concern at the time were material damage to buildings and the deteriorating state of public health caused by air pollution. Amongst the solutions sought for was a proposal to relocate and concentrate enterprises outside the city. Air pollution problems helped to enhance the awareness of other environmental problems, too, such as water and soil pollution.

The issuing of several local statutes regulating the disposal of wastes on land, or in surface water, testifies to the awareness of environmental problems in the Netherlands in the fifteenth century. As the concentration of both people and production is most severe in the cities, they were the first in need of regulations to curb the environmental impacts of productive activities. At that time there was thus a desperate need for legislation in Amsterdam, for example, where health conditions had become notoriously bad because of narrow streets, a humid climate, offensive smells, open sewers, plagues, and streets full of cattle, garbage and vermin.

Then, a large proportion of the environmental conflicts arose, and still arise, from the use of what are called natural 'sinks', such as land, rivers, lakes, etc., for the disposal of man-made wastes. But these sinks are also the sources of water, minerals, oil, agricultural land and other life-sustaining natural resources. And so it was that the irreconcilable economic and political ambitions of the European rulers resulted in an expansionist drift towards Africa, America and Asia to seize and safeguard controlling interests in nature's resources.

At the national level environmental legislation began to become increasingly more important with the onset of industrialisation in the early nineteenth century. Although industrialisation came to the Netherlands a little later, a Royal Decree was already in place in 1824 to control offensive odours from factories. The first edition of the Nuisance Act was made public in 1875, and the transport of dangerous and poisonous materials became regulated by law as a result of the well-known 1807 gunpowder explosion in Leiden.

Environmental issues and problems came to the fore again only after the build-up of western economies in the twentieth century. Typically, it was with a huge body of national legislation that governments reacted to the emerging environmental concerns in the 1960s. For example, a National Environmental Urgency Report was issued in the Netherlands in 1972. It was followed by the coming into force of a large number of Environmental Protection Acts, including the Surface Waters Pollution Act, the Air Pollution Act, the Chemical Wastes Act, the Noise Nuisance Act, the Environmentally Dangerous Substances Act, and the Act on Soil Protection. This huge body of legislation covered all the environmental compartments – air, atmosphere, water and soil – and in principle it enabled elaborate control of emissions aiming at preserving and/or improving the quality of the environment.

An important effect of the growth in environmental legislation has been the increasingly greater use of licences, which has become a source of fierce conflict between industrial enterprises and governmental agencies. Such conflicts arise mainly from governments' dictation of elaborate industrial emissions limits, and from obligations put on companies by governments to report regularly on specific substances emitted and to install pollution-abating equipment of a certain type. All these measures are seen by industry as government interference with their activities at the business level. This microeconomic interference has increased the number and severity of environmental conflicts, because it directly affects private interests by causing delays or even stopping private activities.

Although perhaps somewhat naïve, some of the contemporary western environmental regulations are structured on the basis of what is called *Habermas's Communicative Governance*. Based on reasonableness and public management, Communicative Governance is purported to be able to deal with the phenomenon that learning effects may be realised through the exchange of information whereby support for the resulting strategy will increase (Van Vliet, 1993; Van Asselt, 1989). In this context it is interesting to note the rise in popularity of voluntary initiatives in the 1990s, such as the Business Charter for Sustainable Development, the Responsible Care Programme and the Coatings Care Programme. In fact, industry as a whole is going through a process of change in attitude, as became clear in 1991 at the WICEM (World Industry Conference on Environmental Management) in Rotterdam. There a number of companies ratified the Business Charter for Sustainable Development, which included the following principles:

- An open dialogue with colleagues and the public on the issue of sustainable development.
- Minimisation of waste and emissions from manufacturing processes.
- Product stewardship leading to the eventual development of substitutes.
- Innovative product design and marketing in response to a product's ecological impact and resource conditions.

Charters and Care Programmes are good examples of the effectiveness of voluntary agreements (Buhr, 1991), not least because they lead to closer collaboration among governing bodies, interest groups, and independent companies or groups thereof. Today many of the national and multi-national companies, Shell for example, realise that corporate governance involves social responsibilities, that exhaustion of resources is unacceptable, that constructive solutions to environmental problems must be explored, and that product stewardship is inevitable.

Environmental conflicts are not a new phenomenon, as we have pointed out earlier. However, their origin, subject-matter, and the modalities of their resolution have changed over the course of history. The nature of the effects of human activities on the environment has changed, and so has our knowledge of this causality. The ideas about the value of nature have also changed over time in step with our experience of evolving legislation to resolve environmental conflicts. Ever since the 1980s, there has been a growing willingness amongst the conflicting parties to resolve disputes out of court in some form of organised direct contact. This shows that the values, ideas, and belief systems of authorities, pressure groups and companies towards environmental problems are becoming increasingly more attuned to voluntary initiatives and modalities for dispute resolution.

Large infrastructural projects: an example

Analysis of the management of an environmental conflict implies the identification of the conflict itself, together with how it is to be managed. Clearly, one must first become acquainted with the conflict in question. The focus will then shift to the central issues of the conflict and the actual form of Alternative Environmental Dispute Resolution (AEDR) to be used. An environmental conflict is born the moment the different parties involved fail to come to an agreed decision on any of the pertinent issues. Such a situation commonly occurs when the interests of one or some of the stakeholders are at odds with those of the others, and there is apparently no ground for a compromise. Consider, for example, the construction of an infrastructural project which produces a large quantity of polluted soil. The construction company, whose responsibility it is to pay for cleaning-up the polluted soil, may press for and obtain a postponement of the clean-up operation until such time as when it could be done at least cost. Citizens living in or near the area may, on the other hand, press for immediate but cautious action, whereas environmental pressure groups may prefer some delay in order to improve techniques for reducing the environmental fall-out from the cleaning-up process. As for the national government, in the case of large infrastructural projects it may hold the view that progress demands the rapid expansion of the national infrastructure, even when it is realised at a high cost. Parties without any direct economic stake may, on the other hand, attach a greater value to those alternatives which result in a delayed realisation at lower social and environmental costs. Depending on their situation, some of the industrial interest groups, but not all, may press for delay or postponement in order to acquire new or improved technology, thus confirming that group interests differ from case to case.

In a later section of this chapter we will elaborate on the Mutual Gains Approach (MGA) itself in the specific context of the polluted soil conflict referred to above. An uncomplicated example of such a case is given in Box 23.1. It shows how an alternative environmental dispute resolution was arrived at in the case of a large infrastructural project involving the construction of a dedicated freight railway system from Rotterdam Harbour to Germany. This case is typical of other projects of this kind, such as the expansion of the Schiphol Airport in Amsterdam and the extension of Rotterdam Harbour. These examples may also serve as useful case studies for developing countries.

BOX 23.1 ALTERNATIVE DISPUTE RESOLUTION IN THE CASE OF INFRASTRUCTURAL PROJECTS – EXAMPLE OF THE FREIGHT RAILWAY SYSTEM BETWEEN ROTTERDAM HARBOUR AND GERMANY

Phase 1: problem definition
The problem was defined by the organisation recruited to prepare the Environmental Impact Statement (EIS) under the responsibility of the Dutch National EIS Committee. The EIS comprised the specification of several alternatives to the proposed project. For each alternative it presented an overview of the effects on the environmental compartments such as water and soil. The EIS was tested for such aspects as accuracy, objectivity and completeness before it could be approved by the national EIS Committee.

Phase 2: selection of all possible solutions
Apart from the environmentally preferred alternative, there may be other developed alternative solutions too. Although these solutions may be environmentally sub-optimal, they are nevertheless attractive from the technical and/or economic point of view. For example, several alternatives to the freight railway system were proposed, indicating in each case the trade-offs between the economic, technical, political and environmental aspects. As a result, conflicting opinions were certain to arise.

Phase 3: scope of the problem
In this phase it is most important for all the parties involved to focus on the scope of the problem and to participate in discussions. The forum for this may vary from the local or regional to the national level. In the case of the freight railway system a special 'platform' was constructed to facilitate both discussions and the widest articulation of views.

Phase 4: the preferred strategy
This phase is concerned with finding common solutions. In the case of the freight railway system, the preferred strategy for this was to establish a special committee of the Royal Institute of Engineers.

Phase 5: sustainability of solution
The construction plans must accommodate anticipated future developments and, in addition, should prepare for some compensatory measures. This meant that the freight railway system had to be planned for many decades of full service. Although the plans were subject to a delay of some years, and included several expensive local modifications, they were finally endorsed and supported by all the parties involved.

Aspects of environmental conflicts

The task of finding a generally acceptable and environmentally sound solution to a given environmental conflict is facilitated by a consideration of its various aspects independently. The most important aspects, discussed below, are technological, legal and political, financial, and social.

1 Often the *technological aspect* of an environmental conflict is concerned with the feasibility of the available technology to solve or prevent the environmental problem in question. But the problem here is that the term 'available' is often capable of more than one interpretation. A company may take 'availability' as an indicator of feasibility; that is, technology which has been commercially proven in practice. A pressure group may, on the other hand, interpret it as 'state-of-the-art' technology which is yet to be fully proven by experts based on scientific or professional findings or evidence. Also, because state-of-the-art technology may not be economically feasible yet, such technology usually implies higher risks and costs, although it may bring greater benefits at the end. Consequently, one group may argue for state-of-the-art technology while others may insist on feasible and fully proven technology (Cramer and Zegveld, 1991).

2 The *legal aspect* of an environmental conflict involves the interpretation of the relevant environmental standards by the parties to the conflict. The following variants may occur in this regard: (a) a conflict could relate to the standards incorporated in legislation. For example, when acceptable noise levels are defined as being dependent on the predominant function of an area, among others, then it is important to know whether the area in question is supposed to be

an industrial area, a rural area or a residential area. Clearly, there is potential for conflict until and unless the user-function of the area is defined; (b) the conflict could relate to whether or not the legal protocol of some measurement has been followed; (c) difficult conflicts may arise from whether or not forthcoming standards already carry some legally binding force; and (d) conflicts may arise from the use of non-quantitative standards. It is worth noting in this context that qualitative standards allow for a greater variety of interpretation than quantitative standards (van Dunne, 1991).

3 Environmental conflicts normally acquire a *financial aspect*, too. Typically, pollution conflicts are often concerned with the assignment of financial liability to one or more of the parties to the conflict. The liable party may be a private enterprise or a public authority. Finance is clearly involved when, for example, the premature closure of a factory is called for by the legal authorities, interest groups or by the citizens living nearby. Given the strict provisions of the 'Polluter Pays Principle', a growing number of industries are now becoming interested in insuring themselves against potential environmental claims. It is obviously not in the financial interests of an industrial company to be seen as a polluter, neither is it from its public relations point of view.

4 The *social aspect* of an environmental conflict is pertinent when it is influenced by codes of behaviour. Arrogance and the abuse of power often leads to confrontation and escalation of the conflict, making its resolution even more difficult. This is especially so when the parties to the conflict are not familiar with each other. However, greater socialisation of the parties can contribute much to dispel misunderstanding and distrust, thus creating a climate conducive to an amicable resolution of the conflict through discussions and recognition of mutual interests.

TYPES OF ENVIRONMENTAL CONFLICT RESOLUTION

Broadly defined, environmental conflict management refers to the various forms of direct contact between the parties involved in order to reach an agreement on the disputed environmental issues (Bingham, 1986). In this section we will discuss three different types of environmental conflict resolution in which enlightened self-interest, co-operation and transformation are representative concepts. These three types relate to the three distinct views on dealing with environmental conflicts, as elaborated by Glasbergen (1995).

The concept of self-interest refers to the use of rewards and punishments as external incentives and disincentives, respectively, for changing the demeanour of a stakeholder towards the environment. The concept of reward and punishment is based on the application of outside stimuli for producing a certain type of behavioural response in individuals, groups or organisations. Typically, a punishment might consist of higher levies for excessive pollution, whereas a reward might be a subsidy for investing in new and less polluting processes or technologies. For example, governments may reward companies that invest in clean technologies and processes by allowing them to write off fully such investments against tax in the first year of installation. Clearly, the aim of both punishment and reward in the context of environmental conflicts is to deter and reverse the old attitude of organisations to yield power over resources, exploit our physical environment, and to safeguard it for their economic interests alone.

Enlightened self-interest in environmental conflict resolution, which is a variant of the self-interest referred to above, may often be especially important as a motivating factor in the tactics deployed by a company. In practice, enlightened self-interest entails convincing the parties in dispute that it is in their own long-term interest to take a responsible attitude towards the environment – for example, by switching to more environmentally friendly fuels, materials or processes – because by so doing they would acquire a better public image and, consequently, make greater profits from a larger customer base. At the very least they would benefit from higher resource efficiency, or a reduction in waste generation entailing lower costs of treatment and disposal. Enlightened self-interest need not be limited to enterprises alone. Individuals can also be persuaded to adopt less environmentally harmful lifestyles by applying the concept of enlightened self-interest. For example, one may be persuaded to use energy-efficient light bulbs because they are more economical in the long run. Clearly, changing the behaviour and attitudes of people, groups and organisations in accordance with enlightened self-interest is contingent upon making them aware of the advantages that would be realised through their changed attitude or lifestyle at some future date. But this does not necessarily reflect a change in attitude towards valuing the environment. Even so, the recognition or acknowledgement that caring for the environment would result in private benefits sooner or later, would be beneficial both to the individual and the environment. Enlightened self-interest is thus a mutation of approaches, emphasising the economic aspect with the ultimate aim of preventing the exhaustion of nature's resources.

The concepts of self-interest and enlightened self-interest both represent what is called the 'classical', or the 'first generation', approach to the resolution of environmental conflicts. They convey the view, and are based on the thesis, that rewards and punishments would help to induce necessary changes in economic practices. Emphasis is placed exclusively on power, on one's private interests, and on the legal and financial aspects. Conflicts are either taken to court or subjected to administrative procedures. Although this has been the predominant modality for dealing with environmental conflicts for many centuries, it has lost much of its integrating force in recent environmental conflicts. It is for this reason that it is called the 'classical' or the 'first generation' approach to dealing with environmental conflicts.

The 'second generation' approach to environmental dispute resolution focuses on the concept of co-operation, in line with the recent trend of emphasising the cultural or 'face-to-face' element of environmental conflicts. The roots of environmental conflicts are often to be found in the different values which the parties in dispute attach to the environment. The divergence of the related and often implicitly held values (e.g. prioritising short-term gains versus long-term survival of society) leads to miscommunication, misunderstanding and conflicts. Ever since the 1970s there has been a growing interest amongst those directly involved in environmental conflicts to resolve disputes out of court in some form of organised direct contact. This calls for a certain type of conflict management that is different from the management of non-environmental conflicts. The models of Alternative Environmental Dispute Resolution' (AEDR) offer an alternative perspective on preventing or reducing environmental conflicts through co-operation, especially for conflicts with interdependent behaviour. These models are alternatives to the classical 'first generation' models.

The concept of both self-interest and enlightened self-interest is now giving way to the relatively new and emerging concept of 'transformation' which introduces the 'pursuit of sustainable forms of development as a perspective' (Glasbergen, 1995). Here environmental conflicts are understood as being issue-based and general-interest controversies which affect people beyond those directly involved. Though still in its infancy, the concept of transformation is attracting much attention because it is consistent with sustainable development which is currently high on the environmental agenda (Davis, 1991). Since concern about the quality of the environment motivates people to promote a process of socio-cultural change, it follows therefore that disputes must be resolved by mobilising those who broadly support the objectives of environmental policies. Public administration is seen as the co-ordinator of change, and indeed the change agent, because it organises the environmental problem-solving network of social interests. This new and evolving concept of transformation represents the 'third generation', or the network approach to the resolution of environmental conflicts.

Many, if not most, of the environmental conflicts are site-specific. They reflect contrasting views on, and attitudes to, nature in a given conflict which arises in a particular locality or region. Co-operation implies that conflicts, which are usually multifaceted, are about a 'basket' of benefits that must be shared equitably. However, the benefits accruing from co-operation are often greater than when the parties discuss the environmental problem in question separately and independently from each other. The main difficulty in employing the modality of co-operation for conflict resolution lies in convincing people that it is ultimately to the benefit of all to co-operate. The task of disseminating relevant information to convince people normally lies with the party which stands to gain the larger part of the advantage at the end, or with an independent party charged with the responsibility for solving the environmental conflict in question. Normally some local public authority would be expected to assume that responsibility, because it is the local governments who stand to gain most by controlling the behaviour of polluters, as experience shows. However, as has been shown in several studies (Porter and Van der Linde, 1995; Wubben and Bekkers, 1997), the polluters themselves may also save money, time and resources by cleaning up their acts. Instead of paying compensation for the environmental damage caused, it makes more economic sense for them to co-operate and invest in pollution control technologies or processes that may yield greater profits due to increased energy efficiency, new spin-offs and improved production processes. Realisation and acceptance of this would help to motivate top management to promote and develop an open and co-operative attitude towards external groups, and the public at large, for the resolution of environmental conflicts.

In general, co-operation is more likely to occur, and at a faster pace, when all the parties to a conflict realise that they would gain more or less equally by co-operating. But, while there may be a direct profit to be made by all by co-operating, there is also the potential danger that one or more of the parties may employ tactics to hinder or block progress in order to improve their bargaining position or to secure greater profits for themselves. Actors with mutually obstructive powers seem to be condemned to co-operation, especially when their powers are equal or comparable. But when both the benefits and powers of hindrance are unequal, co-operation can be better achieved through the mediation of a third party which may or may not have a financial interest in initiating co-operation. In such a situation co-operation can be achieved by making it clear to the parties to the conflict that there is a potential advantage, or 'profit', for all in not getting entangled in an environmental conflict. This approach can thus help prevent conflicts before they arise or escalate. It can also help resolve a conflict after it has occurred.

There is no unique recipe for the resolution of environmental disputes. Apart from the variety of approaches, such as negotiation, mediation and consensus-building, there is also a diversity in terms of the subject-matter of conflicts (e.g. from disputes over land use and water resources, via natural resources management to national policies on toxins and air quality). There are various models of Alternative Environmental Dispute Resolution (AEDR), such as 'collaborative problem solving' (Gray, 1989) and 'consensual approaches' (Susskind and Cruickshank, 1987). The Mutual Gains Approach (MGA) is such a consensual alternative approach to environmental conflict management. This approach will take centre stage in the remainder of this chapter because of its practical usefulness and also because it can be explained relatively simply and briefly. We will not discuss the 'third generation' approach to environmental conflict resolution any further because of its relative immaturity. Instead, we will focus on the other two alternative approaches, namely the first and second generation approaches, which feature the Mutual Gains Approach. In particular, whenever environmental dispute resolution is mentioned hereafter, it must be understood to refer to the second generation or alternative forms of environmental dispute resolution.

THE MUTUAL GAINS APPROACH

The Mutual Gains Approach (MGA) sets out the principles and elements necessary for achieving co-operation and, as will be seen, it is superior to the older 'classical' approach. A general overview of the characteristics of the MGA is shown in Figure 23.1. Our purpose here is to discuss the principles of the MGA, classify the principal factors which determine whether or not a given environmental dispute can be successfully managed with the MGA, and to highlight the role of mediators in the practical organisation of environmental conflict management via the MGA.

First principles

The *Mutual Gains Approach* is based on the notion of 'side-by-side' problem-solving rather than 'across-the-table' confrontation via litigation or administrative procedures. Because of dissatisfaction with the results and impacts of the latter, which has been the dominant form of conflict resolution, the Harvard Law School developed a programme for managing public opinion while at the same protecting the reputation of the organisation in dispute. The aim was to develop a more effective and amicable management regime for the resolution of environmental conflicts. Their so-called Public Dispute Programme is based on the Mutual Gains Approach, and it works within the framework of the following rules (Susskind and Cruickshank, 1987):

1 Acknowledge the concerns of the opposing side who believe just as strongly that they are right.
2 Joint fact-finding: a battle between experts may convince the public that anything is true.
3 Propose commitments to minimise likely impacts and promise to compensate for unintended effects.
4 Act in a trustworthy fashion – once lost, trust is very difficult to regain.
5 Accept responsibility, admit mistakes and share power.
6 Focus on maintaining long-term relationships; short-term victories are misleading.

The key factor, which makes the MGA acceptable to the different parties in dispute, turns out to be the participation of those involved in making a realistic estimate of 'the best alternative' to direct contact (i.e. the so-called 'opportunity' costs). This means that there should be more to be gained by all the parties from a negotiated or mediated agreement than could be expected from walking away from the negotiating table. This may be realised through the following arguments that are in favour of participation:

1 Rejection of negotiation may lead to a complete standstill in solving the environmental dispute, and this has the potential to harm all the parties involved.
2 By not negotiating, one or more of the parties may become frustrated and may be tempted to employ obstructive or other detrimental tactics that would be regretted later on.
3 Absence of negotiation may lead to a worsening of the situation. Lack of action on finding a solution, and the continuing environmental damage caused by the polluting activities of the company in the meantime, can have a negative effect on its image, resulting in reduced sales.

Although the formal degree of responsibility of a given party may be difficult to determine, the polluting parties, the supervising authorities and the other parties involved must all acknowledge their responsibility for arriving at an acceptable resolution. In this context the concept of 'common responsibility', which may also be attached to the MGA, may be invoked for finding a solution. For example, if 'common responsibility' is interpreted as 'shared responsibility' – and it would appear eminently reasonable to do so – then this would pave the way to the sharing of the clean-up costs, and, clearly, all the parties would benefit from such a clean-up.

A comparison of approaches

As will be seen from Figure 23.1, the characteristics of the MGA differ fundamentally from those of the 'classical' approach for the settlement of environmental disputes. No litigation or short-term opportunism prevails in the MGA, its focus being on the long term. Also, the financial bias of the classical approach is superseded in the MGA by a cultural or socio-economic orientation. Decision-making in the MGA is not restricted to a closed set of stakeholders who negotiate as hard as possible for preserving or promoting their own interests. Instead, it relies on the free flow of information and dialogue to create a feasible, stable and 'win-win' solution which is

	Classical Approach	*Mutual Gains Approach*
demeanour	defensive	pro-active
orientation	financial, legal	cultural, socio-economic
decision-making	autonomous	participatory
solutions	legalistic	long-term feasibility

Figure 23.1 General characteristics of the 'classical' and the Mutual Gains Approach

acceptable to all the parties as well as beneficial (or at least not damaging) to the environment. The ultimate goal of negotiation, typified by the Harvard Negotiation Process, is to achieve an agreement that is beneficial to all the parties involved. Co-operation, rather than confrontation, is the key to the successful management of conflicts through the MGA.

Figure 23.2 gives an overview of the essentials of the MGA, compared with those of the classical approach. Clearly, the MGA differs fundamentally from the classical approach in all the five phases shown in this figure. The MGA may be characterised by the following example: a chemical company, which makes a declaration of its care and concern for the environment and issues a statement on how it intends to minimise the environmental impacts of its production activities, will enjoy the confidence of the public because the public perception of this would be that the company is not merely complying with legal requirements but is also trying to go a step further proactively to protect the environment (Greeno, 1990). In the classical approach, on the other hand, a company with legal or financial bias would be reluctant to issue any such statement or declaration voluntarily. A probable short-term outcome of this would be less favourable to the company: such an attitude would not please the public, neither would it satisfy the stakeholders, except for the lawyers, while the long-term perspective would be uncertain.

The likelihood of success

There are three main classes of factors which determine the likelihood of success of alternative environmental dispute resolution; namely, the party-related

Classical Approach	*Mutual Gains Approach*
1. Problem definition	
Hire technical experts, lobbyists and press agents	Set up a Community Advisory Group (CAG) to realise joint fact-finding
2. Selection of possible solutions	
Focus on the cheapest solution	Report to CAG on costs and benefits of all relevant solutions
3. Scope of the problem	
Deny any liability beyond direct costs	Acknowledge concern for secondary effects and initiate a thorough investigation Initiate joint site monitoring programmes
4. Preferred strategy	
Impress the public by presentations that are staged and managed by the company Organise media support	Organisation of open community meetings jointly with CAG to reach a consensus
5. Sustainability of solution	
Legalistic and tactical approach to potential liabilities	Safeguard long-term company interests by participating in a joint arbitration panel and in the setting up of a community trust fund

Figure 23.2 The five phases in the 'classical' versus the Mutual Gains Approach

factors, the process- and context-related factors, and the content-related factors (Bingham, 1986).

In the *party-related factors* it is important first to identify all the affected parties. If any of the parties are left out because of not having been identified, or because they were not represented in the negotiation, then the outcome of that negotiation can be challenged later. On the other hand, a dispute may involve so many different parties that it may not be possible to identify them all and to ensure that all are represented in negotiations. Thus the number and types of the affected parties represented in a negotiation is an important determinant of the success of that negotiation.

The *process- and context-related factors* may also influence the likelihood of success of an environmental conflict negotiation. A typical example of such factors is the decision on whether or not the agreement should be in the form of a written statement. If it is *not*, it may be easier for some of the parties involved to reach an agreement, even though others may find it unacceptable. Another example of process-related factors is whether to set a deadline for the agreement in advance. The real sense of urgency, which a deadline set in advance conveys, can be beneficial in the sense that it would oblige the parties to reach an agreement within a given timeframe. Still other process-related factors, such as the frequency with which or the circumstances under which the parties should meet, may also influence the outcome of a negotiation. When chosen carefully, process- and context-related factors can positively influence the outcome of negotiations; but when and how to use them in negotiating environmental conflicts is unlikely to be very different from other types of conflict management. However, an important and characteristic aspect of environmental conflicts is that they are generally fought out in the public arena, and so the parties are more exposed to public scrutiny and criticism about both procedures and context than in other kinds of dispute.

Finally, there are the *content-related factors*. For a successful outcome it is necessary for the parties at least to agree on an agenda of what they will negotiate on before the negotiations begin. If the agenda, or the facts relating to it, are not clear or disputed, there may be confusion and negotiations may be retarded as a result. So the question is how to decide on the facts to be discussed. The principle of joint fact-finding may be employed to reach an agreement on the facts to be discussed. Following this principle, each party writes down its own views and knowledge pertaining to the conflict in question, following which a common document, drafted by an independent person (e.g. an environmental mediator), is agreed upon by all the parties. This document forms the basis of the negotiations that follow. Another example of a content-related factor is the scope of the issues involved. What are the real concerns of the parties involved? Are they only prepared to discuss the conditions under which an activity is to be realised, or are they willing to discuss the activity itself? It is important, therefore, to reach an agreement on the scope of the issues beforehand.

Bingham (1986) gives a useful overview of the success rate of Alternative Environmental Dispute Resolution (AEDR) in her survey of the history and growth of environmental dispute resolution in general, and mediation in particular. The first mediated case in the USA was in 1973 on the long-standing dispute over a flood-control dam on the Snoqualmie river. An agreement was reached and signed by the end of 1974. With the active support of the Ford Foundation and the Hewlett Foundation, the AEDR was developed into an established, well-known, useful and flexible device for resolving environmental disputes. Various universities, such as Harvard and the MIT, run courses in dispute resolution. Some renowned organisations in this field are the Institute for Environmental Negotiation at the University of Virginia, the Conservation Foundation, the Mediation Institute, and the Environmental Resources Management International.

It would be useful to consider the success rate achieved by AEDR to date. In 132 of the 161 documented cases, the objective of the parties was to reach some form of common agreement. Two-thirds of these disputes involved site-specific issues. On the whole agreement was reached in 103 of the cases (or more than 75 per cent), and success rates did not differ significantly between site-specific and policy-level disputes (Bingham, 1986). It follows, therefore, that the likelihood of success of alternative environmental dispute resolution is quite high.

There are, of course, arguments against the usefulness of alternative environmental dispute resolution. First, many of the cases are never brought before a court. It is thus difficult to tell exactly when, or if, a dispute has been resolved to the satisfaction of all the parties involved. Furthermore, power politics and extra-judicial manipulation often make it difficult and even impossible for a court to come to a reasonable resolution of conflicts. Moreover, it is difficult to devise an objective and generally acceptable yardstick with which to evaluate the performance of the various types of conflict management.

Negotiation, mediation and arbitration

Negotiation, mediation and arbitration are the dominant forms of conflict management. A negotiation is really a forum on which the parties to a conflict discuss relevant matters directly with each other with the aim of coming to an agreement. In the case of arbitration or mediation, on the other hand, an independent third party is recruited to act as a mediator or a 'go-between'. A major function of the mediator, which can be an individual or an organisation acceptable to the parties in dispute, is to determine which side has the best case on the basis of arguments advanced and evidence presented. It imposes a binding settlement on the parties instead of expanding their room to manoeuvre. According to Borisof and Victor (1989), 'Mediation is essentially the art of persuasion.' In deference to common practice, we will confine present discussion to external, integrative and independent mediation and not dwell on internal mediation or arbitration (Dorney, 1987).

There have been formal and informal mediators ever since ancient Babylon and China. Mediators have been involved in such diverse fields as arranging marriages, negotiating peace treaties, and resolving marital, labour, industrial and judicial disputes. Conflict management in what are called 'exchange relationships' between human actors involves negotiation and bargaining. Some authors claim that values can be created only by integrative (co-operative) strategies. Habermas's Communicative Governance, which we referred to earlier, is indicative in this respect. In strong contrast to this viewpoint, there are those who foster the view that distributive (competitive) strategies may lead to the creation of values in negotiations, as the parties seek non-adversarial solutions that may have been overlooked (Dworkin, 1990).

The role of mediators in AEDR

The introduction of mediators in AEDR may help to reduce legal costs, prevent conflicts from going through lengthy lawsuits, and may result in conflict resolution sooner than would be possible otherwise. Despite these advantages, many conflicts are still brought before courts because many of the parties prefer to take the classical route to the resolution of environmental conflicts. However, if the parties are really prepared to resolve a conflict without aiming for victory, a mediator may have a useful role to play. Furthermore, when more than two parties are involved, it may be more difficult to bring a case before a court because the case must then be more clearly structured, and its very nature may be changed as a result. Even so, mediation seems to be a pragmatic course of action when the parties are really interested in reaching an agreement rather than winning a victory, especially if more than two parties are involved.

Regarding the organisation of alternative environmental dispute resolution, it should be noted that legal aspects generally play little or no part during the formation of an environmental conflict. Attention focuses on litigation, or administrative hearings, only when the parties become intransigent with regard to their views and perceived rights pertaining to the conflict. Ideally, therefore, an environmental mediator should be called in before arguments begin to be exchanged. Then, by playing a role in setting out the rules for debates and negotiations, the mediator can take into account the factors described earlier (see pp. 445–6), and thus enhance the likelihood of a successful resolution. In particular, the scope of the issues should be defined and the facts behind the conflict clarified. Furthermore, it is important to determine the number of parties involved, and to identify the main players in the conflict.

For example, when the conflict is about the construction of an airport, the parties involved could be those who take the initiative (e.g. a public authority, or a private company, or a partnership of the two), a local municipality (because of the expected noise pollution, or the area to be taken out of its jurisdiction), environmental pressure groups (concerned with the negative environmental impacts of the proposed activity), and of course the central government (Gilbreath, 1984). With such a set of players, it is generally not easy to determine the scope of the issues. Critical questions have to be answered, such as: Should the airport be built at all? Should it be built now, or could it be built in a more environmentally friendly way? etc. The issue of cleaning up the polluted soil arising from the activity is, however, a fundamentally different one. The questions pertaining to it are which of the parties should be responsible for the clean-up, and, if no one is willing to pay, who is going to take the initiative to start the clean-up? Is it going to be the former owner of the site, or the public authority responsible for the health of the local population? One may conclude that the organisation of conflict management will be different in each case, mainly because the number and nature of the parties involved would vary from case to case. And this makes it difficult to formulate the general rules that should apply.

However, as we have pointed out earlier, the key to successful conflict management lies in convincing the parties that discussing pertinent matters in a spirit of co-operation is a better way forward than the adversarial approach. It is certainly more efficient in terms of cost and time, and, above all, it is much more to be recommended than going to court. Indeed, it has been clearly demonstrated to be so in a major project in the Netherlands involving soil pollution. In this instance a number of cases were brought before the court arising from disputes over who was liable for the clean-up costs, even though costly measures had already been taken to prevent further damage to the environment. Such lengthy and costly procedures contribute little or nothing to improve the environment. In fact the contrary prevails, because the companies tend to oppose and delay environmental regulation by highlighting the fixed cost implications of compliance. It is interesting to note for this particular conflict that in the USA 88 per cent of the money, which the insurers paid out between 1986 and 1989 to meet 'environmental' claims, actually went to legal and administrative costs, while a meagre 12 per cent went to actual site clean-ups (Porter and Van der Linde, 1995). It is tempting to speculate on what the result would have been if the AEDR had been applied to the case.

ENVIRONMENTAL CONFLICT MANAGEMENT

Dissemination of information and other tools for environmental conflict management

A particularly important instrument in the successful management of environmental conflicts is the dissemination of relevant information. The success of a negotiation is contingent upon the free flow of relevant information among the various parties involved. In the setting of the Mutual Gains Approach (MGA), the flow of information should be co-ordinated by the mediator. However, control over information and its dissemination gives the mediator enormous powers which he or she could be tempted to use for personal gain. Also, a skilled mediator will emphasise the information which might lead to a successful outcome, or suppress that which could impede the progress of negotiation. Thus it is important to ensure that the mediator is of high personal integrity and has a high level of professional skill in mediation.

The flow of information becomes more complex as the number of parties involved increases. This is because the number of information flows is equal to $N \times (N-1)$, in which N denotes the number of parties involved. There will be 12 information flows, for example, when the number of parties involved (N) is 4, on the assumption that each party will receive information from all the other parties. But in practice this is not always so, because usually all the information flows are not equally important. A decision on their relative importance can be taken by the mediator. In terms of logistics, it is more efficient for the mediator to act as the 'information hub' where all information is received from all the parties and from which it is disseminated to them, rather than all the parties sending information to each other. Under this system each party submits its information to the mediator who then sends the relevant pieces out to all the parties involved, in a summarised form if deemed appropriate.

The resolution of environmental conflicts must be based on open, frank and structured communication between the parties involved. Otherwise the basic knowledge of the different parties on pertinent issues may differ, and this may delay or thwart an agreement. It is possible to begin discussing opinions, based on facts, only when a common factual ground has been established. For example, if a manufacturer starts a new process in an existing factory, it may be that the impacts of the factory on the environment will now be different from what it was before. Whereas the local authority may ask for extra provisions to prevent the new environmental impacts, certain environmental pressure groups may consider the new process to be unacceptable on the ground that the old process was more environmentally friendly. However, it could well be that the new process has different but positive consequences for the environment. Also, the potential impacts of the new process (danger to the environment from leakages in the case of a breakdown, etc.) might be less than in the old process. Of course, the reverse situation can also apply. For example, a third party may demand a specific new process, which was not selected by the initiator of the project, as well as some kind of EIS that could be useful in explaining its environmental credentials. In this situation it may be necessary to provide a complete description of the new process, together with its potential and manifest environmental impacts in order to convince the third parties of its positive aspects. This would call for a kind of *environmental impact assessment* (EIA) or *environmental impact statement* (EIS) (Sadler, 1995). If a simplified EIS is provided for every new project with environmental impacts, instead of only for those above a minimum size, it would be much easier to discuss such impacts with all the parties involved. Clearly, EIS can be a useful tool in environmental conflict management, especially because it constitutes a specific rather than a standardised form of information transfer. There may be other tools too, but it would seem reasonable at present to assume that most of these could be classified as being pseudo-EIS or variants thereof.

The durability of environmental conflict resolution

The durability of the solution to a given environmental conflict depends on the way in which it is resolved. We have already established that environmental conflict management is essentially nothing more or less than bringing the opposing parties to open communication and discussion with each other. But, in order for the solution to be durable, it is necessary to integrate environmental conflict management into general business management. This means that general management should regularly question itself about the impacts of business decisions on the environment, and, at the same time, assess the consequences of current and future environmental issues on business decisions.

In general, environmental pressure groups are more optimistic about the application of a new and more environmentally friendly technology than those who have to apply it, because the latter have to consider other aspects too, such as the economic feasibility of the new technology, its level of safety, and the ease of its implementation. A structured dialogue between the two groups, also involving the manufacturers of the existing production technology, can help prevent a conflict in such a situation. A structured dialogue at the macro-level between the relevant ministry and employers' organisations may also be helpful, assuming that the views of the pressure groups are taken into account by the ministry.

New or emerging environmental conflicts can also be prevented by a structured and permanent dialogue between the parties involved. Potential conflict over whether or not a certain legal provision or standard should be introduced is a typical example of this.

Thus, in order for solutions to environmental conflicts to be durable, it is clear that they should be arrived at through institutional discussions involving the various parties rather than through *ad hoc* decisions. Regular meetings of the parties can help prevent differences in opinions or requirements from escalating into more intractable conflicts. An individual business owner who occasionally meets with local environmental pressure groups, or an organisation of entrepreneurs who discuss the introduction of new environmental standards with the relevant national ministry, are typical examples of this. Under such circumstances environmental conflict management becomes more about the prevention of conflicts by integrating environmental issues into general business management, than about solving them after they have occurred or escalated.

Environmental conflict management: a case study

Several interesting case studies of alternative environmental dispute resolution will be found in Brown *et al.* (1995). To illustrate how the key ideas and strategies in environmental conflict management are applied in practice, a typical case study is given in Box 23.2 which is on the clean-up of contaminated land. In this case a company finds itself responsible for the clean-up of toxic wastes, and therefore has to deal with an angry public. In trying to find a solution to this problem, distinction must be made between the five phases (Figure 23.2) described below. In each phase the company must acknowledge the concerns of the other side, encourage joint fact-finding, and share power. The characteristics of the traditional approach will be demonstrated in each phase, in order to highlight the essentials of the MGA.

The background to this conflict was as follows: A company purchased a plastics manufacturing plant with every intention of operating it. This, however, turned out not to be financially viable. The company therefore sold the land to a real estate developer, who planned to build a shopping centre on it. But the site turned out to be contaminated with a number of toxic chemical compounds. The company agreed, in accordance with the terms of the sale of the land, that it was responsible for the clean-up.

Phase I

There are two ways in which the company can respond to expert findings. The first is the classical approach. In this approach, if followed, the company hires technical experts to ensure that the 'waiver' application is technically correct. Then it hires a

BOX 23.2 AN EXAMPLE OF THE APPLICATION OF THE MGA TO A SOIL POLLUTION PROBLEM

Phase 1
The company faced a clean-up effort lasting at least 6–8 years. It hired a contamination expert to do a preliminary assessment of the risks to human health and safety. The expert reported that the contamination seemed to be substantial (e.g. numerous solvents deemed potentially carcinogenic by the EPA, leaking from underground storage tanks), but that the risks to public health were, in all likelihood, small. As a result, according to the law, the company had the option of applying for a 'waiver', which would mean that elaborate state regulatory review could be suspended, in order to reduce the time it would take to complete the clean-up.

Phase 2
An additional round of studies, required by the state, indicated exactly where and what the contaminants were. The consultants suggested several possible clean-up strategies with their price-tags and inherent characteristics, e.g. that on-site remediation using mobile incinerators was much less expensive, but more controversial, than to dig out the soil, transport it, and clean it up elsewhere by some chemical process.

Phase 3
A resident, living in the area for a long time, announced to the press that there were a 'large number' of families with autistic children living near the site. He or she suggested that there might be a connection between the contamination and the unusually high incidence of autism in the area.

Phase 4
The waiver was granted. Now the company, and not the state, was responsible for managing interactions with the community. So an open community meeting was organised to present the company's risk assessment, remediation strategy, and site management plan.

Phase 5
As a result of the community meeting, the company was pressed to offer various forms of financial compensation to the municipality, as well as to residents living nearby. Some of the complainants pledged not to take legal action if substantial compensation were paid.

lobbyist, who is well-connected to the government to 'grease the skids' for the waiver application, and a press agent to put out positive news items and press releases reporting on the 'low risk' findings of the experts. Finally the company makes an official application for the waiver.

The second approach is the Mutual Gains Approach (MGA). In this the company first sets up a Community Advisory Group (CAG) consisting of public officials and representatives of concerned groups. Then a meeting is convened between the company and the advisory group in which a government official explains the waiver process and what it means. The question of whether or not to apply for the waiver is put to the CAG. It is explained that the waiver might save time and money, and that these funds could be used for cleaning up the site. Next an offer is made to provide funds for the CAG to hire technical experts to prepare a joint fact-finding report.

A comparison of the classical and mutual gains approach in this phase shows that, while the application for a waiver is taken as the starting point in the former, it is seen in the latter as the eventual outcome of a structured dialogue between the parties concerned. If the waiver is not accepted by the CAG in the first round, then a second round of negotiations may be started. The extra time needed for this may be more than compensated for by saving time in the following phases.

Phase 2

The classical approach in this phase (see Figure 23.2) is to prepare a technical report for the government justifying the use of mobile incinerators, to line up experts to explain at public meetings why it is best to use mobile incinerators, and to hire the services of a press agent to put out stories testifying to the benefits of mobile incineration. In contrast, the mutual gains approach involves the presentation of a report to the CAG which sets out the costs and benefits of *each* alternative clean-up option. The members of the CAG then try to reach a consensus on the most appropriate choice. In addition, the company offers to mitigate and compensate for any unintended side effects of mobile incineration, if this particular option is chosen.

Again, a comparison between the traditional and the mutual gains approach reveals fundamental differences. In the former the company chooses the cheapest solution (mobile incinerators), and thereafter seeks to defend this option before the public and the authorities. The MGA, on the other hand, leaves all options open for a decision to be made by the CAG. In addition, whatever option is chosen, the company pledges to compensate for any unintended side effects of the clean-up operation. It appears that the second approach (MGA) is likely to lead to a more durable solution, because it leaves all options open for scrutiny based on merit.

Phase 3

The traditional approach of the company in this phase (Figure 23.2) is to hire some experts to testify that the causes of autism (alleged to be caused by toxic substances) are not well understood and to deny any liability of the company beyond the cost of cleaning-up the contaminants. It is also possible that the company would hire other experts, or influential local residents, to raise doubts about the credibility of concerned residents. By contrast, in the mutual gains approach the seriousness of the residents' concern is acknowledged by the CAG, for example, by sponsoring a one-day seminar on the possible relationship between autism and contamination in which experts from both sides take part. Work then begins (in co-operation with the state public health authority) to assess the incidence of autism in the most professional manner. Finally, a short-term technical solution is proposed to make the site totally secure, proportionate to the level of the community's concern, followed by a joint site monitoring programme.

It is clear from the above that the classical approach is primarily directed at denying any relationship between autism and contamination in order to avoid any liability. The MGA, on the other hand, leaves this question open, to be decided upon later by the CAG and the state public health agency. Meanwhile, it offers a short-term technical solution to meet the current concerns of the community. Thus the MGA provides for greater confidence of the community in the company than the classical approach and, as a result, phases four and five can be dealt with more quickly, easily and cheaply.

Phase 4

The traditional approach involves hiring experts to make a public presentation, staged and managed by the company, followed by a glossy press kit, prepared by a media consultant, and an official statement by the company announcing that a solution to the contamination problem has been found, and that no real risks exist (Figure 23.2). By contrast, public presentations in the MGA are designed jointly by senior officials of the company and the CAG. During presentations all those wishing to speak are given the opportunity to do so. Finally, the company works closely with the CAG to reach a consensus on the best clean-up strategy.

A comparison of the two approaches shows, once again, that the classical approach tries to play down the contamination problem and to impress the public by hired professionals. The MGA, by contrast, is characterised by an open attitude on the part of the company in terms of both the joint organisation of public presentations and its relationship with the CAG. The clean-up strategy is also worked out in collaboration with the CAG. Thus the second approach (MGA) gives a much better guarantee that the solution to the contamination problem will be accepted by all the parties involved.

Phase 5

In the classical approach a company spokesman denies in the first instance all liability beyond that for the clean-up, and indicates the willingness of the company to go to court to prove its innocence. The company may also offer payments to key individuals secretly and without admitting guilt. Finally, it hires the best available lawyers to prepare a legal defence against the remaining opposition, whilst going ahead with its own clean-up plans (Figure 23.2).

In the MGA, on the other hand, a public offer is made to compensate all those who could demonstrate that the contamination had caused them harm. The company would then enlist the services of the CAG to seek consensus on standards and methods of proving the level of harm caused, and agree to participate in a joint arbitration panel to review the claims. The company would also agree to subsidise the cost of long-term health and environmental impacts after the clean-up was completed, and, finally, agree to set up a modest community trust fund to be administered locally for claims arising after ten years.

This last phase reveals most poignantly the differences between the traditional approach and the mutual gains approach in solving environmental conflicts in general, and this conflict in particular. Clearly, the second approach (MGA) emphasises the long-term interests of the company by: (a) promising to subsidise the cost of long-term health and environmental impacts after clean-up; (b) agreeing to set up a community trust fund to meet claims made after ten years; and (c) agreeing to participate in a joint arbitration panel during the next ten years. The classical approach is, on the other hand, adversarial in nature whereby the long-term relationship between the company and the community is determined by lengthy and divisive lawsuits over claims. It goes without saying that this approach does little to enhance or promote the public image of the company which is essential to its profitability.

Lessons from case studies

Perhaps the most important lesson to be learned from this particular case study, and indeed from other case studies, is that the need for acknowledging the interests and concerns of the other parties (e.g. the community, residents living in the neighbourhood, the municipality, the public health agency, etc.) is paramount in environmental conflicts. Also, joint fact-finding should be encouraged to arrive at common standards and methods of proving the extent and level of harm caused, and commitments should be made to minimise impacts or to compensate for unintended effects during and after clean-up operations. And finally, the focus should be on building long-term relationships through consultation before decisions are taken and through the sharing of power. However, although all this may seem very natural and commonsensical, its achievement in practice is often fraught with much difficulty because, the very moment a company is faced with an environmental claim, it often tends to lapse into an adversarial stance and is tempted to use its powers of obstruction. The mutual gains approach, on the other hand, has much to recommend itself, especially for enhancing and promoting the long-term interests of companies *vis-à-vis* the communities they serve.

OBSERVATIONS

Methods and concepts

In many of the environmental conflicts the stakeholders often disagree fundamentally among themselves because their stakes are at odds with each other. While the interest of one of the parties may be to preserve the present state of the environment as perfectly as possible, another may intend to start a new production process, or change an existing one, with untested or doubtful environmental credentials. A third party in the conflict may be the authority, or authorities, responsible for setting and enforcing feasible rules of conduct.

There are many aspects to an environmental conflict, of which the technical, legal and financial aspects merit special attention. Indeed, technology itself can be a source of environmental conflict when there are differing perceptions of whether a specific technology has been proven or not, and also when it is not certain as to who is going to take the possible risks of that technology. An environmental conflict may reduce to a legal dispute when the ambiguity of some standards is at issue and needs to be clarified. Typically, a financial conflict may arise when there is no agreement on how much tax should be paid for a certain amount of pollution created, or what a reasonable compensation for a given environmental damage should be. All such conflicts can, and often do, escalate into court cases.

However, for a variety of reasons, which we have referred to earlier, it is much more efficient and amicable to settle environmental disputes out of court. Such conflicts may also be prevented, or at least alleviated, by efficiently managing the appropriate issues and actions under the provisions of 'enlightened self-interest' and 'responsible care'. The MGA, modelled on these provisions, reflects the concept of 'shared responsibility' which acknowledges the diverse interests of the stakeholders. Thus it makes it possible to come to an agreement that is satisfactory to all the parties to the conflict.

There are three general rules for managing environmental conflicts:

1 Priority should be given to the prevention of environmental conflicts because they are costly to resolve in terms of both time and money, and delay only exacerbates them. To prevent an environmental conflict, it is necessary for all the parties involved to be aware of the advantages of prevention. The concept of *enlightened self-interest* may help to promote this awareness.
2 Temptation to rely on the classical, minimalist concept of *punishment and reward* should be avoided. Resorting to this concept will not convince public environmental groups that they are dealing with a responsible partner; they will see this as the failure of the company to search for creative solutions and as merely its reaction to external stimuli.
3 *Co-operation* is more effective in conflict resolution than confrontation. Parties that co-operate search for solutions which offer benefits to all the active parties. Such solutions can be adapted more easily and quickly than when one of the parties loses out. Thus co-operation is a better approach than fighting over who is right and who is not. Such issues about who is to be blamed are less important than questions on how to cope with the environmental damage and who should take responsibility.

Further consideration of the MGA

The Mutual Gains Approach 'is a practical consequence of the rule of co-operation'. It requires all parties to the conflict to be convinced that negotiations produce better results than walking away from the negotiating table. There are several reasons why the parties may expect more from negotiations, however, apart from avoiding both frustration and a negative impact on the corporate image. Long-term consequences of abandoning negotiations may be that new and more severe conflicts will develop later, resulting in greater frustration which can, in turn, worsen the image further. This negative 'vicious cycle' may be broken by promoting the Mutual Gains Approach. This approach is characterised by the following elements (Susskind, 1987):

1 The mutual gains approach is built on the following four pillars:
   (a) acknowledgement of the interests and concerns of the other parties;
   (b) encouragement of joint fact-finding;
   (c) accepting responsibility, admitting mistakes, and sharing power;
   (d) focusing on the development of long-term relationships.
2 There are five principles of action: (i) taking the initiative; (ii) seeking consensus; (iii) emphasising outcomes; (iv) maintaining credibility; and (v) enhancing legitimacy.
3 There are three arguments with which to convince the parties that serious negotiation produces a better outcome than ignoring the conflict:
   (a) forestalling of negotiation merely shifts the conflict, does not solve it;
   (b) lack of prompt action in the beginning may cause difficulties in resolving the original problem later, because public sentiment may turn against the unwilling party;
   (c) thwarting negotiations may lead to a worsening of the situation brought about by increased environmental damage.
4 Three types of factors determine the success of the MGA:
   (a) party-related factors: for instance, the greater the number of parties involved, the more difficult it would be to resolve the conflict successfully;
   (b) process- and context-related factors: for instance, have deadlines been set for reaching an agreement? Should there be progress reports on negotiations? Do the parties have adequate perceptions of each other?;
   (c) content-related factors: for instance, are the facts of the conflict clear? Is there agreement between the parties on the basic issues of the conflict?
5 The MGA approach may rely on the services of a mediator who is familiar with the environmental issues of the conflict. He or she can convince the parties that direct discussions are the best way to solve the conflict, identify the relevant parties, and to activate the success factors mentioned above. The mediator's most important task is to introduce, explain and implement the Mutual Gains Approach.
6 All parties to the conflict should enjoy free and equal access to information on the present and future environmental impacts of the initiatives under discussion. An environmental impact statement, or a similar report, could be a useful aid for the dissemination of environmental information. The mediator can play a role in co-ordinating the free flow of all relevant information to and from the parties.
7 A lasting and structured dialogue involving all the parties to the dispute is crucial in preventing new conflicts, and usually it leads to durable solutions. It can be organised and co-ordinated by an external mediator, and also by an internal environmental mediator trained for this function. Important tasks for the mediator are: (i) to provide the external parties with a constant flow of information on the environmental management of the organisation, and (ii) to bring together the relevant parties.

Recommendations

1 The Mutual Gains Approach (MGA) should be promoted, and applied wherever possible, for the successful management of environmental conflicts.
2 In order to prevent environmental conflicts, it is advisable to build up long-term relationships with all the stakeholders involved.
3 It is important to acknowledge that the stakeholders in any environmental conflict will have diverse views on the environment as well as different values which they attach to it.

REFERENCES

Asselt, H.Th. van (1989) *A Common Approach: or Maintaining the Status Quo: How to Anticipate Developments in European Environmental Policy*, RSM Management Report Series no. 55.

Bingham, G. (1986) *Resolving Environmental Disputes: A Dcade of Experience*, Washington, DC: The Conservation Foundation.

Brown, V. (1995) *Risk and Opportunities. Managing Environmental Conflict and Change*, London: Earthscan.

Buhr, N. (1991) 'The Environmental Audit: Who Needs It?', *Business Quarterly*, Winter: 27–32.

Cramer, J. and Zegveld, W.C.L. (1991) 'The Future Role of Technology in Environmental Management', *Futures* 23 (5): 451–468.

Davis, J.D. (1991) *Greening Business: Managing for Sustainable Development*, Oxford: Basil Blackwell.

Dorney, R.S. (1987) *The Professional Practice of Environmental Management*, New York: Springer-Verlag.

Dunne, J.M. van (ed.) (1991) *Transboundary Pollution and Liability: The Case of the River Rhine*, Lelystad, The Netherlands: Vermande.

Gilbreath, K. (ed.) (1984) *Business and the Environment: Toward Common Ground* (2nd edition), Washington, DC: The Conservation Foundation.

Glasbergen, P. (ed.) (1995) *Managing Environmental Conflicts. Network Management as an Alternative*, Dordrecht, The Netherlands: Kluwer Academic Press.

Gray, B. (1989) *Collaboration, Finding Common Ground for Multiparty Problems*, London: Jossey-Bass.

Greeno, J. Ladd (1990) 'The Environmental Leadership Strategy: A New Imperative for Management', Presentation at Third German Business Convention, University of Cologne, February, pp. 1–12.

Porter, M.E. and Linde, C. van der (1995) 'Green and Competitive: Ending the Stalemate', *Harvard Business Review* Sept.–Oct.: 120–134.

Sadler, S. (1995) 'Canadian Experience with Environmental Assessment: Recent Changes in Process and Practice', *Australian Journal of Environmental Management* 2: 112–130.

Susskind, L. and Cruickshank, J. (1987) *Breaking the Impasse: Consensual Approaches to Resolving Public Disputes*, New York: UIT, Basic Books.

Winter, G. (1988) *Business and the Environment: A Handbook of Industrial Ecology*, Hamburg: McGraw-Hill Book Company.

SUGGESTED READING

Cairncross, F. (1993) *Costing the Earth: The Challenges for Governments, the Opportunities for Business*, Boston, Mass.: Harvard Business School Press.

Johnson, B.B. (1987) 'Public Concerns and the Public Role in Siting Nuclear and Chemical Waste Facilities', *Environmental Management*, 11 (5): 571–586.

Moore, Chr.W. (1996) *The Mediation Process* (2nd edition), San Francisco, Calif.: Jossey-Bass Publishers.

Neuman, F. (1996) 'The Incorporation of Environmental Elements in Strategic Decision-making Processes in Industry: Government – Corporate Interaction', Dissertation, Erasmus University, Department of Social Studies.

Syne, G. and Eaton, E. (1989) 'Public Involvement as a Negotiation Process', *Journal of Social Studies* 45 (1): 87–107.

SELF-ASSESSMENT QUESTIONS

1 Which of the following statements is most appropriate for explaining the concept of enlightened self-interest? Explain why the concept of self-interest is more effective in preventing environmental conflicts than the concept of reward and punishment.
   (a) The concept of enlightened self-interest makes each party aware of its own responsibility.
   (b) The concept of enlightened self-interest leads to a better mutual understanding of the parties to a conflict.
   (c) The concept of enlightened self-interest gives each party an opportunity to gain competitive advantage by seeking co-operation with the opposing parties.

2 What are the characteristics of the Mutual Gains Approach as compared with the more confrontational types of negotiation? The characteristics are:
   (a) Acknowledgement of the concerns of the other parties.
   (b) Acknowledgement of the concerns of the other parties and joint fact-finding.
   (c) Acknowledgement of the concerns of the other parties and of one's own mistakes, and joint fact-finding.
   (d) Acknowledgement of the concerns of the other parties and of one's own mistakes, joint fact-finding, and transferring responsibility to the other parties.

3 What are the five principles of action needed for implementing the 'mutual gains approach'?
   (a)
   1 taking initiatives;
   2 seeking consensus;
   3 emphasising outcomes;
   4 maintaining credibility;
   5 enhancing legitimacy.

(b)
1 taking initiatives;
2 writing an agreement;
3 emphasising outcomes;
4 maintaining credibility;
5 enhancing legitimacy.
(c)
1 taking initiatives;
2 seeking consensus;
3 emphasising outcomes;
4 maintaining credibility;
5 using an environmental mediator.

4 What are the three main factors which determine the likelihood of success of the Mutual Gains Approach? The three main factors are:
(a)
1 party-related factors;
2 content-related factors;
3 process-related factors.
(b)
1 party-related factors;
2 content-related factors;
3 context-related factors.
(c)
1 party-related factors;
2 content-related factors;
3 process- and context-related factors.

5 What are the tasks and the role of an environmental mediator?
(a) To identify the parties.
(b) To transfer information between the parties.
(c) To explain and implement the Mutual Gains Approach.
(d) To identify the parties, transfer information between parties, and explain and implement the Mutual Gains Approach.
(e) To transfer information between the parties, and explain and implement the Mutual Gains Approach.

24

CURRENT OPTIONS OF THE UN FOR ENVIRONMENTAL CONFLICT MANAGEMENT

Martin Tampier

SUMMARY

International and transnational conflicts, both potential and manifest, in which environmental issues and problems play a major role, have been identified in this chapter together with a description of the existing organisations and institutions of the United Nations which deal with them. The capacity, scope and legal framework of these organisations and institutions, as well as their strengths and weaknesses for settling international and transnational disputes involving environmental problems and issues, have been examined. Suggestions have also been made for making them more effective for settling such disputes, which are almost certain to increase both in number and intensity in the years to come. The political intricacy of environmental diplomacy, against the background of conflicting factors such as national sovereignty, economic interests, political power balance as well as technical and scientific aspects, has also been referred to.

ACADEMIC OBJECTIVES

After completing this chapter the student should have a good working knowledge of the following:

- the concept of a transnational conflict arising from the use of the environment, how it occurs, and a general approach to its management;
- the existing organisations and institutions of the United Nations and their capacity, scope, legal framework, strengths and weaknesses for dealing with international and transnational conflicts;
- the important agreements and conventions currently available to which recourse could be taken for settling or resolving such conflicts;
- possible ways in which both the institutional framework and mechanisms could be improved and made more effective for the settlement or resolution of such conflicts.

DEFINITION OF TRANSNATIONAL ENVIRONMENTAL CONFLICT MANAGEMENT

In order to understand what is meant by Transnational Environmental Conflict Management (TECM), we need to look at every single aspect of what a conflict is and how it arises. To begin with, we note that 'a conflict occurs when the current or planned action of one party is incompatible with the claims of another party or basic principles of conduct'.

This means that a conflict can arise even before the action in question has taken place. In other words, it can arise at the planning stage of an action. A conflict, in the sense used in this chapter, will be said to arise even when there is only *one* obvious party to the 'conflict', where the generally accepted principles (the basic principles of conduct) of equity, justice, fairness, human rights, and in this case also special environmental principles, are disregarded or violated. In this context environmental principles can be based purely on values, or on scientific facts. 'Intergenerational equity' and the 'precautionary principle' are typical examples of value-based principles, while scientific principles would include 'limit values' and 'carrying capacity'. The definition also includes parties to a conflict that are unable to express their desires effectively because of their limited power, or because they are as yet unborn (future generations).

In general, the UN becomes involved in a conflict only when it is *transnational* in nature; that is, when the conflict extends beyond the borders or jurisdiction of only one nation-state, or has the potential to do so. Whereas international conflicts are limited only to the conflicting nation-states, the parties to a transnational conflict can be individuals, corporations, or organisations within different nation-states, or the nation-states themselves.

In order to give an idea of how *environmental* conflicts can be distinguished from other conflicts, Table 24.1 shows a possible classification of environmental conflicts according to five different areas and stages.

The *management* of environmental conflicts primarily comprises their settlement or resolution. The *settlement* of a given conflict does not remove its source, but merely contains its symptoms and includes, if necessary, the measures for its deterrence or repression. Also, whether or not a *settlement* would lead to a timely solution to the problem or issue in dispute depends on the relative power relationship of the adversaries (Burton, 1990). Conflict *resolution*, on the other hand, seeks to eliminate the source of the problem and is meant to prevent future conflicts about the same issue (Burton, 1990). The outcome of a *resolution* contains prevention and avoidance

Table 24.1 Classification of environmental conflicts and some examples

Horizontal classification (areas)	*Vertical classification (stages)*
Transboundary pollution • acid rain • radioactivity • river pollution • oil spills • sea dumping	Unperceived 'conflict' • conflicts between human activities and the eco-system's carrying capacity (scientifically not established) • emission of ozone-depleting substances before the discovery of their harmful effects on the ozone layer in 1970
Environmental resource conflicts • allowable fish catch • allocation of water consumption rights	Conflicts perceived by a minority • emission of ozone-depleting substances in the 1970s • transboundary pollution damage to individuals in country A without standing before the concerned foreign court in country B (origin of pollution) which is not taken up as an issue by the government of country A towards country B
Global conflicts • ozone depletion • greenhouse gases • deforestation • desertification	Acknowledged conflicts • acid rain issue between Sweden and several European countries • conflict between Norway and the International Whaling Commission about whaling
Conflicts relating to nature conservation • rare species • natural area preservation • dolphin-friendly tuna catch methods • allowable fish catch • whaling	Crisis • trade sanctions against Iraq to pay for environmental damage inflicted during the Gulf War • bans on tropical timber imports • boycott of Norwegian products because of continued whaling • 'cod wars' between Great Britain and Iceland
Legal conflicts • national environmental legislation as trade barrier • implementation and interpretation of international agreements • environmental criminality	'Hot' war • civil strife • international armed conflict

measures, and it is lasting in terms of the financial, technical, institutional and ecological aspects of the dispute in question. Other areas of TECM are *preventive* measures in the forefront of an open conflict, and possibly even the *creation* of conflicts (e.g. through research) to reveal activities with potential for violating environmental principles.

GENERAL METHODS FOR SETTLING CONFLICTS

In general, a number of different political and judicial methods are available for settling transnational conflicts. These methods either involve only the parties to the conflict, or also a third party which acts as the mediator or arbitrator. The following methods are recommended in Article 33 of the Charter of the United Nations which also refers to the special role of the UN institutions. A more comprehensive overview of conflict resolution methods will be found in Bilder (1975), Merrills (1991) and Schneider (1979).

Political methods

Direct negotiation between the parties involved, without third-party interference, is generally the first approach to the management and resolution of conflicts. By offering its good offices for negotiations, a third party such as a nation-state or an organisation can provide a channel of communication or facilities without influencing the actual negotiating process. Although environmental disputes may involve only two or a few parties, the impacts of the issue in dispute can be far wider in scale – national, regional or even global; but this is usually not taken into account in the negotiating process. Furthermore, actors other than nation-states can only be represented in negotiations if the government concerned takes the responsibility for both polluters and victims. But there may be a lack of expertise in dealing with such disputes, since only local people (i.e. government representatives) are usually involved. However, in spite of this deficiency, the political process of compromise often proves to be more satisfactory than the 'victory or defeat' solution which results from a judicial process, especially when international issues are at stake.

If negotiations become deadlocked, mediation through a third party suggesting terms of settlement can facilitate further negotiations, although there is usually no commitment on the parties to the dispute to accept the mediator's proposals. Conciliation, which is applied especially to the legal aspects, means a more institutionalised process of mediation which puts greater political pressure on the parties in dispute to find a solution. Generally it involves a special commission, or an international body, which is charged with investigating the dispute and producing a report with proposals for its settlement.

If negotiations or mediation continue fruitlessly for years, as sometimes happens, it may be advisable to split up the conflict into separate aspects to be dealt with individually. Each individual aspect can then be inquired into by an independent commission, or a court, at the request of the parties to the conflict. In such cases the inquiry report is only concerned with facts and does not propose any terms of settlement.

Judicial methods

An inquiry can also be aimed at resolving all the aspects of a dispute. When conducted for such purposes, an inquiry becomes part of an arbitration process which leads to a binding decision. In contrast to arbitration, which is performed through an *ad hoc* commission chosen by the parties to the conflict, a judicial settlement is undertaken by a standing court such as the International Court of Justice (ICJ). The decisions of arbitration, or of judicial processes, are mostly based on international law or on a particular set of rules agreed upon by the parties to the dispute. Such decisions are binding, but not necessarily final because, regardless of what might have been agreed upon by the parties previously, interpretation, revision, rectification, appeal, or even nullification of the decision may be possible subsequently. However, as there is always a risk that a given case might be lost under a judicial or arbitrated settlement, nation-states do not normally use such procedures if they consider dispute resolution to be less important than the outcome itself.

Generally arbitration is seen as being more flexible than judicial processes, because preliminary jurisdictional and choice-of-law questions are settled by the parties themselves. It can also be tailor-made to meet the specific needs of the case, taking into account the interest of the international community with reference to international law. But, on the other hand, it is less pragmatic than political methods, and often very time-consuming too (Schneider, 1979). According to Adede (1987), the latter is due to procedural difficulties and the fact that often the parties to a conflict spend too much time on the appointment of suitable arbitrators.

CURRENT OPTIONS FOR TECM WITHIN THE UN

The Permanent Court of Arbitration

The PCA is a panel of arbitrators available to any nation-state or corporate party seeking to resolve a dispute through an *ad hoc* arbitration tribunal. In this arrangement a Financial Assistance Fund is maintained to help developing countries meet the cost of conflict resolution under the auspices of the PCA. Even so, the PCA has not been very active in the past decades (Trolldalen, 1992).

The International Court of Justice in The Hague

The Statute of the ICJ only allows adjudication between two or more nation-states, and no organisation or body is granted standing. All nations that are members of the UN are also parties to the ICJ Statute, although this does not mean that they are automatically subject to its jurisdiction. A nation-state can submit itself to ICJ jurisdiction by a unilateral declaration under Article 36(2) of the ICJ Statute; or by a submission following its appearance in an action before the ICJ without raising an objection; or by being involved in cases in which international conventions foresee adjudication by the ICJ (EC, 1993: 6). The last procedure allows non-UN members to come under ICJ jurisdiction. A fourth possibility occurs when nation-states mutually consent to submit a particular problem to the Court. This procedure, which is in accordance with the provisions of Article 36(1) of the Statute, is called *special agreement*, so called because it was especially concluded by the parties to a conflict in a certain case (ICJ, 1994: 71). Cases submitted under Article 36(2) can be referred to the Court unilaterally by one party (ICEF, 1992: 82). Under the other procedures, on the other hand, a case can only be taken up by the Court if all parties to the conflict give their consent for it.

Apart from contentious jurisdiction, the Court can also exercise incidental or advisory jurisdiction. The former means a preliminary decision to implement interim protection measures before the final judgement has been pronounced. Advisory jurisdiction, on the other hand, is concerned with any legal question brought to the Court's attention by a body of the UN.

Articles 26–29 of the Statute allow the Court to create chambers composed of fewer judges (mostly five) than the 15 who attend the full Court, in order to facilitate the speedy dispatch of business for particular categories of cases, or to deal with individual cases. An environmental chamber, consisting of seven judges, was created in 1993 (ICJ, 1994), but no case has been submitted to it yet (Arnaldez, 1995).

The parties to a dispute themselves determine the basis on which the Court will take its decision. Court decisions can be taken *ex aequo et bono*; that is, in a way that is completely flexible and meets the specific needs of the case being tried. A trial under this arrangement has never been requested, however, and so cases are generally decided in accordance with the special principles of international law, which again consist of conventions, customary law, general principles of law, or scholars' views as expressed in the literature (Article 38). Resolutions of the UN General Assembly, and of other UN bodies and UN conferences, also contribute to these processes, though their precise significance is disputed (Birnie, 1988).

Cases which the Court has dealt with during the last three decades have mostly taken between one and two years for their resolution (ICJ, 1994). The only 'environmental' case the Court is presently dealing with (actually dealing only with some isolated legal aspects of the case), is the conflict between Hungary and Slovakia about the Gabcikovo-Nagymaros Dam Project on the Danube. Submitted on 23 October 1992 (ICJ, 1994), this case was still to be decided in January 1997.

Proceedings of the Court can be quite expensive, however, even though it covers its procedural costs from the UN budget (ICJ, 1994). For example, in a fishing dispute between the USA and Canada, each side spent approximately 10 million dollars just for preparing briefs, papers, and other documents for the Court (Stein, 1985).

In this respect an important asset of the Court is its Trust Fund (ICJ, 1994) which was established in 1989. The purpose of the Fund is to give financial help to Third World countries, thus enabling them to make use of the Court's adjudication procedures. A secondary purpose of the Fund is the implementation of judgements in case the country in question cannot afford it. Dispersal of the Fund is managed by a panel of experts who examine the merits of each case and make recommendations as to how much should be awarded to the parties to a conflict. In the case of environmental conflicts, it is anticipated that such awards might in future involve very large sums of money. Consequently, this arrangement might lead to a situation, at least theoretically, in which the Fund would help poor countries to implement environmentally sound measures at the expense of the richer countries. The likelihood of this coming to pass is, however, very remote indeed, considering that the Trust Fund is not a part of the regular UN budget and that it derives from voluntary contributions made by nation-states, international organisations and individuals.

Nation-states are legally obliged to comply with the decisions of the Court. If they do not, the UN Security Council can decide upon measures to be taken to enforce the decisions (ICEF, 1992). Even so, history shows that nation-states have been rather reluctant to accept the jurisdiction of the Court:

> Friendly neighbours do not take each other to court.
> (quoted in Bilder, 1975: 173)

By 1994 only 58 of the UN member states, out of a total of 185, had submitted themselves to the Court's jurisdiction under Article 36(2) (ICJ, 1994), many with reservations on the time it takes to settle disputes or on the substance of cases (EC, 1993). Nevertheless, apart from its original purpose and objectives, there are many reasons why the Court is useful (Bilder, 1975): it serves as a 'last resort' for cases which cannot be resolved otherwise; it stimulates the settlement of disputes simply by being available; it clarifies and develops rules and principles; and its sheer existence, together with its functioning dispute resolution machinery, is likely to prevent future disputes. Although third-party involvement is often refused, an impartial judgement would nevertheless be welcome (Merrills, 1991: 238) because:

> Referring a dispute to a court which is demonstrably impartial can be very useful when a government is anxious to settle a dispute, but is under pressure not to make concessions. As the decision is not the government's responsibility, then provided it has been arrived at fairly, a strong case can be made for accepting it even if it is unfavourable.

But even in this respect the Court has been used only to a limited extent to date, mainly because many of the nation-states doubt its competence to deal with important aspects of international problems that could either be political or technical in nature. Indeed, the same argument has also been deployed by environmentalists to make a strong case for a separate 'Environmental Court' (Koskenniemi, 1991). The Court has responded to this charge by establishing a special environmental chamber where environmental cases are heard by judges with special expertise in environmental matters (thus not bringing such cases to the full Court of 15 judges); and by allowing the involvement of expert assessors who can participate in the work of the Court, but cannot vote, in accordance with Article 30 of the Statute (Szasz, 1992). Unfortunately, even this arrangement is handicapped by the difficulty that the environmental chamber is not likely to commend itself politically to all the parties to environmental disputes, because the parties would normally prefer a politically congenial set of judges whose views may not coincide with those of the 'environmental' judges of the chamber (Szasz, 1992).

Another point at issue is the absence of an effective enforcement mechanism (Trolldalen, 1992). Although the Security Council can be called upon to enforce Court decisions, this mechanism can be blocked by the veto of just one permanent member of the Council. Furthermore, many of the transnational conflicts cannot be dealt with by the Court because it has no direct competence over the private entities that are often parties to conflicts. It is for these reasons that a number of nations are still reluctant to accept the authority and legitimacy of the ICJ

(Trolldalen, 1992). Other complaints focus on its lack of speed and flexibility (ibid.).

Finally, a reason why only a few awards could be made is the reluctance of nation-states to include liability provisions in the treaties (Trolldalen, 1992; Rest, 1992), or to develop internationally binding principles of liability. In addition, liability often involves non-state parties, whereas only nation-states have standing before the World Court. This complicates matters even more because legal claims of all non-state parties are not considered by the Court, unless they are represented by their respective governments. The problem, though, is that a non-state party can be represented by its government only if the nation-state in question also regards itself as an injured party (EC, 1993).

The UN Security Council

The primary role of the Security Council is to maintain or re-create peace. This role, however, is limited to serious disputes, 'the continuation of which is likely to endanger the maintenance of international peace and security'. Under Article 36(1) of the UN Charter, the Council can make recommendations as to how a certain conflict should be settled at a given time. Furthermore, Articles 11(3), 35(1) and 99 enable the Council to consider a certain matter at the request of a member state, the General Assembly, or the Secretary-General, even without the consent of the nation-state in question. This distinguishes it from arbitration or conciliation procedures in which a tribunal, or a commission, can intervene only when it is requested to do so by the parties concerned. In addition, the Council is free to consider 'any situation which might lead to international friction'; this means that it can act on matters even before they have escalated to a conflict in the actual sense. Also, unlike other institutions dealing with international conflicts, the Security Council can apply military, economic and political sanctions to enforce its decisions.

As yet no action of the Council has focused on international environmental conflicts in the sense in which we have defined them earlier. The question is, whether environmental problems could start to create conflicts that would endanger international peace and security sufficiently to cause the Security Council to take action. According to many authors this will be the case, and indeed it is already the case in some instances. But general objections have been raised (UNA–USA, 1987: 9), as will be seen from the following:

> It would be far more productive if states sought to use the Council for what it is relatively good at – behind-the-scenes negotiations, consensus-building, providing a face-saving way out when the parties are ready to terminate hostilities, and posing alternatives to the parties for their consideration – rather than for what it does less well – conferring legitimacy, passing resolutions with teeth, and enforcing its judgements.

On the other hand, the Council could be said to have legislative powers under Article 25 and Chapter VII of the Charter to act if certain anti-environmental activities are seen as threatening peace (Szasz, 1992). According to Szasz, the Council is nevertheless completely unsuited to deal with environmental affairs because of its voting system – any one of the permanent members can veto any decision – and because it was established to deal with military threats to peace and not with environmental affairs. In any case, countries that are not members of the Council would resist the idea that it should act in this way in environmental emergencies (French, 1992a; UNGA, 1991).

The UN General Assembly

Under Articles 10–14 of the UN Charter, the Assembly is given powers to discuss any subject within the scope of the Charter. This means that the scope of the General Assembly is much broader than that of the Security Council. It recommends 'measures for peaceful adjustment of any situation, regardless of origin which it deems likely to impair the general welfare or friendly relations among nations', if the Security Council requests it to do so (Article 12(1)). This includes all disputes except those that are 'essentially within the domestic jurisdiction of any state' (Article 2(7)). Officials of the UN General Assembly (UNGA), and other UN officials appointed by UNGA, have already served as mediators in negotiations between nation-states in dispute (Trolldalen, 1992).

Recognising the role that environmental conflicts could play in the future, UNGA has requested a report from UNEP which contains, among other things, proposals and recommendations to strengthen the UN's capacity to monitor, assess and anticipate environmental threats, and to issue early warnings to the international community when such threats become imminent (see UNGA Resolution 44/224 (1990)).

The General Assembly discusses environmental matters every two years in its already overloaded Second Committee (the Economic and Financial Committee), even though the environment is a rather small item on its agenda. Nevertheless, the UNGA has issued a large number of resolutions and other documents concerning environmental matters from its regular sessions which have contributed to the development of international environmental law. They represent norms which a majority of the international community is inclined to accept (Fawcett, 1981) and which eventually facilitate the development of new international binding norms (Broadus, 1995).

The UN Secretary-General

The Secretary-General can intervene in disputes, both through his good offices at the request of the Security Council or the General Assembly, and on behalf of interested parties or even on his own initiative (Articles 98 and 99 of the UN Charter). To this end, he can establish subordinate bodies, such as a commission to investigate and resolve a dispute. In the past activities of the Secretary-General have included investigation and arbitration. His independence, even from the other UN institutions, has proved to be a major factor for the acceptance of his role in dispute settlement. According to Merrills (1991: 189),

> it would be wrong to see the efforts of the Secretary-General in isolation, it is clear that his ability to take the initiative will often have the advantage of enabling a potentially significant influence to be brought to bear on a dispute at a very early stage.

Merrills recommends that the role of the Secretary-General in this regard should be enhanced by providing a larger budget, and by delegating his ceremonial and similar functions to other persons. He could then play a more prominent role, especially in bringing disputes to the attention of the Security Council and the General Assembly at an early stage.

Although the Secretary-General has acted as an impartial mediator in international conflicts in the past, it would nevertheless be improbable for him to play a major role in future TECM; for he will only become involved in conflict settlement when a conflict has already escalated. In any case, given all the conflicts which the world is facing today, the capacity of the UN Secretariat, and of the Secretary-General, would not be sufficient to deal effectively with environmental conflicts as well, notwithstanding the enhanced awareness and seriousness of such conflicts.

The Economic and Social Council

The Economic and Social Council (ECOSOC), which is one of the main UN bodies, has 54 government members. It was established under the UN Charter to co-ordinate economic and social affairs within the UN system and to serve as a forum for the development of common economic and social policies. ECOSOC supervises the legal 'relationship' agreements between the UN and specialised agencies such as the Bretton Woods Institutions (the World Bank and the International Monetary Fund (IMF)). The Regional Economic Commissions, as well as several functional commissions including the Commission on Sustainable Development, also fall under its aegis. ECOSOC receives reports from UNEP and other quasi-autonomous UN bodies and passes them on to the General Assembly.

The NGOs also have a consultative function within the ECOSOC in accordance with Article 71 of the UN Charter (ECOSOC, 1996). ECOSOC is meant to obtain information and advice from organisations with special competence in particular fields, and NGOs can apply for general or special consultative status, or for a listing in the Roster of ECOSOC's Committee for NGOs. Depending on the category to which an NGO is assigned (see Box 24.1), it can play different roles such as participation in UN conferences and sessions including the making of oral statements, or sending written statements to the UN members. The participation of NGOs in the preparation of the UN Conference on Environment and Development has been very intensive, and continues to be so in the work of the

BOX 24.1 DIFFERENT NGO CATEGORIES ADMITTED TO ECOSOC

General consultative status	Organisations which are concerned with most of the activities of the Council and can demonstrate to the satisfaction of the Council that they have marked and sustained contributions to make to the achievement of the objectives of the United Nations (with respect to international economic, social, cultural, educational, health, scientific, technological and related matters and to questions of human rights), and are closely involved with the economic and social life of the peoples of the areas they represent and whose membership, which should be considerable, is broadly representative of major segments of the population in a large number of countries in different regions of the world.
Special consultative status	Organisations which have a special competence in, and are specifically concerned with, only a few of the fields of activity covered by the Council, and which are known internationally within the fields for which they have or seek consultative status.
Roster	Other organisations which can make occasional and useful contributions to the work of the Council or in subsidiary bodies or other UN bodies within their competence.

Source: ECOSOC, 1996

Commission on Sustainable Development (Monahan, 1995).

ECOSOC has been given the role of co-ordinating policies for sustainable development through Agenda-21 (NGO, 1992), but the implications of its role in TECM are not yet clear (Trolldalen, 1992).

Since ECOSOC does not have any standing committee or commission oriented towards environmental matters, it itself has rarely produced any significant work in the environmental field (Szasz, 1992). Furthermore, ECOSOC is too large to represent an effective and intimate forum to deal with environmental matters, and it discourages proposals for using its offices and services for such purposes. It is for these reasons that voices are now being raised for abolishing the Council altogether (Szasz, 1992).

ECOSOC's consulting procedures for NGOs are exemplary within the UN system, but they are impaired by the fact that environmental NGOs such as Greenpeace, the International Union for the Conservation of Nature (IUCN) or Friends of the Earth have no general consultative status, and are therefore denied certain means to become active within the Council.

UN Regional Economic Commissions

These commissions, which are co-ordinated by the ECOSOC, comprise the Economic Commission for Europe (ECE), the Economic Commission for Latin America (ECLA), the Economic and Social Commission for Asia and the Pacific (ESCAP), the Economic Commission for Africa (ECA), and the Economic Commission for Western Asia (ECWA). The ECE, working closely with the CSD, has contributed significantly to the reduction of air pollution in Europe (Trolldalen, 1992).

As has been the case with the ECE, it is possible for regional commissions to co-operate in TECM with other institutions. Although they cannot manage environmental conflicts on their own, the regional approach has many advantages, because (Trolldalen 1992: 24):

> It appears that a regional approach focusing on a particular set of conflicts would be the most effective method of conflict resolution. And, since regions often include nations which share similar cultures, politics, economies, and often languages, many obstacles which typically hinder conflict resolution can be removed.

Future tasks for the regional commissions in the environmental field could be the preparation of the environmental profile of each member state (Trolldalen, 1992). They could then be part of the monitoring procedures in co-operation with UNEP, the CSD, or other newly created institutions.

The Bretton Woods organisations and the Global Environmental Facility

The financial organisations of the UN have been traditionally concerned with technology transfer, and also with policies for development. They can thus play an important role in conflict prevention, especially when setting conditions for possible loans, by financing only those projects which enhance sustainability and take environmental matters fully into account. Also, withholding funds for environmentally doubtful projects can discourage governments from implementing projects that could lead to environmental conflicts. In the past the World Bank has also played the role of a mediator in international environmental conflicts, in the Indus River Basin project in the 1960s for example (Trolldalen, 1992). According to Trolldalen, the Bank 'views resolving international environmental conflicts as part of its traditional role'. In particular, it is able to provide funds to compensate for negative transboundary externalities.

A special role in conflict prevention is played by the Global Environment Facility (GEF). It is concerned with financing projects focusing on the four areas of global warming, ozone depletion, biodiversity, and the protection of international waters. As these are some of the areas in which environmental disputes are most likely to occur, projects and programmes directed at addressing problems in them can obviously help prevent conflict. According to GEF rules, such projects are generally linked with technology transfer, training and other aspects of human resources development, and with the interests of the peoples of the recipient countries (Trolldalen, 1992). Taken together, these constitute an important element of the conflict prevention strategy. Technical and financial assistance could also be co-ordinated by the GEF, since it may turn out to be the future institution for managing environmental trust funds to be established under new conventions (French, 1992a).

While the Bretton Woods institutions work on a weighted voting system which heavily favours the most important financial contributors, the GEF, which works on the 'one-nation-one-vote' decision-making system, has found widespread acceptance, especially because of its mandate to examine transboundary issues with a view to preventing conflicts. Another advantage of the GEF is that its member states can make requests to its Secretariat for assistance in conflict resolution through fact-finding and good offices (Trolldalen, 1995).

Although the Bretton Woods organisations have the possibility to prevent environmental conflicts, unfortunately they have not fulfilled this particular role in the past, at least in so far as can be judged from the attitude and observations of the NGOs (French, 1992b). In 1993 the newly created post of the Vice-President for Environmentally Sustainable Development, together with the restructured high-level Environmental Department, raised hopes for the future integration of environmental concerns into World Bank policies. But the problem was that the degree to which sustainability principles could be implemented depended on the Bank's accountability to those outside it (Werksman, 1993). However, with the mandate to monitor the Bank's activities now being given to the CSD, some development appears to be taking place in that direction. It is also to be noted that, due to the lack of staff equipped with appropriate scientific information and expertise, the role of the Bretton Woods organisations is very much restricted to issues of compensation, and certainly their role as a third party in conflict management should not be overemphasised (Trolldalen, 1992).

During the initial phase of the GEF, criticisms were made on the influence of the World Bank on project selection, dominance of developing countries in making decisions on the development of its operational policy, and the failure to involve NGOs in decision-making (Werksman, 1993). The GEF has certainly been helpful in implementing sustainable projects which could not otherwise be financed by the developing countries themselves. But it remains to be seen whether its budget, which was increased to 2.02 billion US dollars for its second phase until 1996 (IUCN, 1994), is sufficiently large to cover the needs of developing countries to deal effectively with their pressing problems.

Trust funds are crucial in many environmental fields, mainly because they help spread the burden between developing and developed countries. However, GEF's role as the manager of future environmental trust funds might be constrained by the fact that the developing countries would prefer any 'green funds' to be independent of the World Bank which, as they see it, is far too much under the control of the powerful donor nations (French, 1992b; UNGA, 1991; IUCN, 1994).

The World Trade Organisation

The World Trade Organisation (WTO) (previously the GATT agreement), which had 128 members by the end of 1994, is the only organisation which lays down the rules for international trade. At the same time it provides a forum for the discussion and settlement of trade disputes. If the conflicting parties are not able to come to a settlement through bilateral consultations, the Director-General of the WTO offers his good offices, as well as conciliation or mediation services if requested, to the parties in an *ex officio* capacity. Should this also fail, disputes are resolved through a Dispute Settlement Body (DSB) which establishes a panel to deal with each case. The panel, which may be notified about third-party interests, can appoint an expert review group to produce a scientific advisory report. The parties involved may comment on the panel's report before the final report is adopted by the DSB. The entire procedure should not take more than eight months. An appeal to DSB decisions is also possible. A detailed discussion of the WTO and its relevance to the 'environment' is given in Chapter 18 of this volume.

WTO's conflict settlement procedure is to be praised for the range of different options it offers for settling conflicts. The tight time schedule, as set out in its rules of procedure, means that in practice a dispute can be settled within four to seven months. Even when the issue or issues in dispute are extremely complex, the final decision should be given no later than one year after the initiation of the proceedings. Also, third-party interest has been incorporated in an extraordinary manner. Compliance with the decisions of the Dispute Settlement Body is encouraged through public and repeated discussions of the issues during its sessions. Nation-states can even be authorised by the DSB to impose trade sanctions against a non-compliant party.

However, considering that trade liberalisation is the main goal of the WTO, its orientation raises doubts among environmentalists about whether positive outcomes can be expected for the environment from its decisions. Since WTO provisions prohibit discrimination between 'like products on the basis of their method of production', outcomes similar to those in the Tuna–Dolphin dispute between Mexico and the USA (see Box 18.1) are to be expected from its decisions. Indeed, one could go so far as to say, with some justification, that the WTO may even *cause* conflicts because '[t]here is an obvious conflict between an international policy of free trade and a national policy of internalisation of external costs' (Daly, 1992: 8).

Some of the main objections to the WTO are that it has failed to incorporate the principles of sustainability as an objective into its provisions; that it does not embrace the precautionary principle and does not assess the environmental impacts of its decisions; and that it discourages unilateral measures taken beyond or instead of international standards in order to protect the global commons (Cameron, 1993). As discussions on the greening of the WTO continue, it is to be hoped that these deficiencies will be adequately addressed in the foreseeable future.

The United Nations Environment Programme

For a number of years now the United Nations Environment Programme (UNEP) has been involved directly in conflict prevention (Trolldalen, 1992). For example, through its Regional Seas Programme and the Action Plan for Shared Lake and River Basins, it has been engaged in the anticipation and prevention of environmental conflicts.

In addition, international environmental conventions fostered by UNEP have contributed in good measure to the prevention and settlement of conflicts. This has been achieved by creating a set of rules and principles so that breaches can be determined and awards made. It has also established standards, which the parties to a treaty generally comply with, in an attempt to ensure that differences do not escalate into full-blown conflicts (Chayes and Antonia, 1991).

Another function of the treaties is to settle conflicts arising from their interpretation and implementation (see Box 24.2). But a large number of environmental treaties, which depend on common agreement by the parties, do not allow foreseen conflict resolution provisions to be invoked unilaterally (Sand, 1992). However, after examining the provisions concerning the settlement of disputes in 29 out of 83 multilateral Conventions under the auspices of UNEP, Kiss (1983) concluded that there was a clear tendency

BOX 24.2 THE RHINOCEROS CONFLICT

Djibouti, which is not a signatory to the Convention on International Trade in Endangered Species (CITES), serves as an important centre for trade in endangered and rare species in the East African region. This undermines the efforts of many countries, especially Kenya, to restrict poaching. Thanks to its stringent conservation efforts, Kenya has probably prevented the extinction of several endangered species, while at the same time developing a successful tourist industry which is largely based on the country's rich wildlife resources.

Although the problem here is regional in its scope, by May 1989 only Kenya, Somalia and Sudan had become members of CITES. But Sudan, which became a registered CITES member state in 1983, is still an important transit country for the export of rhino horn. This conflict between conservation and short-term maximisation of utility is closely linked to the unstable political environment in the East African region. Because of the fragile political relationship among the countries in the Horn of Africa, little attention is being given to the protection of endangered and rare species.

Source: Trolldalen, 1992

towards obligatory arbitration when other modes of settlement failed. The International Trade Commission of the USA has identified 19 international environmental agreements in which trade sanctions were applied in 1991 (French, 1992b). Sanctions may be severely limited, however, or even prohibited by the WTO, because they may be argued to represent undue restrictions on international free trade. In this context there is a recorded case in which the USA threatened Japan with a trade embargo if it did not comply with the ban on commercial whaling (ibid.).

A convention can also be said to mark the end of a conflict settlement procedure because, after a problem has been identified and brought on to the international political agenda, it is generally considered to have been 'resolved' following the conclusion of a bilateral or multilateral convention.

On the scientific front, UNEP 'has created an impressive corpus of independent, top-quality experts on all aspects of environmental protection and resource use' according to Timoshenko (1992: 413). Considering UNEP's role in fact-finding, either in the form of an inquiry or as an element of conciliation, arbitration or judicial settlement, it is clear that it has developed skills that are crucial to environmental conflict settlement (ibid.). Indeed, UNEP's areas of responsibility (such as in technical assistance, information gathering, monitoring, assessment, exchange of information, and reporting on environmental concerns) have made it the central catalyst, co-ordinator and stimulator in the field of the environment (Tinker, 1990). Although all the functions mentioned above are part of a conflict prevention strategy, they can also be employed in conflict resolution.

Through its regional programmes, UNEP has also been playing a significant role in conflict prevention for some time now. These programmes have been quite successful in identifying both the scope of the conflicts and their possible solutions. The strength of these programmes lies in their ability to focus on the environmental system in question as a whole. On the other hand, they are very much policy-oriented and lack the funds for their effective implementation (Trolldalen, 1992). As a possible solution to this, it has been recommended that UNEP's regional programmes should be implemented in co-operation with regional development banks (ibid.).

In general, the development of international environmental law can be an awesome process requiring diplomacy, much compromise, and, above all, much time. It is not unusual for up to twenty years to elapse before a treaty is finally concluded, as was the case with the Law of the Sea for example. But, clearly, such a protracted time-scale is much too long for the urgent environmental problems facing us today.

Since all environmental agreements are concluded separately, and because there is no mechanism in place to co-ordinate them, common elements in such agreements are not necessarily treated in the same way (Palmer, 1992). Other disadvantages of treaty law are that treaties are commonly concluded on the smallest common denominator because of the unanimous voting procedure (ibid.); that they have to be ratified in order to be binding; and that often there is no enforcement mechanism to ensure compliance. Conventions, on the other hand, mostly leave the responsibility for implementation up to the signatories. These disadvantages of treaty law can be addressed in part by monitoring provisions and follow-up. A good example of this is the Montreal Protocol which established an Implementation Committee. Reporting to the Committee is mandatory; and elements such as environmental auditing, public review, and collective compliance control by the Committee (composed of ten representatives of the member states) have led to 'concurrent enactment of national implementing legislation on such a widespread basis [that] is unprecedented in the history of environmental law' (Trolldalen, 1992: 39).

However, the circumstances under which a convention comes into existence determine if the settlement would be a lasting one, or if the conflict could occur again in the future despite the agreement. Some political scientists claim that such agreements strongly reflect power relationships and are therefore not stable (Vogler, 1992). This means that a new conflict situation can arise if or when power relationships change. If this view holds true for the environmental field, conventions can hardly be counted on as tools for conflict resolution but, rather, as non-sustainable settlement mechanisms. There is, however, another school of thought which holds that lasting regimes can be built in cases where interdependence is high, as in the global environmental problem of ozone depletion for example. According to this school of thought, regimes would be stable (sustainable) as long as there is mutual vulnerability, or, as Chayes has put it, 'the regime will persist if the parties continue to believe that the benefits of membership outweigh the costs' (Chayes and Antonia, 1991: 289).

The Rio Conference on Environment and Development and the Commission for Sustainable Development

No new institution, relating directly to TECM, has been introduced within the UN system following the UN Conference on Environment and Development (UNCED) in 1992. Nevertheless, Agenda-21 gives some insight into the state of discussions on new institutions and institutional change, by calling for governments 'to make international trade and environment policies mutually supportive in favour of sustainable development; and to clarify the role of GATT, UNCTAD and other international organisations in dealing with trade and environment-related issues, including, where relevant, conciliation procedure and dispute settlement' (chapter 2.21). Agenda-21 also defines the following objectives (chapter 38.8):

(a) to ensure and review the implementation of Agenda-21 for achieving sustainable development in all countries;
(b) to enhance the role and functioning of the United Nations system in the field of environment and development.

Chapter 39 sets the goal to 'study and consider the broadening and strengthening of the capacity of mechanisms, *inter alia* in the UN system, to facilitate . . . the identification, avoidance and settlement of international disputes in the field of sustainable development'. This chapter also contains a paragraph on disputes in the field of sustainable development, requesting nation-states to 'study and consider methods to broaden and make more effective the techniques available at present', e.g. the methods outlined in the UN Charter, including the ICJ.

Chapter 38 calls for the establishment of a Commission on Sustainable Development (CSD) to monitor the implementation of Agenda-21. It was established by ECOSOC in 1993 as a high-level body (functional commission) consisting of representatives from 53 different countries. The CSD meets once a year for a period of two to three weeks, and its functions include:

- to give appropriate recommendations to the General Assembly, through the Economic and Social Council, on the basis of an integrated consideration of the reports and issues relating to the implementation of Agenda-21, and to make appropriate recommendations for the review of Agenda-21 in 1997;
- to integrate aspects of sustainable development at regional and sub-regional levels;
- to consider issues related to . . . financial resources, technology transfer, etc.;

- to overview the progress made in the implementation of existing environmental agreements;
- to receive reports from the specialised agencies of the UN, including the World Bank, on their efforts to implement Agenda-21 and to promote sustainable development;
- to receive and analyse information from governments, NGOs, the scientific community, and the private sector concerning the global implementation of Agenda-21, and to ensure that representatives of various parts of the UN system (WTO, regional development banks, etc.), other intergovernmental organisations, and the NGOs can participate effectively in the work of the Commission;
- (to organise) a high-level meeting, with ministerial participation, to have an integrated overview of the implementation of Agenda-21, to consider emerging policy issues and to provide necessary political impetus to the implementation of the decisions of the United Nations Conference on Environment and Development and the commitments contained therein.

A high-level advisory board is proposed for the CSD, consisting of 'eminent persons broadly representative of all regions of the world . . . drawn from relevant scientific disciplines, industry, finance and other major non-governmental constituencies' providing expert advice to the Secretary-General. The possibility of instituting special expert rosters is also mentioned.

As will be seen from the general objectives of Agenda-21, the UN is aware of the need to develop its capacity for resolving environmental conflicts. But, apart from such theoretical statements, the only real institutional outcome of the Rio process, which is tenuously related to these issues, has been the resolution establishing the CSD.

Although the CSD does not have any explicitly defined task in the management of environmental conflicts that may arise in the implementation of Agenda-21, it will certainly be involved in conflicts arising from issues of non-sustainable development. The monitoring tasks, to be carried out by the CSD, will provide a good overview of cases in which issues of non-sustainable development might lead to crisis situations, possibly entailing conflicts on an international scale. It is hoped, however, that with the scientific support of the Advisory Board, and equipped with information received from diverse sources including regional NGOs and the secretariats of other environmental conventions, the CSD would be aware of environmental conflicts at a very early stage.

In pursuance of the provisions of Article 65 of the UN Charter, which authorises ECOSOC to inform and assist the Security Council, it should be possible for the CSD to detect and observe an emerging environmental conflict, and to bring it to the attention of both the General Assembly and the Security Council. In this way the CSD might increasingly take on the role of an 'environmental watchdog' – detecting environmental threats to the international community, warning the relevant UN institutions and requesting them to take appropriate action, and increasing their alertness to emerging conflicts caused by non-compliant nation-states. It is to be noted, however, that this function has already been assumed by UNEP; and, since the CSD has no clear mandate to give early warning of conflicts, its role is expected to be different from that of a 'watchdog' (Monahan, 1995). According to Monahan, the primary role of the CSD is that of a facilitator – for bringing together other more specialised UN bodies for concerted action and for co-ordinating their efforts for the implementation of Agenda-21.

The CSD has not been given any coercive authority for implementing sustainable development policies. Neither has it the power to make changes or decide on amendments to Agenda-21. Even so, it played a crucial role in influencing decisions on the review of Agenda-21 made in a special session of the UNGA in June 1997.

The Law of the Sea Convention

The UN Convention on the Law of the Sea (UNCLOS) came into force on 17 November 1994. It is a highly significant instrument, mainly because it covers an entire environmental sector and can serve as a model for future Conventions dealing with other environmental compartments. It regulates the exploitation of both mineral and living resources of the oceans, and, with a considerable part of it dedicated to environmental issues, it is to date the most comprehensive and complex of all conventions covering environmental aspects. With the inclusion of some amendments to the Convention on deep-sea mining, which incidentally made it acceptable to the OECD countries, the range of ratifying countries has now been extended from almost exclusively Third World countries to several First World countries (e.g. Australia and the Federal Republic of Germany).

There are two different conflict settlement procedures with regard to the Exclusive Economic Zone (EEZ), defined as a zone of the sea enclosed by a boundary 370 km away from the coastline:

1 Conflicts regarding the exploitation of the EEZ shall be subject to an obligatory conciliation procedure without binding results. The only environmental dispute that could be thought of as falling under this provision could occur in the case when a state has manifestly failed to comply with its obligation to ensure that EEZ living resources are not seriously endangered (IUCN, 1995). The reports of the conciliation commission are to be communicated to the relevant international organisations so that the results will most likely influence the development of international law and also impose substantial pressure on the state concerned to comply with its recommendations. International institutions to be established under the Law of the Sea (LOS) Convention would facilitate co-operation for the preservation of highly migratory species for example.
2 Conflicts about the EEZ relating to 'non-resource uses' like ocean research, rights of navigation and over-flight, laying of submarine cables and pipelines, and environmental issues are to be settled by a binding judicial procedure if the parties fail to resolve them otherwise. The environmental disputes covered by this provision are defined as arising from action 'contrary to international rules and standards for the protection and preservation of the marine environment'. As fishing issues (e.g. the allowable catch) are explicitly excluded from falling under the binding settlement procedure (Article 297), environmental disputes under the scope of this procedure can be seen as being restricted to pollution issues, not including threatened species, migratory species, etc. (Franckx, 1995).

Conflicts about fishing outside the EEZ are subject to the compulsory binding settlement procedures. The possibility to settle disputes arising from other fishing agreements under the LOS provisions is also given, either by inclusion of these provisions in the text of the Convention, or by mutual agreement of the parties if they are also parties of UNCLOS. The duty laid down in the LOS to conserve and co-operate in conservation matters can also be enforced against any member of the Convention based on the LOS provisions alone. Such a case might occur, for example, when a nation-state undermines conservation measures agreed upon within a regional agreement through unsustainable harvesting practices (IUCN, 1995).

The basic principle of conflict management under the Law of the Sea Convention is that all nation-states shall settle their disputes by peaceful means in accordance with Article 2(3) of the UN Charter (Article 279). Although the choice of the means for dispute settlement is left to the conflicting parties to make (Article 280), Article 283 obliges them to consult and exchange their views on how the conflict should be settled. Consultation is also obligatory after a settlement, during the implementation phase.

If the method first chosen by the parties in dispute does not lead to a settlement, the Convention foresees either

- the submission of the dispute to a procedure involving a binding decision under another general, regional or bilateral agreement, unless the parties agree otherwise (Article 282),
- or a settlement following the procedure as laid down in later Articles of the Convention, unless the agreement between the parties excludes any further procedure (Article 281).

The latter means that nation-states can still choose to maintain a *status quo* and to live with the impending conflict without being forced to resolve it by any means whatsoever, provided that both parties to the conflict agree on this. Otherwise the conflict will be resolved through a compulsory judicial procedure entailing binding decisions. Nation-states can choose one or more of the four different procedures, described below, by submitting a written declaration to the UN Secretary-General:

- through the International Court of Justice;
- through a new 'International Tribunal of the Law of the Sea' (ITLS), with a special 'Sea-Bed Disputes Chamber' (SBDC) seated in Hamburg (Article 2(1)). The SBDC of the Court, which consists of 11 members chosen by and from among the 21 judges of the ITLS, is set up to deal with all conflicts arising in connection with the complex arrangements of the Convention for the exploration and exploitation of the deep-sea bed;
- through an arbitration tribunal as the preferred method of settlement. It is a common agreed method, based on the declarations of the nation-states in dispute or on compulsory procedures in case no common declaration exists. The members of the tribunal are to be chosen by the Secretary-General from a list with up to four nominations by the nation-states involved (Article 2(1));
- through a special arbitration tribunal for cases involving fishing, protection and preservation of the marine environment, marine scientific research, and navigation including pollution from vessels and dumping at sea. The members of the tribunal can be selected from lists held by specialised UN agencies to which the nation-states in dispute can nominate their own experts (Article 3).

The LOS lays down which procedure should be chosen according to the declarations made by the member states:

3 A State Party, which is a party to a dispute not covered by a declaration in force, shall be deemed to have accepted arbitration in accordance with Annex VII.
4 If the parties to a dispute accept the same procedure for the settlement of the dispute, it may be submitted only to that procedure, unless the parties otherwise agree.
5 If the parties to a dispute do not accept the same procedure for the settlement of the dispute, it may be submitted only to arbitration in accordance with Annex VII, unless the parties otherwise agree.

Article 287 (3–5)

The diversity of procedures reflects the different views expressed during the drafting of the Convention. Special arbitration can involve fact-finding and inquiry, and, if both parties agree, the tribunal may formulate non-binding recommendations to provide the basis for a review by the parties of the questions giving rise to the dispute.

Nation-states are to ensure that recourse to their national judicial systems is possible for compensation or other relief for pollution damage (Article 235.2), including civil proceedings in respect of loss or damage (Article 229) (IUCN, 1995).

For the Sea-Bed Disputes Chamber of the ITLS, the question of granting access to foreign judicial or national persons (corporations) was resolved through the 'local remedies' principle. As a consequence, individuals or other non-state claimants have standing before the Law-of-the-Sea institutions only after they have exhausted the local remedies in the foreign state in question (Adede, 1987). For all other conflicts, including environmental disputes, only state parties have access to the LOS settlement procedures.

The 'International Sea-bed Authority', to be created under the LOS Convention, has considerable powers to intervene in exploitation activities. Apart from developing norms and rules for deep-sea bed mining, including environmental protection issues, it can issue emergency orders to prevent serious harm to the environment and disapprove of exploitation projects in areas where serious environmental damage is to be expected through the mining activities (Article 162.2w, x) (IUCN, 1995).

An interesting feature of the Convention's provisions is that, although most of its environmental obligations are very general, it can include other, more specific agreements by incorporation through reference so that it is indirectly enlarged and specified by those Agreements (IUCN, 1995). Additionally, even if those other Agreements do not contain binding settlement procedures, the LOS provisions become applicable in cases that are covered by these incorporated Conventions (ibid.). Reciprocally, other agreements concerning fishing can incorporate provisions requiring disputes to be resolved through the LOS procedures, and it is also possible for members of the LOS Convention to invoke its procedures for fishing disputes arising under such agreements (ibid.).

Although the LOS Convention represents 'a useful and flexible system of compulsory jurisdiction'

(Merrills, 1991), at least three objections to it have been made:

1 The environmental provisions of Article 297 only deal with violations of *coastal* states. Alleged violations of third states and their vessels should also be subject to the binding and compulsory procedures outlined above (Vukas, 1992).
2 The special arbitration procedure is especially foreseen for a number of environmental issues (fisheries disputes, protection and preservation of the marine environment, and navigation issues including dumping and pollution through vessels (IUCN, 1995)), as follows from Annex VIII of the Convention (Broadus and Vartanov, 1994). This means that, although Article 297 foresees a compulsory settlement procedure with a binding decision, environmental conflicts will at the most be settled without any binding decision.
3 The Convention foresees that a state can be sued for violations of environmental rules, both as contained in the Convention itself and with regard to international rules applicable to the coastal state. Consequently, the state parties to the Convention will not be subjected to the same environmental requirements (Vukas, 1992). While the advantage of these provisions is clear, in that such provisions can now be enforced although no effective enforcement mechanisms had been foreseen in most of the international agreements, unequal treatment of different parties is not acceptable. To address this, common international environmental rules and standards should be developed and applied.

Another problem is concerned with inspection, implementation, and sanctions for breaches of laws on fishing in the EEZ, because

> foreign states are reluctant to provoke a diplomatic controversy by exercising their jurisdiction over ships that pollute or over-fish on the high seas or in other states' jurisdiction unless some treaty specifically approves this . . . States are even more reluctant to resort to sanctions for fear of reciprocal hostile action.
>
> (Birnie, 1988: 112)

At the moment the only country that is not afraid to take enforcement measures is Canada. But clearly, every attempt to provide binding and compulsory procedures, based on complaints made by nation-states, will have to cope with the problem that 'States refuse to apply the mechanism of sanction even if such exists and abstain from claiming an action because of political opportunity' (Rest, 1992: 247).

CONCLUSIONS

The conclusions to be drawn from the foregoing discussion on the current UN system, and its capacity to manage environmental conflicts, are these:

1 The UN has a large set of conflict settlement mechanisms at its disposal. Some are very specialised, such as the WTO which only deals with trade conflicts. All in all, these mechanisms are sufficiently flexible and complement each other so that they can be used to settle a wide range of environmental conflicts.
2 Apart from political settlement procedures, such as good offices and mediation, the applicability and level of acceptance of the existing UN settlement mechanisms is not satisfactory. Too many of the nation-states have reservations, or did not even subscribe to or ratify the constituting documents of the conflict settlement institutions. Also, open and easily accessible mechanisms, such as the Permanent Court of Arbitration, are rarely used.
3 Only the Security Council and the General Assembly are able to react fast to emergency situations. Other conflict settlement procedures take months or even years before a settlement is reached. With regard to political conflict settlement mechanisms, it is to be noted that their application can also take a very long time. A conflict, which cannot be settled through negotiation, may go through the process of good offices, mediation, conciliation and finally arbitration or judicial settlement. This means that many years can elapse before an agreement is finally reached.
4 It is only rarely that a full reimbursement of the costs of using existing settlement procedures can be expected from the parties in dispute.
5 Equal opportunity for all parties, and a consideration of global environmental interests, is only

provided at the highest levels, as in the PCA, the ICJ and the Security Council; but they are also the least used mechanisms.

6 A remarkable conflict settlement procedure is to be found in the WTO. It has many outstanding advantages such as flexibility, acceptance, capacity to make and implement decisions quickly, and to take scientific information into account. But the most regrettable shortcoming of the WTO and other economic treaties is their failure to incorporate environmental principles into their basic rules of operation.

7 The NGOs have no access to the conflict settlement mechanisms, apart from their consultative status within the ECOSOC and related UN commissions. Neither individuals nor NGOs can initiate proceedings, nor can they claim their right to be considered as parties to international conflicts.

8 Apart from the Security Council, the ICJ and the WTO, there are no mechanisms empowered to make binding decisions or to enforce compliance. Most of the present conflict settlement and compliance procedures rely on voluntary compliance.

9 Whether or not an outcome would be lasting often depends rather more on the authority of the institution which makes the decision than on its fairness or a consideration of the wider aspects of the conflict which must also be addressed for its comprehensive settlement. The consideration of scientific evidence is as yet not sufficiently institutionalised to play an important role in the conflict resolution process.

10 There is no institutionalised mechanism in the TECM to take into account the interests of future generations. However, the CSD may be able to implement such aspects, at least in part, through relevant UN bodies such as the General Assembly, the Bretton Woods institutions, and the regional and other commissions. To this extent there is potentially a very important role for the CSD to play.

11 Apart from the LOS Convention, which considers the deep-sea bed and the high seas, there is at present no other mechanism to resolve conflicts arising in the global commons.

12 So far, the UNEP has been playing the main role in conflict prevention. Through its Regional Commissions it has been doing much valuable work in both identifying and dealing with areas of concern such as multinational river basins and transboundary pollution. Significant roles are also being played by the GEF through its financial support for environmental projects, and by UNEP and the General Assembly through their contribution to the development of international environmental rights and obligations. As for the CSD, its future effectiveness will determine the role it will play in implementing the principles of sustainability in international decision-making.

Three areas have to be covered to deal effectively with environmental problems and to prevent and settle conflicts arising from the use of the environment: setting rules, monitoring compliance, and providing an authoritative and binding method of settling conflicts. However, there are still major gaps in the present institutional framework with regard to the co-ordination of environmental activities, integration of environmental concerns into other activities, and the monitoring and assessment of compliance with environmental principles. According to Sir Geoffrey Palmer, Professor of Law at the Victoria University in Wellington, New Zealand, and former Prime Minister and Minister of Environment of New Zealand, the only way to fill this gap is by creating a new international institution.

REFERENCES

Adede, A.O. (1987) *The System for Settlement of Disputes Under the United Nations Convention on the Law of the Sea*, Dordrecht: Martinus Nijhoff Publishers.

Arnaldez, J.-J. (1995) Deputy Registrar of the International Court of Justice, personal communication.

Bilder, R.B. (1975) 'The Settlement of Disputes in the International Law of the Environment', *Collected Courses of the Hague Academy of International Law* 144 (1975–I), pp. 139–239.

Birnie, P. (1988) 'The Role of International Law in Solving Certain Environmental Conflicts', in J.E. Carroll (ed.) *International Environmental Diplomacy*, Cambridge: Cambridge University Press.

Broadus, J.M. and Vartanov, V. (1994) *The Oceans and Environmental Security*, Washington DC: Island Press.

Burton, J.W. (1990) *Conflict: Resolution and Prevention*, London: Macmillan.

Cameron, J. (1993) 'The GATT and the Environment', in Ph. Sands (ed.) *Greening International Law*, London: Earthscan Publications Ltd., pp. 100–121.

Chayes, A. and Antonia, H. (1991) 'Adjustment and Compliance Processes in International Regulatory Regimes', in J. Tuchman Mathews (ed.) *Preserving the Global Environment: The Challenge of Shared Leadership*, New York: W.W. Norton, pp. 280–308.

Daly, H. (1992) Senior Environmental Economist of the World Bank, as cited in *New Scientist*, 30 May 1992, p. 8.

EC (1993) *The Possibility of Establishing an International Environmental Court – Premises, Opportunities and Obstacles: The Position of the European Community and Initiatives it Might Adopt*, Luxembourg: European Commission, DGXII, IV/WIP/93/03/152, 25 August.

ECOSOC (1996) *Non-governmental Organisations – Consultative Relationship Between the United Nations and Non-governmental Organisations*, New York: UN Economic and Social Council, E/1996/L.25 (adopted on 26 July 1996), 17 July 1996.

Fawcett, J.E.S. and Parrey, A. (1981) *Law and International Resource Conflict*, Oxford: Royal Institute of International Affairs.

Franckx, E. (1995) Professor at the Department of Law, Free University of Brussels, personal communication.

French, H. (1992a) 'After the Earth Summit: The Future of Environmental Governance', *Worldwatch Paper 107*, Washington, DC: Worldwatch Institute.

French, H. (1992b) 'From Discord to Accord', *The Federalist* 34 (3): 204–208.

ICEF (1992) 'The History of an Idea', in A. Postiglione (ed.) *Tribunale Internazionale Dell'Ambiente*, Rome: Istituto Poligrafico E Zecca Libreria Dello Stato, pp. 67–97.

ICJ (1994) *Yearbook 1993–1994* (No. 48), The Hague: International Court of Justice.

IUCN (1994) 'A Guide to the Convention on Biological Diversity', *Environmental Law and Policy Paper No. 30*, Gland, Switzerland: The World Conservation Union.

IUCN (1995) *The Law of the Sea: Priorities and Responsibilities in Implementing the Convention*, Gland, Switzerland: The World Conservation Union.

Kiss, A.-C. (1983) 'Le Règlement de Différends dans les Conventions Multilatérales Relatives à la Protection de L'Environnement', in R.-J. Dupuy (ed.) *The Settlement of Disputes on the New Natural Resources*, The Hague: Martinus Nijhoff Publishers.

Koskenniemi, M. (1991) 'Peaceful Settlement of Environmental Disputes', *Nordic Journal of International Law* 60: 82.

Merrills, J.G. (1991) *International Dispute Settlement* (2nd edition), Cambridge: Grotius Publications Limited.

Monahan, M. (1995) UN Department for Policy Coordination and Sustainable Development, Geneva, 28 July (personal communication).

NGO (1992) 'Draft Background Document to the Workshop on Institutional Change', The Hague: Peace Palace.

Palmer, Sir Geoffrey (former Prime Minister of New Zealand) (1992) 'New Ways to Make Environmental Law', *The American Journal of International Law* 86: 259–283.

Rest, A. (1992) 'A New International Court of Justice for the Environment to Implement Environmental Responsibility/Liability Law?', in A. Postiglione (ed.) *Tribunale Internazionale Dell'Ambiente*, Rome: Istituto Poligrafico E Zecca Dello Stato Libreria Dello Stato, pp. 247–260.

Sand, P.H. (1992) *The Effectiveness of International Environmental Agreements*, The United Nations Conference on Environment and Development, Cambridge: Grotius Publications.

Schneider, J. (1979) *World Public Order of the Environment: Towards an International Ecological Law and Organisation*, London: Stevens & Sons.

Stein, R.E. (1985) 'The Settlement of Environmental Disputes: Towards a System of Flexible Dispute Settlement', *Syracuse Journal of International Law and Commerce* 12: 283–298.

Szasz, P.C. (1992) 'Restructuring the International Organizational Framework', in E. Brown-Weiss (ed.) *Environmental Change and International Law: New Challenges and Dimensions*, Tokyo: United Nations University Press.

Timoshenko, A.S. (1992) 'Ecological Security: Response to Global Challenges', in E. Brown-Weiss (ed.) *Environmental Change and International Law: New Challenges and Dimensions*, Tokyo: United Nations University Press.

Tinker, C. (1990) 'Environmental Planet Management by the United Nations: An Idea Whose Time Has Not Yet Come?', *New York University Journal of International Law and Politics* 22 (4): 793–830.

Trolldalen, J.M. (1992) *International Environmental Conflict Resolution: The Role of the United Nations*, Oslo: World Foundation for Environment and Development.

Trolldalen, J.M. (1995) Personal communication, 14 August.

UNA–USA (1993) *A Successor Vision: The UN of Tomorrow*, Final Panel Report of the United Nations Association of the USA, 9/1987, new edition May 1993.

UNGA (1991) 'Preparations for the United Nations Conference on Environment and Development on the Basis of General Assembly Resolution 44/228 and Taking into Account other Relevant Assembly Resolutions – Institutional Proposals', Report by the Secretary-General of the Conference, United Nations General Assembly, A/CONF.151/PC/102, 13 December 1991.

Vogler, J. (1992) 'Regimes and the Global Commons: Space, Atmosphere, and the Oceans', in A. McGrew, P.G. Lewis *et al.* (eds) *Global Politics*, Cambridge: Polity Press, pp. 129–132.

Vukas, B. (1992) 'Settlement of Disputes under the UN Law of the Sea Convention and the Protection of the Marine Environment', in A. Postiglione (ed.) *Tribunale Internazionale Dell'Ambiente*, Rome: Istituto Poligrafico E Zecca Dello Stato Libreria Dello Stato.

Werksman, J. (1993) 'Greening Bretton Woods', in Ph. Sands (ed.) *Greening International Law*, London: Earthscan Publications Ltd., pp. 65–84.

SUGGESTED READING

Calließ, J. (ed.) (1995) 'Treiben Umweltprobleme in Gewaltkonflikte?', *Lokkumer Protokolle* 21/94, Lokkum (Germany): Evangelische Akademie Lokkum.

Libiszewski, S. (1992) 'What is an Environmental Conflict?', *Environment and Conflict Project*, Occasional Paper No. 1, Centre for Security Studies and Conflict Research, Zürich: Swiss Federal Institute of Technology.

Merrills, J.G. (1991) *International Dispute Settlement* (2nd edition), Cambridge: Grotius Publications Limited.

Schmitz, M. (1992) *Les 'conflits verts': aperçu général sur la menace de l'an 2000*, Brussels: GRIP.

Trolldalen, J.M. (1992) *International Environmental Conflict Resolution: The Role of the United Nations*, Oslo: World Foundation for Environment and Development.

UN (1995) *Basic Facts about the United Nations*, New York: Department of Public Information, United Nations.

SELF-ASSESSMENT QUESTIONS

1 Give a definition of transnational environmental conflict.

2 Which of the examples given below are possible areas of transnational environmental conflict?
(a) ozone depletion;
(b) river pollution;
(c) social unrest in the countries of Eastern Europe;
(d) determination of fishing quota;
(e) allocation of fishing quota between two states;
(f) international drug traffic.

3 What are the classical means of conflict abatement?
(a) negotiation, including the provision of good offices;
(b) international votes;
(c) war;
(d) mediation and conciliation;
(e) arbitration and judicial settlement;
(f) consulting through private law firms.

4 Why could the role of institutions like the IMF be crucial in the resolution of transnational environmental conflicts?
(a) because they could offer funds to the governments in question in exchange for compliance with environmental measures;
(b) because many countries do not have the technical and/or financial resources for environmental prevention measures, and the IMF could help to provide funds for such resources;
(c) because the IMF has very good contacts with most of the governments and has much influence on their decisions.

5 What are the areas of TECM in which UNEP has played an important role?
(a) In the legislative area, through the creation of international conventions.
(b) In environment–trade conflicts, through the greening of the GATT/WTO.
(c) In the area of prevention, through its long-term programmes aimed at river basins and other environmentally sensitive areas shared by several states.

6 What could be the future role of the LOS Convention in TECM? Give a brief answer indicating an opportunity and a limiting factor.

7 Based on the discussions of the UN institutions, express your own opinion about the efficacy and adequacy of the present UN system for conflict resolution and suggest possible improvements.

GLOSSARY

Abatement cost: cost of reducing the quantity of residuals being emitted into the environment, or of lowering ambient concentrations.

ABC analysis: instrument to assess an eco-balance. It ranks the considered materials and energy according to their possible risk potential.

Acceptable daily intake (ADI): estimate of the amount of a substance in food or drinking water, expressed on a body mass basis (usually mg/kg_{bw}), which can be ingested over a lifetime by humans without appreciable health risk. To calculate the daily intake per person, a standard body mass of 60 kg is used. The ADI is normally used for food additives, while the tolerable daily intake (TDI) is used for contaminants.

Acceptable risk: a level of risk that arbitrarily is considered tolerable, after consideration of unavoidable risk, such as natural decease or accidental death consequent on a natural disaster (earthquake, floods, lightning, etc.).

ACR: Acute chronic ratio. The inverse of the application factor: LC50 divided by MATC or, for some species, acute EC50 divided by MATC.

Actor model: an assumed structure of human decision-making.

Actors field: a structure of how actors are causally behind other actors.

Acute: rapid onset of condition or arrival at end-point; may also refer to duration of exposure.

Acute toxicity: the adverse effect occurring within a short time of exposure (relative to generation time).

Additivity: the effect whereby the toxicity of two or more chemicals is the sum of the individual toxicities.

Adjudication: the process for the application of legal principles to particular controversies involving two or a reasonably small number of parties, wherein the decision-maker is a judicial official, a jury, or an administrative tribunal.

Adverse effect: change in morphology, physiology, growth, development or lifespan of an organism which results in impairment of its functional capacity or impairment of its capacity to compensate for additional stress or increased susceptibility to the harmful effects of other environmental influences.

Agenda-21: international agreement completed at the UNCED Conference in 1992, without legal, but with moral and inspirational value. It addresses international problems concerning the environment and development from a sustainable point of view. It can be seen as a framework on which national environmental plans can be founded.

Aggregation: simplification of the LCA inventory information by combining parameters to create a convenient number of (impact) scores.

Agro-ecosystem: any ecosystem where crops are grown and/or animals are grazed.

Alarm level: indicates the concentration of a pollutant based on (1) the definition of what effects on human health and the environment the authorities definitely wish to avoid; (2) the definition of the pollutant concentrations generating these effects.

Allocation: a procedure for assigning financial impacts (environmental impacts) to the production steps, sites and products that caused them.

Alternative Environmental Dispute Resolution (AEDR): refers to the second generation of environmental dispute resolution, which stresses the prevention or reduction of environmental conflicts by co-operation. This type of conflict management, e.g. the Mutual Gains Approach, is particularly useful for conflicts with interdependent behaviour. The three main classes of factors that determine the likelihood of its success are party-related factors, process- and context-related factors, and content-related factors.

Alternatives: in EIA, an examination of several locations, different methods, and/or different techniques for a particular project or other activity which leads to a similar result, but in a different way.

Amendment: a change to a treaty or code, often to provide more stringent obligations.

Amenity: a good or service offered by nature without any economic effort by people.

Antagonism: the effect whereby the toxicity of two or more chemicals is less than the sum of the individual toxicities.

Anthroposphere: the human techno-socioeconomic system which exists within the biosphere.

Application factor: for a given substance, the ratio of sublethal toxicity threshold to LC50. The application factor (also termed safety factor) may be used to predict the sublethal threshold of a second substance from its LC50. This term is rarely empirical in nature.

Arbitration: when some specialised tribunal is introduced to determine which party in the conflict has the best case. It imposes a binding settlement upon the parties instead of expanding the room for manoeuvring.

Areawide EIS: EIS prepared for a certain area or region. Areawide EIS methodology was developed by the US Department for Housing and Urban Development.

Arid region: region in an arid climate; a dry region with poor vegetation (steppe and desert).

Autonomy: the degree to which an actor can do, prevent or get done what he would like to do, prevent or get done.

Average life cycle: in LCA represents the 'average' of the life cycles of many similar or homogeneous products, and is used in life-cycle assessments of industrial sectors.

Axiology: science of values.

Behavioural standard: indicates how people should behave to respect and protect the environment.

Bio-accumulation: the net result of the uptake, distribution and elimination of a substance due to all routes of exposure, i.e. exposure to air, water, soil/sediment and food.

Bio-diversity: an ecological concept composed of five key components: species diversity (number of species), genetic diversity (variety of genes within a given population), habitat diversity (variety of biotic and abiotic elements), successional diversity (spectrum of plant communities between early and late ecological succession), landscape diversity (landscape as a mosaic of forests in different succession stages).

Bio-history: historical record of living organisms on earth.

Border Tax Adjustments: a WTO legal mechanism which rebates exporters at the border for taxes paid domestically, and assesses taxes at the border on imports, equivalent to those paid by domestic firms.

Breakdown: abrupt ecosystem degradation.

Bretton Woods Organisation: term for the World Bank group and the International Monetary Fund.

Cancer: disease which results from the development of a malignant tumour and its spread into surrounding tissues.

Cap-and-trade programme: a regulatory framework for environmental quality which allocates a certain amount of allowable emissions (discharges) to each covered source, which can then choose either to reduce emissions (discharges) to that level or purchase the right to emit (discharge) above those limits from other covered sources which are able to reduce their emissions (discharges) below allowable amounts.

Capacity building: refers to efforts and processes initiated with the aim of providing human resources with skills and expertise that will enable them to perform their jobs more efficiently.

Capacity development in the environment: describes the process by which capacity in environment and appropriate institutional structures are enhanced.

Capacity utilisation rate: the proportion of the total capacity of a production process which is utilised at a time.

Carcinogenesis: the development of cancer. Any chemical, physical or biological agent which can cause cancer is said to be carcinogenic.

Carcinogenicity: capacity of chemical, physical or biological agents to induce malignant neoplasms.

Carrying capacity: level of biomass and/or yield that can be sustained in the long term in relation to the productivity of a particular ecosystem or region with regard to the diversity, structure, and trophodynamics of populations and communities within that system.

Causal chain: a string, often branched, of causally interconnected elements or events.

Cause–consequence analysis: a technique combining fault tree analysis and event tree analysis in a single and possibly quantitative evaluation; diagrams in this method tend to be unwieldy, which explains why it is little used.

Characterisation: the Impact Assessment step in which analysis/quantification and, where possible, aggregation of impacts within the given impact categories takes place.

Checklist: a (complete) list for reference and verification.

Chronic: long lasting or continued. Can refer to an effect requiring longer exposure to reach an end-point, or to duration of exposure itself.

Classification: the Impact Assessment step which consists in grouping the data from the inventory analysis into a manageable number of impact categories. Three major categories of impacts should be considered: resource depletion, human health impacts and ecological impacts.

Coase Theorem: the economic theory positing that, assuming no transaction costs, private parties discharging pollutants and those affected by the discharges will bargain towards an economically efficient allocation of pollution control costs and residual damages attributable to the pollution.

Command and control policy: consists of public authorities announcing certain limits on polluters, then enforcing these limits with appropriate enforcement institutions.

Command and control regulation: the legal approach to environmental protection which imposes on regulated entities detailed parameters for design, operation and/or maintenance of systems and processes for pollution control.

Commission for Sustainable Development: Functional Committee of the UN Economic and Social Council; its primary task is to follow up the implementation of Agenda-21.

Commission of the European Union: the Commission is one of the five institutions of the European Union. It is chiefly an executive body whose main powers lie in the field of initiating, implementing and enforcing Community legislation.

Common law: a system of legal rules which are developed and modified over time in the context of particular controversies, by judges responsible for hearing and deciding those disputes.

Comprehensive study: a study comprising or including much or all.

Concentration: the quantifiable amount of a chemical in air, water, food, sediment, tissue, or any other medium.

Context: the context of something is everything outside (but yet related to) that something.

Convention: an agreement, based on international law, between or among states, or between states and international organisations, or between or among international organisations. An international treaty.

Cosmology: total view about the cosmos and its origin.

Countervailing duty: a duty imposed on imports, based on an importing country's assessment that the goods are being traded below their true cost of production.

Cradle to the grave: an approach entailing the environmental aspects over the whole lifetime of a product, from its crude resource and energy demand over the different stages of production and consumption, to its final treatment as waste.

Credibility: the trust and confidence in self-regulatory schemes of the public at large, to be achieved by external oversight by third parties, such as environmental associations.

Critical loading level: a measure of environmental burden, obtained by aggregating across categories by the 'ecopoints approach'. Individual emissions and energy consumption are weighted by means of ecofactors to give 'ecopoints' (units of environmental pollution). This model reduces an inventory table to a single numerical value by totalling the ecopoints.

Critical volume: in LCA a theoretical 'aggregated' emission value, calculated by dividing the amount of each pollutant by their maximum legally tolerated concentrations.

Cultural filter: the cultural constraints on a world-view.

Culture: everything that is transmitted from one generation to the next, by teaching and imitation.

Cumulative Impact Assessment: appraisal of progressive environmental degradation over time arising from a range of activities throughout an area or region, each activity considered in isolation being possibly insignificant in itself.

Decision (as a legal term in the European Union): a decision is one of the binding acts, taken by the Council of the Commission, and shall be binding in its entirety upon those to whom it is addressed (member state(s)/natural or legal person(s)).

Deoxyribonucleic acid (DNA): the constituent of chromosomes which stores the hereditary information of an organism in the form of a sequence of nitrogenous bases. Much of this information relates to the synthesis of proteins.

Deposit-refund system: in buying a good the price is raised by a mark-up that will be refunded (partly) when the good is delivered at a particular place (and in a particular shape).

Design: the selection and combination of potential plan elements ('options') to form the higher system level of a plan.

Determinism: stress on causal links and universal laws.

Dialectical thinking: the quest for a reflexive equilibrium between two extremes.

Directive (as a legal term in the European Union): a directive is one of the binding acts which the Council and the Commission may adopt within the scope of secondary legislation. It shall be binding, as to the result to be achieved, upon each member state to which it is addressed, but shall leave the choice of form and methods to the national authorities.

DNA adducts: modified sections of DNA, generated as the products of irreversible reactions with genotoxic agents.

Dose–effect curve: demonstrates the relationship between dose and the magnitude of a certain effect, either in an individual or in a population. Such curves may take a variety of forms. They may be linear within a given dose range, but more often they are not.

Dose–response assessment: the process of characterising the relationship between the dose of an agent administered or received and the incidence of an adverse health effect in exposed populations.

Draft EIS: a preliminary version or rough form of an Environmental Impact Statement (EIS).

EC50: median effective concentration: the concentration of a substance estimated to have an effect either lethal or non-lethal on half of a group of organisms. The effect as well as the exposure time must be specified. This statistic relates to *individuals* and cannot be applied to effects such as a 50 per cent reduction in the growth of an algal population.

Eco-asset sheet: shows the stock of ecological assets at a specific point of time.

Eco-balance: instrument for environmental management to calculate all flows of material and energy caused by the activities of a company, a production process or a product.

Eco-efficiency (economic-ecological efficiency): the ratio of value added (or £ of economic return) per unit of environmental impact added.

Eco-label: an eco-label can be awarded to products with reduced environmental impact and which highlights more benign alternatives and therefore provides consumers and users with guidance. These products have to be in conformity with health, safety and environmental requirements.

Eco-rational Path Method (EPM): a structured decision process to pilot companies' businesses to eco-efficiency. EPM represents a straightforward practical procedure to integrate two dimensions: an economic dimension and an ecological one. The integration process distinguishes three steps: accounting, judgement and decision.

Ecological accounting: measures the ecological impact of a company on the environment. Its measurements (unlike those of environmentally differentiated traditional accounting) are in physical terms (e.g. kilograms). Internal ecological accounting is designed to collect ecological information for internal managerial purposes. Its purpose is to measure the impact of a company's products and processes on the natural environment, to make possible good management decisions, and to highlight various ways of examining pollution discharges and to calculate eco-efficiency. External ecological accounting is designed to collect data for external environmental reports, thus giving a public stocktaking of the company's environmental impacts.

Ecological assets: are the stock of 'natural capital' at a specific point of time, such as land, forests, etc.

Ecological efficiency: see 'Eco-efficiency'.

Ecological net effect: compares an action of which the costs can be properly passed on to a third party with the consequences of a reaction of a third party.

Ecology cost: any cost caused by the use of ecologically scarce resources and energy and any cost for avoidance, reduction, substitution, recycling and disposal.

Economic instrument: policy measure that uses prices and market mechanisms to induce the desired change in the behaviour of producers and consumers.

Ecosystem: complex of living organisms and their environment.

Ecotoxicity: the toxicity of a sample measured using ecologically relevant end-points.

Effluent charge: compulsory amount of money to be paid on the release of waste water from point sources to surface water.

Emission standard: determines which eject concentrations should not be exceeded either in absolute terms or relating to a particular frequency and/or during a defined period of time.

End-point: the variables (e.g. time, reaction of the test organisms) that indicate the termination of the test and/or the measurements or values derived that characterise the results of the test (e.g. EC50 or NOEC).

Entropy: low energetic quality or potentiality, disorder.

Environment: all the surroundings of an organism, including all subsurface elements, all natural and man-made surface elements, as well as the atmosphere; the conditions for development and growth.

Environmental accounting: involves various methods of estimating income accounts that assign value to resources such as soil, water, forests, minerals and energy and therefore reflect more accurately in the accounting system the depletion of natural resources and the degradation of natural systems.

Environmental assessment: the process of collecting, organising, analysing, interpreting and communicating data that are relevant to a decision.

Environmental Assessment Review Process: former EIA system of Canada, replaced by the Canadian Environmental Assessment Act.

Environmental auditing: a systematic documented periodic evaluation of how well environmental management systems and equipment are performing with the aim of helping to safeguard the environment by (1) facilitating management control of environmental practices; (2) assessing compliance with company policies, which would include meeting regulatory requirements.

Environmental awareness: process of information of individuals with a view to making them realise the impact human beings or human activities have on the environment.

Environmental communication: the provision of information on environmental matters through the systematic use of information/communication instruments such as the mass media.

Environmental conflict: the delay in the decision-making process because of a controversy amongst relevant stakeholders with respect to the (potential) negative environmental impact of current or intended activities. One may distinguish five types of conflict: intrapersonal (within the self), interpersonal (between individuals), intragroup (within a group), intergroup (between groups) and international (between nations).

Environmental education: process of recognising and clarifying values relating to the environment with a view to fostering interest among individuals to take part in the environmental care and conservation process.

Environmental equity (justice): the treatment of individuals in a uniform (equal) fashion with regard to protection of their health and the environmental amenities which they value, where the treatment is not significantly affected by the race, sex, economic status, or human generation in which they find themselves.

Environmental impact added (EA): the measure of all environmental interventions which are assessed according to their relative environmental impact. It is correlated with value added, as no economic activity is without environmental impacts.

Environmental impact assessment: an instrument which is used to aid and improve the decision making process. The objective is to determine the potential environmental, social and health effects of a proposed development project. It attempts to assess these effects in a form that permits a logical and rational decision to be made. Attempts can be made to reduce or mitigate any potential adverse impacts.

Environmental Impact Statement (EIS): a detailed statement describing and analysing the significant environmental effects of a project and discussing ways to mitigate or avoid those effects.

Environmental index: a result of the valuation step in LCA, a parameter representing the perceived harmfulness of a product to the environment obtained by quantitative weighting.

Environmental indicator: a measure which describes the state of, or trends in, the environment and the factors affecting it in terms which are relevant to and readily usable by the decision-maker.

Environmental literacy: state of awareness of facts and phenomena relating to the environment, both on a local and regional scale, which makes people realise how these affect their surroundings and quality of life.

Environmental management: the art of balancing human demands upon the earth with the natural resources on a sustainable basis. The science supporting these activities.

Environmental management system: the organisational structure, responsibilities, practices, procedures, process and resources for determining and implementing environmental policy.

Environmental planning: refers to the sum total of all public environmental measures taken by an environmental authority, or the document (or documents), which are used in guiding operational decision-making concerning public environmental measures.

Environmental problem: a phenomenon including human activities, effects on the environment and impacts on humans (including future generations) and nature, negatively opposed to the normative ('desired') counterparts of these impacts, effects and activities. Because the facts and values making up the environmental problem are interpretations of underlying realities, the formulation of environmental problems is a social construct. This does not mean that problems go away if not formulated, however.

Environmental problem analysis: getting to know the elements of an environmental problem.

Environmental profile: in LCA the outcome of the characterisation step of the impact assessment; the list of all the effect scores for all impact categories.

Environmental quality standard: indicates which amount, concentration or intensity of a harmful agent should not be exceeded in air, water or soil to prevent negative effects on this compartment.

Environmental review: an initial comprehensive analysis of the environmental issues, impact and performance relating to activities at a site.

Environmental space: the total amount of resources and natural absorption capacity used per individual.

Environmentally differentiated accounting: a subset of traditional accounting which reflects the environmentally induced financial impacts on a firm (i.e. differentiated from other financial impacts).

Environmentally friendly product: has minimal impact on the environment, when the whole chain of production, consumption and waste fate is considered.

Equity (intragenerational, intergenerational): the concept that all members of one generation (intragenerational) or even all members of different generations (intergenerational) are treated on the same terms.

Erosion: removal and loss of soil and sediment by the effects of wind, precipitation, etc.

Espoo Convention: convention of the United Nations Economic Commission for Europe on environmental impact assessment in the transboundary context, signed by 27 countries and the European Community at Espoo in Finland on 25 February 1991.

Ethical realism: the conviction that values exist independently from (human) consciousness.

Ethics: general systematisation of what people ought or ought not to do in cases of conflicting values.

Evaluation: an after-action that examines the patterns of modifications, the consequences, and the ensuing decisions among the alternative actions taken.

Event tree analysis (ETA): an inductive reasoning process starting at some initiating event and from there systematically developing sequences of consecutive events, some of which result in accidents.

Ex aequo et bono: judicial term meaning that a verdict on a case is pronounced without precedent, based on a similar case.

Exclusive Economic Zone: 200 natural miles off the shore of coastal states as defined in the Law of the Sea, allowing a state to exploit this zone exclusively.

Expert judgement: a qualitative element of an assessment which relies on individual or consensus interpretation of data on the basis of expertise and previous experience, used particularly when empirical data are insufficient to allow derivation of a full numerical analysis.

Exposure Assessment: an estimation of the exposure of organisms to a chemical or other agent following its release, or relating to its presence, in the environment, taking into account the transport, fate and uptake of the agent and the duration over which exposure occurs.

Exposure standard: determines the maximum amount of an agent to which ecosystems, people or materials can be exposed.

External cost: a cost which is borne by the society or a third party rather than the party which caused it. Characteristic is the non-fulfilment of the polluter pays principle.

Externality (= external effect): an effect on the production or consumption of economic agents caused by activities by other agents without there being a compensating payment.

Extraterritorial: affecting areas or resources outside the territory of the regulator.

Facts: descriptions of events or states of affairs.

Fault tree analysis (FTA): a deductive reasoning process resulting in a graphical representation of various combinations of failures that all produce a specific top event, i.e. the failure mode of interest.

Fertiliser: a substance used for enriching the productivity of soil.

Financial accounting: typically designed to satisfy the information requirements of external stakeholders (i.e. shareholders) of firms with respect to financial impacts.

Formal notice (as a legal term in the European Union): if the Commission considers that a member state has failed to fulfil an obligation it will give the state concerned – by a formal notice – the opportunity to submit its observations, and, as the case may be, to enable the Commission to persuade the state to comply with the requirements of the Treaty.

Framework convention: a type of treaty negotiated to provide a way to address the environmental problem in question through statements of obligation and co-operation, generally without requiring actual abatement measures. This type of convention may set up a process for scientific co-operation and information sharing and is generally designed to lead to negotiation of protocols containing specific obligations once sufficient concern and certainty about the issue exists.

Free-Rider: in the context of MEAs, this is a country which decides not to participate in an agreement, thereby not incurring the costs of compliance while at the same time enjoying the environmental benefits the agreement creates.

Freewill philosophy: stress on human freedom to make conscious choices.

Functional unit of product: the amount of the product necessary to fulfil a well-defined function.

Genome: all the genes carried by a cell.

Genotoxicity: ability to cause damage to genetic material or have an adverse effect in the genome (e.g. mutation, chromosomal damage, etc.) that may lead to a cancer.

Global Environmental Facility (GEF): organisation established by UNEP, UNDP and the World Bank, governing a fund which it gives out for environmental projects focusing on climate change, biodiversity and ozone depletion.

Global Warming Potential: for some substances which contribute to the enhancement of the greenhouse effect, parameters have been developed in the form of Global Warming Potential (GWP). These parameters can be used to express the potential contribution of these substances to the greenhouse effect in a single effect score. The GWP is a relative parameter which uses CO_2 as a reference. In this way atmospheric emissions (in kg) can be converted to CO_2 emissions (in kg) resulting in an equivalent greenhouse effect.

Good laboratory practice (GLP): fundamental rules incorporated in national regulations concerned with the process of effective organisation and the conditions under which laboratory studies are properly planned, performed, monitored, recorded and reported.

Green fund: pool of monetary resources reserved for environmental purposes.

Green tax: includes a variety of measures that would penalise practices that pollute or degrade the environment or encourage excessive waste of a valuable natural resource or service.

Gross domestic product: aggregate value of value added, or the value generated by any unit engaged in a production activity. It is based on the production account. This shows the transactions relating to the production process proper. It is drawn up for institutional sectors and for industries. Its resources include output and its uses include intermediate consumption.

Gross national income: represents total primary income receivable by resident institutional units: compensation of employees, taxes on production and imports less subsidies, property income, operating surplus and mixed income. It equals GDP minus primary income payable by resident units to non-resident units, plus primary income receivable by resident units from the rest of the world. It is conceptually identical with gross national product.

Gross national product: see 'gross national income'.

Guideline: an estimation of the highest allowable concentration which still guarantees a reasonable health or environmental quality condition. A guideline has scientific but no legal value.

Harm: deleterious effects upon the health of living organisms or other interference with the ecological systems of which they form part.

Hazard (toxic): the set of inherent properties of a chemical substance or mixture which makes it capable of causing adverse effects in man or to the environment when a particular level of exposure occurs.

Hazard assessment: determination of the existence of a hazard through comparison of expected or observed environmental concentrations of a contaminant with those eliciting defined end points in toxicity tests.

Hazard identification: identification of the adverse effects which a substance has an inherent capacity to cause, or, in certain cases, the assessment of a particular effect. It also includes the identification of target populations and conditions of exposure.

Head of power: those specific provisions in the US Constitution which authorise action by one or other of the three branches of the national government (Congress, the President, and the Supreme and lower national courts).

Holistic: considering the totality of an event or situation. It is based on the principle that a whole thing or being is more than just a collection of parts added together.

Hookworm: common name of bloodsucking nematodes of the family *Ancyclostomatidae*.

Human carrying capacity (or 'regional support capacity'): the number of people that can live off the land of a certain region sustainably, given a certain farming system and satisfying a certain income standard in a given context of markets, without external subsidies and without external burdening of other regions.

Human development index: an alternative to GNP for measuring the relative socio-economic progress of nations. It enables people and their government to evaluate progress over time and to determine priorities for policy intervention. It also permits instructive comparisons of experiences in different countries.

IC50: inhibition concentration required to inhibit or reduce biological activity such as a 50 per cent reduction in growth of an algal population or a 50 per cent reduction in photosynthesis.

Immission standard: defines which concentration or exposure level should not be exceeded.

Impact assessment: the technical, quantitative, and/or qualitative process to characterise and assess the effects of the environmental burdens identified in the inventory analysis.

Impact profile: see 'Environmental profile'.

Impact score: in LCA the aggregated contribution of a product system to one impact category.

Improvement assessment: the component of LCA in which options for reducing the environmental impacts or burdens of the system(s) under study are identified and evaluated.

Incremental costs: the extra costs of a project to protect the environment.

Incremental unit risk estimate: for an air pollutant, the additional lifetime cancer risk occurring in a hypothetical population in which all individuals are exposed continuously from birth throughout their lifetimes to a concentration of $1 \mu g/m^3$ of the pollutant in the air they breathe.

Index: cardinal and ordinal average to evaluate the environmental conditions.

Indicator (for sustainable development): a value which contains quantitative information that is based on the primary and analysed data describing the environmental, social, economic or institutional aspects of sustainable development.

Input–output account: records material (and energy) flows of a production step, firm, nation, etc.

Institutional capacity building: refers to reinforcing the capacity of an organisation in order to enable it to perform its functions more efficiently.

Integrated environmental management: a philosophy which prescribes a code of practice ensuring that environmental considerations are fully integrated into all stages of the development process in order to achieve a desirable balance between conservation and development.

Inter-generational equity: requires that the needs of the present generation are met without compromising the ability of further generations to meet their own needs.

Internal cost: a cost which is borne by the party which caused the cost and is reflected in its accounting system.

Internalised cost: a cost which is already part of the cost accounting of a company or another organisation.

International Court of Justice: the principal judicial organ of the UN.

International Sea-bed Authority: institution established under the Law of the Sea Convention, authorised to undertake mining operations on the sea-bed and to issue mining rights to private investors.

International Tribunal of the Law of the Sea: international court established under the Law of the Sea Convention, seated in Hamburg, dealing with conflicts arising under the Convention.

Intra-generational equity: requires that people within the present generation have the right to benefit equally from the exploitation of resources and that they have an equal right to a clean and healthy environment.

Inventory analysis: in LCA the technical, data-based process of identification and quantification of energy and raw material requirements, atmospheric emissions, waterborne emissions, solid wastes and other releases of a product, package, process, material, or activity.

Inventory table: a list of environmental interventions caused by processes within a product system.

Joined Time Projection (JTP) models: dynamic material flow models, in which several static flow models of successive time periods are joined together by the input in each model of elements of one or more time-series. Examples of such time-series are demand estimates, projected capacities and changes in technology.

Jurisdiction: the right to pronounce a judicial decision.

LC50: median lethal concentration; the concentration of a substance that is estimated to kill half of a group of organisms. The exposure time must be specified.

LD50: median lethal dose, refers to the amount of toxicant inside an organism as introduced by, *inter alia*, an injection.

Learning-by-doing: a radically cyclical approach in which the implementation of solutions is used to gather experience for a new round of problem analysis, problem explanation, the design of solutions and its implementation.

Least common denominator: the level of regulation that is acceptable to all the states participating in a treaty negotiation; it is usually of a low level because those who resist action are not willing to take difficult or costly measures.

Leopold matrix: impact identification method developed by Leopold in 1971 for the US Geological Survey. Impacts in the matrix are always identified by a series of discrete, two-way linkages between development activities and components.

Lethal: causing death, or sufficient to cause death.

Levy: compulsory, *quid pro quo* payment for the provision of a service good or another asset used in production.

Life-cycle assessment: an objective process to evaluate the environmental burdens associated with a product, process or activity by identifying and quantifying energy and materials used and wastes released to the environment, to assess the impact of the energy and materials released on the environment, and to evaluate and implement opportunities to affect environmental improvements. The assessment includes the entire life-cycle of the product, process, or activity, encompassing extraction and processing of raw materials, manufacturing, transportation and distribution, use/re-use/maintenance, recycling, and final disposal.

Life-cycle assessment with Delta analysis: a method in which one material flow of a life-cycle inventory model is changed with a very small quantity (= Delta), and the effects of the resulting changes in the model analysed as in regular life-cycle assessment.

Life-cycle costing: instrument used to calculate which of two alternatives is the more cost effective, calculated over the whole life-cycle of these products.

Local remedies principle: the rule that redress should be sought at the lowest possible instance (e.g. at local, then regional or national level) before calling upon a higher (e.g. international) institution.

Location alternative: in environmental impact assessment, a different option relating to the site of an operation or activity.

Lowest-observed-effect concentration (LOEC): the lowest concentration that causes a statistically significant adverse effect in a sublethal test.

Management: the creative and error-correcting activity that gives a firm its purpose and cohesion, and assures satisfactory return on investment.

Management system: framework that aims to pull potentially disparate inter-relationships between individuals, tasks and responsibilities in organisations into integrated and organised systems. Such systems cover the totality of organisational operations and help management and workers clearly to see their place in the organisation and to recognise the interdependence of all its aspects.

Managerial accounting: (also called management accounting or cost accounting) is the central monetary information tool and basis of most internal management decisions.

Marginal (cost, productivity, etc.): the first derivative of the function (e.g. marginal cost = change in cost due to a change in output).

Marginal (emission) abatement cost: added cost of achieving a one-unit decrease in emission level.

Marginal land: land of poor quality that is out of use.

Marketable permits: emission quota sold and bought on the stock market.

Mass balance: the mathematical expression in which a summation of all material inputs to a system is equated to a summation of all outputs, accounting for transformation into energy.

Material balance model: a model which supposes a balance between the materials and energy entering and leaving a production or a consumption process.

Material flow model (MFM): a model which analyses man-induced flows and processes of materials and/or products within the anthroposphere, and identifies, measures, and values the material and energy inputs taken by the flow system from the natural environment and the effluent and waste outputs of the system to the environment.

Mathematical model: a set of entities that satisfies all the formulae of a given formal or axiomatic principle.

Matrix: an array of information or data arranged in a rectangle of rows and columns, treated as a single entity.

Maximum acceptable toxicant concentration (MATC): a presumed threshold concentration for effects upon survival, growth and reproduction. For simplicity it is sometimes estimated as the geometric mean of the NOEC and LOEC for the three listed parameters.

Maximum allowable concentration (MAC): regulatory value defining the concentration which, if inhaled daily (for workers: 8h/d over a working week of 40 h; for the general population: 24 h), does not appear capable of causing appreciable harm in the light of present knowledge.

Mediation: the process or action of intervening between parties in dispute to produce agreement or reconciliation.

Mediator: a person who tries to realise the best set of decisions, considering the various interests, by building trust and collaboration.

Mesocosm: an artificial system, designed to represent one or more components of an ecosystem under study, constructed in order to study organism interactions or the fate and effect of contaminants under controlled conditions. Although necessarily simplified representations of the environment, mesocosms provide a greater degree of ecological relevance than standard laboratory conditions and single species tests.

Methodological framework: an interconnected whole of statements of how to do good research; a 'story' of how to discover or design a part of the world for a certain aim. 'Methods' are more specific elements within a methodological framework, speaking, for instance, about how to do a good interview or a statistical test.

Microeconomics: the branch of economics studying the behaviour of private economic agents (consumers, producers) and their relationships through markets.

Mitigation: the action of making something milder or less hostile.

Monitoring: measuring or testing at intervals, especially for the purpose of regulation or control.

Monoculture: a system (forest, agriculture, etc.) in which one species dominates.

Most-Favoured Nation: the MFN is codified in GATT Article I. It states that WTO members must accord, to any product originating in or destined for any other country, the same treatment to all other third members.

Multilateral: undertaken jointly by the international community of countries.

Mutagenesis: introduction of hereditary changes (mutations) in the genotype of a cell as a consequence of genetic alterations or the loss of genes or chromosomes (or parts of them). Any chemical that causes mutations is said to be mutagenic. Some mutagenic chemicals are also carcinogenic.

Mutual gains approach: consists of (1) setting up a Community Advisory Group (CAG) to realise joint fact-finding; (2) reporting to CAG on cost and benefit of all relevant solutions; (3) the acknowledgement of secondary effects, initiating thorough investigation and the setting-up of a joint-site monitoring programme; (4) the organisation of open community meetings jointly with the CAG to reach consensus; and (5) safeguarding long-term company interests by participating in a joint-arbitration panel and the setting-up of a community trust fund. The MGA is based on the notion of 'side-by-side' problem-solving rather than 'across-the-table' confrontation via litigation or administrative procedures. Thereby it is an alternative, i.e. consensual, type of environmental conflict management and involves a systematic endeavour for a long-term and structured dialogue via co-operation to reduce the potential for future conflicts.

National Environmental Policy Act: the Federal law of the USA enacted in 1969 which prescribes the environmental policy of the country and requires EIA to make sure that the environmental policy is taken into account in the decision-making.

National treatment: this principle is codified in GATT Article III, which addresses how a country may treat imports. Any treatment applied to an imported good, such as taxes, regulations or requirements, should be no less favourable than that applied to like domestic goods.

Naturalistic fallacy: the erroneous belief that norms and values can be directly derived from facts.

Neg-entropy: high energetic quality, order.

Negotiation: a process comprising direct interactions between the relevant parties, that seek mutually beneficial solutions, which may have been overlooked. Negotiations may avert both frustration and a negative impact on the corporate image. As it raises co-operative above the confrontational aspects, conflict management by means of negotiation stresses facilitation rather than legislation. For successful negotiation there must be a free flow of information between the various parties involved.

Network: a chain or system of interconnected or intercommunicating immaterial things, points or people.

No-observed-effect concentration (NOEC): the highest concentration in a sublethal test that does not cause a statistically significant adverse effect in comparison to the controls.

Notice-and-comment rule-making: the process of making legally binding administrative regulations, which process requires notice to the public generally regarding the agency's proposal, solicitation of the public's views regarding the proposal, and consideration by the agency of any such comments submitted prior to making a final decision.

Notification (as a legal term in the European Union): a notification means giving a compulsory notice of a measure to the Commission for approval.

Nutrient: chemical element in the soil possessing nourishment for plants.

Objective axiology: the derivation of values and value-priorities from an objective description of facts.

One-hit model: dose–response model of the form $P(d) = 1 - \exp(-bd)$ where P(d) is the probability of cancer death from a continuous dose rate (d) and b is a constant. The one-hit model is based on the concept that a tumour can be induced after a single susceptible target or receptor has been exposed to a single effective unit dose of an agent.

Open-loop recycling: a recycling system in which a product made from one type of material is recycled into a different type of product.

Opinion (as a legal term in the European Union): a statement delivered by the Council or the Commission and having no binding force. An opinion is addressed to member states. It usually expresses a point of view or advice, often in reply to a question put to the institution, or forms an essential procedural step in the decision-making process.

Opt-out provision: a procedure sometimes included in a treaty by which a state can refuse to be bound by a certain part of the treaty or by a decision taken by a regulatory body within the treaty process. This type of provision is almost always present in treaties that allow for regulations to be made without the consensus of all parties to the treaty.

Overlapping consensus: a unity on key issues obtained in spite of, or thanks to, cultural diversity.

Ozone depletion potential: for some substances which contribute to the depletion of the ozone layer, parameters have been developed in the form of ozone depletion potential (ODP). These parameters can be used to express the potential contribution which these substances make to the depletion of the ozone layer in a single effect score. The ODP is a relative parameter which uses CFC-11 as a reference. In this way atmospheric emissions (in kg) can be converted to CFC-11 emissions (in kg) resulting in a equivalent depletion of the ozone layer.

Panel review: an examination of a certain step in the EIA process or an examination of an EIA document by a group of experts and/or other groups of people participating in the EIA process.

Permanent Court of Arbitration: bureau, based in The Hague, the Netherlands, providing a panel of arbitrators and facilities to organise international arbitration.

Pesticide: a substance for destroying pests (eg. insects, weeds, fungi, etc.).

Photosynthesis: production of complex organic materials (e.g. carbohydrates) by green plants from carbon dioxide, water, and nutrients, using sunlight as a source of energy and with the aid of chlorophyll.

Physical model: an illustrative or small-scale copy of an original, constructed to represent the environment (e.g. a visual representation by picture, film or three-dimensional model, or a working model of the environment using wind tunnels or wave chambers).

Planning: a rational methodology for elaborating programmes designed to achieve set goals. It involves the definition of the problem, the establishment of planning objectives, the identification of the assumptions that will form the planning basis, the search for and evaluation of alternative courses of action, and the selection of a particular action to be followed. *Procedural planning* (also called the classical model of the planning process) applies rational procedures of thought and action and consists of a series of integrated steps including the identification of needs, the specification of goals and objectives, the development of alternative means to attain each goal, the estimation of the costs of each alternative, and the selection of the most promising alternative(s). *Strategic planning* is a process for distilling social goals and objectives, and for the designing of alternative scenarios or 'road maps' that become the basis for meaningful collective choice. *Substantive planning* refers to the subject matter of planning or the object of planning.

Policy instrument: a type of social or physical option that may become part of a policy plan.

Polluter-pays-principle: this suggests that the polluter should bear the cost of preventing and controlling pollution. The intent is to force polluters to internalise all of the environmental costs of their activities so that these are fully reflected in the costs of the goods and services they provide.

Post-modernism: a contemporary philosophy challenging some myths of modernity and the Enlightenment.

Post-project analysis: an environmental study undertaken during the operational stage of a project or programme to assess compliance with the terms proposed in the EIS and to consider the quality and possible improvements of environmental management.

Postnormal science: an emerging, extended type of science adapted to face global environmental challenges.

Precautionary principle: holds that where there are threats of serious or irreversible damage, lack of full scientific certainty should not be used as a reason for postponing cost-effective measures to prevent environmental degradation.

Predicted environmental concentration (PEC): the predicted concentration of a toxicant in the environment, taking into account dilution and dispersion.

Predicted no-effect concentration (PNEC): the environmental concentration regarded as below that level at which the balance of probability suggests an unacceptable effect will not occur.

Pressure–state–response model: logical framework developed by the OECD, indicating the relations between the human ecosystem (economic and social activities), the environment (nature and environmental hygiene aspects) and the (nature and policy) responses to changes in these relations.

Price elasticity of demand (own): the percentage change in the demand for a good due to a percentage change in the price of the good.

Probit: a probit, or probability unit, is obtained by modifying the standard variate of the standardised normal distribution by the addition of a constant value of 5 (to avoid negative numbers). Converting a cumulative percentage response to probits followed by plotting it against concentration or dose can provide useful information about the distribution of the response and estimates of the L(E)D50 of L(E)C50 values. This transformation is used for the analysis of dose–response data.

Probit/log transform: the probability unit obtained from the standardised normal distribution plotted against the logarithm of the concentration or dose of a substance when a quantal or graded response has been measured. A linear plot provides evidence that the distribution is lognormal. Estimates of the L(E)C50 and L(E)D50, as well as the standard deviation for the distribution, can then be made.

Process alternative: tries to find different ways to produce something.

Process standard: defines maximally allowed emissions for a particular installation or specifies conditions which need to be taken into account in designing, building or exploiting these facilities.

Product charge: compulsory amount of money to be paid on environmentally harmful products (such as batteries or lubricating oil) to defray the cost of safe treatment.

Product standard: formulates requirements for a particular product.

Production-related standards: standards to address environmental problems inherent in the process or production method (PPM) used in manufacturing the good, or in cases where the production or harvest of the good is itself a problem.

Progressive contextualisation: a research strategy that seeks to explain something by starting out from that thing and then goes 'upstream' along the causal chains that cause the thing in question, led by notions of relevance, and stopping when a sufficient degree of explanation has been reached. Characteristic for this approach is that system assumptions are not needed, and that the elements to be identified are not specified beforehand.

Protectionist: deliberately favouring domestic producers over competing foreign firms.

Protocol: an additional agreement, annexed to a Convention. It defines, modifies and/or supplements (a part of) a principal Convention.

Public good: a good that can be used by several people at once because the use by the one does not lower the utility of the other (non-rivalry).

Public nuisance doctrine: the legal doctrine developed by the courts which allows certain representatives of the public, that is to say, governments, to invoke the aid of the courts in abating threats to the public health and welfare posed by the use of privately owned land.

Public trust doctrine: the legal doctrine developed by the courts which imposes an enforceable duty on the government to act as a trustee in its management and disposition of public lands and resources.

Rapid prototype: a trial version of a plan, made on a basis of almost no analysis, drafted only in order to explore a design situation or to learn how to cross the design threshold, and hence often thrown away after its making.

Realisation alternative: in environmental impact assessment an option in setting up an activity or project.

Reasoned opinion (as a legal term in the European Union): if the Commission considers that a member state has failed to fulfil an obligation, it must deliver a reasoned opinion on the matter after giving the state concerned – by a formal notice – the opportunity to submit its observations. The reasoned opinion determines the precise limits within which the Commission may later bring the matter before the Court of Justice.

Recommendation (as a legal term in the European Union): recommendations made by the Council or the Commission have no binding force. They may be addressed to member states, to individuals, or by the Commission to the Council. They are usually made for the purpose of obtaining a certain course of action or conduct from the addressee, or of laying down procedures and timetables for the achievement of Community objectives.

Regulation (as a legal term in the European Union): a regulation is one of the binding acts which the Council and the Commission may adopt within the scope of secondary legislation. It represents the main form of Community legislation, and has general binding application. It is binding in its entirety and directly applicable in all member states.

Resilience: ability of a healthy ecosystem to withstand disturbances, without dramatically lowering its capacities and functions.

Responsible care: a chemical industry action of self-regulation, initiated in Canada by the CCPA in 1984, by which chemical companies give a signatory commitment, in all aspects of safety, health, and protection of the environment, to seek continuous improvement regarding product use and overall operation. Important elements of the action concern the development and dissemination of environmental performance indicators, and the ongoing process of communication with interested groups outside industry.

Review: a general survey or reconsideration of some subject or thing.

Reward and punishment: the classical, minimalistic concept of motivating individuals by directing their activities via stimulus–response methods appealing to their direct self-interest.

Risk: the probability of an adverse effect on man or the environment resulting from a given exposure to a chemical or mixture. It is the likelihood of harmful effects occurring due to exposure to a risk factor (usually some chemical, physical or biological agent). Risk is usually expressed as the probability of an adverse effect occurring, i.e. the expected ratio between the number of individuals that would experience an adverse effect in a given time and the total number of individuals exposed to the risk factor. The term 'absolute risk' is sometimes expressed per unit dose (or exposure) or for a given dose (exposure).

Risk assessment: a process which entails some or all of the following elements: hazard identification, effects assessment, exposure assessment and risk characterisation. It is the identification and quantification of the risk resulting from a specific use or occurrence of a chemical compound, including the determination of dose–response relationships and the identification of target populations. When little or no quantitative data is available on dose–response relationships for different types of populations, including sensitive groups, such considerations may have to be expressed in more qualitative terms.

Risk characterisation: the estimation of the incidence and severity of the adverse effects likely to occur in a human population or environmental compartment due to actual or predicted exposure to a substance. This may include risk estimation (i.e. quantification of that likelihood). It also serves as a summary and description of the results of a risk analysis for a risk manager, or for the public and other interested parties.

Risk classification: the valuation (or weighting) of risks in order to decide whether risk reduction is required. It is a complex process of determining the significance or value of the identified hazards and estimated risks to those concerned with or affected by the decision. It therefore includes the study of risk perception and the balancing of perceived risks and perceived benefits.

Risk management: a decision making process that entails the consideration of political, social, economic and engineering information – together with risk-related information – in order to develop, analyse and compare the regulatory options and select the appropriate regulatory response to a potential health or environmental hazard.

Risk perception: the subjective perception of the gravity or importance of the risk based on the individual's knowledge of different risks, and the moral and political judgement attached to them and their importance.

Risk–benefit analysis: the process of drawing up a balance sheet of the respective risks and benefits of a proposed risk-reducing action. It is a multidisciplinary task in which the risk manager has to consider not only the risk assessment but also other important aspects such as technical feasibility, economic factors and social and cultural factors as well as legislative and political factors.

Rule-making: the administrative process whereby an agency formulates a (usually detailed) legal instrument designed to control or otherwise affect the behaviour, activities, and other interests of classes of governmental entities and/or private parties.

Safety (toxicological): a high probability that adverse effects will not result from exposure to a substance under specific conditions of quantity and manner of use.

Safety factor: a factor applied to an observed or estimated toxic concentration or dose to arrive at a criterion or standard that is considered safe. 'Safety factor' and 'uncertainty factor' are often used synonymously.

Salinisation: accumulation of salts in the soil, often due to the excess application of irrigation water in areas where evapotranspiration rates are high and drainage poor.

Sanitary and phytosanitary measures: measures which apply specifically to the protection of human or animal life from risks arising from additives, contaminants or disease-causing organisms in their food, or from the spread of pests from imported products.

Schistosomiasis: disease caused by the infection of blood flukes of the genus *Schistosoma*, characterised by chronic symptoms especially of the digestive and urinary systems, and sometimes by fever.

Scoping: process at the beginning of EIA in which a decision is made on the contents of the study. Scoping has to help ensure that an EIA focuses on key issues associated with a proposed activity or development.

Screening: in EIA, a process which results in a decision on the need and extent for the preparation of an EIS.

Social paradigm: model around which the perception of reality is organised.

Socio-economic impact assessment: a subset of EIA, an assessment of the impact on people, society and the economy of major policies, plans, programmes, activities and developments.

Soft law: international measures that states indicate they will undertake, but that are not legally binding.

Sorensen network: impact identification method developed by Sorensen in 1971 which makes it possible to consider explicitly the secondary, tertiary and higher-order impacts that can arise from an initial impact.

Stakeholder: any group or individual who can affect or is affected by the achievement of the firm's objectives.

Standard: the limiting concentration of a chemical (or degree of intensity of some other adverse condition, e.g. pH) which is permitted in an environmental compartment (soil, effluent or waterway). Standards are established for regulatory purposes and are determined on the basis of a judgement of the criteria involved. The standard is dependent on the use (e.g. drinking water or agricultural water for irrigation). Standards are derived from criteria, often by applying safety factors (e.g. quality standards for air, water and soil).

Stare decisis: the practice whereby common law courts continue to apply the same judicially developed rules to controversies as they arise unless there is a very good reason to change the law.

Strategic environmental assessment (SEA): the application of EIA at the level of policies, plans, programmes or other human activities which are more than an individual project.

Strategic management: a continuous iterative process founded in the reality of good management practice which focuses on the present strategy of the firm, supported by the resources and strengths generated in the past, and the design and implementation of a strategy which will carry the organisation successfully into the future.

Strategy: a well co-ordinated set of action programmes aimed at securing a long-term sustainable advantage.

Sublethal: concentration or level not causing death, or an effect which is not directly lethal.

Subsidy: amount of money paid by a public authority to a polluter for every unit of pollution emitted below a fixed bench-mark level. The reverse of a tax or a charge.

Sustainability: the long-term non-impairment of the capacity of the natural system to regenerate raw material inputs and to absorb waste outputs of the human economy.

Sustainable development: an organisation of the affairs of life so that the environment is used in a way to meet the needs and aspirations of this generation, and such that this does not compromise the ability of the future generations to meet their needs.

Sustainable principle: principle in line with the concept of sustainable development.

Sustainable resource: a resource that is replaced in the environment faster than it is being consumed.

SWOT analysis: an instrument to analyse mainly strategically relevant activities in terms of strength, weakness, opportunities and threats.

Synergism: the effect whereby the toxicity of two or more chemicals is more than the sum of the individual toxicities.

System: a bounded and interconnected whole of elements, displaying system-level characteristics. Systems can be 'strong', which means that the elements are highly interdependent and that the system-level characteristics dominate the separate elements; the human body is an example. Other systems are 'weak', i.e. they are loose assemblies of elements that continue to have much life of their own; the ecosystem is an example. Other phenomena are often called 'systems' (or 'community' etc.) because this is romantic and fashionable, while they may in fact not be systems at all. Research should always be systematic, but does not necessarily need to view the world as composed of systems only.

System of national accounts: an internationally compatible accounting framework for a systematic and detailed description of a total economy, its components and its relations with other total economies.

Tariffs: charges assessed at the border on imported goods.

Tax: compulsory, unrequited payment on goods, services, labour, the ownership or use of land, buildings or other assets used in production.

Tax accounting: an accounting system which provides information for the tax agency.

Technical barriers to trade: any sort of restriction placed on imports or exports based on product standards and technical regulations. The former are standards mandating that the product itself meet some requirement: that its characteristics conform to some standard. Any other sort of requirement or procedure mandated by law is a technical requirement.

Technology transfer: the giving or selling (often at low cost) of technology or information to countries in an effort to address environmental problems.

Teratogen: agent which, when administered prenatally to the mother, induces permanent structural malformations or defects in the offspring.

Teratogenesis: the potential or capacity of a substance to cause defects in embryonic and foetal development.

Threshold limit value (TLV): concentration in the air of a substance to which it is believed that most workers can be exposed daily without adverse effect (the threshold between safe and dangerous concentrations). These values are established (and revised annually) by the American Conference of Governmental Industrial Hygienists and are time-weighted concentrations for a 7-h or 8-h working day and a 40-h working week. For most substances the value may be exceeded to a certain extent, provided there are compensating periods of exposure below the value during the working day (or in some cases, the week). For a few substances (mainly those that produce a rapid response) the limit is given as a ceiling concentration (maximum permissible concentration, designated by 'C') that should never be exceeded.

Tiering: refers to the concept of a 'multi-tiered' approach to preparing EISs. The first-tier EIS covers general issues in a broader programme-oriented analysis. Subsequent tiers incorporate by reference the general discussion from the broader EIS, while primarily concentrating on the issues specific to the action being evaluated.

Tolerance: highest concentration of a substance in which test organisms of median sensitivity could live for an indefinite time, i.e. the highest concentration that does not represent a chronic LC50.

Toxic: able to cause injury to a living organism as a result of physico-chemical interaction.

Toxicity: the inherent potential or capacity of a substance to cause adverse effects on a living organism, seriously damaging structure or function or producing death.

Tradable permit: see 'Marketable permits'.

Trade sanctions: trade restrictions on imported goods, but which are not based on concerns over the good in-question. A country might, for example, ban the import of televisions from a country whose fishing practices were environmentally damaging.

Traditional accounting: covers all accounting systems which provide monetary information of the company or organisation.

Tragedy of the commons: a situation in which each actor in a group bears the full positive utility of its action, while the negative utility of the action is shared. Because of this incentive structure actors individually take actions that advantage them in the short run, but that are destructive both collectively and to the individual in the long run. Even though all actors would benefit in the long run from collective restraint, the lack of overarching authority prevents them from achieving this goal.

Trust fund: pool of monetary resources provided by several states, often in the course of international conventions, and made available to parties needing funds to comply with the convention's provisions.

Tuberculosis: a disease characterised by the formation of nodular lesions or tubercles in the tissues.

Ultra vires: 'beyond authority or power', or, otherwise stated, an administrative agency's assertion of authority in violation of the US Constitution, applicable statute, or other law.

UN Economic and Social Council: principal organ of the UN co-ordinating the economic and social work of the UN and the specialised agencies and institutions.

UN Security Council: organ of the UN with 15 state members responsible for the maintenance of international peace and security.

Uncertainty: imperfect knowledge concerning the present or future state of the system under consideration. A component of risk resulting from an imperfect understanding of the degree of hazard or of its spatial and temporal pattern of expression.

Unilateral: undertaken by one country without consultation with, or the consent of, other countries.

User charge: compulsory amount of money to be paid on environmental services as waste, usually levied on households at a flat rate, and on firms by either a flat rate or one that varies with the nature and/or amount of waste taken away.

Valuation: the impact assessment step in which the data of the different specific impact categories are weighted in a quantitative and/or qualitative way so that they can be compared among themselves.

Value: quality of a person, a group, an object or a process, ideally beyond individual or group interests or preferences.

Waste disposal charge: compulsory amount of money to be paid on waste received at disposal sites such as landfills and incinerators.

Waterlogging: saturation with water.

World-view: a provisory, culturally constrained integration of information and beliefs about reality.

Zoning standard: relates environmental quality to the distance of an emission source.

ANSWERS TO SELF-ASSESSMENT QUESTIONS

CHAPTER 1

1 The current world population is nearly 6 billion, and about a quarter million people are added to the world population daily.
2 About 641 kg of food is required per person per year worldwide.
3 To produce food per person, the following resources are approximately required: cropland = 0.27 ha; pastureland = 0.57 ha; water = 0.64 million litres; fossil energy = 1,500 litres of oil equivalents.
4 More than 10 million ha of arable land are being lost each year. Most of this land is being lost because of erosion, but could be protected through the implementation of sound soil conservation practices.
5 About 8 million litres/ha of water are essential to produce a crop. It would be extremely difficult to reduce this minimum amount of water. However, various irrigation practices in use in the world do waste water, and several technologies are available to do a better job of water management.
6 Approximately 80 per cent of all infectious diseases are associated with water. Better management and treatment of human and animal wastes would help reduce the spread of infectious diseases.
7 Most fossil energy resources, like oil, will be depleted in 50 to 100 years.
8 It is doubtful that solar energy technologies can totally replace all the energy needs currently supplied by fossil energy sources.
9 It will be difficult to substitute solar energy for current fossil energy sources because solar energy technologies require enormous amounts of land. Both food and forest production require land and these resources are essential for society.
10 There are approximately 10 million species on earth and estimates are that we may lose 1–2 million by the middle of the next century.
11 Biodiversity is essential for a quality environment and productive agricultural and forestry systems.
12 Population growth is causing increasing environmental degradation and rapidly growing diseases. Approximately 60 per cent of all human deaths are caused by infectious and other environmentally caused diseases, like chemical pollution.
13 Numerous management strategies are available to reduce pollution and environmental degradation. These include improved soil and water conservation and treatment, improved air pollution controls, and employment of sustainable agriculture and forestry management practices.
14 If humans do not control their numbers then starvation, disease, and other environmental disasters will control human population.

CHAPTER 2

1 Using your common sense and the information given in other parts of this book, you will probably have no problem finding the relevant specifications of the final variables here; they pertain to the ecological health of the soil ecosystem, your own health, the intrinsic value of insect-eating birds and so on, to which may be added, if you like, the harmony of your relationship with nature (soil, plants, birds) as it expresses itself in your garden; is the growing of vegetables a matter of dominating nature or of working together with nature? Effect chains run from the problematic action, your applying the pesticides, to these variables, through air, soil and biotic pathways. Here the environmental capacity, the acceptable intensity of the problematic action, is what you allow yourself to burden the environment

with. If you are clever enough, you will have found how the effects and norms chains branch out and weave into the analysis of your vegetable garden problem: what your neighbours do is also relevant for your garden, and the pesticides that you spray have impacts on your neighbours, too. It is important to realise that in this question you are the normative observer and at the same time also the primary actor. The social-scientific contextualisation therefore starts out as a self-analysis: your own options with respect to protecting your plants, the benefits and appropriateness you see attached to these options, and so on. You may apply the 'deeper analysis' schema to yourself, too. (What is my knowledge with respect to other potential options? Is my self-image involved here, e.g. being a macho sprayer or an 'organic' softie?) The secondary actors you may identify in the actors field are, for instance, the pesticide producers, the government which does or does not ban or tax the pesticides, and so on. Would the pesticide producers be tertiary actors putting pressure on the government? Maybe you'd better join Friends of the Earth, so that you become a quaternary actor, putting some pressure on the producers ... (think locally but act globally!)

2 First of all it is crucial that the consultants describe the effects in the relevant terms, i.e. the terms of the policy objectives of your agency and of the national environmental policy. Do not let them get away with just describing effects on arbitrary categories of birds, for instance, but prescribe that they should focus on the rare species, or other categories relevant in normative terms. Generally, you should prescribe that impacts be presented in terms of the final variables, specified so as to fit your specific problem situation. 'Human health and safety' are of course easy to specify here; 'material welfare' can be translated into the economic benefits of the faster flow of goods and services and less wear on vehicles; 'immaterial welfare' can be introduced in terms of landscape impacts, the recreational quality of the area, and so forth. With respect to the evaluation method to be applied in the EIA, you should certainly examine the equity-versus-efficiency issue. One way to operate here is to prescribe that impacts on nature, poor people and future generations should be described separately from the rest of the multi-criteria or cost–benefit analysis, so that they can enter into a separate public discussion. The other way around it is to prescribe that no alternative is allowed to have negative net impacts on nature, poor people or future generations; all the designs should include compensation actions to restore the habitat of endangered species, for instance. A third important issue is the degree to which you prescribe that alternative designs should be truly strategic alternatives ('values-steered'), or allow them to be only technical variants of essentially the same solution. In other words, will the alternatives be alternative ways to support economic growth by means of the smooth flow of goods and services (whatever the technical means), or will they merely be variants of a proposed highway, varying only with respect to the number of lanes or the alignment? Besides these three issues, you will of course have to specify more technical aspects of the EIA procedure: the models to be applied in order to calculate properly the effects chains, the range of possible secondary effects to be included in the EIA (instead of being allowed to simply branch out and remain unevaluated) and so on. In order to do so, it is best to make a kind of rough participatory problem analysis, jointly with the experts but also definitely with inputs from outsiders such as the community and environmental groups (cf. next chapter), in order to agree as far as possible on what are the relevant effect chains and values at stake.

3 What you have to do here is first roughly fill in the PiC picture, thus including some idea of the eutrophication effects chain, the types of impacts on nature, fisheries, etc., the factories as primary actors, their motivations for discharging the pollutant, the secondary actors influencing the factories' options and motivations, and so on. Having done this you can take the list featured in Box 2.2 and try to specify the options for solutions. These start with symptoms-abatement measures such as diverting the pollutant flows away from the lake, and end with (for instance) taking the factory directors on a trip to an unpolluted lake and confronting them with the appearance of the lake they pollute, and the people who suffer by their actions ('show the beauty', etc.). Between these you will find many other options for action. The abstract notion of 'turning macro into micro', for instance, may amount here to trying to establish some lake management authority in which the national government, the factories, the local people, the fisheries organisation, the nature protection agency, etc. participate. This authority may then manage the lake as a commons, moving away from the 'tragedy of free access'. (But take care here that you include not only the local people and organisations, but also the most important actors, e.g. national governments – the core actors when it comes to regulations and the financing of technical measures.)

CHAPTER 3

1 The following should be considered in a comprehensive definition of culture:
   (a) stress the multiple interactions between the cultural filter and transformation of the earth by human activities;
   (b) the impact of social nexus and world-view on human/nature relationships;
   (c) evidence of human history as a subset of all-encompassing bio-history;
   (d) awareness of the growing environmental load exerted by humans, and limits to environmental sustainability.

2 The naturalistic fallacy is the misconception that values and norms can be derived, by simple logical deduction, from a factual description of objective reality. Examples of this fallacy are historical aberrations of world-views dictating scientific descriptions of facts, such as by Galilei, under Nazism and state socialism, and the like.
3 Objectivity of science: the objectivity of scientific experiments and methodology is biased towards a scientific world-view, and it is in fact an important part of the cultural filter of the West. By integrating humans back into nature, and because of concern for culture, ecosystems, the biosphere, human ecology and environmental management are all concerned with both facts and values.
4 The following should be considered in a gradual transition:
   (a) Universality of values and norms: ethical realism; the view that intrinsic values can exist in nature independently of human consciousness offers a good perspective for consensus-building for values and norms;
   (b) Ethical deductions: recent developments in epistemology reveal that ethical deductions are not fundamentally different from factual deductions. The gap between the two depends on differences in complexity and extent of research and development.
5 The strengths and weaknesses of interdisciplinarity are:
   (a) The strengths: reality itself is undivided, and there is a universal psychological need for an integrated world-view which includes the complexity and transboundary nature of environmental management problems, together with an acknowledgement of the interaction between science and society in environmental issues.
   (b) The weaknesses: almost all scientists have a monodisciplinary background. As a result they are ill-equipped to deal with interdisciplinary problems. And, when they participate in such projects or activities, the process becomes complex and time-consuming; in addition, the funding of interdisciplinary projects is usually very difficult.
6 The modalities for promoting public involvement in decision-making should include the following:
   (a) democratic empowerment;
   (b) educational spin-off;
   (c) achievement of objectivity through a confrontational and trial-like procedure;
   (d) political realism;
   (e) an acknowledgement that environmental policy decisions entail more than just scientific facts.
7 The mechanism includes the following:
   (a) hierarchic versus horizontal power distribution;
   (b) individualistic versus collective consciousness;
   (c) quality of life versus performance;
   (d) willingness versus avoidance of risk-taking;
   (e) long-term versus short-term thinking.
8 The purpose can be summarised as follows:
   (a) Intercultural negotiations: with regard to the steadily increasing global environmental problems, and given the prevailing cultural diversity of human societies, it is clear that intercultural negotiations must be undertaken in order to develop and implement a common strategy for survival and development.
   (b) Overlapping consensus: the principle of overlapping consensus holds promise for the realisation of the enormous task of reversing environmental degradation through a realistic articulation of contrasting and incommensurable cultural roots, and this is likely to lead to a possible agreement on a priority platform for urgent action.

CHAPTER 4

1 (c), 2 (c), 3 (b), 4 (a) (c), 5 (a) (c), 6 (a) (b), 7 (a) (c).

CHAPTER 5

The questions listed above do not have 'right' or 'wrong' answers. The following are therefore merely suggestions of the sort of answer you might have given.

1 The concentration of nitrogen dioxide depends on a number of factors. Nitrogen dioxide (NO_2) is a secondary pollutant, produced largely as a result of the oxidation of nitric oxide (NO). Nitric oxide is emitted mainly from road traffic, whereupon it is dispersed through the atmosphere. As a result, levels of nitrogen dioxide generally reflect the volume of road traffic in the city. Weather conditions may also be important, and concentrations tend to build up under stable atmospheric conditions, when there is little wind or turbulence to disperse the gases. Most emissions, however, occur at a low level (i.e. close to the ground). Dispersion is also affected by the structure of the surrounding buildings, and chemical reactions occur as the gases move through the atmosphere. As a result,

large differences in concentrations occur within a given city, and even within a single street. This means that measured concentrations are very dependent on the location of the monitoring site (e.g. how close it is to the road, and the traffic volume on the road). Small differences may also occur depending on the monitoring methods used. Differences in the organisation of the monitoring networks may therefore have a marked effect on the reported statistics. This makes direct comparisons between cities extremely difficult. In this case, for example, it is unlikely that real concentrations in Copenhagen are generally higher than those in London. It is more likely that this reflects differences in monitoring strategies.

Data published by different reports also tend to vary because data from different monitoring stations may be selected for publication, and because different methods may be used to describe the 'average'. Some may deliberately select kerbside sites, for example, while others may select background sites, further away from roads. Some may report the average over all hours of the day, while others may report the average only of the peak hours. Care is therefore needed in comparing statistics from different reports.

2 In each case, four main types of error may be expected: (i) errors in definition or classification; (ii) errors in measurement or survey; (iii) errors relating to location; and (iv) reporting errors. Examples are as follows:

(a) For stream water quality:
- errors/discrepancies in the timing and frequency of sampling;
- errors/discrepancies in the extraction of samples from the stream;
- errors/differences in laboratory analysis of samples;
- errors/inconsistencies in the location of sampling sites;
- errors in extrapolating data from the sampling site to the stream segment;
- mis-coding of results.

(b) for number of mammal species at risk:
- miscounting of the numbers of mammal species present;
- incorrect identification of the species;
- incorrect definition of the critical populations below which the species is truly 'at risk';
- incorrect definition of the range or territory of the species concerned;
- errors in transcription or in the coding of data.

(c) for loss of natural habitats:
- errors/inconsistencies in defining the habitats;
- errors/inconsistencies in defining what constitutes 'lost';
- errors in measuring the area of habitats present or lost;
- errors in defining the geographic extent/location of the habitats;
- mistakes in transcription or in the coding of results.

Note that in most cases the loss of habitats is deduced from a comparison of data for two time periods; errors or inconsistencies in either of these datasets will affect the estimates of loss.

(d) for rates of soil erosion:
- errors in defining 'soil erosion';
- errors in recognising land which meets this definition;
- errors in quantifying the extent of the land area affected;
- errors in attributing the areas of soil erosion to the correct reporting units;
- miscoding or transcription errors.

3 The choice of issues clearly reflects the local character of the area concerned and the use of the report. As an example, however, the following topics were covered by one recent local SoE report:

(a) policy, such as on environmental expenditure, activity by voluntary groups, environmental schemes;
(b) social conditions, such as income, crime, education, health, public attitudes and perceptions;
(c) urban development, such as land use, planning, construction, derelict land, urban green space, public attitudes and perceptions;
(d) industry, such as employment, emissions, pollution events, environmental management, public attitudes and perceptions;
(e) energy, such as energy production, energy consumption, energy conservation, public attitudes and perceptions;
(f) transport, such as transport infrastructure, road traffic, public transport, pedestrians, road maintenance, traffic noise, traffic accidents, public attitudes and perceptions;
(g) waste management, such as waste generation, waste collection, waste disposal, waste recycling, waste minimisation, litter, public attitudes and perceptions;
(h) agriculture, such as employment, land use, farm structure, mechanisation, pesticide use, fertiliser use, grant uptake, public attitudes and perceptions;

(i) forestry, such as employment, area, new planting, clearance, public attitudes and perceptions;
(j) leisure and recreation, such as employment, participation rates, infrastructure, access, public attitudes and perceptions;
(k) air, such as emissions, outdoor concentrations, indoor concentrations, pollution incidents, public attitudes and perceptions;
(l) water, such as water resources, water usage, pollution incidents, stream water quality, drinking water quality, groundwater quality, public attitudes and perceptions;
(m) wildlife and habitats, such as woodlands, grasslands, moorland/heathland, wetlands, field boundaries, free-standing trees, mammals, birds, amphibians/reptiles, fish, invertebrates, plants, protected areas, public attitudes and perceptions;
(n) cultural heritage, such as listed buildings, conservation districts, scheduled monuments, landscape quality, public attitudes and perceptions.

4 Again, the choice of indicators, and the precise way in which they are defined, will depend on the specific area in mind. Examples might include:
(a) For pressures from agriculture on groundwater in Europe:
- the quantities of inorganic pesticides and nitrogenous fertilisers applied per ha of land.

(b) For pressures on the environment from domestic wastes in Europe:
- the volume of domestic waste not collected for recycling per ha of land; that is, the amount generated minus the amount recycled or re-used.

(c) For biodiversity in Europe:
- the area of natural and/or semi-natural habitats (per ha of land)

(d) For air quality in a city of your choice:
- the percentage of days on which air quality exceeds WHO guideline levels.

(e) For health risks from road transport in your country:
- the percentage of the population living within 100 metres of busy major roads.

(f) For actions being taken to reduce reliance on fossil energy sources in your country:
- the amount of energy consumption met by non-fossil fuels (KJ per person).

5 Derivation of sensible compound indicators (question 5a) is notoriously difficult. The components suggested for one such indicator, however, may include:
- nitrogen dioxide concentration in the air;
- sulphur dioxide concentration in the air;
- ozone concentration in the air;
- carbon monoxide concentration in the air;
- particulate concentration in the air;
- bathing water quality;
- stream water quality;
- level of fertiliser usage;
- level of pesticide usage;
- population change;
- number of new dwellings;
- number of tourists;
- traffic volumes;
- waste generation.

CHAPTER 6

1 (c), 2 (a) and (c), 3 (b), 4 (a), 5 (c), 6 (a).

CHAPTER 7

1 (b), 2 (c), 3 (a), 4 (c), 5 (a), 6 (c), 7 (c), 8 (b), 9 (c), 10 (b).

CHAPTER 8

1 Consider a car accident.

(a) Most insurance statistics agree that women drivers are more cautious in traffic, causing fewer severe accidents, whereas very young drivers are less experienced and at times take unreasonable risks. Drunk drivers are a notorious danger both to the public and themselves. Old drivers may develop poor eyesight, react more slowly to a sudden incident, or may be more prone to heart failure. On the other hand, they also drive more slowly and have a lower mileage than average. The authors are not familiar with any link between religious practice and road accident statistics. Old cars are less dependable, and worn-out tyres have a high tendency to skid. Rain reduces visibility and makes roads slippery. Local fog and sleet also create dangerous driving conditions. So, a suggested answer would be that shown in the accident factor table.

These answers should, however, be supported with statistical evidence which can change with time and may be site-specific; for example, in countries with very strict drinking regulations, the frequency of drunk drivers causing accidents may be quite low.

(b) The insurance company should try to obtain information on:

- the yearly mileage travelled by the driver;
- the driving record of the driver showing age, sex, previous accidents, speeding fines, whether or not a drinker, etc., preferably in a controlled manner and without intruding too much into his or her privacy;
- the factors discussed under (a)

(c) A particular driver, or a car, should be banned from the roads if the danger caused is no longer acceptable to society. Reasons for interdiction should be non-discriminatory and may involve the physical condition of the driver, or of the car, or previous driving records. Rational criteria can be developed including physical testing of the driver (general health, eyesight, reaction time, etc.), the car (tyre profile, suspension and direction, etc.) or the number and severity of car accidents which the driver in question has caused in the past.

2 For example, the reasons for *not* playing tennis could be:

The playing of tennis may lead to the development of stiff muscles, a 'tennis elbow', or tendinitis. The frequency and severity of undesirable consequences will vary with factors such as:

- method of warming-up the muscles prior to playing;
- frequency and duration of the game, compared to the physical fitness, resistance and age of the player;
- hardness of the court (hard courts are associated with tendinitis);
- technique, or lack of it, of the player.

Most players have a general idea about the occurrence and gravity of accidents, but this view would be biased:

- favourably towards the frequent and familiar occurrence of small but possibly tenuous physical problems,
- unfavourably towards that of a violent accident, such as a player being hit in the eye, neck or other sensitive areas by a reckless player.

Accident factors (answer 1(a), Ch. 8)

	Factor	*Probability of an accident*	*Severity of an accident*
Driver:	age	+ (older age) ++ young drivers	0/– (older age) ++ (young drivers)
	sex	– (female)	– (female)
	religion	?	?
	drinking	+++	+++
Car:	make, model	?	?
	year	+ (old car)	?
	horsepower	?	+ (high horsepower)
	condition of tyres	+ (old tyres)	+ (old tyres)
Weather:	rain	++	+
	fog	++	+
	sleet	++	++
	snow	?	?
Others		?	?

+ higher 0 neutral – lower ? unknown

Management treatment of accident (answer 4, Ch. 8)

Treatment	*Pros and cons*
Without further interest	Contrary to any sound safety policy No direct cost, maximum long-term cost
In an administrative manner	Red tape treatment leads to reliable statistical data on accidents, but the quality of preventive or curative measures is not ensured. Limited cost, can be expressed in terms of a small number of working hours per accident.
Leads to disciplinary measures	Has undesirable side-effects, such as concealing the frequency and severity of accidents, searching for scapegoats, and an insecure and oppressive work climate. On the other hand, a feeling of responsibility is promoted among the staff and supervisors. This approach is probably better than no action at all. Low cost.
Analysis by an *ad hoc* working group	Ideal approach when based on a sound analysis of the accident and followed by a group discussion on causes and measures to be taken. Takes more time and money, with a sizeable pay-back time.

Figures can usually be gathered from sports insurance records, hospital admissions or interviews with medical doctors and surgeons. Similar examples can also be cited for more hazardous sports such as American football (rugby), skiing, rock climbing, or different kinds of transportation sport such as walking, bicycling, car racing, etc.

3 To some a life-long illness or cancer is less acceptable than sudden death. Such a viewpoint leads to a multiple-criteria approach to risk assessment rather than just the single decision variable of violent death. A problem in considering multiple criteria (aesthetic damage, sound, smell, feeling of insecurity, etc.) is the difficulty in quantifying them and evaluating their impact in financial terms. A general approach to quantification is through posing such questions as 'what is the value to me if that factory is *not* built in my neighbourhood'? To a certain extent an objective value of human life can be estimated by computing, for example, the real value of the future earnings of an individual employee or a worker from a given starting point until the end of his or her professional career. Moral damage is normally evaluated on a common sense basis, for example in a court of law by comparing the (possibly exaggerated) claims of the victim(s) with those of defendant(s).

4 A man fell from a ladder in your enterprise or institution. Management may treat the matter in a variety of ways (see table 'Management treatment of accident').

5 Some obvious limits are:
- the difficulty of forecasting all possible or plausible accidents, and quantifying their probability and impact;
- translating the results of an uncertain computation into generally accepted socio-political recommendations, conclusions or decisions. Even when 'objective' studies confirm the need for constructing a nuclear plant, or a waste incinerator with extensive flue gas cleaning, the result will be challenged by the affected neighbours.

6 Obviously, the various roles proposed require different skills, including:
- *administration*: integrity, fairness, equity and a sense of reality;
- *industry*: technical skills, providence, authority and openness;
- *information*: empathy, fairness and communicative skills.

CHAPTER 9

1 (c), 2 (d), 3 (c), 4 (e), 5 (b), 6 (c), 7 (d), 8 (a).

CHAPTER 10

1 (c), 2 (a), 3 (a), 4 (b), 5 (b).

CHAPTER 11

1 (b), 2 (a), 3 (b), 4 (b), 5 (a), 6 (a), 7 (c), 8 (b), 9 (b).

CHAPTER 12

1 (a), 2 (b), 3 (a) and (b), 4 (b), 5 (b), 6 (c).

CHAPTER 13

1 (b) and (c), 2 (i), 3 (a), 4 (c), 5 (a), 6 (b).

CHAPTER 14

1 (d) and (e).
2 (c).
3 (c).
4 (b).
5 (b) and (c).
6 Calculation of eco-efficiency.
   (a) *Ecological product efficiency of the environmental impact added carriers* (see equation (2) in the text): the efficiency of A is 0.33 product unit per environmental impact added unit, because 2,500 product units cause 7,500 units of environmental impact added. Similarly, the efficiency of B is 0.25, while that of C is 0.33. Efficiencies of A and C are equal, and both are greater than that of B.
   (b) *Ecological function efficiency of the environmental impact added carriers* (see equation (3) in the text): 2.5 g is needed to colour 1 m^2. Therefore, in the case of A, 2,500 g sold would be enough to colour 1,000 m^2; thus, the ecological function efficiency of A = 7,500/1,000 = 7.5 environmental impact added/m^2. Similarly, we can show that the amounts for B and C are 2.0 environmental impact added/m^2 and 9.0 environmental impact added/m^2, respectively. Clearly, B is more efficient than A, which is more efficient than C.
   (c) Eco-efficiency of the environmental impact added carriers
      (c1) *Functional eco-efficiency*: as calculated in (b) above, the ecological function efficiency of A is 7.5 environmental impact added/m^2, and the CM of this is £5.00. Therefore, the functional eco-efficiency of A is 5.00/7.5 = £0.66 per environmental impact added/m^2. The corresponding figures for B and C are 1.00 and 0.33, respectively.
      (c2) *Product eco-efficiency*: for A it is £1.66 per environmental impact added unit per product unit, calculated as £5.00 per product unit × 2,500 units sold/7,500 environmental impact added of sales. The corresponding figures for B and C are 2 and 1, respectively. Clearly, B is better than A, which is better than C.
      (c3) *Eco-efficiency of total sales*: Eco-efficiency is a relative measure. Thus the results are equal to (a), (b), (c1) and (c)2.
   (d) Eco-efficiency cannot be calculated correctly if the environmental impact added carrier does not correspond with the cost carrier.

CHAPTER 15

1 (b) and (c), 2 (a) and (b), 3 (b), 4 (b), 5 (c), 6 (a), 7 (b).

CHAPTER 16

1 (a), 2 (b), 3 (c), 4 (b), 5 (a), 6 (b).

CHAPTER 17

1 See pp. 325–326.

2 See p. 327 and Table 17.1.
3 See p. 329.
4 See p. 329.
5 See p. 330.
6 See pp. 331–334.
7 See pp. 331–334.
8 See pp. 335–337.
9 See pp. 337–338.

CHAPTER 18

1 (b), 2 (c), 3 (c), 4 (a), 5 (b).

CHAPTER 19

1 (a) and (c).
2 (c).
3 States can be persuaded to join international agreements by:
- pressure from their domestic publics;
- scientific evidence that the environmental problem exists, is due to anthropogenic factors within their control, and affects them;
- measures within international agreements to deny them benefits (such as trade in the regulated substances) if they do not join;
- measures within international agreements to give them benefits (such as financial aid or technology transfer to help them meet their obligations) if they join;
- issue linkage;
- threats of trade sanctions or other economic harm.

4 (b) and (c).
5 (a), (c), and (d).
6 (d).
7 (b).
8 (c).
9 (d).
10 (a), (b), and (d).

CHAPTER 20

1 (c), 2 (b), 3 (b), 4 (c), 5 (a), 6 (c), 7 (c), 8 (b), 9 (a), 10 (c).

CHAPTER 21

1 (b), 2 (b), 3 (c), 4 (b), 5 (b), 6 (a).

CHAPTER 22

1 (c), 2 (b), 3 (c), 4 (c), 5 (a), 6 (c), 7 (c), 8 (a), 9 (c), 10 (a).

CHAPTER 23

1 (c), 2 (c), 3 (a), 4 (c), 5 (d).

CHAPTER 24

1 A conflict of a party with the interest of another, or with general environmental principles, which extends beyond the national borders or jurisdiction of a single nation-state, or has the potential to do so.
2 (a), (b), (d).
3 (a), (d), (e).
4 (b).
5 (a), (c).
6 The LOS Convention offers great potential for conflict resolution through a set of mechanisms. In particular, the Special Arbitration procedure allows the involvement of experts who can effectively help solve transnational environmental problems. However, not all outcomes of the possible procedures are binding, and many states are not yet members of the Convention. Whether it would really play an important role in TECM in the future remains to be seen.
7 Individual answer expected.

INDEX

Note: Italic page numbers refer to charts and tables.